Comparative Diagnosis of Viral Diseases

Volume IV

VERTEBRATE ANIMAL AND RELATED VIRUSES
Part B—RNA Viruses

List of Contributors

ARTHUR A. ANDERSEN
MAX J.G. APPEL
EDWARD H. BOHL
M. BRAHIC
LEROY COGGINS
A. J. DELLA-PORTA
BARBARA DETRICK-HOOKS
M. ESSEX
GLYNN H. FRANK
E. PAUL J. GIBBS
A. T. HAASE
JOHN J. HOOKS
MARIAN C. HORZINEK
KARL M. JOHNSON
CHRISTINE KURSTAK
EDOUARD KURSTAK
SAM J. MARTIN
THOMAS P. MONATH
R. MORISSET
J. S. PORTERFIELD
BERT K. RIMA
G. C. SCHILD
DAVID R. SNODGRASS
NEVILLE F. STANLEY
JOHN R. STEPHENSON
WILLIAM P. TAYLOR
VOLKER ter MEULEN
DENNIS W. TRENT
J. van den HURK
WILLIAM G. WINKLER
M. WORLEY

Comparative Diagnosis of Viral Diseases

Volume IV

VERTEBRATE ANIMAL AND RELATED VIRUSES
Part B—RNA Viruses

Edited by

EDOUARD KURSTAK
Groupe de Recherche en Virologie Comparée
Faculté de Médecine, Université de Montréal
Montréal, Canada

CHRISTINE KURSTAK
Laboratoire de Diagnostic des Maladies Virales
Hôtel-Dieu de Montréal, Université de Montréal
Montréal, Canada

1981

ACADEMIC PRESS

A Subsidiary of Harcourt Brace Jovanovich, Publishers

New York London Toronto Sydney San Francisco

ACADEMIC PRESS, INC.
111 Fifth Avenue, New York, New York 10003

United Kingdom Edition published by
ACADEMIC PRESS, INC. (LONDON) LTD.
24/28 Oval Road, London NW1 7DX

Library of Congress Cataloging in Publication Data
Main entry under title:

Vertebrate animal and related viruses.

(Comparative diagnosis of viral diseases ; v. 4)
Includes bibliographies and index.
Contents: --pt. B. RNA viruses.
1. Virus diseases. 2. Vertebrates--Diseases.
3. Viruses, RNA. I. Kurstak, Edouard.
II. Kurstak, Christine. III. Series.
QR302.V47 636.089'60194 81-7951
ISBN 0-12-429704-8(v. 4) AACR2

PRINTED IN THE UNITED STATES OF AMERICA

81 82 83 84 9 8 7 6 5 4 3 2 1

Contents

List of Contributors xiii

Preface xv

Contents of Other Volumes xix

Part I PICORNAVIRIDAE

Chapter 1 **Picornaviruses of Animals: Clinical Observations and Diagnosis**

ARTHUR A. ANDERSEN

I. Introduction 4
II. Description of the Viruses 8
III. Host Range 17
IV. Pathogenesis and Clinical Signs 19
V. Immunity 36
VI. Epizootiology 40
VII. Diagnosis 43
VIII. Preventions 52
IX. Conclusion 54
References 55

Part II REOVIRIDAE

Chapter 2 **Reoviridae: Orbivirus and Reovirus Infections of Mammals and Birds**

NEVILLE F. STANLEY

I. Introduction 67
II. Orbivirus 69

III. Reovirus 94
IV. Provisional Reoviridae 97
V. Conclusion 99
References 99

Chapter 3 Animal Rotaviruses

E. KURSTAK, C. KURSTAK, J. VAN DEN HURK, AND R. MORISSET

I. Introduction 105
II. Incidence 108
III. Morphology and Morphogenesis 108
IV. Physicochemical Properties 112
V. Antigenic Relationships 117
VI. Clinical Features 118
VII. Pathology and Pathogenesis 120
VIII. Laboratory Diagnosis 124
IX. Propagation of Rotaviruses *in Vitro* 128
X. Epidemiology 135
XI. Immunity 137
XII. Prevention, Control, and Treatment 139
XIII. Conclusions 140
References 141

Part III ORTHOMYXOVIRIDAE

Chapter 4 Influenza Infections in Lower Mammals and Birds

G. C. SCHILD

I. Introduction 151
II. Structure and Composition of Influenza Viruses 154
III. Antigenic Classification and Nomenclature of Influenza Viruses 157
IV. Revised Nomenclature System 162
V. Influenza in Lower Mammals and Birds 169
VI. Influenza in Swine 169
VII. Influenza in Horses 171
VIII. Influenza in Birds 172
IX. Influenza in Other Species 178
X. Evidence That Influenza A Viruses from Nonhuman Sources May Be Progenitors of Human Influenza 178
References 180

Part IV PARAMYXOVIRIDAE

Chapter 5 **Paramyxovirus and Pneumovirus Diseases of Animals and Birds: Comparative Aspects and Diagnosis**

GLYNN H. FRANK

I. Introduction 187
II. Family Paramyxoviridae 188
III. Paramyxoviruses of Animals 195
IV. Paramyxoviruses of Birds 211
V. Pneumoviruses of Animals 221
References 229

Chapter 6 **Morbillivirus Diseases of Animals and Man**

MAX J. G. APPEL, E. PAUL J. GIBBS, SAM J. MARTIN, VOLKER ter MEULEN, BERT K. RIMA, JOHN R. STEPHENSON, AND WILLIAM P. TAYLOR

I. Introduction 235
II. Properties of the Virus 237
III. Interactions with Organisms 251
References 287

Part V CORONAVIRIDAE

Chapter 7 **Coronaviruses: Diagnosis of Infections**

EDWARD H. BOHL

I. Introduction 301
II. Characteristics of Coronaviruses 302
III. Antigenic Composition 304
IV. Clinical and Pathological Features of Infections 305
V. Laboratory Diagnosis 311
VI. Prevention 322
References 323

Part VI TOGAVIRIDAE

Chapter 8 Togaviral Diseases of Domestic Animals

THOMAS P. MONATH AND DENNIS W. TRENT

I.	Introduction	332
II.	Morphology, Physicochemistry, and Antigenic Composition of Togaviruses	335
III.	Comparative Biology and Pathogenesis	367
IV.	Epizootiology	395
V.	Comparative Diagnosis	413
VI.	Prevention and Control	421
	References	426

Chapter 9 Nonarbo Togavirus Infections of Animals: Comparative Aspects and Diagnosis

MARIAN C. HORZINEK

I.	Introduction and Classification	442
II.	Comparative Characteristics	443
III.	Pathology and Pathogenesis	451
IV.	Epidemiological Aspects	455
V.	Comparative Diagnosis	459
VI.	Conclusions	467
	References	468

Part VII BUNYAVIRIDAE

Chapter 10 Bunyaviridae: Infections and Diagnosis

J. S. PORTERFIELD AND A. J. DELLA-PORTA

I.	Introduction	479
II.	Akabane Disease	482
III.	Aino and Other Simbu Group Viruses	497
IV.	Rift Valley Fever	499
V.	Bhanja Virus Infections	501
VI.	Nairobi Sheep Disease	502
VII.	Ganjam Virus Infections	503
VIII.	Dugbe Virus Infections	503
IX.	Crimean-Congo Viruses	504
	References	504

Part VIII ARENAVIRIDAE

Chapter 11 Arenaviruses: Diagnosis of Infection in Wild Rodents

KARL M. JOHNSON

I. Introduction 511
II. Virus-Rodent Specificity 512
III. Comparative Biology 515
IV. Comparative Diagnosis 520
V. Conclusions 523
References 524

Part IX RHABDOVIRIDAE

Chapter 12 The Rhabdoviruses

WILLIAM G. WINKLER

I. Introduction 529
II. Morphology 530
III. Physicochemical Properties 532
IV. Antigenic Composition 532
V. Host Range 533
VI. Transmission 535
VII. Incubation Period 536
VIII. Pathogenesis 537
IX. Clinical Illness 539
X. Epidemiology 540
XI. Diagnosis 542
XII. Prevention and Control 546
References 547

Part X RETROVIRIDAE

Chapter 13 Naturally Occurring Retroviruses of Animals and Birds

M. ESSEX AND M. WORLEY

I. Introduction 553
II. Classification of Retroviruses 555

III.	Principal Characteristics of Retroviruses	558
IV.	Comparative Pathobiology	572
V.	Detection of Retroviruses	587
VI.	Conclusions	588
	References	588

Chapter 14 Spumavirinae: Foamy Virus Group Infections: Comparative Aspects and Diagnosis

JOHN J. HOOKS AND BARBARA DETRICK-HOOKS

I.	Introduction	599
II.	Description of the Virion	601
III.	Biological Features	607
IV.	Association with Human Disease	610
V.	Immunity	611
VI.	Epidemiology	612
VII.	Laboratory Diagnosis	613
VIII.	Concluding Remarks	616
	References	616

Chapter 15 Lentivirinae: Maedi/Visna Virus Group Infections

M. BRAHIC AND A. T. HAASE

I.	Introduction	620
II.	Description of Viruses	620
III.	Description of Diseases	626
IV.	Methods	632
V.	Pathogenesis	637
VI.	Conclusion	640
	References	641

Part XI UNCLASSIFIED VIRUSES

Chapter 16 Equine Infectious Anemia

LEROY COGGINS

I.	Introduction	647
II.	Characteristics of EIA Virus	648
III.	Comparative Biology	650
IV.	Immunity	652

V. Epizootiology 653
VI. Comparative Diagnosis 655
VII. Prevention and Control 656
References 657

Chapter 17 **Astroviruses in Diarrhea of Young Animals and Children**

DAVID R. SNODGRASS

I. Introduction 659
II. Description of Astroviruses 660
III. Comparative Biology and Pathogenesis 664
IV. Serology 667
V. Epidemiology 667
VI. Laboratory Diagnosis 667
References 669

Index **671**

List of Contributors

(Numbers in parentheses indicate the pages on which the authors' contributions begin.)

ARTHUR A. ANDERSEN* (1), Plum Island Animal Disease Center, United States Department of Agriculture, Greenport, New York 11944

MAX J. G. APPEL (239), James A. Baker Institute for Animal Health, New York State College of Veterinary Medicine, Cornell University, Ithaca, New York 14853

EDWARD H. BOHL (303), Department of Veterinary Science, Ohio Agricultural Research and Development Center, Wooster, Ohio 44691

M. BRAHIC (591), Department of Medicine, University of California, San Francisco, California 94121

LEROY COGGINS (637), Departments of Microbiology, Pathology, and Parasitology, North Carolina State University, Raleigh, North Carolina 27630

A. J. DELLA-PORTA (477), CSIRO Division of Animal Health, Animal Health Research Laboratory, Parkville, Victoria, 3052 Australia

BARBARA DETRICK-HOOKS (617), Diagnostic Immunology Section, Walter Reed Army Hospital, Washington, D.C.

M. ESSEX (545), Department of Microbiology, Harvard University School of Public Health, Boston, Massachusetts 02115

GLYNN H. FRANK (191), Agricultural Research, North Central Region, National Animal Disease Center, Science and Education Administration, United States Department of Agriculture, Ames, Iowa 50010

E. Paul J. GIBBS (239), College of Veterinary Medicine, University of Florida, Gainesville, Florida 32610

A. T. HAASE (591), Veterans Administration Medical Center, University of California, San Francisco, California 94143

JOHN J. HOOKS (617), Laboratory of Oral Medicine, National Institute of Dental Research, National Institutes of Health, Bethesda, Maryland 20205

*Present address: United States Department of Agriculture, National Program Staff, Beltsville, Maryland 20705

MARIAN C. HORZINEK (441), Veterinary Faculty State University Utrecht, Practicumgebouw, Yalelaan 1, 3584 CL Utrecht, The Netherlands

KARL M. JOHNSON (507), Virology Division, Center for Disease Control, Atlanta, Georgia 30333

CHRISTINE KURSTAK (101), Laboratoire de Diagnostic des Maladies Virales, Service de Microbiologie, Hôtel-Dieu de Montréal, Université de Montréal, Montréal H2W 1T8, Canada

EDOUARD KURSTAK (101), Groupe de Recherche en Virologie Comparée, Faculté de Médecine, Université de Montréal, Montréal H3C 3J7, Canada

SAM J. MARTIN (239), Department of Biochemistry, The Queen's University of Belfast, Belfast BT9 7BL, Northern Ireland

THOMAS P. MONATH (331), Department of Health and Human Sciences, Vector-Borne Disease Division, Center for Disease Control, United States Public Health Service, Fort Collins, Colorado 80522

R. MORISSET (101), Service de Microbiologie, Hôtel-Dieu de Montréal, Université de Montréal, Montréal H2W 1T8, Canada

J. S. PORTERFIELD (477), Sir William Dunn School of Pathology, University of Oxford, Oxford OX1 3RE, England

BERT K. RIMA (239), Department of Biochemistry, The Queen's University of Belfast, Belfast BT9 7BL, Northern Ireland

G. C. SCHILD (157), National Institute for Biological Standards and Control, Holly Hill, Hampstead, London NW3 6RB, England

DAVID R. SNODGRASS (145), Animal Diseases Research Association, Moredun Institute, Edinburgh EH17 7JH, Scotland

NEVILLE F. STANLEY (63), Department of Microbiology, University of Western Australia, Nedlands, 6009 Western Australia

JOHN R. STEPHENSON* (239), Institute für Virologie und Immunobiologie der Universität Würzburg, Federal Republic of Germany

WILLIAM P. TAYLOR (239), The Animal Virus Research Institute, Pirbright Woking,Surrey GU24 ONF, United Kingdom

VOLKER TER MEULEN (239), Institute für Virologie und Immunobiologie der Universität Würzburg, Federal Republic of Germany

DENNIS W. TRENT (331), Vector-Borne Diseases Division, Center for Disease Control, Fort Collins, Colorado 80522

J. VAN DEN HURK (101), Groupe de Recherche en Virologie Compareé, Faculté de Médecine, Université de Montréal, Montréal H3C 3J7, Canada

WILLIAM G. WINKLER (523), Viral Disease Division, Bureau of Epidemiology, Center for Disease Control, Atlanta, Georgia 30333

M. WORLEY (545), Department of Microbiology, Harvard University School of Public Health, Boston, Massachusetts 02115

*Present address: Centre for Applied Microbiology and Research, Porton Down, Salisbury 5P4 DJG, England.

Preface

The first two volumes of the treatise, "Comparative Diagnosis of Viral Diseases" devoted to "Human and Related Viruses," demonstrated the value and interest of a comparative way of looking at virus infections. The unifying concept of comparative virology, on which this multi-volume work is based, is not only well accepted but is of increasing value in basic research, diagnosis, control, and prevention of viral diseases.

Although treatises on fundamental comparative virology are now available, it is notable that among the books devoted to the diagnosis and control of viral infections, only the volumes of this treatise are based on a comparative approach. This principle is shown essential for several groups of viruses infecting animals and man. It is well known that it is difficult to diagnose specifically and rapidly numerous viral diseases without considering the comparative biological, genetic, serological, and physicochemical properties of viruses involved. This treatise demonstrates that comparison of and discrimination among viruses, according to the criteria of classification of the International Committee on Taxonomy of Viruses and to the diseases caused by these viruses, irrespective of the species involved, are essential for their diagnosis and prevention.

The third and fourth volumes of "Comparative Diagnosis of Viral Diseases" are devoted to "Vertebrate Animal and Related Viruses." Volume IV was conceived to cover in separate chapters the infections caused by each RNA virus family, including unclassified viruses. However, in some cases in which a particular interest is manifested for a virus, a group of viruses, or for the disease(s) it induces, a whole chapter is devoted to the subject. This is the case for rotaviruses, morbilliviruses, nonarbo togaviruses, spumaviruses, and lentiviruses, which have elicited great interest in the last few years or which due to their particular characteristics needed a special treatment. Volume III covers all DNA virus families and slow virus diseases inducing infections in vertebrate animals.

These volumes give a comparative description of the principal physicochemical, molecular, structural, genetic, immunological, and biological characteristics of viruses implicated in various diseases, mainly of veterinary importance. With this new concept of comparative diagnosis, the symptoms and the evolution of the diseases are described in detail, as well as the modern methodology for their rapid and specific diagnosis, their control, and prevention. In this respect, Volumes III and IV will interest all virologists and immunologists working in the area of diagnosis and control of animal virus diseases. These two volumes are addressed particularly to the professionals of veterinary sciences working in the field and laboratory and to students of veterinary schools. As numerous animal viruses also can cause severe diseases in man, these volumes are also of interest to all clinical virologists and immunologists, to the professionals of public health, and research workers. The artificial division between the diseases of man and animal is discarded in these volumes.

Each of the contributors to this treatise is well known for his expertise in his field; each has prepared a thoughtful and well-documented treatment of his subject. Personal interpretations and conclusions of the authors, as well as the numerous illustrations and unpublished material, provide a large body of information, which brings into sharp focus current findings and new directions in the comparative diagnosis and prevention of viral diseases.

It is our hope that Volumes III and IV of this treatise will provide a useful tool for all concerned with viral diseases but especially for diagnostic and control centers of animal infectious diseases and for schools of veterinary sciences. These volumes should also serve the formation of veterinary virologists in developing countries where very important economic losses caused by viruses infecting food producing animals are directly responsible for the limited development of human society.

The Second International Conference on the Impact of Virus Diseases on the Development of African and Middle-East Countries, recently held in Nairobi, Kenya, and organized by the International Comparative Virology Organization (ICVO) together with several co-sponsors, concluded that the mortgaging of the world's capacity to produce sufficient food is a result of insufficient interest or effort in the control of viral diseases of animals. There is not a single country exempt from losses due to animal virus infections. These economic losses, especially in developing countries, are estimated at billions of dollars. In the Third World, the substantial reduction in livestock productivity due to virus diseases is directly related to the development. Thus, the continued development of African, Asian, and Latin American countries depends on increasing livestock production to meet the nutritional needs of their population and for bolstering local economic conditions through the exportation of meat and animal products. It is obvious that an international program to control and prevent viral diseases through the creation in the Third World of virology centers, devoted to the formation of virologists and to the rapid diagnosis and control of infections, is an urgent and imperative necessity.

The editors of this volume, published under the advisory sponsorship of ICVO, wish to express their sincere gratitude to the contributors for the effort and care with which they have prepared their chapters. Thanks are due to Professors

R. F. Marsh, F. A. Murphy, and H. Graham Purchase for their very valuable advice; and last, but not least, to the staff of Academic Press for their part in the editing, indexing, proofreading, and other aspects of production of this treatise.

EDOUARD KURSTAK
CHRISTINE KURSTAK

Contents of Other Volumes

Volume I

HUMAN AND RELATED VIRUSES, Part A

Classification of Human and Related Viruses
EDOUARD KURSTAK

Part I DNA Viruses

Parvoviruses. Possible Implications in Human Infections
E. KURSTAK AND P. TIJSSEN
Implication of Papovaviruses in Human Diseases
SILVIA D. GARDNER
Adenoviruses: Diagnosis of Infections
PATRICIA E. TAYLOR
Diagnosis of Herpes Simplex, Varicella, and Zoster Infections
ARIEL C. HOLLINSHEAD AND JOHN J. DOCHERTY
Comparative Diagnosis of Epstein-Barr Virus-Related Diseases: Infectious Mononucleosis, Burkitt's Lymphoma, and Nasopharyngeal Carcinoma
GUY DE THÉ AND GILBERT LENOIR
Comparative Diagnosis of Cytomegaloviruses: New Approach
ENG-SHANG HUANG AND JOSEPH S. PAGANO
Comparative Diagnosis of Poxvirus Diseases
JAMES H. NAKANO

Part II RNA Viruses

Comparative Diagnosis of Picornavirus (Enterovirus and Rhinovirus) Infections
R. GORDAN DOUGLAS, JR.
Diagnosis of Reovirus Infections: Comparative Aspects
NEVILLE F. STANLEY
Rotaviruses: Clinical Observations and Diagnosis of Gastroenteritis
PETER J. MIDDLETON

Orthomyxovirus—Influenza: Comparative Diagnosis Unifying Concept
W. R. DOWDLE, G. R. NOBLE, AND A. P. KENDAL
Paramyxoviruses: Comparative Diagnosis of Parainfluenza, Mumps, Measles, and Respiratory Syncytial Virus Infections
ANDREW E. KELEN AND D. ANGUS MCLEOD
Coronaviruses as Causes of Diseases: Clinical Observation and Diagnosis
KENNETH MCINTOSH
Comparative Diagnosis of Togavirus and Bunyavirus Infections
TELFORD H. WORK AND MARTINE JOZAN
Rubella Virus Infection Diagnosis: Present Status
A. J. RHODES, N. R. PAUL, AND S. IWAKATA
Arenaviruses: Diagnosis of Lymphocytic Choriomeningitis, Lassa, and Other Arenaviral Infections
FREDERICK A. MURPHY
Rhabdoviruses: Rabies and Rabies-Related Viruses
T. J. WIKTOR AND M. A. W. HATTWICK

Volume II

HUMAN AND RELATED VIRUSES, Part B

Part I Unclassified Viruses
Marburg Virus Disease
HERTA WULFF AND J. LYLE CONRAD
Diagnosis of Hepatitis Viral Infections
ARIE J. ZUCKERMAN AND COLIN R. HOWARD
Slow Virus Infections: Comparative Aspects and Diagnosis
RICHARD F. MARSH

Part II Cancer Viruses
Human Proliferative Diseases and Viruses
GABRIEL SEMAN AND LEON DMOCHOWSKI

Part III Control of Viral Diseases: Vaccines and Chemotherapy
Control of Viral Diseases by Vaccines
J. FURESZ, D. W. BOUCHER, AND G. CONTRERAS
Chemotherapy of Viral Diseases: Present Status and Future Prospects
GEORGE J. GALASSO AND FRED J. PAYNE

Part IV Virus Information System
The World Health Organization Virus Information System
FAKHRY ASSAAD AND PAUL BRÈS

Part V Diagnostic Reagents and Newer Methods
Viral Diagnostic Reagents
JOHN R. POLLEY
Immunoperoxidase Technique in Diagnosis Virology and Research: Principles and Applications
E. KURSTAK, P. TIJSSEN, AND C. KURSTAK
Enzyme Immunoassays and Their Potential in Diagnostic Virology
A. VOLLER AND D. E. BIDWELL
Radioimmunoassay in Viral Diagnosis
HARRY DAUGHARTY AND DONALD W. ZIEGLER
Cytohybridization Techniques in Virology
P. TIJSSEN AND E. KURSTAK
Electron and Immunoelectron Microscopic Procedures for Diagnosis of Viral Infections
FRANCES W. DOANE AND NAN ANDERSON

Volume III

VERTEBRATE ANIMAL AND RELATED VIRUSES, Part A—DNA Viruses

Part I Parvoviruses
Animal Parvoviruses. Comparative Aspects and Diagnosis
EDOUARD KURSTAK AND PETER TIJSSEN

Part II Papovaviridae
Papovavirus Infections of Vertebrate Animals
WAYNE D. LANCASTER AND CARL OLSON

Part III Adenoviridae
Adenoviruses of Vertebrate Animals
J. B. MCFERRAN

Part IV Iridoviridae
Comparative Aspects and Diagnosis of the Iridoviruses of Vertebrate Animals
WILLIAM R. HESS

Part V Herpesviridae
Herpesviruses Diseases of Mammals and Birds: Comparative Aspects and Diagnosis
DONALD P. GUSTAFSON

Part VI Poxviridae

Poxviruses of Veterinary Importance: Diagnosis of Infections
DEOKI N. TRIPATHY, LYLE E. HANSON, AND ROBERT A. CRANDELL

Part VII Slow Viral Diseases

Scrapie as a Model Slow Virus Disease: Problems, Progress, and Diagnosis
RICHARD H. KIMBERLIN

Part VIII WHO's Viral Diseases Information System

The World Health Organization's Information Systems for Animal Virus Diseases
K. BÖGEL, A. BETTS, AND V. MILOUCHINE

Part I

PICORNAVIRIDAE

Chapter 1

Picornaviruses of Animals: Clinical Observations and Diagnosis

ARTHUR A. ANDERSEN

I.	Introduction		4
II.	Description of the Viruses		8
	A.	Morphology	8
	B.	Other Physical Characteristics	11
	C.	Chemical Properties	12
	D.	Antigenic Structure	13
III.	Host Range		17
	A.	Aphthoviruses (Foot-and-Mouth Disease)	17
	B.	Rhinoviruses	17
	C.	Cardioviruses	18
	D.	Enteroviruses	18
	E.	Caliciviruses	19
IV.	Pathogenesis and Clinical Signs		19
	A.	Foot-and-Mouth Disease Viruses	20
	B.	Bovine Enteroviruses	26
	C.	Swine Enteroviruses	29
	D.	Bovine Rhinoviruses	34
	E.	Equine Rhinoviruses	35
	F.	Vesicular Exanthema	35
	G.	San Miguel Sea Lion Virus	35
	H.	Feline Caliciviruses	35
V.	Immunity		36
	A.	Aphthoviruses	36
	B.	Rhinoviruses	37
	C.	Cardioviruses	38
	D.	Enteroviruses	38
	E.	Caliciviruses	39

COMPARATIVE DIAGNOSIS
OF VIRAL DISEASES, VOL. IV

ISBN 0-12-429704-8

VI. Epizootiology . . . 40
VII. Diagnosis . . . 43
A. Vesicular Diseases . . . 43
B. Bovine Rhinoviruses . . . 46
C. Equine Rhinoviruses . . . 47
D. Bovine Enteroviruses . . . 47
E. Swine Enteroviruses . . . 48
F. Simian Enteroviruses . . . 50
G. Duck Virus Hepatitis . . . 50
H. Avian Encephalomyelitis . . . 50
I. Cardioviruses . . . 51
J. Caliciviruses . . . 51
VIII. Prevention . . . 52
IX. Conclusion . . . 54
References . . . 55

I. INTRODUCTION

This chapter is limited to the picornaviruses of animals. To facilitate interpretation of the vast quantity of information available, I have used the nomenclature and classification proposed by the Study on Picornaviridae, Vertebrate Virus Subcommittee, International Committee of Taxonomy of Viruses (ICTV), whenever possible. The ICTV subcommittee on classification of picornaviruses has recommended that the family Picornaviridae be limited to four groups—enteroviruses, cardioviruses, rhinoviruses, and aphthoviruses—and that the caliciviruses be placed in a separate family (Cooper *et al.*, 1978). However, historically the caliciviruses were proposed as a separate group of the Picornaviridae (Wildy, 1971) and thus are included in this chapter. The equine rhinoviruses are also of special taxonomic interest because of their wide host range and unusual properties in combination with their virion structure, which is that of a picornavirus (Rueckert, 1976; Newman *et al.*, 1977; Cooper *et al.*, 1978). I intend to give an overview of the properties of this broad group of viruses and the diseases they produce but not to review the literature in detail.

The picornaviruses are ubiquitous; most species of domestic animals are susceptible to one or more (Table I). Although most of the picornaviruses produce little or no morbidity in their host, those causing foot-and-mouth disease (FMD), swine vesicular disease (SVD), and enterovirus encephalomyelitis are of major concern to the livestock industry and are the subject of most of the chapter.

FMD is of historical and international importance. It has had a major impact on livestock production and the economics of nations for centuries. The fear of economic loss from the disease separates nations, as countries free of the disease place embargos against livestock products from countries with the dis-

TABLE I
Picornaviruses of Animals

Enteroviruses
Bovine (serotypes 1–7)
Porcine (serotypes 1–9)
Simian (serotypes 1–18)
Murine encephalomyelitis
Avian encephalomyelitis
Duck virus hepatitis
Cardiovirus (EMC, Col. SK, Mengo, ME, MM)[a]
Rhinoviruses
Bovine (serotypes 1–2)
Aphthoviruses (foot-and-mouth disease)
A
O
C
SAT I
SAT II
SAT III
Asia 1
Unclassified
Equine rhinovirus (serotypes 1–2)
Caliciviruses[b]
Vesicular exanthema of swine
San Miguel sea lion virus
Feline calicivirus

[a] Isolates of a single serotype.
[b] Recommended as a separate family by the Study Group on Picornaviridae of the International Committee on Taxonomy of Viruses.

ease and as countries in which it is epizootic consider such embargos to be unjustified exclusion of their products from the world markets. The first clinical description of the disease was by Fracastorius (1514), who described a 1514 outbreak in Italy. In 1897, Loeffler and Frosch determined that the disease was produced by a nonbacterial filterable agent; this study was the beginning of mammalian virology, as it was the first demonstration that a disease of animals or humans was produced by a filterable agent.

FMD is still a devastating disease of the cloven-hoofed animals—animals vital for the production of food and clothing in many parts of the world. The disease is characterized primarily by vesicular lesions on the feet and in the mouth. Mortality is usually low; however, diseased animals may lose up to 20% of their body weight and stop lactating, resulting in major economic losses (Graves, 1979). The ability of the virus to spread in epizootics across a country, its ability to infect a large number of animal species, and the existence of several serologically distinct types of virus increase the difficulty of control. Countries free of the virus and with sufficient financial means resort to extensive eradication programs when confronted with its introduction. Less affluent developing countries are usually forced to live with the losses and costs associated with FMD.

Enteroviruses have been isolated from mammalian hosts including cats, cattle, deer, fowl, horses, humans, mice, primates, and swine (Rueckert, 1976). The viruses are primarily inhabitants of the gut and are considered nonpathogenic; however, a few have been associated with neurological, enteric, respiratory, and reproductive disorders (Dunne *et al.*, 1974a; Dunne, 1975a,b). The enteroviruses are usually species specific; SVD virus (SVDV) is unusual in its ability to infect both swine and humans (Brown *et al.*, 1973, 1976). The enteroviruses are distributed worldwide. Those of humans, swine, and cattle have received the most attention. Many of the isolates have been obtained from normal-appearing animals. These were formerly called "orphan viruses" because they were considered nonpathogenic; this designation has led to the use of such terms as "ECPO" (enteric cytopathogenic porcine orphan virus) and "ECBO" (bovine cytopathogenic orphan virus) (Dunne *et al.*, 1974a; Dunne, 1975a).

Neurological diseases termed "enterovirus encephalomyelitises" (EE) have been associated with a number of the swine enterovirus strains over the past 50 years. The characteristics of the diseases have varied considerably with the strains of virus, the geographic location, the host susceptibility, and the environment (Jones, 1975; Dunne, 1975a). Through the years the EEs have been given names such as "Teschen disease," "Talfan disease," and "polioencephalomyelitis." A severe form of the disease was reported in the Teschen district of Czechoslovakia in the early 1930s; during World War II it spread to other European countries and to Madagascar (Mills and Nielsen, 1968; Jones, 1975). Milder forms of the disease have been reported from most major swine-producing countries.

The bovine enteroviruses are ubiquitous in the cattle population and can be isolated readily from feces and pharyngeal washings of sick and healthy animals. They have been associated with enteric, respiratory, and reproductive disease; however, their role in the etiology of these diseases is unclear, as they are often isolated simultaneously with other agents (Dunne *et al.*, 1974a). As of late, seven serotypes have been recognized by the WHO/FAO Committee on Comparative Virology; this number is expected to increase with further research because there are reports of more than 600 isolates (Barya *et al.*, 1967), most of which have never been classified.

The cardioviruses, often called "murine" viruses, have a wide host range including humans, hamsters, mice, rats, primates, and swine. They are distinguished from the enteroviruses primarily by their biphasic pH stability in isotonic saline (Speir, 1962; Mak *et al.*, 1970). The five cardiovirus isolates—encephalomyoycarditis (EMC), Maus-Elberfeld (ME), Mengo, MM, and Columbia SK viruses—are all members of a single serotype and are highly infectious for rodents, which probably serve as a reservoir for these viruses in nature (Rueckert, 1976). The cardioviruses multiply readily in most laboratory animals and are usually fatal in mice, hamsters, and cotton rats. Other laboratory animals, though readily infected, usually show no signs of disease. Natural outbreaks, usually limited to rodents and primates, have also been observed in humans and swine (Smadel and Warren, 1947; Murname *et al.*, 1960; Gainer *et al.*, 1968).

Rhinoviruses characteristically are isolated only from the upper respiratory tract, grow best at 33° to 34°C are labile at pH 3, and are highly species specific. Although isolations from the bovines are relatively few, these agents are widespread in the cattle population, as evidenced by serological surveys (Bögel *et al.*, 1962; Mohanty and Lillie, 1968; Ide and Darbyshire, 1972c). The low number of bovine isolates can be attributed to the difficulty of isolating rhinoviruses, their slow growth, and their limitation to low titers in tissue culture.

The equine rhinoviruses have several properties unusual for rhinoviruses. Two serologically distinct serotypes (Ditchfield and Macpherson, 1965) have been reported; they have an unusually wide host range, including monkeys, man, horses, guinea pigs, and rabbits (Plummer, 1963). They also differ from the human and bovine rhinoviruses in sedimentation coefficient, cesium chloride (CsC1), buoyant density, acid stability, and RNA properties (Newman *et al.*, 1977). Equine rhinoviruses, which reproduce primarily in the respiratory tract, have also been isolated from the feces (Plummer, 1963). Because of these unusual properties, they may not be true rhinoviruses, and they are now left unclassified (Cooper *et al.*, 1978).

Caliciviruses obtain their name from their characteristic cuplike surface structure. They include the feline calicivirus (FC), vesicular exanthema of swine (VES) virus, and San Miguel sea lion (SMS) virus. Until 1972, caliciviruses had been isolated only from swine and cats. Since then, a number of isolates have been reported from marine mammals, and possible isolates have been reported from humans and cattle (Madeley and Cosgrove, 1976; Woode and Bridger, 1978). VES virus includes some 13 serotypes that are serologically distinct from the single feline serotype. Since 1972 the VES virus (VESV) and the SMS and other marine animal virus isolates have been found to have the same host range (Madin *et al.*, 1976; Smith *et al.*, 1977). Serological studies indicate that the marine isolates also have antigens in common with VESV and not with the FCs (Burroughs *et al.*, 1978a; Smith *et al.*, 1978; Soergel *et al.*, 1978). It has been suggested that VESV and the marine animal isolates are the same, and that marine animals may serve as a natural reservoir for the viruses (Smith *et al.*, 1973; Sawyer, 1976).

II. DESCRIPTION OF THE VIRUSES

The picornaviruses of animals are presently divided into two families: the Picornaviridae, in which there are four groups of equal-status enteroviruses,—cardioviruses, rhinoviruses, and aphthoviruses (FMD viruses)—and the Caliciviridae (Cooper *et al.*, 1978). The Picornaviridae are similar in physical and chemical properties (Table II); the four groups are distinguished by sensitivity to acid, buoyant density of the virion in CsC1, and clinical manifestations of the infected host.

The caliciviruses have a number of features that distinguish them from the other picornaviruses (Table III), features that have led to the recommendation by the Study Group on Picornaviridae of the International Committee on Taxonomy of Viruses that they be considered a new family (Cooper *et al.*, 1978). They differ from the picornaviruses in both structure and genome strategy.

A. Morphology

Electron microscopy of the picornaviruses reveals a smooth, featureless virion 22–30 nm in diameter (Fig. 1A). Capsomers have rarely, if ever, been observed on the picornaviruses, as their compact virons are essentially impermeable to the electrodense salts used as negative stains. The smooth, featureless nature of the viron has led to considerable past and current disagreement about the number of capsomers. From early studies of X-ray diffraction patterns of crystallized polioviruses, it was concluded that the virus had an intrinsic 5:3:2 sym-

TABLE II
Distinguishing Features of the Four Groups of Picornaviruses

Property	Enterovirus	Cardiovirus	Rhinovirus	Aphthovirus
Morphology	Icosahedral	Icosahedral	Icosahedral	Icosahedral
Diameter (nm)	22–28	24–30	24–30	23–25
Density, CsCl (gm/cm^3)	1.33–1.35	1.34	1.38–1.41	1.43
% RNA	29	31	30	31
Host range	Narrow to wide	Wide	Narrow	Wide
Optimal growth temperature (°C)	36–37	36–37	33–34	36–37
Sensitivity				
Ether	Stable	Stable	Stable	Stable
pH 3	Stable	Stable	Labile	Labile
pH 5–6	Stable	Labile in 0.1 *M* halide	Becoming labile	Stable at high ionic strength

TABLE III
Physicochemical Properties of the Picornaviruses and Caliciviruses

Property	Picornaviruses	Caliciviruses
Electron microscopic morphology	Featureless	Distinctive cuplike
Diameter (nm)	20–30	35–40
Calculated number of structural units	60	180 or 120
UV absorbance (E260/E280)	1.65–1.72	1.48–1.52
Density, CsCl (gm/cm^3)	1.33–1.43	1.36–1.39
Sedimentation coefficient (S value)	140–160	170–180
% RNA	29–32	18–22
RNA molecular weight	2.5–2.6 × 10^6	2.6–2.8 × 10^6
Total molecular weight	8–9 × 10^6	15 × 10^6

metry (Finch and Klug, 1959). By extrapolation it was concluded that the virion could be made up of 60 structural units. During the next few years it became clear that X-ray diffraction patterns were insufficient evidence for determining the number of capsomers; however, the dense, featureless structure of the virion made electron microscopic examination of capsomers exceptionally difficult, if not impossible. Mayor (1964), using electron microscopy and negative staining, interpreted the structure of a poliovirus, a rhinovirus, and three echoviruses in terms of a rhombic pattern. The observations were consistent with a polyhedral structure with icosahedral symmetry and 32 capsomers. Reports on other picornaviruses of both humans and animals confirmed the 32-capsomer construction (Breese *et al.*, 1965; Jamison, 1969; Mattson *et al.*, 1969). There are also two reports of 42 capsomers, which would be consistent with the rhombic structure (Hausen *et al.*, 1963; Agrawal, 1966). Now, with the development of better biochemical techniques, numerous reports list the picornaviruses as containing 60 protomers or subunits. Strobbe (1978) has proposed the existence of an icosahedral virion composed either of 20 capsomers of 3 trimers each or of 60 capsomers as individual trimers; clearly, the morphological analysis of the picornaviruses will be reevaluated.

The larger diameter and the distinct morphology of the calicivirus virion, as determined by electron microscopy (Fig. 1B), clearly distinguishes it from other picornaviruses (Zwillenberg and Burki, 1966; Alemida *et al.*, 1968; Wawrzkiewics *et al.*, 1968). The FC, VESV, and SMSV are all approximately 35–40 nm in diameter. The viruses are spherical. They stain in an unusual pattern, which gives them the distinctive, apparently cuplike surface from which they get

1976). The molecular weight of the RNA is generally agreed to be $2.4–2.6 \times 10^6$ which leaves the molecular weight of the protein at 6×10^6. Sedimentation velocity studies indicate that there are few detectable differences among the picornaviruses and that sedimentation coefficients are 140–160 S, depending on the techniques used.

The caliciviruses share with the enteroviruses an inability to withstand temperatures of 50°C for 1 hour. However, unlike the enteroviruses, the caliciviruses are not stabilized by 1.0 *M* $MgCl_2$ against inactivation by heat (Bürki, 1965). The buoyant density of the caliciviruses is 1.36–1.39 gm/cm^3, which is intermediate between the buoyant densities of the enteroviruses and the rhinoviruses. The reported sedimentation coefficient for the caliciviruses is 160–207 S; 170–180 S is the usual accepted range (Burroughs and Brown, 1974; Cooper *et al.*, 1978). The mass of the RNA is similar in the caliciviruses and the picornaviruses, but the caliciviruses contain only about 18–22% RNA (Ogelsby *et al.*, 1971; Burroughs *et al.*, 1978b); the total mass of the caliciviruses consequently is about 15×10^6 daltons, or about 60% larger than that of the picornaviruses.

C. Chemical Properties

Both the picornaviruses and the caliciviruses lack a lipid-containing envelope and, therefore, are resistant to inactivation by ether or chloroform. The enteroviruses are fairly resistant to ordinary laboratory disinfectants; however, sodium hydroxide, sodium hypochloride, and formaldehyde are quite effective against SVDV (Blackwell *et al.*, 1975). Disinfectants whose activity is based on pH changes, such as sodium hydroxide, sodium carbonate, and acetic acid, are highly effective against the rhinoviruses and aphthoviruses. Disinfectants containing halogen, alkali, organic acid, or phenols are effective against the caliciviruses; however, the iodophors and substituted phenolic-type compounds that inactivate SMSV barely affect VESV (Blackwell, 1978a).

There is general agreement that the protein coat of the picornaviruses contains four major structural polypeptides. Each virion contains about 60 copies each of VP_{gly}, VP_{asp}, VP_{thr}, and VP_4, which, in FMD virus (FMDV), have molecular weights of 34, 30, 26, and 8×10^3, respectively (Bachrach, 1977). The virion also contains one or two copies of VP_0 (molecular weight, 40×10^3), which is the precursor of VP_{asp} and VP_4. A small polypeptide attached to the 5′ end of the RNA has been identified in a number of the picornaviruses.

The caliciviruses—VESV, SMSV, and FC—contain one major polypeptide of molecular weight $60–70 \times 10^3$ (Bachrach and Hess, 1973; Burroughs and Brown, 1974; Schaffer and Soergel, 1976; Burroughs *et al.*, 1978b) and possibly a minor polypeptide of molecular weight 15×10^3, which accounts for less than 2% of the total protein (Burroughs and Brown, 1974; Burroughs *et al.*,

1978b). The infectious RNA of VESV has also been shown to contain a polypeptide that is covalently linked to the RNA (Burroughs and Brown, 1978).

D. Antigenic Structure

Tissue culture fluid from cells infected with FMDV or vesicular fluid from FMD lesions contains a number of antigenic components related to the virus capsid and to the viral infection. The major antigens (140 S, 75 S, 12 S, and virus infection-associated, or VIA) are easily detected by immunodiffusion, complement-fixation (CF), or neutralization tests. The 140, 75, and 12 S antigens are composed of specific viral structural components; their polypeptide structures are described below.

The complete virion (140 S antigen), which is generally considered responsible for the immunogenicity of the FMDV, has a sedimentation rate of about 146 S. The virus contains three similar polypetide chains (VP_{gly}, VP_{asp}, VP_{thr},) with molecular weights of 34, 30, and 26 $\times$ 10^3, respectively, and a fourth smaller polypeptide chain (VP_4) with a molecular weight of 8 $\times$ 10^3 (Bachrach, 1977). A virion may also contain one to two chains of a fifth polypeptide, VP_0, which is the uncleaved precursor of VP_{asp} and VP_4.

The actual immunogenicity of the FMD virion and the attachment of the virus to susceptible cells are due to a trypsin-sensitive polypeptide. The electrophoresis methods used in different laboratories produce different migration patterns; the trypsin-sensitive polypeptide is now identified variously in the literature as VP_1 (Burroughs *et al.*, 1971; La Porte and Lenoir, 1973; Strohmaier and Adam, 1974), VP_2 (Vande Woude *et al.*, 1972), and VP_3 (Bachrach *et al.*, 1975). Strohmaier *et al.* (1978) recommended the adoption of "VP_{thr}" to designate the trypsin-sensitive polypeptide, since it is the only polypeptide with a threonine at the N terminal. To conform to the terminology based on the N terminal amino acid, I use "VP_{gly}," "VP_{asp}," and "VP_{thr}" for "VP_1," "VP_2," and "VP_3," respectively, in this chapter. The VP_{thr} polypeptide chain on the complete virion is cleaved by trypsin into $VP_{thr\text{-}a}$ and $VP_{thr\text{-}b}$; however, the virus retains both segments of the polypeptide, with no change in the sedimentation rate. The cleavage of VP_{thr} alters the antigenicity of the FMD virus, but the effect of cleavage on infectivity depends on the virus type (Bachrach, 1977).

The 75 S antigen is the empty capsid. It occurs naturally in some virus strains and may be produced artificially by treatment of the 140 S virion with EDTA (Rowlands *et al.*, 1957a; Rweyemamu *et al.*, 1979). The 75 S particles produced by these methods are distinctively different. The naturally occurring empty particle contains all the polypeptides present in the intact virion; however, it does not contain RNA, and two of the polypeptides (VP_{asp} and VP_4) are present as uncleaved VP_0. Immunologically, the naturally occurring 75 S antigen is similar to the intact virion. The artificially produced 75 S antigen has a full complement of

VP_{gly}, VP_{asp}, and VP_{thr}, and possibly a small amount of RNA, but does not contain VP_4. The serological properties of the artificially produced 75 S antigen are similar to those of the 12 S subunit, which will be discussed next.

The 12 S antigen is a degradation product of the 140 or 75 S antigen. It can be produced readily by dialysis of the virion against low-ionic-strength buffers at an acidic pH. The 12 S antigen is composed of trimers of the VP_{gly}, VP_{asp}, and VP_{thr} subunits. It contains no VP_4 or RNA. The 12 S antigen is reactive in both immunodiffusion and CF tests. It usually is not considered a significant immunogen; however, it elicits the production of small amounts of neutralizing antibody because it contains VP_{thr} (Cowan, 1973).

Group reactivity with the FMD virus has been associated with two antigenic sites that are exposed following disruption of the virion. One antigenic site, located on the 12 S subunit, gives lines of identity for homotype and heterotype antisera in the immunodiffusion test and fixes complement with both homotype and heterotype antisera (Van Oss *et al.*, 1964; Cowan and Trautman, 1967; Talbot *et al.*, 1973). A second antigenic site, located on the VP_4 polypeptide (the 12 S subunit does not contain VP_4), also exhibits group reactivity, reacting with both homotype and heterotype antigens in CF tests (Talbot *et al.*, 1973).

The VIA antigen appears to be a nonstructural viral antigen that occurs in tissues infected with FMDV. Cowan and Graves (1966) were unable to demonstrate the antigen as a viral component or to demonstrate the antibody in animals immunized with purified, inactivated vaccines. Antibody to the antigen was present in sera of infected cattle; therefore the antigen was thought to be VIA (Cowan and Graves, 1966). The VIA antigen has been found to be specific for FMDV infection but not for virus type.

The location of the VIA antigen has been the subject of a number of subsequent studies. Rowlands *et al.* (1969) suggested that VIA is an internal antigen of the virus. However, studies using electrophoresis on acrylamide gels did not demonstrate a viral polypeptide corresponding to the 57,000 molecular weight of the VIA antigen (Vande Woude *et al.*, 1972). In a review, Bachrach (1977) summarized further evidence: first, that a noncapsid viral protein ($NCVP_5$) with a molecular weight of 57,000 may be identical to the VIA antigen, and second, that $NCVP_5$ is a virus-specified component of FMDV RNA polymerase. Recent vaccine studies have demonstrated that vaccines inactivated by formaldehyde and bivalent acetylethyleneimine elicit a VIA response with repeated vaccination (Dawe and Pinto, 1978; Pinto and Garland, 1979). However, these studies do not preclude the possibility that VIA antigen is in the vaccine as a contaminant rather than as an antigen present in the virus.

There are seven immunologically and serologically distinct types of FMD virus (Bruner and Gillespie, 1973; Callis *et al.*, 1975). The first three types (A, O, and C), which were originally isolated in France and Germany, are known as European types; they have the widest distribution. Three types isolated in the

South African territories are designated "SAT I," "SAT II," and "SAT III." The seventh type was isolated from various parts of Asia and is designated "$Asia_1$." At least 61 subtypes are included under the 7 types—29 type A subtypes; 10 type 0; 5 type C; 7 type SAT I; 3 type SAT II, 4 type SAT III; and 3 type $Asia_1$. The subtype designation is based on CF tests by the FMDV World Reference Laboratories at the Animal Virus Research Institute at Pirbright, Surrey, England, and the Pan American FMD Center at Rio de Janeiro, Brazil (Davie, 1964; Federer *et al.*, 1964; Cottral, 1972).

The enteroviruses, rhinoviruses, and cardioviruses are similar to the aphthoviruses in that they all contain three polypeptides of similar size (VP_1, VP_2, VP_3), a small polypeptide (VP_4), and traces of VP_0 (the precursor of VP_2 and VP_4). Researchers have disagreed about whether all the picornaviruses contain a full complement of VP_4. They agree that some picornaviruses have a full complement of VP_4 chains, and Rueckert (1976) suggests that preparations without this full complement have been damaged or improperly counted.

The antigenic structure of the enteroviruses is described by a different terminology than that used for the FMD viruses. The D (dense) or N (native) antigen refers to the complete virion, and the C (coreless) or H (heated) antigen is the empty capsid (Rueckert, 1976). Again, as with the FMD viruses, the empty capsid of the enteroviruses can occur naturally or be produced from the intact virion. The C antigen of the enterovirus can easily be produced by exposure of the intact virion to heat at 56°C, to ultraviolet light, or to a high pH. The artificially produced C particle contains VP_1, VP_2, and VP_3, but not VP_4. Usually the artificially produced C antigen does not contain RNA; but some preparations contain a significant amount. The C antigen, with or without RNA, has lost the ability to attach to cells.

Breindl (1971) proposed that the VP_4 polypeptide confers the D antigenicity on the full virus particle, i.e., that it is responsible for immunogenicity and cellular attachment in the enterovirus, whereas VP_{thr} is the responsible polypeptide in FMD virus. The naturally occurring C antigen contains all the polypeptide present in the D particle, but VP_0 is not cleaved into VP_2 and VP_4 and RNA is not present. The naturally occurring enterovirus C antigen, unlike the naturally occurring (coreless) FMD antigen, usually does not have the ability to attach to cells. Both the naturally occurring and the artificially produced enterovirus C antigen have group reactivity in the immunodiffusion and CF tests, whereas the D particle is virus specific and responsible for immunogenicity.

Early serological classifications of the bovine enteroviruses have been extensively reviewed (Barya *et al.*, 1967; Dunne *et al.*, 1974a). Serological classifications have been limited to strains isolated by the individual researcher plus those isolated by one or two other workers. As a result, over 63 possible serotypes have been listed (Barya *et al.*, 1967), but there has been no comprehensive serological study involving all the isolates. The most extensive attempt at classification was that of Dunne *et al.* (1974a), who listed eight distinct serotypes from the states of Washington and Pennsylvania. Some of these serotypes are currently recognized by the WHO/FAO Committee on Comparative Virology and are included in the classification of the Western Hemisphere Committee on Animal Virus Characterization (1975).

The procine enteroviruses have been the subject of numerous attempts at classification, and the early classifications have been reviewed (Dunne *et al.*, 1975a). The current classification of the Western Hemisphere Committee on Animal Virus Characterization (1975) lists nine serotypes. The most important of these are type 1, which includes Teschen virus, and type 9, which is SVDV.

The rhinoviruses have antigenic properties similar to those of the FMD viruses in that the naturally occurring empty particle has D as well as C antigenicity and has a limited ability to attach to susceptible cells (Lonberg-Holm and Yin, 1973; Noble and Lonberg-Holm, 1973). The only true rhinovirus known in animals is the bovine rhinovirus (Rueckert, 1976), of which fewer than 20 isolates have been reported. Serological comparison of isolates has been limited by the difficulty of producing specific antiserum and of growing the virus in tissue culture. Only two serotypes are currently recognized by the WHO/FAO Committee on Comparative Virology. More serotypes undoubtedly will be identified as methods of isolation and antiserum production are improved.

The division of caliciviruses (VESV, SMSV, and FC) into three groups is based on the source of the original isolation—swine, sea lions, or felines. The exact number of serotypes of VESV is not known; a number of early isolates were originally thought to be FMDV and were destroyed. There are now 13 recognized, immunologically distinct serotypes of VESV based on serological and cross-immunity tests in swine (Madin *et al.*, 1976).

The SMSV is now receiving considerable attention, both as a relatively newly recognized group of viruses from sea mammals which can infect swine and mink (Smith *et al.*, 1973; Wilder and Dardir, 1978) and because of its similarity to VESV. Eight serotypes have been identified (Smith *et al.*, 1977; Schaffer, cited in Studdert, 1978). Recent studies have demonstrated antigenic similarities among most strains of VESV and SMSV by immunodiffusion, immunoelectron microscopy, and radioimmune precipitation (Burroughs *et al.*, 1978a; Smith *et al.*, 1978; Sorergel *et al.*, 1978). Common antigenicity was demonstrated only between the FC strain F9 and the swine virus strains VESV-D53-G55 by immunoelectron microscopy (Smith *et al.*, 1978).

The FC viruses have a common complement-fixing antigen (Hersey and Maurer, 1961; Kamizono *et al.*, 1968; Tan, 1970) and show considerable cross reactivity in the immunofluorescence test (Gillespie *et al.*, 1971). Early neutralization studies performed with small numbers of viruses demonstrated considerable serological differences among isolates. However, later tests with 46 isolates (Povey, 1974) demonstrated a significant amount of cross reaction among the isolates when cat or goat antisera were used. From these results Povey postulated that all isolates can be regarded as serological variants of a single serotype. Cross-protection studies with cats also showed significant cross relationships among the FCs (Povey and Ingersoll, 1975). The serological differences noted in the earlier studies can be attributed to the antisera used, antisera produced in various species in cross reactivity (Povey, 1974). Olsen *et al.* (1974) also found increased serological cross reactions among strains of FC when he used sera from cats in the convalescent phase of the calicivirus infection.

III. HOST RANGE

A. Aphthoviruses (Foot-and-Mouth Disease)

Natural infections with FMD virus are limited primarily to the cloven-hoofed animals. The domestic animals most often infected are cattle, sheep, swine, and goats. Usually an outbreak on a given farm involves all susceptible species, but occasionally the virus becomes highly adapted to one species and its pathogenicity to other species is reduced. Outbreaks of FMD in swine have been reported in which cattle in close contact with infected swine were not infected. It was difficult to experimentally infect cattle with virus isolated from such outbreaks (Brooksby, 1950; Van Bekkum, 1969). Experimental infection of guinea pigs produces a disease similar to that seen in cattle. However, the virus is not usually spread by contact in guinea pigs, and it may need to be adapted to them by several passages (Gins and Krause, 1924). Suckling mice are extensively used in research because they are highly susceptible to inoculation by most routes (Heatly *et al.*, 1960). The pathogenicity in adult mice is dependent on the strain of virus; some strains produce death (Subak-Sharpe, 1961; Campbell, 1963). Hamsters are highly susceptible to plantar inoculation, which often produces a severe or even fatal infection (Komarov, 1954).

The disease has been reported in a number of wild cloven-hoofed animals; wild animals may be an important reservoir of the virus in some areas (Callis *et al.*, 1975). If infection is indicated by isolation of the virus or detection of the antibody, the virus can be said to infect most animals, with the exception of solipeds (Bruner and Gillespie, 1973; Callis *et al.*, 1975). Human infections are rare and usually mild, even though man is often in close contact with infected animals.

B. Rhinoviruses

Cattle are the only domestic animals from which typical rhinoviruses have been isolated. Rhinoviruses in other species probably have been overlooked because of their low virulence and extreme difficulty of isolation. The bovine rhinoviruses are highly specific for cattle and have a strict affinity for the respiratory tract. These viruses have been isolated only from the upper respiratory tract in natural infections, but pneumonic lung lesions have been reported in experimentally inoculated gnotobiotic and normal calves (Mohanty *et al.*, 1969; Betts *et al.*, 1971).

Although the equine rhinoviruses share many physicochemical properties with the other rhinoviruses, they are atypical in their wide host range. Rabbits, guinea pigs, and monkeys are susceptible to infection with these viruses, if susceptibility is indicated by the production of antibody (Plummer, 1963). Man is also susceptible; antibody surveys demonstrate that up to 25% of stable workers have high serum titers to equine rhinoviruses (Plummer, 1962). Experimental inoculation of a volunteer produced a high fever, a transient viremia that coincided with the fever, and severe pharyngitis (Plummer, 1963).

C. Cardioviruses

The cardioviruses have a wide host range and usually produce an inapparent or mild disease. Antibodies to these viruses have been detected in most animal species and in humans. The isolates Columbia SK, MM, EMC, ME, and Mengo are serologically indistinguishable but have different names because of their isolation histories (Warren, 1965).

The viruses are highly infectious for and can produce a fatal disease in mice, hamsters, and cotton rats. Rodents probably serve as the reservoir for these viruses in nature (Rueckert, 1976; Cooper *et al.*, 1978). The cardioviruses also multiply readily in most laboratory animals, usually without evidence of disease. Natural outbreaks causing extensive losses in swine have been reported (Murnane *et al.*, 1960; Gainer *et al.*, 1968). In humans the natural infection occurs as a sudden, 3-day febrile disease with headache and a stiff neck (Smadel and Warren, 1947).

D. Enteroviruses

Enteroviruses are routinely isolated from mice, fowl, cattle, swine, and monkeys (Andrews and Pereira, 1972). The number of serotypes from each species is not known because classification of the viruses is incomplete. Isolations of enteroviruses from horses, sheep, dogs, and cats have been reported (Andrews and Pereira, 1972), but it is not known whether these viruses are indigenous to the species. The enteroviruses are generally believed to be host specific; however, many animal species have an antibody that neutralizes enteroviruses of another species (Jastrzebski, 1961; McFerran, 1962b). These cross reactions have been considered nonspecific (Dunne, 1975a), although cross reactions between enteroviruses of different species have not been thoroughly studied. A swine enterovirus isolate (Dunne, 1975a) and a bovine enterovirus isolate (Moll, 1964) have caused fetal deaths and abortions.

SVDV is of special interest because it is serologically indistinguishable from the human coxsackievirus B5 (Graves, 1973; Brown and Wild, 1974). Laboratory personnel in contact with SVDV-infected pigs or with SVDV alone have become infected (Brown *et al.*, 1976), with symptoms similar to those produced in humans by the coxsackie group of enteroviruses. SVDV produces a disease in newborn mice that is characterized by nervous signs, progressive paralysis, and death (Nardelli *et al.*, 1968). Mink have been infected by experimental inoculation (Sahu and Dardiri, 1979). Cattle, donkeys, rabbits, guinea pigs, hamsters, and chickens are refractory to the virus (Nardelli *et al.*, 1968; Mowat *et al.*, 1972; Dawe *et al.*, 1973).

Avian encephalomelitis (epidemic tremor) and duck virus hepatitis are two enteroviruses of economic importance, as they can produce extensive losses in chickens and ducks, respectively. The reader is referred to the extensive description of the viruses and the diseases which they produce in the reviews by Luginbuhl and Helmboldt (1978) for avian encephalomelitis and by Levine (1978) for duck hepatitis.

E. Caliciviruses

The caliciviruses all have a limited host range and under natural conditions are restricted to one or a few closely related species. The host range has been studied most extensively for VESV because of the clinical similarity of VES to FMD. Only a limited disease, usually a vesicle or vesicles restricted to the site of inoculation, has been produced in horses, hamsters, and dogs (Madin, 1975). SMSV, which has been isolated from naturally occurring cases in the California sea lion and northern fur seal, has been reported to infect swine, monkeys, mink, and possibly man (Smith *et al.*, 1977, 1978; Wilder and Dardiri, 1978). The relationship between VESV and SMSV is of interest at both the clinical and immunological levels because these may actually be the same virus. SMSV is enzootic in the two species of marine mammals and produces lesions in swine that are indistinguishable from those produced by VESV (Smith *et al.*, 1973). It has been proposed that the original outbreaks of VES in swine were caused by a marine virus, which was introduced into swine through infected garbage (Sawyer, 1976; Sawyer *et al.*, 1978).

IV. PATHOGENESIS AND CLINICAL SIGNS

The picornaviruses are very adaptive viruses; generally they develop a perfect host-parasite relationship—one in which the virus is maintained by the host and the host undergoes little morbidity or mortality. Carrying this relationship further, a number of picornaviruses may persist in a carrier state in seropositive animals, where they are a threat to susceptible animals. Evidence also exists that the viruses can remain dormant for extended periods before producing clinical manifestations upon stress or other unknown conditions (Graves *et al.*, 1971; Andersen, 1978b).

Each group of picornaviruses has a classic site of infection in the host. The enteroviruses infect the gastrointestinal tract, producing mild to severe diarrhea. Rhinoviruses infect the upper respiratory tract, producing rhinitis, and the cardioviruses infect the central nervous system (CNS) and heart, often producing a fatal disease in rodents. The aphthoviruses or FMD viruses produce a general infection, with vesicular lesions primarily around the hoofs and mouth. The pH and temperature ranges of each virus tend to limit it to a specific organ system; however, in studying these viruses, it is soon apparent that individual strains especially of enteroviruses can become adapted to other organ systems, often producing a severe or fatal disease. Teschen and SVDV can produce fatal CNS involvement; swine enteroviruses can produce reproductive complications and fetal death. Despite the challenging juxtaposition of ubiquity and the potential for infection of uncharacteristic tissues, little is known about the involvement of the picornaviruses in disease. This is because picornaviruses are routinely isolated from both normal and morbid animals, both with and without other viruses. The role of the apparently avirulent picornaviruses in causing or eliciting disease, by themselves or as contributing agents, needs review and further study.

A. Foot-and-Mouth Disease Viruses

FMDV can produce highly variable types of disease, depending on the method of inoculation, the strain of virus, and the general health of the host. The typical textbook picture of FMD in cattle is of a highly contagious disease characterized by depression, fever, drooling, and vesicles (Bruner and Gillespie, 1973). The vesicles soon rupture, producing large, denuded, ulcerative lesions (Figs. 2, 3, 4). The mortality from FMD is usually low, but the disease is extremely important because of morbidity losses. Affected animals often become nonproductive for long periods; extreme soreness of the mouth and feet severely reduces both meat and milk production.

The virus is spread by contact by contaminated animal products, by mechanical transfer on humans, by wild animals and birds, by vehicles and fomites, and by the airborne route (Hyslop, 1970). The virus was first believed to enter the animal by virus-contaminated material puncturing the epithelium of the mouth or feet; in early laboratory studies, therefore, the virus was inoculated intradermal-lingually. Intradermal–lingual inoculation produces a primary vesicular reaction at the site of inoculation (Graves *et al.*, 1971; McVicar, 1977). A viremia then

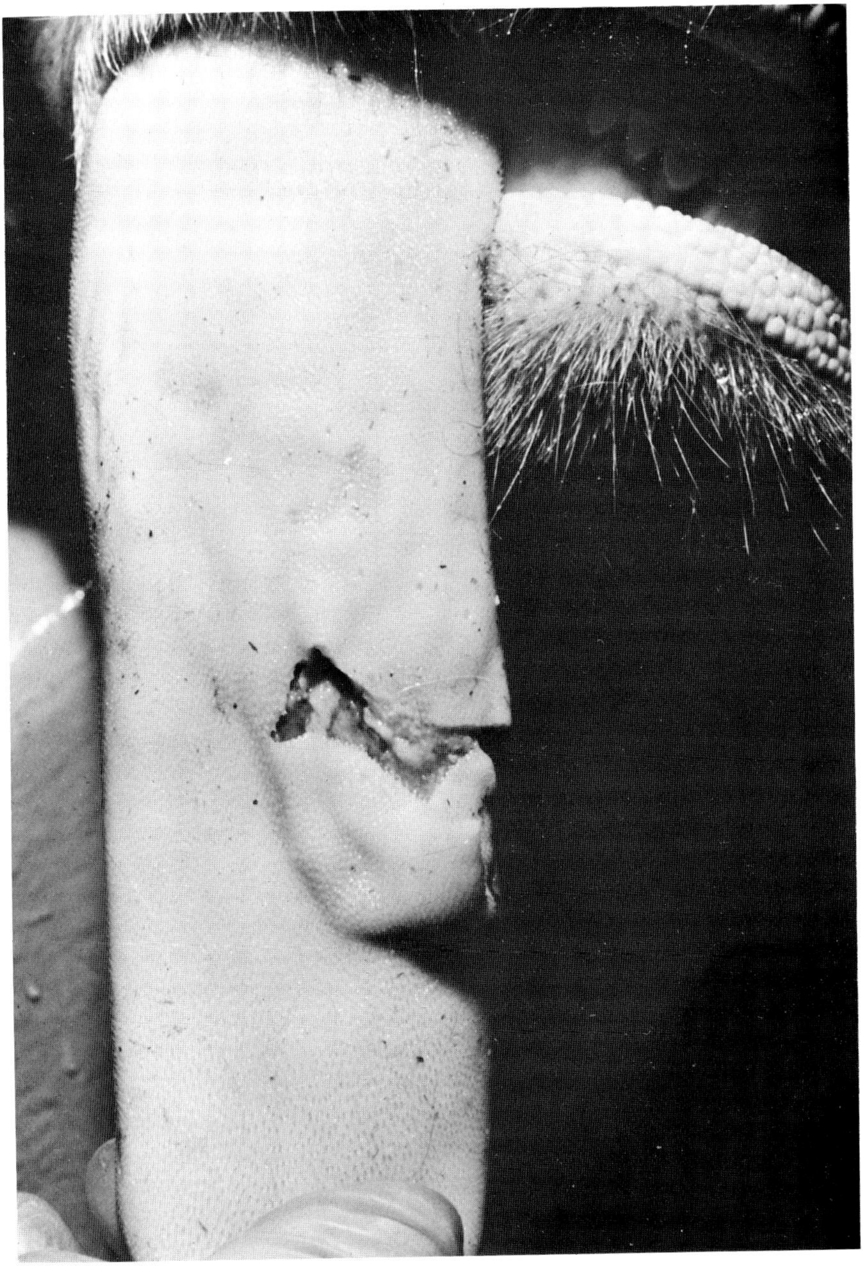

Fig. 2. Foot-and-mouth disease, early tongue lesions in the bovine. (Courtesy of Dr. D. A. Gregg.)

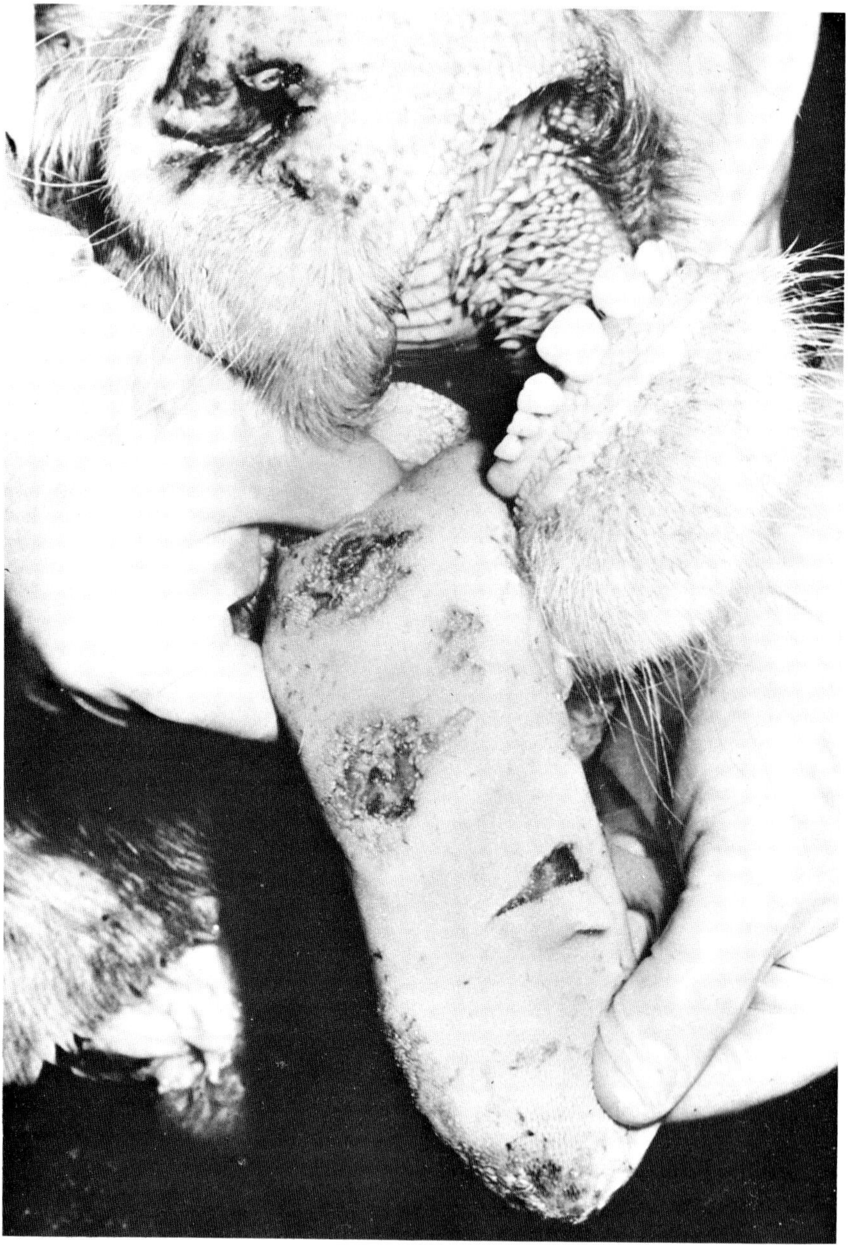

Fig. 3. Foot-and-mouth disease, late tongue lesions in the bovine. (Courtesy of Dr. D. A. Gregg.)

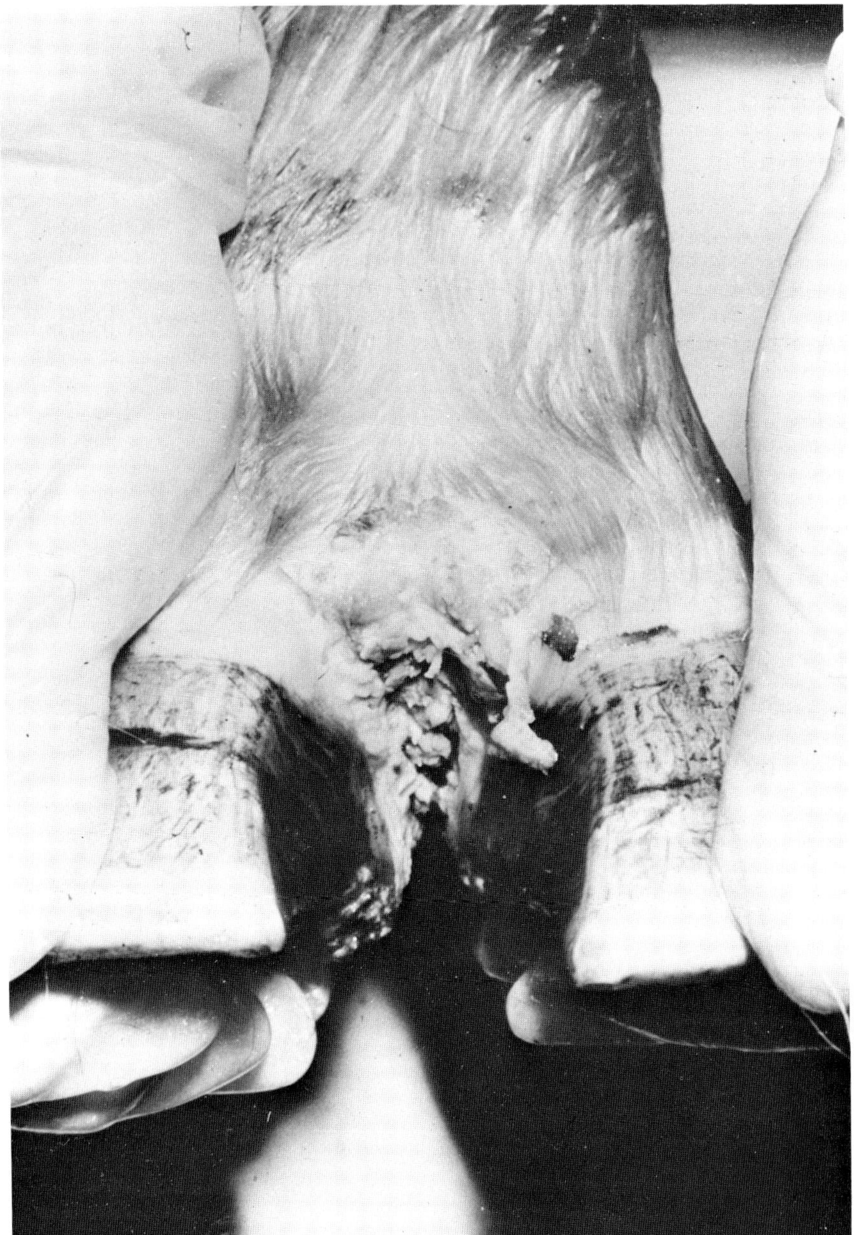

Fig. 4. Foot-and-mouth disease, late hoof lesion in the bovine. (Courtesy of Dr. D. A. Gregg.)

develops with subsequent spread and production of vesicles at susceptible secondary sites. However, it was soon observed that the first sign of infection after contact with infected animals was an upper respiratory infection, with growth of the virus in cells of the pharyngeal area. The viremia that soon developed then spread the virus to the secondary sites, including the muscosa of the tongue and buccal cavity, where vesicles are often seen.

FMDV is now generally considered to be spread primarily as an aerosol and to enter the animal through the respiratory tract (McVicar, 1977). Animals in the prodromal period and during frank infection have high titers of virus in most body secretions, thus supplying large amounts of virus for aerosolization. Large quantities of virus have been detected in the air both before and after clinical signs were observed (Hyslop, 1965b; Sellers and Parker, 1969; Sellers, 1971).

A recent study with susceptible, vaccinated, and recovered pigs, cattle, and sheep, all of which were exposed to FMDV by aerosol, demonstrated two periods of virus aerosolization (Sellers *et al.*, 1977). The first period, which occurred with all animals, was from 30 minutes to 22 hours after exposure and was attributed to virus trapped on the animal during exposure. The second period was from 2 to 7 days after exposure in susceptible animals, vaccinated animals that developed clinical disease, and a number of vaccinated and recovered animals that did not develop clinical disease. The airborne virus found during the second period was believed to have resulted from limited multiplication in the respiratory tract.

The site at which the virus enters the respiratory tract walls and produces the primary infection may depend on the level of aerosol exposure. It has been suggested that 90% of the airborne virus in the vicinity of the infected animal is carried in particles of the right size to lodge in the upper respiratory tract or bronchus (Sellers and Parker, 1969). The remaining 10% of the airborne virus is contained in droplets small enough to penetrate to the alveoli, where the virus can readily pass into the bloodstream.

Experimental data indicate that the virus rapidly enters the blood and is disseminated to the upper digestive tract. The patterns of virus isolation following intravenous or contact exposure are strikingly similar in that the virus is first recovered from oral swab samples and not from tissue samples (McVicar, 1977). Tracheal intubation, permitting direct, controlled aerosol exposure of the lungs, and intramammary infusion give similar results, with virus rapidly appearing in the oral fluids (Burrows *et al.*, 1971; Sutmoller and McVicar, 1976). Virus is soon detected in tissue samples, mostly from epithelial predilection sites and lymphoid tissues. Virus is isolated more frequently from the upper digestive tract than from the respiratory tract.

By dissemination through the blood, FMDV reaches all parts of the body and multiplies in a number of locations. The mammary gland is an important multiplication site; virus appears in the milk before clinical signs develop (Burrows *et al.*, 1971). High levels of virus also have been recovered from, and virus is believed to multiply in, the skin, pituitary gland, pancreas, and muscle tissue, especially the heart (Potel, 1958; Barboni and Manocchio, 1962; Scott *et al.*, 1965; Gailiunas and Cottral, 1966). Virus has also been isolated from the lymph nodes and kidneys (Hess *et al.*, 1960; Cottral *et al.*, 1963).

In cattle, FMD is characterized by fever, depression, drooling, lameness, and the appearance of vesicles filled with clear serous fluid. The vesicles are located primarily in the mucous membranes of the mouth, tongue, cheeks, gums, and dental pad and in the skin of the muzzle, the interdigital space, and around the top of the hoof. Vesicles can occur on the teats and occasionally on the surface of the udder, and may be seen around the base of the horn and in the pharynx, larynx, trachea, esophagus, and wall of the rumen (Bruner and Gillespie, 1973).

Seibold (1963) described the pathological changes in the epithelium of the tongue and feet. The lesions are characterized by intercellular edema and necrosis of the epithelial cells in the stratum spinosum. A circumscribed, slightly elevated, blanched area termed the "initial lesion" develops in the lingual mucosa. The initial lesion can proceed to separation of the mucosa from the underlying tissue, forming a full-blown vesicle. If the initial lesion does not form a vesicle, the edematous fluid probably seeps through cracks in the stratum corneum, producing characteristic discolored, desiccating necrotic mucosa. Failure of the mucosa to separate and form a vesicle is attributed to the firm attachment of the thick bovine lingual mucosa by numerous well-developed conical papillae. Lesions in the interdigital skin are similar in initial development, but failure to vesiculate is rare. The large size of vesicles on the hoofs is attributed to the stress and motion to which the interdigital skin is subjected and to the ease with which the edematous skin is detached from the dermal papillae.

The vesicles in the mouth rupture within a few hours after they develop, leaving large flaps of whitish, detached epithelium over raw, bleeding surfaces. Often large areas of the tongue are denuded. Tongue and mouth lesions heal promptly with little or no residual effect, and the soreness usually disappears within a week. Denuded areas between the claws usually become secondarily infected with bacteria, causing necrosis of tissue and suppuration that frequently undermine the claws. The claws are often lost, producing permanent deformities of the hoof and lameness. Secondary bacterial infections also often cause pneumonia or severe mastitis, adding to the long-term unproductiveness of the animal.

Both gross and microscopic degenerative lesions have been reported in muscle tissue, particularly in the heart (Potel, 1958). The degenerative lesions are characterized by yellowish streaks and foci of parenchymatous degeneration. Severe heart lesions are seen most often in young calves and pigs; these lesions may account for the higher mortality in young stock. Some strains of FMDV produced greater heart damage with increased mortality in both adults and young. In this malignant form of FMD, deaths occur early in the course of the infection and are attributed to the specific action of the virus.

The pathogenesis in the endocrine system has received some attention because of the potential for using the effect of FMD on the pancreas as a model for the study of diabetes. Barboni and Manocchio (1962) hypothesized that the growth of the virus in the pancreas of cattle leads to the disappearance of β-cells and the development of a diabetes-like syndrome. Bhalla and Sharma (1968) also observed in goats the marked degeneration of the pancreas, including the islets of Langerhans, following FMD. The pancreas is regularly involved in FMD in adult guinea pigs (Platt, 1958) and in young and adult mice (Platt, 1956; Seibold, 1960). In the pregnant mouse, the levels of virus were higher and persisted for a day longer in the pancreas than in other tissues, making the pancreas a reliable indicator of viral infection (Andersen and Campbell, 1976). Other glands of the endocrine system have been implicated in aspects of the symptomatology of FMD. Bhalla and Sharma (1967) considered the panting and the decrease in milk yield in cattle with FMD to be aftereffects attributable to degenerative lesions in the thyroid and adrenal glands.

Abortion is a recognized complication of FMD (Graham, 1959; Andrievskii and Redkin 1967; Hyslop, 1970; Callis *et al.*, 1975); however, little research has been reported on the effects of the virus on reproduction (Andersen and Campbell, 1976). Andrievskii and Redkin (1967), in a survey of reproductive problems in two large herds of cattle after enzootics of FMD, found high frequencies of abortion, stillbirth, placental retention, and associated reproductive problems. These authors also reported field observations of newborn calves with FMD lesions. In a study on the placental transfer of FMD virus in mice, the virus was detected in the maternal placenta for 6 days but was rarely isolated from the fetus (Andersen and Campbell, 1976). Mortality in the young increased only when the dam was infected late in gestation, and this increase was associated with increased morbidity in the dams. It was concluded that the placenta is an active site of infection with FMDV in pregnant mice, but that the fetus is relatively resistant to infection. Experiments with pregnant cattle are necessary to determine the precise role of FMD in reproductive losses.

B. Bovine Enteroviruses

Few if any enteroviruses have been implicated in the major bovine diseases. Consequently research on these viruses, especially on their method of transmission and their pathogenesis, is limited. The viruses have been associated with and isolated from animals with respiratory, enteric, and reproductive system signs and lesions. They are commonly isolated from pharyngeal washings and feces and have been isolated from nasal washings, vaginal mucosa, semen, and aborted fetuses. The unusually wide distribution of the viruses in various secretions and their stability permit transmission by most known routes, including contact, aerosol, contaminated feed and equipment, and breeding. The method and site of entry of virus into the host, and perhaps the disease manifestations as well, depend upon the method of transmission.

Enteroviruses, classically considered to be pathogens of the gastrointestinal tract, appear to have only limited significance in the production of enteric disease of cattle. The viruses are often isolated from feces of both normal cattle and cattle with diarrhea. Researchers have attempted to reproduce enteric signs by experimentally infecting calves and adult cattle, with mixed results; most animals showed only mild or no clinical signs (McFerran, 1962a; Niederman *et al.*, 1963; Afshar *et al.*, 1964; Van Der Maaten and Packer 1967; Dunne *et el.*, 1974a).

Attempts to isolate enteroviruses from cattle with winter dysentery have yielded a limited number of isolates; however, only one, from France, has been reported to reproduce the disease (Charton *et al.*, 1963). Virus from the eighth tissue culture passage of the isolate was injected intravenously into two heifers, which developed a painful catarrhal stomatitis and diarrhea on days 4 to 6. Recovery was complete by day 9 postinfection. However, animal inoculation and a serological survey in the United States failed to link the enteroviruses with winter dysentery (Andersen and Scott, 1976).

The most critical age for infection with enteroviruses may be the first week of life. Dunne *et al.*, (1974b) experimentally inoculated 10 colostrum-fed and 9 colostrum-deprived calves at ages 1 day to 8 weeks. Experimental inoculation was associated with the development of fever (13/19), diarrhea (8/19), and leukopenia (6/19). Four of the calves died following infection, but the virus may only have been a contributing factor, since three of the calves were colostrum deprived and the death of the fourth was attributed to a septicemia. In a study by Van Der Maaten and Packer (1967), two of four inoculated calves developed diarrhea, which indicates that the agent is involved in enteric diseases of young calves. However, the virus was difficult to reisolate from the calves with diarrhea, which indicates that the agent is involved in enteric diseases of young calves. However, the virus was difficult to reisolate from the calves with diarrhea, except in fecal samples collected during the first few days of illness. McFerran (1962a) inoculated six calves orally with enterovirus strain VG/5/27, producing in three a marked diarrhea that began within 60 hours of infection and lasted for about 36 hours. However, the animals appeared normal, gave no other clinical response, and did not even stop feeding. The results of these experimental inoculations indicate that enteroviruses, if involved in epizootics of calf diarrhea under field conditions, probably are initial invaders to be accompanied later by a secondary infection.

The involvement of the enteroviruses in bovine respiratory infections has been explored, with mixed results. There have been numerous isolations from cattle with clinical signs of respiratory tract disease (Moll and Ulrich, 1963; Huck and Cartwright, 1964; Moll, 1964; Woods *et al.*, 1970; Mattson and Reed, 1974; Kurogi *et al.*, 1976). Moll and Davis (1959) isolated enteroviruses from cattle in several herds with respiratory disease. A rise in antibody to the enterovirus was demonstrated in the cattle—evidence that supports the relationship of the enterovirus to the respiratory syndrome. Dunne *et al.* (1974a) inoculated a number of enterovirus isolates into calves and produced pleuritis and pneumonia in a calf that died following infection. Knowledge of the true role of enteroviruses in respiratory tract infections requires a study of pathogenesis in gnotobiotic calves, because mixed viral infections are common and the enteroviruses may act synergistically with other disease-producing agents.

The bovine enteroviruses may affect the cattle industry most significantly with regard to reproduction in both the male and the female. Enteroviruses appear to be closely associated with disease syndrome of infertility, abortion, stillbirth, and weakness and death of calves shortly after birth. Enteroviruses have been isolated from the semen (Weldon *et al.*, 1978), vaginal mucus (Afshar *et al.*, 1964; Straub and Bohm 1964; Straub, 1965), placenta (Huck and Cartwright, 1964) and fetus (Moll and Finalayson, 1957; Huck and Cartwright, 1964). In a serological survey, Dunne *et al.* (1973) detected antibody to enteroviruses in 41% of aborted and stillborn fetuses; in 13% of the fetuses, enterovirus antibody was the only antibody detected. The researchers concluded that the antibody was of fetal origin, for a number of reasons: the antibody was predominantly IgM; the fetal tissues had antibody-containing cells; the dam's serum contained antibodies to more viruses than did the fetal serum; occasionally, the fetal serum had higher titers than the dam's serum; and incidence of antibody to bovine viral diarrhea and infectious bovine rhinotracheitis was lower in the fetuses than in the dams. Also, fetuses had no antibody when abortion occurred before the antibody-producing system of the fetus was functional. Huck and Cartwright (1964) reported the isolation primarily of enterovirus strain F266a from fetuses and from herds with reproductive problems. The F266a strain was determined by serological neutralization to be similar to the G-VP virus, which was originally isolated from the genitalia of bulls. Dunne *et al.* (1973, 1974a) found that antibodies in the fetus were predominantly to the virus strain BES 6, with BES 2 the second most common strain. The serological relationship of F266a to BES 6 and BES 2 has not been determined.

During the 1950s at least two strains of enteroviruses were involved in outbreaks of catarrhal vaginitis in cattle (McKercher and Kendrick, 1958; Straub and Bohm 1964; McKercher, 1979). The vaginitis was mild to severe, with occasional cervical involvement; it was characterized by the accumulation of a heavy, tenacious, mucoid exudate on the floor of the vagina. The incubation period was 1–2 days and the disease course was 7–10 days. The only known effect on reproduction was the delay of conception when estrus occurred during the acute phase. The viruses were transmitted by coitus; however, occurrence of the disease in calves and virgin heifers indicated that it could also be transmitted by other methods. Other enteroviruses have been isolated from the vaginal mucosa of normal cattle and cattle with reproductive problems (Afshar *et al.*, 1964; Straub, 1965; Odegaard and Straub, 1970). Attempts at reproducing clinical vaginitis by vaginal inoculation have resulted only in mild disorders.

The limited studies on the pathogenicity of enteroviruses for the fetus have involved primarily animal model systems. Enteroviruses are known to cause stunting and death of chicken embryos. Changes in livers of the embryos are readily identifiable as lesions caused by the virus (Kunin and Minuse, 1958; Dunne *et al.*, 1974a). Intracardial injection of pregnant guinea pigs with bovine enteroviruses caused abortion and stillbirths (Moll, 1964; Van Der Maaten and Packer, 1967).

The placenta may be the primary site of infection for the bovine enteroviruses. Moll (1964) isolated the virus from more placentas than fetuses from inoculated guinea pigs. Limited studies with bovine enteroviruses in pregnant mice, as with FMDV, have supported the belief that placental infection is important; viruses were rarely recovered from fetuses but were recovered readily and in high titers from the placenta (Andersen and Campbell, 1976a; A. A. Andersen, unpublished data). Afshar *et al.* (1964) also suggested that the F46 strain of bovine enterovirus may have an affinity for the placenta, because pregnant cattle developed higher serum antibody titers than nonpregnant animals. The F46 virus, originally isolated from the vaginal mucosa, did not produce abortion.

In the bull, infertility, decreased libido, and testicular lesions have been associated with enterovirus infections of the genital tract. Enteroviruses have been isolated from testicles of bulls with orchitis (Bögel *et al.*, 1963; Bouters, 1963). Reinoculation of the enterovirus into the prepuce of young bulls decreased semen quality and caused histological abnormalities in the genital epithelium (Bouters, 1964; Bouters and Vandeplassche, 1964). A bovine enterovirus serologically classified as BES 1 was isolated from both semen and fecal samples of a bull with orchitis, testicular degeneration, aspermatogenesis, and loss of libido (Weldon *et al.*, 1978). Enteroviruses have also been isolated from samples of semen from artificial insemination centers. Some of the isolates were from clinically normal bulls; others were from bulls with various reproductive disorders (Branny and Zembala, 1971).

C. Swine Enteroviruses

The swine enteroviruses cause a number of highly significant swine diseases. Individual strains are tropistic for different tissues; consequently, their method of transmission and their spread within the body may vary. The viruses can be spread through contaminated feces, by contact, or by material contaminated by or originating from infected animals (Dunne,1975b). Enteroviruses are probably introduced into the farrowing house by carrier sows or by newborn piglets infected *in utero*. A common method of introduction of the viruses to new geographic areas is through the feeding of uncooked garbage (Graves and McKercher, 1975), a practice which has been made illegal in many countries. The swine enteroviruses are known to enter through the mouth and nose (Betts and Jennings, 1960; Mayr and Hecke, 1960; Sibalin and Lannek, 1960) and usually produce a primary infection in the intestinal tract. Whether the viruses can first infect the respiratory tract by aerosol and then be transmitted to the intestine by the blood, or whether intestinal tract infection is by ingestion only, has not been adequately established. A viremia soon develops, with spread of the virus to various target organs, depending on the strain of enterovirus.

1. Enterovirus Encephalomyelitis

The most elaborate pathogenesis studies have been with Teschen virus, a highly virulent strain of encephalomyelitis virus. The virus is found in large amounts in the tonsils and adjacent lymph nodes within 24 hours of oral inoculation. The virus then rapidly spreads to the mesenteric lymph nodes and the epithelium of the colon, where it persists for 6–8 days (Hecke, 1958). A transient viremia develops at 5–6 days postinfection, spreading the virus to most body tissues. The viremic stage may occur up to 10–12 days before the onset of clinical signs (Horstmann, 1952).

The pathogenetic process usually stops at this stage of development, resulting in a subclinical infection with continued propagation of the virus in the colon and shedding of the virus in the feces for up to 7 weeks (Hecke, 1958; Mayr and Hecke, 1960). However, a paralytic stage occasionally may develop. Apparently during the transient viremic stage the virus may pass the blood–brain barrier to cause the characteristic clinical signs and histological lesions of Teschen disease. The reason that the paralytic stage develops in only some animals is not understood. This stage usually follows an incubation period of 10–12 days, during which the virus is not detected in the CNS until just before the onset of clinical signs. Viral levels in the CNS peak during the late incubation period and early paralytic phase (Dardiri *et al.*, 1966). Higher viral titers in the CNS are associated with shorter incubation periods, rapid onset of clinical paralysis, and increased histological changes (Kötsche, 1958).

EE affects pigs of all ages; however, once enzootic in an area, it affects only the newborn and recently introduced animals. Signs of the disease often appear simultaneously in a number of baby pigs. The first signs, which often go unrecognized, are mild fever and depression. A second stage is characterized by mild ataxia of the hind limbs and general weakness, followed by a period of nervous excitement with muscle tremors precipitated by the stimulus of noise or touch. Periodic muscle tremors, nystagmus, opisthotonos, stiffness of limbs, and tonic-clonic convulsions may occur. Recovery usually takes 2–3 weeks, but some pigs, usually adults, have neurological deficiencies for longer periods. In severely affected animals the disease may progress to paresis, followed by paralysis beginning posteriorly and involving all four limbs, the neck, and the head. Coma and death usually follow, although recovery has been reported.

Swine EE throughout the world varies greatly in severity, even though the causative viruses are serologically the same (Mills and Nielsen, 1968). Signs of the disease in Denmark include high fever, diminished control of hind limbs, and slight ataxia. Many of the pigs are anorectic and incoordinated and often remain lying down. Usually recovery is complete and occurs within a few days; however, a few animals develop ascending paralysis initially affecting the hind limbs so that the pig assumes a dog's sitting position. Severely affected pigs soon recover and grow as rapidly as their unaffected littermates. The disease in England is similar, with low morbidity and mortality (Harding *et al.,* 1957). In contrast to the Danish and English forms, EE in North America and Australia resembles the classical Teschen disease, with high mortality in pigs under 2 weeks of age (Alexander *et al.,* 1959; Gardiner, 1962; Koestner *et al.,* 1962).

In all types of the disease, the gross lesions at necropsy are nonspecific, and significant microscopic lesions are limited to the CNS. The type and intensity of the neurological lesions vary, but typically the lesions are a nonsuppurative inflammation involving primarily the gray matter. Perivascular cuffing is a prominent feature in the microscopic picture, but it is neither specific to nor limited to this disease. The inflammatory cells are predominantly lymphocytes, plasma cells, and glial cells; polymorphonuclear leukocytes are absent. Neuronal necrosis that may accompany the inflammation primarily involves the dorsal root ganglia, ventral columns of the spinal cord, cerebellar cortex, diencephalon, mesencephalon, and thalamus. For a more extensive description of the microscopic lesions and variations associated with different isolates, see reviews by Mills and Nielsen (1968) and Jones (1975).

2. *Swine Vesicular Disease*

SVD is a moderately contagious disease of swine. Often it is first detected by the sudden appearance of lameness in several animals in a herd. Lameness and extreme reluctance to rise often precede the appearance of vesicles and are indirectly responsible for a decrease in food consumption. Temperatures of 1–2°C above normal are often seen. A viremia coincides with the febrile response and the appearance of vesicles. The vesicles are of various sizes. They appear first about the coronary band and then in the interdigital spaces and on the soles of the feet. Vesicles may also appear in the epithelium of the buccal cavity and the tongue, on the snout, and on the teats. The vesicles soon rupture, leaving a raw, denuded, bleeding ulcer. In severe cases, the ulcerative skin lesions on the feet can extend to the metacarpus and metatarsus, with loosening of the sole pad and loss of the hoof or claw. Clinically the disease is indistinguishable from FMD, vesicular stomatitis (VS), and VES (Graves and McKercher, 1975).

There are two other forms of SVD: a subclinical form, which usually goes undetected (Burrows *et al.*, 1974), and a neurological form. The neurological form of the disease has been reported both in field outbreaks and after experimental infection of swine (Monlux *et al.*, 1974, 1975; Lai *et al.*, 1979). Signs of the neurological form are a very unsteady gait, shivering, and chorea-type movements involving the legs.

The incubation period varies with the route of infection or inoculation. Lesions can be detected as early as 30 hours after intravenous inoculation of the virus and from 3–11 days after the animal has eaten contaminated food. The timing of lesion development in susceptible pigs in contact with SVD virus-infected pigs is variable, but most susceptible pigs show signs in 2–7 days. The subclinical form of SVD has followed experimental inoculation of low levels of virus in the nose and mouth, and there is evidence that it also occurs in the field (Burrows *et al.*, 1974; Graves and McKercher, 1975).

Recovery from SVD is rapid, with most pigs returning to normal within 3 weeks; however, secondary infection of foot lesions can lead to chronic lameness, especially when hooves have been lost. In field outbreaks, morbidity is usually moderate and death is rare (Zoletto *et al.*, 1973). The infection can be highly fatal to baby pigs; experimental inoculation of a sow within a few hours of farrowing caused a severe infection, fatal in all the baby pigs (Graves and McKercher, 1975).

The gross lesions of SVD are limited to vesicular lesions that are both grossly and microscopically the same as those described for FMD (Graves and McKercher, 1975). The presence of microscopic neurological lesions is a differentiating feature. Following experimental inoculation and contact exposure, both the Hong Kong and UKG strains produce a diffuse encephalomyelitis of mild to moderate intensity (Monlux *et al.*, 1974, 1975). With the UKG strain, perivascular cuffing with lymphocytes and formation of neuroglia cell foci were most prominent in the telencephalon, diencephalon, and mesencephalon. With the Hong Kong strain, the lesions were most severe in the diencephalon, mesencephalon, metencephalon, and myelencephalon.

3. Embryonic and Fetal Death, Stillbirth, and Infertility

The term "SMEDI" originally designated the swine enteroviruses known to cause stillbirth (S), mummification of fetuses (M), embryonic death (ED), and infertility (I) (Dunne *et al.*, 1965). However, it has been shown that other viruses, including parvoviruses and reoviruses, can cause similar reproductive problems; these viruses are often considered to be in the SMEDI group. The original SMEDI (entero) viruses were isolated from dead fetuses following hysterectomy, from stillborn pigs, and from baby pigs that died shortly after birth.

The viruses have also been isolated from herds with histories of stillbirth, mummification, embryonic deaths, infertility, and small litter size. Four enteroviruses (serogroups 1, 3, 6, and 8) have been associated with reproduction problems (Dunne, 1975b). Antibodies to serogroups 4, 5, and 7 have been detected in fetuses and correlated with the production of small litters (Dunne, 1975b; Cropper *et al.*, 1976); antibody to enteroviruses was detected in young in 11% of 127 litters with four or fewer live pigs, but not in any of 46 litters with nine or more live pigs.

The effect of the enteroviruses on litter size has been demonstrated by experimental inoculation of susceptible pregnant gilts at different stages of gestation (Dunne, 1975b). Inoculation at 21–27 days of gestation had the greatest effect on reproduction—an average decrease of 2.4 live pigs per litter, with a comparable increase in mummified fetuses. Sows with antibody to the particular enterovirus had litter sizes comparable to those of control gilts, a minimum number of mummified fetuses, a low rate of embryonic death, and a high litter survival rate at 5 days postpartum.

Gilts inoculated early in pregnancy had a high rate of return to estrus at irregular estrus cycles of 25–28 days (Dunne, 1975b). Such a pattern would be expected if the fetuses died early, because sows with fewer than four live embryos at 15 days of gestation generally return to estrus (Dhindsa and Dziuk, 1968). In some sows all the fetuses died after 35 days of gestation. These sows failed to return to estrus and were permanently sterile (Dunne, 1975b).

The viruses have little or no pathological effect on the sow. In a few neonatal pigs, perivascular cuffing with round cells in the brain stem and mild focal gliosis are observed. However, most pigs dying in the perinatal period have no lesions that can be attributed to virus infection. Death of the newborn is usually attributed to bacterial infection, which may be a result of decreased resistance following the *in utero* viral infection (Dunne, 1975b).

4. *Pneumonia, Diarrhea, and Pericarditis*

Swine enteroviruses have been implicated in numerous other signs, including pneumonia, diarrhea, and pericarditis. The evidence that the enteroviruses can cause these conditions is quite convincing; however, attempts to reproduce the specific signs by inoculation in the laboratory have been inconclusive, usually producing only a mild disease in a few animals. The failure of laboratory confirmation could be due to (1) attenuation of the virus through laboratory passage; (2) the absence of stress and other environmental factors; or (3) a requirement for synergistic action of the virus with other viral or bacterial agents. The field and experimental results were reviewed by Dunne (1975a).

Evidence that the swine enteroviruses are etiological agents of pneumonia is based primarily on results from pigs experimentally inoculated by the intranasal or oral route. The studies have usually involved normal or gnotobiotic pigs less than 6 weeks of age. Experimental inoculation with serogroups 2 and 3 has produced pneumonia (Sibalin and Lannek, 1960; Singh *et al.*, 1964). The pneumonic lesions are usually generalized throughout the lung and are characterized histologically as an interstitial pneumonia with mild damage to bronchial epithelium. Serogroups 2 and 8 have been isolated from lung tissue of pigs with pneumonia (Dunne *et al.*, 1965; Veznikova *et al.*, 1969).

Pericardial and myocardial lesions have been observed in pigs experimentally inoculated with serogroups 2, 3, and 9 (SVD) swine enteroviruses (Sibalin and Lannek, 1960; Long *et al.*, 1966; Lai *et al.*, 1979). Serogroup 3 may contain the most significant pathogens; 5-day-old germ-free pigs experimentally inoculated by Long *et al.* (1966) all developed a serofibrinous pericarditis by the tenth day postinoculation. Enterovirus 3 also has been isolated from the pericardial fluid of two pigs with mulberry heart disease in two separate herds (Dunne *et al.*, 1967).

Diarrhea has been produced in suckling pigs by inoculation with serogroups 3 and 8 (Yamanouchi *et al.*, 1964; Long *et al.*, 1966), which were originally isolated from pigs with diarrhea. Other enterovirus isolates have been obtained from herds in which diarrhea was a common sign (Hancock *et al.*, 1959; Moscovici *et al.*, 1959; Bohl *et al.*, 1960); however, their role in the etiology of the diarrhea is still uncertain. It is possible that some enteroviruses can act as primary pathogens in the production of this disease, but enteroviruses may also act synergistically with other weak pathogens or act as a predisposing factor for pathogens such as *Escherichia coli*.

D. Bovine Rhinoviruses

Bovine rhinoviruses have been isolated from normal calves and from calves with respiratory infection (Rosenquist, 1972; Kurogi *et al.*, 1974, 1975; Mattson and Reed, 1974; Rosenquist and Dobson, 1974). Frequently other respiratory viruses were isolated simultaneously. The rhinoviruses are limited to the respiratory tract and are assumed to be spread by the aerosol route. Their role in the production of disease is not well understood. Tests of pathogenicity in normal and gnotobiotic calves give mixed results, with bold clinical and histological examination indicating that the virus produces only a mild disease, if any. The lesions and signs produced depend on the strain and may include one or more of the following: severe focal rhinitis, diphasic temperature response, necrosis of epithelial cells of turbinates and trachea, and mild pneumonic lesions (Mohanty *et al.*, 1969; Betts *et al.*, 1971; Ide and Darbyshire, 1972b). The frequent isolation of rhinoviruses in combination with other pathogens may indicate that a synergist is required to induce disease. A synergistic effect has been demonstrated *in vitro:* A bovine rhinovirus enhanced the growth of mycoplasma in tracheal organ cultures (Reed, 1972).

E. Equine Rhinoviruses

Equine rhinoviruses are common viral agents that circulate in horses, especially at large stables and racetracks. Serological surveys indicate that 60–80% of these horses have antibody to type 1 virus, and a lesser percentage have antibody to type 2 virus (Hofer *et al.*, 1973; Moraillon *et al.*, 1973; Powell *et al.*, 1974; Rose *et al.*, 1974). Newly introduced susceptible horses usually become rapidly infected (Hofer *et al.*, 1973; Powell, 1975). Infected horses shed the virus for long periods of time, thus serving as a reservoir for the virus (Powell, 1975). Subclinical infections are common and may be important in maintaining the virus in a stable (Rose *et al.*, 1974; Coggins and Kemen, 1975). The virus produces a mild upper respiratory tract infection that lasts 7–14 days (Kemen, 1975). the clinical signs are variable, commonly including a copious serous nasal discharge that later becomes mucopurulent. Other common signs are low-grade fever, pharyngitis, and a mild cough that may persist (Plummer and Kerry, 1962; Moraillon *et al.*, 1973; Kemen, 1975; Powell, 1975).

F. Vesicular Exanthema

VES is a highly contagious disease of swine. The incubation period is 24–72 hours. The disease is characterized by fever and the formation of vesicles on the snout, lips, gums, tongue, and feet. The vesicles are similar to those produced by FMD; clinically, the two diseases cannot be differentiated. The vesicles contain a clear serous fluid with large quantities of virus. The vesicles or blisters rupture in 24–48 hours, leaving a raw, eroded area. These vesicles heal in 5–7 days in uncomplicated cases. For more extensive information on VES, see reviews by Bankowski (1965), Madin (1975), and Studdert (1978).

G. San Miguel Sea Lion Virus

SMS, a newly recognized virus, shares many properties, including host range, with VES virus (Madin *et al.*, 1976). Inoculation of swine produces a disease that is clinically indistinguishable from VES (Madin *et al.*, 1976; Sawyer, 1976). Little is known about the pathogenesis of the virus in marine mammals; it is believed to be responsible primarily for reproductive losses in California sea lions and possibly in northern fur seals (Odell, 1970). The virus has been isolated from aborted fetuses; from nasal, throat, and rectal swabs, and from ground products of fur seal carcasses (Smith *et al.*, 1973). Vesicular lesions on the flippers of northern fur seals and California sea lions have been observed annually in approximately 2% of the seals (Sawyer, 1976).

H. Feline Caliciviruses

The FC produce a wide range of diseases from inapparent infection to rhinitis, conjunctivitis, glossitis, and pneumonia. An excellent review of the disease was given by Gillespie and Scott (1973). The severity of the disease is influenced by the strain of virus, the level of preexisting immunity, the presence of concurrent bacterial and viral infections, and the husbandry practices used to maintain the cat colony.

The FC produce primarily an upper respiratory tract infection and are thought to be disseminated from carrier animals to kittens by the aerosol route. Following aerosol infection, most kittens develop a diphasic febrile response, with a brief initial febrile response 1 day after exposure. In some kittens a second febrile response occurs between days 4 and 7 postinfection. Severe conjunctivitis and a mild ocular involvement develop early during the infection and last for up to 13 days postexposure. During the first 10 days, kittens also have a rhinitis, with a serous to mucoid to mucopurulent discharge accompanied by sneezing. Ulcerative lesions are observed on the tongue and the mucosa of the hard palate in approximately 25% of the affected kittens. Mortality among infected kittens varies greatly with the strain of virus, the age of the kittens, and the husbandry methods. Most deaths occur during the first 5 days, and the rest before 2 weeks postexposure (Gillespie and Scott, 1973).

The effect of the strain on the type of disease is shown by a study in which specific pathogen-free cats were tested with virus strain 255 (Hoover and Kahn, 1973). This strain produces primarily a viral pneumonia, sparing the nasal mucosa and conjunctiva. The pneumonia began as a multifocal exudative pneumonia and progressed rapidly to interstitial pneumonia characterized by marked adenomatoid proliferation of pneumocytes. Also, small, sharply delineated ulcers were common on the anterior dorsolateral margin of the tongue, the hard palate lateral to the midline, or both.

V. IMMUNITY

The picornaviruses all elicit an immune response which protects the animal against a second infection with the homologous virus for 1–2 years, after which it can again develop typical signs from the same virus. There are numerous serotypes of each virus, and immunity to one serotype usually does not protect against another. The serological response can be domonstrated by most standard serological tests, including serum neutralization, agar gel precipitation, and complement fixation.

A. Aphthoviruses

A natural or experimentally induced FMDV infection produces a type-specific immunity and, to a degree, a subtype-specific immunity (Davie, 1964; Federer *et al.*, 1964). The antibody response with local or histogenic immunity develops as early as 3 days at and around the site of the lesion. Humoral antibody detectable by either the single radial immunodiffusion test or the serum-neutralization test is also present at about day 3 postinoculation. The CF antibody does not appear until days 7–8 postinfection. Titers rise rapidly, peaking at 10–14 days (Cowan, 1973).

The sequential change in the physicochemical characteristics of antibodies to FMDV infection was first demonstrated by Brown and Graves (1959) and Brown (1960). They showed that antibody present at 7 days after infection had the electrophoretic characteristics of β-globulin and that antibody present after 14 days had the slow mobility γ-globulin. Later studies identified the β-globulin as immunoglobulin M (IgM) or 19 S antibody and the γ-globulin as immunoglobulin G (IgG) or 7 S antibody (Brown *et al.*, 1964; Graves *et al.*, 1964; Cowan and Trautman, 1965; Cowan, 1966). Secretory antibody following FMD was reported by Kaaden and Matthaeus (1970), who subsequently identified it as immunoglobulin A (IgA). Matsumoto *et al.* (1978) demonstrated that the IgA was first present in saliva at 30 days postinfection even though high-serum antibody titers were present at 10 days postinfection. The IgA titers peaked at about 90 days. Matsumoto *et al.* (1978) reported that small amounts of IgA were still present for at least 6 months.

The length of immunity following natural infection is difficult to assess. Generally, cattle which have recovered from FMD have adequate immunity to protect them from the same virus type for a year or more, but the resistance is not lifelong (Cunliffe, 1964; Bruner and Gillespie, 1973). Such animals may, however, immediately be infected with one of the other types of virus and may then exhibit typical signs. Little is known about the duration of immunity in swine except that it is shorter than that in cattle (Cunliffe, 1962; McKercher and Farris, 1967; McKercher and Giordano, 1967). Whether the drop in protection is due to a decrease in antibody level, a change in antibody avidity, or an antigenic shift in the virus is not known; however, the drop in immunity following natural infection illustrates the difficulty that is faced in obtaining and maintaining protection by vaccination.

Passive transfer of antibody through ingestion of colostrum protects against FMDV for 6–12 weeks. The passively immunized calf will not respond to vaccination until the serum antibody reaches low levels, whereas actively immunized calves of the same age produce antibody in response to vaccination (Graves, 1963).

B. Rhinoviruses

The bovine rhinoviruses elicit an immune response in cattle, but little research has been done to determine the expected level of response, the duration of immunity, or the specificity of the antibody. Mohanty *et al.* (1969) found the immune response of calves to be poor even after they had been given a high dose of the CO-7 bovine rhinovirus. A few calves had a low titer to the isolate at the start of the experiment, but the titer failed to protect them from infection. Ide and Darbyshire (1972b) also found that a low serological response was present for at least 5 months following inoculation with the RS3X strain of bovine rhinovirus. However, a high-titer amnestic response followed a second inoculation by the intravenous route.

The two equine rhinoviruses (type 1 and type 2) are serologically distinct, both eliciting a fairly rapid neutralizing-antibody response. Although the relationship between antibody levels and protection against disease has not been established, immunity is thought to be complete and to provide long-term protection against the homologous viruses (Burrows, 1970; Coggins and Kemen, 1975).

C. Cardioviruses

The cardioviruses elicit an immune response in surviving animals. The immunity is presumed to be long-lived, although data on its duration and specificity are lacking.

D. Enteroviruses

The bovine enteroviruses usually produce an inapparent infection that stimulates the production of specific neutralizing antibody, but little is known about the duration, avidity, and protectivity of the antibody. Research is needed on the effectiveness of antibody to the enteroviruses in controlling infection, for the carrier state can last 9 months or more in calves (Klein and Earley, 1957). It is not known whether the carrier state is due to a failure of virus to elicit an immune response during low-level infection in the gastrointestinal tract or to a failure of the antibody to neutralize the virus.

Neutralizing antibodies to the bovine enteroviruses have been detected in a number of aborted fetuses. Dunne *et al.* (1973) showed that the antibody is of fetal origin and is produced after intrauterine infection of the calf. Antibody can also be transferred to the calf through the colostrum and is believed to protect the newborn calf from enterovirus infection for the first weeks of life (Cliver and Bohl, 1962a,b).

The swine enteroviruses, like the bovine enteroviruses, produce a specific serum-neutralizing antibody and a local gastrointestinal antibody. The antibody titers are assumed to be long-lived, with swine having antibody to four or more serogroups (Rodeffer *et al.*, 1975; Cropper *et al.*, 1976). After oral inoculation, antibody is present in the serum by 4 to 6 days and reaches high levels by 9 days. Hazlett and Derbyshire (1976) first detected gastrointestinal-neutralizing activity by 14 days after oral dosing; this response coincided with a fall in virus titer. The gastrointestinal-neutralizing response peaked at 23 days and dropped markedly by 36 days. This local neutralizing activity was found predominantly in the IgA class, although some activity was present in the IgM and IgG classes (Hazlett and Derbyshire, 1977a,b). The serum-neutralizing activity was present in all three antibody classes but predominated in the IgG class.

The presence of neutralizing antibody in aborted and stillborn piglets has received attention because it indicates the existence of intrauterine infections (Dunne *et al.*, 1974b). Wang *et al.* (1973) determined that swine fetuses inoculated *in utero* at days 68, 82, and 96 of gestation developed antibodies within 1 week of inoculation. The antibody was type specific and was of fetal origin.

Antibody against SVDV is found in sera of infected pigs by day 7 after infection, and peaks by day 28 (Burrows *et al.*, 1973, 1974). Neutralizing antibody persists in recovered pigs for at least 4 months. Subclinical infections elicit much lower antibody levels, which decline fairly quickly (Burrows *et al.*, 1973). Maternal antibody does not cross the placenta; however, antibody transferred through the colostrum can protect the baby pigs (Gourreau *et al.*, 1975).

Following oral feeding of enterovirus encephalomyelitis (Teschen) virus, serum-neutralizing antibody was detected as early as 5 to 9 days before the onset of paralytic signs (Mayr, 1962); the antibody titer rose during the clinical syndrome and persisted for at least 9 months after recovery. Precipitating antibodies have been demonstrated on the second or third day of illness and have persisted as long as the neutralizing antibodies (Mayr, 1962). Two different forms of serological response may develop after an inapparent infection. In the first form, the antibody response, which is characterized by the production of neutralizing and precipitating antibodies, is the same as it is in the diseased animal. In the second form, low levels of neutralizing antibody are produced, but precipitating antibody is not demonstrable. Mayr and Wittmann (1959) postulated that in the second case the infection was probably confined to the intestinal tract and corresponding lymph nodes.

E. Caliciviruses

Only limited data are available on the duration of immunity following infection with the caliciviruses. Swine that have recovered from an experimentally induced infection with VESV are immune to the homologous strain for as long as 30 months (Mott, 1957; Bankowski, 1965). The duration of immunity appears to depend on the infecting strain. Little or no protection is provided against heterologous strains of VESV (Crawford, 1937). The development of vaccines has been limited; originally, the disease was controlled by slaughter. More recently, control has consisted of restricting the movement of swine products, decontaminating stockyards and livestock-handling equipment, and enforcement of garbage-cooking regulations.

The FC viruses elicit the production of neutralizing, complement-fixing, and precipitating antibodies in the domestic cat. Early reports gave little encouragement for the development of vaccines; immunity after infection appeared limited to the prevention of clinical illness caused by homologous or closely related strains. Recent studies in which live virus of low-virulence strains were used as vaccines have demonstrated significant protection against exposure to highly virulent virus strains (*Kahn et al.*, 1975; Povey and Ingersoll, 1975). The protection may be enhanced by a latent infection with the avirulent strain; Kahn *et al.* (1975) were able to isolate the avirulent strain from four of six animals at 35 days postexposure. Encouraged by these findings, a number of manufacturers have developed FC vaccines. The vaccines have been formulated in two ways—from low-virulence strains further attenuated by passage at a low temperature and given in two intramuscular doses 27 days apart, or from highly avirulent, naturally occurring strains inoculated into the conjunctival sacs and nostrils in a single dose.

VI. EPIZOOTIOLOGY

The epizootiology of FMDV has been thoroughly studied because of the need to control or eradicate FMD in specific localities and to prevent its spread from enzootic areas to FMD-free areas. The topic is well covered in a review by Hyslop (1970). FMD has been a major threat to the health of livestock for centuries. It has been eradicated from several of the more highly developed countries; however, it is virtually unchecked and uncontrolled throughout large areas of Asia, Africa, and some parts of South America.

Infected cattle produce large quantities of virus that can be present in all tissues and secretions. During the early febrile period, virus is abundant in the saliva, feces, milk, semen, and vaginal and urethral secretions (Hyslop, 1965a; Scott *et al.*, 1966; Burrows, 1968; Cottral *et al.*, 1968; Hedger, 1970; Burrows *et al.*, 1971; Parker, 1971). During frank disease, nasal discharge, urine, and vesicular epithelium and fluid should be added to the list. Virus has also been detected in the air surrounding infected cattle, pigs, and sheep (Hyslop, 1965b; Sellers and Parker, 1969).

The virus, though readily inactivated by drying, can remain infectious for extended periods in most livestock products. Because the virus is stabilized, it can readily be transmitted from one farm to the next or from one country to the next by livestock, by products from infected animals, or by fomites and improperly cleaned equipment. Evidence exists that the standard methods of pasteurizing milk and processing of some milk products do not destroy all the virus, but the significance of the residual virus in transmission of FMD is not known (Hyde *et al.*, 1975; Blackwell, 1976, 1978b,c; Blackwell and Hyde, 1976; Cunliffe and Blackwell, 1977).

antigens in common (Burroughs *et al.*, 1978a; Smith *et al.*, 1978; Soergel *et al.*, 1978) and produce similar lesions in swine (Sawyer, 1976).

FC is ubiquitous in the cat population; 90–100% of adult cats have neutralizing antibody to FC in their serum (Studdert, 1978). The virus is maintained in the cat population by carrier animals. The frequency of carrier-status development after infection is not known; however, it may be a usual sequel of infection. The disease is most often seen in cats under 1 year of age after they have been brought together in pet shops, veterinary hospitals, breeding establishments, and such.

VII. DIAGNOSIS

The vesicular diseases all produce similar clinical signs, so the signs are of little value in differentiating them. Because of the economic importance of the vesicular diseases in many parts of the world, a means for rapid and definitive diagnosis should be sought. Whenever a vesicular disease is suspected, it should be reported immediately to authorities who act promptly to obtain a definitive laboratory diagnosis. Most countries have state or national laboratories capable of performing such tests. International vesicular diagnostic centers are located at Pirbright, England, and Rio de Janeiro, Brazil. The World Reference Center for Vesicular Diseases at Pirbright is responsible for the typing and subtyping of FMD viruses.

Diagnosis of nonvesicular diseases produced by the picornaviruses also requires laboratory confirmation. Clinical signs can aid in the diagnosis only of diseases such as EE in swine and FC disease. Other diseases produced by the picornaviruses are nonspecific; similar signs are often produced by a variety of other viruses. Definitive diagnosis requires isolation and identification of the viral agent as well as results from paired serological tests.

The rapid diagnosis of the vesicular diseases—FMD, VES, SVD, and VS—is of major concern; a number of countries are free of one or more of the diseases. The ease of differentiating a naturally occurring vesicular disease depends greatly on the species involved. When a disease occurs in horses, the diagnosis is clear, because horses are naturally susceptible only to VSV. When a disease occurs in cattle, both FMD and VS must be considered. However, swine may become infected with all vesicular diseases; further, when only swine are involved in the field, one cannot eliminate FMDV even though other species are present, because an occasional field strain of FMDV may be relatively species specific (Brooksby, 1950).

A. Vesicular Diseases

Specimens for the diagnosis of the vesicular diseases should be collected from two to three animals and include the following: (1) vesicular fluid, (2) epithelial tissue attached to the edge of lesions and covering vesicles, (3) esophageal-pharyngeal fluid from ruminants diluted immediately with equal volumes of tissue culture medium, (4) blood with anticoagulant, and (5) serum (Cottral, 1978). Fecal samples are also required for swine vesicular disease suspects. With the exception of serum samples, it is important that all samples be promptly frozen.

The viruses causing vesicular diseases in livestock can be differentiated by a variety of laboratory procedures (Table IV) such as CF, virus neutralization, agar gel diffusion, differential growth in tissue culture, and measurement of physicochemical properties. The CF and virus-neutralization tests are the most specific, and of these the CF test is the more rapid. The CF test can give a preliminary diagnosis within 3 hours when vesicular fluid or lesion material contains adequate amounts of virus.

The CF antibodies were first reported in the sera of guinea pigs that had been infected with FMDV (Ciuca, 1929). The test has since been modified, expanded (Brooksby, 1952; Graves, 1960; Cowan and Trautman, 1967), and adapted for the diagnosis of the other vesicular diseases (Callis *et al.*, 1975; Madin, 1975). Field viruses are usually identified in vesicular lesion material. The material can be preserved in buffered glycerin solution and kept at 4–7°C, or it can be frozen. It is used as antigen in the presence of specific hyperimmune guinea pig sera prepared against the various vesicular-disease viruses.

The agar gel diffusion test can be used to identify both virus and antibodies (Cowan, 1966; Cowan and Graves, 1966, 1968; McKercher and Giordano, 1967;

TABLE IV
Optimal Isolation Methods and Serological Tests for Vesicular Diseases of Livestock

Virus	Tissue culture	Animal inoculation	Serological tests[a]
Foot-and-mouth disease	Bovine kidney Bovine thyroid Swine kidney	Natural host, suckling mice, guinea pigs	CF, VN, AD, FA
Swine vesicular disease	Swine kidney	Swine, suckling mice	CF, VN, AD, FA
Vesicular exanthema of swine	Swine kidney Tissue cultures of swine origin	Swine	CF, VN, AD
Vesicular stomatitis	Embryonating eggs Mammalian and avian cell cultures	Horse, suckling mice, guinea pigs	CF, VN

[a]CF, complement fixation; VN, virus neutralization; AD, agar gel diffusion; FA, fluorescent antibody.

Graves *et al.*, 1968). However, caution is necessary in interpreting the results of this test, because a number of normal sera can give false-positive reactions to the FMD viruses (Andersen 1975, 1977a,b, 1978a; Meloen, 1978).

The agar gel diffusion test for VIA antibody can be used to determine whether antibody in a given animal is due to actual infection or to immunization by a killed-virus vaccine. Generally, immunized animals do not develop VIA antibody, but recent reports indicate that low levels of VIA antibody can be detected in animals that have been repeatedly immunized (Dawe and Pinto, 1978; Pinto and Garland, 1979). The VIA test also has potential as a screening test, since the VIA antigen is not type specific and may react equally well with heterologous and homologous sera (Cowan and Graves, 1966). The test is performed using the agar gel double-diffusion technique. The VIA antigen is prepared from the supernatant of the FMDV-infected tissue culture (Cowan and Graves, 1966).

The virus-neutralization test can be used to identify antibodies or virus. Sera from convalescent or recovered animals are tested for virus-neutralizing antibodies. Virus-specific antibodies inhibit the multiplication of FMDV in tissue culture (Bachrach *et al.*, 1955; Sellers, 1955; McVicar *et al.*, 1974) or reduce viral infectivity when mixtures of serum and virus are injected into suckling mice (Skinner, 1953). Care must be taken in interpreting the results because a number of normal sera have low levels of virus-neutralization activity (Andersen, 1975, 1977a,b, 1978a; Meloen, 1978). The cross activity is usually to IgM antibody, which likely is produced in response to subclinical infections with cross-reacting agents (Andersen, 1977b, 1978b).

Caution must also be exercised in interpreting the plaque-reduction neutralization results, because IgG produced against some strains of FMDV can be relatively ineffective in preventing the formation of plaques (Rweyemamu *et al.*, 1977). The same antibody neutralizes the viruses and prevents infection when injected into susceptible animals. The reason that the antibody fails to neutralize the virus in tissue culture is not understood; however, the antibody combines with the virus, and the addition of species-specific anti-IgG prevents the formation of plaques. Research is needed on the mechanism of neutralization of the FMD viruses by antibody and on the reason for the failure of this mechanism in tissue culture. Until the mechanism and extent of nonneutralization are known and understood, the addition of species-specific anti-IgG in neutralization tests performed on tissue culture should be considered.

The serum-protection test is used primarily as an indication of protective immunity after vaccination. It is similar to the virus-neutralization test, except that constant amounts of antibody are inoculated into several susceptible animals and the virus is titrated in the animals 1 hour later (Cunha and Honigman, 1963). Trautman and Bennett (1979) obtained more consistent results when the virus was inoculated 6 to 24 hours after the antibody.

FMDV is usually isolated by inoculation of susceptible animals or tissue cultures. The preferred method for detection of minute amounts of FMDV is by intradermal-lingual inoculation of cattle (Hyde *et al.*, 1975). This method can isolate virus not demonstrable in tissue culture. The preferred cell cultures are primary calf thyroid and primary calf kidney cell cultures. The FMD viruses also grow in baby hamster kidney BHK_{21} and swine kidney cells.

The method of field diagnosis by inoculation of animals was developed by Traum (1936). He recommended that at least two cows be inoculated—one in the mucosa of the tongue, lip, or dental pads and the other either intravenously or intramuscularly. He also recommended that a horse be inoculated, because horses are susceptible only to VS, and that guinea pigs be inoculated in the footpad. The inoculation of swine was not recommended, because swine are susceptible to all the vesicular diseases. The expected results of inoculating various animals with the different viruses are listed in Table V. It should be noted that the diseases—SVD, VES, and SMS—that could occur only in swine would not be differentiated by this method; however, SVD is the only one known to occur naturally in swine at this time

B. Bovine Rhinoviruses

Definitive diagnosis of clinical disease produced by a bovine rhinovirus would require the isolation and identification of the causal agent, and the preferable course would be to reproduce the disease by reinoculation of animals under controlled conditions to establish the pathogenicity of the virus. Isolation is usually in bovine embryonic kidney (BEK) or calf cells. The viruses prefer a low concentration of bicarbonate (0.035 or 0.07%), a pH around neutrality, and a temperature of 33–34°C (Ide and Darbyshire, 1972a; Rosenquist, 1972). The use of roller cultures instead of stationary cultures may enhance the cytopathic effect

TABLE V
Susceptibility of Various Species to Experimental Infection with the Vesicular Disease Viruses

		Susceptibility to indicated virus[a]				
Species	Route of inoculation	FMD	VS	SVD	VES	SMS
Horse	Intradermal-lingual	–	+	–	–	–
Cow	Intradermal-lingual	+	+	–	–	–
Cow	Intravenous or intramuscular	+	–	–	–	–
Guinea pig	Intradermal-foot pad	+	+	–	–	–
Pig	Intradermal-intradigital	+	+	+	+	+

[a]FMD, foot-and-mouth disease; VS, vesicular stomatitis; SVD, swine vesicular disease; VES, vesicular exanthema of swine; SMS, San Miguel sea lion virus.

(CPE). The initial CPE upon BEK cells consists of one or more foci of a few rounded or oval refractive cells on what appears to be a normal monolayer. The CPE usually progresses slowly, compared with that of the enteroviruses, and seldom completely destroys the monolayer even after prolonged incubation. In some cases, the initial CPE may regress. The virus isolate is identified as a rhinovirus on the basis of its typical rhinovirus CPE in BEK-cell monolayers and its small size, acid lability, RNA content, and ether resistance (Rosenquist, 1972).

Serological diagnosis of rhinovirus infection with paired bovine serum specimens is still impractical and unreliable because of the current uncertainty about the number of serotypes. If further research on the rhinoviruses shows that serotypes are few, the neutralization tests with selected strains may be valuable (Rosenquist, 1972).

C. Equine Rhinoviruses

Confirmation of the occurrence of equine rhinovirus disease depends upon the isolation and identification of the virus and the demonstration of a significant rise of specific antibody in the serum. The virus can readily be isolated from nasal secretions and feces with tissue cultures prepared from several animal species. The equine rhinovirus differs from the rhinoviruses of other species in that it grows in tissue cultures of other species and does not need low bicarbonate concentrations and lowered temperatures (Ditchfield and MacPherson, 1965; Coggins and Kemen, 1975). The detection of the rise in specific antibody titer is necessary because virus can often be isolated from horses long after clinical or inapparent infection (Plummer and Kerry, 1962; Powell, 1975). According to Dobbertin *et al.* (1974), the neutralization test is preferred for detection of antibody against the equine rhinovirus; these researchers found that this test yielded higher titers and seemed more useful than the immune-adherence and CF tests.

D. Bovine Enteroviruses

Diagnosis of the bovine enteroviruses requires the isolation and identification of the viral agent and preferably the determination that the isolate reproduces signs of the disease when inoculated into susceptible animals. Often these viruses produce no clinical symptoms; they are routinely isolated from both normal and diseased animals. However, the isolation of enteroviruses from semen of bulls with low fertility, from bulls with orchitis, and from aborted fetuses is generally accepted as a valid demonstration of their involvement in decreased fertility. It is difficult to determine that an enterovirus has affected the fetus to produce abortion, stillbirth, or a weak calf, because the virus has usually been eliminated by the time of parturition. Thus, this diagnosis must be based on serology. The fetus produces antibody to the viral agent when infection occurs after 60 days of gestation. Neonatal serum samples must be collected before nursing begins if the introduction of colostrum antibody is to be prevented. Antibodies to enteroviruses have often been detected in stillborn and aborted fetuses, alone or with antibody to other viral agents, but the enterovirus has not been unequivocally demonstrated as the causative agent.

Primary isolation of the bovine enteroviruses is usually done in primary bovine kidney or calf testicle cells. After isolation, the viruses can often be adopted to multiply in a wide range of cell cultures. The bovine enteroviruses produce a CPE consisting of rounding of cells, with some of them clumping and others floating free from the cell sheet. Complete CPE may occur within 24 hours but usually takes 3 to 4 days (Cottral, 1978). Some strains require a subpassage following initial isolation before CPE is seen. To confirm an isolate as an enterovirus, the following test should be performed: (1) exposure to chloroform or ether at room temperature for 10 minutes, (2) exposure to pH 3 at room temperature for 3 hours, (3) exposure to 1 M $MgC1_2$ for 1 hour at 50°C, and (4) filtration through a 50-nm-pore filter (Cottral, 1978). Virus size determination can be made by electron microscopy.

Serological tests are usually limited to virus-neutralization tests performed in tissue culture and with viruses of each of the known serotypes. The plaque-reduction neutralization test has been shown to be more sensitive than tube neutralization and is the method of choice (Cottral, 1978). Either 100 $TCID_{50}$ or 100 plaque-forming units of virus with variable amounts of serum is most often used. The serum–virus mixture should be incubated for 30 minutes at 37°C prior to inoculation of the tissue culture. Hyperimmune serum for identification of the BEV has been successfully produced in chickens, goats, rabbits, and guinea pigs.

E. Swine Enteroviruses

Enterovirus encephalomyelitis of swine should be considered when swine have fever in addition to signs referable to CNS lesions. A presumptive diagnosis can be made for a herd upon the appearance of fever, irritability, and convulsions, followed by progressive spinal paralysis. The paralysis spreads rapidly through the herd, especially attacking the young animals. The condition must be differentiated from a number of other diseases, such as pseudorabies, hog cholera, African swine fever, rabies, gut edema, and sodium salt poisoning (Mills and Nielsen, 1968). Diagnosis must be confirmed by laboratory procedures. Microscopic lesions in the CNS are particularly valuable in the diagnosis of EE. Neuronal necrosis and neuronophagic and cellular nodules are prominent features in EE. These features are minimal or absent in African swine fever and hog cholera (Jones, 1975). Isolations of virus from aseptically collected samples of spinal cord and brain are also useful in differenting EE from other diseases. A 10% suspension of the tissue in physiological saline or tissue culture media is inoculated intracerebrally into baby pigs. Production of the characteristic signs and demonstration of typical lesions in the CNS are used as diagnostic criteria (Jones, 1975). Virus can also be isolated in tissue cultures of swine origin with subsequent serological confirmation.

Virus neutralization in tissue culture is a reliable method of detecting antibody to EE (Gasparini and Nani, 1955). Serum-neutralizing antibody is usually detectable 5–7 days before paralytic signs (Mayr and Wittmann, 1959). The CF test with infected tissue-culture fluid as an antigen has been used in some countries (Darbyshire and Dawson, 1963). The agar gel double-diffusion technique has also been useful in detecting antibody. Precipitating antibody appears at about the third day of clinical illness and persists as long as neutralizing antibody (Mayr, 1962). Determination of the presence of neutralizing antibody but not precipitating antibody before or at the onset of paralysis, along with the development of precipitating antibody following the onset of paralysis, is sufficient for the serological diagnosis of EE.

Diagnosis of the remaining swine enteroviruses is difficult because the viruses usually produce no clinical signs. Confirmation of cardiac, respiratory, or gastrointestinal disease produced by an enterovirus requires the isolation and identification of the virus, followed by the production of the disease in swine under laboratory conditions. This second step is necessary because enteroviruses are often isolated from both diseased and healthy animals and may be present only as secondary viral agents.

Diagnosis of reproductive complications from SMEDI viruses is based on clinical signs with serological confirmation (Dunne, 1975b). The presence of mummified fetuses in litters of fewer than 10 live pigs is considered evidence of virus infiltration because only subclinical viral infections are known to kill embryos or fetuses selectively without producing abortion. Litters of four or fewer piglets also indicate subclinical viral infections of the sow, in this case during early gestation, because litters originally of this size do not implant (Dhindsa and Dziuk, 1968). A presumptive serological diagnosis is usually all that can be obtained, because viral agents are usually no longer present at the time of farrowing. Baby pigs exposed to viruses beyond the first third of gestation develop specific antibody. The diagnosis is based on detection of antibody to the specific enterovirus serogroups in body fluids of stillborn pigs or sera of newborn pigs collected before ingestion of colostrum (Wang *et al.*, 1973; Dunne *et al.*, 1974b; Cropper *et al.*, 1976). The antibody is of fetal origin, as antibody does not cross the placenta.

F. Simian Enteroviruses

The simian enteroviruses are most readily isolated in primary kidney cultures of the primate species yielding the fecal specimen. Primary rhesus kidney cell cultures are the most useful cell cultures; however, they do not support growth of all the chimpanzee isolates and are of doubtful use with New World species isolates (Cottral, 1978). A chimpanzee liver culture is useful in working with the chimpanzee isolates (Douglas *et al.*, 1966). Human diploid cells will support the growth of a number of chimpanzee isolates, but none of the simian enteroviruses will grow in cell cultures of human origin (Cottral, 1978). All cultures infected with simian enteroviruses produce a typical CPE and when stained with H&E produce a typical eosinophilic mass in the cytoplasm, causing the nucleus to be pushed to one side with indentation (Heberling and Cheever, 1965).

The preferred serological test for serotyping the simian enteroviruses is the virus-neutralization test, but the poor antigenicity of the viruses and the frequency of cross reactions make serotyping difficult (Herberling and Cheever, 1965). This test is usually done only by investigators specifically interested in these viruses.

G. Duck Virus Hepatitis

Diagnosis of duck virus hepatitis is made by the isolation of the virus in chicken embryos and the identification by virus neutralization using specific immune serum (Levine, 1978). The inoculation of infectious liver suspensions or blood into the allontoic sac of 9-day-old chicken embryos produces stunted and edematous embryos that die on the fifth or sixth day. The amniotic sac contains an excess of fluid, and the yolk sac is reduced in size. Often a greenish discoloration of the embryonic fluid, yolk sac, and liver is seen. Frequently the parenchyma of the liver contains whitish-yellow necrotic foci. Usually identification of the virus is made by virus neutralization; however, the fluorescent antibody technique developed to study the early stages of experimental infection has been recommended for diagnostic use (Vertinskii *et al.*, 1968).

H. Avian Encephalomyelitis

Avian encephalomyelitis (AE) is predominantly a disease of 2–3-week-old chicks which can frequently be diagnosed by histopathological examination of typical specimens along with a complete history of the flock. Histological evidence of gliosis, a lymphocytic perivascular infiltration axonal type of neuronal degeneration in the CNS, and hyperplasia of the lymphoid follicles in visceral tissue usually are sufficient evidence for a positive diagnosis (Luginbuhl and Helmboldt, 1978). However, AE must be differentiated from other avian diseases producing similar clinical signs, such as Newcastle disease, equine encephalomyelitis infection, Marek's disease, and nutritional disturbances.

Virus isolation and identification give the most specific diagnoses. Brain tissue is the preferred source of virus for isolation and is inoculated via the yolk sac into 5–7-day-old chick embryos obtained from a susceptible flock. The eggs are allowed to hatch, and the chicks are observed for signs of the disease during the first 10 days of life. When clinical signs appear, the brain, pro ventriculus, and pancreas are examined histologically. Fluorescent antibody procedures are often used to identify the virus (Luginbuhl and Helmboldt, 1978).

Flock immunity tests based on the number of eggs with maternal antibody have been developed to determine which breeding flocks have experienced the infection (Sumner *et al.*, 1957). Fertile eggs are selected from the flock and incubated. Each embryo is inoculated via the yolk sac with 100EID_{50} of the egg-adopted virus at 6 days of age. The embryos are examined for 10–12 days for characteristic lesions. The flock is considered susceptible when 100% of the embryos are infected.

The virus-neutralization test and the agar gel diffusion test have been of only limited use. The virus-neutralization test performed in eggs is more specific; however, a number of chickens fail to produce detectable antibody (Calnek and Jehnich, 1959). The test's acceptance has also been deterred because of the inconsistent supply of susceptible eggs. The agar gel test developed by Lukert and Davis (1971) has been confirmed as reliable.

I. Cardioviruses

Cardiovirus diagnosis requires the isolation and serological identification of the virus. For virus isolation from swine, specimens should be collected from the heart ventricle, liver, spleen, kidney, pancreas, brain, blood, and feces (Littlejohns and Acland, 1975). For rodents or other animals with CNS signs, the brain specimen is most important, but samples from the heart, kidney, and pancreas are also useful (Cottral, 1978). Fetal mouse fibroblast or swine embryo kidney cells are recommended; however, the virus will grow in numerous other cell cultures or embryonating eggs. Suckling mice and hamsters are also useful for initial virus isolations. The virus is readily identified by standard serological tests such as the virus-neutralization, CF, fluorescent antibody, or agar gel diffusion tests (Warren, 1965).

J. Caliciviruses

The caliciviruses can be isolated readily during the acute viral infection by inoculation of appropriate cell cultures. The preferred tissues for isolation of SMS and VES viruses are the lesions and the vesicular fluids, as described under the vesicular diseases. The FC virus can be isolated from nasal excretions, pharyngeal swab samples, and conjunctival scrapings (Gillespie and Scott, 1973). Primary feline kidney cell cultures are preferred for initial isolation. The virus produces a characteristic CPE where the infected cells develop fibrillar cytoplasmic processes and then become rounded (Cottral, 1978). The distinctive fibrillar cytoplasmic processes are also a pronounced feature of the CPE when the fetal diploid line of feline tongue cells (FC_3Tg) or the Crandell feline kidney (CRFK) cell line are used (Cottral, 1978). A tissue-culture isolate is easily identified as an FC by use of physicochemical properties and immunofluorescent tests. However, identification of a specific serotype is cumbersome, time-consuming, and expensive, and is usually done only by investigators specifically interested in research on these viruses (Gillespie and Scott, 1973).

VIII. PREVENTION

The methods used for prevention and control of the vesicular diseases depend on the disease and the country involved. The initial effort is to prevent the introduction of SVD and FMD into countries now free of these diseases. To this end, the importation of live animals, and of meat or meat products and milk or milk products not properly heated, from countries in which the diseases have been identified is contraindicated. Following the introduction of either virus, most developed countries that are currently free of the disease would attempt eradication by quarantine and slaughter. Countries in which FMD is enzootic or in which an outbreak is too extensive to control by slaughter may use vaccines to control the viruses. Vaccines for the control of SVD are still under development; however, the prospect of lasting protection through the use of killed virus vaccine is promising.

Vaccine control of FMD requires for optimal effectiveness a killed-virus vaccine produced from the field strain or from a closely related subtype (Bachrach, 1968). Modified live-virus vaccines for FMD have been developed (Martin *et al.*, 1962; Mowat and Prydie, 1962; Goldsmit, 1964; Pay and Bracewell, 1966), but their use is limited since they frequently cause disease in swine or in cattle under stress (Bachrach, 1968). Consequently, only limited amounts of attenuated-virus vaccines are used today. The killed-virus vaccines have been found both safe and effective in most countries in which FMD is enzootic (Cottral, 1975).

The first effective inactivated virus FMD vaccine for cattle was produced from vesicular fluid and tongue epithelial tissue of cattle (Schmidt, 1938). This method of virus production was both expensive and inconvenient, and was largely replaced by the Frenkel technique of propagating FMDV *in vitro* in fragments of bovine tongue epithelium (Frenkel, 1951). The tissue cultures first used for the large-scale production of FMDV were primary calf kidney cell cultures (Ubertini *et al.*, 1960), which later were replaced with the BHK_{21} cell line (Mowat and Chapman, 1962).

Dilute formaldehyde solution was commonly used to inactivate the virus in the early FMD vaccines (Graves, 1963). However, the complete inactivation of FMD virus was a prolonged process; often some residual live virus was present after 72 hours of incubation with 0.05% formaldehyde solution (Wesslen and Dinter, 1957). Brown and Crick (1959) introduced *N*-acetylethylenimine (AEI), an ethyleneimine derivative, as an inactivating agent. They found that inactivation was complete and rapid, and that the resulting vaccine was highly effective in guinea pigs. Since then, AEI has been widely used as an inactivating agent in FMD vaccine production. Recently ethylenimine (EI) has been found to be a more desirable inactivant for FMDV (Cunliffe, 1973; Warrington *et al.*, 1973; Bahneman, 1975).

The formulation of an effective FMD vaccine requires the use of adjuvants. An inexpensive adjuvant currently widely used is aluminum hydroxide gel. The gel releases the antigen at a rather high rate, giving a peak response in 2–3 weeks; however, it has the disadvantage of providing only brief immunity, so that three doses of vaccine per year must be given to maintain adequate protection. Recent tests with oil adjuvants (incomplete Freund's adjuvant) have demonstrated that the bovine can be protected with one or two doses per year (McKercher and Graves, 1977). The oil adjuvant allows a slower release of antigen, resulting in increased titers that remain high for 4–6 months.

The immunization of swine against FMD is more difficult than the immunization of cattle and requires the use of large quantities of antigen. However, properly formulated vaccines are highly effective and, with the use of oil adjuvant, give immunity lasting three times longer than the immunity from natural exposure.

Vaccines are available in some countries for swine EE (Jones, 1975), duck virus hepatitis (Levine, 1978), and AE (Luginbuhl and Helmboldt, 1978). Excellent vaccines could be produced for SVD (Mowat *et al.*, 1974; Mitev *et al.*, 1978) and the equine rhinoviruses (Kemen, 1975) since the antigens produce excellent, long-lasting immune responses.

Hazlett and Derbyshire (1977a,b) have compared the protection conferred by swine enteroviruses inoculated by different methods with that conferred by the live virus. The best protection, measured by inhibition of growth of virus in the gastrointestinal tract following challenge, was produced by the live-virus vaccines. Usable vaccines are needed for the control of serotypes associated with reproductive losses and baby pig losses. Dunne (1975a,b) gives guidelines for the planned exposure of pigs before breeding and farrowing. The preferred methods involve the use of contact exposure or feeding of feces from young pigs to breeding-age gilts 3–4 weeks prior to breeding. Boars newly introduced into breeding herds should be quarantined for 30 days, and controlled exposure of sows to the flora of the new boars prior to breeding should be considered.

The bovine rhinoviruses and enteroviruses at present are not conclusively identified with any significant disease syndrome, so no control programs have been advanced. If in the future the enteroviruses are associated with extensive reproductive losses, vaccines and control programs may be justified.

The control and eradication of VESV in the United States can serve as an example of sound livestock management practices. The virus spread from California to 42 other states during 1952 and 1953. The rapid spread forced legislatures to enact a series of hygiene laws, the most important of which required the cooking of garbage. This had been proposed in California in 1943 but was not implemented. During the 1952–1953 outbreak, 46 states enacted laws requiring the cooking of garbage, restricting the movement of infected swine, and requiring cleaning and disinfection of contaminated equipment. The control measures were exceptionally effective; the last case of VES was reported in New Jersey in 1956 (Mills and Nielsen, 1968).

Early research indicated that FC comprised many distinct serotypes, and that the prospects for a vaccine consequently were poor (Gillespie and Scott, 1973). Nevertheless, Povey and Ingersoll (1975) and Kahn *et al.* (1975) described experiments that demonstrated a degree of heterotype protection in the cat. Following these *in vivo* studies, several manufacturers developed FC vaccines. The vaccines contain either virus attenuated by passage at low temperature and inoculated intramuscularly or unmodified, naturally occurring virus of low virulence (Bittle and Rubic, 1976; Davis and Beckenhauer, 1976; Scott, 1977). It is too early to evaluate the FC vaccines; however, the results are encouraging, and it is likely that vaccination will provide a method of controlling FC.

IX. CONCLUSION

The picornaviruses are an exceptionally diverse group of viruses capable of producing lesions in most organ systems. A few of the viruses have been extensively researched because of their economic importance, either through morbidity and death losses or through importation embargos.

The vast majority of the picornaviruses are much more elusive. They are constantly isolated from both diseased and normal animals, and they often do not cause disease upon experimental inoculation. Whether these viruses produce morbidity when acting synergistically with other agents or when the animal is under stress is not known.

The viruses often produce subclinical infections, which in the past have been of little concern to scientists. Now evidence is accumulating that subclinical infections may be responsible for reproductive failures in both the male and the female. Abortions, stillbirths, breeding failure, or delayed breeding and other reproductive failures represent the largest economic loss for the livestock producer. These widespread and undramatic losses for years have been considered normal. It is only when epizootics or unusual losses occur that interest in etiology is fostered. A silent viral agent affecting either the reproductive organs or the fetus usually disappears long before its effects on reproductive efficiency are seen; determination of etiology then presents a far more difficult problem than is encountered in more obvious epizootics. The enteroviruses have increasingly been implicated, through serological studies and a limited number of virus isolations from fetuses, as one of these silent viral infections affecting reproduction in both the bovine and the swine. The evidence, though intriguing, is far from conclusive, and extensive research involving both epizootiology and animal inoculations is needed to evaluate the effects of subclinical viral infections.

REFERENCES

Afshar, A., Huck, R. A., Millar, P. G., and Gitter, M. (1964). *J. Comp. Pathol.* **74,** 500–513.

Agrawal, H. O. (1966). *Arch. Gesamte Virusforsch.* **19,** 365–372.

Alexander, T. J. L., Richards, W. P. C., and Roe, C. K. (1959). *Can. J. Comp. Med.* **23,** 216–319.

Almeida, J. D., Waterson, A. P., Prydie, J., and Fletcher, E. W. L. (1968). *Arch. Gesamte Virusforsch.* **25,** 105–114.

Andersen, A. A. (1975). *Am. J. Vet. Res.* **36,** 979–983.

Andersen, A. A. (1977a). *Am. J. Vet. Res.* **38,** 1757–1759.

Andersen, A. A. (1977b). *Proc. 81st Annu. Meet. U. S. Anim. Health Assoc.* pp. 264–269.

Andersen, A. A. (1978a). *Am. J. Vet. Res.* **39,** 59–63.

Andersen, A. A. (1978b). *Am. J. Vet. Res.* **39,** 603–606.

Andersen, A. A., and Campbell, C. H. (1976). *Am. J. Vet. Res.* **37,** 585–589.

Andersen, A. A., and Scott, F. W. (1976). *Cornell Vet.* **66**(2), 232–239.

Andrews, C., and Pereira, H. G. (1972). "Viruses of Vertebrates," 3rd ed. p. 451. Baillière, London.

Andrievskii, V. Y., and Redkin, I. P. (1967). *Veterinariya (Moscow)* **43,** 84–86.

Bachrach, H. L. (1968). *Annu. Rev. Microbiol.* **22,** 201–244.

Bachrach, H. L. (1977). *Beltsville Symp. Agric. Res.* **1,** 3–32.

Bachrach, H. L., and Hess, W. R. (1973). *Biochem. Biophys. Res. Commu.* **55,** 141–149.

Bachrach, H. L., Hess, W. R., and Callis, J. J. (1955). *Science* **122,** 1269–1270.

Bachrach, H. L., Moore, D. M., McKercher, P. D., and Polatnick, J. (1975). *J. Immunol.* **115,** 1636–1641.

Bahneman, H. G. (1975). *Arch. Virol.* **47,** 47–56.

Bankowski, R. A. (1965). *Adv. Vet. Sci.* **10,** 23–64.

Barboni, E., and Manocchio, L. (1962). *Arch. Vet. Ital.* **13,** 477–489.

Barya, M. A., Moll, T., and Mattson, D. E. (1967). *Am. J. Vet. Res.* **28,** 1283–1294.

Betts, A. O., and Jennings, A. R. (1960). *Res. Vet. Sci.* **1,** 160–171.

Betts, A. O., Edington, J., Jennings, A. R., and Reed, S. E. (1971). *J. Comp. Pathol.* **81,** 41–48.

Bhalla, R. C., and Sharma, G. L. (1967). *Indian J. Vet. Sci.* **37,** 287–297.

Bhalla, R. C., and Sharma, G. L. (1968). *Indian J. Vet. Sci.* **38,** 60–66.

Bittle, J. L., and Rubic, W. J. (1976). *Am. J. Vet. Res.* **37,** 275–278.
Blackwell, J. H. (1976). *J. Dairy Sci.* **59,** 1574–1579.
Blackwell, J. H. (1978a). *Res. Vet. Sci.* **25,** 25–28.
Blackwell, J. H. (1978b). *J. Dairy Res.* **45,** 283–285.
Blackwell, J. H. (1978c). *J. Food Prot.* **41,** 631–633.
Blackwell, J. H., and Hyde, J. L. (1976). *J. Hyg.* **77,** 77–83.
Blackwell, J. H., Graves, J. H., and McKercher, P. D. (1975). *Br. Vet. J.* **131,** 317–323.
Bögel, K., Korn, G., and Lorenz, R. J. (1962). *Monatsh. Tierheilkd.* **14,** 227–287.
Bögel, K., Straub, O. C., and Dinter, Z. (1963). *Zentralbl. Bakteriol., Parasitenkd., Infektionskr. Hyg., Abt. 1: Orig.* **190,** 1–6.
Bohl, E. H., Singh, K. V., Hancock, B. B., and Kasza, L. (1960). *Am. J. Vet. Res.* **21,** 99–103.
Bouters, R. (1963). *Med. Veearts. Sch. Rijksunis Gent.* **7**(3), 173–174.
Bouters, R. (1964). *Nature (London)* **201,** 217–218.
Bouters, R., and Vandeplassche, M. (1964). *Tieraerztl. Wochenschr.* **77,** 87–90.
Branny, J., and Zembala, M. (1971). *Br. Vet. J.* **127,** 88–92.
Breese, S. S., Jr., Trautman, R., and Bachrach, H. L. (1965). *Science* **150,** 1303–1305.
Breindl, M. (1971). *Virology* **46,** 962–964.
Brooksby, J. B. (1950). *J. Hyg.* **48,** 184–195.
Brooksby, J. B. (1952). *Agric. Res. Counc., Rep.* **12,** 1–40.
Brown, F. (1960). *J. Immunol.* **85,** 298–303.
Brown, F., and Crick, J. (1959). *J. Immunol.* **82,** 444–447.
Brown, F., and Graves, J. H. (1959). *Nature (London)* **183,** 1688–1689.
Brown, F., and Wild, F. (1974). *Intervirology* **3,** 125–128.
Brown, F., Cartwright, B., and Newman, J. F. E. (1964). *J. Immunol.* **93,** 397–402.
Brown, F., Talbot, P., and Burrows, R. (1973). *Nature (London)* **245,** 315–316.
Brown, F., Wild, T. F., Rowe, L. W., Underwood, B. O., and Harris, F. J. R. (1976). *J. Gen. Virol.* **31,** 231–237.
Bruner, D. W., and Gillespie, J. H. (1973). "Hagan's Infectious Diseases of Domestic Animals," pp. 1205–1231. Cornell Univ. Press, Ithaca, New York.
Bulloch, W. (1927). *J. Comp. Pathol. Ther.* **40,** 75–76.
Bürki, F. (1965). *Arch. Gesamte Virusforsch* **15,** 690–696.
Burroughs, J. N., and Brown, F. (1974). *J. Gen Virol.* **22,** 281–286.
Burroughs, J. N., and Brown, F. (1978). *J. Gen. Virol.* **41,** 443–446.
Burroughs, J. N., Rowland, D. J., Sangar, D. V., Talbot, P., and Brown, F. (1971). *J. Gen. Virol.* **13,** 73–84.
Burroughs, J. N., Doel, T., and Brown, F. (1978a). *Intervirology* **10,** 51–59.
Burroughs, J. N., Doel, T., Smale, C. J., and Brown, F. (1978b). *J. Gen. Virol.* **40,** 161–174.
Burrows, R. (1966). *J. Hyg.* **64,** 81–90.
Burrows, R. (1968). *Vet. Rec.* **82,** 387–388.
Burrows, R. (1970). *Proc. Int. Conf. Equine Infect. Dis., 2nd, 1969,* pp. 154–164.
Burrows, R., Mann, J. A., Greig, A., Chapman, W. G., and Goodridge, D. (1971). *J. Hyg.* **69,** 307–321.
Burrows, R., Greig, A., and Goodridge, D. (1973). *Res. Vet. Sci.* **15,** 141–144.
Burrows, R., Mann, J. A., and Goodridge, D. (1974). *J. Hyg.* **72,** 135–143.
Callis, J. J., McKercher, P. D., and Shahan, M. S. (1975). *In* "Diseases of Swine" (H. W. Dunne and A. D. Leman, eds.), 4th ed., pp. 325–345. Iowa State Univ. Press, Ames.
Calnek, B. W., and Jehnich, H. (1959). *Avian Dis.* **3,** 225–239.
Campbell, C. H. (1963). *J. Bacteriol.* **86,** 593–597.

Charton, A., Faye, P., Lecoanet, J., Desbrosse, H., and Lelayec, C. (1963). *Recl. Med. Vet.* **139,** 897-908.

Ciuca, A. (1929). *J. Hyg.* **28,** 325-339.

Cliver, D. O., and Bohl, E. H. (1962a). *J. Dairy Sci.* **45,** 921-925.

Cliver, D. O., and Bohl, E. H. (1962b). *J. Dairy Sci.* **45,** 926-932.

Coggins, L., and Kemen, M. J. (1975). *J. Am. Vet. Med. Assoc.* **166,** 80-83.

Coombs, G. P. (1974). *Proc. 77th Annu. Meet. U.S. Anim. Health* pp. 332-335.

Cooper, P. D., Agol, V. I., Bachrach, H. L., Brown, F., Ghendon, Y., Gibbs, A. J. Gillespie, J. H., Lonberg-Holm, K., Mandel, B., Melnick, J. L., Mohanty, S. B., Povey, R. C., Rueckert, R. R., Schaffer, F. L., and Tyrrell, D. A. J. (1978). *Intervirology* **10,** 165-180.

Cottral, G. E. (1972). *J. Am. Vet. Med. Assoc.* **161,** 1293-1298.

Cottral, G. E. (1975). *In* "Foreign Animal Diseases," 3rd ed., pp. 109-128. U.S. Anim. Health Assoc., Richmond, Virginia.

Cottral, G. E. (1978). "Manual of Standardized Methods for Veterinary Microbiology," pp. 248-272. Cornell Univ. Press, Ithaca, New York.

Cottral, G. E., Gailiunas, P., and Campion, R. I. (1963). *U.S. Livestock Sanit. Assoc. Proc.* **67,** 463-472.

Cottral, G. E., Gailiunas, P., and Cox, B. F. (1968). *Arch. Gesamte Virusforch.* **23,** 362-377.

Cowan, K. M. (1966). *Am. J. Vet. Res.* **27,** 1217-1227.

Cowan, K. M. (1973). *Adv. Immunol.* **17,** 195-253.

Cowan, K. M., and Graves, J. H. (1966). *Virology* **30,** 528-540.

Cowan, K. M., and Graves, J. H. (1968). *Virology* **34,** 544-548.

Cowan, K. M., and Trautman, R. (1965). *J. Immunol.* **94,** 858-867.

Cowan, K. M., and Trautman, R. (1967). *J. Immunol.* **99,** 729-736.

Crawford, A. B. (1937). *J. Am. Vet. Med. Assoc.* **90,** 380-395.

Cropper, M., Dunne, H. W., Leman, A. D., Starkey, A. L., and Hoefling, D. C. (1976). *J. Am. Vet. Med. Assoc.* **168,** 233-235.

Cunha, R. G., and Honigman, M. N. (1963). *Am. J. Vet. Res.* **24,** 371-375.

Cunliffe, H. R. (1962). *Can. J. Comp. Med. Vet. Sci.* **26,** 182-185.

Cunliffe, H. R. (1964). *Cornell Vet.* **54,** 501-510.

Cunliffe, H. R. (1973). *Appl. Microbiol.* **26,** 747, 750.

Cunliffe, H. R., and Blackwell, J. H. (1977). *J. Food Prot.* **40,** 389-392.

Darbyshire, J. H., and Dawson, P. S. (1963). *Res. Vet. Sci.* **4,** 48-55.

Dardiri, A. H., Seibold, H. R., and DeLay, P. D. (1966). *Can. J. Comp. Med. Vet. Sci.* **30,** 71-81.

Davie, J. (1964). *J. Hyg.* **62,** 401-411.

Davis, E. V., and Beckenhauer, W. H. (1976). *VM/SAC, Vet. Med. Small Anim. Clin.* **71,** 1405-1410.

Dawe, P. S., and Pinto, A. A. (1978). *Br. Vet. J.* **134,** 504-511.

Dawe, P. S., Forman, A. J., and Smale, C. J. (1973). *Nature (London)* **241,** 540-542.

Derbyshire, J. B., Clarke, M. C., and Jessett, D. M. (1966). *Vet. Rec.* **79,** 595-599.

Dhindsa, O. S., and Dziuk, P. J. (1968). *J. Anim. Sci.* **28,** 668-672.

Dimmock, N. J., and Tyrrell, D. A. J. (1964). *Br. J. Exp. Pathol.* **45,** 271-280.

Ditchfield, J., and MacPherson, L. W. (1965). *Cornell Vet.* **55,** 181-189.

Dobbertin, S., Teufel, P., and Wernery, R. (1974). *Berl. Muench. Tieraerztl. Wochenschr.* **87,** 350-352.

Douglas, J. D., Vassington, P. J., and Noel, J. K. (1966). *Proc. Soc. Exp. Biol. Med.* **121,** 824-829.

Dunne, H. W. (1975a). *In* "Diseases of Swine" (H. W. Dunne, and A. D. Leman, eds.), 4th ed., pp. 353-368. Iowa State Univ. Press, Ames.

Dunne, H. W. (1975b). *In* "Diseases of Swine" (H. W. Dunne and A. D. Leman, eds.), 4th ed., pp. 918-952. Iowa State Univ. Press, Ames.
Dunne, H. W., Gobble, J. L., Hokanson, J. F., Kradel, D. C., and Bubash, G. R. (1965). *Am. J. Vet. Res.* **26,** 1284-1297.
Dunne, H. W., Kradel, D. C., Clark, C. D., Bubash, G. R., and Ammerman, E. (1967). *Am. J. Vet. Res.* **28,** 557-568.
Dunne, H. W., Ajinkya, S. M., Bubash, G. R., and Griel, L. C. (1973). *Am. J. Vet. Res.* **34,** 1121-1126.
Dunne, H. W., Huang, C. M., and Lin, W. J. (1974a). *J. Am. Vet. Med. Assoc.* **164,** 290-294.
Dunne, H. W., Wang, J. T., and Huang, C. M. (1974b). *Am. J. Vet. Res.* **35,** 1479-1481.
Federer, K. E., Saille, J., and Gomes, I. (1964). *Bull. Off. Int. Epizoot.* **61,** 1563-1578.
Finch, J. T., and Klug, A. (1959). *Nature (London)* **183,** 1709-1714.
Fracastorius (1514). Libre iii. Published in Venice, 1546. cited by Bulloch W. (1927).
Frenkel, H. S. (1951). *Am. J. Vet. Res.* **12,** 187-190.
Gailiunas, P., and Cottral, G. E. (1966). *J. Bacteriol.* **91,** 2333-2338.
Gainer, J. H., Sandefur, J. R., and Bigler, W. J. (1968). *Cornell Vet.* **58,** 31-47.
Gardiner, M. R. (1962). *Aust. Vet. J.* **38,** 24-26.
Gasparini, G., and Nani, S., (1955). *G. Microbiol.* **1,** 170-175.
Gillespie, J. H., and Scott, F. W. (1973). *Adv. Vet. Sci.* **17,** 163-200.
Gillespie, J. H., Judkins, A. B., and Kahn, D. E. (1971). *Cornell Vet.* **61,** 172-179.
Gins, H. A., and Krause, C. (1924). *Ergeb. All. Pathol. Pathol. Anat. Menschen Tiere* **20,** 805-807.
Goldsmit, L. (1964). *Bull. Off. Int. Epizoot.* **61,** 1177-1182.
Gourreau, J. M., Fremont, A., Bourlier, A., Bruder, C., and Langneau, F. (1975). *Rev. Med. Vet. (Toulouse)* **126**(3), 357-364; Abst.: *Foot Mouth Dis. Bull.* **14**(6), 57(1975/1980).
Graham, A. M. (1959). *Vet. Rec.* **71,** 383-387.
Graves, J. H. (1960). *Am. J. Vet. Res.* **21,** 687-690.
Graves, J. H. (1963). *J. Immunol.* **91,** 251-256.
Graves, J. H. (1973). *Nature (London)* **245,** 314-315.
Graves, J. H. (1979). *J. Am. Vet. Med. Assoc.* **174,** 174-176.
Graves, J. H., and McKercher, P. D. (1975). *In* "Diseases of Swine" (H. W. Dunne, and A. D. Leman, eds.), 4th ed., pp. 346-352. Iowa State Univ. Press, Ames.
Graves, J. H., Cowan, K. M., and Trautman, R. (1964). *J. Immunol.* **92,** 501-506.
Graves, J. H., Cowan, K. M., and Trautman, R. (1968). *Virology* **34,** 269-274.
Graves, J. H., McVicar, J. W., Sutmoller, P., Trautman, R., and Wagner, G. G. (1971). *J. Infect. Dis.* **124,** 270-276.
Hancock, B. B., Bohl, E. H., and Birkeland, J. M. (1959). *Am. J. Vet. Res.* **20,** 127-132.
Harding, J. D. J., Done, J. T., and Kershaw, G. F. (1957). *Vet. Rec.* **69,** 824-832.
Hausen, P., Hausen, H., Rott, R., Scholtissek, C., and Schafer, W. (1963). *In* "Viruses, Nucleic Acids, and Cancer," 17th Symposium on Fundamental Cancer Res., pp. 282-295. Williams & Wilkins, Baltimore, Maryland.
Hazlett, D. T. G., and Derbyshire, J. B. (1976). *Can. J. Comp. Med.* **40,** 370-379.
Hazlett, D. T. G., and Derbyshire, J. B. (1977a). *Can. J. Comp. Med.* **41,** 257-263.
Hazlett, D. T. G., and Derbyshire, J. B. (1977b). *Can. J. Comp. Med.* **41,** 264-273.
Heatley, W., Skinner, H. H., and Subak-Sharpe, H. (1960). *Nature (London)* **186,** 909-911.
Heberling, R. L., and Cheever, F. S. (1965). *Am. J. Epidemiol.* **81,** 106-123.
Hecke, F. (1958). *Monatsh. Tierheilkd.* **10,** 197-217.
Hedger, R. S. (1970). *Vet. Rec.* **87,** 180-188.
Hersey, F. D., and Maurer, F. D. (1961). *Proc. Soc. Exp. Biol. Med.* **107,** 645-646.
Hess, W. R., Bachrach, H. L., and Callis, J. J. (1960). *Am. J. Vet. Res.* **21,** 1104-1108.
Hofer, B., Steck, F., Gerber, H., Lohrer, J., Nicolet, J., and Paccaud, M. F. (1973). *In* "Equine

Infectious Diseases III" (J. T. Bryans and H. Gerber, eds.), Vol. III, pp. 527-545. Karger, Basel.
Hoover, E. A., and Kahn, D. E. (1973). *Vet. Pathol.* **10,** 307-322.
Horstmann, D. M. (1952). *J. Immunol.* **69,** 379-394.
Huck, R. A., and Cartwright, S. F. (1964). *J. Comp. Pathol.* **74,** 346-365.
Hyde, J. L., Blackwell, J. H., and Callis, J. J. (1975). *Can. J. Comp. Med.* **39,** 305-309.
Hyslop, N. St. G. (1965a). *J. Comp. Pathol.* **75,** 111-117.
Hyslop, N. St. G. (1965b). *J. Comp. Pathol.* **75,** 119-126.
Hyslop, N. St. G. (1970). *Adv. Vet. Sci. Comp. Med.* **14,** 261-307.
Ide, P. R., and Darbyshire, J. H. (1972a). *Arch. Gesamte Virusforsch.* **36,** 166-176.
Ide, P. R., and Darbyshire, J. H. (1972b). *Arch. Gesamte Virusforsch.* **36,** 335-342.
Ide, P. R., and Darbyshire, J. H. (1972c). *Arch. Gesamte Virusforsch.* **36,** 343-350.
Jamison, R. M. (1969). *J. Virol.* **4,** 904-906.
Jastrzebski, T. (1961). *Arch. Exp. Veterinaermed.* **15,** 408-423.
Jones, T. C. (1975). *In* "Diseases of Swine" (H. W. Dunne and A. D. Leman, eds.), 4th ed., pp. 369-384. Iowa State Univ. Press, Ames.
Kaaden, O., and Matthaeus, W. (1970). *Arch. Gesamte. Virusforsch.* **30,** 263-266.
Kahn, D. E., Hoover, E. A., and Bittle, J. L. (1975). *Infect. Immun.* **11,** 1003-1009.
Kamizono, M., Knoishi, S., Ozata, M., and Korbori, S. (1968). *Jpn. J. Vet. Sci.* **30,** 197-206.
Kemen, M. J. (1975). *J. Am. Vet. Med. Assoc.* **166,** 85-88.
Kilham, L., Mason, P., and Davies, J. N. P. (1956). *Am. J. Trop. Med. Hyg.* **5,** 655-663.
Klein, M., and Earley, E. (1957). *Bacteriol. Proc.* **57,** 73.
Koestner, A., Long, J. F., and Kasya, L. (1962). *J. Am. Vet. Med. Assoc.* **140,** 811-814.
Komarov, A. (1954). *Refu. Vet.* **11,** 239-241.
Kötsche, W. (1958). *Acta Virol. (Engl. Ed.)* **2,** 103-112.
Kunin, C. M., and Minuse, E. (1958). *J. Immunol.* **80,** 1-11.
Kurogi, H., Inaba, Y., Goto, Y., Takahashi, E., Sato, K., Omori, T., and Matumoto, M. (1974). *Arch. Gesamte Virusforsch.* **44,** 215-226.
Kurogi, H., Inaba, Y., Takahashi, E., Sato, K., Goto, Y., and Omori, T. (1975). *Natl. Inst. Anim. Health Q.* **15,** 201-202.
Kurogi, H., Inaba, Y., Takahashi, E., Sato, K., and Omori, T. (1976). *Natl. Inst. Anim. Health Q.* **16,** 49-58.
Lai, S. S., McKercher, P. D., Moore, D. M., and Gillespie, J. H. (1979). *Am. J. Vet. Res.* **40,** 463-468.
La Porte, J., and Lenoir, G. (1973). *J. Gen. Virol.* **20,** 161-168.
Lee, K. M., and Gillespie, J. H. (1973). *Infect. Immun.* **7,** 678-679.
Levine, P. P. (1978). *In* "Diseases of Poultry" (M. S. Hofstad, ed.), 7th ed., pp. 611-619. Iowa State Univ. Press, Ames.
Littlejohns, I. R., and Acland, H. M. (1975). *Aust. Vet. J.* **51,** 416-422.
Loeffler, P. (1909). *Dtsch. Med. Wochenschr.* p. 2097; cited by Hyslop (1970).
Loeffler, P., and Frosch, P. (1897). *Zentralbl. Bakteriol., Parasitenkd. Infektionskr., Abt. I* **22,** 257-259.
Lonberg-Holm, K., and Yin, F. H. (1973). *J. Virol.* **12,** 114-123.
Long, J. F., Koestner, A., and Kasza, L. (1966). *Am. J. Vet. Res.* **27,** 274-279.
Luginbuhl, R. E., and Helmboldt, C. F. (1978). *In* "Diseases of Poultry" (M. S. Hofstad, ed.), 7th ed., pp. 537-547. Iowa State Univ. Press, Ames.
Lukert, P. D., and Davis, R. B. (1971). *Avian Dis.* **15,** 935-938.
McFerran, J. B. (1962a). *Ann. N. Y. Acad. Sci.* **101,** 436-443.
McFerran, J. B. (1962b). *J. Pathol. Bacteriol.* **83,** 73-83.
McKercher, D. G. (1977). *Beltsville Symp. Agric. Res.* **1,** 83-100.

McKercher, P. D., and Farris, H. E. (1967). *Arch. Gesamte Virusforsch.* **22,** 451-461.
McKercher, P. D., and Giordano, A. R. (1967). *Arch. Gesamte Virusforsch.* **20,** 39-53.
McKercher, P. D., and Graves, J. H. (1977). *Dev. Biol. Stand.* **35,** 107-112.
McKercher, D. G., and Kendrick, J. W. (1958). *In* "Reproduction and Infectivity" (F. X. Gassner, ed.), pp. 29-37. Pergamon, Oxford.
McVicar, J. W. (1977). *Bol. Cient. Panam. Fiebre Aftosa* **26,** 9-14.
McVicar, J. W., Stumoller, P., and Andersen, A. A (1974). *Arch. Gesamte Virusforsch.* **44,** 168-172.
Madeley, C. R., and Cosgrove, B. P. (1976). *Lancet* **1,** 199.
Madin, S. H. (1975). *In* "Diseases of Swine" (H. W. Dunne and A. D. Leman, eds.), 4th ed., pp. 286-307. Iowa State Univ. Press, Ames.
Madin, S. H., Smith, A. W., and Akers, T. G. (1976). *In* "Wildlife Diseases" (L. A. Page, ed.), pp. 197-204. Plenum, New York.
Mak, T. W., O'Callaghan, D. J., and Colter, J. S. (1970). *Virology* **40,** 565-571.
Martin, W. B., Davies, E. B., and Smith, I. M. (1962). *Res. Vet. Sci.* **3,** 357-367.
Matsumoto, M., McKercher, P. D., and Nusbaum, K. E. (1978). *Am. J. Vet. Res.* **39,** 1081-1087.
Mattson, D. E., Moll, T., and Balya, M. A. (1969). *Am. J. Vet. Res.* **30,** 1577-1585.
Mattson, J. M., and Reed, D. E. (1974). *Am. J. Vet. Res.* **35,** 1337-1341.
Mayor, H. D. (1964). *Virology* **22,** 156-160.
Mayr, A. (1962). *Ann. N. Y. Acad. Sci.* **101,** 423-427.
Mayr, A., and Hecke, F. (1960). *Bull. Off. Int. Epizoot.* **54,** 445-448.
Mayr, A., and Wittmann, G. (1959). *Z. Immunitaetsforsch.* **117,** 45-52.
Meloen, R. H. (1978). *Arch. Virol.* **58,** 35-43.
Mills, J. H. L., and Nielsen, S. E. (1968). *Adv. Vet. Sci.* **12,** 33-104.
Mitev, G., Tekerlekov, P., Dilovsky, M., Ognianev, D., and Nikolova, E. (1978). *Arch. Exp. Veterinaermed.* **32,** 29-33.
Mohanty, S. B., and Lillie, M. G. (1968). *Proc. Soc. Exp. Biol. Med.* **128,** 850-852.
Mohanty, S. B., Lillie, M. G., Albert, T. F., and Sass, B. (1969). *Am. J. Vet. Res.* **30,** 1105-1111.
Moll, T. (1964). *Am. J. Vet. Res.* **25,** 1757-1762.
Moll, T., and Davis, A. D. (1959). *Am. J. Vet. Res.* **20,** 27-32.
Moll, T., and Finlayson, A. V. (1957). *Science* **126,** 401-402.
Moll, T., and Ulrich, M. I. (1963). *Am. J. Vet. Res.* **24,** 545-550.
Monlux, W. S., Graves, J. H., and McKercher, P. D. (1974). *Am. J. Vet. Res.* **35,** 615-617.
Monlux, W. S., McKercher, P. D., and Graves, J. H. (1975). *Am. J. Vet. Res.* **36,** 1745-1749.
Moraillon, A., Moraillon, R., and Brion, A. (1973). *Ann. Rech. Vet.* **4,** 293-304.
Moscovici, C., Ginevri, A., and Mazzarocchio, V. (1959). *Am. J. Vet. Res.* **20,** 625-626.
Mott, L. O. (1957). *In* "Proceedings of The Symposium on Vesicular Disease," ARS 45-1 1957, pp. 74-83. Plum Island Anim. Dis. Lab. Greenport, New York.
Mowat, G. N., and Chapman, W. G. (1962). *Nature (London)* **194,** 253-255.
Mowat, G. N., and Prydie, J. (1962). *Res. Vet. Sci.* **3,** 368-381.
Mowat, G. N., Darbyshire, J. H., and Huntley, J. F. (1972). *Vet. Rec.* **90,** 618-621.
Mowat, G. N., Prince, M. J., Spier, R. E., and Staple, R. F. (1974). *Arch. Gesamte Virusforsch.* **44,** 350-360.
Murnane, T. G., Craighead, J. E., Mondragon, H., and Shelokov, A. (1960). *Science* **131,** 498-499.
Nardelli, L., Lodetti, E., Gualandi, G. L., Burrows, R., Goodridge, D., Brown, F., and Cartwright, G. (1968). *Nature (London)* **219,** 1275-1276.
Newman, J. F. E., Rowlands, D. J., Brown, F., Goodridge, D., Burrows, R., and Steck, F. (1977). *Intervirology* **8,** 145-154.
Niederman, R. A., Luginbuhl, R. E., and Helmboldt, C. F. (1963). *Cornell Vet.* **53,** 550-560.

Noble, J., and Lonberg-Holm K. (1973). *Virology* **51,** 270–278.

Odegaard, O. A., and Straub, O. C. (1970). *Acta Vet. Scand.* **11,** 536–544.

Odell, D. K. (1970). *Proc. 7th Annu. Conf. Biol. Sonar Diving Mammals* p. 185.

Ogelsby, A. S., Schaffer, F. L., and Madin, S. H. (1971). *Virology* **44,** 329–341.

Olsen, R. G., Kahn, D. E., Hoover, E. A., Saxe, N. J., and Yohn, D. S. (1974). *Infect. Immun.* **10,** 375–380.

Parker, J. (1971). *Vet. Rec.* **88,** 659–662.

Pay, T. W. F., and Bracewell, C. D. (1966). *Bull. Off. Int. Epizoot.* **65,** 313–332.

Phillips, R. M., Foley, C. W., and Lukert, P. D. (1972). *J. Am. Vet. Med. Assoc.* **161,** 1306–1316.

Pinto, A. A., and Garland, A. J. M. (1979). *J. Hyg.* **82,** 41–50.

Platt, H. (1956). *J. Pathol. Bacteriol.* **72**(1), 299–312.

Platt, H. (1958). *J. Pathol. Bacteriol.* **76**(1), 119–131.

Plummer, G. (1962). *Nature (London)* **195,** 519–520.

Plummer, G. (1963). *Arch. Gesamte Virusforsch.* **12,** 694–700.

Plummer, G., and Kerry, J. B. (1962). *Vet. Rec.* **74,** 967–970.

Pope, J. H., and Scott, W. (1960). *Aust. J. Exp. Biol.* **38,** 447–450.

Potel, K. (1958). *Monatsh. Veterinaermed.* **13,** 401–405.

Povey, R. C. (1974). *Infect. Immun.* **10,** 1307–1314.

Povey, C., and Ingersoll, J. (1975). *Infect. Immun.* **11,** 877–885.

Powell, D. G. (1975). *Vet. Rec.* **96,** 30–34.

Powell, D. G., Burrows, R., and Goodridge, D. (1974). *Equine Vet.* **6,** 19–24.

Reed, S. E. (1972). *J. Comp. Pathol.* **82,** 267–278.

Rodeffer, H. E., Leman, A. E., Dunne, H. W., Cropper, M., and Sprecher, D. J. (1975). *J. Am. Vet. Med. Assoc.* **166,** 991–992.

Rose, M. A., Hopes, R., Rossdale, P. D., and Beveridge, W. I. B. (1974). *Vet. Res.* **95,** 484–488.

Rosenquist, B. D. (1972). *Proc. 76th Annu. Meet. U.S. Anim. Health Assoc.* pp. 724–735.

Rosenquist, B. D., and Dobson, A. W., (1974). *Am. J. Vet. Res.* **35,** 363–365.

Rowlands, D. J., Cartwright, B., and Brown, R. (1969). *J. Gen. Virol.* **4,** 479–487.

Rowlands, D. J., Sangar, D. V., and Brown, F. (1975a). *J. Gen. Virol.* **26,** 227–238.

Rowlands, D. J., Shirley, M. W., Sangar, D. W., and Brown, F. (1975b). *J. Gen. Virol.* **29,** 223–234.

Rueckert, R. R. (1971). *In* "Comparative Virology" (K. Maramorosch and E. Kurstack, eds.), pp. 255–306. Academic Press, New York.

Rueckert, R. R. (1976). *In* "Comprehensive Virology" (H. Fraenkel-Conrat and R. R. Wagner, eds.), Vol. 6, pp. 131–213. Plenum, New York.

Rweyemamu, M. M., Booth, J. C., and Pay, T. W. F. (1977). *J. Hyg.* **78,** 99–112.

Rweyemamu, M. M., Terry, G., and Pay, T. W. F. (1979). *Arch. Virol.* **59,** 69–79.

Sahu, S. P., and Dardiri, A. H. (1979). *J. Wildl. Dis.* **15,** 489–494.

Sawyer, J. C. (1976). *J. Am. Vet. Med. Assoc.* **169,** 707–709.

Sawyer, J. C., Madin, S. H., and Skilling, D. E. (1978). *Am. J. Vet. Res.* **39,** 137–139.

Schaffer, F. L., and Soergel. M. E. (1976). *J. Virol.* **19,** 925–931.

Schmidt, S. (1938). *Z. Immunitaetsforsch.* **92,** 392–409.

Scott, F. W. (1977). *Am. J. Vet. Res.* **38,** 229–234.

Scott, F. W., Cottral, G. E., and Gailiunas, P. (1965). *Proc. 69th Annu. Meet. U.S. Livestock Sanit. Assoc.* pp. 87–93.

Scott, F. W., Cottral, G. E., and Gailiunas, P. (1966). *Am. J. Vet. Res.* **27,** 1531–1536.

Seibold, H. R. (1960). *Am. J. Vet. Res.* **21,** 870–877.

Seibold, H. R. (1963). *Am. J. Vet. Res.* **24,** 1123–1130.

Sellers, R. F. (1955). *Nature (London)* **176,** 547–549.

Sellers, R. F. (1971). *Vet. Bull.* **41,** 431–439.

Sellers, R. F., and Herniman, K. A. J. (1974). *J. Hyg.* **72,** 61–65.
Sellers, R. F., and Parker, J. (1969). *J. Hyg.* **67,** 671–677.
Sellers, R. F., Herniman, K. A. J., and Gumm, I. D. (1977). *Res. Vet. Sci.* **23,** 70–75.
Sibalin, M., and Lannek, N. (1960). *Arch. Gesamte Virusforsch.* **10,** 31–45.
Singh, K. V., Bohl, E. H., and Sauger, V. L. (1964). *Cornell Vet.* **54,** 612–628.
Skinner, H. H. (1953). *Proc. Int. Vet. Congr., 15th, 1953,* Vol. 1, pp. 195–200.
Smadel, J. E., and Warren, J. (1947). *J. Clin. Invest.* **26,** 1197.
Smith, A. W., and Akers, T. G. (1976). *J. Am. Vet. Med. Assoc.* **169,** 700–703.
Smith, A. W., Akers, T. G., Madin, S. H., and Vedros, N. A. (1973). *Nature (London)* **244,** 108–109.
Smith, A. W., Prato, C. M., and Skilling, D. E. (1977). *Intervirology* **8,** 30–36.
Smith, A. W., Skilling, D. E., and Ritchie, A. E. (1978). *Am. J. Vet. Res.* **39,** 1531–1533.
Soergel, M. E., Schaffer, F. L., Sawyer, J. C., and Prato, C. M. (1978). *Arch. Virol.* **57,** 271–282.
Speir, R. W. (1962). *Virology* **17,** 588–592.
Straub, O. C. (1965). *Dtsch. Tieraerztl. Wochenschr.* **72,** 54–55.
Straub, O. C., and Bohm, H. O. (1964). *Arch. Gesamte Virusforsch.* **14,** 272–275.
Strobbe, R. (1978). *Ann. Med. Vet.* **122**(4), 257–270.
Strohmaier, K., and Adam, K. H. (1974). *J. Gen. Virol.* **22,** 105–114.
Strohmaier, K., Kaaden, O. R., Adam, K. H., and Wittmann-Liebold, B. (1978). *In* "Report of the Session of the Research Group of the Standing Technical Committee of the European Commission for the Control of Foot-and-Mouth-Disease, Food and Agriculture Organization of the United Nations," Appendix IV, pp. 55–66.
Studdert, M. J. (1978). *Arch. Virol.* **58,** 157–191.
Subak-Sharpe, H. (1961). *Arch. Gesamte Virusforsch.* **11,** 373–399.
Summer, F. W., Luginbuhl, R. E., and Jungherr, E. L. (1957). *Am. J. Vet. Res.* **18,** 720–723.
Sutmoller, P., and Gaggero, C. A. (1965). *Vet. Rec.* **77,** 968–969.
Sutmoller, P., and McVicar, J. W. (1976). *J. Hyg.* **77,** 245–253.
Talbot, P., Rowlands, D. J., Burroughs, J. N., Sangar, D. V., and Brown, F. (1973). *J. Gen. Virol.* **19,** 369–380.
Tan, R. J. S. (1970). *J. Med. Sci. Biol.* **23,** 419–424.
Traum, J. (1936). *J. Am. Vet. Med. Assoc.* **88,** 316–327.
Trautman, R., and Bennett, C. E. (1979). *J. Gen. Virol.* **42,** 457–466.
Ubertini, B., Nardelli, L., and Panina, G. (1960). *J. Biochem. Microbiol. Technol. Eng.* **2,** 327–333.
Van Bekkum, J. G. (1969). *Proc. Congr. Int. Pig Vet. Soc. Camb. 1st, 1969,* p. 45.
Van Bekkum, J. G., Frenkel, H. S., Frederiks, H. H., and Frenkel, S. (1959). *Tijdschr. Diergeneeskd.* **84,** 1159–1167.
Van Der Maaten, M. J., and Packer, R. A. (1967). *Am. J. Vet. Res.* **28,** 677–684.
Vande Woude, G. F., Swaney, J. B., and Bachrach. H. L. (1972). *Biochem. Biophys. Res. Commun.* **48,** 1222–1229.
Van Oss, C. O., Dhennin, L., and Dhennin, L. (1964). *Virology* **22**(3), 428–430.
Vertinskii, K. I., Bessarabov, B. F., Kurilenko, A. N., Strelnikov, A. P., and Makhno, P. M. (1968). *Veterinariya (Moscow)* **7,** 27–30; *Vet. Bull.* **39,** 2074 (abstr.) (1969).
Veznikova, D., Gois, M., Cerny, M. (1969). *Arch. Exp. Veterinaermed.* **23,** 59–64; *Vet. Bull.* **39,** 781 (abstr.) (1970).
Wallis, C., and Melnick, J. L. (1962). *Virology* **16,** 504–506.
Wang, J. T., Dunne, H.W., Griel, L. C., Hokanson, J. F., and Murphy, D. M. (1973). *Am. J. Vet. Res.* **34,** 785–791.
Warren, J. (1965). *In* "Viral and Rickettsial Infections of Man" (F. L. Horsfall and I. Tamm, eds.), 4th ed., pp. 526–568. Lippincott, Philadelphia, Pennsylvania.

Warrington, R. E., Cunliffe, H. R., and Bachrach, H. L. (1973). *Am. J. Vet. Res.* **34,** 1087–1091.
Wawrzkiewics, J., Samle, C. J., and Brown, F. (1968). *Arch. Gesamte Virusforsch.* **25,** 337–351.
Weldon, S. L., Blue, J. L., Wooley, R. E., and Lukert, P. D. (1978). *J. Am. Vet. Med. Assoc.* **174,** 168–169.
Wesslen, T., and Dinter, Z. (1957). *Arch. Gesamte Virusforsch.* **7,** 394–402.
Western Hemisphere Committee on Animal Virus Characterization (1975). *Am. J. Vet. Res.* **36,** 861–872.
Wilder, F. W., and Dardiri, A. H. (1978). *Can. J. Comp. Med.* **42,** 200–204.
Wildy, P. (1971). *Monogr. Virol.* **5,** 1–81.
Woode, G. N., and Bridger, J. V. (1978). *J. Med. Microbiol.* **11,** 441–452.
Woods, G. T., Watrach, A. M., Fearino, P., and Zingilieta, M. (1970). *Can. J. Comp. Med.* **34,** 122–125.
Yamanouchi, K., Bankowski, R. A., Howarth, J. A., and Huck, R. A. (1964). *Am. J. Vet. Res.* **25,** 609–612.
Zoletto, R., Carlotto, F., Stilas, B., and Carcelloti, F. (1973). *Vet. Ital.* **24,** 310–316.
Zwillenberg, L. O., and Bürki, F. (1966). *Arch. Gesamte Virusforsch.* **19,** 373–384.

Part II

REOVIRIDAE

Chapter 2

Reoviridae: Orbivirus and Reovirus Infections of Mammals and Birds

NEVILLE F. STANLEY

I. Introduction 67
II. Orbivirus 69
A. Characterization and Antigenic Grouping 69
B. Morphology and Structure 73
C. African Horse Sickness 80
D. Bluetongue Group 81
E. Colorado Tick Fever 85
F. Kemerovo Group 87
G. Other Groups 88
H. Laboratory Diagnosis 92
I. Problems of Characterization 92
III. Reovirus 94
A. Comparative Aspects of Mammalian Types 94
B. Importance of Analysis of the Double-Stranded RNA Genome . 95
C. Comparative Aspects of Avian Types 96
IV. Provisional Reoviridae 97
A. Infectious Pancreatic Necrosis Virus 98
B. Infectious Bursal Disease Virus 98
C. Other Possible Members 98
V. Conclusion 99
References 99

I. INTRODUCTION

Viruses in the family Reoviridae show great ubiquity in infecting plants, arthropods, and vertebrates. For this reason and the paucity of data on recent

COMPARATIVE DIAGNOSIS
OF VIRAL DISEASES, VOL. IV

ISBN 0-12-429704-8

isolates, taxonomic problems have yet to be resolved. The International Committee on Taxonomy of Viruses met in 1978 to discuss the Reoviridae, and it was accepted that the family comprises five named genera: *Reovirus, Orbivirus, Rotavirus, Phytoreovirus,* and *Fijivirus*. No genus name has yet been agreed on for the cytoplasmic polyhedrosis viruses (see Fig. 1; Matthews, 1979).

Reoviridae attract interest for three main reasons: (1) molecular biologists study them because they all have a double-stranded RNA genome subject to experimental manipulations; (2) students of infectious disease are involved because some of them produce infections with or without disease in man, animals, plants, and insects; (3) they offer interesting models for the study of antigenic configuration, disease pathogenesis, and virus evolution.

Genera within the family have common properties which differentiate them from all other viruses (see Fig. 1; Joklik, 1974; Stanley, 1974, 1978). The orbiviruses and plant viruses have arthropod transmission, whereas the rotaviruses and reoviruses have not yet been shown to be naturally transmitted this way. Considered from the viewpoint of transmission and replication in arthropods, orbiviruses are classified as among the arboviruses, but in addition to their double-stranded RNA, they differ from typical togaviruses by their relative resistance to lipid solvents and desoxycholate (Borden *et al.*, 1971). This relates to the absence of a true envelope like that possessed by the togaviruses. Most of the arboviruses currently fall into the following families and genera (Porterfield, 1975):

Togaviridae	*Alphavirus*
	Flavivirus
Rhabdoviridae	Unidentified genera
Bunyaviridae	*Bunyavirus*
	Unidentified genera
Reoviridae	*Orbivirus*

A notable characteristic of the Reoviridae is their similarity in base composition in spite of their replication in widely different hosts, including arthropods, plants, and vertebrates.

Although the number of viruses characterized in this family exceeds 100, only 11 are known to infect man: the 3 mammalian reoviruses, the human rotaviruses, Colorado tick fever (CTF), and specific isolates from the Changuinola, Corriparta, Kemerovo, and ungrouped orbiviruses. Those infecting humans are marked in the tables. As the *Rotavirus, Phytoreovirus,* and *Fijivirus* genera are considered in other chapters of this volume, and as the reoviruses were examined in Chapter 10 of the first volume of this series (Stanley, 1978), this chapter will be primarily concerned with orbiviruses and to a lesser extent with current and relevant aspects of animal and avian reoviruses not discussed in Chapter 10, Volume I. With the exception of laboratory diagnosis, the properties of the viruses, their pathogenesis, epidemiology, epizootiology, prevention, and control will be discussed where appropriate with individual groups or viruses.

One of the problems of presenting useful guidelines for the study of Reoviridae relates to the increasingly large number of isolates in the genera *Orbivirus* and *Rotavirus*. Many of these were initially referred to as "reovirus-like" and have yet to be characterized. Some of the "provisional" Reoviridae may well be placed in one of the genera seen in Fig. 1. In his 1970 review, Verwoerd divides the Reoviridae into two groups: (1) the reoviruses, rice dwarf and CTF, which possess a double-layered capsid with 92 morphological units in the outer layer and (2) those viruses of the bluetongue/African horse sickness group that have a single-layered capsid of 32 capsomers. These difficulties and interpretations of morphology and structure are discussed later. It would appear that effective characterization of viruses fitting the description of the main and basic properties of Reoviridae listed at the top of Fig. 1 will depend not only on further studies of morphology, antigenic configuration, and sensitivity to pH and lipid solvents but on the information revealed by analysis of the double-stranded RNA by such techniques as polyacrylamide gel electrophoresis (PAGE). The value of this appraisal is now being appreciated in the characterization of viruses within the *Orbivirus* genus. Furthermore, the *Orbivirus* RNA pattern is distinct from that of other genera of the Reoviridae family such as *Reovirus*, *Rotavirus*, and *Phytoreovirus*, and the size distribution of the RNA segments of a new isolate could therefore be useful in preliminary characterization.

II. ORBIVIRUS

[R/2:Σ12/*:S/S:I,V/Ve/Ac,Di]

A. Characterization and Antigenic Grouping

Viruses in this genus replicate in both insects and vertebrates and are therefore grouped with arboviruses. The first clear recognition of this taxonomic group was anticipated by Verwoerd (1970), who suggested the name "Diplornavirus" for those viruses known to have double-stranded RNA genomes. Borden *et al.* (1971) and Murphy *et al.* (1971), in their attempts to differentiate 27 members from togaviruses, suggested the name "Orbivirus" (ring or circle), as it reflected the large doughnut-shaped capsomers seen in negative contrast preparations. Bluetongue virus (BTV) (with 32 morphological units and an icosahedral structure) was accepted as the prototype virus but is now known to represent only one

*Information inadequate.

TABLE I
Characteristics of Reovirus (Mammalian Type 1) and Orbivirus (Bluetongue)[a]

Virus	Capsomers	Lipid solvent	pH 3	Natural arthropod cycle
Reovirus	92[b]	Resistant	Resistant	−
Bluetongue	32[c]	Partially sensitive	Sensitive	+

[a] Modified after Murphy *et al.* (1971).
[b] Outer capsid.
[c] Inner capsid layer. All orbiviruses have a double-layered capsid composed of seven polypeptides (Verwoerd *et al.*, 1979; see also text).

serogroup of the orbiviruses. Further studies, in addition to electron microscopy, showed different properties for orbiviruses and reoviruses (see Table I). It was shown that the RNA from BTV consisted of 10 fragments resembling in some ways the double-stranded RNA fragments of the reoviruses (Verwoerd, 1969; Verwoerd *et al.*, 1970). Morphology and structure still appear to be the prime criteria on which an initial characterization is made, and the factors to be considered are summarized in Table II.

The loss of infectivity by exposure of BTV to a pH of 3 has been shown to apply to many other orbiviruses and clearly demonstrated with six Australian orbivirus isolates (Corriparta, Wallal, Warrego, Mitchell River, Eubenangee, and D'Aguilar) by Gorman (1978). The mechanism of inactivation by lipid solvents is not yet fully understood, and there are variations with individual types such as ether for Wallal and chloroform for Corriparta. Some strains of orbivirus (CTF and Chenuda) are far more sensitive to deoxycholate (and presumably to lipid solvents) than most other isolates tested (R. E. Shope, personal communication, 1979). Generally, though, the degree of sensitivity is significantly different from the absolute resistance shown by reoviruses (see Table I). Table I also emphasizes the difference in acid sensitivity exhibited by orbiviruses and reoviruses (Stanley, 1967).

It has been possible to divide the 90 accepted orbiviruses into groups depending on their antigenic configuration (see Table III). Because many of these viruses have been isolated during arthropod surveillance studies for arboviruses, it has been customary to give the appropriate geographic name to a virus when it has been recognized as a distinct arbovirus by reference centers and confirmed by the World Reference Center for Arthropod-borne Viruses at the Yale Arbovirus Research Unit, New Haven, Connecticut. These names are registered in the "International Catalogue of Arboviruses" (Berge, 1975), and an outline for one of these viruses (D'Aguilar of the Palyam antigenic group) is reproduced here with permission (pp. 70–74). This outline of information required for each virus

TABLE II
Characteristics of Analogous Layers of Virus Particles Representative of Three Genera of the Reoviridae Family[a]

Genus	Virus particle layers		
	Outer	Middle	Inner
Rotavirus	Sharply defined margin forming a continuous covering over the main capsid	Well-defined capsomers arrayed with $T = 3$ primary symmetry and $T = 9$ secondary symmetry	Particles without resolved subunits but with 5-3-2 symmetry characteristic of an icosahedron
Reovirus (type 3)	Featureless layer located either over or among capsomers of the main capsid	Capsomers difficult to define; $T = 3$ symmetry resolved only with rotational image enhancement	1. Particles with subunits considered to be capsomers 2. Particles with capsomers and projections at vertices of an icosahedron
Orbivirus (bluetongue serotype 10)	Fuzzy, indistinct layer covering the main capsid	Well-defined capsomers arrayed with $T = 3$ symmetry, with secondary symmetry similar to that of *Rotavirus*	Particles without resolved subunits, but with 5-3-2 symmetry characteristic of an icosahedron

[a] After Palmer *et al.* (1977), published with permission of the authors.

clearly shows the data required before a virus can be characterized and accepted. This constitutes a delay which is, of course, necessary, but it does add to the difficulties of orbivirus characterization. This fact has been clearly appreciated by those research workers involved. Gorman (1979) has proposed an additional system to describe the orientation and clustering of the established antigenic groups by use of letters A, B, C, etc., which are referred to as the "Gorman grouping" in Table III. Like most other major virus groups (e.g., Reovirus, Adenovirus) which share a complement-fixing antigen, individual types may be differentiated by neutralization tests (in cell cultures or newborn mice). This technique, however, has limitations, and the ultimate division into "cluster groups" will almost certainly depend on an analysis of the double-stranded RNAs and their interactions. The capacity to interact genetically would appear to be a more useful and meaningful approach to virus relationships. Reference to these studies using PAGE, RNA–RNA hybridization, recombination between serotypes, and polypeptide analysis in relation to antigenic configuration will be considered, where results are available, with individual cluster groups. PAGE

TABLE III
Viruses Accepted as Orbiviruses

Antigenic group prototype	Gorman grouping[a]	Number of serotypes or antigenically related viruses	Vertebrate host	Vector
African horse sickness	A	9 serotypes	Mammals	*Culicoides*
Bluetongue	B	20 serotypes		
		Epizootic hemorrhagic disease of deer[b] (7)		*Culicoides*
		Eubenangee[b]	Mammals,	
		Pata[b]	including	*Culicoides*
		Tilligerry[b]	marsupials	and
		Ibaraki[b]		Dermacentor
Colorado tick fever	C	Colorado tick fever	Mammals[c]	
		Eyach		
Palyam	D	Palyam	Cattle	Mosquitoes
		Kasba	and	and
		Vellore	sheep	*Culicoides*
		D'Aguilar		
		Abadina		
		Nyabira		
Changuinola	E	Changuinola	Mammals[c] and	Phlebotamines
		Irituia (Be An 28873)	marsupials	
		Be Ar 35646		
		Be Ar 41067		
		Pan D50		
Corriparta	F	Corriparta	Mammals[c]	Mosquitoes
		Acado	and	
		Bambari	birds	
Kemerovo	G	20 viruses (see Table VI)	Mammals[c] and birds	Ticks
Warrego	H	Warrego	Marsupials	*Culicoides*
		Mitchell River	Mammals	
Wallal	I	Wallal	Marsupials	*Culicoides*
		Mudjinbarry	Marsupials	*Culicoides*
Equine encephalosis	J	5 serotypes	Horses	
Ungrouped		Lebombo	Rodents[c]	Culicines
		Orungo	Mammals[c]	Culicines
		Japanaut	Bats	Culicines
		Umatilla	Birds	*Culex* spp.
		Paroo River	?	Mosquitoes

[a] See Gorman (1979).
[b] For antigenic relationships see, text.
[c] Evidence of human infection.

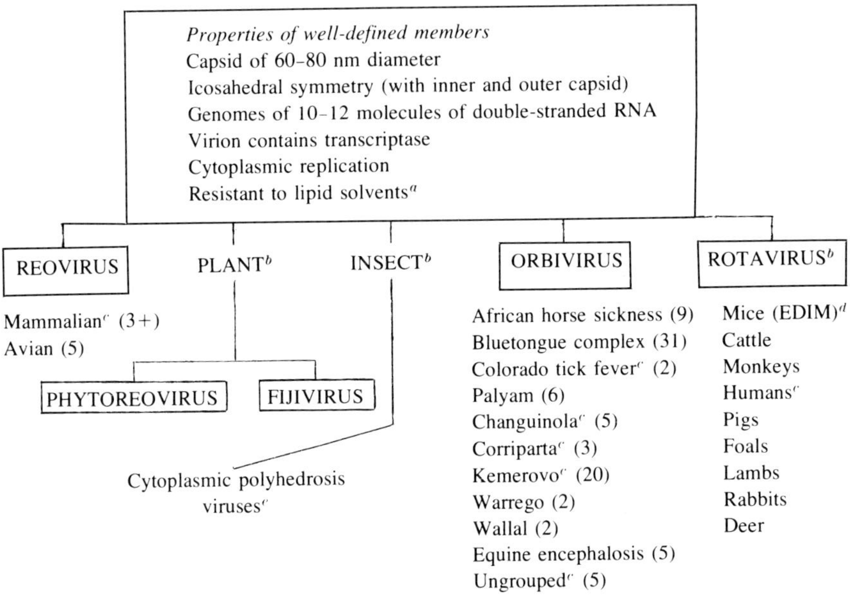

[a]There is variation in sensitivity to lipid solvents between reovirus and orbivirus and also within the genus *Orbivirus* (for details see text).
[b]Rotaviruses and the plant–insect viruses are not considered in this chapter (see Chapters).
[c]Infects humans (isolation and/or positive serology).
[d]Epidemic diarrhea of infant mice.
[e]A name has not yet been approved for this group (Matthews, 1979); CPV has also been observed in the freshwater crustacean *Simocephalus expinosus* (Federici and Hazard, 1975).
( ) Numbers refer to serotypes.
Note: There are five other viruses that may be considered as provisional Reoviridae (see text).

Fig. 1. Outline of the Reoviridae.

has been generally useful not only in comparing the protein components of orbiviruses such as BTV, African horse sickness, and epizootic hemorrhagic disease of deer (EHD), which are similar, but also in demonstrating marked differences from and between reoviruses, rotaviruses, and cytoplasmic polyhedrosis viruses (Rodger *et al.*, 1975; Payne and Rivers, 1976). The presence of different-sized polypeptides in the diffuse outer layer of orbiviruses may well be associated with the characterization of the different types recorded in Tables III, IV, VI, VII, and VIII.

B. Morphology and Structure

Although a wide range of sizes and variations in morphology have been recorded, interpretation must be careful and frequently accepted with some re-

D'AGUILAR (DAG)

I. Virus Name and/or Number: D'AGUILAR (B8112)

Information from: R. L. Doherty Day 1 Mo. Feb. Yr. 1972

Address: Queensland Institute of Medical Research, Brisbane.

II. Antigenic: Group Palyam Ungrouped: (Give details 2nd page)

III. Original Source: Isolated by Doherty *et al.* (1) at Brisbane

Genus and species: *Culicoides brevitaraia* Kieffer Sentinel Animal

Age or stage: adult Sex: female

Isolate from: Whole blood ... Clot: ... Serum or plasma .. Other fluids

Organs and Tissues:

Signs and symptoms of illness:

Time of collection: Day 2 Month April Year 1968

Method of collection: aspirated from cattle

Place collected: Bunya, Old. Lat.: 27°22′S Long: 152°57′E

Macrohabitat: east coastal plain south Queensland

Microhabitat: cattle stud, sucalypt forest partly cleared for pasture

Method of storage until inoculated: overnight at 5°C, sorted to species on metal trays cooled with dry ice, then stored at −60°C

IV. Method of Isolation in Laboratory: Inoculation: Day 26 Month April Year 1968

Animal used: mouse Age: 1 day Embryonated egg:

Diluent: 10% rabbit-serum-saline Inoc.: 0.015 ml. Route inoc.: intracerebral

Cell Culture: Type of Cell Primary: Continuous Line:

Diploid: Medium composition:

Manner of Recognition: CPE Plaques: Other:

V. Validity of Isolation: Reisolation: Yes No X Not Tried

Homologous antibody formation by original source animal: Yes ... No ...

Not tested

Tests used: .... Other reasons supporting validity of isolation: further isolates in same areas in 1970; widespread antibody in cattle

VI. Properties of Virus

Physical: Filtered Yes Type(s) Filter Millipore

Size: <100 mμ How estimated: passed filter APD 100nm, not APD 50 nm

Envelope: Present X Diameter: 100 ± 5 mμ Absent:

Capsid: X Diameter: 66 $\pm_{41}$ 70 ± 3* mμ Shape spherical: polygonal with obvious capsomers* No. of Capsomers: uncertain

Other information: core 36± 3nm. *results from thin section and negative contrast in that order. See reference 5 for electron microscopy.

Chemical: RNA DNA Other information

Resistance to chemicals:

Ether: Dil. 50% final After treatment titer: $10^{5.0}$ LD_{60}/0.015 m Control titer: $10^{5.4}$ LD_{50}/0.015ml

Sod. deoxycholate: Dil. 1/1000 fio after treatment titer: ($10^{4.6}$ LD_{50}/0.015 ml (3P)

Control titer ($10^{2.3}$ LD_{50}/0.015ml$^{(3P)}$

Other chemicals: ($10^{4.75}$$LD_{50}$/0.015ml(11P) ($10^{4.5}$$LD_{50}$/0.015ml $^{(11P)}$

Antigenic: hemagglutinin produced: Yes No X Not Tried

Source material used: brain and blood of infected mice

Methods employed: sucrose-acetone extraction followed by treatment with protamine, sonication or trypsin

Source of erythrocytes: goose

pH: Range 6.0-7.6 Optimal Temp: Range Optimal

Remarks: Orbivirus taxon

Antigenic relationship and lack of relationship to other viruses:

Studies at Queensland Institute of Medical Research:

No antigenic relationship by complement-fixation and neutralization tests to any arbovirus or suspected arbovirus isolated or available at this laboratory: Group A (Sindbis, Ross River, Getah,

Bebaru), Group B (Murray Valley encephalitis, Kunjin, Kokobera, Edge Hill, Stratford, Alfuy, CBE, SLE, dengue types 1–4), Koongel Group (Koongol, Wongal), Mapputta Group (Mapputta, Trubanaman, Mk7532), Simbu Group (Akabane, Samford), Quaranfil Group (Abal), Corriparta Group (Corriparta), Eubenanges Group (Eubenanges), Warrego Group (Warrego, Mitchell River), ungrouped (Kowanyama, Almpiwar, Upolu, ephemeral fever, Belmont, Charleville, Wallal, MRM13443, MRM14556).

Studies at Yale Arbovirus Research Unit:

C. G. Carley and R. E. Shope found B8112 antigen non-reactive by CF test to immune ascitic fluids to 13 Groups (A, B, C, Guama, Capim, Simbu, Bunyamwera, vesicular stomatitis, Anopheles A, Turlock, california, phlebotomus fever and Tacaribe) and 103 other arboviruses.

R. E. Shope subsequently found B8112 related to but recognizably distinct from members of the Palyam group:

	B8112 Antigen				B8112 Immune Ascitic Fluid		
	CF		NT		CF		NT
Immune Ascitic Fluids	H+ Ho	Ratio	H+ Ho	Antigens	H+ Ho	Ratio	H+ Ho
Palyam	16/32	1/2	0/>2.4	Palyam	64/128	1/2	0.3/>4.0
Kasba (G15534)	512/1024	1/2	0/>2.8	Kasba (G15534)	64/128	1/2	1.4/>4.0
Vellore (68886)	512/1024	1/2	1.2/>3.3	Vellore (68886)	64/128	1/2	3.0/>4.0

These complement-fixation results, and tests against 22 other solvent-resistant arboviruses, have been published[2].

Methods found most useful for investigating antigenic characteristics: CF

Biologic characteristics:

Virus recovered from (all isolations from vertebrates): Blood (M) ... (LV) ... Cerebro spinal fluid (M) . (LV) . CNS(M) . (LV) . Heart(M) .. (LV) .. Lung(M) .. (LV) .. Liver(M) .. (LV) ... Spleen(M) . (LV) .. Kidney(M) .. (LV) .. Skin lesions(M) .. (LV) .. Nasopharyngeal(M) .. (LV) .. Milk(M) .. (LV) .. Urine(M) ... (LV) .. Feces(M) .. (LV) .. Salivary gland(M) .. (LV) .... Mammary gland(M) .. (LV) . Lymph node(M) .. (LV) .. Skeletal muscles(M) ... (LV) ....

Other ....

Laboratory methods of virus recovery (all isolations): Suckling mice X ......, Weanling mice

Baby chick . Chick embryo . Hamster . Rabbit . . Guinea pig . Primate . . Other animals . . . .

. . . .

Cell culture: Human(P) . . (L) . . Other primate(P) . (L) . . Other mammal(P) . . (L) . Avian(P) . . . .

(L) Other . . . .

Natural Host Range

Vertebrate and Arthropod	No. Isolations/ No. Tested	Country and Region	No. with antibody/No. Tested Test Used	Country and Region
Culicoides brevitarsis[1]	1/17907	South-east Queensland, 1968		
	1/2755	South-east Queensland, 1970		
Cattle[1]			104/132	Queensland
Sheep[1]			9/23	Western Queensland
Various other vertebrates[1]			1/244	Queensland

Susceptibility to Experimental Infection: Vertebrates, Embryos, Arthropods, and Cell Cultures (TC)

Experimental Host	Passage History Strain Used	Age of Animal or Egg	Inoculation Route ml	Evidence of Infection	AST Days	Titer (Logs) per ml
Mice	3,11	1–4 day	ic	Death	4–5 at 10^{-2}	6.7, 7.4LD_{50}
Mice	3,11	1–4 day	ip	No overt sign of illness		<3.5, <3.5LD_{50}
Mice	3,11	3–4 week	ic	No overt sign of illness		<3.5, <3.5LD_{50}
Mice	11	3–4 week	ip	Antibody formation	NOT TITRATED	
Cell culture:						
PS (pig kidney)[3]	11		0.1	Plaques under agar		6.5 × 10^6PFU
Experimentally inoculated mosquitoes:						
Aedes aegypti[1]	11		intrathoracic, 0.0006 ml = $10^{1.9}LD_{50}$ mosquito	titration of whole mosquitoes at intervals 1–20 days after inoculation	Virus undetectable (<20LD_{50} per mosquito) 0.5 and 1 days, then rise to >10^4LD_{50} per mosquito days 12–20	

Histopathology: Character of lesions . . . .

. . . .

. . . .

Inclusion bodies: Cytoplasmic (M) . . (LV) . . Intranuclear(M) . . (LV) . . . .

Organs and tissues significantly affected: Brain(M) .. (LV) .. Spinal cord(M) .. (LV) ....

Lungs(M) .. (LV) .. Liver(M) .. (LV) .. Spleen(M) .. (LV) .. Kidney(M) .. (LV) .. Heart(M) .. (LV)

Blood vessels(M) .. (LV) .. Marrow(M) .. (LV) .. Skeletal muscles(M) .. (LV) ....

Secretory glands(M) .. (LV) ....

Species of lower vertebrates (LV) used in study ....

Category of tropism ....

Human disease: In nature: (R) .. (S) . Death (R) ... (S) .. Residue(R) .. (S) ....

Laboratory infection: Subclinical(R) .. (S) .. Overt disease(R) . (S) ....

Clinical Manifestations: Fever(R) .. (S) .. Headache(R) .. (S) . Prostration(R) .. (S) .. Conjunctival

inflammation(R) .. (S) .. Stiff neck(R) ... (S) .. Myalgia(R) .. (S) .. Arthralgia(R) ... (S) .. CNS signs

(including encephalitis(R) ... (S) .. Hemorrhagic signs(R) .. (S) .. Respiratory involvement(R) .. (S)

Leukopenia(R) .. (S) .. CNS Pleocytosia(R) .. (S) .. Rash(R) .. (S) .. Lymphadenopathy(R) .. (S) ..

Jaundice(R) .. (S) .. Vomiting(R) .. (S) .. Other significant symptoms ....

Category: (see instructions reverse side) 1.)Febrile illness 2.)Febrile Illness with rash .. 3.)Hem- .. orrhagic fever .. 4.)Encephalitis ....

Number of cases observed: ....

Known geographic distribution: Virus Isolation: southeast Queensland

Suspected geographic distribution: (from serological surveys) wide area of eastern and northern Australia and New Guinea

References:

1. Doherty, R. L., Carley, J. G., Standfast, H. A., Dyce, A. L., Snowdon, W. A. Virus strains isolated from arthropods during an epizootic of bovine ephemeral fever in Queensland. *Aust. vet. J.*, 1972, 48:81.
2. Borden, E. C., Shope, R. E., Murphy, F. A. (1971). Physicochemical and morphological relationships of some arthropod-borne viruses to bluetongue virus—a new taxonomic group. Physicochemical and serological studies. *J. gen. Virol.*, 13, 261–271.
3. *Westaway, E. G. (1966). Assessment and application of a cell line from pig kidney for plaque assay and neutralization tests with twelve Group B arboviruses. *Am. J. Epidemiol.*, 84, 439–456.
4. Carley, J. G., Standfast, H. A., Kay, B. H. Multiplication of viruses isolated from arthropods and vertebrates in Australia in experimentally-infected mosquitoes, in preparation.

* Reference to cell line, not to results with D'Aguilar virus.

5. Schnagl, R. D., Holmes, I. H. (1971). A study of Australian arboviruses resembling bluetongue virus. *Aust. J. biol. Sci.*, 24, 1151.

Remarks: SEAS rating: Possible arbovirus

servation. With most of the orbiviruses that have been adequately examined, the capsid symmetry is icosahedral, with the inner capsid comprising 32 capsomers (I. H. Holmes, personal communication, 1979). It is now clear that the capsids of reovirus and BTV are similar both in size and in the possession of a double layer (Verwoerd *et al.*, 1972; Martin and Zweerink, 1972), but the assembly of the viruses appears to be different. It is possible, as Gorman (1979) has suggested, that nucleocapsids rather than complete virions may have been seen in the reported structure of many orbiviruses prior to 1972. An example of a typical orbivirus (Australian isolate of BTV) morphology is shown in Fig. 2.

Morphology has been the main characteristic for the placement of all orbiviruses listed in Table III, and this feature distinguishes them from reoviruses. The diameters of most of the members are between 60 and 70 nm, as determined by positive staining techniques of sections of infected cells. The discrepancies in size recorded in negatively stained preparations and purification procedures are now explained by the demonstration of a double-layered capsid (Verwoerd *et al.*, 1972, 1979). The existence of a structureless outer capsid layer is characteristic of all orbiviruses, and it is this that distinguishes them from reoviruses and rotaviruses, with the exception of CTF. For a comparative study of structure, readers are referred to Verwoerd *et al.* (1979), where both core particles and complete virions are demonstrated for selected orbiviruses, as well as to the comparative morphological studies of rotavirus, reovirus 3, and bluetongue presented by Palmer *et al.* (1977).

During virus replication, tubules have been described with reoviruses

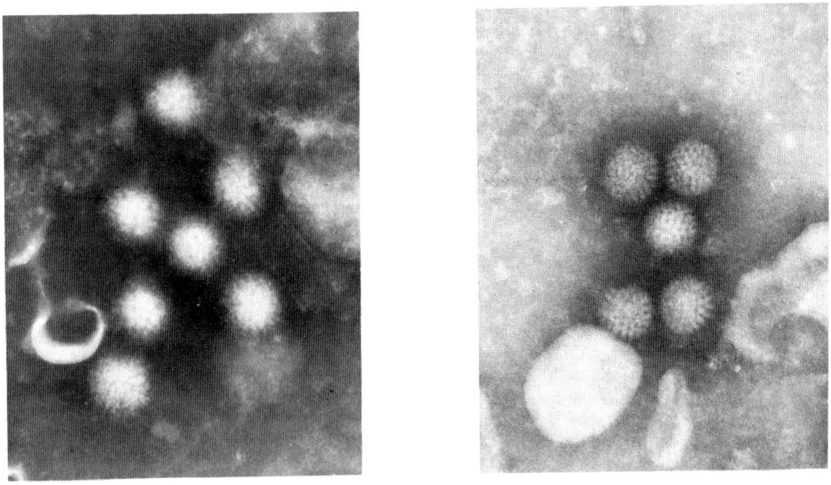

Fig. 2. Electron micrographs of bluetongue virus type 20 stained with 3% phosphotungstic acid, pH 6. × 188,800. C. J. Smale and B. M. Gorman, unpublished micrographs.

(Spendlove *et al.*, 1963; Dales *et al.*, 1965), rotaviruses (Holmes *et al.*, 1975; Kimura and Murakami, 1977), and orbiviruses that have been adequately studied (Verwoerd *et al.*, 1979). With BTV, in addition to the seven polypeptides, two polypeptides (P5a and P6a) are synthesized (Huismans, 1978). Polypeptide P5a is produced in large quantities early in the infectious cycle. Huismans and Els (1978) showed that most of the P5a is associated with a complex of hollow tubular structures with a diameter of 68 nm that vary in length. Tubules of African horse sickness virus, EHD virus, and BTV have almost the same molecular weight, but the diameter is characteristic for each virus, being 68 nm for BTV, 54 nm for EHD virus, and 18 nm for African horse sickness virus (Verwoerd *et al.*, 1979).

C. African Horse Sickness

1. Characterization and Pathogenesis

The nine antigenic types show a common complement-fixing antigen but may be differentiated by neutralization and hemagglutination inhibition (HI) tests (Howell, 1962; Pavri and Anderson, 1963). The protein and nucleic acid components have been studied by gel electrophoresis (Bremer, 1976).

Yolk-sac inoculation of fertile chick embryos supports the growth of all nine types (Goldsmit, 1967), but most virus strains grow well in cell cultures from mammals (Hopkins *et al.*, 1966) and in one mosquito cell line (Mirchamsy *et al.*, 1970).

The virus causes natural disease in horses, mules, and donkeys, with death from pulmonary edema in severe cases. Viremia is a frequent and significant event, necessary for transmission. Experimentally, rodents may be infected after intracerebral (i.c.) inoculation, and continuous passage results in attenuation for the natural host (horses), although viremia occurs. It has been suggested that elephants and zebras are possible reservoir hosts for this virus (Davies and Otieno, 1977).

2. Transmission and Control

Transmission is mainly by nocturnal biting insects, with *Culicoides pallidipennis* being the main vector in which the virus replicates (Mellor *et al.*, 1975). The disease is endemic in western, central, southern, and eastern Africa. As with BTV, African horse sickness virus may be spread over relatively vast distances by wind-blown infected *Culicoides* (Sellers *et al.*, 1977; Sellers, 1978). This interesting spread must indeed merit serious consideration, as the movement of infected *Culicoides* on the wind is believed to be responsible for spread of the virus across the sea from Morocco to Spain (1966), from Turkey to Cyprus (1960), and from Senegal to the Cape Verde Islands in 1943 (Sellers *et al.*, 1977). Sellers (1978) suggests a flight endurance of up to 20 hours and a flight range of up to 700 km at temperatures between 15 and 40°C. If this is correct, then it could be an important factor deserving analysis in the control of spread. A similar situation has been examined in Australia, where the spread of bovine ephemeral fever has been associated with the distribution of *C. brevitarsis* (St. George *et al.*, 1977). Synoptic weather charts could yield information of value in reducing the risk of spread from endemic areas by wind dispersal. Overall control includes restriction of the movement of horses, vaccination, and the use of insecticides.

D. Bluetongue Group

The BTV complex currently comprises three clusters of viruses (see Table III) based on antigenic structure and not significantly on disease production or geographic location. All are *Culicoides* transmitted. The most recent BTV serotype was isolated in Australia (strain CSIRO 19), a continent believed to be free from the virus (St. George *et al.*, 1978). On first isolation, the relationship of some of the isolates in the subgroup to BTV was not established. Further studies confirm cluster groups comprising 31 types:

B1-20	BTV (20 types)
B21-28	EHD (7 types)—Ibaraki
B29-31	Eubenangee
	Pata
	Tilligerry

Studies by agar gel precipitation, indirect fluorescent antibody, and neutralization tests have shown that Ibaraki shares antigens with some types of EHD but not with BTV (Campbell *et al.*, 1978). EHD also shares antigens with BTV, Pata, and Eubenangee. Obviously an antigenic spectrum exists, as shown in Fig. 3, and this requires further studies for examining the suggested EHD virus link in this cluster group. Because of antigenic heterogeneity, protection from disease by vaccination is complex, and earlier investigators were aware of but could do little about the antigenic complexity or the variations in virulence. The BTV types have a common complement-fixing antigen but may be differentiated by neutralization tests (Howell, 1970).

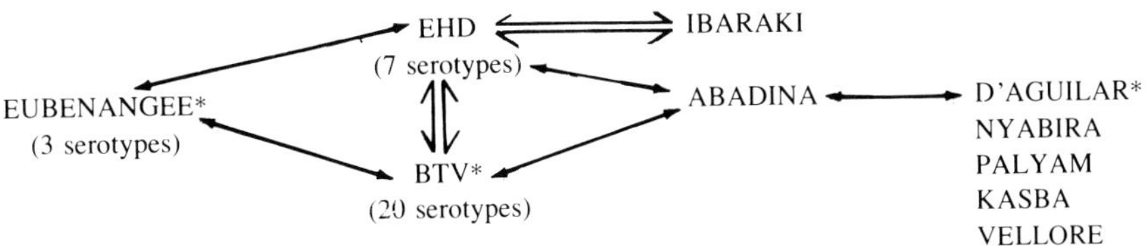

Fig. 3. An antigenic spectrum. *Australian isolates. ⇆, neutralization; ↔, complement fixation; ⟷, agar gel precipitation.

The Australian CSIRO 19 isolate has been recognized as a new serotype—serotype 20—and this virus appears to grow in cell cultures and chick embryos better than other BTV types. Where this virus came from, how it arose, and how long it has been on the Australian continent are questions which have yet to be answered. However, there is some evidence (A. J. Della-Porta, personal communication, 1979) that it is related to BTV-17 and that it could have arisen by gene reassortment.

Verwoerd *et al.* (1972) showed that BTV, like reovirus, had polypeptide molecular weights which correlated closely with the sizes of the genome segments (each one acts as a gene for a polypeptide). The virion transcriptase differs from that of reovirus in its dependence on Mg^{2+} ions as well as in its low optimum temperature of 28°C. There is no cross hybridization with reovirus (Verwoerd, 1970). Temperature-sensitive (ts) mutants have been isolated and classified into six genetic recombination groups (Shipham and de La Rey, 1976). This approach is likely to be of value in vaccine production.

Ibaraki virus was initially isolated from cattle (Omori *et al.*, 1969), but Australian isolates from this group have come from mosquitoes (Doherty, 1968, 1970). In Australia, marsupials as well as cattle appear to be involved. Ibaraki virus is antigenically related to EHD virus (Campbell *et al.*, 1975, 1977, 1978) and produces clinical bluetongue in cattle but not in sheep. Pata (Africa), Eubenangee (Australia), and EHD virus share complement-fixing antigens (Borden *et al.*, 1971), as do Eubenangee, Tilligerry, and BTV (Gorman, 1979).

Although BTV is naturally transmitted by *Culicoides* spp. (see Section II,D,2), it has been possible to demonstrate BTV replication (without CPE) in mosquito cell lines from *Aedes albopictus* and *A. pseudoscutellaris* (Jennings and Boorman, 1979). Both EHD virus and African horse sickness virus have also been shown to replicate in mosquito cell lines (Mirchamsy *et al.*, 1970; Willis and Campbell, 1973).

The importance of the heterogeneity of the RNA genome segments in generating antigenic diversity in the BTV group has recently been demonstrated by Gorman and Taylor (1978). The RNA which they extracted from Tilligerry virus separated into 10 segments (using PAGE) where molecular weights differed from the corresponding segments of Eubenangee virus, a member of the same subgroup of BTV (see Table III).

Mosquitoes have provided the primary source of virus isolates of the Eubenangee subgroup (see Table IV).

1. Clinically Recognizable Disease

This has been recorded for BTV, EHD, and Ibaraki viruses, but not for Pata, Tilligerry, or Eubenangee viruses.

Bluetongue has had clinical recognition as a disease of ruminants for over a century (Erasmus, 1975a). Its earliest recognition was in Africa, but later the

TABLE IV
Eubenangee Subgroup of Bluetongue (Gorman Group B)

Name	Arthropod	Locality	Vertebrate host association
Eubenangee[a]	Mosquitoes (11 spp.) *Culicoides*[d]	Innisfail, Queensland, Australia	Cattle, marsupials
Pata[b]	*Aedes palpalis*	Pata, Central African Republic, Africa	?
Tilligerry[c]	*Anopheles annulipes*	Nelson Bay, N.S.W., Australia	?

[a] Doherty (1968).
[b] Digoutte and Pajot (1968).
[c] Gard *et al.* (1973).
[d] A. J. Della-Porta (personal communication, 1979); T. D. St. George, (personal communication, 1979).

disease was recognized in Mediterranean countries, India, Pakistan, and Japan. As with African horse sickness, it is possible that BTV may have reached Portugal from Morocco in 1956, and Cyprus from Turkey and Syria in 1968 and 1977, by wind-blown BTV-infected *C. pallidipennis*. Serological studies confirm the presence of BTV in Egypt, Jordan, Iran, and Iraq (Hafez *et al.*, 1978). It was first observed in North America in 1948 (Hardy and Price, 1952). Although the disease has not occurred in Australia, and that continent was believed to be free from the agent, the virus (CSIRO 19 isolate) was recently isolated in the Northern Territory (St. George *et al.*, 1978) from *Culicoides* collected in 1975 (Shope, 1977). This very interesting observation has not only stimulated BTV research in Australia but has emphasized the presence of a potential pathogen that has not yet been shown to produce significant disease in Australian cattle or sheep. St. George and McCaughan (1979) have reported that the clinical picture in sheep, following inoculation of the CSIRO 19 virus, was consistent with bluetongue.

In countries where the disease is established, it is usual to find that lambs are seriously affected and frequently present with fever, erosions, and crusting and cyanosis around the mouth. There may be muscle damage, and edema of the head and neck is more frequently observed than pulmonary edema. The mortality varies in different geographic areas (from about 5 to 30%), and this may be associated with variations in the virulence of virus strains. Cattle are frequently infected, as detected by virus isolation and by serological techniques (Hourrigan and Klingsporn, 1975a). Although complement fixation (CF) may be used for the detection of antibody to the group antigen, Thomas *et al.* (1976) showed good correlation between CF tests, agar gel precipitation, and plaque-neutralization tests. In their hands the plaque-neutralization tests proved to be the most sensitive and detected antibody early in the course of infection. Another serological technique which may be used for the group antigen is a hemolytic plaque assay using sheep erythrocytes with BTV adsorbed to their surface (Oellerman *et al.*, 1976). Infected cattle may serve as a source of infection for sheep (Bourne, 1971). BTV also infects and produces disease in deer (Stair *et al.*, 1968; Frank and Willis, 1975) which is clinically similar to EHD (Shope *et al.*, 1960; Tables I and III). Ibaraki virus produces a bluetongue-like disease in cattle but not in sheep (Omori *et al.*, 1969).

2. *Ecology and Transmission*

Natural transmission of BTV is mainly through *Culicoides* spp.—in South Africa by *C. pallidipennis* (du Toit, 1944; Erasmus, 1975b) and in North America by *C. variipennis* (Price and Hardy, 1954). The virus replicates in the vector, which carries the virus for life (several weeks). The vector can fly or be wind-blown or transported greater distances by ground vehicle or aircraft. Transovarial transmission has not been proven, although overwintering has been observed (Hourrigan and Klingsporn, 1975b). The viremic state in a calf born of a dam infected while pregnant has persisted for up to 3 years (Hourrigan and Klingsporn, 1975a,b), but this is unusual. It is not known whether such persistence is the result of the development of immunological tolerance, defective interfering particles, or some other variation in virus–host interaction.

The maintenance of an arbovirus/orbivirus in one geographic environment is not completely understood, but there are obviously different mechanisms operating. Some of these could be transovarial transmission in the arthropod, overwintering in a poikilothermic vertebrate, or persistent viremia in which the virus may be protected from antibody by an intracellular existence, as in CTF. The persistence of BTV in cattle and the persistent viremia need further study. For example, Luedke *et al.* (1977) demonstrated BTV latency in a bull over a 5-year period with activation to the viremic state by the bites of *C. variipennis*. This natural stress factor caused "showering" of virus into the blood, and the BTV was presumably within red cells, as the serum contained antibody. The cell site of BTV latency was not clear, but experiments similar to those of Emmons and his colleagues for CTF could yield further information.

C. variipennis is also the vector for EHD (Foster *et al.*, 1977). Jones and Foster (1978) used a standard colonized population of *C. variipennis* for comparing infection rate responses of different vector field populations to four serotypes of BTV. There were marked differences in susceptibility and infection rates.

3. *Control of Bluetongue in Enzootic and Epizootic Situations*

This matter is now of considerable relevance to the Australian continent since the isolation of BTV from *Culicoides* spp. in the Northern Territory was reported in 1978. The view is currently held that once BTV has entered a large country, it is difficult to eliminate (Geering, 1975). Slaughter and vaccination remain accepted procedures for disease control and have been moderately successful, as in Portugal and Spain (Geering, 1975).

However, the Australian situation again demonstrates that the virus can be present without obvious clinically recognizable disease—a phenomenon shown by other arboviruses and often related to their transmission and ecology. Transmission is directly related to the biology of *Culicoides* spp. where breeding is related to warmth and moisture. Successful control really depends on early diagnosis of the disease, disposal or confinement of viremic animals, disinfection of the infected area and, in Australia, the maintenance of a large quarantine zone separating infected cattle from susceptible sheep. In addition, the Australian plans have considered both ring and mass vaccination, and contingent plans were developed before the recognition of the CSIRO 19 strain as a new serotype of BTV. If there is only one serotype associated with disease, then mass vaccination with an egg-attenuated vaccine is somewhat easier than with multiple serotypes.

E. Colorado Tick Fever

This virus and its human disease have been selected for discussion because our epizootiological knowledge is relatively more extensive than with some other orbivirus subgroups and because of the interesting pathogenesis of the virus, even though its distribution is confined to the northwestern United States. The virus is transmitted to man by the bite of the wood tick *Dermacentor andersoni*. For many years only one virus type of CTF was recognized, although there is a report indicating a serological relationship to an arbovirus (Eyach) isolated from *Ixodes ricinus* ticks (Rahse-Küpper *et al.*, 1976). In man the disease is a noncontagious, self-limiting, febrile illness, but there has been no recognition of clinical disease in the natural vertebrate host or the tick vector.

1. Ecology and Transmission

The virus is maintained in small, medium, and large mammals and the blood-feeding ticks which parasitize them. This probably reflects a long and successful adaptation to a stable ecological niche (R. W. Emmons, personal communication, 1978). It appears that the virus is naturally maintained in a cycle between immature stages of the tick and the mammalian host. Nymphal ticks assist in the overwintering of the virus, but rodents may have prolonged viremias. In this situation, feeding larval ticks may become infected prior to molting to nymphs (Burgdorfer, 1977). Transovarial transmission of CTF virus in female ticks is equivocal (Eklund *et al.*, 1961; Florio *et al.*, 1950). As no other form of transmission has been defined, the natural biocenose indicates that large populations of mammals and ticks would be required to maintain infective CTF virus. The extensive period of natural viremia is confirmed by experimental infections of squirrels, chipmunks, deer, mice, porcupines, and golden-mantled ground squirrels (Emmons and Lennette, 1966).

The adult tick (*D. andersoni*) with CTF virus in its salivary glands is the usual source of infection for man. Very rarely human infections have been recorded without tick intervention, and both have been iatrogenic from infected mammalian blood (laboratory-acquired infection or a blood transfusion) (Randall *et al.*, 1975).

With many mammals serving as hosts for immature ticks (see Table V for those known to be naturally infected with CTF virus), one would expect different biocenoses with geographic location, and this indeed does occur (Burgdorfer, 1977).

2. *Pathogenesis in Man and Experimental Animals*

As with African horse sickness, one of the interesting aspects of CTF infections is the persistent viremia. The virus persists only transiently in serum but remains for much longer periods in the cellular component of blood. Emmons *et al.* (1972) have shown in a series of very clear studies, using immunofluorescence, electron microscopy, and virus isolation techniques, that the persistent viremia is associated with the presence of virus in circulating erythrocytes. They suggested in 1972 that this may have been the result of CTF virus infection of erythrocyte precursor cells in the bone marrow. These cells were subsequently released into the bloodstream, but the virus was protected from the humoral and cellular host immune responses because of its intraerythrocytic location. Recent studies have convincingly confirmed this interesting infection (Oshiro *et al.*,

TABLE V
Mammals Naturally Infected with Colorado Tick Fever Virus

Common name	Generic name
Porcupine	*Erethizon dorsatum epixanthum*[a]
Wood rat	*Neotoma cinerea cinerea*[b]
Golden-mantled ground squirrel	*Spermophilus lateralis tescurum*[c,d,e]
Columbian ground squirrel	*S.c. columbianus*,[c,d]
Pine squirrel	*Tamiasciurus hudsonieus richardsoni*[c]
Chipmunk	*Eutamias* spp.[c,d]
Meadow vole	*Microtus* spp.[c]
Red-backed vole	*Clethrionomys* spp.[b]
Deer mouse	*Peromyscus maniculatus*[c,e]

[a] Eklund *et al.* (1958).
[b] Burgdorfer (1977).
[c] Burgdorfer and Eklund (1959).
[d] Burgdorfer and Eklund (1960).
[e] Clark *et al.* (1970).

1978). The Berkeley studies of Emmons and Oshiro (cited above) of experimental mouse inoculation support the thesis of virus replication within infected erythropoietic cells at the time of cell differentiation. This results in the presence of virions within erythrocytes, with consequent protection from antibody. This is not only an attractive hypothesis, but the model offers opportunities to improve diagnostic procedures by immunofluorescence techniques and to explore the mechanisms of erythrocyte infection that may occur with other orbiviruses and arboviruses.

Hamsters, mice, and rhesus monkeys can be experimentally infected and develop a viremia. If mice are infected during pregnancy, abnormalities may appear in the offspring (Harris *et al.*, 1975).

3. *Laboratory Diagnosis*

CTF may be diagnosed by virus isolation in infant mice (blood or tick suspensions) or by direct immunofluorescence of smears thought to contain the virus (Emmons and Lennette, 1966). Serological diagnosis seeking a significant rise in antibody may employ CF, immunofluorescence, or neutralization (plaque-reduction) tests. The techniques were analyzed by Emmons *et al.* (1969), and their results favored an indirect fluorescent antibody technique as being the most simple and sensitive.

F. Kemerovo Group

This group of 20 viruses (see Table VI) derives from international collaboration between laboratories whose investigations are obviously geographically expanding, but originated from the Russian–Czechoslovak expedition to a known natural focus of tick-borne encephalitis in the Kemerovo region of western Siberia in 1962 (Libikova and Casals, 1971). As may be seen in Table VI, they are all tick-borne—either Argasid or Ixodid—with the predominant vertebrate host association being seabirds. Many of the serotypes distinguished by neutralization tests share antigens when tested by CF (Main, 1978), and clustering on antigenic configuration may be recognized (Libikova and Buckley, 1971), as shown with the bluetongue group. There is strong evidence that Kemerovo virus infects man and produces disease (virus isolated in CNS disease), which may also be true of Tribec and Lipovnik (Casals, 1975). Kemerovo virus did induce an outbreak of disease in the Soviet Union in 1962, and isolations were made from the CSF of patients who developed neutralizing antibodies. Prior to the observation of human disease in Siberia, a strain of Kemerovo virus was isolated from a migrating common redstart (*Phoenicurus phoenicurus*) by Schmidt and Shope (1971; see also Table VI). This is an example of the movement of arboviruses around the planet by migrating birds and could represent a significant link between eastern Siberia and Australia, where seabirds have annual and well-defined mi-

TABLE VI
Kemerovo Group (Gorman Group G)

Name	Tick	Locality of first isolates	Vertebrate host association
Baku (1)	*Ornithodoros capensis*	Azerbaijan, Caspian Sea, USSR	Birds
Bauline (2)	*Ixodes uriae*	Great Island, Newfoundland, Canada	Birds
Cape Wrath (3)	*Ixodes uriae*	Clo Mor, Cape Wrath, Scotland	—
Chenuda (4)	*Argas reflexus hermanni*	Chenuda Village, Nile Delta, Egypt	Birds
Great Island (2)	*Ixodes uriae*	Great Island, Newfoundland, Canada	Birds
Huacho (5)	*Ornithodoros amblus*	Punta Salinas, Huacho, Peru	Birds
Kemerovo (6)	*Ixodes persulcatus*	Kemerovo, Siberia, USSR[a]	Birds, horses, cattle, small mammals, man[b]
Kenai[c] (7)	*Ixodes signatus*	Gull Island, Alaska	Birds
Lipovnik (8)	*Ixodes ricinus*	Lipovnik, eastern Slovakia, Czechoslovakia	?Man
Mono Lake (5)	*Argas cooleyi*	Mono County, California	Birds
Mykines (9)	*Ixodes uriae*	Faeroe Islands, North Atlantic Ocean	Birds
Nuggett (10)	*Ixodes uriae*	Macquarie Island, Southern Ocean, Australia	—
Okhotskiy (11)	*Ixodes (Ceratixodes) putus*	Tyuleniy Island, Okhotsk, USSR	Birds
Poovoot (7)	*Ixodes uriae*	St. Laurence Island, Alaska	Birds
Seletar (12)	*Boophilus microplus*	Seletar district, Singapore	Cattle

grations. Yunker (1975) described the geographic distribution of some members (including Kenai and Poovoot as provisional ones) on the North American continent. Doherty *et al.* (1975) isolated Nugget virus from *Ixodes uriae* at Macquarie Island in the Southern Ocean. This observations show the extent of the subarctic/subantarctic distribution.

G. Other Groups

Space does not permit individual detailed consideration of the eight groups remaining in Table III. The following account selects those facets of epizootiol-

TABLE VI—*Continued*

Name	Tick	Locality of first isolates	Vertebrate host association
Sixgun City (13)	*Argas cooleyi*	Sunday Canyon, Texas	Birds
Tindholmur (9)	*Ixodes uriae*	Faeroe Islands, North Atlantic Ocean	Birds
Tribec (14)	*Ixodes ricinus*	Tribec Mountains, southwestern Slovakia, Czechoslovakia (also in Italy) (15)	Mice, goats, voles, cattle, man
Wad Medani (4)	*Rhipicephalus sanguineus*	Wad Medani, Sudan	—
Yaquina Head (16)	*Ixodes uriae*	Lincoln County, Oregon	—

[a] Also isolated from a migrating common redstart captured in Egypt (Schmidt and Shope, 1971).
[b] See text.
[c] ?Relation to Great Island (Ritter and Feltz, 1974).
(1) Berge (1975).
(2) Main (1973).
(3) Main *et al.* (1976).
(4) Taylor *et al.* (1966).
(5) Johnson and Casals (1972).
(6) Chumakov *et al.* (1963).
(7) Yunker (1975).
(8) Libikova *et al.* (1964).
(9) Main (1978).
(10) Doherty *et al.* (1975).
(11) Lvov *et al.* (1973).
(12) Rudnick *et al.* (1967).
(13) Yunker *et al.* (1972).
(14) Gresikova *et al.* (1965).
(15) Verani *et al.* (1978).
(16) Yunker (1973).

ogy, virus characterization, and diagnosis that are thought to be useful for understanding the problems of orbiviruses and to illustrate the need for further research.

1. Palyam Group

C. brevitarsis and *Culex* spp. are so far the only known vectors, and cattle and sheep the only vertebrates with detectable antibody (see Table VII). Nyabira and D'Aguilar have been isolated from vertebrates, but it has not yet been established that Nyabira has arthropod transmission or that it causes abortion in cattle (Swanepoel and Blackburn, 1976). D'Aguilar virus has been isolated from a cow (St. George and Dimmock, 1976), and it has been regularly isolated from

TABLE VII
Palyam Group (Gorman Group D)

Name	Arthropod	Locality	Vertebrate host association
D'Aguilar (1)	*Culicoides brevitarsis*	Bunya, Queensland, Australia	Cattle[a], sheep
Kasba (2)	*Culex "vishnui"*	Tamil Nadu, India	?
Palyam (2)	*Culex "vishnui"*	Tamil Nadu, India	?
Vellore (3)	*Culex pseudovishnui*	Tamil Nadu, India	Cattle
Abadina (4)	*Culicoides* spp. Mosquitoes	Nigeria	?
Nyabira (5)	?	Nyabira, Rhodesia	Cattle[a]

[a]Virus isolation.
(1) Doherty *et al.* (1973).
(2) Dandawate *et al.* (1969).
(3) Myers *et al.* (1971).
(4) Lee *et al.* (1974).
(5) Swanepoel and Blackburn (1976).

bovine blood (B. M. Gorman, personal communication, 1979; T. D. St. George, personal communication, 1979).

2. *Changuinola, Corriparta, Warrego, and Wallal Groups*

Table VIII provides limited information currently available on vector, vertebrate host, and geographic distribution. It is worth noting that the Wallal groups have been detected only in northern Australia in a biocenose probably involving *Culicoides* spp. and marsupials (Doherty *et al.*, 1973, 1977). The Warrego group has a similar geographic association, but antibodies have been found in cattle as well as marsupials and Warrego has been isolated from *Anopheles* mosquitoes (Doherty *et al.*, 1973, 1977). Viruses of the Corriparta group have been isolated from *Culex* spp. in Ethiopia and Australia and from birds in Australia (Carley and Standfast, 1969). The Changuinola viruses have so far been located only in Central and South America.

3. *Ungrouped Orbiviruses*

Table IX summarizes the pertinent available data on five ungrouped orbiviruses. Two (Orungo and Lebombo) infect man. As Orungo virus has been isolated only from man and mosquitoes, the possibility exists that man may be the natural host (Tomori and Fabiyi, 1976; Tomori, 1978). Further study of this orbivirus is required regarding the resistance of its outer coat layer (Tomori *et al.*, 1976). Although the other viruses are accepted as orbiviruses, comment is not warranted until further data are available.

TABLE VIII
Orbiviruses: Gorman Groups E, F, H, and I

Group	Name	Arthropod	Locality	Vertebrate host association
E Changuinola	Changuinola (1)	*Phlebotomus* spp.	Bocas del Toro, Canal Zone, Central America	Arboreal mammals, opossums, man[a]
	Irituia (Be An 28873) (2)	?*Phlebotomus* spp.	Belém, Pará, Brazil	*Oryzomys* spp.[b]
	Be Ar 35646 (2)	*Phlebotomus* spp.	Belém, Pará, Brazil	*Lutzomyia* spp.[b]
	Be Ar 41067 (2)	*Phlebotomus* spp.	Belém, Pará, Brazil	*Lutzomyia* spp.[b]
	Pan D50 (2)	*Culex nigripalpus*	Belém, Pará, Brazil	?
F Corriparta	Corriparta	*Culex annulirostris* (3)	Queensland, Australia	Man,[a] birds[b] (5), cattle
		A. catasticta (4)	Western Australia, Australia	Marsupials, horses
	Acado (6)	*C. antennatus* and *C. univittatus neavi*	Baro River, Ethiopia	—
	Bambari (7)	*Culex* spp.	Bambari, Central African Republic	Birds
H Warrego	Warrego (8)	*Culicoides* spp.	Charleville, Queensland, Australia	Marsupials
		Anopheles meraukensis		Cattle
	Mitchell River (8)	*Culicoides* spp.	Mitchell River, Queensland, Australia	Cattle, marsupials
I Wallal	Wallal (8)	*Culicoides dycei*	Charleville, Queensland, Australia	Marsupials
		Culicoides marksi		
	Mudjinbarry (8)	*Culicoides marksi*	Northern Territory, Australia	Marsupials

[a] Neutralizing antibody.
[b] Virus located. (1) Borden *et al.* (1971). (2) Woodall (1967); Theiler and Downs (1973). (3) Doherty (1970). (4) Liehne *et al.* (1976). (5) Whitehead *et al.* (1968). (6) Berge (1975). (7) Annual Report (1973). (8) Doherty *et al.* (1973).

TABLE IX
Ungrouped Orbiviruses

Name	Arthropod	Locality	Vertebrate host association
Orungo (1, 2)	*Aedes* *Anopheles*	Nigeria	Man, sheep, cows, monkeys
Lebombo (3)	*Aedes circumluteolus* *Mansonia africana* *Scipio aulacodi*	Natal, Nigeria	Man, rodents
Japanaut (3, 4)	Culicines	Sepik River, Papua, New Guinea	Bats
Umatilla (3)	*Culex* spp.	Oregon, U.S.A.	?
Paroo River (5, 6) (Aus. GG 668)	Mosquitoes	Queensland, Australia	?

(1) Tomori and Fabiyi (1976).
(2) Tomori (1978).
(3) Berge (1975).
(4) I. Marshall (personal communication, 1979).
(5) I. Marshall (personal communication, 1979).
(6) Annual Report, YARU (1977).

H. Laboratory Diagnosis

Although immunofluorescence and electron microscopy have been used in the diagnosis of individual infections, natural or experimental (e.g., CTF), it is clear that most of the 88 orbiviruses described have been isolated primarily from arthropods and rarely from vertebrates during surveys for arthropod-borne disease. Most have been isolated by i.c. and/or intraperitoneal (i.p.) inoculation of newborn mice and occasionally by cell culture. On isolation of a presumed virus, the procedures adopted for characterization are as outlined earlier (Section I,II, A,B) and involve electron microscopy, detection of double-strandedness of RNA, sensitivity to lipid solvents and pH, and ability to replicate in arthropods, arthropod cell cultures, or mammalian cell cultures. Provisional identity is then determined by the use of CF and neutralization tests on which the present characterization is based (see also Table X). The limitations of this approach have become obvious and will now be discussed.

I. Problems of Characterization

Virologists have successfully employed variations of the CF, HI, and neutralization tests to characterize the major genera and antigenic types within these genera. This is universally accepted as successful and critical in many areas such as evolution, diversity, and disease prevention. The orbiviruses so far described

TABLE X
Techniques Employed in the Laboratory Diagnosis of Some Orbivirus and Reovirus Infections[a]

	Virus	Source	Isolation/detection	Antibody
Reovirus	Mammalian	V	NBM, CC, IF, IP, EM, RIA, LS, CAM, F	CF, HI, N, IF, IP, GD
		A	NBM	—
	Avian	V	CC, CE, CAM, EM, IF, LS, CAM	CF, N, GD
Orbivirus	African horse sickness	V	CC, NBM, CE, EM, F	CF, N, GD, HI
		A	CC, NBM, EM, F	—
	Bluetongue	V	NBM, CC, CE, LS, EM, F, GD	CF, N, GD, HPA, IF
		A	NBM, CC, F	—
	Colorado tick fever	V	NBM, CC, IF, EM, CE	CF, N, IF
		A	NBM, CE	—

[a]It is worth noting that although ELISA has been used for rotavirus diagnosis, this method has not been used for orbivirus or reovirus characterization.

V, vertebrate; A, arthropod; NBM, newborn mice; CC, cell cultures; IF, immunofluorescence; IP, immunoperoxidase; EM, electron microscopy; RIA, radioimmunoassay; LS, lipid sensitivity; CAM, chorioallantoic membrane; F, filtration; CF, complement fixation; HI, hemagglutination inhibition; N, neturalization (including variations like plague reduction); GD, gel diffusion; HPA, hemolytic plaque assay; CE, chick embryo (other than CAM).

do, however, illustrate the weakness of this approach in establishing diversity. This is being overcome to a great extent by utilizing the knowledge that these viruses have 10 genome segments, any one of which may provide the genetic information for one protein (antigen) (Huismans, 1979). The demonstration of genetic reassortment between reoviruses (Sharpe *et al.*, 1978) and orbiviruses (Gorman *et al.*, 1977a,b) indicates that a clear division into antigenic groups with both genera is unlikely.

Gorman (1979) has examined the correlation of antigenic differences between orbiviruses and the molecular structure, and readers are referred to this review for details. Gorman's analysis of the relative value of RNA–RNA hybridization, PAGE, and genetic reassortment as mechanisms for variation in orbiviruses clearly points the way for future studies. It is most interesting to note that the reovirus studies of Fields and Joklik and their colleagues (see Section III,B) show that an identical situation with reoviruses exists.

Wallal and Mudjinbarry viruses (see Table VIII) may be distinguished by neutralization but not by CF tests (Doherty *et al.*, 1977). Gorman *et al.* (1978) record a high-frequency combination between ts mutants of Wallal and Mudjinbarry, the genomes of which comprise distinct segments of double-stranded RNA. These results favor reassortment of genome segments as the most likely mechanism. Gorman (1979) has summed up the situation concisely by stating, "Comparative analysis of each of the 10 genes and their function is needed to understand the genetic basis of diversity within orbiviruses." In addition, Knudson *et al.* (1977) report on the use of PAGE for analysis of the RNAs of 27 viruses of the Kemerovo group in the hope that it will clarify the closeness of relationship of multiple serotypes occurring in the same tick populations; this could lead to the characterization of useful cluster groups.

III. REOVIRUS

[R/2:15/15:S/S:V/O]

A. Comparative Aspects of Mammalian Types

The comparative aspects, properties, and diagnosis of three mammalian and five avian reoviruses have been discussed in some detail in Chapter 10, Volume I, of this series (see Stanley, pp. 385–421), as well as in a number of reviews (Hassan and Cochran, 1966; Stanley, 1967, 1974, 1978; Rosen, 1968; Spendlove, 1970). Readers are referred to these reports for details of comparative diagnosis. There appears to be no antigenic overlap between the mammalian and avian groups, but types within the groups may be differentiated by neutralization tests. The Nelson Bay virus (Gard and Marshall, 1973) isolated from the blood of the fruit bat (*Pteropus poliocephalus*) shows antigenic overlap with the mammalian reoviruses. Two bovine serotypes have been isolated from cattle and may be distinguished from each other and from the three human types (with which they share complement-fixing antigens) by neutralization and HI tests (Kurogi *et al.*, 1974). The three feline isolates described by Csiza (1974) and the three isolates associated with cloacal pasting in chicks are also antigenically related to the three mammalian types (Deshmukh *et al.*, 1969). Possibly other types may be characterized.

Reoviruses are characterized by great ubiquity, and the three human mammalian types have been isolated from cattle, dogs, cats, mice, marsupials, pigs, wild birds, lambs, sheep, chimpanzees, *Macaca* spp., *Cercopithecus* spp., insects (Stanley, 1978), horses (Theim and Harth, 1976), and chickens (Jones, 1976). Serological studies reveal even greater ubiquity, with Stanley and Leak (1963) finding evidence of antibody in all vertebrates tested except sperm whales. This ubiquity creates a diagnostic problem which requires the antigenic type of reovirus isolated to be characterized. This has not been done in many of the avain infections, and in many instances it is not known whether the reovirus is a mammalian, an avian, or some other member of the Reoviridae.

As opposed to those orbiviruses which are arthropod-transmitted, the epidemiological pattern and experimental studies indicate that the fecal–oral route is the common route of transmission with reoviruses, with some spread also via the respiratory tract. It is likely that animal strains infect man, and human strains may infect animals. Their physical properties enable the mammalian reoviruses to survive in the environment and contaminate it. Stanley (1978) suggested that reoviruses could be used as a marker for bird, animal, or human fecal pollution of water, as all types have been frequently isolated from this source and in high concentration. This problem is now relevant to the recycling of water for human use and the need for detection of small numbers of enteric viruses in large volumes of water. Since Stanley reported on this problem in 1977, there has been improvement in virus detection methods in the United States and the confirmation of the presence of reoviruses (Sharp *et al.*, 1976; Floyd and Sharp, 1977; Hurst *et al.*, 1978; Sobsey *et al.*, 1978; Vaughn *et al.*, 1978).

Diagnostic procedures have not changed within the last 2 years, and an infection may be revealed by (1) the detection and identification of virus by cell culture, newborn mouse inoculation, immunofluorescence, immunoperoxidase, or electron microscopy, and (2) the detection of antibody by HI, CF, neturalization, or immunodiffusion tests (see Stanley, 1978, for details). A radioimmunofluorescent antibody technique has now been developed for the detection of reovirus antigens in culture (McCammon, 1976).

B. Importance of Analysis of the Double-Stranded RNA Genome

The significance of studies on orbivirus RNA in relation to characterization of virus isolates has been referred to in Section II, A. A similar situation applies to the three mammalian reoviruses where there has been a genetic approach to both virulence and antigenic characterization.

The reovirus genome comprises 10 segments of double-stranded RNA. As with the orbiviruses, genetic reassortment is almost certainly the mechanism of recombination (Cross and Fields, 1976; Sharpe *et al.*, 1978). The RNAs and the polypeptides are distinguished by PAGE (Ramig *et al.*, 1977). Using PAGE to analyze the RNA patterns of recombinants between serotypes has permitted the construction of a genetic map which correlates segments between the serotypes (Sharpe *et al.*, 1978; Ramig *et al.*, 1978; Mustoe *et al.*, 1978). The technique used permitted ts mutants to be mapped onto specific segments of the genome and proteins to be correlated with the genome segments in which they are encoded. Using hybrid recombinant clones, it has been possible to determine the gene segregating with and responsible for neutralization and type specificity of reoviruses (Weiner and Fields, 1977). This approach has distinct advantages for the study of the biological properties of the virions and virus–host interactions. Different patterns of cell tropism and virulence are exhibited by the three types of reovirus (Stanley, 1974). Using recombinant clones derived from types 1 and 3, Weiner *et al.* (1977) showed that the SI genome segment is responsible for the differing cell tropisms of the serotypes and is certainly the determinant of neurovirulence. This approach offers a molecular basis for studying virus pathogenesis and cell specificity and relates directly to the well-defined murine experimental models (Stanley, 1974) as well as to naturally occurring virulent and avirulent strains. This approach with reoviruses was first put forward by Fields (1972) and may well be applicable to other viruses producing more serious disease in man and animals. It could also be usefully employed in examining the nature of virus–cell interactions in the murine experimental induction of diabetes mellitus by reovirus type 3 after passage in pancreatic β-cell cultures (Onodera *et al.*, 1978). An identical approach, using PAGE, was used by Payne and Rivers (1976) to examine the genome segments of 33 isolates of cytoplasmic polyhedrosis viruses (see Fig. 1). Major differences offered the opportunity for a provisional characterization.

C. Comparative Aspects of Avian Types

Confusion due to lack of adequate characterization persists in this area, and readers are referred to Stanley (1978), Chapter IV, Volume I, of this series for discussion of the problem.

It has become clear that the five avian serotypes described by Kawamura *et al.* (1965) are sufficiently distinct in their antigenicity and biological properties to be considered as a separate subgroup. The three mammalian types have been isolated from birds, and serological studies suggest widespread infection with or without disease. In addition, the isolates of Deshmukh *et al.* (1969), Lee *et al.* (1973), and Spradbrow and Bains (1974) are clearly closely related antigenically and biologically to the three mammalian types. It is perhaps unfortunate that these are referred to as avian reoviruses, as the observations seem to reflect the ubiquity of mammalian types and their ability to infect and produce disease in poultry. It is also unfortunate that the isolates from birds have not been satisfactorily checked against an appropriate range of specific sera by neturalization, CF, or HI tests even though all have been provisionally characterized as reoviruses on physicochemical grounds.

Irrespective of whether a reovirus is mammalian or avian in its major properties, it is clear that it can produce disease in poultry, as indicated below.

Arthritis (Walker *et al.*, 1972; Olson and Khan, 1972; Glass *et al.*, 1973)
Nephrosis (Mandelli *et al.*, 1969)
Cloacal pasting (Deshmukh and Pomeroy, 1969a,b)
Enteritis (Gerhsowitz and Wooley, 1973)
Chronic respiratory disease (Fahey and Crawley, 1954; Petek *et al.*, 1967)
Myocarditis (Bains *et al.*, 1974; Spradbrow and Bains, 1974)

With the limited information available, it is still possible to divide the reoviruses into two subgroups—mammalian and avian. These appear to be antigenically distinct, but each subgroup has its own common complement-fixing antigen. The types within each group (five avian and three mammalian) are defined by neutralization tests. On primary isolation, the following properties are frequently described.

Avian	Mammalian
Pocks on chorioallantoic membrane	No pocks on chorioallantoic membrane
Chick embryo death	No chick embryo death
No disease in newborn mice	Severe disease in newborn mice
Do not hemagglutinate	Hemagglutinate
Syncytia in cell cultures	Characteristic perinuclear replication in cell cultures

Two avian viruses (2207/68 and Wi) isolated from chickens with infectious bursitis were shown to be typical reoviruses by Nick *et al.* (1975). PAGE analysis showed that the double-stranded RNA genome of both strains consisted of 10 segments, but there were marked differences in the RNA patterns which differentiated them from the mammalian reoviruses. There was no antigenic relationship between reovirus 3 and these viruses when tested by neutralization. Unfortunately serological tests were not done with reovirus types 1 or 2, nor with the avian reoviruses of Kawamura *et al.* (1965). Later Nick *et al.* (1976) reported that these same two viruses had no structural or biological similarities to avian reovirus. Similar results were obtained by Spandidos and Graham (1976) with an avian reovirus (S1133), but again the serological testing was inadequate. Although these viruses closely resemble the mammalian reoviruses in their properties, all 10 double-stranded RNA segments of the avian isolate may be distinguished by PAGE from those of reovirus type 3. In view of the worldwide association of reoviruses with poultry and some poultry diseases, adequate characterization of the reoviruses infecting poultry is required.

IV. PROVISIONAL REOVIRIDAE

The following five viruses are looking for a home and, at the moment, the Reoviridae appears to be the most acceptable one.

Infectious pancreatic necrosis virus
Infectious bursal disease virus
Infectious myocarditis of goslings
Bluecomb
Syncytial virus of rabbits

A. Infectious Pancreatic Necrosis Virus*

Infectious pancreatic necrosis virus (IPNV) causes a highly contagious disease of salmonids, frequently with high mortality in the younger fish (Wolfe *et al.*, 1960). Although much of the earlier work on the RNA and virus structure was controversial, it is now generally recognized that this virus morphologically resembles reovirus and has a double-stranded RNA (Cohen *et al.*, 1973; MacDonald and Yamamoto, 1977; MacDonald *et al.*, 1977). The double-stranded RNA has been examined by PAGE analysis, which reveals two RNA species of differing molecular weights (2.3 and 2.5×10^6 daltons). It has been shown that some preparations contain defective interfering particles (MacDonald and Yamamoto, 1978). Almost identical viruses have been isolated from zebra fish, *Brachydanio rerio* (Seeley *et al.*, 1977), and from the bivalve mollusc, *Tellina tenuis* (Underwood *et al.*, 1977). The viruses can be isolated in cell cultures from a variety of fish, fish embroyos, and gonads; immunofluorescence, precipitation, and neutralization tests are currently employed (Malsberger and Cerini, 1965; Dobos and Rowe, 1977). IPNV has been isolated in Europe, Japan, and North America.

B. Infectious Bursal Disease Virus†

Infectious bursal disease virus (IBDV) produces a contagious disease in poultry; chickens affected show a profuse diarrhea with a mortality up to 15% (Farragher *et al.*, 1974). At postmortem examination there are lesions in the kidney and the bursa of Fabricius, with the latter organ frequently showing inflammation, necrosis, and atrophy. Such infections with IBDV may interfere with the avian responses to vaccines such as NDV (Farragher *et al.*, 1972), and infected birds had significantly lower antibody titers when given *Mycoplasma synoviae*, infectious bronchitis virus, or NDV (Giambrone *et al.*, 1977). It has been possible to provide protection against the virulent virus by infecting chickens subclinically with a small plaque variant (Cursiefen *et al.*, 1979). IBDV is of interest not only because of its economic importance to the poultry industry but also because it can destroy the bursa of Fabricius, the prime organ for avian B lymphocyte-associated immunity.

C. Other Possible Members

Bluecomb (avian diarrhea), infectious myocarditis of goslings, and syncytial virus of rabbits all share some properties which suggest they could be placed

*Since completing this chapter, it has come to my attention that this virus will be excluded from the Reoviridae and probably placed in a new family (F. A. Murphy, personal communication, 1979).

†Since completing this chapter, it has come to my attention that this virus will be excluded from the Reoviridae and probably placed in a new family (F. A. Murphy, personal communication, 1979).

within the Reoviridae. Unfortunately they have been insufficiently characterized for useful comment and will not be discussed in this chapter. Readers are referred to Andrewes *et al.* (1978) for further information.

V. CONCLUSION

The Reoviridae comprise five genera: *Reovirus, Orbivirus, Rotavirus, Phytoreovirus,* and *Fijivirus,* plus the cytoplasmic polyhedrosis viruses. All share common properties but may be clearly differentiated and characterized. Orbiviruses are arthropod-transmitted and reoviruses are not. Only a few of the orbiviruses produce diseases in mammals, but those disease are of serious economic consequence. Characterization within the genera *Reovirus* and *Orbivirus* has been outlined from the results of serological tests such as CF, HI, neturalization, immunofluorescence, immunoperoxidase, gel diffusion, and radioimmunoassay, as well as from morphological, ecological, and epizootiological information. These tests are inadequate for the taxonomic characterization of the large numbers of isolates now accumulating. The viral genome, consisting of 10 segments of double-stranded RNA, facilitates genetic reassortment. This exchange of genetic information probably relates to the evolution of many closely related orbiviruses. It is suggested that analysis of the double-stranded RNA genome through techniques such as PAGE is required if effective progress on characterization is to be made. Although the numerous names are delightful and of geographic interest, it is suggested that the Gorman grouping be adopted for orbiviruses and that a similar approach be used for reoviruses and other genera of the Reoviridae. As the study of the Reoviridiae at the molecular level is now recognized and accepted, it is clear that solution of these virological problems could significantly contribute to our understanding of evolution, host specificity, and pathogenicity in this ubiquitous family.

REFERENCES

Andrewes, C., Pereira, H. G., and Wildy, P. (1978). "Viruses of Vertebrates," 4th Ed., pp. 42–66. Baillière, London.

Annual Report (1973). "WHO Regional Arbovirus Center for West Africa." Inst. Pasteur, Dakar, Senegal.

Bains, B. S., MacKenzie, M., and Spradbrow, P. B. (1974). *Avian Dis.* **18,** 472–476.

Borden, E. C., Shope, R. E., and Murphy, F. A. (1971). *J. Gen. Virol.* **13,** 261–271.

Bourne, J. G. (1971). *Adv. Vet. Sci. Comp. Med.* **15,** 1–9.

Bremer, C. W. (1976). *Onderstepoort J. Vet. Res.* **43,** 193.

Burgdorfer, W. (1977). *Acta Trop.* **34,** 103–126.

Burgdorfer, W., and Eklund, C. M. (1959). *Am. J. Hyg.* **69,** 127–137.

Burgdorfer, W., and Eklund, C. M. (1960). *J. Infect. Dis.* **107,** 379–383.

Campbell, C. H., Barber, T. L., and Jochim, M. M. (1978). *Vet. Microbiol.* **3,** 15–22.
Carley, J. G., and Standfast, H. A. (1969). *Am. J. Epidemiol.* **89,** 583–592.
Casals, J. (1975). *Med. Biol.* **53,** 249–258.
Chumakov, M. P., Karpovich, L. G., Sarmonova, E. S., Sergeeva, G. I., Bychkova, M. B., Tapupeve, V. O., Libikova, H., Mayer, V., Rehacek, J., Kozuch, O., and Ernek, E. (1963). *Acta Virol.* **7,** 82–83.
Clark, G. M., Clifford, C. M., Fadness, L., and Jones, E. K. (1970). *J. Med. Entomol.* **7,** 189–197.
Cohen, J., Poinsard, A., and Scherrer, R. (1973). *J. Gen. Virol.* **21,** 485–498.
Cross, R. K., and Fields, B. N. (1976). *Virology* **74,** 345–362.
Csiza, C. K. (1974). *Infect. Immun.* **9,** 159–166.
Cursiefen, D., Kaufer, I., and Becht, H. (1979). *Arch. Virol.* **59,** 39–46.
Dales, S., Gomatos, P. J., and Hsu, K. C. (1965). *Virology* **25,** 193–211.
Dandawate, C. N., Rajagopolan, P. K., Pavri, K. M., and Work, T. M. (1969). *Indian J. Med. Res.* **57,** 1420–1426.
Davies, F. G., and Otieno, S. (1977). *Vet. Rec.* **100,** 291–292.
Deshmukh, D. R., and Pomeroy, B. S. (1969a). *Avian Dis.* **13,** 239–240.
Deshmukh, D. R., and Pomeroy, B. S. (1969b). *Avian Dis.* **13,** 427–439.
Deshmukh, D. R., Sayed, H. I., and Pomeroy, B. S. (1969). *Avian Dis.* **13,** 16–22.
Dobos, P., and Rowe, D. (1977). *J. Virol.* **24,** 805–820.
Doherty, R. L. (1968). *Trans. R. Soc. Trop. Med. Hyg.* **62,** 862–867.
Doherty, R. L. (1970). *Trans. R. Soc. Trop. Med. Hyg.* **64,** 748–753.
Doherty, R. L., Carley, J. G., Standfast, H. A., Dyce, A. L., Kay, B. H., and Snowdon, W. A. (1973). *Trans. R. Soc. Trop. Med. Hyg.* **67,** 536–543.
Doherty, R. L., Carley, J. G., Murray, M. D., Main, A. J., Kay, B. H., and Domrow, R. (1975). *Am. J. Trop. Med. Hyg.* **24,** 521–526.
Doherty, R. L., Standfast, H. A., Dyce, A. L., Carley, J. G., Gorman, B. M., Filippich, C., and Kay, B. H. (1977). *Aust. J. Biol. Sci.* **31,** 97–103.
du Toit, R. M. (1944). *Onderstepoort J. Vet. Sci. Anim. Ind.* **19,** 7–11.
Edgar, S. A. (1976). *Dev. Biol. Stand.* **33,** 349–356.
Eklund, C. M., Kohls, G. M., and Jellison, W. L. (1958). *Science* **128,** 413.
Eklund, C. M., Kohls, G. M., and Kennedy, R. C. (1961). *Czech. Acad. Sci.* p. 401.
Els, H. L., and Verwoerd, D. V. (1969). *Virology* **38,** 213–219.
Emmons, R. W. (1966). *Am. J. Trop. Med. Hyg.* **15,** 428–433.
Emmons, R. W., and Lennette, E. H. (1966). *J. Lab. Clin. Med.* **68,** 923–929.
Emmons, R. W., Dondero, D. V., Devlin, V., and Lennette, E. H. (1969). *Am. J. Trop. Med. Hyg.* **18,** 796–802.
Emmons, R. W., Oshiro, L. S., Johnson, H. N., and Lennette, E. H. (1972). *J. Gen. Virol.* **17,** 185–195.
Erasmus, B. J. (1975a). *Aust. Vet. J.* **51,** 165–170.
Erasmus, B. J. (1975b). *Aust. Vet. J.* **51,** 196–198.
Erasmus, B. J. (1975c). *Aust. Vet. J.* **51,** 209–210.
Fahey, J. E., and Crawley, J. F. (1954). *Can. J. Comp. Med.* **18,** 13–21.
Farragher, J. T., Allan, W. H., and Cullen, G. A. (1972). *Nature (London), New Biol.* **237,** 119.
Farragher, J. T., Allan, W. H., and Wyeth, P. J. (1974). *Vet. Rec.* **95,** 385.
Federici, B. A., and Hazard, E. I. (1975). *Nature (London)* **254,** 327.
Fields, B. N. (1972). *N. Engl. J. Med.* **287,** 1026–1033.
Florio, L., Miller, M. S., and Mugrage, E. R. (1950). *J. Immunol.* **64,** 257–263.
Floyd, R., and Sharp, D. G. (1977). *Appl. Environ. Microbiol.* **33,** 159–167.
Foster, N. M., Breckon, R. D., Luedke, A. J., Jones, R. H., and Metcalfe, H. E. (1977). *J. Wildl. Dis.* **13,** 9–16.

Frank, J. F., and Willis, N. G. (1975). *Aust. Vet. J.* **51,** 174–177.

Gard, G. P., and Marshall, I. D. (1973). *Arch. Gesamte Virusforsch.* **43,** 34–42.

Gard, G. P., Marshall, I. D., and Woodroofe, G. M. (1973). *Am. J. Trop. Med. Hyg.* **22,** 551–560.

Geering, W. A. (1975). *Aust. Vet. J.* **51,** 220–224.

Gershowitz, A., and Wooley, R. E. (1973). *Avian Dis.* **17,** 406–414.

Giambrone, I. J., Eidson, C. S., and Kleven, S. H. (1977). *Am. J. Vet. Res.* **38,** 251–253.

Glass, S. E., Nagi, S. A., Hall, C. F., and Kerr, K. M. (1973). *Avian Dis.* **17,** 415–424.

Goldsmit, L. (1967). *Am. J. Vet. Res.* **28,** 19–24.

Gorman, B. M. (1978). *Aust. J. Exp. Biol. Med. Sci.* **56,** 359–367.

Gorman, B. M. (1979). *J. Gen. Virol.* **44,** 1–15.

Gorman, B. M., and Taylor, J. (1978). *Aust. J. Exp. Biol. Med. Sci.* **56,** 369–371.

Gorman, B. M., Taylor, J., Brown, K., and Melzer, A. J. (1977a). *32nd Annu. Rep. Queens. Inst. Med. Res.* pp. 15–16.

Gorman, B. M., Walker, P. J., and Taylor, J. (1977b). *Arch. Virol.* **54,** 153–158.

Gresikova, M., Nosek, J., Kozuch, O., Ernek, E., and Lichard, M. (1965). *Acta Virol.* **9,** 83–88.

Hafez, S. M., Pollis, E. G., and Mustafa, S. A. (1978). *Trop. Anim. Health Prod.* **10,** 95–98.

Harkness, J. W., Alexander, D. J., Pattison, M., and Scott, A. C. (1975). *Arch. Virol.* **48,** 63–73.

Harris, R. E., Morahan, P., and Coleman, P. (1975). *J. Infect. Dis.* **131,** 397–402.

Hassan, S. A., and Cochran, K. W. (1966). *Bacteriol. Rev.* **42,** 115.

Holmes, I. H., Ruck, B. J., Bishop, R. F., and Davidson, G. P. (1975). *J. Virol.* **16,** 937–943.

Hopkins, J. G., Hazrati, A., and Ozawa, Y. (1966). *Am. J. Vet. Res.* **27,** 96–105.

Hourrigan, J. L., and Klingsporn, A. L. (1975a). *Aust. Vet. J.* **51,** 170–174.

Hourrigan, J. L., and Klingsporn, A. L. (1975b). *Aust. Vet. J.* **51,** 203–208.

Howell, P. G. (1962). *Onderstepoort J. Vet. Res.* **29,** 139–149.

Howell, P. G. (1970). *J. S. Afr. Vet. Med. Assoc.* **41,** 215–223.

Huismans, H. (1978). *Virology* **92,** 385–396.

Huismans, H. (1979). *Virology* **94,** 417–429.

Huismans, H., and Els, H. J. (1978). *Virology* **92,** 397–406.

Hurst, C. J., Farrah, S. R., Gerba, C. P., and Melnick, J. L. (1978). *Appl. Environ. Microbiol.* **36,** 81–89.

Jennings, M., and Boorman, J. (1979). *Arch. Virol.* **59,** 121–126.

Johnson, H. N., and Casals, J. (1972). *In* "Transcontinental Connections of Migratory Birds and their Role in the Distribution of Arboviruses" (A. Cherepanov *et al.*, eds.), Publishing House, Siberian Branch, Novosibirsk.

Joklik, W. K. (1974). *In* "Comprehensive Virology" (H. Fraenkel-Conrat and R. R. Wagner, eds.), Vol. 2, pp. 231–344, Plenum, New York.

Jones, R. C. (1976). *Vet. Rec.* **99,** 458.

Jones, R. H., and Foster, N. M. (1978). *Am. J. Trop. Med. Hyg.* **27,** 178–183.

Kawamura, H., Shimizu, F., Maeda, M., and Tsubahara, H. (1965). *Natl. Inst. Anim. Health Q.* **5,** 115–124.

Kimura, T., and Murakami, T. (1977). *Infect. Immun.* **17,** 157–160.

Knudson, D., Shope, R., and Main, A. (1977). "Annual Report," p. 62. Yale Arbovirus Research Unit, New Haven, Connecticut.

Kurogi, H., Inaba, Y., Takahashi, E., Sato, K., Goto, Y., Omori, T., and Matumoto, M. (1974). *Arch. Gesamte Virusforsch.* **45,** 157–160.

Lee, L. F., Nazerian, K., and Burmester, B. R. (1973). *Avian Dis.* **17,** 559–567.

Lee, V. H., Causey, O. R., and Moore, D. L. (1974). *Am. J. Vet. Res.* **35,** 1105–1008.

Libikova, H., and Buckley, S. M. (1971). *Acta Virol.* **15,** 79–86.

Libikova, H., and Casals, J. (1971). *Acta Virol.* **15,** 65–78.

Libikova, H., Rehacek, J., Gresikova, M., Kozuch, O., Somogyiova, J., and Ernek, E. (1964). *Acta Virol.* **8,** 96.

Liehne, C. G., Leivers, S., Stanley, N. F., Alpers, M. P., Paul, S., Liehne, P. F. S., and Chan, K. H. (1976). *Aust. J. Exp. Biol. Med. Sci.* **54,** 499–504.

Luedke, A. J., Jones, R. H., and Waltn, T. E. (1977). *Am. J. Trop. Med. Hyg.* **26,** 313–325.

Lvov, D. K., Timofeeva, A. A., and Gromashevski, V. L. (1973). *Arch. Gesamte Virusforsch.* **41,** 160–164.

McCammon, J. R. (1976). *Infect. Immun.* **14,** 811–815.

MacDonald, R. D., and Yamamoto, T. (1977). *J. Gen. Virol.* **34,** 235–247.

MacDonald, R. D., and Yamamoto, T. (1978). *Arch. Virol.* **57,** 77–89.

MacDonald, R. D., Roy, K. L., Yamamoto, T., and Chang, N. (1977). *Arch. Virol.* **54,** 373–377.

Main, A. J. (1973). *J. Med. Entomol.* **10,** 229–235.

Main, A. J. (1978). *J. Med. Entomol.* **15,** 11–14.

Main, A. J., Shope, R. E., and Wallis, R. C. (1976). *J. Med. Entomol.* **13,** 304–308.

Malsberger, R. G., and Cerini, C. P. (1965). *Ann. N. Y. Acad. Sci.* **126,** 555–565.

Mandelli, G., Rinaldi, A., Cessi, D., Cervio, G., Pasencci, S., and Valeri, A. (1969). *Atti Soc. Ital. Sci. Vet.* **23,** 1–8.

Martin, S. A., and Zweerink, H. J. (1972). *Virology* **50,** 495.

Matthews, R. E. F. (1979). *Intervirology* **11,** 133–135.

Mellor, P. S., Boorman, J., and Jennings, M. (1975). *Arch. Virol.* **47,** 351–356.

Mirchamsy, H., Hazrati, A., Bahrami, S., and Shapyi, A. (1970). *Am. J. Vet. Res.* **31,** 1755–1760.

Murphy, F. A., Borden, E. C., Shope, R. E., and Harrison, A. (1971). *J. Gen. Virol.* **13,** 273–288.

Mustoe, T. A., Ramig, R. F., Sharpe, A. H., and Fields, B. N. (1978). *Virology* **85,** 545–556.

Myers, R. M., Carey, D. E., Reuben, R., Jesudass, E. S., and Shope, R. E. (1971). *Indian J. Med. Res.* **59,** 1209–1213.

Nawathe, D. R., Onunkwo, O., and Smith, J. M. (1978). *Vet. Rec.* **102,** 444.

Nick, H., Cursiefen, D., and Becht, H. (1975). *Arch. Virol.* **48,** 261–269.

Nick, H., Cursiefen, D., and Becht, H. (1976). *J. Virol.* **18,** 227–234.

Oellermann, R. A., Carter, P., and Marx, M. J. (1976). *Infect. Immun.* **13,** 1321–1324.

Olson, N. O., and Khan, M. A. (1972). *Avian Dis.* **16,** 1073–1078.

Omori, T., Inaba, Y., Morimoto, T., Tanaka, Y., Ishitani, R., Kurogi, H., Munakata, K., Matsuda, K., and Matumoto, M. (1969). *Jpn. J. Microbiol.* **13,** 139–157.

Onodera, T., Jenson, A. B., Yoon, J. W., and Notkins, A. L. (1978). *Science* **201,** 529–531.

Oshiro, L. S., Dondero, D. V., Emmons, R. W., and Lennette, E. H. (1978). *J. Gen. Virol.* **39,** 73–79.

Palmer, E. L., Martin, M. L., and Murphy, F. A. (1977). *J. Gen. Virol.* **35,** 403–414.

Pavri, K. M., and Anderson, C. R. (1963). *Indian J. Vet. Sci.* **33,** 113–117.

Payne, C. C., and Rivers, C. F. (1976). *J. Gen. Virol.* **53,** 71–85.

Petek, M., and Mandelli, G. (1968). *Atti Soc. Ital. Sci. Vet.* **22,** 875–879.

Petek, M., Felluga, B., Borghi, G., and Baroni, A. (1967). *Arch. Gesamte Virusforsch.* **21,** 413–423.

Porterfield, J. S. (1975). *Med. Biol.* **53,** 400–405.

Price, D. A., and Handy, W. T. (1954). *J. Am. Vet. Med. Assoc.* **124,** 255–258.

Rahse-Küpper, B., Casals, J., Rehse, E., and Ackerman, R. (1976). *Acta Virol.* **20,** 339–342.

Ramig, R. F., Cross, R. K., and Fields, B. N. (1977). *J. Virol.***22,** 726–733.

Ramig, R. F., Mustoe, T. A., Sharpe, A. H., and Fields, B. N. (1978). *Virology* **85,** 531–544.

Randall, W. H., Simmons, J., Casper, E. H., and Philip, R. N. (1975). CDC, Atlanta, Georgia, **24,** No. 50, 422–423.

Ritter, D. G., and Feltz, E. T. (1974). *Can J. Microbiol.* **20,** 1359–1366.

Rodger, S. M., Schnagl, R. D., and Holmes, I. H. (1975). *J. Virol.* **16,** 1229–1235.

Rosen, L. (1968). *Monogr. Virol.* **1,** 73–107.
Rudnick, A., Marchette, N. J., and Garcia, R. (1967). *Abstr., Southeast Asian Reg. Semin. Trop. Med., 1st, 1967* pp. 40–41.
St. George, T. D., and Dimmock, C. K. (1976). *Aust. Vet. J.* **62,** 598.
St. George, T. D., and McCaughan, C. I. (1979). *Aust. Vet. J.* **55,** 198–199.
St. George, T. D., Standfast, H. A., Christie, D. G., Knott, S. G., and Morgan, I. R. (1977). *Aust. Vet. J.* **53,** 17–28.
St. George, T. D., Standfast, H. A., Cybinski, D. H., Dyce, A. L., Doherty, R. L., and Carley, J. G. (1978). *Aust. Vet. J.* **54,** 153–154.
Schmidt, J. R., and Shope, R. E. (1971). *Acta Virol.* **15,** 112.
Seeley, R. J., Perlmutter, A., and Seeley, V. A. (1977). *Appl. Environ. Microbiol.* **34,** 50–55.
Sellers, R. F. (1978). *Abstr. Int. Congr. Virol., 4th, 1978* p. 289.
Sellers, R. F., Pedgley, D. E., and Tucker, M. R. (1977). *J. Hyg.* **79,** 279–298.
Sharp, D. G., Floyd, R., and Johnson, J. D. (1976). *Appl. Environ. Microbiol.* **31,** 173–181.
Sharpe, A. H., Ramig, R. F., Mustoe, T. A., and Fields, B. N. (1978). *Virology* **84,** 63–74.
Shipham, S. O., and de La Ray, M. (1976). *Onderstepoort J. Vet. Res.* **43,** 189–192.
Shope, R. E. (1977). "Annual Report," pp. 29–30. "Yale Arbovirus Research Unit, New Haven, Connecticut.
Shope, R. E., MacNamara, L. G., and Mangold, R. (1960). *J. Exp. Med.* **111,** 155–170.
Sobsey, M. D., Carrick, R. J., and Jensen, H. R. (1978). *Appl. Environ. Microbiol.* **36,** 121–128.
Spandidos, D. A., and Graham, A. F. (1976). *J. Virol.* **19,** 968–976.
Spendlove, R. S. (1970). *Prog. Med. Virol.* **12,** 161–191.
Spendlove, R. S., Lennette, E. H., and John, A. C. (1963). *J. Immunol.* **90,** 554–560.
Spradbrow, P. B. and Bains, B. S. (1974). *Aust. Vet. J.* **50,** 179.
Stair, E. L., Robinson, R. M., and Jones, L. P. (1968). *Pathol. Vet.* **5,** 164–171.
Stanley, N. F. (1967). *Br. Med. Bull.* **23,** 150–155.
Stanley, N. F. (1974). *Prog. Med. Virol.* **18,** 257–272.
Stanley, N. F. (1978). *In* "Comparative Diagnosis of Viral Diseases" (E. Kurstak and C. Kurstak, eds.), Vol. 1, pp. 385–421. Academic Press, New York.
Stanley, N. F., and Leak, P. J. (1963). *Am. J. Hyg.* **78,** 82–88.
Swanepoel, R., and Blackburn, N. K. (1976). *Vet. Rec.* **99,** 360.
Taylor, R. M., Horlbut, H. S., Work, T. H., Kingston, J. R., and Hoogstraal, H. (1966). *Am. J. Trop. Med. Hyg.* **15,** 76–86.
Theiler, M., and Downs, W. G. (1973). "The Arthropod-borne Viruses of Vertebrates: An Account of the Rockefeller Foundation Virus Program 1951–1970." Yale Univ. Press, New Haven, Connecticut.
Theim, P., and Harth, G. (1976). *Zentralbl. Veterinaermed., Reihe B.* **23,** 698–701.
Thomas, F. C., Girard, A., Boulanger, P., and Ruckerbauer, G. (1976). *Can. J. Comp. Med.* **40,** 291–297.
Thornton, D. H. (1976). *Dev. Biol. Stand.* **33,** 343–348.
Tomori, O. (1978). *Br. Vet. J.* **134,** 108–112.
Tomori, O., and Fabiyi, A. (1976). *Niger, Med. J.* **7,** 5.
Tomori, O., Fabiyi, A., and Murphy, F. (1976). *Arch. Virol.* **51,** 285–298.
Underwood, B. O., Smale, C. J., and Brown, F. (1977). *J. Gen. Virol.* **36,** 93–109.
Vaughn, J. M., Landry, E. F., Baranosky, L. J., Beckwith, C. A., Dahl, M. C., and Delihas, N. C. (1978). *Appl. Environ. Microbiol.* **36,** 47–51.
Verani, P., Balducci, M., and Lopes, C. (1978). *Acta Virol.* **22,** 170.
Verwoerd, D. W. (1969). *Virology* **38,** 203–212.
Verwoerd, D. W. (1970). *Prog. Med. Virol.* **12,** 192–210.
Verwoerd, D. W., Louw, H., and Oellermann, R. A. (1970). *J. Virol.* **5,** 1–7.

Verwoerd, D. W., Els, H. J., De Villiers, E. -M., and Huismans, H. (1972). *J. Virol.* **10,** 783–794.
Verwoerd, D. W., Huismans, H., and Erasmus, B. J. (1979). *In* "Comprehensive Virology" (H. Fraenkel-Conrat and R. R. Wagner, eds.), Vol. 14, p. 285–345. Plenum, New York.
Walker, E. R., Friedman, M. H., and Olson, N. O. (1972). *J. Ultrastruct. Res.* **41,** 67–79.
Weiner, H. L., and Fields, B. N. (1977). *J. Exp. Med.* **146,** 1305–1310.
Weiner, H. L., Drayna, D., Averill, D. R., and Fields, B. N. (1977). *Proc. Natl. Acad. Sci. U.S.A.* **74,** 5744–5748.
Whitehead, R. H., Doherty, R. L., Domrow, R., Standfast, H. A., and Wetters, E. J. (1968). *Trans. R. Soc. Trop. Med. Hyg.* **62,** 439–445.
Willis, N. G., and Campbell, J. B. (1973). *Proc. Int. Colloq. Invertebr. Tissue Cult., 3rd, 1971* pp. 347–366.
Wolfe, K., Snieszko, S. F., Dunbar, C. E., and Pyle, E. (1960). *Proc. Soc. Exp. Biol. Med.* **104,** 105–108.
Woodall, J. P. (1967). *Atas Simp. Biota Amazon.* **6,** 31–63.
Yunker, C. E. (1973). *J. Med. Entomol.* **10,** 264–269.
Yunker, C. E. (1975). *Med. Biol.* **53,** 302–311.
Yunker, C. E., Clifford, C. M., Thomas, L. A., Cory, J., and George, J. E. (1972). *Acta Virol.* **16,** 415–421.

Chapter 3

Animal Rotaviruses

E. KURSTAK, C. KURSTAK, J. VAN DEN HURK,
AND R. MORISSET

I. Introduction 105
II. Incidence 108
III. Morphology and Morphogenesis 108
IV. Physicochemical Properties 112
V. Antigenic Relationships 117
VI. Clinical Features 118
VII. Pathology and Pathogenesis 120
VIII. Laboratory Diagnosis 124
IX. Propagation of Rotaviruses *in Vitro* 128
X. Epidemiology 135
XI. Immunity 137
XII. Prevention, Control, and Treatment 139
XIII. Conclusions 140
References 141

I. INTRODUCTION

Acute infectious gastroenteritis is a very common disease of young mammals, including children. The major symptoms of this self-limited disease are diarrhea and vomiting. In addition, nausea, abdominal pain and cramps, low-grade fever, and malaise may occur (Steinhoff, 1978). At times, severe outbreaks of the disease in domestic animals, often with a high mortality, cause enormous economic losses. In addition to the mortality, which can reach 80%, veterinary costs and the disruption of normal farm management cause difficulties. Gastroenteritis may also increase the time necessary to reach slaughter, weight of calves

COMPARATIVE DIAGNOSIS
OF VIRAL DISEASES, VOL. IV

ISBN 0-12-429704-8

(Woode, 1976), causing reduced productivity. Morbidity among children caused by acute gastroenteritis in the developed countries is rather high. For example, 20,000 cases are admitted to hospitals annually in England and Wales (Flewett, 1977). As 6% of the deaths in the first year of childhood in England and Wales are caused by gastroenteritis (Thornton and Zuckerman, 1975), the mortality is not negligible. However, in the developing countries, childhood mortality caused by acute gastroenteritis is much higher: approximately 5 to 18 million deaths per year in Asia, Africa, and Latin America (Elliot and Knight, 1976; Kurstak, 1977; Agarwal, 1979; Acha and Szyfres, 1980). In India alone, 1.4 million children die from noncholera diarrheal diseases each year (Editorial, 1975). It is certain that *Salmonella, Shigella* and, above all, *Escherichia coli* are agents which may produce diarrhea in young animals and children (Cramblett *et al.*, 1971; Woode, 1976; Kurstak, 1977; House, 1978) but, as was suspected by many virologists for some time, most outbreaks of gastroenteritis are caused by viral infection. At the moment, rotaviruses, parvoviruses, coronaviruses, astroviruses, caliciviruses, and "fuzzy-wuzzys" have been found to cause gastroenteritis (Flewett, 1978). In this chapter, the rotaviruses, which are very important etiological agents of gastroenteritis in animals (Woode, 1976; Flewett, 1977) and the major cause of diarrhea in children (DuPont *et al.*, 1977; Middleton, 1977), will be discussed.

The epizootic diarrhea of infant mice (EDIM) virus was the first rotavirus recognized to be the causative agent of diarrhea in suckling mice (Kraft, 1957, 1966; Adams and Kraft, 1963). The virus was found to be very infectious, resistant to heat and ether, and 65–75 nm in diameter. It was further characterized by Much and Zajac (1972), who established that it was rather acid resistant, stable on storage, and contained RNA. Electron microscopic (EM) studies of sections of infected mouse gut were performed by Adams and Kraft (1967) and Banfield *et al.* (1968). It was possible to cultivate EDIM virus in organ cultures of intestinal epithelium, but propagation in cell culture was not yet achieved (Rubinstein *et al.*, 1971).

The simian virus SA.11 was isolated from a vervet monkey and the "O" agent from abattoir waste (Malherbe and Strickland-Cholmley, 1967). Both grow well in vervet monkey kidney (VMK) cells, in which they produce round or oval cytoplasmic inclusions. SA.11 virus is resistant to ether and chloroform, but the O agent is not. Both agents are morphologically identical and stable at pH 4 but are slowly inactivated at pH3 (Lecatsas, 1972). They are double-shelled, 72 ± 0.5 nm in diameter (Els and Lecatsas, 1972), and contain RNA (Kalica *et al.*, 1978a). Other biochemical characteristics were described by Rodger *et al.* (1977) and Kalica *et al.* (1978a,b).

A very important discovery was made by Mebus *et al.* (1969), who showed that diarrhea could be caused by inoculating calves with bacterium-free filtrates of diarrheal feces. Further studies revealed virus particles 65 nm in diameter in feces of calves with gastroenteritis (Fernelius *et al.*, 1972). Since diarrhea in calves causes large economic losses, much research was carried out on this neonatal calf diarrhea virus (NCDV) following these reports. Physiochemical studies were reported by Welch and Thompson (1973), Newman *et al.* (1975), Bridger and Woode (1976), Cohen (1977), Verly and Cohen (1977), and others and revealed similarities to the EDIM virus, the SA.11 virus, and the O agent. The pathology of the calf rotavirus was studied in calves experimentally infected with NCDV (Mebus *et al.*, 1971b; Stair *et al.*, 1973; Mebus and Newman, 1977). Unlike many other rotaviruses, calf rotavirus can be adapted to serial passage in continuous cultures of calf kidney cells (Mebus *et al.*, 1971a; Fernelius *et al.*, 1972; Welch and Twiehaus, 1973; McNulty *et al.*, 1977).

The human rotavirus or infantile gastroenteritis virus (IGV) was discovered a few years later in different parts of the world. Bishop *et al.* (1973) demonstrated the presence of numerous virus particles by thin section EM in duodenal biopsies from children with acute gastroenteritis in Melbourne, Australia. These virus particles resembled the EDIM virus, but at that time they were called "orbiviruses." Shortly after this report, Flewett *et al.* (1973) in Birmingham, England, found reovirus-like particles in stool suspensions from young children using negative contrast EM. Then, in Toronto, Canada, Middleton *et al.* (1974) discovered similar viruses in the feces and duodenal biopsies of gastroenteritis patients. In Washington, in the United States, Kapikian *et al.* (1974) not only showed reovirus-like particles in feces of children with gastroenteritis but also found elevated antibody levels during the course of the disease. In addition, they described a morphological similarity and an antigenic relationship of this reovirus-like agent to EDIM virus and NCDV.

More recently, virus particles morphologically indistinguishable from the viruses described above have been found in a large number of animal species.

Early in the history of rotaviruses, they were subject to some taxonomic confusion because of their resemblance to reoviruses on the one hand and orbiviruses on the other. Thus, they were given a number of names, such as "reovirus," "reovirus-like" (Banfield *et al.*, 1968; Fernelius *et al.*, 1972; Lecce *et al.*, 1976), and "orbivirus" (Middleton *et al.*, 1974). Later, more information about their morphological, biochemical, and biophysical properties became available, thus making a classification possible. The International Committee on Taxonomy of Viruses (ICTV) placed these agents into the family of Reoviridae, although in a separate genus because of the apparent differences from reoviruses and orbiviruses (Fenner, 1976). The generic names "rotavirus" (*rota*, L. = wheel), because of the resemblance of the virions to a spoked wheel (Flewett *et al.*, 1974), and "duovirus," because of the double-layered capsid and the duodenal origin of the virus (Davidson *et al.*, 1975a), have been proposed. Although no definite decision has been made yet, it is anticipated that the widely used name "rotavirus" will be accepted by the ICTV as the generic name (Derbyshire and Woode, 1978). Recently, the reoviridae working team, established in 1975 under the WHO/FAO Comparative Virology Program for animal viruses, chose the Nebraska calf diarrheal virus strain of bovine rotavirus as the reference virus, because it is the best characterized of the genus (Derbyshire and Woode, 1978). All species names of the various gastroenteritis-causing viruses in this genus still have to be decided upon; as much as possible, the most currently used names will be used in this chapter.

II. INCIDENCE

Among the early reports of rotavirus infections were those with EDIM virus (Adams and Kraft, 1967; Banfield *et al.*, 1968; Much and Zajac, 1972), SA.11 virus in simians (Malherbe and Strickland-Cholmley, 1967; Els and Lecatsas, 1972), and the O agent, found in intestinal washings of cattle and sheep (Malherbe and Strickland-Cholmley, 1967). Later, rotaviruses were detected in calves (Mebus *et al.*, 1969, 1971b; Turner *et al.*, 1973; Meyling, 1974; Newman *et al.*, 1975; Woode and Bridger, 1975; McNulty *et al.*, 1976d) apparently with a worldwide incidence. The same also holds for the human rotavirus, which was described somewhat later (Flewett *et al.*, 1973; Bishop *et al.*, 1974; Holmes *et al.*, 1974; Ørstavik *et al.*, 1974; Sexton *et al.*, 1974; White *et al.*, 1974; Davidson *et al.*, 1975a) and extensively investigated (Petric *et al.*, 1975; Rodger *et al.*, 1975a; Kalica *et al.*, 1976; Kapikian *et al.*, 1976a,b; Schnagl and Holmes, 1976). In addition, rotaviruses have been found in pigs (Rodger *et al.*, 1975b; Lecce *et al.*, 1976; McNulty *et al.*, 1976b; Woode *et al.*, 1976b), lambs (McNulty *et al.*, 1976a; Snodgrass *et al.*, 1976a,b), foals (Flewett *et al.*, 1975a), deer (Tzipori *et al.*, 1976), rabbits (Bryden *et al.*, 1976), pronghorn antelope (Flewett and Woode, 1978), and avian species (McNulty *et al.*, 1978c). Woode *et al.* (1975) reported serological evidence for rotavirus infections in guinea pigs and goats. Virus-specific antibodies have been found in brown bears, dogs, and cats (Gustafson *et al.*, 1978; McNulty *et al.*, 1978b).

III. MORPHOLOGY AND MORPHOGENESIS

After the reports of rotavirus infections by Bishop *et al.* (1973) and Flewett *et al.* (1973), the agent was tentatively called either "orbivirus," based on its size, morphology, and location in the endoplasmic reticulum, or "reovirus," because of its double-capsid structure. However, later studies showed that this agent is a new genus, called "rotavirus."

The virions of rotaviruses of man, calves, pigs, lambs, foals, and mice are morphologically identical (McNulty *et al.*, 1976a; Woode *et al.*, 1976a) (Fig. 1). Moreover, on the basis of morphological characteristics, Kapikian *et al.* (1976b) also included in this genus SA.11 virus and the O agent. The virions of rotaviruses consist of an electron-dense core about 38 nm in diameter. From this

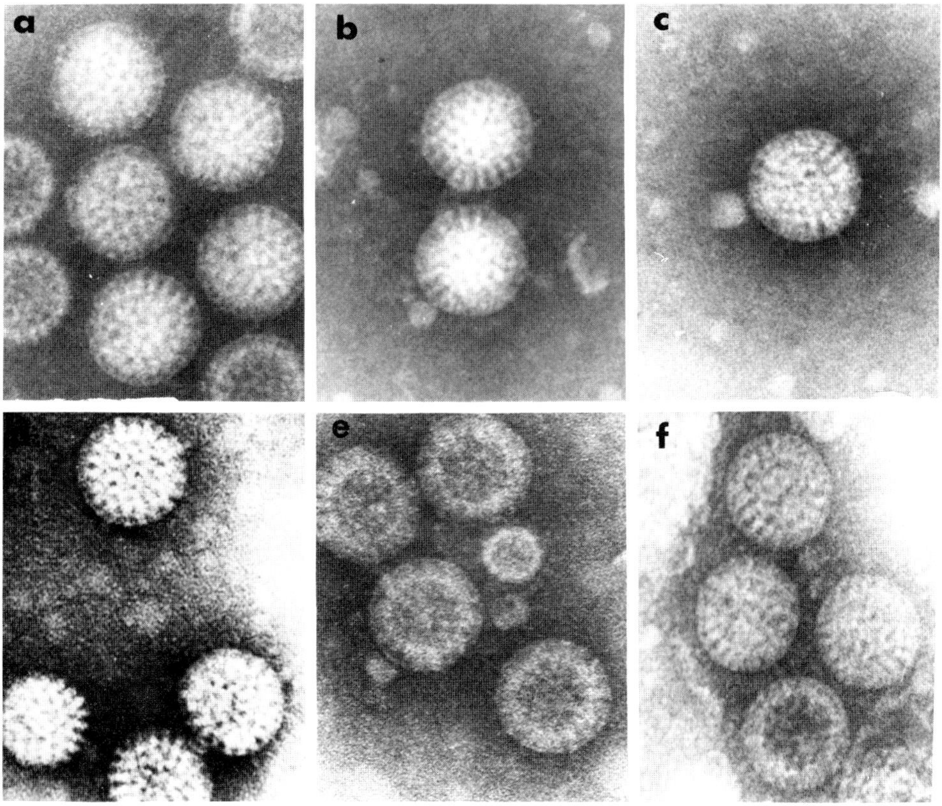

Fig. 1. Negatively stained, intact, smooth rotavirus particles from (a) man, (b) calf, (c) pig, (d) lamb, (e) foal, and (f) mouse. Courtesy of Drs. T. H. Flewett, M. C. McNulty, and M. E. Bégin.

central core, short cylindrical capsomers radiate outward like the spokes of a wheel, and an additional layer of capsid subunits is attached to the tops of these capsomers which gives the virion its smooth surface. The intact, double-shelled particles measure approximately 71 nm in diameter (Table I). When the outer layer is removed from the virions, the particles acquire a rough surface and become single-shelled, about 57 nm in diameter (Fig. 2; Table I).

Middleton (1977) described four different particle types isolated from fecal preparations of human rotavirus. Particles of larger diameter with entire circumferences and nonempty cores (density of 1.36 gm/cm^3 in CsCl gradient), particles of smaller diameter with scalloped circumferences and nonempty cores (1.38 gm/cm^3), particles of larger diameter with entire circumferences and empty cores (1.28 gm/cm^3), and empty cores (1.30 gm/cm^3). Rodger *et al.* (1975a) reported earlier about the same results for NCDV. Later, Chasey (1977) demon-

TABLE I
Morphological Properties of Rotaviruses

Virus	Diameter of D.S.[a] particle (nm)	Diameter of S.S.[b] particle (nm)	Diameter of core (nm)
Infantile gastroenteritis virus (IGV)	70–75[c] 70[b]	60[c,d]	33[c] 45[d]
Neonatal calf diarrhea virus (NCDV)	66 ± 0.4[g]	55 ± 0.4[e]	38–40[e]
Epizootic diarrhea of infant mice (EDIM) virus	65–75[k]	54 ± 0.2[f]	
Simian virus SA.11	72 ± 0.5[g]	58 ± 0.5[g]	
Lamb rotavirus	68–70[h]	58–60[h]	40[h]
O agent	72 ± 0.5[g]	55 ± 0.5[g]	
Pig rotavirus	77[g]	50–55[i]	31–38[i]
Foal rotavirus		55[j]	

[a] D.S., double-shelled. [b] S.S., single-shelled. [c] Holmes *et al.* (1975a). [d] Konno *et al.* (1975a). [e] Bridger and Woode (1976). [f] Much and Zajac (1972). [g] Els and Lecatsas (1972). [h] Snodgrass *et al.* (1976a). [i] Saif *et al.* (1978). [j] Eugster and Whitford (1978). [k] Adams and Kraft (1963).

strated five morphologically different types of particles, three of which predominate in tissue culture and in *in vivo* systems and two of which are less common. Type I particles, 26 to 35 nm in diameter, were associated with dense inclusions; type II particles, 70–80 nm in diameter and consisting of a dense core and a well-defined membrane, were found within dilated regions of rough endoplasmic reticulum; and type III particles, somewhat smaller in diameter and with indistinct outlines, were also seen in the endoplasmic reticulum. These groups of particles represent the characteristic normal development of infectious virus. Type IV particles were circular, with a diameter of 50 to 65 nm, usually evenly electron-dense without clear substructural details, and were found within large vacuoles. Type V particles were 50 nm in diameter and consisted of a dense central core surrounded by an electron-lucent region, enclosed again by a moderately dense profile. They were found in large vacuoles and were connected by thin filaments. These two forms may correspond to the noninfectious form of the virus. McNulty *et al.* (1976c) suggested that the enveloped (type II) and uneveloped (type III) forms correspond to the negatively stained double- and single-shelled particles, respectively. However, according to Chasey *et al.* (1977), the unenveloped type III particles are the double-capsid structures, and

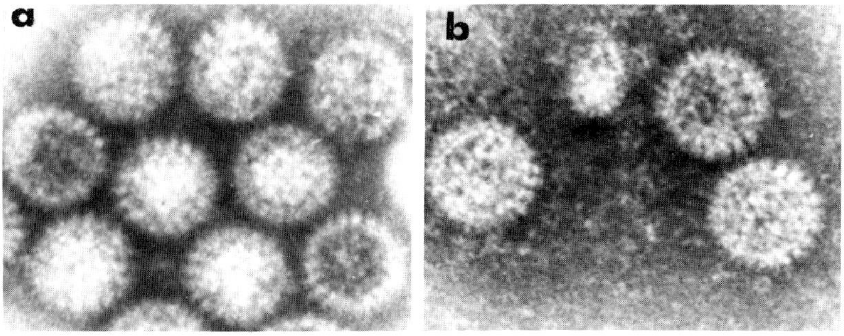

Fig. 2. Negatively stained, incomplete, rough rotavirus particles from (a) pig and (b) mouse. Courtesy of Dr. T. H. Flewett.

the type IV particles represent the single-capsid forms. This interpretation conforms with the association of infectivity with the double-capsid particles (Bridger and Woode, 1976).

Palmer *et al.* (1977) described the arrangement of the central core, which has a 5-3-2 symmetry characteristic of an icosahedron. Apparently, there is no unanimity yet about the structure of the two layers of the capsid. Martin *et al.* (1975) did not even observe the outer layer and suggested that the virus looked double-shelled by superimposition of subjacent capsomers. They also reported that the capsid layer was composed of 32 capsomers which were clustered conforming to a $T = 3$ morphology. This is a structure similar to that of orbiviruses. These capsomers were composed of six wedge-shaped subunits. As each of these was again a trimer, the capsomers conform to a $T = 9$ icosadeltahedron on the level of structural units. Such a structure had been proposed earlier by Casper and Klug (1963). Thus, the virus seems to possess 540 structural units ($20 \times 6 \times 3 + 12 \times 5 \times 3$) on its surface. However, in 1977 Stannard and Schoub postulated that the inner capsid of the virion has an icosahedral symmetry and consists of 180 morphological units arranged in an open lattice formation, with the space at each of the 12 apices surrounded by 5 capsomers and the other 80 spaces in the lattice surrounded by 6 capsomers. Furthermore, the outer layer of the capsid was clearly shown to have a honeycomb-like lattice which corresponds to the lattice arrangement of the inner capsid and is attached by short radial septa to the projections of the inner layer. A similar model was proposed by Esparza and Gil (1978), though with one major difference.

This structure of the inner capsid of rotaviruses resembles that of the outer surface of reoviruses, although the honeycomb-like structure of the outer surface of rotaviruses has not been previously described.

Among the first to describe the morphogenesis of a rotavirus were Adams and Kraft (1967) and Banfield *et al.* (1968), who studied by light microscopy and EM the intestinal epithelium of infant mice infected with EDIM virus. Later reports about the morphogenesis of calf (Stair *et al.*, 1973; Hall *et al.*, 1976), human (Holmes *et al.*, 1975a), pig (Chasey and Lucas, 1977), and lamb (McNulty *et al.*, 1976a) rotaviruses gave similar results. Studies with SA.11 virus and the O agent (Lecatsas, 1972) in cell cultures have also been reported.

In general, intracellular viruses are seen exclusively in the cytoplasm of the infected cells. In the cytoplasmic matrix of such cells, a non-membrane-bound viroplasm containing virus cores is found. After replication in the matrix, the virus particles appear to enter cisternae of the granular endoplasmic reticulum, so that a big part of the endoplasmic reticulum of the cells is dilated, containing numerous virus particles and virus-associated lipids. The virus particles acquire outer shells by budding through either matrices of granular electron-dense viroplasm or membranes of the distended rough endoplasmic reticulum. Crystalline arrays of virus were not observed, and no association of the virus with mitotic spindle tubules was found. The most severely infected cells were found to rupture and liberate large numbers of virus particles into the intestinal lumen.

Different exceptions to this sequence of general morphogenesis have also been found. Crystalline arrays of virus have been demonstrated in cells infected with the O agent; in this case, viroplasm was not present (Lecatsas, 1972). Lecatsas (1972) also described nuclear and cytoplasmic membranous elements and nuclear viroplasm in cells infected with SA.11 virus. Although unlike orbiviruses most rotaviruses have not been observed to form tubules, an exception is the EDIM virus, which forms tubules in the nucleus and cytoplasm of infected mouse epithelial cells (Banfield *et al.*, 1968). Identical tubules were observed in intestinal epithelial cells of pigs experimentally infected with pig rotavirus (McNulty, 1978) and in the nuclei of a few porcine kidney cells infected with porcine rotavirus (Saif *et al.*, 1978).

IV. PHYSICOCHEMICAL PROPERTIES

The sedimentation coefficient of calf rotavirus purified from calf kidney cell cultures, which had been infected with a strain of tissue culture-adapted virus, was 500 S (Newman *et al.*, 1975). Petric *et al.* (1975) found a sedimentation coefficient of 520 to 530 S for the human rotavirus.

Rotaviruses have a slightly varying stability against different inactivating agents. The infectivity of calf rotavirus is stable at pH 3, whereas the O agent and SA.11 virus are more acid labile (Malherbe and Strickland-Cholmley, 1967; Welch and Thompson, 1973). Calf rotavirus, EDIM virus, and SA.11 virus are stable in ether and chloroform (Much and Zajac, 1972; Welch and Thompson, 1973), but the infectivity of the O agent is partially destroyed by these lipid solvents (Malherbe and Strickland-Cholmley, 1967). All viruses mentioned are

TABLE II
Physicochemical Properties of Rotaviruses

Virus	Sedimentation coefficient of virion (S)	Density in CsCl of D.S.[a] particle (gm/cm³)	Density in CSCl of S.S.[b] particle (gm/cm³)	Number and molecular weight range of RNA segments (10^6)	Number and molecular weight range of proteins (10^3)
Infantile gastro-enteritis virus (IGV)	520–530[c]	1.36[c,i] 1.35–1.37[b]	1.38[c,i]	11[c,o] 0.23–2.04[c]	8[f] 21–127[d]
Neonatal calf diarrhea virus (NCDV)	500[g]	1.36[g,h]	1.38[g,h]	11–12[f,g] 0.24–2.2[f,g]	8/9[f] 14.5–131[f]
Epizootic diarrhea of infant mice (EDIM) virus					
Simian virus SA.11		1.36[j]	1.38[j]	11[k]	9[j] 14–133[j]
Lamb rotavirus		1.35[l]		11–12[l] 0.3–2.1[l]	10[l] 30–128[j]
O agent					
Pig rotavirus		1.36[m]		11–12[n] 0.3–2.1[n]	

[a] D.S., double-shelled. [b] S.S., single-shelled. [c] Petric *et al.* (1975). [d] Kapikian *et al.* (1975). [e] Schnagl and Holmes (1976). [f] Rodger *et al.* (1975a). [g] Newman *et al.* (1975). [h] Bridger and Woode, 1976. [i] Elias (1977b). [j] Rodger *et al.* (1977). [k] Kalica *et al.* (1978a). [l] Todd and McNulty (1977). [m] Middleton *et al.* (1975). [n] Todd and McNulty (1976). [o] Kalica *et al.* (1976).

thermolabile at 50°C in the presence of $MgCl_2$ (Malherbe and Strickland-Cholmley, 1967; Welch and Thompson, 1973).

Intact double-shelled rotavirus particles may be separated from single-shelled or empty particles by density gradient centrifugation. The buoyant density of double-shelled particles in cesium chloride is 1.36 gm/cm^3, whereas the single-shelled and empty particles band at 1.38 gm/cm^3 and 1.30–1.32 gm/cm^3, respectively (Middleton *et al.*, 1975; Newman *et al.*, 1975; Petric *et al.*, 1975; Rodger *et al.*, 1975a, 1977; Todd and McNulty, 1977; Table II). The particle/infectivity ratio of preparations of single-shelled particles is estimated to be more than 1000-fold greater than that of double-shelled particles (Bridger and Woode, 1976; Elias, 1977b).

Early work on the characterization of rotaviruses showed that calf rotavirus and EDIM virus both contained RNA (Welch, 1971; Much and Zajac, 1972), which was supposedly double-stranded (Welch and Thompson, 1973). Rodger *et al.* (1975a) confirmed the double-stranded nature of the RNA using the orcinol and diphenylamine reactions and determined the thermal denaturation characteristics of the nucleic acid of calf rotavirus. Newman *et al.* (1975) and Rodger *et al.* (1975a) demonstrated that the genome consisted of 11 or 12 double-stranded segments of RNA, ranging in molecular weight from 0.24×10^6 to 2.2×10^6. The total molecular weight of the viral RNA was $11–12 \times 10^6$. The human rotavirus genome was found to consist of 11 double-stranded segments of RNA with a molecular weight ranging from 0.23×10^6 to 2.04×10^6 and a total molecular weight of $10–11 \times 10^6$ (Holmes *et al.*, 1975b; Kalica *et al.*, 1976;

TABLE III
RNA Composition of Human, Calf, Lamb, and Pig Rotavirus

	Human[a]		Calf[b]		Lamb[c]		Pig[d]	
Band number	M.W.[e] $\times 10^6$	Molar ratio	M.W.[e] $\times 10^6$	Molar ratio	M.W.[e] $\times 10^6$	Molar ratio	M.W.[e] $\times 10^6$	Molar ratio
1	2.04	0.62	2.2	—	2.1	1.1	2.1	—
2	1.58	1.61	1.9**	—	1.7	1.0	1.7	—
3	1.40	0.99	1.7	—	1.6	1.0	1.6	—
4	0.81	1.09	0.9	—	1.5	0.9	1.5	—
5	0.75	1.06	0.8	—	0.9	1.0	0.9	—
6	0.50	3.38	0.5[f]	—	0.8	0.9	0.8	—
7	0.28	0.99	0.3	—	0.5	3.7	0.5[f]	—
8	0.23	1.14	0.2	—	0.3	1.0	0.3	—
9					0.3	1.1	0.3	—

[a]Schnagl and Holmes (1976). [b]Newman *et al.* (1975). [c]Todd and McNulty (1977). [d]Todd and McNulty (1976). [e]Molecular weight. [f]The molar ratio of band 2 and band 6 of calf rotavirus indicates that they consist of 2 and 3 or 4 species, respectively. The molar ratio of band 7 of Pig rotavirus indicates that it consists of 3 or 4 species.

Schnagl and Holmes, 1976; Table II). Similar results were obtained for the RNA of pig (Todd and McNulty, 1976) and lamb (McNulty *et al.*, 1976a) rotaviruses and for that of the SA.11 virus and the O agent (Kalica *et al.*, 1976). However, as can be seen in Table III, the RNAs of rotaviruses from different animal species differed with respect to the electrophoretic mobility of one or more genome segments on polyacrylamide gels (Newman *et al.*, 1975; Schnagl and Holmes, 1976; Todd and McNulty, 1976, 1977; Rodger and Holmes, 1979). A size variation of the RNA segments for different isolates of human (Schnagl and Holmes, 1976) and calf (Verly and Cohen, 1977; Rodger and Holmes, 1979) rotaviruses has also been found (Table IV).

The protein composition of the human rotavirus has been studied by Rodger *et al.* (1975a, 1977) and Obijeski *et al.* (1977). The calf (Newman *et al.*, 1975; Bridger and Woode, 1976), lamb (Todd and McNulty, 1977), and SA.11 (Rodger *et al.*, 1977) rotaviruses are also well known with regard to their protein profiles (Table II). Preparations of single-shelled rotavirus particles contain two major and three or four minor proteins (Newman *et al.*, 1975; Bridger and Woode, 1976; Rodger *et al.*, 1977; Todd and McNulty, 1977; Table V). Three or four proteins are associated with the outer shell (Rodger *et al.*, 1977; Todd and McNulty, 1977; Table V). On the basis of the components of the inner shell, the four rotaviruses are indistinguishable from each other. However, a comparison of the outer shell components reveals variations among the low-molecular-weight proteins. This supports the concept that the group- and species-specific antigens of the virus are located on the inner and outer capsid layers, respec-

TABLE IV
RNA Composition of Different Isolates of Calf Rotavirus[a]

Band number	Wild rotavirus isolate M.W.[b] × 10^6	Cell culture-adapted rotavirus M.W. × 10^6
1	2.2	2.2
2	1.8	1.85
3	1.73	1.70
4	1.60	1.55
5	0.94	1.00
6	0.78	0.82
7	0.53	0.51
8	0.50	0.51
9	0.50	0.51
10	0.26	0.26
11	0.20	0.20

[a] Verly and Cohen (1977).
[b] Molecular weight.

TABLE V
Molecular Weights and Location of Viral Proteins in the Double-Shelled Virion of Human, Calf, Simian, and Lamb Rotaviruses

Human[a]		Calf[a]		Simian[a]		Lamb[b]	
Inner shell	Outer shell	Inner shell	Outer shell	Inner shell	Outer shell	Inner shell	Outer shell
127,000		131,000		133,000		128,000	
103,000		103,000		102,000		108,000	
97,000		97,000		99,000		99,000	
88,000		92,000		92,000		96,000	
	58,000		58,000		58,000		72,000
32,000		32,000		31,000		50,000	
	26,000		22,000		23,000	48,000	
	21,000		16,500		18,000		45,000
			14,500		14,000		37,000
							30,000

[a]Rodger *et al.* (1977).
[b]Todd and McNulty (1977).

tively. Rodger *et al.* (1977) also found that major components of the outer shells of human, calf, and simian rotaviruses are glycosylated. This observation supports the postulate of Holmes *et al.* (1976) that rotaviruses contain a lactase or β-galactosidase substrate on their surface, while the intestinal brush border enzyme, lactase, acts as the host cell receptor for the virus. The presence of glycoproteins in the outer shell also explains the resistance of rotaviruses to many proteases (Rodger *et al.*, 1977). In contrast, human rotavirus may be degraded to the single-shelled form by α-chymotrypsin and calf rotavirus by β-galactosidase. The uncoated particles prepared *in vitro* are indistinguishable from naturally occurring single-shelled particles.

More data on the proteins of rotaviruses have become available. Kalica and Theodore (1979) studied SA.11 virus and found eight polypeptides whose molecular weight ranged from 48,000 to 128,000. Five proteins were found in the inner shell, whereas three proteins were present in the outer shell. The same composition of the viral proteins was found for calf rotavirus by Matsuno and Mukoyama (1979). However, a comparative study on viral proteins of calf, pig, lamb, mouse, foal, rabbit, and human rotaviruses by Thouless (1979) showed four polypeptides in the inner shell, four in the outer shell, and three nonstructural polypeptides.

An RNA-dependent RNA polymerase has been detected in purified calf rotavirus particles (Cohen, 1977). Optimum polymerase activity was found between 45° and 50°C, at pH 8, and in the presence of 10 mM Mg ions. The enzyme was activated by EDTA treatment of intact particles or by heat shock; herewith the density of the particles in CsCl changed from 1.369 to 1.378 gm/cm^3, i.e., to that of single-shelled particles. The product was highly sensitive to pancreatic RNase (97%) in low or high salt concentration, and this single-stranded product hybridized completely (100%) with double-stranded virion RNA.

V. ANTIGENIC RELATIONSHIPS

Absolutely no serological relationship between calf rotavirus and reovirus or bluetongue virus, either by neutralization (Neu) tests (Fernelius *et al.*, 1972) or by the immunofluorescence (IF) technique (Welch and Twiehaus, 1973), could be found. Kapikian *et al.* (1974, 1975, 1976b) developed a complement fixation (CF) test for human and calf rotaviruses, EDIM virus, SA.11 virus, and the O agent and subsequently showed a lack of antigenic relationship of these five rotaviruses to reovirus types 1, 2, and 3 and 20 orbiviruses, including bluetongue virus. At the time, this was strong evidence for the existence of the new genus "rotavirus."

Although until 1977 the human rotavirus supposedly represented a single serotype (Middleton, 1977), two serotypes were distinguished by CF (Zissis and Lambert, 1978), EM (Zizzis and Lambert, 1978), and IF (Thouless *et al.*, 1978). These findings were supported by Fonteyne *et al.* (1978), who described two consecutive epidemics of rotavirus gastroenteritis with a difference in the severity of the symptoms. Rodriguez *et al.* (1978) also described two serotypes of the human rotavirus which were responsible for two sequential infections in the same children. Polyacrylamide gel electrophoresis of RNA extracted from both serotypes revealed a variation in the molecular weight of individual RNA segments, and the two serotypes reacted with different antibodies in the enzyme-linked immunosorbent assay (ELISA). This technique was subsequently used by Yolken *et al.* (1978d) to study the epidemiology of both types 1 and 2. Of 414 rotavirus isolates, 77% were type 2 and the remainder were type 1, with the serotype distribution being similar in different parts of the world. As immunity to one serotype apparently does not provide resistance to illness caused by the other serotype, a vaccine must induce resistance to both serotypes to be completely effective. For the other known rotaviruses, only one serotype has been so far found. However, Verly and Cohen (1977) and Rodger and Holmes (1979) demonstrated a size variation of the RNA segments for different isolates of calf rotavirus, which could also imply differences in their biological properties.

Flewett *et al.* (1974) were the first to demonstrate a serological relationship between calf and human rotavirus using immunoelectron microscopy (IEM) and IF. They suggested the existence of a serologically similar internal capsid protein. Subsequently, it was shown by Kapikian *et al.* (1975, 1976b) that human and calf rotavirus, SA.11 virus, EDIM virus, and the O agent share a group of antigens, which was demonstrated by CF. Woode *et al.* (1976a) and Thouless *et al.* (1977a) extended these findings to pig, lamb, foal, and rabbit rotavirus using the CF, IF, and gel diffusion (GD) techniques. IEM studies (Woode *et al.*, 1976a; Schoub *et al.*, 1977) revealed that single-shelled rotavirus particles were always agglutinated by all convalescent sera, either homologous or heterologous, and that the antibodies adhered to the inner capsid layer. In addition, Fauvel *et al.* (1978) demonstrated that the CF activity was more strongly associated with the single-shelled particles banding at a density of 1.38 gm/cm^3. Consequently, the general concept today is that the group-specific antigens are located on the inner capsid of rotaviruses.

Contrary to the CF, IF, GD, and IEM studies, Neu tests detect species-specific rotavirus antigens (Woode *et al.*, 1976a; Bridger, 1978). Any virus is neutralized at a much higher dilution of the homologous convalescent serum than with any heterologous serum, and very little cross reaction is observed. However, sera with a high neutralizing titer for the homolous virus also produce proportionally higher titers against heterologous viruses (Thouless *et al.*, 1977a). Calf rotavirus and human rotavirus were both shown to carry hemagglutinin (Inaba *et al.*, 1977; Fauvel *et al.*, 1978; Matsuno and Nagayoshi, 1978), which was found to be associated with intact virions (density 1.36 gm/cm^3), but not with virions lacking the outer capsid layer (density 1.38 gm/cm^3) (Fauvel *et al.*, 1978). Inaba *et al.* (1977) developed a hemagglutination adherence inhibition test (HAI) for calf rotavirus, and the HAI titers were closely related to the Neu titers of the sera. Quantitative estimation of human rotavirus antigens was also carried out with a hemagglutination adherence (HA) test (Matsuno and Nagayoshi, 1978), which gave a good correlation between HA titers and EM counts. These results support the concept that the species-specific antigens are located on the outer capsid of rotaviruses.

VI. CLINICAL FEATURES

Rotavirus infections show a wide clinical spectrum, i.e., subclinical, clinical, with different degrees of severity, and fatal outcomes, both in infants and young children, as well as in young animals. Evidence of rotavirus infections in adults has also been reported (Gomez-Baretto *et al.*, 1976; Zissis *et al.*, 1976; Bolivar *et al.*, 1978; von Bonsdorff *et al.*, 1978), but generally with no or less severe symptoms. The prevalent clinical signs are anorexia, depression, diarrhea, and dehydration (Acha and Szyfres, 1980).

In colostrum-deprived or gnotobiotic calves, diarrhea develops within 18 to 24 hours following infection with NCDV (Mebus *et al.*, 1969, 1971b; Woode and

Bridger, 1975), the feces being brilliant yellow to white if they are on a milk diet and watery, brown-gray or light green with blood and mucus if they are on a solid diet. Most of the calves become depressed and lose weight. Severely ill calves, however, can recover if fed with a mixture of glucose and saline instead of milk (Woode and Crouch, 1978). Clinically severe epizootics may be caused by an excess intake of food when the appetite recovers. If the diarrhea is prolonged at this stage, the animal can die of dehydration after 4 to 7 days of diarrhea. In addition, pneumonia may also contribute to the death of the calves 2 to 3 weeks after the beginning of the illness (Woode and Crouch, 1978). Human rotaviruses, when inoculated orally into conventionally reared calves, provoke similar symptoms (Light and Hodes, 1949).

Woode *et al.* (1976b) described similar clinical signs in naturally infected piglets and in gnotobiotic piglets experimentally infected with two pig virus isolates or with the calf virus. Within 18–24 hours of infection the piglets showed depression, anorexia, and a reluctance to move, followed by vomiting after 18–24 hours and diarrhea a few hours later. The experimental piglets on a milk diet shed yellow diarrhea, frequently with floccules floating in the liquid, whereas the conventionally reared piglets on a solid diet showed diarrhea of a yellow to dark gray color. Both groups of piglets lost weight rapidly. After a period of anorexia 24–72 hours after infection, the piglets recovered their appetite and clinical signs regressed 4–6 days after infection. Diarrhea sometimes persisted for 7–14 days, but recovery was quicker when water replaced milk for 24 hours. Similar symptoms were described by Pearson and McNulty (1977), with no fever observed; daily temperatures were 37.0° to 38.8°C.

When three 1-day-old lambs were infected with lamb rotavirus, all of them developed diarrhea 11 to 18 hours after the infection and two showed abdominal tension. Anorexia was apparent, but no fever could be observed. By 48 hours after infection, the lambs recovered except for the diarrhea, which lasted for 3 to 4 days. Very mild diarrhea was also observed in 12-day-old lambs after infection with the rotavirus (Snodgrass *et al.*, 1976b).

Similar clinical features were described, although less extensively, for rotavirus infections in other economically important animals, such as foals (Eugster and Whitford, 1978) and turkeys (McNulty *et al.*, 1978c). Furthermore, a relationship between diarrhea and rotavirus infection has been observed in mice, simians, deer, rabbits, and pronghorn antelopes.

In man, rotavirus infections sometimes produce only a mild disease or even no disease at all (Murphy *et al.*, 1975; Chrystie *et al.*, 1975; Flewett *et al.*, 1975b; Shepherd *et al.*, 1975). When symptoms follow infection, the incubation period is 48–72 hours (Shepherd *et al.*, 1975). Some children vomit during the first 1–2 days of the illness. The vomiting can be projectile following feedings, but usually it is a regurgitation. Diarrhea develops after a variable time, i.e., watery green or yellow stools, free from blood or mucus (Tallett *et al.*, 1976). Often there is accompanying fever; severely ill children may have temperatures of 38° to 39°C, rising to 40°C if dehydration is advanced. In very severe cases, hypernatremia may also appear (Parrott, 1976). With standard treatment, the illness is uncomplicated; the diarrhea usually lasts for 5 to 8 days, rarely less than 2 or more than 15 days. However, Middleton (1977) reported 16 fatal cases in which there was a 2- to 4-day history of vomiting and diarrhea, followed by dehydration and hypernatremia.

As evidenced by the high incidence of seropositivity in man and animals, subclinical infection with rotavirus is quite common. Under two different conditions a rotaviral infection has been reported to be subclinical. When rotaviruses isolated from children, foals, or lambs were inoculated into gnotobiotic piglets, these animals showed an antibody response but no symptoms at all (Bridger *et al.*, 1975; Woode *et al.*, 1976a). This observation indicates a varying virulence of rotaviruses for different animal species. An infection which would normally be virulent can become subclinical if the calf, lamb, or pig involved is fed with colostrum which contains antibodies (Woode *et al.*, 1975; Snodgrass and Wells, 1976; Snodgrass *et al.*, 1977b).

It is not yet clear whether other viruses or bacteria are an additional cause of the death of animals or children infected with a rotavirus. Certainly, parvoviruses, coronaviruses, adenoviruses, astroviruses, caliciviruses, and "fuzzy-wuzzys" have been isolated from a variety of animals and children, and bacterial infections such as colibacillosis and salmonellosis have also been reported to cause diarrhea (Madeley and Cosgrove, 1975; Woode and Bridger, 1975; Morin *et al.*, 1976; Whitelaw *et al.*, 1977; Flewett, 1978; Woode *et al.*, 1978a).

VII. PATHOLOGY AND PATHOGENESIS

Pathological changes caused by rotavirus infections have been studied in man (Barnes and Townley, 1973; Bishop *et al.*, 1973; Davidson *et al.*, 1975b; Middleton, 1976), calves (Mebus *et al.*, 1969, 1971b; Woode *et al.*, 1974; Morin *et al.*, 1976; Mebus and Newman, 1977), pigs (Hall *et al.*, 1976; Lecce *et al.*, 1976; Torres-Medina *et al.*, 1976; Woode *et al.*, 1976b; Pearson and McNulty, 1977; Lecce and King, 1978), lambs (Snodgrass *et al.*, 1977c), and mice (Adams and Kraft, 1967).

Although very little is known about the histopathology of naturally occurring rotavirus infections, solid information is available on the course of the disease in experimentally infected animals. Usually, rotaviral infection occurs in the small intestine. However, calf rotavirus was also found in the lungs and mesenteric lymph nodes of an infected calf (Mebus *et al.*, 1971b). Snodgrass *et al.* (1977a) detected virus antigens by IF in the colon and cecum of infected lambs and

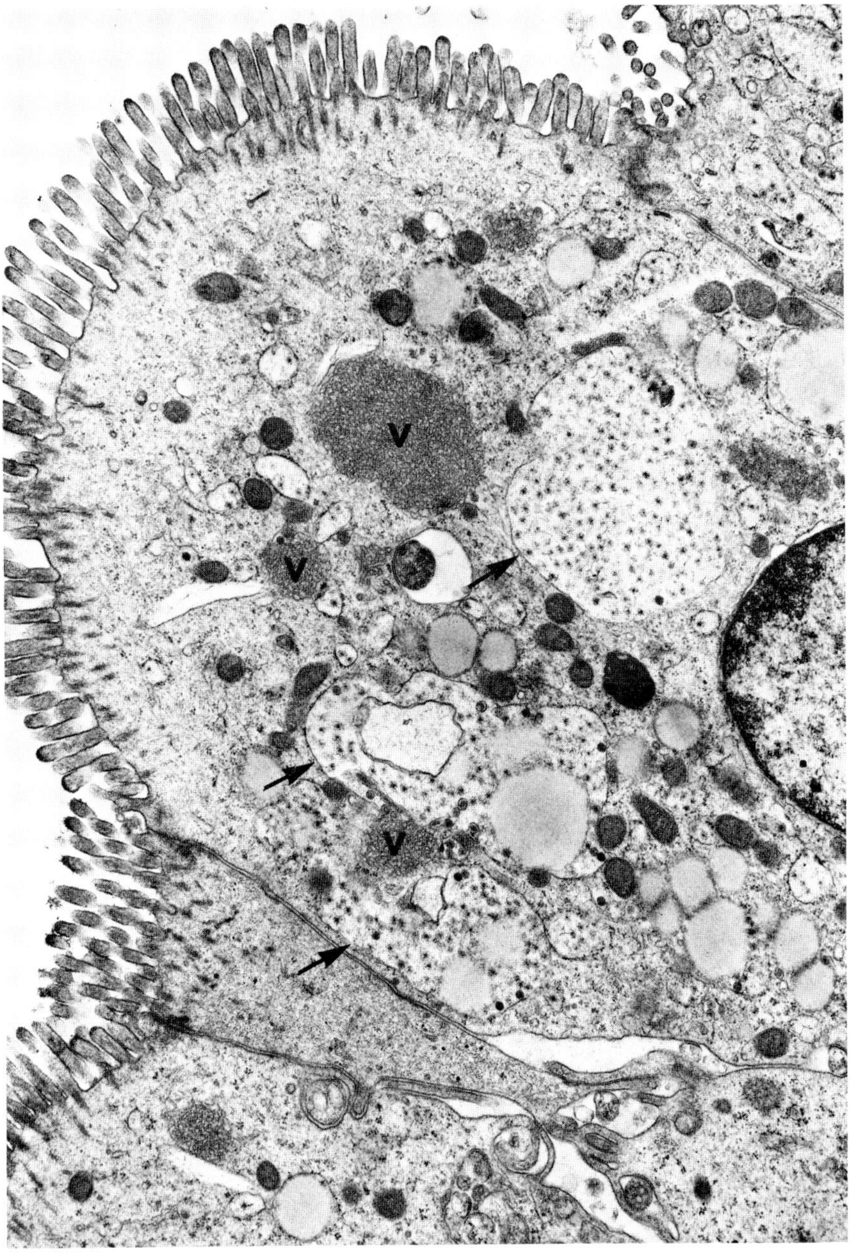

Fig. 3. Villous epithelial cell infected with rotaviruses, 18 hours postinfection. Virus particles are found in cisternae of the endoplasmic reticulum, which is often distended to form large vesicles (arrowed). Coarsely granular viroplasm (v) can be seen close to these cisternae. From Snodgrass *et al.* (1977c); courtesy of Dr. D. R. Snodgrass.

observed hepatic and pulmonary changes which could not be related to rotavirus infections. Histopathological changes, although mild, were observed in the stomach and lungs of infected pigs, but these changes were not associated with the presence of viral antigens (Hall *et al.*, 1976; Pearson and McNulty, 1977).

Rotaviruses infect the epithelial cells of the absorptive portion of the villi but not the crypt cells. In the epithelial cells, virus particles were found in the cytoplasmic vesicles by EM (Fig. 3) and IF. The villi are shortened or stunted (Figs. 4, 5), and the columnar brush-bordered epithelium is lost from their tips and replaced by cuoidal or squamous cells which lack a brush border. An increased number of round cells is seen in the lamina propria. As the infected cells migrate to the tip of each villus, they may be replaced by noninfected cuboidal cells from the crypt, lacking rotavirus receptors (Snodgrass *et al.*, 1977c). This observation would support the suggestion of a wave of infection, progressing posteriorly from the anterior small intestine (Mebus *et al.*, 1971b). However, no

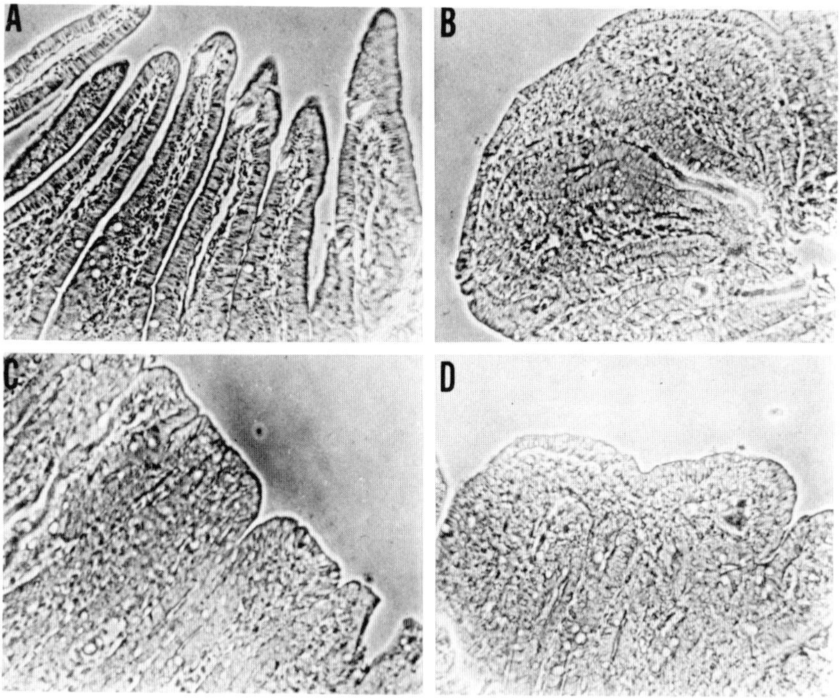

Fig. 4. Phase contrast photomicrographs of the mid-jejunum. (A) elongated, thin villi from a normal 3-week-old colostrum-deprived piglet reared in isolation; (B, C) blunted, shortened, and fused villi from a newly weaned 24-day-old piglet naturally infected with rotavirus; (D) blunted, shortened, and fused villi from a 12-day-old colostrum-deprived piglet 3 days after inoculation with rotavirus. From Lecce and King (1978); courtesy of Dr. J. G. Lecce.

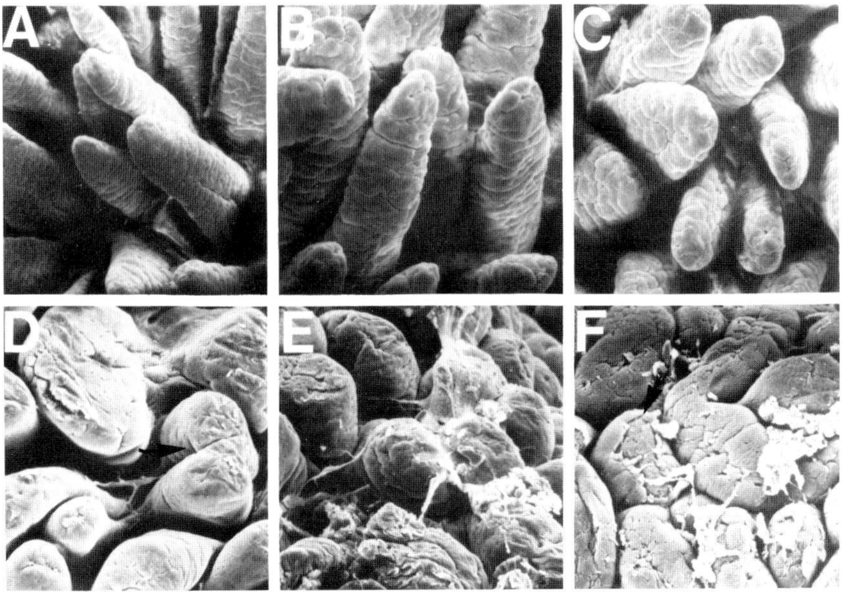

Fig. 5. Scanning electron micrographs of the mid-jejunum. (A, B, C) 3-week-old normal colostrum-deprived piglets reared in isolation. Villi are thin and elongated; (D) newly weaned 24-day-old piglet naturally infected with rotavirus. Villi are blunted, shortened, and fused at the point of the arrow; (E) newly weaned 24-day-old piglet naturally infected with rotavirus, showing more damage to the villi; (F) severely damaged villi from a 12-day-old colostrum-deprived piglet 3 days after inoculation with rotavirus. A villus is denuded at the point of the arrow. From Lecce and King (1978); courtesy of Dr. J. G. Lecce.

further evidence for this concept has been reported. In a later stage, stunning and adhesion of the villi produce a flat, avillous mucosa, but the lesions are not uniform throughout the small intestine (Pearson and McNulty, 1977; Lecce and King, 1978).

In the naturally occurring infections in man, Bishop *et al.* (1973) and Barnes and Townley (1973) showed shortened and blunted villi of duodenal epithelial cells and an increased number of reticulum-like cells in the lamina propria. The epithelial cells were more cuboidal and irregular. Virus antigens were detected in the cytoplasm of enterocytes in the crypts and villi from the duodenum to the ileum (Middleton, 1976). Similar results were obtained by Morin *et al.* (1976) for calves.

Different factors are important in the pathogenesis of rotavirus infections. As damaged epithelial cells are rapidly replaced by cuboidal immature cells which have migrated from the crypt, the villous epithelium shows characteristics of the crypt epithelium during the diarrhea phase. Thus, the villous tip cells contain diminished disaccharase levels, which results in a decreased ability to utilize dietary lactose. The lactose further accumulates in the small intestine, where it prevents absorption of water from the feces, causing a high osmotic pressure and even a positive dehydrating effect (Flewett, 1977). In isolated jejunal villous enterocytes of infected piglets, thymidine kinase and sucrase activities were typical of normal crypt cells, i.e., increased and decreased, respectively (Davidson *et al.,* 1977). *In vitro* experiments showed that in tissues from piglets infected with rotavirus the net Na^+ flux-to-glucose was blunted, which caused an impaired glucose-coupled Na^+ transport (Davidson *et al.,* 1977). These findings are similar to those in a coronavirus enteritis of pigs but differ from those seen with enterotoxigenic diarrhea (Middleton, 1978). Furthermore, the loss of the absorptive cells of the small intestine is assumed to cause another malabsorption syndrome. D-xylose malabsorption has been described both in children with rotaviral infection (Mavromichalis *et al.,* 1977) and in gnotobiotic calves inoculated with rotavirus (Woode *et al.,* 1978b). Although destruction of the brush-bordered epithelium is not complete, there still is an inflammatory response to the infection which tends to increase the peristaltic activity. This results in a faster passage of food, so that the time for digestion and water absorption is reduced and the animal is even more predisposed to diarrhea (McNulty, 1978). Other microorganisms, particularly *E. coli,* may contribute to the severity of rotavirus infection (Woode and Crouch, 1978). Thus, a synergistic effect of *E. coli* and rotavirus was demonstrated in neonatal calf diarrhea (Gouet *et al.,* 1978; Dubourguier *et al.,* 1978). Perhaps fermentation of lactose by coliforms in the large bowel to 6 moles of short-chain acid for each mole of lactose, giving a highly hypertonic solution, can exacerbate the dehydrating effect caused by rotavirus infection Flewett, 1977).

VIII. LABORATORY DIAGNOSIS

A variety of techniques have been used for the diagnosis of rotavirus infections, the criteria being the demonstration of the agent in a clinical specimen and/or the proof of a serological response to the infection.

The first report on rotavirus infection (Bishop *et al.,* 1973) showed the presence of virus particles in epithelial cells of duodenal mucosa by thin section EM. In subsequent studies the virus was concentrated by differential centrifugation and precipitation with polyethylene glycol 6000 (Bishop *et al.,* 1974) or by ultracentrifugation (Flewett *et al.,* 1974) and negatively stained prior to examination in the electron microscope. Middleton *et al.* (1974) simply mixed stool and distilled water, placed a drop of this suspension on an electron microscope grid, and stained with phosphotungstic acid. A simple and rapid pseudoreplica technique was described by Gomez-Baretto *et al.* (1976). IEM was used by Fernelius *et al.* (1972), Flewett *et al.* (1974), Kapikian *et al.* (1974), Bridger and Woode

(1975), and Saif *et al.* (1977). In this technique virus samples are mixed with homologous convalescent antiserum, and the aggregates coated with specific antibodies are observed in the electron microscope. Advantages of the negative staining contrast technique for EM are that it yields clear images with many details, it is relatively unaffected by impurities, only small amounts of virus are required, and after a relatively simple preparation of specimens for examination, a diagnosis can be reached in less than 2 minutes. One of the main disadvantages of this technique is that a high concentration of virus particles is necessary, i.e., at least 10^5 particles/ml and 10^8 particles/ml for a rapid diagnosis (Flewett, 1978). However, IEM provides a higher sensitivity, because aggregates instead of single virus particles are detected. Another important disadvantage is that it required expensive equipment and expertise and is not suitable for large-scale testing.

The IF technique was later developed for the detection of rotavirus infection. Middleton *et al.* (1974) detected rotavirus antigens in necropsy specimens from patients who died from gastroenteritis. Cells from the duodenal and jejunal mucosa were incubated with an antiviral antiserum prepared in a guinea pig and with a fluorescein-conjugated antiguinea pig immunoglobulin which revealed the presence of viral antigens in the columnar cells. This indirect immunofluorescence (IIF) technique was also used by Davidson *et al.* (1975b), who demonstrated rotaviral antigens in the cytoplasm of epithelial cells of the villi of the duodenal mucosa from children with gastroenteritis. They also determined the immunoglobulin class of the specific antibodies shortly after infection. Immunofluorescent staining of fecal smears (Mebus *et al.*, 1969; McNulty *et al.*, 1976e) is a very convenient diagnostic technique because of its rapidity and simplicity. Bridger and Woode (1975) examined cell cultures infected with fecal filtrates containing NCDV, using the IIF technique with gnotobiotic calf antiserum to NCDV, and found individual cells with brightly fluorescing cytoplasm. A similar immunofluorescent cell assay with NCDV was also successful with virus present in fecal samples or produced in cell culture (Barnett *et al.*, 1975). The sensitivity of this cell culture IF method has been increased by depositing the infective agent on the cells with low-speed centrifugation (Banatlava *et al.*, 1975; Bryden *et al.*, 1977) or by inoculating freshly trypsinized cells in suspension with fecal filtrates (McNulty *et al.*, 1976a). Foster *et al.* (1975) developed a fluorescent virus precipitin (FVP) test in which free virus present in fecal samples from scouring calves is reacted with anti-NCDV conjugate and the antigen–antibody complexes are detected by IF. This technique was later applied to detect antigen–antibody complexes formed between viral antigens in stools of young children and serum containing anti-IGV antibodies (Yolken *et al.*, 1977b). Several investigators reported that IF and EM examinations gave similar results. However, although less specialized equipment and less expertise are required for IF techniques, the difficulty of handling large numbers of samples is still present.

Among other methods for the detection of virus or viral antigens in feces is the CF technique, although it is not used frequently for this purpose. It is less sensitive and less specific than EM or IF because of the possible anticomplementary activity found in many feces specimens (Spence *et al.*, 1975; Middleton *et al.*, 1976; Acha and Szyfres, 1980).

There is no agreement yet concerning the value of counterimmunoelectrophoresis (CIEP) as a diagnostic technique. This technique forms precipitates of identity by reacting antigen moving toward the anode electrophoretically with antibody moving toward the cathode by electroendosmosis. Under well-defined electrophoretic conditions, the antigen and antibody meet in optimal proportions so that a precipitin line is formed. In the CIEP assay, the quality and purity of the aragose are major factors for good and reproducible results. Middleton *et al.* (1976) used CIEP for the detection of infantile gastroenteritis virus (IGV) in stool and found its sensitivity lower than that of EM. However, it appeared to be a rapid, convenient, and inexpensive method for identifying rotavirus. In addition, in subsequent reports the CIEP method was shown to be as sensitive as EM (Tufvesson and Johnson, 1976) or even more sensitive (Grauballe *et al.*, 1977; Bégin *et al.*, 1978).

One of the more recently developed techniques for the detection of rotavirus is the solid-phase radioimmunoassay (RIA). Basically, all RIA methods measure the extent of antigen–antibody reactions by determining the amount of unbound antibody or antigen after the reaction. In this solid-phase RIA, the specific antibody is attached to a solid support, e.g., polystyrene tubes. The antigen is then added, so that an antigen–antibody complex is formed, and this antigen is detected by another specific antibody labeled with a radioactive iodine, ^{125}I or ^{133}I. Middleton *et al.* (1977) adapted this assay for the detection of rotavirus in the stool or as purified virus preparation and obtained very satisfactory results. Compared with EM, the assay was up to 10 times more sensitive for detection of virus in stool and 128 times more sensitive for the detection of the purified virus antigen. In addition, the results were obtained on the same day and a high-volume operation was possible. A micro-RIA described by Kalica *et al.* (1977) had the additional advantage of being well suited for very small amounts of crude specimens with a specificity and sensitivity equal to those obtained with EM. Obvious disadvantages of the technique are the need for labile radioactive reagents and expensive counting equipment.

The recently described enzyme-linked immunosorbent assay (ELISA) for the detection of rotavirus is very promising. In principle, the ELISA is similar to the RIA but utilizes an enzyme instead of a radioactive isotope as the marker for the immunoglobulin. This assay was developed for the detection of IGV in human stools (Yolken *et al.* 1977a) and of calf rotavirus in fecal extracts (Ellens and de Leeuw, 1977; Scherrer and Bernard, 1977). Specific antibody is coated on a solid phase, which may be wells of a microtiter plate or polystyrene cuvettes.

The sample of stool filtrate or fecal extract is then added to this antibody-coated phase. After washing, antirotavirus immunoglobulin labeled with an enzyme, such as alkaline phosphatase or peroxidase, is added. The amount of antibody bound is determined by colorimetry after the addition of the substrate, which is converted by the enzyme into a colored product. This assay was successfully modified for the detection of rotavirus serotypes and for the study of the epidemiology of these viruses (Yolken *et al.,* 1978a). ELISA has multiple advantages in comparison with other types of assays. It provides a reproducible, rapid, and quantitative method for the detection of rotavirus antigens, and it is as sensitive as RIA and more sensitive than EM, IF, CP, or CIEP (Ellens *et al.,* 1978). In addition, ELISA utilizes reagents which are both stable and nonradioactive, and it can be accurately performed in large numbers in a short time. Furthermore, no sophisticated equipment is required, and the tests may be interpreted by a colorimeter or even by the naked eye (Voller and Bidwell, 1977).

Another method recently used for the detection of rotavirus antigens and antibodies is the immunoperoxidase (IP) technique. The IP technique resembles the IF and RIA methods but uses an enzyme (horseradish peroxidase) coupled to antibodies instead of a fluorochrome or radioactive label. Applications of the IP technique are reported by Kurstak 1971; Kurstak *et al.,* 1977). The IP methods generally used in the assays are the labeled antibody techniques (direct, indirect, and anticomplement IP methods) and the noncovalent labeled peroxidase antiperoxidase (PAP) method (Sternberger, 1974). Rotavirus antigens were demonstrated in BS-C-1 cells infected with trypsinized NCDV with the direct IP method (E. Kurstak *et al.,* unpublished data) and in MA 104 cells infected with SA.11 virus with the PAP technique (Graham and Estes, 1979). The direct IP method was as sensitive as the direct IF technique for the detection of rotavirus antigens (E. Kurstak *et al.,* unpublished data). Advantages of the IP method are: (i) it is a simple, highly sensitive technique; (ii) for detection only a standard light microscope is required; (iii) preparations may be kept for later observation or reference; (iv) virus detection on the ultrastructural level is also possible (Kurstak *et al.,* 1978; van den Hurk and Kurstak, 1980).

Serological methods for diagnosis of rotavirus infections comprise IEM (Flewett *et al.,* 1974; Kapikian *et al.,* 1974), IF (Woode *et al.,* 1976a; Ørstavik *et al.,* 1976a; Elias, 1977a), CF tests (Kapikian *et al.,* 1974, 1975, 1976b; Elias, 1977b; Thouless *et al.,* 1977a), CIEP (Middleton *et al.,* 1976), Neu tests (Elias, 1977a; Thouless *et al.,* 1977a), RIA (Middleton *et al.,* 1976; Acres and Babiuk, 1978), HA tests (Inaba *et al.,* 1977; Shinozaki *et al.,* 1978; Kalica *et al.,* 1978b; Fauvel *et al.,* 1978), and the ELISA (Yolken *et al.,* 1978a,b,c). Kapikian *et al.* (1974) demonstrated an antibody response by CF to IGV in patients who had rotavirus particles in their stools. IGV appeared to be antigenically related to EDIM virus and NCDV. These results were further extended (Kapikain *et al.,* 1975) when NCDV produced in cell culture was used as antigen, instead of the human stool filtrate that is not always readily available. Moreover, IGV, NCDV, EDIM virus, SA.11 virus, and the O agent were found to be similar in reciprocal CF tests. Thus the SA.11, O, and EDIM viruses can be used as substitute antigens. The O agent was about as efficient as IGV and certainly more efficient than the NCDV for detecting seroresponses (Kapikian *et al.,* 1976b). In addition, Thouless *et al.* (1977a) showed that human, calf, piglet, mouse, and foal rotaviruses reacted with human, calf, foal, and lamb convalescent serum by CF, so that it was not possible to distinguish the different rotaviruses by CF. Elias (1977a) used the CF test for a serological survey to establish the distribution and titers of antirotavirus antibodies in different age groups, using tissue culture-adapted calf rotavirus as antigen. Neu tests were used to detect species-specific antigens (Thouless *et al.,* 1977a). In this study it was found that any virus was neutralized by homologous species convalescent serum at a much higher dilution than by any heterologous serum. With the exception of the mouse virus, there was almost no cross reaction. As rotaviruses can infect individuals of species other than their host, this test can be used to determine the identity of the original host. Calf, human, and simian rotaviruses were shown to carry hemagglutinin (Inaba *et al.,* 1977; Matsuno and Nagayoshi, 1978; Kalica *et al.,* 1978b), found in the intact virion but not in virions lacking the outer capsid layer (Fauvel *et al.,* 1978; Kalica *et al.,* 1978b). This finding led to the development of two assays: the HA and HAI tests. The former is able to detect viral antigens in a sample of fecal extract, whereas the latter may be used for the demonstration of antiviral antibodies. As well as the Neu test, the HAI method seems to detect type-specific rather than group-specific antibodies (Fauvel *et al.,* 1978; Shinozaki *et al.,* 1978). Indeed, the HAI titers were very similar to the Neu test titers in the experiments of Inaba *et al.* (1977).

IX. PROPAGATION OF ROTAVIRUSES *IN VITRO*

Since tissue culture techniques generally offer many advantages over *in vivo* methods for the isolation and characterization of viruses, the development of suitable serological tests, and a possible production of vaccine, numerous efforts have been made to adapt rotaviruses to serial passage in cell cultures. The first propagation of a rotavirus in cell culture has been achieved with SA.11 virus and the O agent in velvet monkey kidney (VMK) cells (Malherbe and Strickland-Cholmley, 1967). Subsequently, the calf rotavirus was adapted to serial passage in primary fetal bovine kidney (FBK) cells (Mebus *et al.,* 1971a; Welch and Twiehaus, 1973; Bridger and Woode, 1975; L'Haridon and Scherrer, 1976; Chasey, 1977). Propagation of a cytopathic calf rotavirus in secondary and continuous cultures of calf kidney cells was achieved later (McNulty *et al.,* 1976a, 1977). Fernelius *et al.* (1972) adapted and propagated NCDV in several

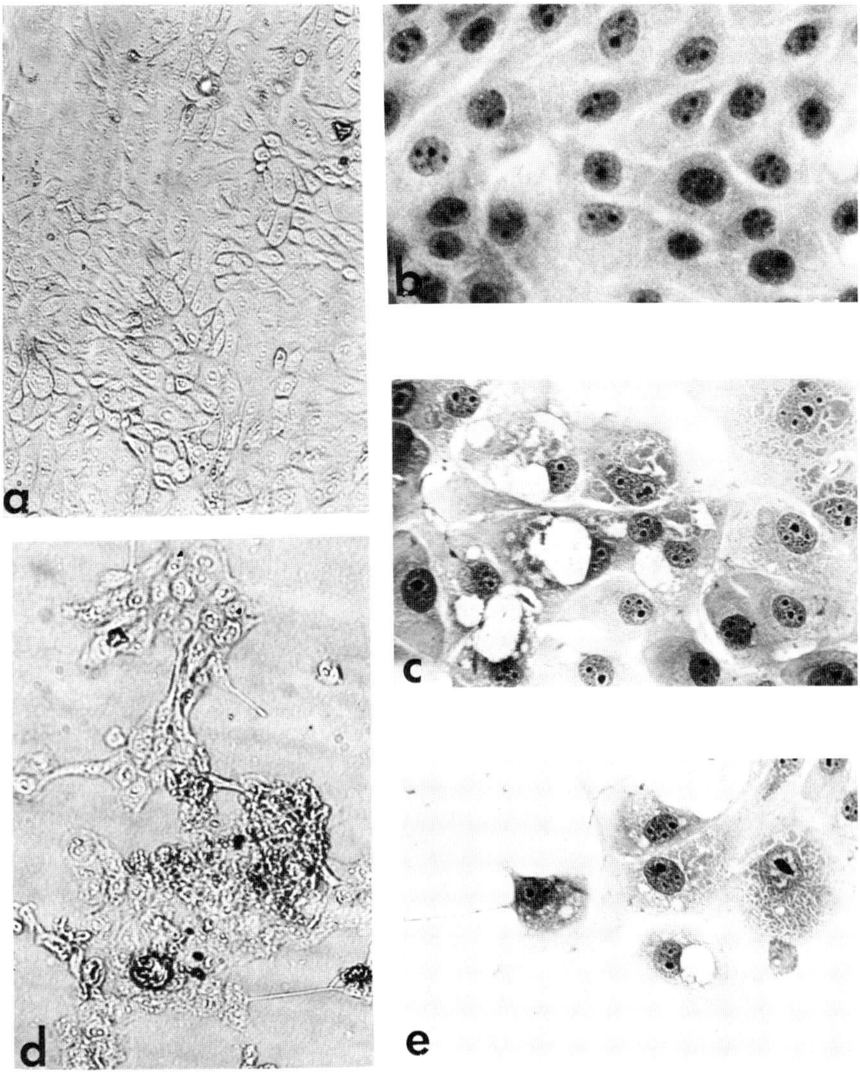

Fig. 6. Cytopathic effect of calf rotavirus 75-447 in MDBK cells. (a, b) uninfected control cultures; (c) 24 hours postinfection; note vacuolation and intracytoplasmic inclusions; (d, e) 48 hours postinfection; note loss of cells from the monolayer and cytoplasmic stranding. From McNulty *et al.* (1977); courtesy of Dr. M. S. McNulty.

culture systems and obtained a cytopathic effect in infected cultures of PK-15, embryonic bovine trachea, lamb kidney, and L cells. More recently, Banatlava *et al.* (1975) found that the human virus could be made to infect monolayer cultures of a pig kidney cell line, and Purdham *et al.* (1975) detected human rotavirus in infected monolayer cultures of human embryonic gut (HEG). In addition, Wyatt *et al.* (1976a) obtained growth of the human rotavirus over 14 passages in human embryonic kidney (HEK) cells. A strain of type 2 human rotavirus (Wa) has been adapted for propagation *in vitro* in primary cultures of African green monkey kidney cells, after the virus had been passaged 11 times serially in newborn gnotobiotic piglets, which grow to high titer (Wyatt *et al.*, 1980). Both murine and human rotaviruses have been grown in intestinal epithelial organ cultures (Rubinstein *et al.*, 1971; Wyatt *et al.*, 1974). However, even samples containing many virus particles were not highly infective. The rate of infection could be increased by centrifuging the specimen onto the cells (Banatlava *et al.*, 1975; Bryden *et al.*, 1977).

The cytopathic effect of the calf rotavirus on primary FBK cells and continuous Madin-Darby bovine kidney (MDBK) cells consisted of cytoplasmic vacuolation, some eosinophilic cytoplasmic inclusions and degeneration of cells, accompanied by their detachment from the monolayer (Fig. 6) (Welch and Twiehaus, 1973; McNulty *et al.*, 1977). Immunofluorescent staining showed the presence of the rotavirus antigen only in the cytoplasm of the infected cells (Fig. 7; McNulty *et al.*, 1977).

Similar results were demonstrated with the immunoperoxidase (IP) technique in BS-C-1 cells infected with NCDV. The cytoplasm of the infected cells was strongly colored after enzymatic revelation of the immune-peroxidase complexes (Fig. 8). BS-C-1 cells infected with decreasing NCDV concentrations showed a decreasing number of positively colored cells (Fig. 9) (E. Kurstak *et al.*, unpublished data).

The morphogenesis of pig rotativus in pig kidney (PK) cell cultures was described by Saif *et al.* (1978) and appeared to be similar to that observed *in vivo* in porcine epithelial cells. Cytoplasmic, nonmembrane bound viroplasm, and accumulation of virus particles within cisternae of the rough endoplasmic reticulum were observed. In infected cells, double-shelled particles about 77 nm in diameter, single-shelled particles 50–55 nm in diameter, and electron-dense nucleoids 31–38 nm in diameter were noted. Virus particles acquired outer shells by a budding process through matrices of granular electron-dense viroplasm or membranes of the rough endoplasmic reticulum. Tubules were observed in the nuclei of a few infected cells. Virus was released from the cells by discharge through breaks in the plasma membrane of the heavily infected cells. Observations on the morphogenesis of a calf rotavirus in MDBK cells showed similar results (McNulty *et al.*, 1976d). However, when PK-15 cells were infected with

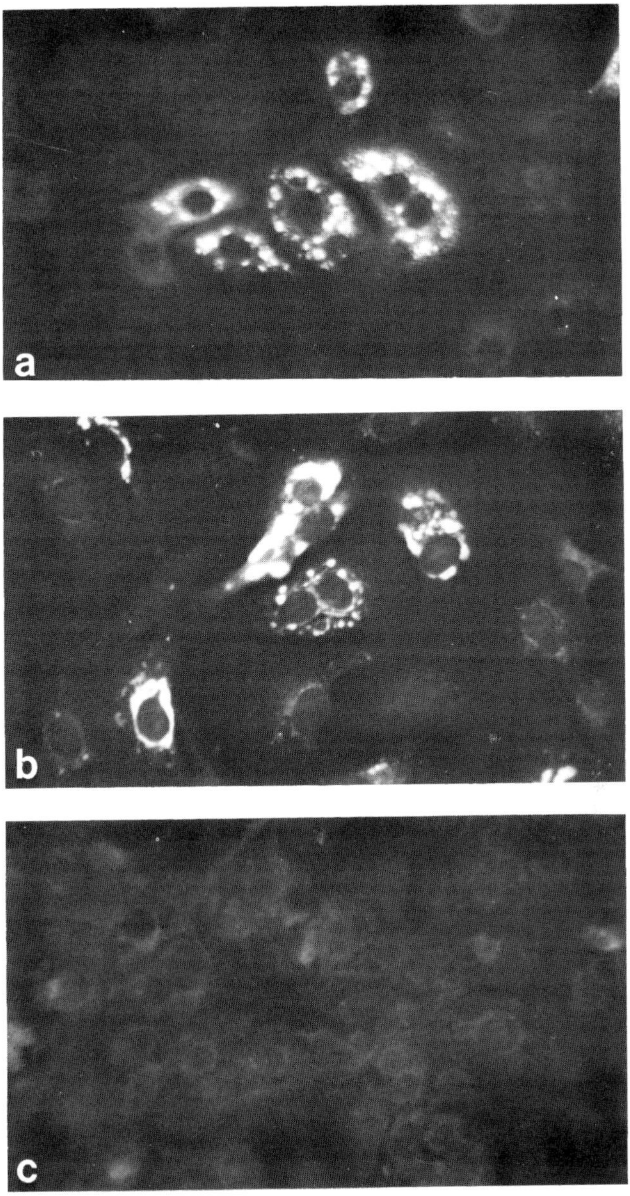

Fig. 7. Immunofluorescence. (a) MDBK cells infected with calf rotavirus 75-447 and stained with rotavirus conjugate; note cytoplasmic fluorescence; (b) VERO cells infected with mammalian reovirus type II and stained with mammalian reovirus type II conjugate; note cytoplasmic fluorescence; (c) MDBK cells infected with calf rotavirus 75-447 and stained with mammalian reovirus type II conjugate; no fluorescence. From McNulty *et al.* (1977); courtesy of Dr. M. S. McNulty.

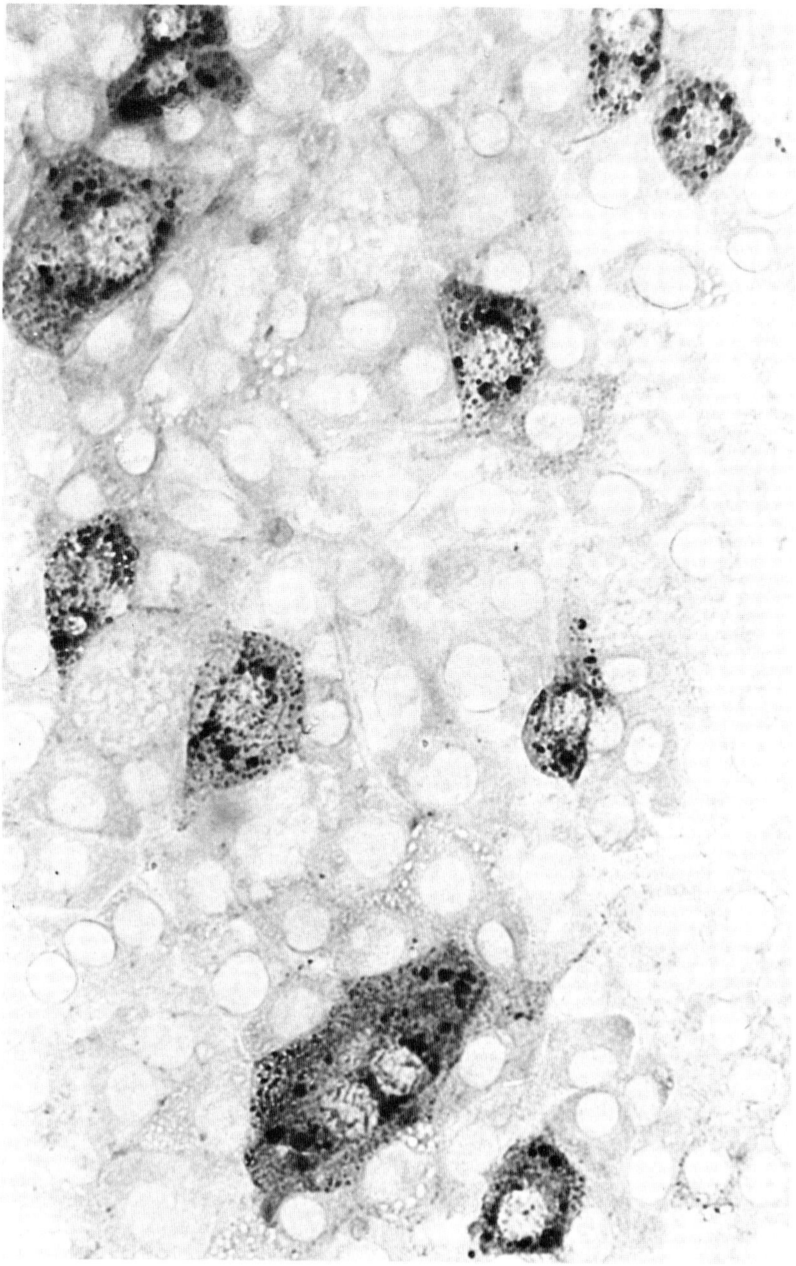

Fig. 8. Calf rotavirus antigen detection by the direct immunoperoxidase technique (serum dilution 1:250) in BS-C-1 cells infected with NCDV (1 PFU per cell).

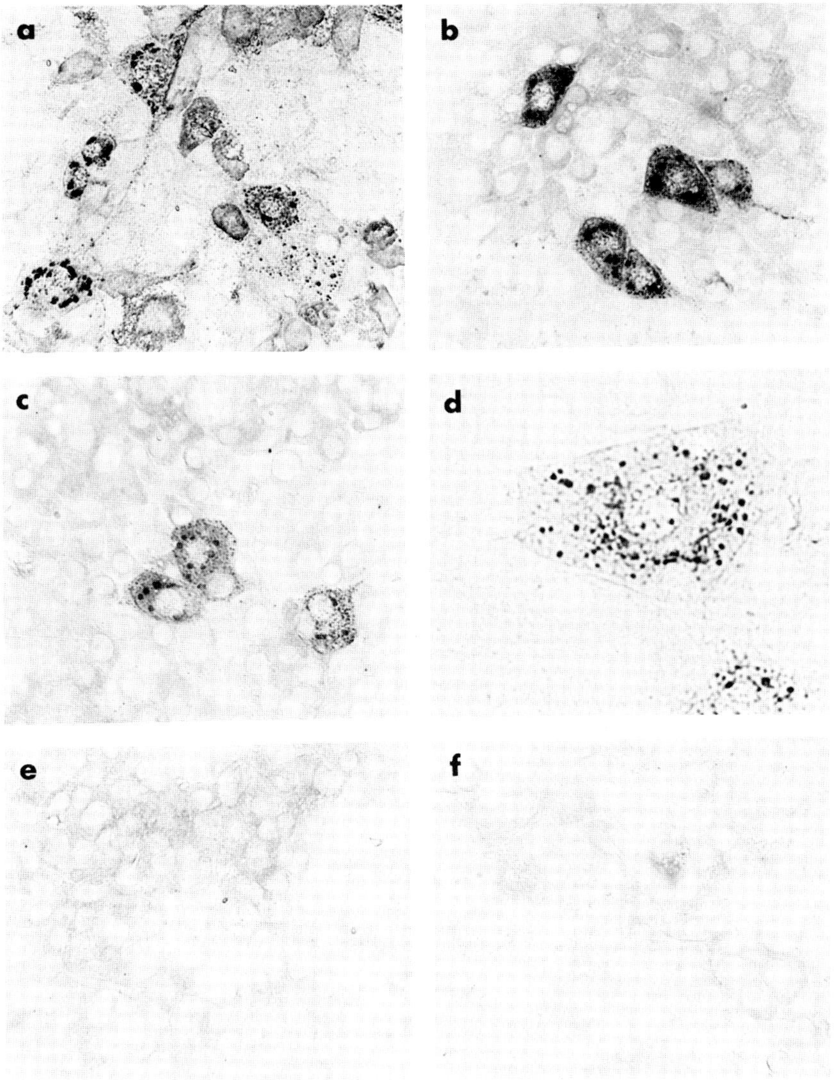

Fig. 9. Demonstration of rotavirus antigens by the direct immunoperoxidase method in BS-C-1 cells infected with NCDV at various multiplicities. (a, c) 5 PFU; (b) 1 PFU; (c, d) 0.1 PFU; (e) uninfected BS-C-1 cells. Calf antibovine rotavirus serum conjugated with peroxidase was used in a dilution of 1:250 (a, b, c, d, f) and peroxidase-conjugated control serum in a dilution of 1:20 (e). Note the decreasing number of positively colored cells in (a) to (c), the granular staining in (d) (higher magnification), and the absence of staining in the controls (c, f).

pig and lamb rotavirus strains which were not adapted to serial growth in cell cultures, a difference between virus morphogenesis *in vitro* and *in vivo* was observed which manifested itself by the generation of large numbers of coreless virus particles in PK-15 cells (McNulty *et al.*, 1978a). The number of these particles increased with increasing multiplicity of infection.

As the adaptation of rotavirus to its *in vitro* growth is a difficult task, several studies were carried out to understand these difficulties. As *in vivo* rotaviruses replicate in the small intestinal villous epithelium, it was considered that rotaviruses might replicate only in differentiated cells (Rubinstein *et al.*, 1971). It was also suggested that the lactase present in the brush border of the intestinal epithelial cells might act as a combined receptor and uncoating enzyme for the rotavirus (Holmes *et al.*, 1976). On the other hand, McNulty *et al.* (1978a) demonstrated that PK-15 cells could be infected with pig and lamb rotaviruses which were not adapted to propagation *in vitro*. As a result, large numbers of coreless and probably defective virus particles appeared in the cells. Therefore, it was postulated that not only lack of appropriate receptors but also nonpermissiveness of the cells to subsequent events in the replication made propagation of rotaviruses *in vitro* a fortuitous event (McNulty *et al.*, 1978a). Theil *et al.* (1978) postulated that pancreatic enzymes might be essential for rotavirus infection. These enzymes might convert rotaviruses from a noninfective to an infective form early in the infection, presumably during the initial interactions between the virion and the cell membrane. Thus, initially, fecal samples with some virus particles already activated to an infective form by endogenous enzymes would be able to infect the cell's culture, but progeny virus particles would have to be exposed to pancreatic enzymes to become infectious. This theory could find support in some earlier reports. Pancreatic enzymes were found to be essential for infectivity of pig rotavirus for primary PK cell cultures (Theil *et al.*, 1977). Trypsin, one of the three enzymes in pancreatic extracts, was at least partially responsible for the observed effect on the serial passage *in vitro*. In addition, it was reported on several occasions that freshly trypsinized cells in suspension were more susceptible to rotaviral infections than nontrypsinized cells (Woode *et al.*, 1976a; McNulty *et al.*, 1976a). Plaque formation was obtained with NCDV in monolayers of macacus rhesus monkey kidney (Ma-104) cells, when DEAE-dextran and trypsin were included in the overlay medium (Matsuno *et al.*, 1977). A procedure for the isolation and serial propagation of calf rotavirus in cell cultures in the presence of trypsin was also described by Babiuk *et al.* (1977). Furthermore, when trypsin was incorporated in the maintenance medium throughout the growth cycle, 1000 times more calf rotavirus was obtained in primary calf kidney cells and several continuous and semicontinuous cell lines (Almeida *et al.*, 1978). Apparently, these results might prove to be very important for the successful propagation of rotavirus in cell culture.

X. EPIDEMIOLOGY

In all parts of the world, rotaviruses have been implicated as the major cause of diarrhea in infants and young animals. Calf rotaviruses have been reported in the United States (Mebus *et al.*, 1969), Australia (Turner *et al.*, 1973), Canada (Morin *et al.*, 1974). Denmark (Meyling, 1974), England (Newman *et al.*, 1975; Woode and Bridger, 1975), and Ireland (McNulty *et al.*, 1976d). In pigs rotaviruses have been found in Australia (Rodger *et al.*, 1975b), the United States (Lecce *et al.*, 1976), England (Woode *et al.*, 1976b), and Ireland (McNulty *et al.*, 1976b). In Scotland (Snodgrass *et al.*, 1976a,b) and Ireland (McNulty *et al.*, 1976a) rotaviruses were observed in lambs. Areas where human rotaviruses have been related to diarrhea include England (Flewett *et al.*, 1973, 1974), Canada (Middleton *et al.*, 1974), Australia (Bishop *et al.*, 1974), New Zealand (Barnes, 1975), the United States (Kapikian *et al.*, 1974; Gomez-Baretto *et al.*, 1976), Argentina (Lombardi *et al.*, 1975), South Africa (Schoub *et al.*, 1975; 1978), Rhodesia (Cruickshank *et al.*, 1975), India (Holmes *et al.*, 1974), and Japan (Konno *et al.*, 1975a). Few reports exist about the incidence of other rotaviruses in mice, simians, foals, deer, rabbits, pronghorn antelopes, and avian species.

In man, a seasonal incidence for rotavirus gastroenteritis has been reported by different investigators. In January and February, 73 and 64% of gastroenteritis patients admitted to a hospital in Toronto excreted rotaviruses, whereas in June and July these percentages dropped to 8 and 7%, respectively (Middleton *et al.*, 1974; 1977). Similar results have been reported for the incidence of rotavirus infections in England (Bryden *et al.*, 1975; Chrystie *et al.*, 1978), France (Pouget *et al.*, 1978), and Japan (Konno *et al.*, 1978). The situation in the other hemisphere is the same, Davidson *et al.* (1975a) described a peak frequency of children with rotavirus gastroenteritis of 73% during the winter, but during the hottest month of the year the frequency was 23%. A comparison of human rotavirus disease in tropical (San Jose) and temperate (Dallas) localities showed that in both settings rotavirus accounted for 50 to 60% of the gastroenteritis episodes from December through February. This period is the winter in Dallas and the dry season in San Jose. During the rest of the observed year rotavirus was not found in any of the Dallas patients, but 30 to 40% of the San Jose patients had rotavirus in every month but May (Hieber *et al.*, 1978). Since this is the only report about this aspect of the epidemiology of rotavirus gastroenteritis, with the exception of the report from India by Maiya *et al.* (1977), more information is essential for a proper concept. Almost nothing has been reported about a possible seasonal incidence of rotavirus infections in animals.

Although rotavirus infections occur in young animals and children, there is a difference between the situation of these two groups. In mice (Cheever and

Mueller, 1947), calves (Woode and Bridger, 1975), foals (Flewett *et al.*, 1975a), pigs (Lecce *et al.*, 1976), lambs (Snodgrass *et al.*, 1976a), deer (Tzipori *et al.*, 1976), rabbits (Bryden *et al.*, 1976), pronghorn antelopes (Flewett and Woode, 1978), and turkeys (McNulty *et al.*, 1978a), rotavirus-associated diarrhea occurs in the first few weeks of life. In older animals the disease is less severe (Bridger and Woode, 1975; Lecce *et al.*, 1976; Snodgrass *et al.*, 1976b). Some cases of infection of adult animals have also been reported (Bridger and Woode, 1975; McNulty *et al.*, 1976b). Children up to 5 years of age are most susceptible to infection by the human rotavirus, and their susceptibility varies with age (Davidson *et al.*, 1975a). Infections of neonates have been reported, but these are often very mild or symptomless (Murphy *et al.*, 1975, 1977; Chrystie *et al.*, 1978). Later, the susceptibility increases until 1 year of age and then decreases again (Davidson *et al.*, 1975a; Middleton, 1977). Children are most likely to be infected between 6 months and 1 year of age (Davidson *et al.*, 1975a; Madeley *et al.*, 1977; Middleton, 1978). The male/female ratio of rotavirus gastroenteritis was 60 to 40% for all reported cases (Middleton, 1977) and 57 to 43% for 21 fatal cases (Carlson *et al.*, 1978) in Toronto. Rotavirus infections of school children (6–12 years old) have also been noted, but rarely (Hara *et al.*, 1976). In addition, natural rotavirus infections in adults exposed to infected children have been described by several investigators (Kapikian *et al.*, 1976c; Ørstavik *et al.*, 1976b; von Bonsdorff *et al.*, 1976, 1978; Zissis *et al.*, 1976; Kim *et al.*, 1977; Bolivar *et al.*, 1978). Symptoms in the adults varied from severe diarrhea, sometimes with high fever, to none at all.

Infection of animals and man can be caused by contact with infected individuals or by a contaminated environment. Recurrent yearly outbreaks of diarrhea in herds are common. These may be caused by repeated infection and excretion of virus in immune animals. On the other hand, rotaviruses are very stable in feces at average temperatures of 18 to 20°C (Woode and Bridger, 1975) and rather stable to desinfectants, such as Lysol, sodium hypochlorite, and formol-saline (Snodgrass and Herring, 1977). Thus, once infection has been introduced into farm buildings, the rotavirus can probably survive for many years outside the animals, perhaps in dried dung, and be a continuous source of infection which is difficult to eliminate. Another source of infection may be the dam which carries rotaviruses; it has been suggested that rotaviruses can cross the placenta and infect the fetus *in utero* (McNulty *et al.*, 1976f). No evidence for latent infections could be detected when lambs were treated with an immunosuppressant 4 months after recovery (Snodgrass *et al.*, 1976b). In hospitals, the same problem as in animal houses can arise. Different reports of nosocomial rotavirus infection (Chrystie *et al.*, 1975; Davidson *et al.*, 1975a; Flewett *et al.*, 1975b; Murphy *et al.*, 1975; Middleton, 1977; Tallett *et al.*, 1977) provide support for the expected persistence of the human rotavirus in the environment.

Recently, a genetic diversity of the bovine rotavirus genome was demonstrated by polyacrylamide gel electrophoresis (Verly and Cohen, 1977; Rodger and Holmes, 1979), with distinct epidemiological implications (Rodger and Holmes, 1979). Thus, different electropherotypes can cause successive outbreaks of rotaviral diarrhea in calves and can coexist on one farm during an outbreak of the disease. The genome profiles of rotavirus isolates also was analysed by ethidium bromide staining after fractionation on polyacrylamide gels (Todd *et al.*, 1980) and by 3′ terminal labelling of dsRNA extracted from isolates with T_4 RNA ligase (Clarke and McCrae, 1981). These methods were tentatively used for molecular epidemiological study of genome profiles of wild rotavirus isolates.

Cross infection of rotaviruses of different species has been observed. Thus, it was possible to infect experimentally pigs (Bridger *et al.*, 1975; Middleton *et al.*, 1975; Torres-Medina *et al.*, 1976), calves (Mebus *et al.*, 1976), monkeys (Lambeth and Mitchell, 1975; Wyatt *et al.*, 1976b), and lambs (Snodgrass *et al.*, 1977b) with the human rotavirus. Infection of piglets with the calf rotavirus has also been described (Hall *et al.*, 1976; Woode *et al.*, 1976a). These findings may have important implications for the epidemiology of rotaviruses under field circumstances.

XI. IMMUNITY

In man the level of antibodies to rotavirus fluctuates considerably for different age groups. At birth, 73–80% of babies possess serum antibodies to rotavirus (Middleton, 1977), but at 3 months antibodies are almost undetectable by IF (Elias, 1977a). Antibodies then appear very rapidly and by the age of 6 years over 90% of children have antibodies to rotavirus (Kapikian *et al.*, 1975; Blacklow *et al.*, 1976; Middleton *et al.*, 1976). Antibodies are detectable in the serum as early as 2 days after the onset of symptoms by IF (Davidson *et al.*, 1975b) and after 2 to 4 days by IEM (Flewett *et al.*, 1974). IgM is found first, which is soon replaced by IgG (Konno *et al.*, 1975b; Ørstavik and Haug, 1976). CF antibodies diminish fast after infection (Elias, 1977a), but antibodies detected by neutralization or by IF reach their highest level at 6 years of age and then diminish gradually until they are almost undetectable in sera of individuals over 70 years (Elias, 1977a). In addition to IgM and IgG antibodies, Watanabe *et al.* (1978) found IgA antibodies in feces, colostra, and sera, probably bound to the virus. It was, therefore, postulated that IgA antibodies might play a major role in the intestinal resistance to rotavirus infection. Only four instances of a second attack have been reported thus far (Davidson *et al.*, 1975a; Ørstavik *et al.*, 1976b; Middleton, 1977), two of which occurred in adults in contact with infected children. Their antibody levels were boosted by their reinfection.

Rotavirus infection can cause clinical illness in calves at all ages, but it is most severe at the age of less than 7 weeks. Most often, it occurs within the first 7 days of life (Flewett and Woode, 1978). In England, two herds of cattle in which diarrhea had been observed were studied with respect to their antibody response (Woode, 1976); 10% of yearlings and adults had serum Neu titers of 40 or less and 58% had titers of 320 or more. Thus, many animals, including adult animals. are susceptible to rotavirus infection. Mebus *et al.* (1973) suggested that local immunity in the epithelium of the small intestine was more important than circulating antibodies in providing resistance to infection. In the absence of reinfection and subsequent increase of the antibody levels in the cows, the adults do not develop sufficient levels of antibodies in their colostrum and milk to protect the young calves. In Canada, 79% of cattle have antibodies to rotavirus (Flewett and Woode, 1978). In a survey of sheep in Scotland, antibodies to rotavirus were detected in the sera of 38% of adult sheep and 56% of neonatal lambs (Snodgrass *et al.*, 1977a).

In cattle and sheep, antibodies to rotavirus are secreted in large amounts in first-day colostrum, but they decrease within 3 days after parturition (Woode *et al.*, 1975; Snodgrass and Wells, 1978a). Thus, although neonatal animals generally have circulating antirotaviral antibodies, they still can become infected by the virus (Woode *et al.*, 1975; Snodgrass *et al.*, 1977b). It is clear now that passive protection by colostrally derived antibodies has limited value. However, subsequent studies have indicated that serum antibody was not the essential agent for protection against rotaviral infection, but rather the presence of colostral antibody or immune serum in the intestinal lumen. This situation is found to be true for calves (Bridger and Woode, 1975; McNulty *et al.*, 1976f), pigs (Lecce *et al.*, 1976), and lambs (Snodgrass and Wells, 1976, 1978a) and is similar to the "lactogenic immunity" of transmissible gastroenteritis infections. Important for the protective effect of colostrum is its antibody titer, the volume ingested, and the period of feeding (McNulty *et al.*, 1976f; Snodgrass and Wells, 1978a). Ingestion of large amounts of colostrum for a short period gives protection only for 48 hours, whereas continuous feeding of smaller amounts of colostrum can protect for as long as it is continued. Immune serum was found to protect by the presence of specific rotaviral antibodies, probably IgG (Lecce *et al.*, 1976; Snodgrass *et al.*, 1977b), and it is quite possible that normal colostrum protects the same way. Inoculation of the dam with inactivated rotavirus was shown to increase the amounts of rotaviral antibodies in the colostrum and milk, resulting in a reduced incidence of diarrhea in the neonates (Mebus *et al.*, 1973; Snodgrass and Wells, 1978a).

In man, the presence of preexisting circulating serum antibody to rotavirus does not prevent infection in children or in adults (Kapikian *et al.*, 1976b; Kim *et al.*, 1977). Whether or not antibodies in the mother's milk provide protection to infants is not yet sure. Cruickshank *et al.* (1974) found rotavirus infections in African children who were exclusively breast-fed. An increased opportunity for spreading infection by bottle feeding, in comparison with breast feeding, was suggested by Schoub *et al.* (1977), thus explaining that breast-fed infants were less susceptible to rotavirus infection than bottle-fed infants. However, this phenomenon, which was also observed by Chrystie *et al.* (1975), as well as the high proportion of symptomless infections in human neonates (Madeley and Cosgrove, 1975; Murphy *et al.*, 1975), may very well be explained by the presence of antiviral activity other than antibodies in human and cow milk (Matthews *et al.*, 1976; Thouless *et al.*, 1977b). This activity is probably caused by nonspecific viral inhibitors. Such inhibitors seem to be relatively heat-stable macromolecules in the glycoprotein fraction.

Antibodies have also been detected in human colostrum (Thouless *et al.*, 1977b; Simhon and Mata, 1978). These antibodies are probably of the secretory IgA type, supposedly most important in the intestinal resistance to rotavirus infection.

XII. PREVENTION, CONTROL, AND TREATMENT

Prevention of rotavirus infection by hygienic measures is extremely difficult, because the viruses are very stable (Woode, 1976; McNulty, 1978) and are present in high concentration in feces of man and animals with rotaviral diarrhea. For example, 10^{11} calf rotavirus virions were found per gram of feces (Flewett, 1977). Exposure of susceptible animals to sources of rotavirus infection should be avoided; special care should be taken for neonates, which are very susceptible to infection. Stables should be cleaned and disinfected after outbreaks of rotavirus infection (Gustafson *et al.*, 1978). Unfortunately, not much information of rotavirus disinfectants is yet available. Inactivation of lamb rotaviruses in stools by short treatments (10 seconds) with formol-saline (10%) and Lysol (5%) resulted in partial to nearly no virus inactivation; longer exposure (2 hours) demonstrated that both agents are effective disinfectants (Snodgrass and Herring, 1977).

Active immunization of neonates is one possibility for the control of rotaviral diseases. A tissue culture-attenuated calf rotavirus strain is commercially available and was tested, with controversial results (Mebus *et al.*, 1973; Newman *et al.*, 1973; Acres and Radostits, 1976). Thurber *et al.* (1977) reported that protection could be achieved only after vaccination of all calves in the field trials. In an (Kapikian *et al.*, 1976b; Kim *et al.*, 1977), calves (Mebus *et al.*, 1973), and lambs (Snodgrass and Wells, 1976, 1978a), the presence of antibody to rotavirus was not found to correlate with the resistance to infection. Another possibility for the control of rotavirus diarrhea is passive immunity. Passive protection to rotavirus was achieved in the neonates of lambs (Snodgrass and Wells, 1976, 1978a), pigs (Lecce *et al.*, 1976), and calves (Mebus *et al.*, 1973) after oral administration of colostrum and serum containing antibodies to rotaviruses. The feeding of first-day colostrum as part of the diet throughout the period of greatest risk as an effective protection against lamb rotavirus infections was suggested by Snodgrass and Wells (1978a). Exposure of lambs to rotavirus fed with colostrum or antibodies to rotavirus may also immunize lambs against subsequent infections and thus play an important role in the prophylaxis of lamb rotaviral infections (Snodgrass and Wells, 1976, 1978a; Snodgrass *et al.*, 1977b). Passive immunity may also be obtained by immunization or hyperimmunization of the dam, which provides the offspring with higher amounts of antibodies in the colostrum and milk. Inoculation of cows with inactivated calf rotavirus and of ewes with inactivated lamb rotavirus showed a reduced incidence of diarrhea in calves (Mebus *et al.*, 1973) and lambs (Snodgrass and Wells, 1978a,b), respectively.

A rapid recovery of calves infected with calf rotavirus was obtained by feeding the calves with water instead of milk for 30 hours at the onset of diarrhea (Woode, 1976). Oral electrolyte solutions containing glucose were recommended as a successful therapy for dehydration in children and adults having rotavirus-induced diarrhea (Sack *et al.*, 1978). Glucose was found to be preferable to sucrose (Sack *et al.*, 1978), and the results obtained after oral therapy with glucose-electrolyte solution did not differ from those obtained after intravenous hydration (Nalin *et al.*, 1978).

XIII. CONCLUSIONS

Recently, the rotaviruses have been placed into the family of the Reoviridae by the ICTV, on the basis of the characteristics so far available. With respect to their generic name, no decision has been made yet, but in all probability the generally used name "rotavirus" will be accepted in the near future. More biochemical and biophysical data are definitely required to differentiate between the rotavirus species found in man and animals.

The etiological role of rotaviruses in gastroenteritis has to be established in connection with other enteropathogenic viruses and bacteria. In order to determine the potency of rotaviruses as a source of infection, their presence should be studied in a larger range of animals. The interesting question of why the incidence of rotavirus infection is much higher in winter than in summer in temperate localities in both hemispheres and in the rainy season than in the dry season in tropical settings still awaits an answer. The reason for the wide clinical spectrum—subclinical, clinical, or fatal—that is characteristic of rotavirus infections should also be investigated in relation to host resistance and virulence of the strain.

The ELISA technique has been used for the detection of rotaviral antigens and antibodies, as well as for the determination of the immunoglobulin class, IgM, IgG or IgA, of the antirotaviral antibodies. In addition, this assay has been used for the differentiation of serotype-specific rotavirus antigens and antibodies. Application of this reproducible, rapid, simple, and inexpensive method should be further extended.

The presence and protective role of the different types of immunoglobulins associated with rotaviral infections and the methods to stimulate the production of these antibodies should be further investigated. More data are needed on the mechanism of rotaviral infection and on the replication of rotaviruses in order to overcome the difficulties encountered in their propagation in cell cultures. The number of rotaviral serotypes for each mammalian species has to be established. For the development of effective vaccines, it is of crucial importance to classify the viral strains isolated from the same and different species and to establish rotavirus strains attenuated in tissue culture.

ACKNOWLEDGMENT

The authors wish to acknowledge Drs. M. E. Bégin, T. H. Flewett, J. C. Lecce, M. S. McNulty, and D. R. Snodgrass for the photographs they have kindly provided.

REFERENCES

Acha, P. N., and Szyfres, B. (1980). Zoonoses and Communicable Diseases Common to Man and Animals. Pan American Health Organization, Washington, D.C., pp. 1–700.

Acres, S. D., and Babiuk, L. A. (1978). *J. Am. Vet. Med. Assoc.* **173,** 555–559.

Acres, S. D., and Radostits, O. M. (1976). *Can. Vet. J.* **17,** 197–212.

Adams, W. R., and Kraft, L. M. (1963). *Science* **141,** 359–360.

Adams, W. R., and Kraft, L. M. (1967). *Am. J. Pathol.* **51,** 39–60.

Agarwal, A. (1979). *Nature* **278,** 389.

Almeida, J. D., Hall, T., Banatlava, J. E., Totterdell, B. M., and Chrystie, I. L. (1978). *J. Gen. Virol.* **40,** 213–218.

Babiuk, L. A., Mohammed, K., Spence, L., Fauvel, M., and Petro, R. (1977). *J. Clin. Microbiol.* **6,** 610–617.

Banatlava, J. E., Totterdell, B., Chrystie, I. L., and Woode, G. M. (1975). *Lancet* **2,** 821.

Banfield, W. G., Kasnic, G., and Blackwell, J. H. (1968). *Virology* **36,** 411–421.

Barnes, G. L. (1975). *Lancet* **1,** 1192.

Barnes, G. L., and Townley, R. R. W. T. (1973). *Arch. Dis. Child.* **48,** 343–349.

Barnett, B. B., Spendlove, R. S., Peterson, M. W., Hsu, L. Y., Lasalle, V. A., and Egbert, L. N. (1975). *Can J. Comp. Med.* **39,** 462–465.

Bégin, M. E., Dea, S., Dagenais, L., and Roy, R. S. (1978). *Proc. Int. Symp. Neonatal Diarrhea, 2nd 1978* pp. 273–285.

Bishop, R. F., Davidson, G. P., Holmes, I. H., and Ruck, B. J. (1973). *Lancet* **2,** 1281–1283.

Bishop, R. F., Davidson, G. P., Holmes, I. H., and Ruck, B. J. (1974). *Lancet* **1,** 149–151.

Blacklow, N. R., Echeverria, P., and Smith, D. H. (1976). *Infect. Immun.* **13,** 1563–1566.

Bolivar, R., Conklin, R. H., Vollet, J. J., Pickering, L. K., DuPont, H. L., Walters, D. L., and Kohl, S. (1978). *J. Infect. Dis.* **137,** 324–327.

Bridger, J. C. (1978). *J. Clin. Microbiol.* **8,** 625–628.

Bridger, J. C., and Woode, G. N. (1975). *Br. Vet. J.* **131,** 528–535.

Bridger, J. C., and Woode, G. N. (1976). *J. Gen. Virol.* **31,** 245–250.

Bridger, J. C., Woode, G. N., Jones, J. M., Flewett, T. H., Bryden, A. S., and Davies, H. A. (1975). *J. Med. Microbiol.* **8,** 565–569.

Bryden, A. S., Davies, H. A., Hadley, R. E., Flewett, T. H., Morris, C. A., and Oliver, P. (1975). *Lancet* **2,** 241–243.

Bryden, A. S., Thouless, M. E., and Flewett, T. H. (1976). *Vet. Rec.* **99,** 323.

Bryden, A. S., Davies, H. A., Thouless, M. E., and Flewett, T. H. (1977). *J. Med. Microbiol.* **10,** 121–125.

Carlson, J. A. K., Middleton, P. J., Szymanski, M. T., Huber, J., and Petric, M. (1978). *Am. J. Dis. Child.* **132,** 477–479.

Casper, D. L. D., and Klug, A. (1963). *Collect. Pap. Annu. Symp. Fundam. Cancer Res.* **17,** 27–39.

Chasey, D. (1977). *J. Gen. Virol.* **37,** 443–451.

Chasey, D., and Lucas, M. (1977). *Res. Vet. Sci.* **22,** 124–125.

Cheever, F. A., and Mueller, J. H. (1947). *J. Exp. Med.* **85,** 405–416.

Chrystie, I. L., Totterdell, B., Baker, M. J., Scopes, J. W., and Banatlava, J. E. (1975). *Lancet* **2,** 79.

Chrystie, I. L., Totterdell, B., and Banatlava, J. E. (1978). *Lancet* **1,** 1176–1178.

Clarke, I. N., and McCrae, M. A. (1981). *J. Virol. Methods* **2,** 203–209.

Cohen, J. (1977). *J. Gen. Virol.* **36,** 395–402.

Cramblett, H. G., Azini, P., and Haynes, R. F. (1971). *Ann. N. Y. Acad. Sci.* **176,** 80–92.

Cruickshank, J. G., Axton, J. H. M., and Webster, U. F. (1974). *Lancet* **1,** 1353.

Cruickshank, J. G., Zilberg, B., and Axton, J. H. M. (1975). *S. Afr. Med. J.* **49,** 859–863.

Davidson, G. P., Bishop, R. F., Townley, R. R. W., Holmes, I. H., and Ruck, B. J. (1975a). *Lancet* **1,** 242–246.

Davidson, G. P., Goller, I., Bishop, R. F., Townley, R. R. W., Holmes, I. H., and Ruck, B. J. (1975b). *J. Clin. Pathol.* **28,** 263–266.

Davidson, G. P., Gall, D. G., Petric, M., Butler, D. G., and Hamilton, J. R. (1977). *J. Clin. Invest.* **60,** 1402–1409.

Derbyshire, J. B., and Woode, G. H. (1978). *J. Am. Vet. Med. Assoc.* **173,** 519–521.

Dubourguier, H. C., Gouet, P., Mandard, O., Contrepois, M., and Bachelerie, C. (1978). *Ann. Rech. Vet.* **9,** 441–451.

DuPont, H. L., Portnoy, B. L., and Conklin, R. H. (1977). *Annu. Rev. Med.* **28,** 167–177.

Editorial (1975). *Lancet* **1,** 257–259.

Elias, M. M. (1977a). *J. Hyg.* **79,** 365–372.

Elias, M. M. (1977b). *J. Gen. Virol.* **37,** 191–194.

Ellens, D. J., and de Leeuw, P. W. (1977). *J. Clin. Microbiol.* **6,** 530–532.

Ellens, D. J., de Leeuw, P. W., Straver, P. J., and van Balken, J. A. M. (1978). *Med. Microbiol. Immunol.* **166,** 157–163.

Elliot, K. M., and Knight, J., eds. (1976). *Ciba Found. Symp.* [*N.S.*] **42,** 341.

Els, H. J., and Lecatsas, G. (1972). *J. Gen. Virol.* **17,** 129–132.

Esparza, Y., and Gil, F. (1978). *Virology* **91,** 141–150.

Eugster, A. K., and Whitford, H. W. (1978). *J. Am. Vet. Med. Assoc.* **173,** 857–858.

Fauvel, M., Spence, L., Babiuk, L. A., Petro, R., and Bloch, S. (1978). *Intervirology* **9,** 95–105.

Fenner, F. (1976). *Intervirology* **7,** 34–36.

Fernelius, A. L., Ritchie, A. E., Classick, L. G., Norman, J. O., and Mebus, C. A. (1972). *Arch. Gesamte Virusforsch.* **37,** 114–130.

Flewett, T. H. (1977). *Recent Adv. Clin. Virol.* **1,** 151–169.

Flewett, T. H. (1978). *J. Am. Vet. Med. Assoc.* **173,** 538–543.

Flewett, T. H., and Woode, G. N. (1978). *Arch. Virol.* **57,** 1-23.

Flewett, T. H., Bryden, A. S., and Davies, H. (1973). *Lancet* **2,** 1497.

Flewett, T. H., Bryden, A. S., Davies, H., Woode, G. N., Bridger, J. C., and Derrick, J. M. (1974). *Lancet* **1,** 61.

Flewett, T. H., Bryden, A. S., and Davies, H. (1975a). *Vet. Rec.* **96,** 477.

Flewett, T. H., Bryden, A. S., Davies, H., and Morris, C. A. (1975b). *Lancet* **1,** 4-5.

Fonteyne, J., Zissis, G., and Lambert, J. P. (1978). *Lancet* **1,** 983.

Foster, L. G., Peterson, M. W., and Spendlove, R. S. (1975). *Proc. Soc. Exp. Biol. Med.* **150,** 155-160.

Gomez-Baretto, J., Palmer, E. L., Nahmias, A. J., and Hatch, M. H. (1976). *J. Am. Med. Assoc.* **253,** 1857-1860.

Gouet, P., Contrepois, M., Dubourguier, H. C., Riou, Y., Scherrer, R., Laporte, J., Vautherot, J. F., Cohen, J., and L'Haridon, R. (1978). *Ann. Rech. Vet.* **9,** 433-440.

Graham, D. Y., and Estes, M. K. (1979). *Infect. Immun.* **26,** 686-689.

Grauballe, P. C., Genner, J., Meyling, A., and Hornsleth, A. (1977). *J. Gen. Virol.* **35,** 203-218.

Gustafson, D. P., Byrnes, R. J., Casselberry, M. H., Clark, A. S., Gale, C., Gillespie, J. H., Mebus, C. A., Mengeling, W. L., Olander, H. J., Peacock, G. V., Scherba, G., Timoney, J. F., West, R. L., and Wyatt, R. G. (1978). *J. Am. Vet. Med. Assoc.* **173,** 515-518.

Hall, G. A., Bridger, J. C., Chandler, R. L., and Woode, G. N. (1976). *Vet. Pathol.* **13,** 197-210.

Hara, M., Mukoyama, J., Tsuhuhara, T., Saito, Y., and Tagaya, I. (1976). *Lancet* **1,** 311.

Hieber, J. P., Shelton, S., Nelson, J. D., Leon, J., and Mohs, E. (1978). *Am. J. Dis. Child.* **132,** 853-858.

Holmes, I. H., Mathan, M., Bhat, P., Albert, M. J., Swaminathan, S. P., Maiya, P. P., Pereira, S. M., and Baker, S. J. (1974). *Lancet* **2,** 58.

Holmes, I. H., Ruck, B. J., Bishop, R. F., and Davidson, G. P. (1975a). *J. Virol.* **16,** 937-943.

Holmes, I. H., Moritz, N. A., Rodger, S. M., Ruck, B. J., Schnagl, P. D., Gust, I. D., Bishop, R. F., Davidson, G. P., and Cameron, D. (1975b). *Abstr., Int. Congr. Virol., 3rd, 1975* p. 70.

Holmes, I. H., Rodger, S. M., Schnagl, R. D., Ruch, B. J., Gust, I. D., Bishop, R. F., and Barnes, I. D. (1976). *Lancet* **1,** 1387-1389.

House, J. A. (1978). *J. Am. Vet. Med. Assoc.* **173,** 573-576.

Inaba, I., Sato, K., Takahashi, E., Kurogi, H., Satoda, K., Omori, T., and Matumoto, M. (1977). *Microbiol. Immunol.* **21,** 531-534.

Kalica, A. R., and Theodore, T. S. (1979). *J. Gen. Virol.* **43,** 463-466.

Kalica, A. R., Garon, C. F., Wyatt, R. G., Mebus, C. A., VanKirk, D. M., Chanock, R. M., and Kapikian, A. Z. (1976). *Virology* **74,** 86-92.

Kalica, A. R., Purcell, R. H., Serano, M. M., Wyatt, R. G., Kin, H. W., Chanock, R. M., and Kapikian, A. Z. (1977). *J. Immunol.* **118,** 1275-1279.

Kalica, A. R., Sereno, M. M., Wyatt, R. G., Mebus, C. A., Chanock, R. M., and Kapikian, A. Z. (1978a). *Virology* **87,** 247-255.

Kalica, A. R., Harvey, D., James, H. D., Jr., and Kapikian, A. Z. (1978b). *J. Clin. Microbiol.* **7,** 314-315.

Kapikian, A. Z., Kim, H. W., Wyatt, R. G., Rodriguez, W. J., Ross, S., Cline, W. L., Parrott, R. H., and Chanock, R. M. (1974). *Science* **185,** 1049-1053.

Kapikian, A. Z., Cline, W. L., Mebus, C. A., Wyatt, R. G., Kalica, A. R., James, H. D., Jr., VanKirk, D., Chanock, R. M., and Kim, H. W. (1975). *Lancet* **1,** 1056-1061.

Kapikian, A. Z., Kalica, A. R., Wai-Kuo Shih, J., Cline, W. L., Thornhill, T. W., Wyatt, R. G., Chanock, R. M., Kim, H. W., and Gerin, J. L. (1976a). *Virology* **70,** 564-569.

Kapikian, A. Z., Cline, W. L., Kim, H. W., Kalica, A. R., Wyatt, R. G., VanKirk, D. H., Chanock, R. M., James, H. D., Jr., and Vaughn, A. L. (1976b). *Proc. Soc. Exp. Biol. Med.* **152,** 535-539.

Kapikian, A. Z., Kim, H. W., Wyatt, R. G., Cline, W. L., Arrobio, J. D., Brandt, C. D., Rodriguez, W. J., Sack, D. A., Chanock, R. M., and Parrott, R. H. (1976c). *N. Engl. J. Med.* **294**, 965-972.
Kim, H. W., Brandt, C. D., Kapikian, A. Z., Wyatt, R. G., Arrobio, J. D., Rodriguez, W. J., Chanock, R. M., and Parrott, R. H. (1977). *J. Am. Med. Assoc.* **238**, 404-407.
Konno, T., Suzuki, H., and Ishida, H. (1975a). *Lancet* **1**, 918-919.
Konno, T., Imai, A., Suzuki, H., and Ishada, H. (1975b). *Lancet* **2**, 1312.
Konno, T., Suzuki, H., Imai, A., Kutsuzawa, T., Ishida, H., Katsushima, N., Sakamoto, M., Kitaoka, S., Tsusoi, R., and Adachi, M. (1978). *J. Infect. Dis.* **138**, 569-576.
Kraft, L. M. (1957). *J. Exp. Med.* **106**, 743-755.
Kraft, L. M. (1966). *Natl. Cancer Inst. Monogr.* **20**, 55-61.
Kurstak, E. (1971). *In* "Methods in Virology," (K. Maramorosch and H. Koprowski, eds.) Vol. V, pp. 423-444. Academic Press, New York.
Kurstak, E. (1977). *Comp. Virol. Symp. Abstracts,* 8-14.
Kurstak, E., Tijssen, P. and Kurstak, C. (1977). *In* "Comparative Diagnosis of Viral Diseases" (E. Kurstak and C. Kurstak, eds.) Vol. II, Part B, pp. 403-448. Academic Press, New York.
Kurstak, E., de Thé, G., van den Hurk, J., Charpentier, G., Kurstak, C., Tijssen, P., and Morisset, R. (1978). *J. Med. Virol.* **2**, 189-200.
Lambeth, L., and Mitchell, J. D. (1975). *Aust. Paediatr. J.* **11**, 127-128.
Lecatsas, G. (1972). *Onderstepoort J. Vet. Res.* **39**, 133-137.
Lecce, J. G., and King, M. W. (1978). *J. Clin. Microbiol.* **8**, 454-458.
Lecce, J. G., King, M. W., and Mock, R. (1976). *Infec. Immun.* **14**, 816-825.
L'Haridon, R., and Scherrer, R. (1976). *Ann. Rech. Vet.* **7**, 373-381.
Light, J. S., and Hodes, H. L. (1949). *J. Exp. Med.* **90**, 113-115.
Lombardi, G. H., Roseto, A. M., Stamboulian, D., and Dro, J. G. B. (1975). *Lancet* **2**, 1311.
McNulty, M. S. (1978). *J. Gen. Virol.* **40**, 1-18.
McNulty, M. S., Allan, G. M., Pearson, G. R., McFerran, J. B., Curran, W. L., and McCracken, R. M. (1976a). *Infect. Immun.* **14**, 1332-1338.
McNulty, M. S., Pearson, G. R., McFerran, J. B., Collins, D. S., and Allan, G. M. (1976b). *Vet. Microbiol.* **1**, 55-63.
McNulty, M. S., Curran, W. L., and McFerran, J. B. (1976c). *J. Gen. Virol.* **33**, 503-508.
McNulty, M. S., Allan, G. M., and McFerran, J. B. (1976d). *Res. Vet. Sci.* **21**, 114-115.
McNulty, M. S., Allan, G. M., Curran, W. L., and McFerran, J. B. (1976e). *Vet. Rec.* **98**, 463-464.
McNulty, M. S., McFerran, J. B., Bryson, D. G., Logan, E. F., and Curran, W. L. (1976f). *Vet. Rec.* **99**, 229-230.
McNulty, M. S., Allan, G. M., and McFerran, J. B. (1977). *Arch. Virol.* **54**, 201-209.
McNulty, M. S., Curran, W. L., Allan, G. M., and McFerran, J. B. (1978a). *Arch. Virol.* **58**, 193-202.
McNulty, M. S., Allan, G. M., Thomson, D. W., and O'Boyle, J. D. (1978b). *Vet. Rec.* **102**, 534-535.
McNulty, M. S., Allan, G. M., and Stuart, J. C. (1978c). *Vet. Rec.* **103**, 319-320.
Madeley, C. R., and Cosgrove, B. P. (1975). *Lancet,* **2**, 451-452.
Madeley, C. R., Cosgrove, B. P., Bell, E. J., and Fallon, R. J. (1977). *J. Hyg.* **78**, 261-273.
Maiya, P. P., Pereira, S. M., Mathan, M., Bhat, P., Albert, M. J., and Baker, S. J. (1977). *Arch. Dis. Child.* **52**, 482-485.
Malherbe, H. H., and Strickland-Cholmley, M. (1967). *Arch. Gesamte Virusforsch.* **22**, 235-245.
Martin, M. L., Palmer, E. L., and Middleton, P. J. (1975). *Virology* **68**, 146-153.
Matsuno, S., and Mukoyama, A. (1979). *J. Gen. Virol.* **43**, 309-316.
Matsuno, S., and Nagayoshi, S. (1978). *J. Clin. Microbiol.* **7**, 310-311.
Matsuno, S., Inouye, S., and Kono, R. (1977). *J. Clin. Microbiol.* **5**, 1-4.

Matthews, T. H. J., Nair, C. D. G., Lawrence, M. K., and Tyrrell, D. A. J. (1976). *Lancet* **2,** 1387–1389.
Mavromichalis, J., Evans, N., McNeish, A. S., Bryden, A. S., Davies, H. A., and Flewett, T. H. (1977). *Arch. Dis. Child.* **52,** 589–591.
Mebus, C. A., and Newman, L. E. (1977). *Am. J. Vet. Res.* **38,** 553–558.
Mebus, C. A., Underdahl, N. R., Rhodes, M. B., and Twiehaus, M. J. (1969). *Res. Bull. Nebr., Agric. Exp. Stn.* **233,** 1–16.
Mebus, C. A., Kono, M., Underdahl, N. R., and Twiehaus, M. J. (1971a). *Can. Vet. J.* **12,** 69–72.
Mebus, C. A., Stair, E. L., Underdahl, N. R., and Twiehaus, M. J. (1971b). *Vet. Pathol.* **8,** 490–505.
Mebus, C. A., White, R. G., Bass, E. P., and Twiehaus, M. J. (1973). *J. Am. Vet. Med. Assoc.* **163,** 880–883.
Mebus, C. A., Wyatt, R. G., Sharpee, R. L., Sereno, M. M., Kalica, A. R., Kapikian, A. Z., and Twiehaus, M. J. (1976). *Infect. Immun.* **14,** 471–474.
Meyling, A. (1974). *Acta Vet. Scand.* **15,** 457–459.
Middleton, P. J. (1976). *Symp. Gastroenteritis, 1976* p. 132–139.
Middleton, P. J. (1977). *In* "Comparative Diagnosis of Viral Diseases" (E. Kurstak and C. Kurstak, eds.) Vol. 1, Part A, pp. 423–445. Academic Press, New York.
Middleton, P. J. (1978). *J. Am. Vet. Med. Assoc.* **173,** 544–546.
Middleton, P. J., Szymanski, M. T., Abbott, G. D., Bortolussi, R., and Hamilton, J. R. (1974). *Lancet* **1,** 1241–1244.
Middleton, P. J., Petric, M., and Szymanski, M. T. (1975). *Infect. Immun.* **12,** 1276–1280.
Middleton, P. J., Petric, M., Hewitt, C. M., Szymanski, M. T., and Tam, J. S. (1976). *J. Clin. Pathol.* **29,** 191–197.
Middleton, P. J., Holdaway, M. D., Petric, M., Szymanski, M. T., and Tam, J. S. (1977). *Infect. Immun.* **16,** 439–444.
Morin, M., Lamothe, P., Gagnon, A., and Malo, R. (1974). *Can J. Comp. Med.* **38,** 236–242.
Morin, M., Lanviere, S., and Lallier, R. (1976). *Can J. Comp. Med.* **40,** 228.
Much, D. H., and Zajac, I. (1972). *Infect. Immun.* **6,** 1019–1024.
Murphy, A. M., Albrey, M. B., and Hay, P. J. (1975). *Lancet* **2,** 452–453.
Murphy, A. M., Albrey, M. B., and Crewe, E. B. (1977). *Lancet* **2,** 1149–1150.
Nalin, D. R., Mata, L., Vargas, W., Loria, A. R., Levine, M. M., de Cespedes, C., Lizano, C., Simhon, A., and Mohs, E. (1978). *Lancet* **2,** 277–279.
Newman, F. S., Myers, L. L., Firehammer, B. D., and Catlin, J. E. (1973). *Proc. 77th Annu. Meet. U.S. Anim. Health Assoc.* pp. 59–64.
Newman, F. S., Brown, F., Bridger, J. C., and Woode, G. H. (1975). *Nature (London)* **258,** 631–633.
Obijeski, J. F., Palmer, E. L., and Martin, M. L. (1977). *J. Gen. Virol.* **34,** 485–497.
Ørstavik, I. and Haug, K. W. (1976). *Scand. J. Infect. Dis.* **8,** 237–240.
Ørstavik, I., Figenschau, K. J., and Ulstrup, J. C. (1974). *Lancet* **2,** 1083.
Ørstavik, I., Figenschau, K. J., Haug, K. W., and Ulstrup, J. C. (1976a). *Scand. J. Infect. Dis.* **8,** 1–5.
Ørstavik, I., Haug, K. W., and Søvde, A. (1976b). *Scand. J. Infect. Dis.* **8,** 277–278.
Palmer, E. L., Martin, M. L., and Murphey, F. A. (1977). *J. Gen. Virol.* **35,** 403–414.
Parrott, R. H. (1976). *Symp. Gastroenteritis, 1976* p. 264–271.
Pearson, G. R., and McNulty, M. S. (1977). *J. Comp. Pathol.* **87,** 363–375.
Petric, M., Szymanski, M. T., and Middleton, P. J. (1975). *Intervirology* **5,** 233–238.
Pouget, M. M., de Micco, P., and Tamalet, J. (1978). *Med. Mal. Infect.* **8,** 519–521.
Purdham, D. R., Purdham, P. A., Evans, M., and McNeish, A. S. (1975). *Lancet* **2,** 977.
Rodger, S. M., and Holmes, I. H. (1979). *J. Virol.* **30,** 839–846.

Rodger, S. M., Schnagl, R. D., and Holmes, I. H. (1975a). *J. Virol.* **16,** 1229–1235.
Rodger, S. M., Craven, J. A., and Williams, I. (1975b). *Aust. Vet. J.* **51,** 536.
Rodger, S. M., Schnagl, R. D., and Holmes, I. H. (1977). *J. Virol.* **24,** 91–98.
Rodriguez, W. J., Kim, H. W., Brandt, C. D., Yolken, R. H., Arrobio, J. D., Kapikian, A. Z., Chanock, R. M., and Parrott, R. H. (1978). *Lancet* **1,** 37.
Rubinstein, D., Milne, R. G., Buckland, R., and Tyrrell, D. A. J. (1971). *Br. J. Exp. Pathol.* **52,** 442–445.
Sack, D. A., Eusof, A., Merson, M. H., Black, R. E., Chowdhury, A. M. A. K., Akbar Ali, M., Islam, S., and Brown, K. H. (1978). *Lancet* **2,** 280–283.
Saif, L. J., Bohl, E. H., Kohler, E. M., and Hughes, J. H. (1977). *Am. J. Vet. Res.* **38,** 13–20.
Saif, L. J., Theil, K. W., and Bohl, E. H. (1978). *J. Gen. Virol.* **39,** 205–217.
Scherrer, R., and Bernard, S. (1977). *Ann. Microbiol. Inst. Pasteur* **128A,** 499–510.
Schnagl, R. D., and Holmes, I. H. (1976). *J. Virol.* **19,** 267–270.
Schoub, B. D., Koornhof, H. J., Lecatsas, G., Prozesky, O. W., Freiman, I., Hartman, E., and Kassel, H. (1975). *Lancet* **1,** 1093–1094.
Schoub, B. D., Lecatsas, G., and Prozesky, O. W. (1977). *J. Med. Microbiol.* **10,** 1–16.
Schoub, B. D., Prozesky, O. W., Lecatsas, G., and Oosthuizen, R. (1978). *J. Med. Microbiol.* **11,** 25–31.
Sexton, M., Davidson, G. P., Bishop, R. F., Tounley, R. R. W., Holmes, I. H., and Ruck, B. J. (1974). *Lancet* **2,** 355.
Shepherd, R. W., Truslow, S., and Walker-Smith, J. A. (1975). *Lancet* **2,** 1082–1084.
Shinozaki, T., Fuji, R., Sato, K., Takahashi, E., Ito, Y., and Inaba, Y. (1978). *Lancet* **1,** 877–878.
Simhon, A., and Mata, L. (1978). *Lancet* **1,** 39–40.
Snodgrass, D. R., and Herring, J. A. (1977). *Vet. Rec.* **101,** 81.
Snodgrass, D. R., and Wells, P. W. (1976). *Arch. Virol.* **52,** 201–205.
Snodgrass, D. R., and Wells, R. W. (1978a). *J. Am. Vet. Med. Assoc.* **173,** 565–568.
Snodgrass, D. R., and Wells, R. W. (1978b). *Vet. Rec.* **102,** 146–148.
Snodgrass, D. R., Smith, W., Gray, E. W., and Herring, J. A. (1976a). *Res. Vet. Sci.* **20,** 113–114.
Snodgrass, D. R., Herring, J. A., and Gray, E. W. (1976b). *J. Comp. Pathol.* **86,** 637–642.
Snodgrass, D. R., Herring, J. A., Linklater, K. A., and Dyson, D. A. (1977a). *Vet. Rec.* **100,** 341.
Snodgrass, D. R., Madeley, C. R., Wells, P. W., and Angus, K. W. (1977b). *Infect. Immun.* **16,** 268–270.
Snodgrass, D. R., Angus, K. W., and Gray, E. W. (1977c). *Arch. Virol.* **55,** 263–274.
Spence, L., Fauvel, M., Bouchard, S., Babiuk, L., and Saunders, J. R. (1975). *Lancet* **2,** 322.
Stair, E. L., Mebus, C. A., Twiehaus, M. J., and Underdahl, M. R. (1973). *Vet. Pathol.* **10,** 155–170.
Stannard, L. M., and Schoub, B. D. (1977). *J. Gen. Virol.* **37,** 436–439.
Steinhoff, M. C. (1978). *Am. J. Dis. Child.* **132,** 302–307.
Sternberger, L. A. (1974). *In* "Immunocytochemistry" (L. A. Sternberger, ed.), pp. 129–171. Prentice-Hall, Englewood Cliffs, New Jersey.
Tallett, S. E., MacKenzie, C., Middleton, P. J., Kerzner, B., and Hamilton, J. R. (1976). *Pediatr. Res.* **10,** 350 (abstr.)
Tallett, S. E., MacKenzie, C., Middleton, P. J., Kerzner, B., and Hamilton, J. R. (1977). *Pediatrics* 60, 217–222.
Theil, K. W., Bohl, E. H., and Agnes, A. G. (1977). *Am. J. Vet. Res.* **38,** 1765–1768.
Theil, K. W., Bohl, E. H., and Saif, L. J. (1978). *J. Am. Vet. Med. Assoc.* **173,** 548–551.
Thornton, A., and Zuckerman, A. J. (1975). *Nature (London)* **254,** 557–558.
Thouless, M. E. (1979). *J. Gen. Virol.* **44,** 187–197.
Thouless, M. E., Bryden, A. S., Flewett, T. H., Woode, G. N., Bridger, J. C., Snodgrass, D. R., and Herring, J. A. (1977a). *Arch. Virol.* **53,** 287–294.

Thouless, M. E., Bryden, A. S., and Flewett, T. H. (1977b). *Br. Med. J.* **2,** 1390.
Thouless, M. E., Bryden, A. S., and Flewett, T. H. (1978). *Lancet* **1,** 39.
Thurber, E. T., Bass, E. P., and Beckenhauer, W. H. (1977). *Can. J. Comp. Med.* **41,** 131-136.
Todd, D., and McNulty, M. S. (1976). *J. Gen. Virol.* **33,** 147-150.
Todd, D., and McNulty, M. S. (1977). *J. Virol. 21,* 1215-1218.
Todd, D., McNulty, M. S., and Allan, G. M. (1980). *Archiv. Virol.* **63,** 87-94.
Torres-Medina, A., Wyatt, R. G., Mebus, C. A., Underdahl, M. R., and Kapikian, A. Z. (1976). *J. Infect. Dis.* **133,** 22-27.
Tufvesson, B., and Johnson, T. (1976). *Acta Pathol. Microbiol. Scand., Sect. B: Microbiol.* **84,** 225-228.
Turner, A. J., Caple, I. W., Craven, J. A., and Reingahum, C. (1973). *Aust. Vet. J.* **49,** 544.
Tzipori, S., Caple, I. W., and Butler, R. (1976). *Vet. Rec.* **99,** 398.
van den Hurk, J., and Kurstak, E. (1980). *J. Virol. Methods* **1,** 11-26.
Verly, E., and Cohen, J. (1977). *J. Gen. Virol.* **35,** 583-586.
Voller, A., and Bidwell, D. E. (1977). *In* "Comparative Diagnosis of Viral Diseases" (E. Kurstak and C. Kurstak, eds.) Vol. II, Part B, pp. 449-457. Academic Press, New York.
von Bonsdorff, C. -M., Hovi, T., Måkelå, P., Hovi, L., and Tevalvoto-Aarnio, M. (1976). *Lancet* **2,** 423.
von Bonsdorff, C. -M., Hovi, T., Måkelå, P., and Mőrtlinen, A. (1978). *J. Med. Virol.* **2,** 21-28.
Watanabe, H., Gust, I. D., and Holmes, I. H. (1978). *J. Clin. Microbiol.* **7,** 405-409.
Welch, A. B. (1971). *Can. J. Comp. Med.* **35,** 195-202.
Welch, A. B., and Thompson, T. L. (1973). *Can. J. Comp. Med.* **37,** 295-301.
Welch, A. B., and Twiehaus, M. J. (1973). *Can. J. Comp. Med.* **37,** 287-294.
White, G. B. B., Ashton, C. I., Roberts, C., and Parry, H. E. (1974). *Lancet* **2,** 726.
Whitelaw, A., Davies, H., and Parry, J. (1977). *Lancet* **1,** 361.
Woode, G. N. (1976). *Vet. Annu.* **16,** 30-34.
Woode, G. N., and Bridger, J. C. (1975). *Vet. Rec.* **96,** 85-88.
Woode, G. N., and Crouch, C. F. (1978). *J. Am. Vet. Med. Assoc.* **173,** 522-526.
Woode, G. N., Bridger, J. C., Hall, G. A., and Dennis, M. J. (1974). *Res. Vet. Sci.* **16,** 102-104.
Woode, G. N., Jones, J., and Bridger, J. C. (1975). *Vet. Rec.* **97,** 148-149.
Woode, G. N., Bridger, J. C., Jones, J. M., Flewett, T. H., Bryden, A. S., Davies, H. A., and White, G. B. B. (1976a). *Infect. Immun.* **14,** 804-810.
Woode, G. N., Bridger, J. C., Hall, G. A., Jones, J. M., and Jackson, G. (1976b). *J. Med. Microbiol.* **9,** 203-209.
Woode, G. N., Bridger, J. C., and Meyling, A. (1978a). *Vet. Rec.* **102,** 15-16.
Woode, G. N., Smith, C., and Dennis, M. J. (1978b). *Vet. Rec.* **102,** 340-341.
Wyatt, R. G., Kapikian, A. Z., Thornhill, T. S., Sereno, M. M., Kim, N. W., and Chanock, R. M. (1974). *J. Infect. Dis.* **130,** 523-528.
Wyatt, R. G., Gill, V. W., Sereno, M. M., Kalica, A. R., VanKirk, D. H., Chanock, R. M., and Kapikian, A. Z. (1976a). *Lancet* **1,** 98-99.
Wyatt, R. G., Sly, D. L., London, W. T., Palmer, A. E., Kalica, A. R., VanKirk, D. H., Chanock, R. M., and Kapikian, A. Z. (1976b). *Arch. Virol.* **50,** 17-27.
Wyatt, R. G., James, W. D., Bohl, E. H., Theil, K. W., Saif, L. J., Kalica, A. R., Greenberg, H. B., Kapikian, A. Z., and Chanock, R. M. (1980). *Science* **207,** 189-191.
Yolken, R. H., Kim, H. W., Clem, T., Wyatt, R. G., Kalica, A. R., Chanock, R. M., and Kapikian, A. Z. (1977a). *Lancet* **2,** 263-266.
Yolken, R. H., Wyatt, R. G., Kalica, A. R., Kim, H. W., Brandt, C. D., Parrott, R. H., Kapikian, A. Z., and Chanock, R. M. (1977b). *Infect. Immun.* **16,** 467-470.
Yolken, R. H., Wyatt, R. G., Kim, H. W., Kapikian, A. Z., and Chanock, R. M. (1978a). *Infect. Immun.* **19,** 540-546.

Yolken, R. H., Barbour, B., Wyatt, R. G., and Kalica, A. R. (1978b). *Science* **201,** 259–262.

Yolken, R. H., Barbour, B., Wyatt, R. G., and Kapikian, A. Z. (1978c). *J. Am. Vet. Med. Assoc.* **173,** 552–554.

Yolken, R. H., Wyatt, R. G., Zissis, G., Brandt, C. D., Rodriguez, W. J., Kim, H. W., Parrott, R. H., Urrutia, J. J., Mata, L., Greenburg, H. B., Kapikian, A. Z., and Chanock, R. M. (1978d). *N. Engl. J. Med.* **299,** 1156–1161.

Zissis, G., and Lambert, J. P. (1978). *Lancet* **1,** 38–39.

Zissis, G., Lambert, J. P., Fonteyne, J., and de Kegel, D. (1976). *Lancet* **1,** 96.

Part III

ORTHOMYXOVIRIDAE

Chapter 4

Influenza Infections in Lower Mammals and Birds

G. C. SCHILD

I.	Introduction	151
	Historical Perspectives	153
II.	Structure and Composition of Influenza Viruses	154
	A. Genome	154
	B. Polypeptides and Antigens	155
III.	Antigenic Classification and Nomenclature of Influenza Viruses	157
	Other Considerations for Taxonomy	161
IV.	Revised Nomenclature System	162
V.	Influenza in Lower Mammals and Birds	169
VI.	Influenza in Swine	169
VII.	Influenza in Horses	171
VIII.	Influenza in Birds	172
IX.	Influenza in Other Species	178
X.	Evidence That Influenza A Virus from Nonhuman Sources May Be Progenitors of Human Influenza	178
	Diagnostic and Serological Studies	180
	References	180

I. INTRODUCTION

Epizootics among horses described as influenza were reported commonly in the eighteenth and nineteenth centuries, while fowl plague, a fatal infection of chickens caused by influenza A virus, was described over 100 years ago. Historically the isolation of influenza viruses from nonhuman hosts predates the first successful isolation of influenza viruses from man in 1932, and the past two

COMPARATIVE DIAGNOSIS
OF VIRAL DISEASES, VOL. IV

ISBN 0-12-429704-8

decades have witnessed an explosion of information on the influenza A viruses of nonhuman hosts. The wide antigenic and biological spectrum of the influenza A viruses in animals and birds has led to an abundance of material for the study of the biological, pathological, epidemiological, and molecular aspects of the virology of influenza. Indeed the extensive reservoir of influenza A virus strains which exists in nature is considered by some virologists to present a permanent threat to mankind as a potential source of viruses, or at least of their genetic components, for future human pandemics (see Kaplan and Webster, 1977). The animal viruses merit serious consideration *per se* as causes of zoonoses of veterinary and economic significance, but their main attraction is that their study may lead to a fuller understanding of the nature and the conquest of human influenza.

In man, influenza has been described as the "last great plague" (Beveridge, 1977). The pandemic of 1918–1919, which probably killed tens of millions of persons, must qualify as one of the most devastating episodes of epidemic disease in human history. Historic accounts over the past three centuries suggest that there may have been about a score of major pandemics. Only two of these have been clearly defined by laboratory studies, those caused by influenza A viruses of the "Asian" (H2N2) subtype in 1957–1958 and by the Hong Kong (H3N2) subtype in 1968–1969. In addition to these major events, almost annual influenza outbreaks of lesser epidemiological impact and restricted geographic distribution occur.

It is because of the possibility that influenza viruses of lower mammals and birds might be involved in the origins of pandemic influenza of man (Pereira, 1969; Laver and Webster, 1979) that the World Health Organization in 1957 developed a coordinated research program on the ecology of influenza in animals involving many laboratories throughout the world (Kaplan and Beveridge, 1972). This program has done much to foster the collection of information on the zoonotic aspects of influenza. Much attention has been paid to the study of antigenic and genetic comparisons between influenza A virus of human and nonhuman sources. The conclusions of these studies will be discussed in this chapter.

So far only influenza A strains have been unequivocally demonstrated in nonhuman mammals and in birds. Evidence of infection by influenza B viruses in these hosts remains uncertain. Epidemics of disease caused by influenza A viruses occur frequently in swine and horses. Of considerable interest was an outbreak of fatal illness in the common seal on the coast of the United States in 1980 which was associated with influenza A virus (see Section IX). However, by far the largest spectrum of antigenically and biologically different influenza A viruses exists in a wide range of domestic and wild birds, including chickens, ducks, turkeys, quail, pheasants, geese, and several species of seabirds. The present review cannot attempt to cover the whole field of animal influenza and is selective in treating the areas of major biological significance.

Historical Perspectives

In the early history of influenza there is much anecdotal information (reviewed by Beveridge, 1977) on epidemics of influenza-like disease occurring among nonhuman species, usually domestic or agricultural animals such as dogs, cats, horses, and cattle. Such episodes are sometimes reported to have coincided with epidemics of influenza in humans. Thompson (1890), in an historical survey of influenza, mentions frequent reports of "influenza" among horses in the seventeenth to nineteenth centuries, which were apparently coincident with human epidemics. Creighton (1894) propounded the "generally held belief" that influenza epidemics in man and horses were epidemiologically related events. The significance of such reports has not been established, and it is clear that they should be regarded with much caution. Nevertheless the principle that certain influenza A viruses are potentially capable of transmitting across species barriers from lower animals to man was confirmed unequivocally when, in 1976, influenza A viruses of swine (H1N1* subtype) spread from pigs in the United States to produce a limited outbreak of illness among military recruits (Topp and Russell, 1977; see Section X).

The beginning of the modern history of the virology of influenza relied heavily on the potential of human influenza viruses to infect animals. In the autumn of 1932 in London, Wilson Smith and his colleagues attempted to transmit the disease to various laboratory animals using garglings from patients with influenza. Earlier, Shope (1931) had reported the successful transmission of swine influenza to healthy pigs using filtrates of respiratory secretions inoculated intranasally. Rabbits and guinea pigs appeared refractory to the human viruses, but inoculated ferrets developed a febrile respiratory disease (Smith *et al.*, 1933) which was contagious to their cagemates. Later it was found that infection could be artificially transmitted from ferrets to mice, producing a sometimes fatal pneumonia (Andrewes *et al.*, 1934). These findings led to a rapid development of knowledge of the immunological and pathological properties of the influenza virus. In 1940, a second virus serologically distinct from the 1933 isolates was recovered by Francis and Magill in New York (Francis, 1940). This virus was later designated "influenza B" to distinguish it from that of Smith *et al.* (1933), which became known as "influenza A." A third type, "influenza C," was isolated in 1949 (Taylor, 1949). Alternatives to experimental animals for the isolation and cultivation of influenza viruses became available later when Burnet (1940, 1941) discovered that the virus would grow in the amniotic and allantoic cavities of the developing chick embryo. This method also led to the discovery of the hemagglutinating properties of the virus (Hirst, 1941) and to the use of

*Previous designation Hsw1N1. Antigenic designations of influenza viruses used in this chapter are those recommended by the World Health Organization (1980).

hemagglutination inhibition (HI) techniques for the serological study of the virus. These two techniques have probably contributed more to the knowledge of the influenza virus than any others. Later biochemical work (Gottschalk, 1954, 1957) provided the observation that influenza viruses contain the enzyme neuraminidase which, with the hemagglutinin, is a major antigenic component of the virus particle.

II. STRUCTURE AND COMPOSITION OF INFLUENZA VIRUSES

The true influenza viruses, classified as orthomyxoviruses (Melnick, 1973), form a group of agents related by their morphological, molecular, and replicative characteristics. The morphology, molecular composition, and replication of influenza viruses have been extensively reviewed elsewhere (Choppin and Compans, 1975; Schultz, 1975; Stuart-Harris and Schild, 1976; Wrigley, 1980; Palese *et al.*, 1980; Lamb and Choppin, 1980; Skehel *et al.*, 1980; Laver and Air, 1980), and a detailed description is not intended here.

Negatively stained preparations of influenza viruses reveal irregular spherical or kidney-shaped particles some 80–120 nm in diameter. In some preparations, particularly those of recently isolated virus strains with few laboratory passages, filamentous particles 1000 nm or more in length may be abundant. The particles are covered with densely arranged, radial projections—"spikes"—of two distinct morphologies (Laver and Valentine, 1969), corresponding to the hemagglutinin (HA) and neuraminidase (NA) proteins of the virus. The surface projections are inserted in a lipid membrane that encloses the virus genome and core proteins.

A. Genome

The single-stranded RNA genome of the virus is made up of eight distinct species of RNA of molecular weights ranging from 3×10^5 to 1×10^6 daltons (Skehel, 1971). The virion RNA is not functional as message (negative-stranded) and is complementary to virus-specific RNA molecules present in the polysomal fraction of infected cells. The assignment of coding function for the eight species of RNA has been established (Table I). The RNA molecules are monocistronic, coding for one viral protein in the case of the seven largest RNA segments (genes 1–7) and for two different nonstructural (NS) proteins of the virus (NS1 and NS2) in the case of the smallest RNA segment (gene 8). The three largest RNA segments (genes 1–3) each code for a different one of the three largest internal proteins of the virus particle (P1, P2, and P3) of molecular weight 90,000–100,000, which are thought to be involved in the RNA polymerase activity of the virus. Three intermediate-size RNA segments code for the two surface glycoproteins, HA and NA, respectively, and for the nucleoprotein (NP), of molecular weight 60,000, which is a major protein and antigen of the virus, closely associated with the RNA molecules in the virus core. Of the two smallest RNA segments, gene 7 codes for the M protein, which is the smallest and most abundant protein of the virus, of molecular weight 25,000 and may have potential to code for an additional smaller protein (Allen *et al.*, 1980). Gene 8 codes the two NS proteins, NS1 and NS2, of molecular weight approximately 20,000 and 10,000 (Lamb and Choppin, 1980), which appear to result from transcription of the single gene in two different reading frames.

At present, very rapid progress is being made in cloning the influenza virus genes into bacterial plasmids and determining their primary structure by restriction enzyme mapping and sequencing. The complete sequence of the HA genes of at least three distinct influenza A virus HA subtypes, including that of fowl plague virus (H7N7)* and two human subtypes H3N2 and H2N2, has been established (Porter *et al.*, 1979; Min Jou *et al.*, 1980; Sleigh *et al.*, 1980). The sequences of representatives of most of the eight gene fragments of the influenza A and B viruses are likely to be known in the near future. This will enable rapid progress to be made in the determination of genetic relationships between influenza viruses and in the elucidation of primary structure of their functional and antigenic proteins.

B. Polypeptides and Antigens

Analyses of the proteins present in extracts of virus-infected cells indicate that virus-specific proteins can be detected even before the eventual inhibition of cellular protein synthesis becomes established. A total of 10 distinct polypeptides, specific for infected cells, can be resolved by polyacrylamide gel electrophoresis of the detergent-solubilized cell extracts. The molecular weight of these polypeptides shows a clear correspondence with the molecular weight of the genome RNA segments (see Table I).

The large-molecular-weight internal proteins P1, P2, and P3 are present in small quantities in the virus core; their antigenic characteristics and relationships between strains are unknown. It is likely that these proteins exercise complex functions in controlling transcription and replication of viral RNA.

The influenza NP is abundant in the virus particle and in infected cells. Together with the virus RNA, it forms the nucleocapsid of the virus, which is often visualized as a helically coiled structure. This is a major type-specific antigen forming the basis of the classification of influenza viruses into type A, B, and C, depending on the ability of their NPs to cross-react in serological tests such as complement fixation or double-immunodiffusion. It has been recognized (Schild *et al.*, 1979; G. C. Schild, unpublished observations) that minor dif-

*Former designation Hav1Neq1.

TABLE I
Assignment of Coding Capacity for the Eight RNA Gene Components of the Influenza Virus

RNA gene		Polypeptide products		
Number	Molecular weight ($\times 10^{-5}$)	Designation	Molecular weight ($\times 10^{-4}$)	Product function and location
1	9–10	P1	9.4	Internal virion possible Polymerase involvement
2	9–10	P2	8.5	Internal virion possible Polymerase involvement
3	9–10	P3	8.1	Internal virion possible Polymerase involvement
4	7.5	HA[a]	7.5	Envelope glycoprotein, hemagglutinin[a]
5	6.0	NP	5.3	Internal virion nucleocapsid
6	6.5	NA	6.8	Envelope glycoprotein, neuraminidase
7	3.5	M	2.5	Internal virion matrix protein
8	3.0	NS1	2.3	Nonstructural protein
		NS2	1.0	Nonstructural protein

[a]The haemagglutinin is cleaved into two polypeptides with a molecular weight of approximately 5.5 and 2.8 $\times 10^4$ in many host cell systems.

ferences in antigenic specificity exist in the NP antigens of human influenza A strains isolated during different periods. Early strains isolated between 1933 and 1942 and more recently isolated strains could be readily distinguished on the basis of their NP antigens. Distinctions could also be observed between the NP antigens of human influenza A viruses and those from avian sources (see Figures 10 and 11). The NP antigen is thus potentially of value in ecological studies since its antigenic specificity is apparently related to the host of origin of the virus.

The matrix protein is abundant in the virus and appears to be important for the stability of the virus particle as well as in the assembly and maturation process of virions in the infected cell. The matrix protein, like the NP, is a type-specific antigen (Schild, 1972).

The HA and NA subunits are glycoproteins; their covalently attached carbohydrate chains are of host cell origin. Both components are major antigens of the virus and are highly variable, undergoing antigenic "drift" and "shift" (Schild and Dowdle, 1975). The HA and NA form the basis for the serological classification of influenza A viruses into subtypes and strains (see Section III). Antibodies to both HA and NA contribute to immunity to influenza in experimental animals and man (Stuart-Harris and Schild, 1976; Schild, 1979; Virelizier *et al.*, 1979; Potter and Oxford, 1979). However, there are fundamental differences in the biological properties of antibodies to these two antigens (Schild and Dowdle, 1975). Antibody to HA is capable of direct and efficient neutralization of virus infectivity on contact with infectious virus. In contrast, antibody to NA neutralizes infectivity only when present in high titer, but it has the property of specifically neutralizing enzyme activity and limiting the dissemination of infectious virus in cultures or infected hosts. Antibody rises to HA and NA are of diagnostic significance (Schild and Dowdle, 1975) and are useful as a serological index of immunity following infection or vaccination (Potter and Oxford, 1979; Tyrrell and Smith, 1979). HA polypeptide is a single gene product but is cleaved into two components (HA1 and HA2) when the virus replicates in certain cell types. Such cleavage is essential for the generation of infectious virus. Each functional HA spike contains three HA1 and three HA2 polypeptides, the HA1 and HA2 molecules being linked by disulfide bonds. There is strong evidence that efficiency of cleavage of the HA is related to pathogenicity (Rott, 1980). For avian influenza A viruses there is a direct correlation between the potential of the HA to be cleaved when the virus replicates in a wide range of cell types of a given host and the pathogenicity of the virus for that host (Rott, 1980). In strains of fowl plague viruses that are highly pathogenic for chickens, such cleavage is highly efficient.

NA is less abundant than HA on the surface of the virus. This protein possesses enzymic activity responsible for the destruction of receptors associated with glygoproteins and glycolipids on host cells to which influenza virus HA binds. Like HA, NA is synthesized as a single polypeptide chain, but the functional enzyme spike comprises four polypeptides (Wrigley *et al.*, 1973).

III. ANTIGENIC CLASSIFICATION AND NOMENCLATURE OF INFLUENZA VIRUSES

In the system of nomenclature for influenza viruses in use from 1971 to 1980 (World Health Organization, 1971), the viruses were classified into types (A, B, and C) on the basis of the antigenic character of the NP. Type A viruses were further divided into subtypes based on the antigenic character of their HA and NA antigens. The HA antigen subtypes of human influenza A viruses were designated H0, H1, H2, and H3, representative of the human influenza A viruses prevalent from 1932 to 1946 (H0N1), 1947 to 1957 (H1N1), 1957 to 1968 (H2N2, Asian), and 1968 to the present (H3N2, Hong Kong). There was one HA subtype of swine influenza virus (Hsw1), two subtypes of equine virus (Heq1 and Heq2), and eight HA antigen subtypes of avian influenza viruses (Hav1 to Hav8). Since 1971, however, it has become apparent that antigenic relationships exist between certain viruses which were classified into different HA antigen subtypes (Schild, 1970; Baker *et al.*, 1973; Laver and Webster, 1973; Schild *et al.*, 1980). These relationships have been demonstrated by a variety of methods but particularly by immuno-double-diffusion (IDD) tests in gels with potent, specific antisera to isolated HA antigens. Such tests have been found to be more broadly reactive than conventional HI tests (Schild and Dowdle, 1975). Evidence is now available to support the designation of two additional HA antigen subtypes among recently characterized avian influenza A viruses (Webster *et al.*, 1976a; Hinshaw and Webster, 1979). The NA antigens were likewise divided into antigenic subtypes. NA inhibition tests (Aymard *et al.*, 1973) have been widely employed for the study of neuraminidase antigens, but IDD tests with NA-specific sera are of relatively broad specificity and are of value for the identification of subtypes (Schild and Dowdle, 1975). Among human influenza A viruses there were two NA antigen subtypes, N1 and N2, represented by the viruses prevalent from 1932 to 1956 (H0N1 and H1N1) and from 1957 to the present (H2N2 and H3N2). Among swine influenza viruses there was one subtype (N1), closely related to the human N1 subtype. Among equine influenza viruses two distinct NA antigen subtypes, designated Neq1 and Neq2, were described. For avian influenza A strains there were eight subtypes of NA antigen; two of these (N1 and N2) were shared with human influenza A viruses, two (Neq1 and Neq2) were shared with equine viruses, and four (Nav1 to Nav4) were unique to viruses of avian origin. Since 1971 evidence for two additional NA subtypes among avian subtypes has become available (Downie and Laver, 1973; Webster *et al.*, 1976b).

A comprehensive antigenic analysis (Schild *et al.*, 1980) has recently been completed of a large collection of prototype strains of influenza A virus of human, swine, equine, and avian origin, employing IDD tests with HA- and NA-specific antisera. Representative IDD reactions are illustrated in Figs. 1–9. This and other new information obtained since 1971 on their immunology, biochemistry, genetics, and ecology has permitted a re-evaluation of the nomenclature of the influenza A viruses (World Health Organization, 1979, 1980). It has been proposed (World Health Organization, 1980) that several of the subtypes described in the 1971 system be merged. In the revised system for in-

---

Plates 1 and 2. Immuno-double-diffusion reactions in agarose gels of antisera to purified H or N antigens of influenza A viruses. Undiluted rabbit, sheep, or goat sera (see Table I) were added to the central position in each pattern of wells, and purified, concentrated influenza virus antigen preparations, disrupted with sarcosyl detergent, were added to the peripheral wells. The precipitin reactions were photographed unstained. Cross reactions between the antigens of different virus strains, as revealed by shared precipitin lines, indicate the possession of common antigenic determinants and may be considered evidence for designation of the viruses into a common H or N antigen subtype. Antigenic differences between H and N antigens of viruses within a common subtype are revealed by the formation of "spurs" between the precipitin reactions which produced homologous and heterologous virus antigens tested in adjacent wells. (See p. 160 for Figs. 6–9.)

Figs. 1–5. Reactions of antisera to H antigens. Hav3 was found to be antigenically unique, i.e., not cross-reacting with other subtypes (Fig. 1). In contrast, subtypes HO, Hl, and Hswl (Fig. 2), H3, Heq2, and Hav7 (Fig. 3), and Heq1 and Hav1 (Fig. 4) were found to be antigenically related. Fig. 5 shows that A/turkey/Wisconsin/56 virus, formerly classified as Hav6, does not cross-react with other Hav6 viruses.

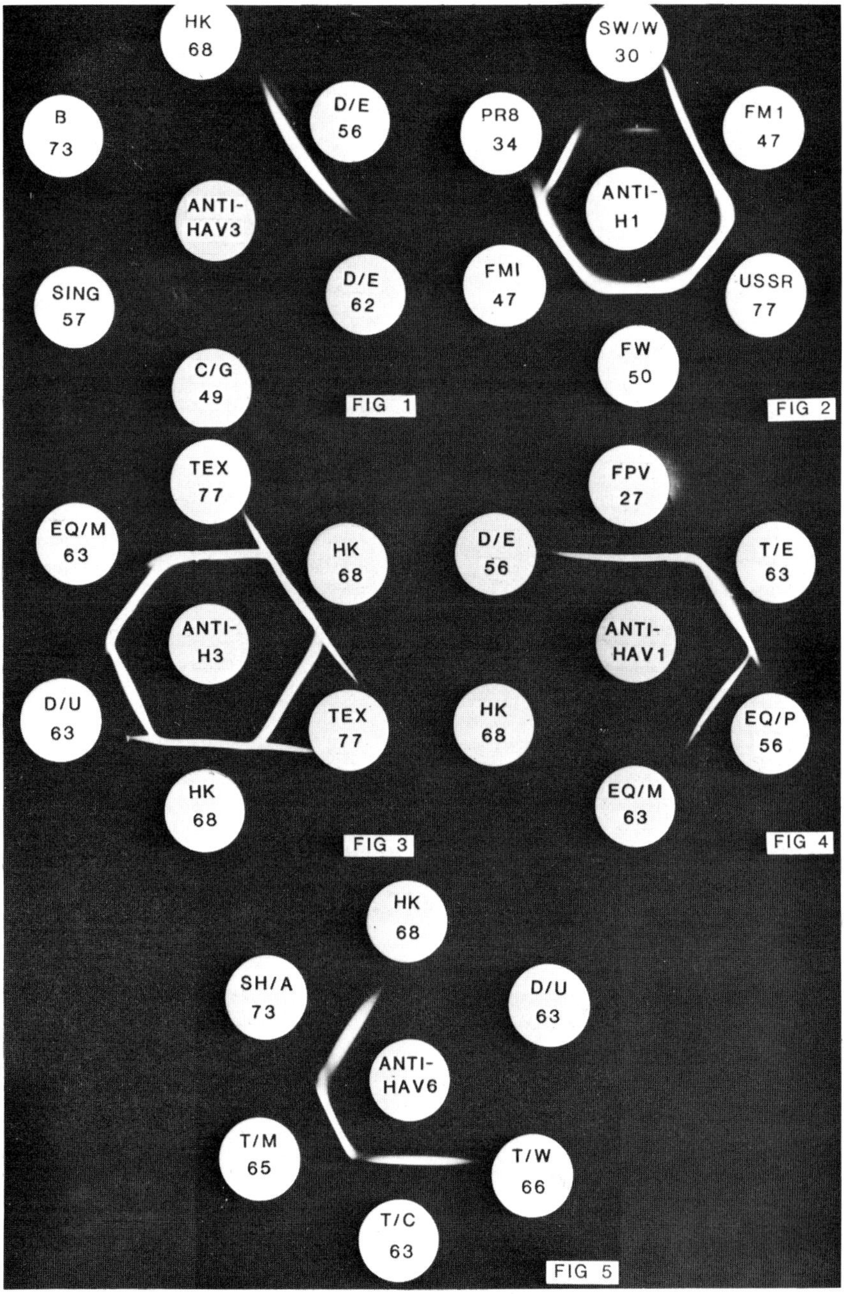

Plate 1 *(continued)*

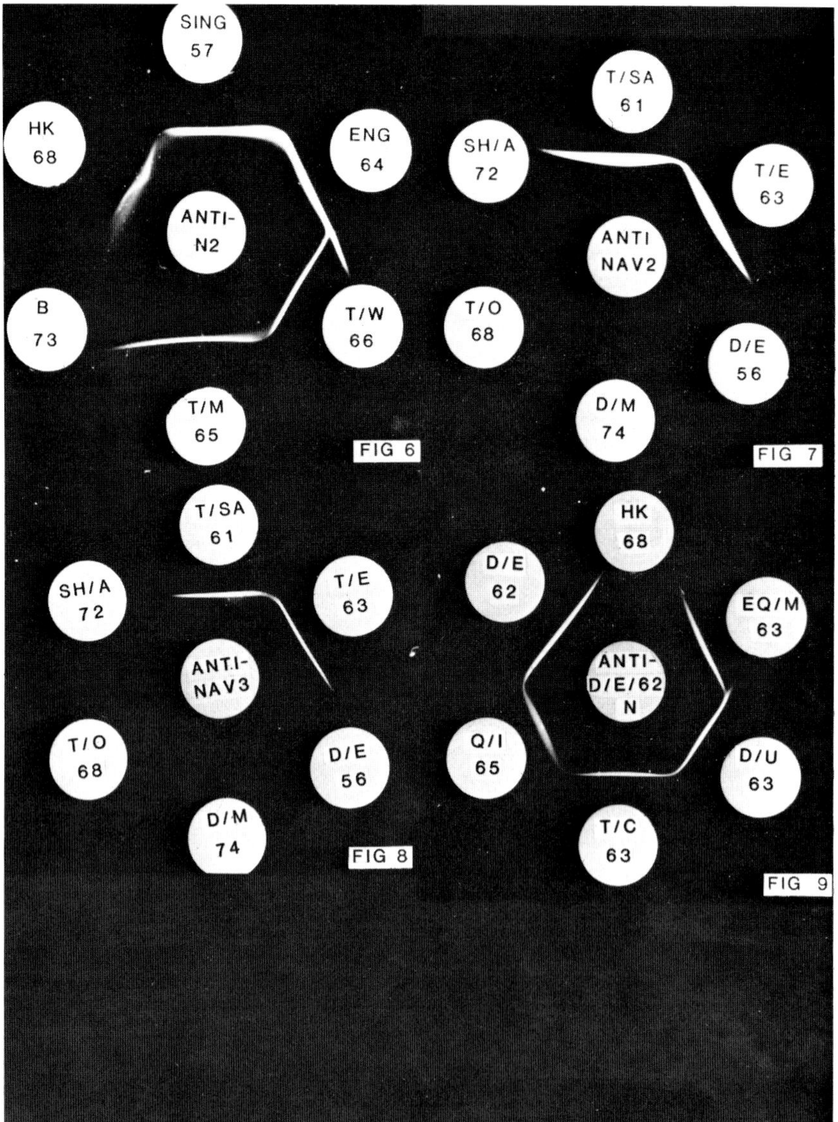

Plate 2

Figs. 6–9. Reactions of antisera to N antigens. Certain N antigens, e.g., N2, are shared between viruses of human and nonhuman origin (Fig. 6). Figs. 7 and 8 show that viruses A/tern/South Africa/61 and A/turkey/England/63, formerly classified into different N antigen subtypes (Nav2 and Nav3), contain related neuraminidase and may be regarded as a single subtype. The N antigen of A/duck/England/62, formerly classified as Navl, was found to be related to that of A/equine/Miami/63 (Neq2) (Fig. 9).

fluenza A viruses from all species, the HA antigens have been regrouped into 12 subtypes and the NA antigens into 9 subtypes.

Other Considerations for Taxonomy

Antigenic similarities between the surface antigens HA and NA of influenza A viruses of different subtypes have been observed even in the absence of demonstrable cross reactions in IDD tests. These similarities include evidence of relationships based on cross protection or of cell-mediated immunity (Doherty *et al.*, 1977; Russel and Liew, 1979). However, where these relationships are not supported by serological (IDD) cross reactions between the surface antigens, they have not been considered relevant to the subtyping of virus strains.

The RNAs of several influenza virus strains and their recombinants have been characterized by various methods, including RNA–RNA (Scholtissek, 1978) and RNA–DNA hybridization, oligonucleotide analysis (Young and Palese, 1979), and nucleotide sequence determinations (Porter *et al.*, 1979). Results obtained using RNA–RNA hybridization techniques have indicated that the genes coding for the HA antigens of viruses of the previously (World Health Organization, 1971) designated Hsw1, HO, and H1 subtypes were closely related (Scholtissek, 1978), thus supporting the inclusion of these antigens in a single subtype H1 in the revised (World Health Organization, 1980) nomenclature system. The genes coding for the HA antigens of viruses previously designated H3, Hav7, and Heq2 subtypes in the 1971 system also exhibited a high base-sequence homology, as did those coding for Heql and Hav1. These antigens have likewise been allocated to common subtypes, H3 and H7, respectively. The gene coding for Hav5 appeared to be unrelated to the HA genes of other subtypes (P. Palese and C. Scholtissek, unpublished observations). The results of the analyses of genes coding for the Nav2 and Nav3 proteins supported the inclusion of these antigens in a single group. Based on similar analyses, Nav6 appeared to constitute a separate subtype (Scholtissek, 1978).

Genes coding for corresponding nonsurface proteins of different influenza A viruses have been compared using similar methods, and it was found that influenza A viruses could be separated into two groups on the basis of the genes coding for the NS protein (Scholtissek, 1978; see Table II). However, the groupings do not appear to have immediate taxonomic or epidemiological significance.

Knowledge of the primary structure of influenza virus proteins, obtained by tryptic peptide, nucleic acid sequence, and amino acid sequence analyses, is at present restricted to the HA molecule. The results of comparative tryptic peptide analyses, particularly of the HA2 components of the HAs so far obtained, are consistent with the revised (World Health Organization, 1980) nomenclature system. The nucleotide sequences that have been determined indicate that the HAs within a subtype are much more closely related to each other than to those of

TABLE II
Grouping of Influenza Subtypes According to the Genetic Relatedness of RNA Segment 8 (NS Gene)[a]

Influenza virus strain	Subtype	Group
A/fowl plaque virus/Rostock/34	(H7N7)	1
A/chicken/Germany/N/49	(H10N7)	2
A/turkey/Canada/63	(H6N8)	2
A/turkey/Oregon/71	(H7N3)	2
A/duck/Ukraine/1/63	(H3N8)	1
A/turkey/England/63	(H7N3)	1
A/Puerto Rico/8/34	(H0N1)	1
A/Fort Monmouth/1/47	(H1N1)	1
A/Singapore/1/57	(H2N2)	1
A/Hong Kong/1/68	(H3N2)	1
A/swine/1976/31	(H1N1)	1
A/equine/Miami/1/63	(H3N8)	1
A/equine/Prague/1/56	(H7N7)	1
A/duck/England/56	(H11N6)	1
A/duck/Czechoslovakia/63	(H4N6)	1
A/turkey/Ontario/7732/66	(H5N9)	1
A/duck/Germany/1868/68	(H6N1)	1
A/turkey/Ontario/6118/68	(H8N4)	1
A/chicken/Scotland/59	(H5N1)	1
A/duck/Memphis/546/74	(H11N9)	1
A/fowl plague virus/Dutch/27	(H7N7)	1
A/heron/Chabarovsk/700/73	(H3N8)	2
A/duck/Chabarovsk/698/73	(H3N8)	2
A/duck/Chabarovsk/1610/72	(H3N8)	2

[a]After Scholtissek (1978).

other subtypes (Porter *et al.*, 1979; Min Jou *et al.*, 1980; Gething *et al.*, 1980; Sleigh *et al.*, 1980).

Monoclonal antibodies, when used in addition to conventional serological reagents, have proved to be useful in antigenic analysis of influenza viruses (Webster *et al.*, 1979). Although monoclonal antibodies to the HA or NA antigens of influenza viruses are valuable for discriminating between very closely related antigenic variants within a subtype, they are often too specific for routine serological typing and subtyping of influenza viruses.

IV. REVISED NOMENCLATURE SYSTEM

In the revised system of nomenclature recommended by the World Health Organization for use from 1980, the strain designation for influenza viruses contains the following information:

1. A description of the antigenic type of the virus based on the antigenic specificity of the NP antigen (type A, B, or C). Since 1971, a further type-specific internal antigen of the influenza A and B viruses, the matrix (M) protein, has been described (Schild, 1972). Typing of influenza A and B viruses based on the M protein is consistent with the results obtained with NP antigen (Dowdle *et al.*, 1974).

2. The host of origin is not indicated for strains isolated from human sources, but it is indicated for all strains isolated from nonhuman hosts, e.g., swine, horse (equine), chicken, and turkey. For viruses from nonhuman species, both the Latin binomial nomenclature and the common name of the host of origin should be recorded in the original publication describing the virus isolate, e.g., *Anas acuta* (pintail duck). Thereafter, the common name of the species should be used for the strain, e.g., A/duck/USSR/695/76 (H2N3). When viruses are isolated from nonliving material, the nature of the material should be specified e.g., A/lake water/Wisconsin/1/79.

3. Geographical origin.

4. Strain number.

5. Year of isolation.

For influenza A viruses the antigenic description includes:

1. An index describing the antigenic character of the HA, i.e., H1, H2, H3,

TABLE III
Proposed Subtypes of HA Antigens of Influenza A Viruses

Proposed subtypes (from 1980)	Previous subtypes (1971 system)
H1[a]	H0, H1 Hsw1
H2	H2
H3	H3, Heq2, Hav7
H4	Hav4
H5[a]	Hav5
H6	Hav6
H7	Heq1, Hav1
H8	Hav8
H9	Hav9
H10	Hav2
H11	Hav3
H12	Hav10

[a]Minor antigenic cross reactions have been detected in some laboratories between the HA antigenic subtypes H5 (A/chick/Scotland/49 and A/tern/South Africa/61) and H1 (A/swine/Iowa/15/30).

TABLE IV
Proposed Subtypes of NA Antigens of Influenza A Viruses

Proposed subtypes from 180	Previous subtypes (1971 system)
N1	N1
N2	N2
N3	Nav2, Nav3
N4	Nav4
N5	Nav5
N6	Nav1
N7	Neq1
N8	Neq2
N9	Nav6

TABLE V
Examples of Reference Strains for the Proposed Subtypes of HA and NA Antigens of Influenza A Viruses Isolated from Man

HA and NA subtypes	Reference strains
H1N1	A/PR/8/34 (H1N1)
	A/Weiss/43 (H1N1)
	A/FM1/47 (H1N1)
	A/England/1/51 (H1N1)
	A/Denver/1/57 (H1N1)
	A/New Jersey/8/76 (H1N1)
	A/USSR/90/77 (H1N1)
H2N2	A/Singapore/1/57 (H2N2)
	A/Japan/305/57 (H2N2)
	A/England/12/64 (H2N2)
	A/Tokyo/3/67 (H2N2)
H3N2	A/Hong Kong/1/68 (H3N2)
	A/England/42/72 (H3N2)
	A/Port Chalmers/1/73 (H3N2)
	A/Victoria/3/75 (H3N2)
	A/Texas/1/77 (H3N2)

TABLE VI
Examples of Reference Strains for Proposed Subtypes of HA and NA Antigens of Influenza A Viruses Isolated from Swine

HA and NA subtypes	Reference strains
H1N1	A/swine/Iowa/15/30 (H1N1)
	A/swine/Wisconsin/67 (H1N1)
H3N2	A/swine/Taiwan/1/70 (H3N2)

H4, etc. The numbering of subtypes is a simple sequential system that applies uniformly to influenza viruses from all sources (Table III).

2. An index describing the antigenic character of the NA, i.e., N1, N2, N3, N4, etc., applied uniformly to all influenza A viruses (Table IV).

It is implicit that a given HA or NA subtype designation will encompass strains exhibiting some antigenic variation within the subtype (antigenic drift). The exact antigenic character of an influenza virus variant may be defined by indicating similarities to designated reference strains.

The 1980 nomenclature system does not provide for the description of distinct subtypes of influenza B and C viruses because of the lack of clear-cut antigenic distinctiveness between members of these types. The description of these viruses is therefore limited to strain designation, e.g., B/England/5/66, C/Paris/1/67.

Examples of reference strains of influenza A viruses of human, swine, equine, and avian influenza viruses and their new subtype designations are given in Tables V–IX.

The 1980 nomenclature system was not designed to provide information on the host range or virulence of influenza viruses. The isolation from different hosts of antigenically similar influenza A viruses is well established, and examples in which one subtype of the surface antigens has been found in influenza viruses from different species are numerous (Hinshaw *et al.*, 1980; see Table X). Rep-

TABLE VII
Examples of Reference Strains for Proposed Subtypes of HA and NA Antigens of Influenza A Viruses Isolated from Horses

HA and NA subtypes	Reference strains
H7N7	A/equine/Prague/1/56 (H7N7)
H3N8	A/equine/Miami/1/63 (H3N8)

TABLE VIII
Examples of Reference Strains for Proposed Subtypes of HA Antigens of Influenza A Viruses Isolated from Avian Species

Proposed subtype designation	Previous subtype designation	Reference strains	Other strains with related antigens
H1	Hsw1	A/duck/Alberta/35/76 (H1N1)	A/duck/Alberta/97/77 (H1N8)
H2	H2	A/duck/Germany/1215/73 (H2N3)	A/duck/Germany/1/72 (H2N9)
H3	Hav7	A/duck/Ukraine/1/63 (H3N8)	A/duck/England/62 (H3N8)
			A/turkey/England/69 (H3N2)
H4	Hav4	A/duck/Czechoslovakia/56 (H4N6)	A/duck/Alberta/300/77 (H4N3)
H5	Hav5	A/tern/South Africa/61 (H5N3)	A/turkey/Ontario/7732/66 (H5N9)
			A/chick/Scotland/59 (H5N1)
H6	Hav6	A/turkey/Massachusetts/3740/65 (H6N2)	A/turkey/Canada/63 (H6N8)
			A/shearwater/Australia/72 (H6N5)
			A/duck/Germany/1868/68 (H6N1)
H7	Hav1	A/fowl plague virus/Dutch/27 (H7N7)	A/chick/Brescia/1902 (H7N1)
			A/Turkey/England/63 (H7N3)
			A/fowl plague virus/Rostock/34 (H7N1)
H8	Hav8	A/turkey/Ontario/6118/68 (H8N4)	—
H9	Hav9	A/turkey/Wisconsin/1/66 (H9N2)	A/duck/Hong Kong/147/77 (H9N6)
H10	Hav2	A/chick/Germany/N/49 (H10N7)	A/quail/Italy/1117/65 (H10N8)
H11	Hav3	A/duck/England/56 (H11N6)	A/duck/Memphis/546/74 (H11N9)
H12	Hav10	A/duck/Alberta/60/76 (H12N5)	—

TABLE IX
Examples of Reference Strains for Proposed Subtypes of NA Antigens of Influenza A Viruses Isolated from Avian Species

Proposed new subtype designation	Previous grouping	Reference strains	Other strains with related N antigens
N1	N1	A/chick/Scotland/59 (H5N1)	A/duck/Alberta/35/76 (H1N1) A/duck/Germany/1868/68 (H6N1)
N2	N2	A/turkey/Massachusetts/3740/65 (H6N2)	A/turkey/Wiscònsin/66 (H9N2) A/turkey/England/69 (H3N2)
N3	Nav2	A/tern/South Africa/61 (H5N3)	A/duck/Germany/1215/73 (H2N3)
	Nav3	A/turkey/England/63 (H7N3)	
N4	Nav4	A/turkey/Ontario/6118/68 (H8N4)	A/duck/Wisconsin/6/74 (H6N4)
N5	Nav5	A/shearwater/Australia/1/72 (H6N5)	A/duck/Alberta/60/76 (H12N5)
N6	Nav1	A/duck/Czechoslovakia/56 (H4N6) A/duck/England/56 (H11N6)	—
N7	Neq1	A/fowl plague virus/Dutch/27 (H7N7)	A/chick/Germany/N/49 (H10N7)
N8	Neq2	A/quail/Italy/1117/65 (H10N8)	A/turkey/Canada/63 (H6N8) A/duck/England/62 (H3N8)
N9	Nav6	A/duck/Memphis/546/74 (H11N9)	A/turkey/Ontario/7732/66 (H5N9)

TABLE X
Tabulation of Antigenic Subtypes of Influenza A Viruses Isolated from Different Species[a,b]

HA Subtype	N1	N2	N7
H1	Sw/Iowa/15/30 Human/PR/8/34 Dk/Alb/35/76		Whale/Pacific/76 Dk/Miss/77
H2	Dk/Can/76	Human/Sing/57 Dk/604/78	Dk/Germ/1215/73
H3	Dk/Alb/25	Human/HK/1/68 Tk/Eng/69 Sw/Taiwan/70 Ck/Kamchatka/12/71 71 (32) Calf/Duschanbe/ 55/71 Dk/Wis/10/74 Crow/Kazan/20/72 Guillemot/USSR/74	Dk/Can/1516/77
H4	Dk/Ger/210/67	Dk/Ont/4134/67	Dk/HK/174/77
H5	Ck/Scot/59 Tk/Ont/6213/66	Tk/Ont/5265/66 Pheasant/Quebec/647/74	
H6	Dk/Pa/69 Tk/Ont/4689/67 Dk/Ger/1868/68 Goose/Tx/2/75	Dk/S.D./2/75 Tk/Ont/5050/64 Dk/HK/110/69 Tern/Pechora/105/72 Goose/Tx/1/75 Lake/Alb/1/78 Feces/Alb/1/78	
H7	Ck/"FPV"/Brescia/ 1902 Dk/Ca/72 (48) Parrot/Ulster/1/73 Pond/HK/78	Dk/HK/47/76 Pond/HK/298/78	Ck/Dutch/27 Eq/Pr/1/56 Tk/Aust/75 Seal/Mass/80
H8			
H9		Tk/Wi/1/66[c] Dk/HK/86/76	
H10	Dk/HK/35/76	Dk/It/574/66	Ck/"N" Ger/49 Dk/Manitoba/53

[a] Abbreviations used in virus name: Sw, swine; Dk, duck; Tk, turkey; Ck, chicken; Eq, horse.
[b] The earliest recorded viruses with the designated subtypes isolated from mammals or birds are presented
[c] The earliest characterized virus with the designated subtype.

resentatives of each of the NA antigen subtypes of influenza viruses from man, pigs, and horses (N1, N2, N7, and N8) have also been isolated from birds. Similarly, representatives of each of the HA subtypes of man, pigs, and horses (H1, H2, H3, and H7) have been isolated from avian species.

V. INFLUENZA IN LOWER MAMMALS AND BIRDS

Although the earliest isolations of influenza from man were not achieved until 1932 (Smith *et al.*, 1933; Laidlaw, 1935), viruses had been isolated earlier from nonhuman sources which only subsequently were shown to be influenza A strains. The first clinical descriptions of fowl plague were reported in Italy in 1878 (cited by Easterday and Tumova, 1972), and the disease was shown in 1900 by Centanni and Savonuzzi (see Stubbs, 1965), using experimental transmission studies, to be due to a "filterable virus." Fowl plague remained of little interest to influenza research workers for over 50 years until Schäfer (1955) showed that the agent possessed the antigenic and morphological attributes of influenza A viruses isolated from man—and was indeed a true influenza virus. Similarly, swine influenza virus, first transmitted from pig to pig with bacteria-free filtrates (Shope, 1931), was shown (Andrewes *et al.*, 1935; Francis and Shope, 1936) to have antigenic similarities to human influenza A viruses. Since these early observations numerous isolations of type A influenza viruses from nonhuman sources have been recorded, and much has been learned of the antigenic and biological spectrum of influenza A viruses in lower mammals and birds and their relationships with human influenza A viruses.

Evidence that influenza B (or C) viruses commonly infect nonhuman hosts is lacking. Although on a few occasions evidence of influenza B infections in animals has been reported, these findings have not generally been confirmed, and it is likely that the viruses are infrequent or even absent in nonhuman species.

VI. INFLUENZA IN SWINE

In 1918, an American veterinarian registered the sudden onset of epizootics of respiratory disease among swine herds in Iowa (Koen, 1919). These epizootics appeared to coincide with the onset of the great pandemic of human influenza in the midwestern United States. The clinical similarities between the diseases of swine and man lead to the description "swine influenza." Koen emphasized that the swine influenza in 1919 was apparently a new clinical entity. Although the primary infectious agent was undoubtedly a virus (Shope, 1931), a bacterial commensal (*Haemophilus influenza suis* appeared to be important in the pathogenesis of the disease. Retrospective serological studies in man (Davenport *et al.*, 1953) and subsequent investigations (reviewed by Stuart-Harris and Schild, 1976) provided strong circumstantial evidence that the Shope strain was antigenically close to the virus responsible for the 1918–1919 human pandemic.

Since 1918–1919 outbreaks of influenza associated with classical swine (H1N1)* influenza A virus have occurred in pigs almost annually in the United States, and recent intensive surveillance in that country has indicated that the virus is widespread throughout the states and may circulate at all times of the year (Hinshaw *et al.*, 1978). Epizootic influenza in commercial pig herds in the United States is probably of considerable economic importance.

International surveys carried out since 1976 suggest that pigs in the United States may be the major reservoir of swine (H1N1) viruses in the world. Investigations of pigs in other countries have indicated a low prevalence of infection in some areas, including Japan, Italy, Israel, and Central Europe, but these areas may have acquired the viruses from the United States. Evidence of infection in the United Kingdom has not been apparent since 1939, and pigs in several other European countries appear to be free of the disease. Swine influenza virus is probably endemic in pigs from Hong Kong, Singapore, and the People's Republic of China (Shortridge and Webster, 1979). Detailed serological studies (Schild *et al.*, 1972; Kendal *et al.*, 1977) have shown that a moderate degree of antigenic drift has occurred in the prevalent swine (H1N1) influenza viruses between 1930 and 1977. Genetic studies of swine influenza viruses isolated from pigs have revealed considerable heterogeneity in the RNAs of the viruses (Hinshaw *et al.*, 1978). The swine (H1N1) viruses from different farms were usually distinctive, suggesting the restricted transmission of viruses between farms.

There is widespread evidence of infection of pigs with human Hong Kong (H3N2) viruses. A/Hong Kong/68 (H3N2)-like viruses were first isolated in 1970 from pigs in Taiwan (Kundin, 1970), and subsequently several of the later antigenic variants of human H3N2 virus have been detected in pigs in almost every country where studies have been done. Although in some cases the studies have indicated a very high frequency of infection (Schild *et al.*, 1972; Harkness *et al.*, 1972), it appears that viruses of the H3N2 subtype cause no significant disease in pigs. It is thought that infection is commonly transmitted from man to pigs. However, viruses resembling closely the 1968 prototype strain of A/HK/68 (H3N2) were isolated from pigs in Hong Kong in 1976 (Shortridge *et al.*, 1977a), several years after their disappearance from man. Thus, it is possible that pigs may serve as a repository of viruses of past human pandemics as well as a possible source of genetic information for the production of recombinant viruses between human and porcine strains of influenza A virus (Kaplan and Webster, 1977). Serological studies of pigs in the United Kingdom in 1980 (D. H. Roberts, unpublished observation) have provided evidence of infection of pigs

*Previous designation Hsw1N1.

with H1N1 viruses resembling A/USSR/77 (H1N1) strains which circulated in humans since 1977.

There is evidence that human H3N2 strains may spread from man to species other than swine. In the Soviet Union, isolations of viruses (e.g., A/duck/Kamchatka/71) antigenically resembling human A/Hong Kong/68 (H3N2) have been made from domestic chickens with respiratory disease in the USSR. In addition, H3N2 viruses have been isolated from calves with pneumonia in the Soviet Union (A/calf/Duschaube/7; Cambell *et al.*, 1977) and from calves in Hungary.

VII. INFLUENZA IN HORSES

Although epizootics among horses referred to as "influenza" were common in the seventeenth to nineteenth centuries, the specific viral etiology of equine influenza was not established until 1956. Initially this was based on the demonstration of serological evidence of infection in horses in Sweden (Heller *et al.*, 1956) and subsequently on the isolation of type A influenza virus from an epizootic of respiratory disease in horses in Czechoslovakia (Sovinova *et al.*, 1958). In the following years, virus isolations and serological evidence indicated that infection by the same agent occurred among horses in other Central European countries, followed by the United States and Western Europe (McQueen *et al.*, 1968). The 1956 isolate was found, on its subsequent antigenic characterization, to possess HA and NA antigens unrelated to those of human influenza A viruses. It is designated in the World Health Organization (1980) nomenclature system as A/equine/Prague/1/56 (H7N7).* This strain is also commonly referred to as the equine-1 virus, being the first subtype of equine influenza virus to be isolated. Equine-1 virus persists up to the present and still produces outbreaks in horses in many areas of the world. The most recently documented outbreak was in England (Powell *et al.*, 1974) and was shown to be caused by a virus exhibiting a minor degree of antigenic difference from the 1956 isolate, suggesting that the equine-1 virus has undergone only a small degree of antigenic drift over the past 18 years. Tumova and Pereira (1968) found minor antigenic relationships between equine-1 virus and certain avian influenza viruses. It is now clearly established that there are indeed antigenic relationships between the NAs of equine-1 and several avian influenza viruses, and they are classified in a single NA subtype (N7). Similarly, Schild *et al.* (1979) showed that the HA antigen of equine-1 virus is related to that of fowl plague virus, and both are classified as H7N7 in the 1980 nomenclature system. This relationship was demonstrated by IDD tests with antibody to purified HA antigens (Schild *et al.*, 1979) (see Fig.

*Previous designation Heq1 Neq1.

1), and also by the fact that chickens immunized with equine-1 influenza virus are immune to the normally lethal effects of infection with fowl plague virus.

In 1963 a second subtype of equine influenza was isolated (Waddell *et al.*, 1963) from extensive outbreaks of influenza in horses in the United States. Later this virus was detected in several South American countries and in Europe. Serological evidence of infection has been detected among the large populations of horses in Mongolia. The new isolate designated A/equine/Miami/1/63 (H3N8)* possesses HA and NA antigens unrelated to those of the equine-1 virus and has been termed the "equine-2 subtype." The disease produced by equine-2 is similar to that of equine-1 virus, being essentially an acute upper respiratory tract infection, but epidemics tend to be more severe and widespread. Infected horses may be off form for several months, and the infection may severely compromise the performance of competition horses. It has been observed by some workers (Tumova *et al.*, 1972; Jensen, 1973) that in some outbreaks both equine-1 and equine-2 viruses may be involved, and furthermore, simultaneous infection with both viruses may occur in individual animals.

Up to 1969, isolates of equine-2 virus were antigenically close to the prototype 1963 isolate. In 1969 equine-2 isolates from Brazil (Pereira *et al.*, 1972) were found to have a significant degree of antigenic difference from the 1963 strain. Later, severe epidemics among racehorses in Tokyo yielded equine-2 strains showing even greater differences (Kono *et al.*, 1972).

Several workers have shown antigenic relationships between the HA antigen of equine-2 virus and that of human A/Hong Kong/1/68 virus, and both are classified as H3 in the 1980 nomenclature. It has also been clearly demonstrated (Tumova and Schild, 1972; Schild *et al.*, 1980) that the NA antigen of equine-2 virus is of the same subtype (N8) as the NAs of several avian influenza strains.

Although all breeds and ages of horses, as well as donkeys and mules, are susceptible to infection with equine influenza viruses, racehorses appear to be particularly affected. Routine vaccination of horses with inactivated equine influenza virus vaccines is practiced in several countries, particularly among racing animals and breeding stock.

VIII. INFLUENZA IN BIRDS

Avian species are the most abundant source of influenza viruses from nonhuman sources, both numerically and in terms of agents of widely different biological and antigenic characteristics. Some of these agents, like fowl plague virus, are responsible for rapidly fatal infections associated with viremia and pantropic

*Previous designation Heq^2 Neq^2.

dissemination of virus. In other cases, infection is apparently without pathology. The pathology of influenza infections in birds has been reviewed extensively by Easterday and Tumova (1972, 1975). Table XI summarizes the origins and clinical symptoms produced by representative numbers of avian influenza virus strains that have been isolated and characterized in several parts of the world. Fowl plague was shown as long ago as the late nineteenth century to be of viral origin (Stubbs, 1965). Later, other agents isolated from ducks with acute sinusitis in Czechoslovakia (Koppel *et al.*, 1956) and in England (Andrewes and Worthington, 1959; Roberts, 1964) were subsequently identified as being type A influenza.

Since these early findings, influenza viruses have been isolated in many countries from a very wide range of domestic birds, including chickens, ducks, turkeys, quail, pheasant, seabirds, and passerines (Pereira, 1969; Easterday and Tumova, 1972, 1975; Easterday, 1975, 1980; Stuart-Harris and Schild, 1976; Laver and Webster, 1979; Hinshaw *et al.*, 1980). More recently isolations have been made in the United States and Western Europe from several species of imported exotic aviary birds including myna birds, parrots, parakeets, cockatoo, and weaver birds (Slemans *et al.*, 1973). The majority of strains were antigenically related to the duck influenza isolate A/duck/England/62 (H3N8), a strain originally described by Roberts (1964). It is not certain whether infection was acquired in their natural habitat or had become endemic in the aviary.

Becker (1963) isolated influenza virus, A/tern/South Africa/61 (H5N3), from an outbreak of disease with a high fatality rate which occurred in the common tern (*Sterna hirundo*) off Cape Town. In contrast, an influenza virus (A/Shearwater/E.Australia/72 (H6N5) isolated off the Australian Barrier Reef from shearwater (*Puffinus pacificus*), a seabird pelagic in the Pacific Ocean (Downie and Laver, 1973), was apparently avirulent in its natural host.

Although most earlier investigations of avian influenza involved studies of domestic or cage birds, it is apparent from recent ecological investigations that wild bird populations harbor a wide range of antigenic varieties of type A influenza virus. Antibodies to influenza antigens have been found in wild geese in North America (Winkler *et al.*, 1972), in various seabirds of the Australian Barrier Reef (Dasen and Laver, 1970), and in a wide variety of migratory birds in the far east regions of the Soviet Union (Slepuskin *et al.*, 1972; Zakstelshaya *et al.*, 1972). During a study of migratory wild duck populations in California, Slemans *et al.* (1974) isolated 43 strains of avian influenza A virus from the respiratory tract in several thousand wild duck.

A wide range of different influenza A viruses have also been isolated from fecal specimens of ducks in Hong Kong and on the Mississippi (Shortridge *et al.*, 1977b; Hinshaw *et al.*, 1980). It has been shown (Webster *et al.*, 1978) that avian strains of influenza virus replicate in the lungs and also in the cells lining the intestinal tract of ducks. The viruses reach the lower intestinal tract despite

TABLE XI
Origin, Host, and Disease Pattern of Some Avian Influenza A Viruses

Virus strain, year of isolation, and antigenic designation	Host of origin	Country of origin	Nature of disease	Reference
1. Fowl plague virus 1927 (Dutch strain, H7N7)	Domestic poultry and wild birds	Indonesia	Generalized, rapidly fatal infection CNS involvement, high mortality	Stubbs (1965)
2. Fowl plague 1934 (Rostock strain, H7N1)	Domestic poultry	Germany	Generalized, rapidly fatal infection, CNS involvement, high mortality	—
3. Virus "N" (Dinter strain N/Germany/49 (H10N7)	Chickens	Germany	Generalized infection, 20% mortality	Dunter (1949)
4. duck/Czech/56 (H4N6)	Ducklings	Czechoslovakia	Sinusitis, high morbidity, high mortality	Koppel *et al.* (1956)
5. duck England/56 (H11N6)	Ducklings	England	Respiratory disease, sinusitis	Andrewes and Worthington (1959)
6. chicken/Scotland/59 (H5N1) (Smith strain)	Adult chickens	Scotland	Fowl-plague-like	Wilson (1960) (unpublished)

7. tern/S. Africa/61 (H5N3)	Common tern *Sterna hirundo*	South Africa	Fowl-plague-like	Wilson (1960, unpublished)
8. duck/Ukraine/1/63 (H3N8)	Ducklings	USSR	Sinusitis	Isachenko, USSR (unpublished)
9. turkey/Canada/63 (H6N8)	Turkeys	Canada	Mild respiratory infection	Lang *et al.* (1965)
10. turkey/England/63 (H7N6)	Turkeys	England	Severe fowl-plague-like	Wells (1963)
11. turkey/Massachusetts/65 (H6N2)	Turkeys	USA	Respiratory disease, low morbidity, low mortality	Pereira *et al.* (1966)
12. quail/Italy/1117/65 (H10N8)	Quail	Italy	Respiratory disease, high morbidity, moderate mortality	Pereira *et al.* (1967)
13. pheasant/Italy/647/66 (H10N2)	Pheasant	Italy	Respiratory disease, high morbidity, low mortality	Rinaldi (unpublished, 1966)
14. parrot/England/70 (H4N6)	Parrot in aviary	England	Fatal infection	Chu (unpublished, 1970)
15. shearwater/Australia/1/72[a] (H6N5)	Shearwater *Puffinus pacificus*	Australian Barrier Reef	Avirulent	Downie and Laver (1973)

[a]Isolates from wild birds.

the low pH of the gizzard and are shed in high concentrations in the feces. Human influenza A viruses replicate in ducks, but only in the upper respiratory tract, not in the intestinal tract (Webster *et al.*, 1978).

In some studies on wild ducks in North America, a very high proportion of healthy birds have yielded influenza A viruses from cloacal samples. Among the influenza viruses isolated from ducks are strains possessing many of the possible combinations of HA and NA antigen subtypes (Hinshaw *et al.*, 1979). Several viruses from duck are as yet unclassified and may constitute new subtypes (Hinshaw *et al.*, 1980). Influenza A viruses have been isolated from water samples collected from lakes frequented by ducks in Canada at the time when the birds were assembling before migration, and viruses were also isolated from fecal samples collected on the shores of these lakes (Hinshaw *et al.*, 1980). The high concentration of duck influenza viruses in fecal material, the isolation of influenza viruses in fecal material, and the isolation of influenza A viruses from lake water offer a possible mechanism for transmission of avian influenza viruses from feral ducks to domestic avian and mammalian species.

As indicated above in the section on nomenclature, the influenza A viruses from avian species so far characterized are divided into 12 HA antigen subtypes (H1 to H12) and 9 NA antigen subtypes (N1 to N9). Of the HA subtypes, four (H1, H2, H3, H7) also have counterparts among influenza A viruses of human, swine, or equine origin, and the remaining eight are so far unique to avian species. For the NA subtypes, four (N1, N2, N7, and N8) are represented among human, swine, or equine viruses, while five are unique to viruses from avian species.

An important finding in regard to avian influenza viruses, and quite unlike the general situation with human, swine, or equine viruses, is that a given HA antigen is not always associated with the same NA antigen (see Table X). Thus among strains possessing a common NA antigen may be two or more subtypes of HA; for example, the classic fowl plague virus has the antigenic composition H7N7, whereas chicken/Germany/N/49 virus is H10N7. Strains that possess common HA antigens may have different NA antigens (Table X). It is striking that there is no correlation between antigenic character and host of origin, and antigenically identical strains have been isolated from several different species of bird (Pereira, 1969; Hinshaw *et al.*, 1980). A further extremely variable feature of avian influenza viruses is their virulence, which again is not necessarily associated with antigenic character. Recently strains of virus antigenically close to the classic and highly virulent fowl plague virus (H7N7) have been found to be completely avirulent. Besides fowl plague, other viruses produce rapidly fatal infections in birds. Among these are A/tern/South Africa/61 (H5N3), A/turkey/England/63 (H7N3), and A/turkey/Ontario/7732/66 (H5N9). Both A/tern/South Africa/61 and A/turkey/England/63 are highly virulent for laboratory mice and hamsters (Uyo and Becker, 1967).

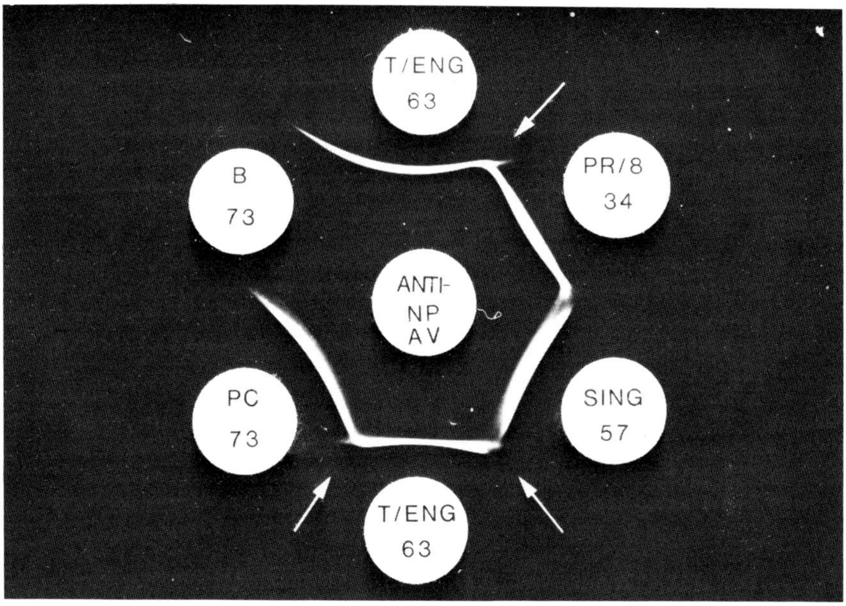

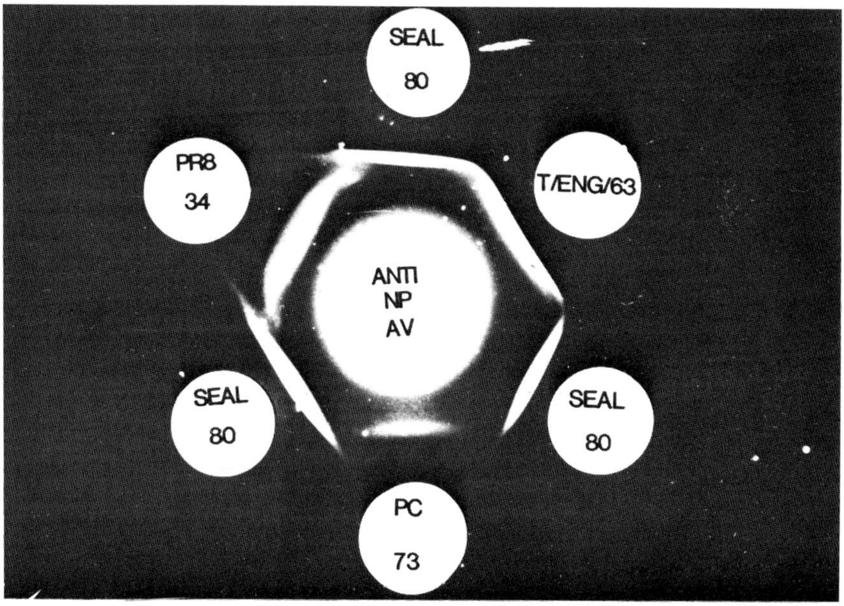

Figs. 10–11. IDD reactions of antiserum (anti-NP AV) to the purified NP antigen of avian (A/turkey/England/63) influenza virus. Fig. 10 shows antigenic differences (spurs marked by arrow) between the NP antigens of A/turkey/England/63 (H7N3) and those of various human influenza viruses A/PR8/34 (H1N1) A/Singapore/1/75 (H2N2) and A/Port Chalmers/1/73 (H2N2). Fig. 11 shows that the NP antigen of A/seal/Massachusetts/80 (H7N7) virus is closely similar to that of avian influenza virus (A/turkey/England/63) and distinct from that of human viruses A/PR8/34 (H1N1) and A/Port Chalmers/73 (H3N2). This provides supportive evidence for the origin of the seal virus from avian sources.

IX. INFLUENZA IN OTHER SPECIES

Influenza infections in nonhuman species other than swine, horses, and birds have been infrequently confirmed. However, a few observations are worthy of note here.

The most striking of these concerns epizootic pneumonia in seals. Between November 1979 and June 1980 (R. G. Webster, personal communication) approximately 500 common seals (*Phoca vitulina*) were found to have died on the New England coast of the United States. Apparently up to 20% mortality occurred in the seals, and the fatalities peaked in January 1980. Influenza A virus designated as A/seal/Massachusetts/1/80 (H7N7*), antigenically close to the Dutch strain of fowl plague virus (Tables VIII and IX), was isolated from several individuals and was present in high titers in the lungs together with a number of bacterial species, including *Mycoplasma*. Isolates were also made from brain tissue. Pathological examination of the lungs of dead animals showed evidence of primary viral pneumonia. The virus was found to be avirulent for domestic avian species, including chickens, ducks, and turkeys. Examination of the NP antigen of the A/seal/Massachusetts/80 virus by IDD tests indicated that it was antigenically identical to the NP of avian influenza viruses distinct from that of human strains. This finding (G. C. Schild, unpublished data; see also Figs. 10, 11) suggests that the seal virus may be of avian origin. The outbreak represents the first convincing evidence of an apparently avian influenza A virus infection in a mammalian species, but its origin from avian sources requires confirmation. Whether seals have previously been infected with H7N7 viruses remains to be established, as does the contribution to the severity of the disease produced by nonviral agents, bacteria, and *Mycoplasma*.

Other evidence of the involvement of large sea mammals is provided by the isolation of an H1N7 virus from a whale in the Pacific Ocean (A/whale/Pacific Ocean/76; Lvov *et al.*, 1978). This agent, however, may have come from avian sources, and its isolation remains an interesting observation of uncertain significance.

X. EVIDENCE THAT INFLUENZA A VIRUSES FROM NONHUMAN SOURCES MAY BE PROGENITORS OF HUMAN INFLUENZA

This fascinating area of influenza virology has been reviewed in detail elsewhere (see Stuart-Harris and Schild, 1976; Laver and Webster, 1979). Since 1918 three pandemic viruses have circulated in man: H1N1 virus probably from

* Previous designation Hav1 Neq1.

1918 to 1956, Asian (H2N2) virus from 1957 to 1968, and Hong Kong (H3N2) virus from 1968. In 1977 H1N1 viruses closely resembling antigenically and genetically human H1N1 strains which previously had circulated around 1950 (Nakajima *et al.*, 1978; Young and Palese, 1979) reemerged as epidemic viruses of children and young adults and circulated for the next several years (Pereira, 1979). The source of the H1N1 virus in 1977 is uncertain, but several possible explanations for the preservation of its genome in nature have been proposed (Laver and Webster, 1979). These explanations include its escape from laboratory sources, its persistence in infectious form in animals or birds, or its possible latency in a human or nonhuman host, or even in a parasite of man.

Evidence that the 1918–1919 pandemic virus was related to swine (H1N1) virus is described above in this chapter and by Stuart-Harris and Schild (1976). In 1976 evidence of the potential of H1N1 swine (H1N1) virus from pigs to infect man was obtained. Viruses antigenically close to swine (H1N1) virus were isolated from young military recruits at Fort Dix, New Jersey (Topp and Russell, 1977). There is strong circumstantial evidence that the source of infection was pigs. The outbreak was limited, involving some 500 recruits with serological evidence of infection, of whom a small proportion developed clinical influenza and one fatal case occurred. Unequivocal evidence for the infection of humans by swine (H1N1) viruses from porcine sources has been obtained by the isolation of genetically identical "swine" viruses from pigs and man on a farm in Wisconsin (Hinshaw *et al.*, 1978).

Concerning the Asian (H2N2) pandemic virus, there is evidence (reviewed by Laver and Webster, 1979) that it possessed four genes probably derived from the former human H1N1 virus, i.e., genes 1, 5, 7, and 8, coding for polypeptides P1 NP, MP, and the NS proteins (see Table I). The remaining genes 2, 3, 4, and 6, coding for P2, P3, HA, and NA, were probably derived from an unidentified influenza A virus possessing H2N2 surface antigens. Recent studies have shown that the genes for H2 and N2 are present in certain avian influenza strains, including viruses from duck isolated in Hong Kong (K. F. Shortridge, personal communication). There is thus strong circumstantial evidence that the pandemic Asian virus arose as a recombinant between the former human H1N1 virus and a possibly avian virus possessing H2N2 genes.

In 1968 the Hong Kong (H3N2) virus was found to possess an antigenically distinct HA antigen (H3), but its NA antigen was close to that of the human H2N2 viruses of 1967. Serological evidence and biochemical analyses have shown that H3 HA is closely related to the HA of certain avian influenza viruses, A/duck/Ukraine/1/63 (H3N7), and also to that of equine-2 virus, i.e., A/equine/Miami/1/63 (H3N8). Other studies suggest that the human Hong Kong (H3N2) virus contains seven RNA genes closely similar to the corresponding genes of the human H2N2 virus; only gene 4, coding for HA, was not homologous for the two viruses. It is thus possible that the H3N2 virus arose as a result of recombination between the human H2N2 virus (which donated seven of the eight genes) and a virus from another source (which donated the H3 gene). This finding provides probably the strongest circumstantial evidence yet available that lower mammals or birds may have a role in the origin of human influenza pandemics. However, it is clear that the mechanism of genetic recombination cannot be invoked as a universal explanation for the emergence of new human viruses since the reemerged human H1N1 virus of 1977 clearly did not originate in this manner.

Diagnostic and Serological Studies

Detailed technical accounts of the methodology for the laboratory diagnosis of influenza in mammals and man have been published elsewhere (World Health Organization, 1959; Easterday and Tumova, 1972, 1975; Aymard *et al.*, 1973; Palmer *et al.*, 1975; Schild and Dowdle, 1975; Stuart-Harris and Schild, 1976). It should be emphasized, however, that the isolation of viruses from clinical material in fertile eggs or cell cultures should be attempted wherever possible. Serology provides valuable confirmatory evidence of the etiology of influenza infections when used diagnostically or in serosurveys, but serological evidence alone has often been found to be of limited value.

REFERENCES

Allen, H., McCauley, J., Waterfield, M., and Gelhing, M. -J. (1980). *Virology* **107,** 548–551.

Andrewes, C. H., and Worthington, G. (1959). *Bull. W. H. O.* **20,** 435–440.

Andrewes, C. H., Laidlaw, P. P., and Smith, W. (1934). *Lancet* **2,** 859–861.

Andrewes, C. H., Laidlaw, P. P., and Smith, W. (1935). *Br. J. Exp. Pathol.* **16,** 566–582.

Aymard, M., Coleman, M. T., Dowdle, W. R., Laver, R. G., Silver, G. C., and Webster, R. G. (1973). *Bull. W. H. O.* **48,** 199–205.

Baker, N., Store, H. O., and Webster, R. G. (1973). *J. Virol.* **11,** 137–142.

Becker, W. B. (1963). *Virology* **20,** 318–321.

Beveridge, W. I. B. (1977). "Influenza: The Last Great Plague." Heinemann, London.

Burnet, F. M. (1940). *J. Exp. Pathol.* **21,** 147–152.

Burnet, F. M. (1941). *J. Exp. Biol. Med. Sci.* **19,** 291–295.

Cambell, C. H., Easterday, B. C., and Webster, R. G. (1977). *J. Infect. Dis.* **135,** 678–680.

Choppin, P. W., and Compans, R. W. (1975). *In* "The Influenza Viruses and Influenza" (E. D. Kilbourne, ed.), pp. 15–51. Academic Press. New York.

Creighton, C. (1894). "A History of Epidemics in Britain," Vol. 2. Frank Cass, London (2nd ed., 1965).

Dasen, C. A., and Laver, W. G. (1970). *Bull. W. H. O.* **42,** 885–891.

Davenport, F. M., Hennessy, A. V., and Francis, T., Jr. (1953). *J. Exp. Med.* **98,** 641–650.

Doherty, P. C., Effros, Rita B., and Bennink, J. (1977). *Proc. Nat. Acad. Sci. U.S.A.* **74,** 1209–1213.

Dowdle, W. R., Galphin, J. C., Coleman, M. T., and Schild, G. C. (1974). *Bull. W. H. O.* **51,** 213–218.

Downie, J. C., and Laver, W. G. (1973). *Virology* **51,** 259–269.

Dinter, Z. (1949). *Tieraerztl. Umsch.* **4,** 185–190.

Easterday, B. C. (1975). *In* "The Influenza Viruses and Influenza" (E. D. Kilbourne, ed.), pp. 449–477. Academic Press, New York.

Easterday, B. C. (1980). *In* "Influenza," Eds. H. G. Pereira and D. A. J. Tyrrell, pp. 145–151. Royal Society, London.

Easterday, B. C., and Tumova, B. (1972). *In* "Diseases of Poultry" (M. S. Hofstad, ed.), 6th ed., pp. 670–700. Baillière, London.

Easterday, B. C., and Tumova, B. (1978). *In* "Diseases of Poultry" (M. S. Hofstad, ed.), 7th ed., pp. 549–576. Baillière, London.

Francis, T., Jr. (1940). *Science* **92,** 405–409.

Francis, T., Jr., and Shope, R. E. (1936). *J. Exp. Med.* **63,** 645–652.

Gething, M. J., Bye, Jackie, Skekel, J., and Waterfield, M. (1980). *In* "Structure and Variation in Influenza Virus" (W. G. Laver and G. Air, eds.), pp. 1–10. Elsevier, Amsterdam.

Gottschalk, A. (1954). *Yale J. Biol. Med.* **26,** 352.

Gottschalk, A. (1957). *Physiol. Rev.* **37,** 66–69.

Harkness, J. W., Schild, C. C., Lamart, P. B., and Brand, C. M. (1972). *Bull. W. H. O.* **46,** 709–719.

Heller, L., Espmark, A., and Vriden, P. (1956). *Arch. Gesamte Virusforsch.* **7,** 120–130.

Hinshaw, V. S., and Webster, R. G. (1979). *J. Gen. Virol.* **45,** 751–754.

Hinshaw, V. S., Bean, W. J., Jr., Webster, R. G., and Easterday, B. C. (1978). *Virology* **84,** 51–62.

Hinshaw, V. S., Webster, R. G., and Turner, B. (1979). *Intervirology* **11,** 66–68.

Hinshaw, V. S., Webster, R. G., and Rodriguez, R. J. (1980). *Arch. Virol.* **63,** 165–170.

Hirst, G. K. (1941). *Science* **94,** 22–24.

Jensen, K. (1973). *Acta Vet. Scand.* **14,** 205–210.

Kaplan, M. M., and Beveridge, W. I. B. (1972). *Bull. W. H. O.* **47,** 439–443.

Kaplan, M. M., and Webster, R. G. (1977). *Sci. Am.* **237,** 88–106.

Kendal, A. P., Noble, G. R., and Dowdle, W. R. (1977). *Virology* **82,** 111–121.

Koen, J. S. (1919). *J. Vet. Med.* **14,** 468–472.

Kono, Y., Ishikawa, K., Fukunaya, Y., and Fujino, M. (1972). *Natl. Inst. Anim. Health Q.* **12,** 183–190.

Koppel, Z. J., Vriak, V. M., and Spiesz, S. (1956). *Veterinarstvi* **6,** 267–276.

Kundin, W. D. (1970). *Nature (London)* **228,** 857–859.

Laidlaw, P. P. (1935). *Lancet* **1,** 1118–1121.

Lamb, R. A., and Choppin, P. W. (1980). *In* "Influenza," pp. 39–49. Royal Society, London.

Lang, G., Furgusson, A. E., Connell, M. C., and Wills, C. G. (1965). *Avian Dis.* **9,** 495.

Laver, W. G., and Air, G., eds. (1980). "Structure and Variation in Influenza Virus." Elsevier, Amsterdam.

Laver, W. G., and Valentine, R. C. (1969). *Virology* **38,** 105–110.

Laver, W. G., and Webster, R. G. (1973). *Virology* **51,** 383–391.

Laver, W. G., and Webster, R. G. (1979). *Br. Med. Bull.* **35,** 29–34.

Lvov, D. D., Zhdanov, V. M., Sazonov, A. A., Braude, N. A., Pysina, T. V., and Osherovich, A. M. (1978). *Prob. Virol. (USSR)* **2,** 151–156.

McQueen, J. L., Steele, J. H., and Robinson, R. Q. (1968). *Adv. Vet. Sci.* **12,** 285–290.

Melnick, J. (1973). *Prog. Med. Virol.* **15,** 380.

Min Jou, W. *et al.* (1980). *Cell* **19,** 683–696.

Nakajima, K., Desselberger, U., and Palese, P. (1978). *Nature (London)* **274,** 334–339.

Palese, P., Racaniello, V. R., Desselberger, U., Young, J., and Baez, M. (1980). *In* "Influenza," pp. 11–17. Royal Society, London.

Palmer, D. F., Dowdle, W. R., Coleman, M. T., and Schild, G. C. (1975). "Advanced Laboratory

Techniques for Influenza Diagnosis," Immunol. Ser. No. 6. Centre for Disease Control, U.S. Dept. of Health, Education and Welfare, Atlanta, Georgia.
Pereira, H. G. (1969). *Prog. Med. Virol.* **11,** 46–79.
Pereira, H. G., Lang, G., Olesiuk, O. M., Snoeybus, G. H., Roberts, D. H., and Easterday, B. C. (1966). *Bull. W. H. O.* **35,** 799–805.
Pereira, H. G., Tumova, B., and Webster, R. G. (1967a). *Nature (London)* **215,** 982–986.
Pereira, H. G., Renald, A., and Nordelli, L. (1967b). *Bull. W. H. O.* **37,** 553–558.
Pereira, H. G., Takimoto, S., Piegas, N. S., and Ribeiro Do Valle (1972). *Bull. W. H. O.* **47,** 465–470.
Pereira, M. S. (1979). *Br. Med. Bull.* **35,** 9–14.
Porter, A. G. *et al.* (1979). *Nature (London)* **284,** 471–477.
Potter, C. W., and Oxford, J. S. (1979). *Br. Med. Bull.* **35,** 69–76.
Powell, D. G., Thompson, G. R., Plowright, W., Burrows, R., and Schild, G. C. (1974). *Vet. Rec.* **94,** 282–290.
Proc. Nat. Acad. Sci. U.S.A. **76,** 6547–6551.
Roberts, D. H. (1964). *Vet. Rec.* **76,** 470–476.
Rott, R. (1980). *In* "Influenza," pp. 105–112. Royal Society, London.
Russell, S. M., and Liew, F. Y. (1979). *Nature (London)* **280,** 147–148.
Schäfer, W. (1955). *Z. Naturforsch.* **10,** 81–85.
Schild, G. C. (1970). *J. Gen. Virol.* **9,** 191–200.
Schild, G. C. (1972). *J. Gen. Virol.* **15,** 99–103.
Schild, G. C. (1979). *Postgrad. Med. J.* **55,** 87–97.
Schild, G. C., and Dowdle, R. W. (1975). *In* "The Influenza Viruses and Influenza" (E. D. Kilbourne, ed.), pp. 316–368. Academic Press, New York.
Schild, G. C., Brand, C. M., Harkness, J. W., and Lamont, P. B. (1972). *Bull. W. H. O.* **46,** 720–730.
Schild, G. C., Oxford, J. S., and Newman, R. W. (1979). *Virology* **93,** 569–573.
Schild, G. C., Newman, R. W., Webster, R. G., Major, D., and Hinshaw, V. S. (1980). *Arch. Virol.* **63,** 171–184.
Scholtissek, C. (1978). *Curr. Top. Microbiol. Immunol.* **80,** 139–169.
Schultz, I. T. (1975). *In* "The Influenza Viruses and Influenza" (E. D. Kilbourne, ed.), pp. 53–79. Academic Press, New York.
Shope, R. E. (1931). *J. Exp. Med.* **54,** 349–359.
Shortridge, K. F., and Webster, R. G. (1979). *Intervirology* **11,** 9–15.
Shortridge, K. F., Webster, R. G., Butterfield, W. K., and Cambell, C. H. (1977a). *Science* **196,** 1454–1455.
Shortridge, K. F., Butterfield, W. K., Webster, R. G., and Cambell, C. H. (1977b). *Bull. W. H. O.* **55,** 15–20.
Skehel, J. J. (1971). *J. Gen. Virol.* **11,** 103–107.
Skehel, J. J., Waterfield, M. D., McCauley, J. W., Elder, K., and Wiley, D. C. (1980). *In* "Influenza," pp. 47–53. Royal Society, London.
Sleigh, M. J. *et al.* (1980). *In* "Structure and Variation in Influenza Virus" (W. G. Laver and G. Air, eds.), pp. 69–80. Elsevier, Amsterdam.
Slemans, R. D., Cooper, R. S., and Osborn, J. S. (1973). *Avian Dis.* **17,** 746–755.
Slemans, R. D., Johnson, D. C., Osborn, J. S., and Hayes, F. (1974). *Avian Dis.* **18,** 119–122.
Slepuskin, A. N., Rysina, T. V., Gonsovsky, F. K., Sazanov, A. A., Isachenko, V. A., Sokolova, N. N., and Lvov, D. D. (1972). *Bull. W. H. O.* **47,** 527–531.
Smith, W., Andrewes, C. H., and Laidlaw, P. P. (1933). *Lancet* **2,** 66–68.
Sovinova, O., Tumova, B., Pouska, F., and Nemec, J. (1958). *Acta Virol.* **2,** 52–60.
Stuart-Harris, C. H., and Schild, G. C. (1976). "Influenza: The Viruses and the Diseases." Arnold, London.

Stubbs, E. L. (1965). *In* "Diseases of Poultry" (H. E. Biester and D. D. Schwarte, eds.), 5th ed., pp. 813–822. Iowa State Univ. Press, Ames.
Taylor, R. M. (1949). *Am. J. Public Health* **39,** 171–175.
Thompson, E. S. (1890). "Influenza: An Historical Survey of Past Epidemics in Great Britain from 1510 to 1890." Percival, London.
Topp, F. H., Jr., and Russell, P. K. (1977). *J. Infect. Dis.* **136,** S376–S380.
Tumova, B. (1976). *J. Gen Virol.* **32,** 217–225.
Tumova, B., and Pereira, H. G. (1968). *Virology* **27,** 253–260.
Tumova, B., and Schild, G. C. (1972). *Bull. W. H. O.* **47,** 453–460.
Tumova, B., Easterday, B. C., and Stumpa, A. (1972). *Am. J. Epidemiol.* **95,** 80–86.
Tyrrell, D. A. J., and Smith, J. W. G. (1979). *Br. Med. Bull.* **35,** 77–86.
Uyo, C. J., and Becker, W. B. (1967). *J. Comp. Pathol. Ther.* **77,** 167–180.
Virelizier, T. L., Allison, A. G., and Schild, G. C. (1979). *Br. Med. Bull.* **35,** 65–69.
Waddell, G. H., Teigland, M. B., and Siegel, M. M. (1963). *J. Am. Vet. Assoc.* **143,** 587–590.
Webster, R. G., Tumova, B., Hinshaw, V. S., and Lang, G. (1976a), *Bull. W. H. O. 54,* 555–560.
Webster, R. G., Yakhno, M., Hinshaw, V. S., Bean, W. J., and Murti, K. G. (1978). *Virology* **84,** 268–278.
Webster, R. G. *et al.* (1979). *Virology* **96,** 258–265.
Webster, R. G., Maurita, M., and Pridgen, C.
Wells, R. J. H. (1963). *Vet. Rec.* **75,** 783–786.
Winkler, W. G., Trainer, D. O., and Easterday, B. C. (1972). *Bull. W. H. O.* **47,** 507–512.
World Health Organization (1959). *W. H. O. Tech. Rep. Ser.* **170.**
World Health Organization (1971). *Bull. W. H. O.* **45,** 119–126.
World Health Organization (1979). *Bull. W. H. O.* **57,** 227–233.
World Health Organization (1980). *Bull. W. H. O.* **58,** (in press).
Wrigley, N. G. (1980). *Br. Med. Bull.* **35,** 35–39.
Wrigley, N. G., Skehel, J. J., Charlwood, P. A., and Brand, C. M. (1973). *Virology* **51,** 525–531.
Young, J. F., and Palese, P. (1979).
Zakstelshaya, L. Y., Isachenko, V. A., Ozidze, N. G., Timofeeva, C. C., Slepuskin, A. N., and Sokdova, N. N. (1972). *Bull. W. H. O.* **4,** 497–501.

Part IV

PARAMYXOVIRIDAE

Chapter 5

Paramyxovirus and Pneumovirus Diseases of Animals and Birds: Comparative Aspects and Diagnosis

GLYNN H. FRANK

I. Introduction ... 187
II. Family Paramyxoviridae ... 188
A. Genus *Paramyxovirus* ... 194
B. Genus *Pneumovirus* ... 195
III. Paramyxoviruses of Animals ... 195
A. Parainfluenza-3 Virus ... 195
B. Canine SV-5 Virus ... 201
C. Paramyxoviruses of Rodents ... 204
IV. Paramyxoviruses of Birds ... 211
A. Newcastle Disease Virus ... 211
B. Other Paramyxoviruses of Birds ... 218
V. Pneumoviruses of Animals ... 221
A. Bovine Respiratory Syncytial Virus ... 221
B. Pneumonia Virus of Mice ... 226
References ... 229

I. INTRODUCTION

Paramyxoviruses and pneumoviruses infect man and a broad spectrum of animals and birds. Those that primarily infect humans are covered by Kelen and McLeod in the first volume of this series and will not be considered in this

COMPARATIVE DIAGNOSIS
OF VIRAL DISEASES, VOL. IV
ISBN 0-12-429704-8

chapter. Of those that infect animals and birds, only the more virulent strains of Newcastle disease virus cause a definite, recognizable disease in their host, the chicken. The others are involved in various nondescript respiratory diseases, which can vary from asymptomatic infections to severe clinical conditions. Usually, experimental infection of the susceptible host results in a very mild respiratory disease of short duration and of little consequence. In most cases of natural respiratory disease involving paramyxoviruses and pneumoviruses, other etiological agents are isolated along with the virus, making the exact roles of each difficult to assess. Even though the exact roles of the viruses are not known, they are an integral part of the disease complexes in which they are involved.

Antibodies to most of the paramyxoviruses and pneumoviruses are commonly found in their hosts, indicating that the viruses are highly infectious and easily transmissible. It is also common for host animals in colonys or herds to seroconvert, even in the absence of a history of respiratory disease, indicating that the viruses are usually of little consequence in the healthy, unstressed host.

II. FAMILY PARAMYXOVIRIDAE

For many years, certain of the paramyxoviruses and pneumoviruses have been studied both in their relationship to disease and as models for basic studies in virology and immunology. Some, such as Newcastle disease virus, Sendai virus, and SV-5 virus, have been studied quite extensively, while others have scarcely been investigated. The available data have been extensively reviewed by the Paramyxovirus Study Group of the International Committee on Taxonomy of Viruses (ICTV) and other investigators (Fenner, 1977; Kingsbury *et al.*, 1978). A simplified classification scheme, based on the ICTV's recommendations, is presented in Table I. It is by no means complete because there are viruses for which comparative data are not available. This is especially true with some of the newly or rarely isolated avian paramyxoviruses. To make matters more confusing, certain of the paramyxoviruses were first isolated from a naturally infected laboratory host animal being used for attempted virus isolation from another species. Such was the case with both Sendai virus and pneumonia virus of mice. Paramyxoviridae have been isolated from nonhuman primates which are highly susceptible to a number of human Paramyxoviridae. The origin of such infections is questionable. The true origin of newly isolated Paramyxoviridae must be carefully determined, since the members of the family infect a broad spectrum of species.

Information on classification and characteristics presented here is from the detailed work of many virologists as presented in several reviews (Fenner, 1977; Kingsbury, 1977; Andrews *et al.*, 1978; Kingsbury *et al.*, 1978).

Members of the family Paramyxoviridae are pleomorphic in shape, although

TABLE I
Classification of the Paramyxoviridae

Taxonomic status: Family	Approved name: Paramyxoviridae	Common name
Genus	*Paramyxovirus*	Paramyxovirus group
Members		Natural host
Newcastle disease virus		Avian
Mumps		Human
Parainfluenza-1 (Sendai)		Human, murine
Parainfluenza-2 (SV-5)		Human, canine
Parainfluenza-3		Human, bovine, ovine
Parainfluenza-4		Human
Yucaipa		Avian
Parainfluenza turkey/Ontario		Avian
(Other avian paramyxoviruses)		
Genus	*Morbillivirus*	Measles-distemper group
Members		Natural host
Measles virus		Human
Canine distemper virus		Canine
Rinderpest virus		Ungulates
Pest des petits ruminants		Caprine, ovine
Genus	*Pneumovirus*	Respiratory syncytial virus group
Members		Natural host
Respiratory syncytial virus		human
Bovine respiratory syncytial virus		bovine, ovine
Pneumonia virus of mice		murine

usually roughly spherical, and approximately 100–200 nm in diameter (Fig. 1). Very large aberrant forms as well as filamentous forms up to several micrometers in length are common (Fig. 2). The nucleocapsid ("herringbone") has a helical symmetry with a diameter of 12–17 nm, depending upon the genus (Fig. 3). Virions are studded with surface projections 8–12 nm in length and usually 8–10 nm apart unless they are disarranged by surface changes in the virion's envelope (Fig. 4).

The viral genome is one continuous single-stranded RNA molecule with a molecular weight of $5\text{–}8 \times 10^6$. Replication takes place in the cytoplasm of the infected cell, and maturation occurs by budding through the virus-modified cell surface membrane.

Three genera are included in the family Paramyxoviridae. They are *Paramyxovirus* (the type species is Newcastle disease virus), *Morbillivirus* (the

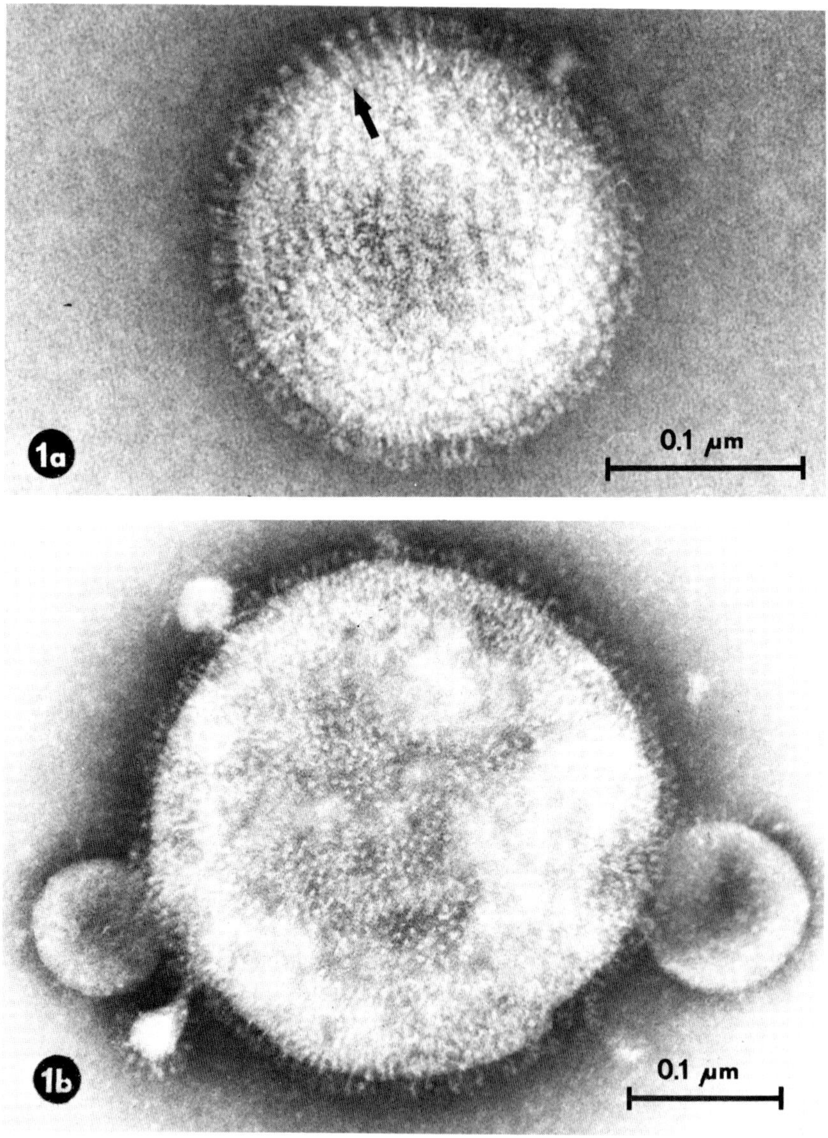

Fig. 1. (a) Newcastle disease virion with spherical symmetry and close spacing of its surface projections (arrow); (b) respiratory syncytial virion illustrating common, but minor, blebbing of the envelope. (Courtesy of A. E. Ritchie.)

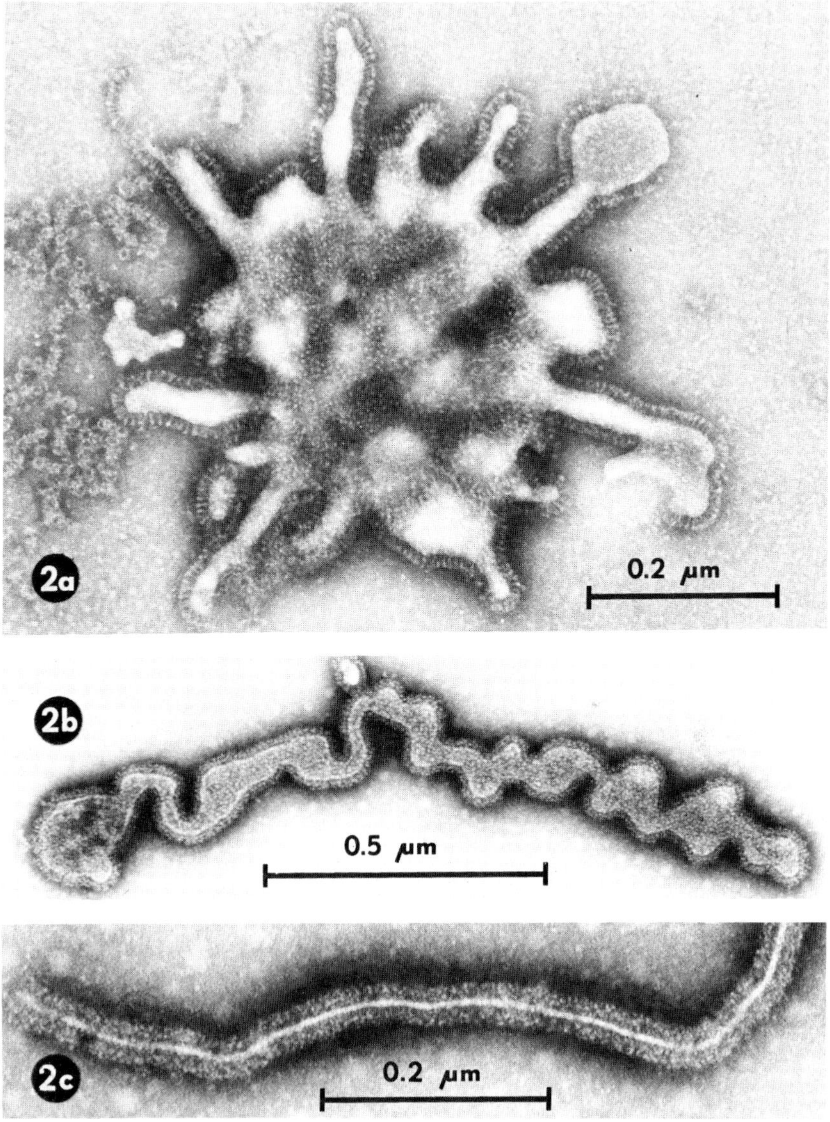

Fig. 2. (a) Respiratory syncytial virus illustrating aberrant pleomorphic distension; (b) parainfluenza-3 virus illustrating filamentous pleomorphism; (c) parainfluenza-3 virus illustrating a limiting form of tubular distension. (Courtesy of A. E. Ritchie.)

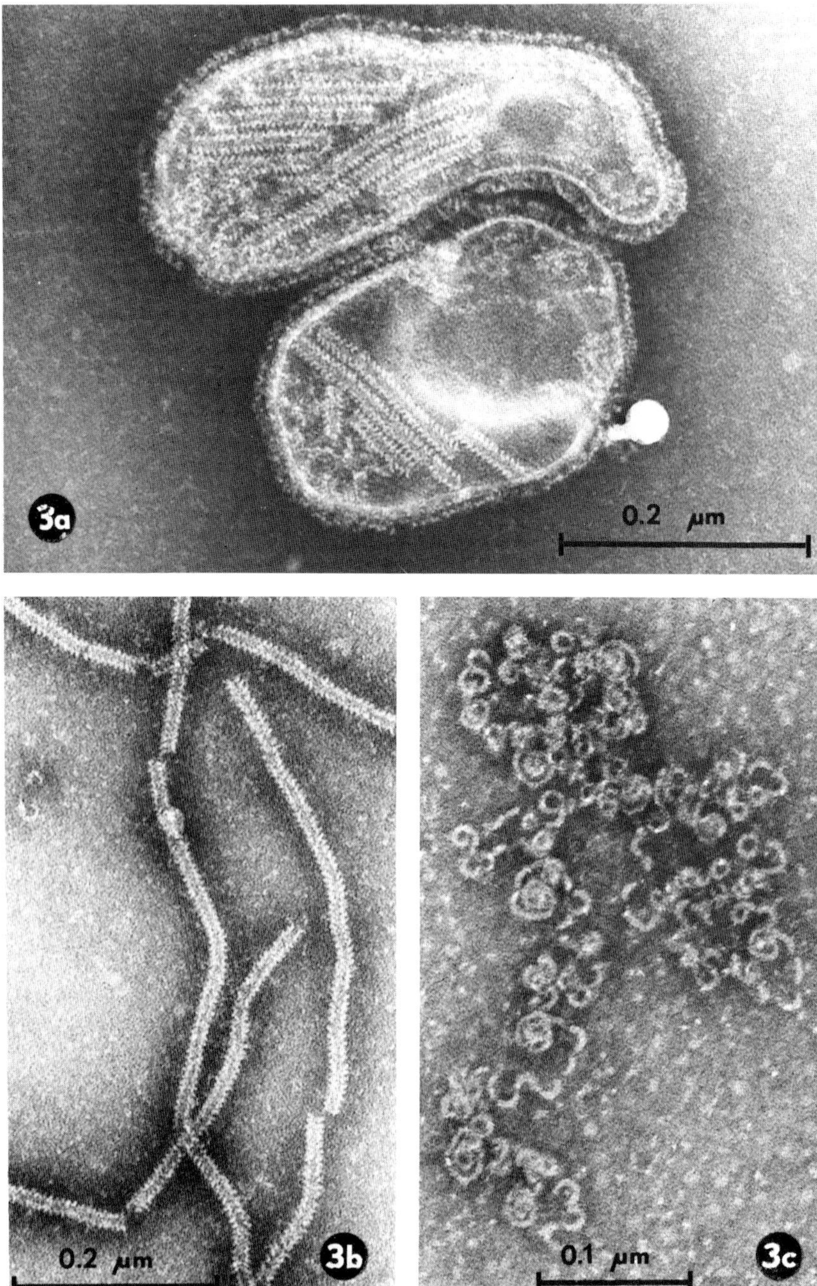

Fig. 3. (a) Newcastle disease virions with internal fragments of "herringbone," presumably nucleocapsid structure; (b) parainfluenza-3 virus "herringbone" not associated with a virion, illustrating fractures induced by handling; (c) respiratory syncytial virus nucleocapsid undergoing typical uncoiling. (Courtesy of A. E. Ritchie.)

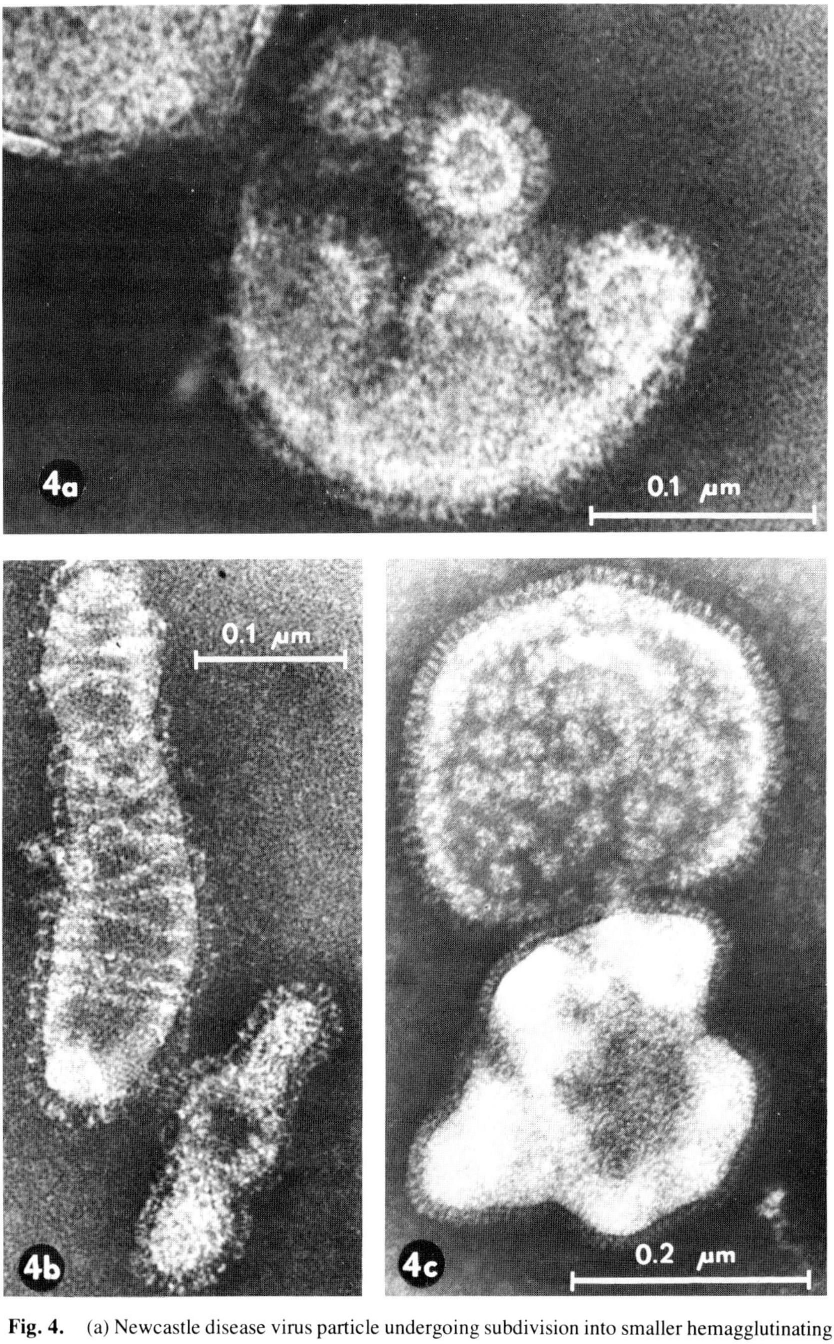

Fig. 4. (a) Newcastle disease virus particle undergoing subdivision into smaller hemagglutinating units without major alteration in the packing of the surface projections; (b) Newcastle disease virus particles illustrating packing differences of the surface projections. Note their linear association on the larger particle; (c) parainfluenza-3 virus with distended surface bleb, illustrating clusters of the disarranged surface projections. (Courtesy of A. E. Ritchie.)

type species is measles virus), and *Pneumovirus* (the type species is respiratory syncytial virus). Only members of genus *Paramyxovirus* and genus *Pneumovirus* which naturally infect animals and birds will be considered in this chapter.

Ample biochemical and physical data for classification purposes and knowledge of the process of replication are known for a few species of the genus *Paramyxovirus*. Those most studied have been Sendai virus, SV-5 virus, and Newcastle disease virus. Much less is known about other Paramyxoviridae, and especially members of the genus *Pneumovirus*.

A. Genus *Paramyxovirus*

1. Morphology

Members of the genus *Paramyxovirus* are pleomorphic, usually roughly spherical viruses, approximately 120–300 nm in diameter. The nucleocapsid is 17–18 nm in diameter, arranged in a helical symmetry, and covered with a lipid-containing envelope studded with projections 8 nm in length and 8–10 nm apart.

2. Nucleocapsid

The nucleocapsid contains a genome consisting of a single species of single-stranded 50 S RNA with a molecular weight of $5\text{-}6 \times 10^6$. The RNA is covered with protein subunits of 60,000 molecular weight arranged as a rod-shaped structure held in a helical symmetry. An RNA-dependent RNA polymerase (transcriptase) which synthesizes viral mRNA is part of the nucleocapsid.

3. Viral Envelope

The viral envelope consists of a modified segment of cell surface membrane containing virus-specified proteins. Biological activities associated with the proteins of the viral envelope are hemagglutination, neuraminidase, hemolysin, and cell fusion. Hemagglutination and neuraminidase activities are both associated with a common glycosylated polypeptide which are present as projections on the envelope surface and have a molecular weight of 70,000. Hemolysin and cell fusion are associated with another glycosylated polypeptide present as a surface projection of 50,000 molecular weight. A nonglycosylated protein of 40,000 molecular weight is present and is probably internal to the envelope.

4. Physicochemical Properties

Weight and sedimentation coefficients of paramyxoviruses (PMV) vary according to their size and the stability of the population being measured. PMVs are at least 5×10^8 in molecular weight, and the sedimentation coefficient is at least 1000 S. Their buoyant density ranges from 1.18 to 1.25 gm/cm^3. The virions are sensitive to lipid solvents, nonionic detergents, formaldehyde, and oxidizing agents.

5. *Replication*

PMVs gain entry into the cell by fusion of the viral envelope with that of the cell surface membrane. The RNA genome is a negative stand, considering the mRNA or viral message as being the positive strand. The intact viral nucleocapsid is the template for transcription of a spectrum of mRNA species and contains transcriptase which synthesizes the mRNA. The nucleocapsid is also the template for complete transcript of the viral genome, which in turn serves as the template on which replicas of the viral genome are produced. Synthesis of a new protein is necessary before the genome is replicated. Nucleocapsids containing the viral genome travel to an area beneath the cell surface membrane where viral envelope proteins have been inserted. Maturation occurs when the nucleocapsid is enclosed by budding through the virus-modified cell surface membrane.

B. Genus *Pneumovirus*

Pneumoviruses are pleomorphic, roughly spherical viruses approximately 80–500 nm in diameter, and sometimes filaments 60–110 nm in diameter and up to 5 μm in length are seen. The nucleocapsid is 12–15 nm in diameter, smaller in diameter than that of PMV. The nucleocapsid is covered with a lipid-containing envelope studded with projections 10–12 nm in length. Intact virions have a buoyant density of 1.15–1.26 depending on the virus preparation and buoyant medium.

The genus *Pneumovirus* differs from the genus *Paramyxovirus* primarily in the size of the nucleocapsid and the lack of neuraminidase. Hemagglutinating activity has been described for pneumonia virus of mice, but not for respiratory syncytial virus.

III. PARAMYXOVIRUSES OF ANIMALS

The PMVs of animals covered in this chapter are parainfluenza-3 (PI-3) virus of cattle and sheep, parainfluenza-1 (Sendai) virus and other PMVs of rodents, and parainfluenza-2 (SV-5) virus of dogs.

A. Parainfluenza-3 Virus

1. *History*

Parainfluenza-3 (PI-3) virus was first isolated from cattle in the United States and characterized by Reisinger *et al.* (1959). Abinanti and Huebner (1959) reported that the virus had the general properties of type 1 hemadsorption virus isolated from children with respiratory disease as described by Chanock *et al.* (1958). Numerous reports from other countries indicated that PI-3 virus was worldwide in distribution. In nearly all cases, PI-3 virus was isolated from cattle suffering from respiratory disease. The virus was isolated from sheep with respiratory disease (Hore, 1966; Hore *et al.*, 1968), from horses with acute respiratory infections (Ditchfield *et al.*, 1963; Sibinovic *et al.*, 1965), and from Egyptian water buffalo calves with pneumoenteritis (Singh and Baz, 1967). Serological surveys revealed PI-3 virus antibody in some wild deer in the United States and in India (Shah *et al.*, 1965).

Bovine strains of PI-3 virus were distinguished serologically from the human HA-1 type (Abinanti *et al.*, 1961; Kelter *et al.*, 1961). Thus far, the bovine isolates as well as those from other animals fall into one serological group.

2. *Properties*

a. Host Tissue Range. Bovine strains of PI-3 virus grow well and cause a cytopathic effect (CPE) in a variety of primary cells and established cell lines from a number of animal species. Kidney cells used for PI-3 virus isolation and propagation include those of sheep, water buffalo, camel, goat, monkey, dog, pig, and cat.

The CPE usually consists of syncytia formation and development of intracytoplasmic and intranuclear inclusion bodies (Reisinger *et al.*, 1959; Inaba *et al.*, 1963; Dawson, 1964; Singh and Baz, 1967; Singh and El Cicy, 1967; Kita and Gillespie, 1968). Churchill (1963) found that the bovine strain of PI-3 virus produced eosinophilic intracytoplasmic and intranuclear inclusions in syncytia formed in HEp2 cell cultures and in primary monkey kidney cells, but was unable to demonstrate intranuclear inclusions with a human strain of PI-3 virus.

In embryonic bovine kidney cells, two types of CPE have been observed (Fig. 5). One is a rounding of individual cells scattered throughout the monolayer, and the other consists mainly of large syncytia (Dinter *et al.*, 1960; Inaba *et al.*, 1963; Dawson, 1964; Frank, 1970). Frank (1970) associated the two types of CPE with different plaque types but found that serial passage of progeny from either plaque type resulted in a mixture of the two after a few passages. A viral morphogenesis study showing sequential development and maturation of PI-3 virus in embryonic bovine kidney cells was conducted by Tsai and Thompson (1975a).

Bovine PI-3 virus was grown in chicken embryos inoculated by the amniotic route (Marshall, 1964). One isolate was passaged in suckling mice by intracerebral inoculation (Schiøtt and Jensen, 1962).

b. Hemagglutination. Bovine PI-3 virus hemagglutinates the erythrocytes (RBC) of many species, including cow, guinea pig, human O, chicken, turkey, sheep, pig, rabbit, goose, goat, pigeon, and mouse RBC. Various PI-3 virus isolates have differed in their hemagglutination (HA) spectrums, and the HA

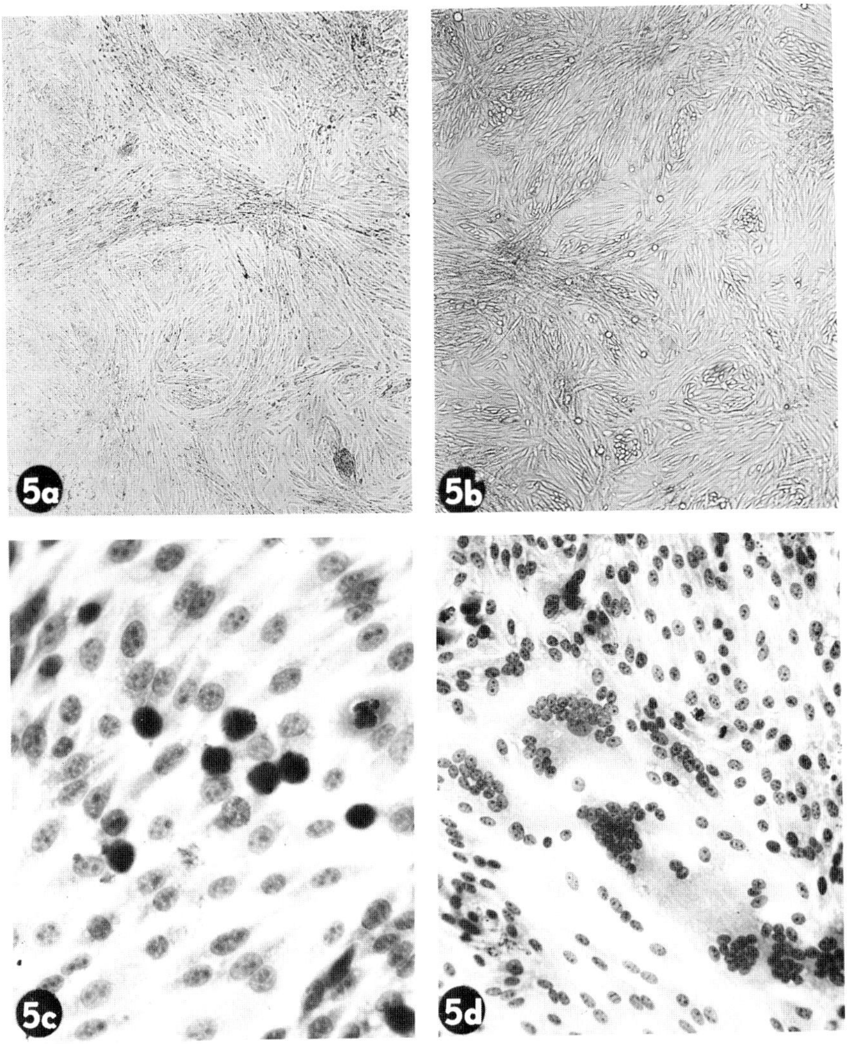

Fig. 5. Embryonic bovine kidney cells infected with bovine PI-3 virus. (a) Uninfected monolayer, unstained. × 63; (b) infected monolayer showing scattered, rounded cells, unstained. × 63; (c) rounded, pyknotic cells, MGG stain. × 450; (d) syncytium formation, MGG stain. × 250.

spectrum has been used as a strain marker. Frank (1970) found the HA spectrum not to be a stable characteristic of a PI-3 virus population since it changed with passage of the virus.

3. *Disease*

a. Natural. Under field conditions, PI-3 virus is often isolated from cattle with respiratory diseaee, and in the United States it has been associated with the respiratory disease complex known as "shipping fever," which causes serious economic losses. The condition usually occurs after cattle are removed from their home farms, mixed with cattle from other sources, and then transported to feedyards for fattening.

Usually within 10 days after arrival at the feedyard, cattle may develop respiratory disease characterized by fever, coughing, dyspnea, and a mucopurulent nasal discharge. Those dying of acute respiratory disease usually have an extensive fibrinous bronchopneumonia.

Agents commonly isolated from the respiratory disease complex along with PI-3 virus include *Pasteurella* spp., *Mycoplasma* spp., and other bovine respiratory disease viruses. The association of PI-3 virus with respiratory disease in cattle has been the subject of several reviews (Reisinger, 1962; Sinha and Abinanti, 1962; Hamdy *et al.*, 1965; Nugyen-Ba-Vy, 1967a,b; Woods, 1968; Gale, 1970; Frank and Marshall, 1973).

Other isolations of PI-3 virus from naturally infected cattle include isolation from the testicles of an infertile bull (Deas *et al.*, 1966) and from the milk and nasal secretions of cattle with acute respiratory illness (Kawakami *et al.*, 1966a,b). The virus has occasionally been recovered from bovine and ovine fetuses, as reviewed by Swift (1973).

As in cattle, PI-3 virus is involved in respiratory diseases in sheep. The virus was isolated from newly weaned lambs which had been moved into close quarters with adult sheep (Hore, 1966). It was isolated from older lambs undergoing a natural outbreak of respiratory disease in which *Pasteurella haemolytica* as well as *Mycoplasma* spp. were involved (Hore *et al.*, 1968).

Natural PI-3 virus infections, uncomplicated by other active disease agents and stress factors, usually result in a rise in serum antibody titer, with no history of noticeable signs of respiratory disease (Frank and Marshall, 1973). Antibody to PI-3 virus is very common in cattle in many parts of the world. The same seems to be true in sheep, since serological surveys have shown antibody to be commonly present (Fischman, 1965; Hore, 1966). In a flock of sheep monitored for seroconversion to PI-3 virus antibody, 9-month-old seronegative lambs seroconverted within a 4-month period. No PI-3 virus was isolated by periodic sampling and no respiratory disease occurred (Fischman, 1967). Antibody was found in a serological survey of horses (Sibinovic *et al.*, 1965).

b. Experimental. Infection of cattle with PI-3 virus under experimental conditions has been reviewed by Frank and Marshall (1973). For best results in causing an experimental disease, a large amount of virus must be administered in such a manner as to reach the lungs, and the calf must be fully susceptible. Exposure of colostrum-deprived seronegative calves to an aerosol of 3μ droplets generated from stocks containing 10^8 to 10^9 plaque forming units (PFU) of PI-3 virus per milliliter resulted in fevers beginning on the second day and lasting for 7 to 10 days. Respirations were shallow and rapid, and viral shedding occurred for 8 to 10 days. Most calves recovered completely in 10 days (Baldwin *et al.*, 1967; Frank and Marshall, 1971). Exposure to less virus in the same manner resulted in viral shedding but no clinical illness (Frank and Marshall, 1971).

Many workers have exposed seronegative cattle to PI-3 virus by various routes and methods. Generally, those producing a clinical disease describe it as a mild respiratory involvement with fever, nasal discharge, and viral shedding. Others report viral shedding in the absence of clinical illness.

Lesions resulting from experimental infection of calves with PI-3 virus have been described (Dawson *et al.*, 1965; Woods *et al.*, 1965b; Omar *et al.*, 1966; Stevenson and Hore, 1970). Grossly, lung lesions are usually in the ventral portions of the cranial lobes and vary from congestion to consolidation, with congestion of the tracheal mucosa and nasal mucous membranes. Histopathologic features include hypertrophy and hyperplasia of the bronchiolar and alveolar epithelium, along with multinucleated epithelial cells. Intracytoplasmic and intranuclear inclusion bodies occur in the epithelial and syncytial cells. Epithelial hyperplasia and hypertrophy, viral inclusions, and syncytial cells in the lungs are the most distinctive morphological features of PI-3 viral pneumonia (Omar *et al.*, 1966). An ultrastructural study of viral pathogenesis and bovine respiratory tissue responses to PI-3 virus infection was conducted by Tsai and Thomson (1975b).

After experimental exposure by the respiratory route, virus was usually isolated from the lung, trachea, larynx, and nasal turbinates, and less frequently from the tonsil, retropharyngeal lymph nodes, and bronchial lymph nodes in calves killed 3 to 9 days after infection (Omar *et al.*, 1966; Van Der Maaten, 1969). The virus can be isolated from the lungs for a longer period of time than from other tissues.

The experimental disease produced in sheep is a mild upper respiratory tract disease, much like that produced in calves, with the same viral shedding pattern. Lesions in experimentally infected sheep were much like those described in calves (Woods *et al.*, 1965a; Hore and Stevenson 1967, 1969; Faye and Charton, 1967; Stevenson and Hore, 1970). However, unlike the lung lesions in calves, intranuclear inclusion bodies were not observed in lung lesions in sheep (Hore and Stevenson, 1967, 1969).

Experimental infections of calves and sheep with other disease agents (especially *Pasteurella haemolytica*) in combination with PI-3 virus have usually resulted in a more severe disease and more severe lesions than those produced by either agent alone (Baldwin *et al.*, 1967; Biberstein *et al.*, 1971; Davies *et al.*, 1977).

4. Diagnosis

a. Virus Isolation. Virus can be isolated from nasal swabs of infected cattle and sheep during the course of clinical respiratory disease. Usually, viral isolation attempts are not successful from normal or recovering animals. Cattle and sheep experimentally exposed by the respiratory route shed virus for approximately 8–10 days. Susceptible cells, usually bovine or ovine kidney cells, are inoculated with the virus-containing material, then observed for the expected CPE. The necessity of cell passage for virus detection will depend upon the amount of virus in the inoculum. Usually, it is not necessary. Hemadsorption with bovine or guinea pig RBC can be a useful technique.

b. Serology. Since the time of virus isolation is critical, one may look for an increase in serum antibody titer during convalescence. Paired serum samples are essential since most cattle will have a serum antibody titer before the disease occurs.

Standard hemagglutination inhibition (HI) or serum neutralization (SN) procedures are usually used. HI tests are usually run with bovine or guinea pig RBC. Some bovine sera contain nonspecific HA inhibitors. In high-titered sera, nonspecific HA inhibitors are diluted out, but in low-titered or negative sera, they can lead to confusion (Frank, 1966). Procedures to remove this inhibitor include heat treatment at 56°C for 30 minutes, kaolin adsorption, and treatment with receptor-destroying enzyme. Some sera must be absorbed with RBC to remove nonspecific hemagglutinins. Frank (1966) reported that a combination of heat treatment (56°C for 30 min.) and kaolin adsorption was necessary to remove nonspecific HA inhibitors from sera of colostrum-deprived, seronegative calves.

5. *Immunity*

a. Naturally Acquired Antibody. Under natural conditions, young calves are very likely to have maternal antibody to PI-3 virus. Maternal antibody lessens the severity of or prevents clinical illness in calves after an aerosol challenge exposure (Dawson *et al.*, 1965; Marshall and Frank, 1975). Even though calves with maternal antibody can become infected and shed virus, their serum antibody levels do not increase during convalescence (Marshall and Frank, 1975).

Older calves which have undergone a natural PI-3 virus infection are likely to have local antibody in the respiratory secretions as well as serum antibody. Frank and Marshall (1971) found that the nasal secretion antibody level better reflected resistance to illness than the serum antibody level. Cattle with diminished serum antibody levels and undetectable nasal secretion antibody levels can undergo clinical illness as a result of aerosol challenge exposure (Woods *et al.*, 1965a; Frank and Marshall, 1971).

b. Vaccination. A number of vaccines containing modified live or killed PI-3 virus are available for cattle. While such products may afford some protection against experimental challenge exposure, their ability to lessen the severity of or prevent natural respiratory disease involving PI-3 virus is still questionable. One may question the need for routine vaccination when antibody to the virus is commonly found in most cattle.

B. Canine SV-5 Virus

1. History

Morris *et al.* (1956) found antibody to mumps virus in canine sera and demonstrated specific antibodies in dogs after experimental infection. In a serological survey for antibody to human PMV in canine sera, HI antibodies against PI-1, PI-2, PI-3, DA, and mumps viruses were found (Cuadrado, 1965).

Thus far, only one PMV has been isolated from naturally infected dogs. It was isolated from throat swabs of canine distemper and infectious canine hepatitis-vaccinated dogs with an upper respiratory tract illness (Binn *et al.*, 1967). The isolate was very closely serologically related to SV-5 virus. Other reports of isolating canine SV-5 viruses from dogs with upper respiratory tract disease soon followed (Binn *et al.*, 1968; Crandell, et al., 1968; Appel and Percy, 1970; Appel *et al.*, 1970; Binn *et al.*, 1970; Saona Black and Lee, 1970; Cornwell *et al.*, 1976; Emery *et al.*, 1976; McCandlish *et al.*, 1978).

2. Properties

a. Host Cell Range. Canine SV-5 viruses have usually been isolated in canine kidney cells, both primary and established cell lines (Binn *et al.*, 1967, 1968; Crandell *et al.*, 1968; Saona Black and Lee, 1970; Cornwell *et al.*, 1976). Other successfully used cells include primary green monkey kidney (Binn *et al.*, 1967, 1968), rhesus monkey kidney (Crandell *et al.*, 1968), feline kidney (Saona Black and Lee, 1970), and a human foreskin cell line (Crandell *et al.*, 1968). The virus was also isolated by inoculating 8- to 9-day-old chicken embryos in the amniotic cavity. No embryonic deaths occurred in 5 days, but hemagglutinins were present in the allanto-amniotic fluids (Crandell *et al.*, 1968).

Canine SV-5 viruses have been propagated in primary dog kidney (Saona Black and Lee, 1970; Emery *et al.*, 1976), feline kidney (Saona Black and Lee, 1970), and canine kidney cell lines (Rosenberg *et al.*, 1971; Cornwell *et al.*, 1976).

b. Hemagglutination. Canine SV-5 viruses hemagglutinate RBC from a number of species, but guinea pig RBC are most frequently used. Positive HA occurs with human O, chicken, guinea pig, rat, rabbit, dog, cat, and sheep RBC at 4 and 24°C. Guinea pig RBC were found to be the most sensitive (Crandell *et al.*, 1968).

c. Antigenic Relationship to Other PMV. Several serological techniques using various types of reference antisera have been used to determine the antigenic relationship of canine SV-5 viruses to other PMV. They include SN using rabbit antisera in canine and human embryo kidney cell lines (Binn *et al.*, 1967), reciprocal HI tests with ferret reference antisera (Binn *et al.*, 1968), guinea pig, and rabbit antisera (Crandell *et al.*, 1968), reciprocal HI, SN, and complement fixation (CF) tests with guinea pig antisera (Saona Black and Lee, 1970), reciprocal HI and SN tests with ferret reference antisera (Lazar *et al.*, 1970), reciprocal HI and hemadsorption neutralization tests with guinea pig antisera (Rosenberg *et al.*, 1971), and reciprocal HI tests with known commercial SV-5 reference sera (Cornwell *et al.*, 1976).

Generally, the results with various canine SV-5 virus isolates show a close antigenic relationship to simian SV-5 virus isolates and to each other. They cannot be distinguished from simian SV5 virus isolates serologically since the antigenic differences between canine and simian isolates are not as great as those among some simian isolates (Lazar *et al.*, 1970).

3. *Disease*

a. Natural. Canine SV-5 virus is one of the disease agents involved in the kennel cough complex in dogs. Kennel cough frequently occurs in dogs soon after they are congregated from different sources. It is characterized by a dry cough which lasts for 1 to 2 weeks. The morbidity is high, but the mortality is very low. Other agents commonly isolated from dogs with kennel cough include *Bordatella bronchiseptica, Mycoplasma* spp., canine adenovirus types 1 and 2, canine herpesvirus, and reovirus type 1. Review articles on kennel cough have been published (Appel *et al.*, 1970; Thompson *et al.*, 1975; Appel and Bemis, 1978). Occasionally, in a few individual dogs, the canine SV-5 virus has been the only agent isolated.

Lesions in dogs with naturally occurring kennel cough (Cornwell *et al.*, 1976; McCandlish *et al.*, 1978) and in dogs experimentally infected with canine SV-5 virus alone (Appel and Percy, 1970; Rosenberg *et al.*, 1971) have been described.

In naturally occurring kennel cough, pulmonary lesions have varied from slight edema and focal congestion to broncho-pneumonia and consolidation. Tracheobronchitis is the most common finding and is characterized by mucosal congestion and a mucoid or mucopurulent exudate. Tonsils and regional lymph nodes may be enlarged and congested. Focal tracheobronchitis and intra-alveolar hemorrhage have been described in dogs from which only canine SV-5 virus was isolated.

In dogs experimentally inoculated with canine SV-5 virus, petechial hemorrhages hve been observed in the lungs 3 to 4 days after exposure. Microscopic examination has revealed some inflammatory responses in the upper and lower respiratory tract.

b. Experimental. Infection of dogs with canine SV-5 virus isolates by the respiratory route has resulted in only very mild respiratory disease. Clinical signs have ranged from none (Saona Black and Lee, 1970) to swollen, reddened tonsils for 1 to 2 weeks (Crandell *et al.*, 1968), with a slight serous nasal discharge lasting for 1 to 2 weeks (Crandell *et al.*, 1968; Appel and Percy, 1970; Lazar *et al.*, 1970) and a mild fever lasting for 2 to 3 days (Appel and Percy, 1970; Emery *et al.*, 1976). The virus can usually be isolated from nasal or throat swabs for approximately 1 week. The virus is readily transmitted to contact dogs during this time (Appel and Percy, 1970; Lazar *et al.*, 1970; Rosenberg *et al.*, 1971). Parenteral injection of the virus has not resulted in clinical illness or viral shedding (Appel and Percy, 1970; Emery *et al.*, 1976). Intranasal instillation of the virus in cats (Saona Black and Lee, 1970) and in ferrets (Lazar *et al.*, 1970) has not resulted in signs of disease.

4. Diagnosis

a. Virus Isolation. The virus is isolated by infecting primary or established cultures of canine kidney cells with throat or nasal swab material from live dogs. Tissues from which virus is commonly isolated include lung and in some cases tonsil, spleen, kidney, and liver, although lung is the tissue of choice (Binn *et al.*, 1967; Cornwell *et al.*, 1976; McCandlish *et al.*, 1978).

After inoculation, cells are observed for CPE for 1 to 2 weeks, then checked for hemadsorption with guinea pig RBC (Binn *et al.*, 1967; Crandell *et al.*, 1968). Usually there is little or no evident CPE. Some swollen, refractile cells and multinucleate cells have been seen, some of which contain eosinophilic cytoplasmic inclusion bodies (Cornwell *et al.*, 1976). Syncytia and eosinophilic cytoplasmic inclusions have developed in primary canine, feline, and embryonic bovine kidney cells and in a feline tongue cell line (Saona Black and Lee, 1970), and after passage in primary rhesus monkey kidney cells (Crandell *et al.*, 1968).

The canine SV-5 virus is identified by reciprocal serological procedures with known reference antisera. Usually the HI procedure with guinea pig RBC is used.

b. Serology. Serum antibody to canine SV-5 virus is measured by routine serological procedures, including SN (Binn *et al.*, 1967, 1968; Crandell *et al.*, 1968; Saona Black and Lee, 1970; Emery *et al.*, 1976), neutralization of hemadsorption (Rosenberg *et al.*, 1971; Fulton *et al.*, 1974), HI (Binn *et al.*, 1967; Saona Black and Lee, 1970; Rosenberg *et al.*, 1971), and CF (Saona Black and Lee, 1970). A conglutinating-complement-adsorption procedure was used to measure antibody in feline sera (Saona Black and Lee, 1970).

5. Immunity

Dogs with serum antibody titers are at least partially immune to canine SV-5 virus. The status of local antibody usually is not known, but dogs developing a serum antibody response after a natural or experimental aerosol exposure probably develop a local antibody response in the respiratory tract as well. In one study, dogs with serum antibody did not shed virus after experimental exposure (Lazar *et al.*, 1970), but in another study, dogs given a second aerosol exposure 3 weeks after the first did shed virus for a short period of time (Appel and Percy, 1970). Intramuscular vaccination with canine SV-5 virus resulted in a serum antibody response. Upon aerosol challenge exposure, the dogs did not exhibit signs of disease and shed virus for a shorter period of time than did nonvaccinates (Emery *et al.*, 1976). Vaccination against canine SV-5 virus may be desirable when young dogs will possibly be mixed with dogs from other sources.

C. Paramyxoviruses of Rodents

1. History

Known PMV which are found in rodents include Sendai virus (murine parainfluenza type 1), Peromyscus virus, Nariva virus, and J virus. Sendai virus was first isolated in Japan from mice used for passage of virus isolation materials from humans suffering from upper respiratory tract diseases. Fukumi *et al.* (1954) isolated the virus from naturally infected mice, described it, and recognized it as a virus of murine origin. Sendai virus was first named "hemagglutinating virus of Japan" by the Society of Japanese Virologists. Serological surveys demonstrated the prevalence of Sendai virus antibody in mouse colonies in Japan (Fukumi *et al.*, 1962).

A mouse breeding colony in the United States was found to have a high incidence of antibody to Sendai virus (Parker *et al.*, 1964a). A survey of other mouse colonies in the United States resulted in isolation of Sendai virus and confirmed its prevalence (Parker *et al.*, 1964b).

Natural and experimental disease caused by Sendai virus in mice and other laboratory animals and the persistence of the virus in laboratory animal colonies have been studied. Sendai virus infection of mice has been used as a virus-host model to study viral pneumonias and pneumonias caused by combined viral and bacterial infections.

Comparatively little is known about the other PMV of rodents, since little has been done beyond their isolation, description, and limited host range studies. Peromyscus virus was isolated in Virginia from healthy white-footed mice (*Peromyscus leucopus*) which had no gross lesions (Morris *et al.*, 1963). Nariva virus was isolated from a forest rodent (*Zygodontomys brevicauda*) in Trinidad (Tikasingh *et al.*, 1966). The J virus was isolated from moribund wild mice in North Queensland, Australia (Jun *et al.*, 1977). Of these, only the J virus had been associated with disease in its host.

2. Properties

a. Host Cell Range. Sendai virus readily replicates in primary monkey kidney cells (Fukumi and Nishikawa, 1961; Parker *et al.*, 1964b) and HeLa cells (Fukumi and Nishikawa, 1961). Growth medium for primary monkey kidney cells sometimes includes 0.2% anti-SV-5 virus serum (Parker *et al.*, 1964b). Baby hamster kidney (BHK-21) cells were reported to be as sensitive as primary monkey kidney cells (Bhatt Jonas, 1974). Sendai virus grew in mouse peritoneal macrophages, causing no CPE, but addition of an excessive amount of Sendai virus to the infected monolayer resulted in syncytia formation (Mims and Murphy, 1973). The virus is usually detected by hemadsorption after a period of observation for CPE.

Embryonating chicken eggs are also used for virus isolation as well as for propagation. Fukumi *et al.* (1954) showed both the amniotic and allantoic routes of inoculation to be effective, but the amniotic route to be a little more sensitive. Later, it was determined that the most rapid growth and highest hemagglutinating titers were obtained by inoculating 8-day-old chicken embryos by the amniotic route (Fukumi *et al.*, 1962). After 4 days of incubation, the amniotic and allantoic fluids contain hemagglutinins (Fukumi *et al.*, 1962; Parker *et al.*, 1964b).

Sendai virus has been propagated in several systems for use in mouse studies. It was grown in embryonating eggs (Parker *et al.*, 1966), in mice (Van Nunen and Van Der Veen, 1967), and in primary monkey kidney cells (Parker and Reynolds, 1968).

b. Hemagglutination. Sendai virus hemagglutinates the RBC of many species. Positive HA occurs with chicken, human O, guinea pig, mouse, rat, hamster, rabbit, cow, and sheep RBC, but not with horse RBC (Fukumi *et al.*, 1954).

c. Antigenic Relationship to other PMV. Sendai virus was found to share CF antigens with HA-2 (human parainfluenza 1) virus, but guinea pig anti-Sendai virus serum did not inhibit HA by HA-2 virus. Rabbit anti-Sendai virus serum neutralized HA-2 virus to a much lesser extent than it neutralized Sendai virus, and the reverse was true with rabbit anti-HA-2 virus serum (Chanock *et al.*, 1958). Although Andrews *et al.* (1959) found the viruses to be very similar in their biological characteristics, Fukumi and Nishikawa (1961) described several characteristics which distinguish one from the other. HA-2 virus was more heat and acid pH stable than Sendai virus. Sendai virus grew more rapidly and to a higher titer in HeLa cells than did HA-2 virus. Sendai virus produced CPE in HeLa cells, whereas HA-2 virus did not. One strain of HA-2 virus tested did not grow well in embryonating chicken eggs.

3. Disease

a. Natural. In a 13-year survey in the United States, more than 60% of the mouse, rat, and hamster colonies and 44% of the guinea pig colonies tested were seropositive for Sendai virus (Parker *et al.*, 1968). In seropositive mouse colonies, over 50% of the individual mice were seropositive (Parker *et al.*, 1966; Parker and Reynolds, 1968). Of 400 sera from wild mice, however, none were positive (Parker *et al.*, 1966; Parker and Reynolds, 1968).

Two types of Sendai virus infection patterns in mouse colonies were described by Parker and Reynolds (1968). The most common was an enzootic type in which nearly all susceptible mice became infected between 1 and 2 months of age, then remained antibody positive for life. The other, an epizootic type, occurred as a sudden infection in a mouse colony, persisted for 7 months, then disappeared. No illness was attributable to the epizootic. Epidemic periods of infection lasting 3–4 months with no outward signs of illness were described in mouse colonies in Japan by Fukumi *et al.* (1962).

Mice which become infected with Sendai virus under natural conditions usually show no signs of disease (Parker and Reynolds, 1968; Van Der Veen *et al.*, 1970). However, Bhatt and Jonas (1974) described an epizootic in a closed colony of mice which consisted of morbidity and mortality in suckling mice. Litters of mice 8 to 14 days old underwent respiratory distress and died in a few hours. Litter mortality peaked at 2 weeks, then declined. Natural outbreaks sometimes occur after mice are shipped. Two outbreaks were described in which 21-day-old mice became ill within a week of arrival and lost weight (Ward, 1974).

Generally, Sendai virus seems to have the same relationship with other laboratory animals as with mice. Hamsters have developed Sendai virus antibody without showing signs of disease. The virus has been isolated from the lungs of apparently normal newborn hamsters (Profeta *et al.*, 1969) and from rats (Parker *et al.*, 1966). Sendai virus is transmissible among rats but causes no signs of disease (Tyrrell and Coid, 1970).

b. Experimental. For study, mice and other laboratory animals have been inoculated intranasally, by aerosol, by contact exposure, intracranially, intravenously, and intraperitoneally with Sendai virus. Results of infection depend upon the origin and dosage of the virus inoculum and upon host susceptibility as well.

Mice have been killed by aerosol or intranasal exposures to Sendai virus, and the lethal dose depended upon whether the virus was egg-propagated or mouse-propagated. Mouse lethal doses for egg-propagated Sendai virus have been reported as being $5\times 10^{6.5}$ EID_{50} for day-old mice (Fukumi *et al.*, 1962), 2–4×10^4 PFU (Van Nunen and Van Der Veen, 1967), and 10^6–10^7 EID_{50} (Mims and Murphy, 1973). Lesser doses have been infectious but nonlethal. However, mouse lethal doses for mouse lung-propagated virus have been reported as 50 EID_{50} for day-old mice (Fukumi *et al.*, 1962) and 2×10^3 PFU (Van Nunen and Van Der Veen, 1967). Aerosol exposure to infectious doses of virus can result in development of lesions in the absence of signs of systemic disease (Appell *et al.*, 1971). Deaths from lethal aerosol or intranasal doses of Sendai virus usually occur 6–10 days after exposure (Fukumi *et al.*, 1962; Mims and Murphy, 1973).

Contact exposure to experimentally infected mice led to infection with no clinical disease or mortality (Parker and Reynolds, 1968; Van Der Veen, *et al.*, 1970; Iida, 1972). Intravenous or intraperitoneal injection of 10^7 EID_{50} of virus did not kill mice, but they developed a highly viscous peritoneal exudate. Intracranial inoculation with 10^7 EID_{50} caused illness and convulsions in 1 to 2 days and some deaths by 5 days. Others recovered by 7 days (Mims and Murphy, 1973).

Host factors influencing the course of Sendai virus infection in mice include differences in genetic susceptibility and the immunological capacity of the mouse. Parker *et al.* (1978) found the occurrence of clinical disease and mortality to be directly related to mouse strain, whereas permissiveness to infection was not. In genetically nude mice, naturally occurring Sendai virus infection resulted in a chronic wasting disease, with virus persisting in the host for weeks (Ward *et al.*, 1976). Attempts to reactivate Sendai virus infection in preexposed and recovered mice by intranasal instillation of broth and cyclophosphamide and hydrocortisone treatments were not successful (Van Der Veen *et al.*, 1974). The presence of other pathogens will influence the course of disease. Mice exposed by aerosol to 10^4 $TCID_{50}$ of egg-passaged Sendai virus, followed by an exposure to *Pasteurella pneumotropica* 6 days later, died with massive lung consolidation within 2 to 4 days (Jakab, 1974).

Lung lesions were produced in rats by intranasal exposure to 10^4 EID_{50} of egg-passaged virus. Clinical signs included respiratory distress and a staring hair coat (Tyrrell and Coid, 1970).

Virus isolation studies have been conducted to determine the location of virus in host tissues after infection. After intranasal exposure, virus was recovered from the lungs of 1-day-old mice from 1 to 14 days and from the lungs of 4-week-old mice for up to 8 days (Sawicki, 1962). In contact-exposed mice, virus was isolated from the lungs and saliva for 5 to 11 days and could be infrequently isolated from the kidney and liver (Parker and Reynolds, 1968). Van Der Veen *et al.* (1970) isolated Sendai virus from nasal washings and from lungs by 5 days in contact-exposed mice and found the peak virus titers on days 6 to 9. In aerosol-exposed mice, virus was recovered from nasal washings by day 2. Nasal wash viral titers peaked at day 5, were markedly reduced by day 9, and were negative by day 12. The same general pattern occurred in the lung, with virus recovery beginning on day 1 and peak titer occurring on day 6 (Appell *et al.*, 1971).

Blandford and Heath (1972) observed the Sendai virus distribution in mice at intervals after intranasal exposure by immunofluorescent staining. Viral antigen was on the luminal surface of bronchial mucosal cells by day 1 and in the cytoplasm by day 2, and antigen-containing sloughed cells were found in the bronchial lumen by day 3. Most viral antigen in the lungs was seen at days 3 and 4. Heaviest infection was in the trachea and bronchi. Some viral antigen was seen in alveolar macrophages and in the spleen.

Lesions in intranasally infected mice were first described by Fukumi *et al.* (1954). They involved primarily the bronchiolar epithelial cells and began at 2 days after infection. Robinson *et al.* (1968) infected mice intranasally with egg-passaged Sendai virus and gave a detailed report on development of lesions.

Generally, changes were first observed in the cells of the bronchial mucosa at 24 hours and progressed to severe degenerative changes by the fifth day. They were accompanied by cellular debris appearing in the bronchial lumen. The changes were marked with peribronchial and periarterial cuffing of inflammatory cells. Signs of systemic disease were not observed.

Appell *et al.* (1971) reported adhesions between the serosa of the lung and pleura 2 days after aerosol exposure of mice to Sendai virus. Microscopically, lesions were rhinitis, tracheitis, bronchitis, and interstitial pneumonitis. Small eosinophilic intracytoplasmic inclusions were seen in the tracheal and bronchial epithelium. Tissue lesions varied from mild to severe and resolved rapidly.

Mims and Murphy (1973) described the pathology in mice infected by various routes. After intranasal instillation of Sendai virus into adult mice, lesions were largely confined to the epithelial lining of the larger airways. In suckling mice, lesions extended into the terminal air spaces.

Bhatt and Jonas (1974) reported a Sendai virus epizootic with high morbidity and mortality in a closed mouse colony. Significant lesions were restricted to the lung with mild to moderate bronchial epithelial necrosis, with degeneration and sloughing of cells into the lumen.

Ward (1974) described the lesions of pneumonia in two natural clinical outbreaks of Sendai virus infection in mice after arrival from a commercial breeding laboratory. Upon necropsy, the mice had variable degrees of pneumonia. Histologically, there was hyperplasia of the bronchiolar epithelium, with necrosis and ulceration.

Parker *et al.* (1978) described the sequential development of lesions in intranasally exposed mice of two strains, one (129/J) which developed clinical disease and the other (SJL/J) which remained clinically normal. Qualitatively, the microscopic lesions were the same, but there were quantitative and chronological differences. Generally, the lesions included hypertrophy to massive sloughing of the bronchial epithelial cells and loss of alveolar architecture at the time of most severe damage. The lesions appeared earlier, were more extensive, and persisted longer in the more sensitive strain of mice. Lung lesions in a young adult mouse killed 5 days after intranasal inoculation with Sendai virus are shown in Fig. 6.

Shimokata *et al.* (1977) produced an infection of the middle ear of newborn mice by inoculating them intracranially with 10^6 EID_{50} of egg-propagated Sendai virus. No remarkable inflammatory changes were noted, but viral envelope and nucleocapsid antigens were present at days 1 and 2, after which the amount of antigen decreased.

Tyrrell and Coid (1970) infected weanling rats intranasally and produced lung lesions which were resolved within 2 weeks.

4. *Diagnosis*

a. Virus Isolation. From live mice, Sendai virus can be isolated from oral swabs or from nasal washings. The tissue commonly used for isolation is the

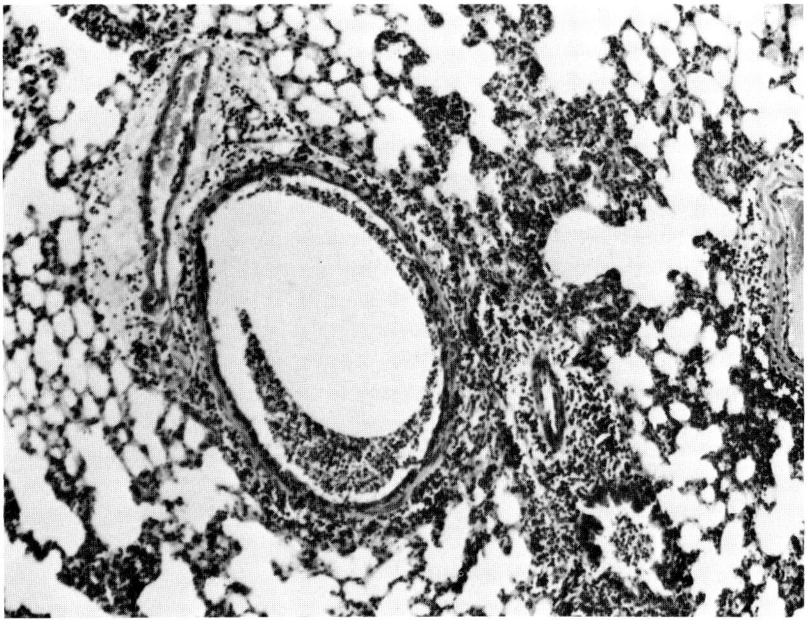

Fig. 6. Lung lesion in young adult mouse killed 5 days after intranasal exposure to Sendai virus, showing bronchiolitis, peribronchial accumulation of lymphoid cells, and alveolar thickening. H&E stain. (Courtesy of N. F. Cheville.)

lung, usually prepared as a 10% homogenate. Cells used for isolation are usually primary monkey kidney maintained in medium containing 0.2% anti-SV-5 serum. After inoculation, the cells should be observed for CPE for 1 to 2 weeks, then hemadsorbed (Parker *et al.*, 1964b; Parker and Reynolds, 1968). Bhatt and Jonas (1974) found BHK-21 cells to be as sensitive as primary monkey kidney cells.

For Sendai virus isolation in eggs, Fukumi *et al.* (1954) found the amniotic route to be more sensitive than the allantoic route. After incubation, allantoic fluids contain hemagglutinins.

A very sensitive test for the presence of viral antigen is the mouse antibody production (MAP) test. Mice are inoculated intranasally with lung homogenate, then bled 4 weeks later to check for Sendai virus antibody (Van Der Veen *et al.*, 1974).

Sendai virus isolates are usually identified by the HI technique with known reference antisera.

b. Serology. Antibody to Sendai virus is usually measured by HI or CF procedures. For HI, Fukumi *et al.* (1962) found nonspecific HA inhibitors in mouse sera to be no problem at 1:16 dilutions. However, some sera were found to have natural hemagglutinins for chick RBC, making preadsorption of the serum with RBC necessary. Sera have been treated by various methods, including incubation at 56°C for 30 min. and treatment with RDE. Profeta *et al.* (1969) used a hemadsorption neutralization technique, using primary monkey kidney cells.

5. *Immunity*

In an infected colony, suckling mice had low titers of antibody and were negative for virus isolation (Parker and Reynolds, 1968). Several workers reported that they could not isolate Sendai virus from antibody-positive mice after experimental infection. Mice exposed to Sendai virus by the respiratory route were not reinfected by a second exposure by the respiratory route (Van Der Veen *et al.*, 1974; Jakab, 1974).

Fukumi and Takeuchi (1975) reported that a vaccine made from parainfluenza-1 virus with a mineral oil–Aracil adjuvant raised HI titers in mice and guinea pigs. The newborn of vaccinated pregnant mice had increased antibody levels for 8 weeks. Vaccinated 4-week-old mice and guinea pigs were resistant to natural infection for about 20 weeks.

6. *Other Paramyxoviruses of Rodents*

a. Peromyscus Agent. This PMV was isolated and described by Morris *et al.* (1963), who inoculated the yolk sacs of 7-day-old chicken embryos with tissue homogenates of healthy white-footed mice. On rhesus monkey heart cells and kidney cell lines, it caused syncytia formation beginning at 48 hours. On primary rhesus monkey kidney cells, it caused ballooning and vacuolization beginning at 2 to 3 days. Guinea pig RBC were adsorbed diffusely to primary cells and only on syncytia on cell lines. The Peromyscus agent killed suckling hamsters, Swiss mice, and baby white-footed mice in 7 to 12 days when injected intracranially and suckling hamsters when injected intraperitoneally. No antigenic relationships were shown with other known PMV by CF and hemadsorption inhibition procedures, using guinea pig reference antisera. The Peromyscus agent causes no known disease, and antibody has been found only in wild white-footed mice.

b. Nariva Virus. Tikasingh *et al.* (1966) isolated Nariva virus by inoculating 2-day-old Swiss mice intracranially with tissue homogenates from healthy forest rodents (*Zygodontomys b. brevicauda*) of Trinidad. The virus was studied in more detail by Karabatsos *et al.* (1969). It caused syncytia formation in BHK-21 cells and Vero cells, which became hemadsorption positive to guinea pig RBC. Hemagglutinin could be demonstrated only after infected cells and fluids were sonicated. Reference sera of other PMV did not inhibit HA by the Nariva virus.

c. J Virus. This PMV was isolated by Jun *et al.* (1977) by kidney autoculture of moribund mice in North Queensland. The virus replicated in primary mouse kidney and lung cells and in mouse kidney, pig kidney, monkey kidney, MRC 5, and HEp-2 cell lines. The CPE consisted of vacuolated syncytia formation, resulting in monolayer destruction. Eosinophilic cytoplasmic inclusions were seen. No HA or hemadsorption occurred with mouse, rat, guinea pig, rabbit, chicken, goose, pig, sheep, cow, horse, or human O RBC at 4°, 25°, or 37°C. Intranasal or subcutaneous infection of rats and mice resulted in a viremia and hemorrhagic interstitial pneumonia. Antibody was found in wild mice, rats, and pigs. No antigenic relationships were demonstrated with other PMV by reciprocal CF procedures.

IV. PARAMYXOVIRUSES OF BIRDS

Paramyxoviruses of birds include Newcastle disease virus (NDV), the type species of the genus *Paramyxovirus,* Yucaipa virus, and turkey paramyxovirus. Other PMVs have been isolated from birds, but sufficiently detailed comparisons have not been made among them for definite classification purposes. NDV is of particular interest because of its economic impact on the poultry industry and its many properties, which make it a good model for laboratory studies. The literature covering NDV and Newcastle disease (ND) covers all aspects of virology, immunology, pathogenesis, and epidemiology. Fortunately, the literature has been reviewed periodically. Beginning in 1966, the newest ND panzootic appeared in many parts of the world. It caused great concern because it produced disease in vacinnated chickens that were capable of resisting viruses previously prevalent (Hanson *et al.,* 1973).

Hemagglutinating viruses, possibly PMVs, which were not neutralized with either influenza or NDV hyperimmune sera, have been collected during surveys for avian influenza viruses. Such isolates were usually stored for study at a future date.

A. Newcastle Disease Virus

1. History

After the peracute form of ND was first described in 1926, it soon became evident that the disease was present in all poultry-producing areas of the world, and that the disease had more than one clinical form. The history of ND and modes of its spread have been periodically reviewed (Hanson, 1964; Lancaster, 1966, 1977; Lancaster and Alexander, 1975; Hanson, 1978).

2. Properties

NDV strains are differentiated from other PMV by serological methods and from each other by biotyping. The virus can hemagglutinate the RBC of most reptiles, amphibians, and birds and those of several mammals. The virus contains neuraminidase which will elute it from the RBC. It can also hemolyse RBC.

a. Antigenic Types. Some minor antigenic differences are found among NDV strains, but strains are essentially serologically indistinguishable. Strains of NDV were found to consist of heterogenous mixtures of plaque mutants (Schloer, 1974). By neutralization kinetics procedures, Schloer (1974) demonstrated some differences in plaque mutants originating from different strains as well as from the same strain. However, all antigenic variations were within the range of differences found within a serotype. Amino acid analysis of three major polypeptides from three strains of NDV revealed differences among the proteins of the different strains (Moore and Burke, 1974). Pennington (1978) found differences in neutralization rate constants among egg-propagated NDV strains of different virulence and different places of origin. Antigenic differences were also shown with immunodiffusion procedures, using Tween 80 and ether-treated virus. No antigenic differences were found with the HI procedure.

b. Strains of NDV. A strain of NDV is a culture that has been recovered from the infected host by inoculation of a suitable laboratory host system such as embryonated chicken eggs or cell cultures (Hanson, 1978). There are many NDV strains, and they are indistinguishable by serological techniques. However, they do differ markedly in their virulence for the chicken. Therefore, NDV strains have been grouped into three classes of virulence for the chicken (Hanson and Brandly, 1955; Hanson, 1978). Velogenic strains produce severe lesions and death, regardless of the route of exposure. Mesogenic strains cause mild disease symptoms and rarely kill chickens when administered by peripheral routes, but they do cause severe disease and death when inoculated into the central nervous system. Lentogenic strains cause a mild or inapparent disease, regardless of the route of exposure (Hanson, 1978).

Actually, NDV strains fall somewhere along a virulence gradient and in some instances fall between classes of virulence. New NDV strains are therefore classified by a biotyping system to establish their pathotype. The system includes four tests which are standardized and described in detail in a publication, ''Isolation and Identification of Avian Pathogens,'' by the American Association of Avian Pathologists (Hanson, 1975).

The four tests are, briefly:

1. Mean death time of the minimum lethal dose (MDT/MLD), which establishes the time in hours it takes a minimum lethal dose of virus to kill the chicken embryo. Lentogenic strains take over 100 hours, while mesogenic and velogenic strains take 60 hours or less (Hanson, 1978).

2. Pathogenicity for 8-week-old chickens. In this test, susceptible chickens are inoculated by swabbing NDV onto the conjunctiva and into the cloaca. Clinical signs of disease, lesions, and time of death are noted.

3. Ability to plaque on chicken embryo fibroblast monolayers with and with-

out DEAE and Mg^{2+} ions. In this test, plaque morphology and size are also observed.

4. Intracerebral pathogenicity for day-old chicks in which a survival index is determined by scoring the death times and clinical symptoms of day-old chicks inoculated intracerebrally with NDV.

After the four tests are completed, the NDV strain can be pathotyped by keying the test results. The MDT/MLD is a measure of virulence, but the relative virulence of a strain may vary with the test used. Therefore, for a more complete pathotyping, results of the four tests should be considered.

Other markers unrelated to virulence are used to differentiate strains. They are rate of elution from chicken RBC (Spalatin *et al.*, 1970), thermostability of hemagglutinin (Hanson and Spalatin, 1978), and agglutination of equine RBC (Hanson, 1975).

c. Effect on Chicken Embryos. All NDV strains will kill chicken embryos, with the yolk sac or intravenous routes of inoculation causing the most rapid death (Hanson, 1978). Inoculation of velogenic NDV strains into the allantoic cavity of 10-day-old chicken embryos causes necrosis and hemorrhage of the chorioallantoic membranes, as shown in Fig. 7 (Cheville *et al.*, 1972). High

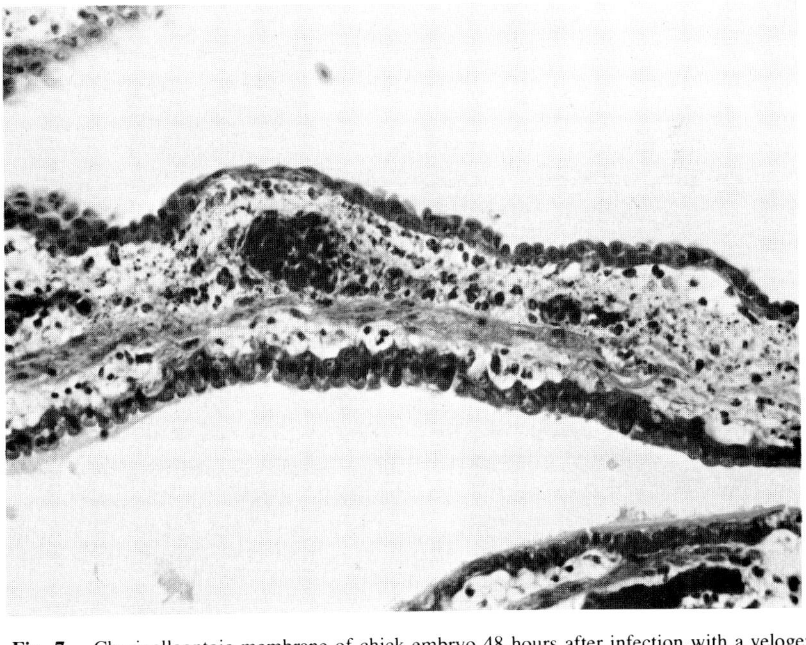

Fig. 7. Chorioallaontoic membrane of chick embryo 48 hours after infection with a velogenic strain of NDV, showing hyperemia, edema, necrosis of the mesenchymal tissue, and degeneration and necrosis of epithelial cells. H&E stain. (Courtesy of N. F. Cheville.)

titers of virus are present in the extraembryonic fluids 24 to 48 hours before death of the embryos. As previously mentioned, the time of death varies with the strain of NDV.

d. Cytopathic Effect. For basic virological studies, NDV has been propagated in many kinds of cells and cell lines. The subject has been reviewed (Bankowski, 1964; Lancaster, 1966; Lancaster and Alexander, 1975). A wide range of cells from various species are susceptible to NDV, and factors such as viral replication, type of CPE, and behavior of the virus with passage depend upon both the cell type and the strain of NDV. The two primary types of CPE are necrosis and syncytia formation.

NDV will form plaques in monolayers of various chicken embryo cells and in some human cell lines. The plaque morphology of 14 strains of NDV on primary chicken kidney monolayers was reported by Schloer and Hanson (1968a). Virulence of NDV was associated with large plaques, red plaques, and strains which cause a great amount cell fusion (Hanson, 1978). Large plaque size and the presence of large plaques in a strain of NDV were related to virulence, while the absence of large plaques within a strain was related to reduction of virulence in the chicken embryo (Schloer and Hanson, 1968b).

3. Disease

Episodes of severe ND occur in some parts of the world, but in the United States, the severe form is not common. The usual situation in the United States has been a flare-up of respiratory disease in immunologically depressed or stressed chickens. In such instances, the NDV can act in combination with other respiratory disease agents to cause a respiratory disease complex. The disease can also be noncharacteristic in chickens which have a low level of immunity. The pure disease in susceptible chickens has several clinical forms. The form of ND in other birds may have no relationship to that in chickens.

a. Symptoms and Lesions. The spectrum of ND was grouped into four forms of disease (Hanson, 1978):

1. Doyle's form: This form is acute and lethal and may occur so suddenly that chickens die without clinical signs or lesions. Usually the chickens become listless and weak, breathe rapidly, become prostrate, and die in 4 to 8 days. There are usually hemorrhagic lesions in the digestive tract. Sometimes there is edema of the tissues surrounding the eye and a greenish diarrhea. Mortality is very high. Surviving birds may have CNS involvement. This form is caused by velogenic strains.

2. Beach's form: This form is also acute and lethal. Chickens exhibit respiratory distress with coughing and gasping, anorexia, a drop of cessation in egg production, and appearance of CNS involvement. Lesions develop in the respiratory and nervous systems. The mortality is variable but can be very high in immature chickens. This form is also caused by velogenic strains.

3. Beaudette's form: This is an acute respiratory disease with coughing in adult chickens. There is a loss of appetite along with a drop in egg production and egg quality. This form is caused by mesogenic strains.

4. Hitchner's form: This form is a very mild respiratory infection and may be inapparent. There is very low if any mortality. It is caused by lentogenic strains.

NDV is usually more virulent in young chickens. Natural routes of exposure usually result in respiratory disease, while parenteral routes of inoculation result in neurovirulence (Hanson, 1978).

In Doyle's form, primary lesions are hemorrhages in the intestinal wall. In other forms, primary lesions are in the respiratory tract. They include serous or catarrhal exudate in the upper respiratory tract, and the air sacs may be thickened and may contain exudate (Hanson, 1978).

Beard and Easterday (1967b) studied tracheal lesions in young chickens aerosol-exposed in lentogenic and velogenic strains of NDV. Lesions produced by both strains were similar but varied in intensity. Changes were first detected by histopathology 2 days after aerosol exposure and were characterized by loss of cilia, engorgement of capillaries, and flattened tracheal epithelial cells. Lesions were maximal at days 3 and 4, then began regressing on days 5 and 6. On day 10 there were dense aggregates of mononuclear cells in the tracheal epithelium.

Lung changes in acute disease are chiefly exudative, with hyperplasia and hypertrophy of the cells of the alveolar walls. Fluids and cells accumulate in the air spaces and air passages. Inflammatory changes may occur in the air sacs (Hanson, 1978). Cheville *et al.* (1972) reported severe hyperemia, edema, degeneration, and necrosis of the parabronchi in chickens with extensive exudation of fibrin and albumin into the lumen and the air spaces after intranasal exposure with certain viscerotrophic, velogenic strains of NDV (Fig. 8). Encephalitis, when present, consists of glial foci, neuronal degeneration, perivascular lymphocytic infiltration, and hypertrophy of the endothelial cells (Hanson, 1978).

b. Transmission. The primary mode of transmission of NDV is by aerosol. The infected chicken can shed virus for at least 2 days before clinical symptoms are evident (Hanson, 1978). Modes of transmission and spread are reviewed (Lancaster, 1963a, 1966, 1977; Lancaster and Alexander, 1975). Generally, ND is spread by movement of infected birds and poultry products. Occasionally, NDV has been found as a contaminating virus in other vaccines.

Asymptomatic birds can shed NDV. The role of carrier chickens with circulating antibody in transmitting NDV has not been determined. Tracheal organ cultures started from aerosol-exposed birds at 4 to 7 days after exposure persistently produced NDV, even though the same culture fluids neutralized NDV (Heuschele and Easterday, 1970a). By the fluorescent antibody technique, NDV antigen was shown to be present for at least 120 days in such organ cultures. The antigen was confined to the histiocytes in the tracheal mucosa (Heuschele and Easterday, 1970b).

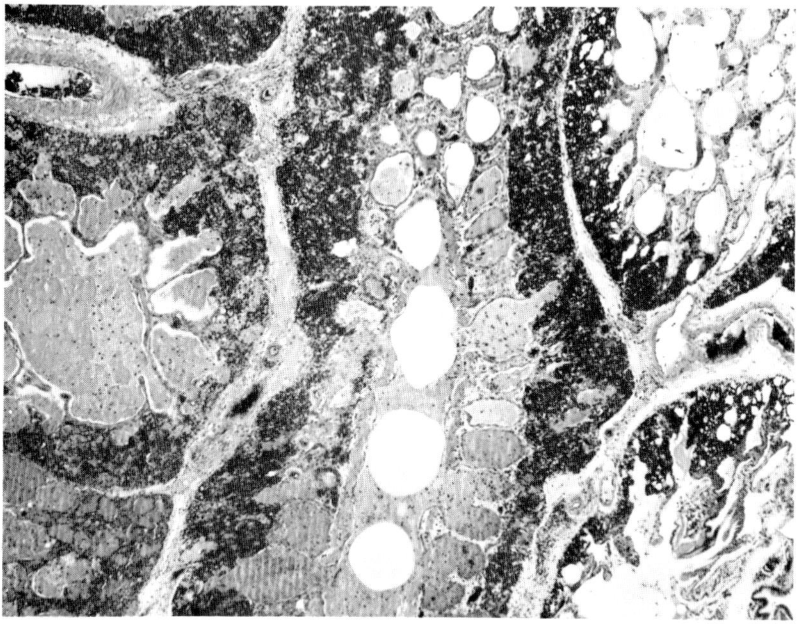

Fig. 8. Lung of chicken which died 7 days after aerosol inoculation with the Largo strain of NDV, showing edema, degeneration and necrosis of the parabronchi, and extensive exudation of fibrin and albumin into the lumen of the parabronchi. H&E stain. (From Cheville *et al.*, 1972, with permission.)

c. Newcastle Disease in Other Birds. ND primarily attacks chickens and turkeys, although other birds can be infected with the virus. It can produce a mild and sometimes fatal disease in various wild birds. The subject of ND in other birds has been reviewed (Lancaster, 1963a, 1966, 1977; Lancaster and Alexander, 1975).

Waterfowl are more resistant to NDV than are chickens and turkeys. Infection of adult geese with velogenic strains of NDV has usually resulted in an inapparent disease. The virus has been isolated from migratory ducks (Spalatin and Hanson, 1974; Bahl *et al.*, 1977), and antibody has been found in Canada geese, wild ducks, and domestic geese (Spalatin and Hanson, 1974).

Erickson *et al.* (1977) determined the effect of aerosol exposure with velogenic, viscerotrophic NDV in six pet bird species. Some had clinical signs of disease, some shed NDV from the cloaca for long periods of time, and none developed visceral lesions as found in chickens. Parrots and budgerigars developed encephalitis after aerosol or contact exposure, whereas canaries developed no lesions (Proctor *et al.*, 1974).

Warm- and cold-blooded animals have been infected with NDV by various means. The subject is reviewed by Lancaster (1966).

4. *Diagnosis*

a. Virus Isolation. Since ND has various clinical forms in its hosts, virus isolation is the best method of diagnosis. Diagnostic procedures have been reviewed (Lancaster, 1963b, 1966) and presented in detail (Hanson, 1975, 1978). For virus isolation, it is best to collect tracheal and cloacal swabs from birds in the incubative or early stages of disease. Tissues of choice for NDV isolation are the lung, brain, spleen, liver, and bone marrow (Hanson, 1975).

The virus-containing material is inoculated into the allantoic fluid of 9- to 11-day-old chicken embryos. The embryos are observed for viability by candling for 5 days. Those that die the first day are suspected as dying from other causes. After the period of incubation or embryonic death, the allantoic fluids should be clear and bacteria-free and hemagglutinate of 1% suspension of chicken RBC. The hemagglutinating agent can be identified as NDV by the HI procedure.

For cell culture techniques, chicken embryo fibroblasts or kidney cells are usually used. Plaques will form in 2 to 7 days, depending on the NDV strain (Schloer and Hanson, 1968a). The plaque type may indicate the possible pathotype of NDV isolated.

b. Serology. For determining the involvement of NDV in a respiratory disease outbreak in the absence of viral isolation, serum samples from both the acute and convalescent stages of the disease are necessary. For most purposes, however, serological techniques are used to evaluate a flock's response to vaccination, to monitor specific pathogen-free flocks, and for serological surveys to detect NDV antibody in other hosts.

The HI procedure is most commonly used. There are many variations of the HI procedure, and results will vary depending upon reagents and procedures used. A microtiter technique using a standardized antigen has recently been published in detail (Carbrey *et al.*, 1974). The antigen was an egg-propagated, formalin-inactivated, propylene glycol-concentrated, glycerine-stabilized lentogenic strain of NDV (Carbrey *et al.*, 1974; Beard *et al.*, 1975). A standardized, stable, inactivated HA antigen will allow workers in different laboratories to compare titer results more closely.

Virus neutralization procedures in embryonated eggs, cell cultures, and by plaque reduction are outlined by Hanson (1975).

5. *Immunity*

Attempts to prevent and control ND include programs to prevent contact with the virus and vaccination. A combination of both may be necessary in certain areas. In the poultry industry, chickens are raised for various uses (egg production, breeding stock, and meat production), and reasons and methods for vaccination will vary accordingly. Methods will depend on the economics of protecting certain birds, the duration of immunity needed, the age at which the birds need to be protected, and their immune status at the time of vaccination. Other programs may include raising birds in filtered-air, positive-pressure houses, space isolation, and local slaughter and quarantine programs.

a. Naturally Acquired Antibody. Chicks less than 3 weeks of age from immune hens will have a maternal antibody titer to NDV. This fact, plus immaturity of the immune mechanism at that age, presents special problems for effective immunization by vaccination (Hanson, 1978). Actively or passively acquired antibody inhibits development of immunity by parenteral vaccination. However, there is conflicting evidence on the effect of serum antibody on vaccination by natural routes. Beard and Easterday (1967a) reported that after intravenous administration of hyperimmune serum or IgG, chickens were protected from an intramuscular viral challenge and did not respond with an increased HI titer. However, such chickens shed virus after an aerosol challenge exposure. Those with the lowest HI titers at the time of aerosol challenge exposure had the greatest HI titer response. The same relationship between passively acquired HI titer level and immune response was demonstrated by Stone *et al.* (1975), who challenged chicks from vaccinated hens intraocularly at 1 and at 34 days of age. The 1-day-old chicks had high HI titers at challenge, excreted virus for 60 days, while HI titers decreased. The 34-day-old chicks had low HI titers at challenge, excreted virus for 10 days, and developed high HI titers.

b. Vaccination. An enormous amount of work has been done on vaccination of chickens against NDV. The subject has been reviewed (Lancaster, 1964, 1966; Hanson, 1976, 1978). Subjects of study have included NDV strains, methods of inactivation, adjuvants, routes and methods of exposure, dosage, and various other factors influencing the degree of protection.

Most vaccination is done with live lentogenic NDV strains given by methods of mass administration, such as spraying, dusting, and administration in the drinking water. The problems encountered in vaccinating chicks by these methods are reviewed by Hanson (1976). Many problems can arise from mishandling and abuses in vaccine production and administration, especially in the case of live virus vaccines, which will have little or no effect if the virus is inactivated.

B. Other Paramyxoviruses of Birds

Hemagglutinating viruses have been isolated from various birds and propagated in chicken embryos or tissue cultures. Some of these have been identified as PMVs, serologically distinct from NDV. At least two non-NDV PMVs are well documented, Yucaipa virus and turkey parainfluenza virus. Partially characterized PMVs have been isolated from various birds. Some that are partially serologically related to Yucaipa virus have been isolated from finches, turkeys, parrots, and free flying birds. Others with no known serological relationship to other PMVs have been isolated from budgerigars. Undoubtedly, more have been isolated and scarcely studied. There is a lack of sufficiently detailed comparative information at present to classify these isolates.

1. Yucaipa Virus

a. History. A PMV was isolated along with infectious laryngotracheitis virus from tracheas of 3-week-old chicks suffering from unusually severe laryngotracheitis (Bankowski *et al.*, 1960). Recovered birds had no HI antibody for NDV and were not resistant to NDV when challenged. The PMV was isolated in embryonating eggs, and the hemagglutinins in the allantoic fluids were not inhibited with NDV antiserum. The virus was not related to NDV or to other PMV by HI and SN procedures. The virus was designated "myxovirus strain Yucaipa" (Bankowski and Corstvet, 1961).

b. Host Range. Chicken embryos were not killed in 6 days when inoculated by the chorioallantoic membrane or allantoic routes, but the allantoic fluid contained hemagglutinins. The ability to kill chicken embryos developed with passage of the virus (Bankowski *et al.*, 1960).

Yucaipa virus grows in pig kidney and in HeLa cells, causing a mild CPE. Infected cells hemadsorb guinea pig, chicken, bovine, and equine RBC (Bankowski *et al.*, 1960). Dinter *et al.* (1964) reported that the virus grew in chicken kidney and monkey kidney cells without causing a definite CPE, and that the virus did not grow in calf kidney, pig kidney, HeLa, KB, and BHK cells. The Yucaipa virus was characterized as a PMV by Dinter *et al.* (1964). Kawamura and Tsubahara (1968) reported that the Yucaipa virus caused a CPE consisting of vacuolization and syncytia formation in chick kidney cells. Yucaipa virus also caused formation of eosinophilic cytoplasmic inclusion bodies in chick kidney cells (Kawamura and Tsubahara, 1968) and in monkey kidney cells (Dinter *et al.*, 1964).

c. Properties. Yucaipa virus hemagglutinates chicken, guinea pig, cow, and human O RBC. It was reported to agglutinate (Bankowski and Corstvet, 1961) and not agglutinate (Dinter *et al.*, 1964) horse RBC.

d. Disease. Intramuscular or intrabursal inoculation of 37-day-old chicks caused no signs of disease, whereas intratracheal inoculation caused mild rales at 4 to 6 days and the Yucaipa virus was recovered from the trachea on day 4 (Bankowski *et al.*, 1960). Serological surveys of chicken and turkey flocks in the United States showed the Yucaipa virus antibody to be rare in chickens and more common in turkey flocks (Bankowski *et al.*, 1968).

e. Yucaipa-Related Virus Isolates. A PMV, designated the "Bangor isolate," serologically related to Yucaipa virus by HI and SN procedures, was isolated from tissue pools of imported finches (*Ureginthus angolensis*) which died of disease in Ireland (McFerran *et al.*, 1973, 1974). The hemagglutinins of Yucaipa virus and the Bangor isolate were antigenically related to each other, but the neuraminidases were serologically distinct (Alexander, 1974a). In comparative studies, both the Yucaipa virus and the Bangor isolate hemagglutinated chicken, turkey, pigeon, human O, and guinea pig RBC, but not calf, sheep, and horse RBC. Also, the hemagglutinin of the Bangor isolate was much more heat resistant than that of the Yucaipa virus (McFerren *et al.*, 1974). Both isolates grew in lamb and calf kidney, BHK, and Vero cells. In chick kidney cells, both isolates caused vacuolization, syncytia formation, and rounding and detachment of cells (McFerran *et al.*, 1974). No illness occurred in chicks infected orally or intracranially with the Bangor isolate (McFerran *et al.*, 1974). No antibody to the Bangor isolate was found in serological surveys of poultry farms in Ireland (McFerran *et al.*, 1973).

A PMV serologically related to Yucaipa virus by HI and SN procedures was isolated from turkeys with a severe respiratory disease and sinusitis in Canada (Lang *et al.*, 1975). The virus was isolated from the trachea. In two other instances, it was isolated from tissue pools from young turkeys with respiratory disease. In all three cases, the virus was isolated along with other pathogens known to cause respiratory disease in turkeys.

Another PMV, designated "0121," serologically related to the Yucaipa and Bangor isolates by HI, was isolated from the tissues of an African gray parrot which died of respiratory disease (Collings and Fitton, 1975). Neuraminidase of the 0121 isolate was closely antigenically related to that of both the Yucaipa and Bangor viruses by a neuraminidase inhibition procedure, even though neuraminidases of Yucaipa and Bangor viruses were antigenically distinct from each other. The hemagglutinin of the 0121 isolate was more heat stable than that of the Bangor isolate, which in turn was more stable than that of the Yucaipa virus. Exposure of 6-week-old chicks by aerosol, intramuscular, or intratracheal routes caused no signs of disease.

Other Yucaipa-related PMVs were isolated from free-flying birds in Germany (Nymadava *et al.*, 1977) and from the feces of a wild bird in Senegal (Fleury, 1978). The German isolate was serologically related to Yucaipa virus, Bangor virus, and had some cross reactivity to bovine PI-3 virus by HI procedures (Starke *et al.*, 1977).

3. Turkey Parainfluenza Virus

Parainfluenza viruses serologically related to each other have been isolated from turkeys in Wisconsin and Ontario (unpublished work referred to by Alexander, 1974b). The viruses were not related to Yucaipa, Bangor, or NDV by reciprocal HI procedures (McFerran *et al.*, 1974).

4. Parainfluenza Viruses from Budgerigars

A PMV was isolated from budgerigars suffering from an acute enteritis with a high mortality in Queensland (Mustaffa-Babjee *et al.*, 1974).It was isolated from a tissue pool in chick embryo fibroblast cultures. Infected cell culture fluids and allantoic fluids did not hemagglutinate chicken, goose, turkey, guinea pig, or rabbit RBC. The virus was not related to other avian PMVs by SN procedures. In chick embryo fibroblasts, the CPE was observed at 24 hours but was nonprogressive. In Vero cells, the CPE was observed at 48 hours, then developed into syncytia with intranuclear and intracytoplasmic inclusion bodies. Infected cells hemadsorbed chicken RBC. The virus was nonlethal to chicken embryos. Intranasal, oral, or intra-abdominal inoculation of budgerigars caused an acute fatal enteritis in the immature birds, with hemorrhages in the digestive tract. Mature birds were ill for 4 to 7 days, then recovered. The virus caused no disease in pigeons and chicks.

Another PMV, designated the "Kunitachi virus," was isolated from the lung tissues of dead budgerigars in Japan (Nerome *et al.*, 1978). The virus grew in chicken embryos inoculated by the amniotic route, but not by the allantoic route, and grew well in primary chick embryo cells. Amniotic fluid containing the Kunitachi virus did not hemagglutinate RBC, but virus purified by density gradient centrifugation hemagglutinated chicken, duck, goose, guinea pig, and human O RBC at 4° and 25°C. It was not serologically related to the other avian PMV by reciprocal HI procedures. However, it was not compared with the virus isolated by Mustaffa-Babjee *et al.* (1974). The virus caused depression, dyspnea, diarrhea, and death in budgerigars.

V. PNEUMOVIRUSES OF ANIMALS

The pneumoviruses infecting animals are bovine respiratory syncytial virus (RSV) and pneumonia virus of mice (PVM). As with the PMVs in animals, the pneumoviruses are associated with respiratory disease conditions, from which other etiological agents are commonly isolated. Antibody to the pneumoviruses are commonly found in their hosts.

A. Bovine Respiratory Syncytial Virus

1. History

An inhibitor to human RSV was demonstrated in sera of normal cattle by an SN technique. It was found to be antibody, and titers to the virus increased in cattle after some episodes of respiratory disease (Doggett *et al.*, 1968). The authors speculated that a bovine RSV could be associated with respiratory disease in cattle. A bovine RSV was first isolated in Switzerland from cattle with an acute respiratory disease consisting of fever, nasal discharge, coughing, and signs of bronchopneumonia (Paccaud and Jacquier, 1970). Since then, the bovine RSV has been isolated in different areas of the world. Serological surveys have indicated a widespread incidence of antibody in cattle. Antibody to RSV was found to be prevalent in sheep (Berthiaume *et al.*, 1973; Smith *et al.*, 1975). The subject of bovine RSV has been reviewed (Woods, 1974).

2. *Properties*

Bovine RSV isolates have the morphological and physical characteristics of human RSV isolates. They are closely related antigenically but are distinguishable by serological techniques. They also vary somewhat in their host cell ranges.

a. Host Tissue Range. Bovine RSV upon first isolation is usually slow to produce a CPE, but does so more rapidly after one or more passages. There are conflicting reports on whether bovine RSV grows in certain types of cells. Paccaud and Jacquier (1970) first isolated the virus in embryonic calf kidney cells, where it caused small foci of CPE in 12 to 14 days. After passage, the virus caused syncytion formation in 7 days. Their virus caused foci of ballooned cells and syncytia in embryonic calf kidney and lung cells, and formation of eosinophilic intracytoplasmic inclusion bodies, but no intranuclear inclusions. Their virus did not grow in human embryo fibroblast, KB, HeLa, BHK-21, secondary rhesus monkey kidney, and primary embryonic sheep kidney cells. Other workers (Inaba *et al.*, 1970a; Wellemans *et al.*, 1970) also reported syncytia formation with eosinophilic cytoplasmic inclusions in bovine kidney and testicle cells.

Matumoto *et al.* (1974) studied the growth of a bovine and a human RSV isoalte in a variety of bovine cells and other cells at 34°C. The bovine cells included kidney, testicle, duodenum, rectum, and thyroid cells. Bovine RSV grew in all, but human RSV did not grow as well and did not grow at all in the bovine rectum and thyroid cells. Of the other cells, both viruses grew in BHK-21, HEp-2, HeLa, HEL, and Vero cells. Only the bovine isolate grew in embryonic swine kidney and hamster lung cells, and only the human isolate grew in HEK cells. Neither isolate grew in chick embryo kidney or in chick embryo cells. Lehmkuhl and Cutlip (1979) reported that bovine RSV grew well in fetal ovine kidney and lung cells. Embryonic bovine lung cells infected with bovine RSV are shown in Fig. 9.

The effect of bovine RSV on fetal bovine tracheal organ culture was studied by Thomas *et al.* (1976). The virus grew but had no effect on ciliary activity. It flattened cuboidal cells in patchy areas, and most viral antigen was found in the connective tissue surrounding the trachea.

The virus was passaged intracranially in suckling mice but caused no clinical symptoms. Older mice inoculated by various routes did not experience disease or gross lesions or produce serum antibody by 4 weeks. (Matumoto *et al.*, 1974). Inoculation of 300-gm guinea pigs by various routes did not cause signs of disease, but neutralizing antibody was produced in 2 to 3 weeks, except in those inoculated subcutaneously (Matumoto *et al.*, 1974).

b. Physicochemical Properties. Properties of bovine RSV are the same as

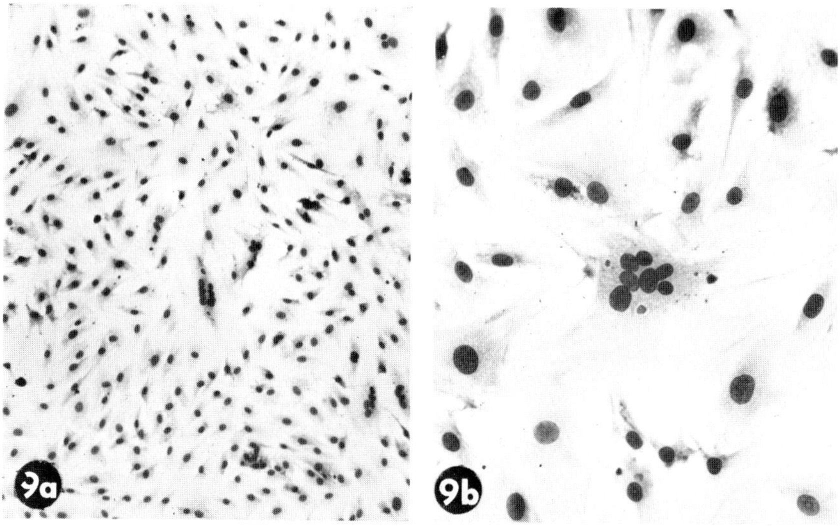

Fig. 9. Embryonic bovine lung cells infected with bovine RSV, showing syncytial formation and intracytoplasmic inclusions. MGG stain. (a) × 100; (b) × 250. (Courtesy of H. D. Lehmkuhl.)

those of the human isolates. They are morphologically the same, chloroform and ether sensitive, as well as sensitive to sodium deoxycholate, trypsin, and acid treatment (Inaba *et al.,* 1973). Bovine isolates do not hemagglutinate RBC of any species tested thus far.

c. Propagation of Bovine RSV. Bovine RSV has been propagated for use in various bovine cells, usually kidney, lung, and testicle cells, by routine methods. Rossi and Kiesel (1978) reported that including 40 μg of diethylaminoethyldextran per ml of viral inoculum resulted in a faster onset of CPE, more infected cells, and a higher viral titer in bovine embryonic lung cells. They also determined that inoculation of cells by passage of infected cells rather than infectious supernatant fluid alone resulted in consistently higher viral titers.

3. Disease

a. Natural. The natural disease involving bovine RSV has been described by investigators from many areas of the world (Inaba *et al.,* 1970b, 1972; Paccaud and Jacquier, 1970; Jacobs and Edington, 1971; Rosenquist, 1974; Smith *et al.,* 1975; Wellemans and Leunen, 1975; Wellemans, 1977). In all cases, the disease was respiratory, usually characterized by fever, respiratory distress, coughing, nasal discharge, and sometimes lacrimation. Reports varied as to duration and severity. Virus isolations from nasal secretions were usually sporadic and only from a low percentage of the samples collected. Usually the mortality was very low, although Wellemans and Leunen (1975) reported that some individuals developed profound dyspnea and died within a few hours with widespread pulmonary emphysema.

b. Experimental. Calves have been experimentally infected with bovine RSV by the respiratory route and by aerosol (Inaba *et al.*, 1972; Jacobs and Edington, 1975; Mohanty *et al.*, 1975; Smith *et al.*, 1975; Moteane *et al.*, 1978). The resulting clinical disease was mild, characterized by fever, depression, and sometimes a serous nasal discharge and a mild cough. The disease was of short duration, and not all exposed calves had signs of systemic disease. Viruses were isolated from nasal secretions sporadically, usually from 5 to 10 days after exposure. Viral-neutralizing antibody usually was present by 2 weeks after exposure. Moteane *et al.* (1978) found no lesions attributable to RSV infection in calves killed at 3-day intervals after exposure, even though the calves had signs of disease and shed virus. Jacobs and Edington (1975) found no gross lesions in calves killed 7 to 8 days postexposure, but they did find some focal areas of proliferation and necrosis in the nasal and tracheal mucosa and proliferation and desquamation of epithelium and syncytia formation in the bronchial mucosa, with some peribronchial congestion. No lesions were found in calves killed at days 3 and 10. Mohanty *et al.* (1975) found areas of consolidation in the lungs, with mononuclear infiltration of the alveolar walls, syncytia in the alveolar lumen, peribronchiolitis, and pseudoepithelization.

Lehmkuhl and Cutlip (1979) infected colostrum-deprived lambs by the respiratory route with bovine RSV and killed them at intervals. Mild clinical signs of disease were observed and were characterized by slight depression, increased respiratory rate between 5 and 9 days postexposure, and fluctuating febrile responses between days 3 and 9. Virus was isolated from one nasal swab collected on day 5 and from the tracheal fluids and lung tissue of those killed on days 7 and 9. Basic lesions in those killed on days 7, 9, and 11 were multiple foci of interstitial pneumonitis with mononuclear cells and other cell debris in the septa and air spaces (Fig. 10). Serum antibody response was first detected by plaque reduction techniques within 1 week.

4. Diagnosis

Determining the role of bovine RSV in a respiratory disease situation is difficult. The virus is isolated sporadically if at all. One may rely upon demonstration of viral antigen by fluorescent antibody (FA) techniques or associate a rise in serum antibody titer with the disease. Diagnostic procedures have been reviewed by Wellemans (1977).

a. Virus Isolation. From natural cases of respiratory disease in cattle involving bovine RSV, virus isolation has been sporadic and difficult. Various workers reported very low isolation rates from nasal secretions of sick cattle, such as 2 of 12 (Paccaud and Jacquier, 1970), 1 of 14 (Jacobs and Edington, 1971), 2 of 12 (Ito *et al.*, 1973), 2 of 9 (Rosenquist, 1974), and 2 of 5 (Smith *et al.*, 1975). Lehmkuhl *et al.* (1979) isolated the virus from the nasal secretions of 2 of 7 calves with very early signs or no signs of disease and none

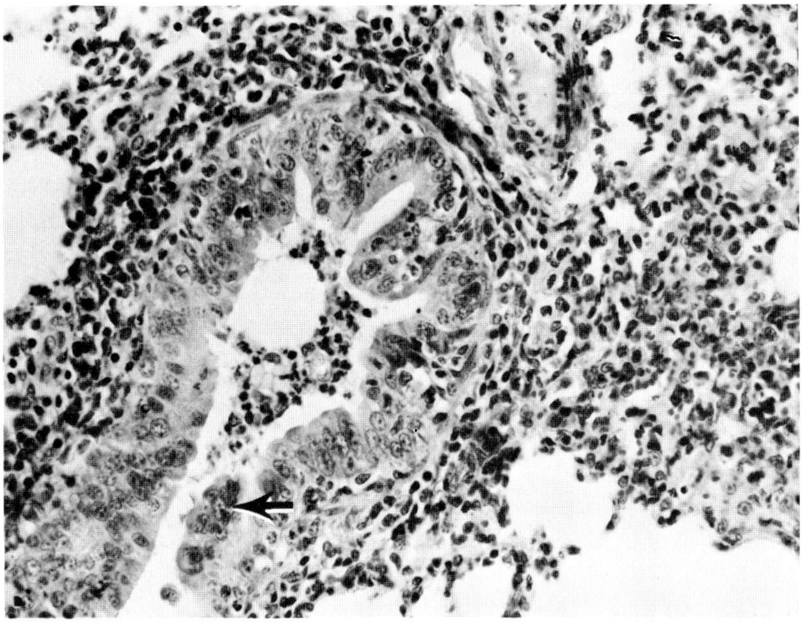

Fig. 10. Interstitial pneumonitis and bronchiolitis in a lamb killed 7 days after intratracheal inoculation with bovine RSV. There is accumulation of mononuclear cells in the interstitium and degeneration of the bronchiolar epithelium. The bronchiole has a syncytium in the epithelium (arrow) and cellular debris in the lumen. H&E stain. (Courtesy of R. C. Cutlip.)

from 8 acutely ill calves in a herd during an epizootic. The importance of early isolation from nasal secretions was reported by Wellemans and Leunen (1975) and Wellemans (1977). As mentioned earlier, isolations from experimentally infected cattle are also sporadic.

Techniques of virus isolation from secretions or tissues involve long periods of incubation and blind passage of cells or supernatant fluids. Cells of bovine origin, embryonic bovine kidney, testicle, embryonic bovine lung, and bovine turbinate cells have usually been used. Viral CPE may become evident after blind passages. Virus isolation may be hampered by the presence of local antibody in the specimen, especially since the virus is very slow to develop CPE.

Wellemans (1977) described a method of demonstrating viral antigen in ultrathin sections by an FA procedure. The antigen was usually demonstrated in the mucus membranes of the bronchioles.

b. Serology. Antibody to bovine RSV has been demonstrated in convalescent sera by CF, SN, immunodiffusion, and indirect FA procedures. Antibody, when present, can be demonstrated by all three procedures (Wellemans and Leunen, 1975). In sera from experimentally infected cattle, CF and SN titer rises correlated and there was a linear relationship between the titers obtained by both procedures (Takahashi *et al.*, 1975). Indirect FA procedures for demonstrating antibody to bovine RSV were described and reported as being reliable and sensitive (Potgieter and Aldridge, 1977; Wellemans, 1977).

An enzyme-linked immunosorbent assay was described for detecting antibody to human RSV in sera (Richardson *et al.*, 1978). The assay was found to be specific, reproducible, and more sensitive than other procedures.

5. *Immunity*

a. Natural Antibody. Calves with maternal antibody to bovine RSV became mildly ill and shed virus after exposure to the virus by the respiratory route (Jacobs and Edington, 1975). Mohanty *et al.* (1976) reported that calves had no clinical signs of disease, but they did shed virus and had increased serum antibody titers after a second exposure by the respiratory route.

b. Vaccination. Wellemans and Leunen (1975) reported that vaccination with a formalin-inactivated bovine RSV with Freund's adjuvant resulted in a good serum antibody response. However, 6 days after experimental challenge, a more severe localized pneumonia was present in the vaccinated than in the nonvaccinated cattle. At present, vaccination of cattle against bovine RSV is at an experimental stage.

B. Pneumonia Virus of Mice

1. *History*

Pneumonia virus of mice (PVM) was discovered by Horsfall and Hahn (1939, 1940) while mouse-passaging human nasal washing material from noninfluenza respiratory diseases. They noticed areas of lung consolidation that increased in frequency and severity with mouse passage. The same phenomena occurred when lung material from uninoculated normal mice was passaged. The causative agent was found in mice originating from three of eight sources. In a survey a sera, neutralizing antibody was found in chimpanzee, cotton rat, guinea pig, hamster, monkey, rabbit, and human serum, as well as in mouse serum when mice were inoculated to detect the unneutralized virus (Horsfall and Curnen, 1946). Later, Tennant *et al.* (1966) surveyed sera from 10 species of animals and found HI antibody only in some rat and hamster sera. No antibody was found in sera of persons with prolonged laboratory contact with the virus.

2. *Properties*

a. Host Tissue Range. PVM was grown in suckling hamster kidney cells (Tennant and Ward, 1962). Viral replication was detected by the presence of hemagglutinins in the culture fluid, CPE, and hemadsorption with mouse RBC.

Focal destruction of cells began at 7 days. PVM could not be detected by CPE or hemagglutinin production in mouse kidney and lung cells, human embryo skin and muscle cells, HeLa cells, chick embryo, embryonic bovine kidney, and mouse fibroblast cells (Tennant and Ward, 1962). Harter and Choppin (1967) also grew PVM in BHK-21 cells and reported scattered rounding of cells and detachment from the surface at 48 hours. There were also eosinophilic cytoplasmic inclusions which became large and numerous. Viral antigen was present in the cytoplasm by 24 hours, reaching a maximum concentration at 72 hours. Hemagglutinin production was good, and infectivity titer was maintained during serial passage. After 10 serial passages, the virus was antigenically the same as the mouse-passaged virus, although the infectivity titer for mice was slightly lower. PVM was also grown in Vero cells in which the CPE was slow and progressive. Eosinophilic intracytoplasmic inclusion bodies were formed, but syncytia were not (Berthiaume *et al.*, 1974). Gallapsy *et al.* (1978) established a persistent infection in BHK-21 cells by passaging the cells 72 hours after infection. The cells hemadsorbed mouse RBC, and the hemadsorption was blocked with specific antiserum. Viral antigens were demonstrated by FA techniques and by immune electron microscopy, but no infectious virus was detected.

Horsfall and Hahn (1940) were unable to infect ferrets, rabbits, guinea pigs, rhesus monkeys, voles, deer mice, skunks, woodchucks, opossums, and hamsters with PVM by intranasal instillation of the virus.

b. Physicochemical Properties. Horsfall and Hahn (1940) found the infectivity of lung suspension PVM to be heat sensitive, even becoming rapidly inactivated at room temperature. Mills and Dochez (1944) observed that blood seeping from freshly cut lung lesions agglutinated spontaneously, but that extracts prepared from lung suspensions did not agglutinate RBC. However, by heating the lung to 80°C, they were able to prepare extract which agglutinated mouse RBC. A simplified method of preparing the hemagglutinating lung extract was to grind the lung in saline, then heat the suspension to 75° to 80°C for 10 minutes, and then centrifuge out the lung debris (Mills and Dochez, 1945). Such preparations agglutinated mouse and hamster RBC but not human O, fowl, guinea pig, rat, cotton rat, sheep, cat, dog, or ferret RBC. Curnen *et al.* (1947) obtained a hemagglutinating extract by perfusing the lungs with saline before removal from the mouse to remove many RBC. The lung extract hemagglutinated without prior heat treatment.

Tennant and Ward (1962) found infected primary hamster kidney cell culture fluids hemagglutinated without heat treatment. The HA activity was not increased by heat treatment, but the virus was able to combine with the HI component from mouse lung (Harter and Choppin, 1967). The HA activity of PVM was not prevented by pretreatment of the RBC with neuraminidase (Harter and Choppin, 1967).

3. *Disease*

Mice inoculated intranasally with PVM-containing lung suspension remained healthy for 5 to 7 days, then became less active and emaciated, had ruffled hair, slow, deep respirations, cyanosis of the ears and tail, and some died in 12 to 13 days. Other routes of inoculation, and also contact exposure, had no effect (Horsfall and Hahn, 1940). The dosage of PVM-containing lung material to confer immunity was 10^{-4} of a lethal dose or 10^{-2} of an infectious dose, although the dosages varied with the source of mice. Lesions were like those of influenza virus in mice. Bronchial epithelium was hyperplastic, and bronchi were surrounded by mononuclear cells. Alveolar septa were thickened, and alveoli contained cellular exudate (Horsfall and Hahn, 1940).

Parker *et al.* (1966) found HI antibody to PVM in 22 of 34 mouse colonies, but less than 50% of the individuals in the positive colonies had antibody. In one colony, antibody was present in 2-month-old mice, and the seroconversion rate increased to 90% by 6 months. In a seronegative colony, 50% of the individuals suddenly seroconverted at 7 months of age. In both colonies the number of individuals with antibody decreased after 6 to 7 months. Tennant *et al.* (1966) found that less than 25% of the mice in 13 seropositive colonies had HI antibody to PVM, which indicates a low rate of infectivity or of transmissibility.

PVM is not passed transplacentally, is not harbored in young mice before they produce antibody, and does not persist in the presence of antibody, but probably causes acute enzootic infections which remain focal in the colony because of the low infectivity of the virus (Tennant *et al.*, 1966). Virus was present in the lungs of intranasally inoculated mice 3 to 10 days postinfection but was gone by day 20. Serum CF and HI antibody titers to PVM were present by 9 to 10 days. The CF antibody titer declined at 3 weeks and was gone in 3 months, while the HI titer persisted.

4. *Immunity*

Mills and Dochez (1945) found that injection of a heat-inactivated hemagglutinating extract prepared from PVM-infected lung material conferred immunity in mice to intranasal PVM challenge, as well as stimulating an HI antibody response. Live PVM injected at 14-day intervals also conferred protection against an intranasal challenge 14 days after the second injection (Horsfall and Hahn, 1940).

ACKNOWLEDGMENT

I thank A. E. Ritchie for kindly supplying and interpreting his electron micrographs depicting the morphological variations of the Paramyxoviridae.

REFERENCES

Abinanti, F. R., and Huebner, R. J. (1959). *Virology* **8,** 391–394.

Abinanti, F. R., Chanock, R. M., Cook, M. K., Wong, D., and Warfield, M. (1961). *Proc. Soc. Exp. Biol. Med.* **106,** 466–469.

Alexander, D. J. (1974a). *Arch. Gesamte Virusforsch.* **44,** 28–34.

Alexander, D. J. (1974b). *Arch. Gesamte Virusforsch.* **46,** 291–301.

Andrews, C., Periera, H. G., and Wildy, P. (1978). *In* "Viruses of Vertebrates," 4th ed., pp. 221–254. Bailliere, London.

Andrews, C. H., Bang, F. B., Chanock, R. M., and Zhdanov, V. M. (1959). *Virology,* **8,** 129–130.

Appel, M., and Bemis, D. A. (1978). *Cornell Vet.* **68,** Suppl. 7, 70–75.

Appel, M. J. G., and Percy, D. H. (1970). *J. Am. Vet. Med. Assoc.* **156,** 1778–1781.

Appel, M., Pickerill, P. H., Menegus, M., Percy, D. H., Parsonson, I. M., and Sheffy, B. E. (1970). *Proc. Gaines Vet. Symp.,* Oct., 1970, pp. 15–23.

Appell, L. H., Kovatch, R. M., Reddecliff, J. M., and Gerone, P. J. (1971). *Am. J. Vet. Res.* **32,** 1835–1841.

Bahl, A. K., Pomeroy, B. S., Mangundimedjo, S., and Easterday, B. C. (1977). *J. Am. Vet. Med. Assoc.* **171,** 949–951.

Baldwin, D. E., Marshall, R. G., and Wessman, G. E. (1967). *Am. J. Vet. Res.* **28,** 1773–1782.

Bankowski, R. A. (1964). *In* "Newcastle Disease Virus: An Evolving Pathogen" (R. P. Hanson, ed.), pp. 231–246. Univ. of Wisconsin Press, Madison.

Bankowski, R. A., and Corstvet, R. (1961). *Avian Dis.* **5,** 253–269.

Bankowski, R. A., Corstvet, R. E., and Clark, G. T. (1960). *Science* **132,** 292–293.

Bankowski, R. A., Conrad, R. D., and Reynolds, B. (1968). *Avian Dis.* **12,** 259–278.

Beard, C. W., and Easterday, B. C. (1967a). *J. Infect. Dis.* **117,** 62–65.

Beard, C. W., and Easterday, B. C. (1967b). *J. Infect. Dis.* **117,** 66–70.

Beard, C. W., Hopkins, S. R., and Hammond, J. (1975). *Avian Dis.* **19,** 692–699.

Berthiaume, L., Joncas, J., Boulay, G., and Pavilanis, V. (1973). *Vet. Rec.* **93,** 337–338.

Berthiaume, L., Joncas, J., and Pavilanis, V. (1974). *Arch. Gesamte Virusforsch.* **45,** 39–51.

Bhatt, P. N., and Jonas, A. M. (1974). *Am. J. Epidemiol.* **100,** 222–229.

Biberstein, E. L., Shreeve, B. J., Angus, K. W., and Thompson, D. A. (1971). *J. Comp. Pathol.* **81,** 339–351.

Binn, L. N., Eddy, G. A., Lazar, E. C., Helms, J., and Murnane, T. (1967). *Proc. Soc. Exp. Biol. Med.* **126,** 140–145.

Binn, L. N., Lazar, E. C., Rogul, M., Shepler, V. M., Swango, L. J., Claypoole, T., Hubbard, D. W., Asbill, S. G., and Alexander, A. D. (1968). *Am. J. Vet. Res.* **29,** 1809–1815.

Binn, L. N., Lazar, E. C., Helms, J., and Cross, R. E. (1970). *Am. J. Vet. Res.* **31,** 697–702.

Blandford, G., and Heath, R. B. (1972). *Immunology* **22,** 637–649.

Carbrey, E. A., Beard, C., Cooper, R., Hansen, R. P., and Pomeroy, B. S. (1974). *Proc. 17th Annu. Meet. Am. Assoc. Vet. Lab. Diagnosticians* pp. 1–6.

Chanock, R. M., Parrott, R. H., Cook, K., Andrews, B. E., Bell, J. A., Reichelderfer, T., Kapikian, A. Z., Mastrota, F. M., and Huebner, R. J. (1958). *N. Engl. J. Med.* **258,** 207–213.

Cheville, N. F., Stone, H., Riley, J., and Ritchie, A. E. (1972). *J. Am. Vet. Med. Assoc.* **161,** 169–179.

Churchill, A. E. (1963). *Nature (London)* **197,** 409.

Collings, D. F., and Fitton, J. (1975). *Res. Vet. Sci.* **19,** 219–221.

Cornwell, H. J. C., McCandlish, I. A. P., Thompson, H., Laird, H. M., and Wright, N. G. (1976). *Vet. Rec.* **98,** 301–302.

Crandell, R. A., Brumlow, W. B., and Davison, V. E. (1968). *Am. J. Vet. Res.* **29,** 2141–2147.

Cuadrado, R. R. (1965). *Bull. W.H.O.* **33,** 803–808.
Curnen, E. C., Pickels, E. G., and Horsfall, F. L., Jr. (1947). *J. Exp. Med.* **85,** 23–38.
Davies, D. H., Dungworth, D. L., Humphreys, S., and Johnson, A. J. (1977). *N. Z. Vet. J.* **25,** 263–265.
Dawson, P. S. (1964). *Res. Vet. Sci.* **5,** 81–88.
Dawson, P. S., Darbyshire, J. H., and Lamont, P. H. (1965). *Res. Vet. Sci.* **6,** 108–113.
Deas, D. W., Johnston, W. S., and Vantsis, J. T. (1966). *Vet. Rec.* **78,** 739–740.
Ditchfield, J., Zbitnew, A., and Macpherson, L. W. (1963). *Can. Vet. J.* **4,** 175–180.
Dinter, Z., Hermodsson, S., and Bakos, K. (1960). *Acta Pathol. Microbiol. Scand.* **49,** 485–492.
Dinter, Z.. Hermodsson, S., and Hermodsson, L. (1964). *Virology* **22,** 297–304.
Doggett, J. E., Taylor-Robinson, D., and Gallop, R. G. C. (1968). *Arch. Gesamte Virusforsch.* **23,** 126–137.
Emery, J. B., House, J. A., Bittle, J. L., and Spotts, A. M. (1976). *Am. J. Vet. Res.* **37,** 1323–1327.
Erickson, G. A., Mare, C. J., Gustafson, G. A., Miller, L. D., Proctor, S. J., and Carbrey, E. A. (1977). *Avian Dis.* **21,** 642–654.
Faye, P., and Charton, A. (1967). *Recl. Med. Vet.* **143,** 1225–1229.
Fenner, F. (1977). *Intervirology* **7,** 1–115.
Fischman, H. R. (1965). *Proc. Soc. Exp. Biol. Med.* **118,** 725–727.
Fischman, H. R. (1967). *Am. J. Epidemiol.* **85,** 272–281.
Fleury, H. J. A. (1978). *Avian Dis.* **22,** 196–197.
Frank, G. H. (1966). *Proc. U.S. Livestock Sanit. Assoc., 70th Annu. Meet.* pp. 59–66.
Frank, G. H. (1970). *Am. J. Vet. Res.* **31,** 1085–1091.
Frank, G. H., and Marshall, R. G. (1971). *Am. J. Vet. Res.* **32,** 1707–1713.
Frank, G. H., and Marshall, R. G. (1973). *J. Am. Vet. Med. Assoc.* **163,** 858–860.
Fukumi, H., and Nishikawa, F. (1961). *Jpn. J. Med. Sci. Biol.* **14,** 109–120.
Fukumi, H., and Takeuchi, Y. (1975). *Dev. Biol. Stand.* **28,** 477–481.
Fukumi, H., Nichikawa, F., and Kitayama, T. (1954). *Jpn. J. Med. Sci. Biol.* **7,** 345–363.
Fukumi, H., Mizutani, H., and Takeuchi, Y. (1962). *Jpn. J. Med. Sci. Biol.* **15,** 153–163.
Fulton, R. W., Ott, R. L., Duenwald, J. C., and Gorham, J. R. (1974). *Am. J. Vet. Res.* **35,** 853–855.
Gale, C. (1970). *J. Dairy Sci.* **53,** 621–625.
Gallaspy, S. E., Coward, J. E., and Howe, C. (1978). *J. Virol.* **26,** 110–114.
Hamdy, A. H., Morrill, C. C., and Hoyt, H. H. (1965). *Ohio Agric. Res. Dev. Cent., Res. Bull.* **975.**
Hanson, R. P., ed. (1964). "Newcastle Disease Virus: An Evolving Pathogen." Univ. of Wisconsin Press, Madison.
Hanson, R. P. (1975). *In* "Isolation and Identification of Avian Pathogens" (S. B. Hitchner, C. H. Domermuth, H. G. Purchase, and J. E. Williams, eds.), pp. 160–173. Arnold Printing Corp., Ithaca, New York.
Hanson, R. P. (1976). *Dev. Biol. Stand.* **33,** 297–301.
Hanson, R. P. (1978). *In* "Diseases of Poultry" (M. S. Hofstad, ed.), 7th ed., pp. 513–535. Iowa State Univ. Press, Ames.
Hanson, R. P., and Brandly, C. A. (1955). *Science* **122,** 156–157.
Hanson, R. P., and Spalatin, J. (1978). *Avian Dis.* **22,** 659–665.
Hanson, R. P., Spalatin, J., and Jacobson, G. S. (1973). *Avian Dis.* **17,** 354–361.
Harter, D. H., and Choppin, P. W. (1967). *J. Exp. Med.* **126,** 251–266.
Heuschele, W. P., and Easterday, B. C. (1970a). *J. Infect. Dis.* **121,** 486–496.
Heuschele, W. P., and Easterday, B. C. (1970b). *J. Infect. Dis.* **121,** 497–504.
Hore, D. E. (1966). *Vet. Rec.* **79,** 466–467.

Hore, D. E., and Stevenson, R. G. (1967). *Vet. Rec.* **80,** 26-27.
Hore, D. E., and Stevenson, R. G. (1969). *Res. Vet. Sci.* **10,** 342-350.
Hore, D. E., and Stevenson, R. G., Gilmour, N. J. L., Vantsis, J. T., and Thompson, D. A. (1968). *J. Comp. Pathol.* **78,** 259-265.
Horsfall, F. L., Jr., and Curnen, E. C. (1946). *J. Exp. Med.* **83,** 43-64.
Horsfall, F. L., Jr., and Hahn, R. G. (1939). *Proc. Soc. Exp. Biol. Med.* **40,** 684-686.
Horsfall, F. L., Jr., and Hahn, R. G. (1940). *J. Exp. Med.* **71,** 391-408.
Iida, T. (1972). *J. Gen. Virol.* **14,** 69-75.
Inaba, Y., Omori, T., Kono, M., and Matumoto, M. (1963). *Jpn. J. Exp. Med.* **33,** 313-329.
Inaba, Y., Tanaka, Y., Sato, K., Omori, T., and Matumoto, M. (1970a). *Jpn. J. Exp. Med.* **40,** 473-474.
Inaba, Y., Tanaka, Y., Sato, K., Ito, H., Omori, T., and Matumoto, M. (1970b). *Jpn. J. Microbiol.* **14,** 246-248.
Inaba, Y., Tanaka, Y., Sato, K., Omori, T., and Matumoto, M. (1972). *Jpn. J. Microbiol.* **16,** 373-383.
Inaba, Y., Tanaka, Y., Omori, T., and Matumoto, M. (1973). *Jpn. J. Microbiol.* **17,** 211-216.
Ito, Y., Tanaka, Y., Inaba, Y., and Omori, T. (1973). *Arch. Gesamte Virusforsch.* **40,** 198-204.
Jacobs, J. W., and Edington, N. (1971). *Vet. Rec.* **88,** 694.
Jacobs, J. W., and Edington, N. (1975). *Res. Vet. Sci.* **18,** 299-306.
Jakab, G. J. (1974). *J. Am. Vet. Med. Assoc.* **164,** 723-728.
Jun, M. H., Karabatsos, N., and Johnson, R. H. (1977). *Aust. J. Exp. Biol. Med. Sci.* **55,** 645-647.
Karabatsos, N., Buckley, S. M., and Ardoin, P. (1969). *Proc. Soc. Exp. Biol. Med.* **130,** 888-892.
Kawakami, Y., Kaji, T., Kume, T., Omuro, M., Hiramune, T., Murase, N., and Matumoto, M. (1966a). *Jpn. J. Microbiol.* **10,** 159-169.
Kawakami, Y., Kaji, T., Omuro, M., Maruyama, Y., Hiramune, T., Murase, N., and Matumoto, M. (1966b). *Jpn. J. Microbiol* **10,** 171-182.
Kawamura, H., and Tsubahara, H. (1968). *Natl. Inst. Anim. Health Q.* **8,** 1-7.
Kelter, A., Hamparian, V. V., and Hilleman, M. R. (1961). *J. Immunol.* **87,** 126-133.
Kingsbury, D. W. (1977). *In* "The Molecular Biology of Animal Viruses" (D. P. Nayak, ed.), Vol. 1, pp. 349-382. Dekker, New York.
Kingsbury, D. W., Bratt, M. A., Choppin, P. W., Hanson, R. P., Hosaka, Y., TerMeulen, V., Norrby, E., Plowright, W., Rott, R., and Wunner, W. H. (1978). *Intervirology* **10,** 137-152.
Kita, J., and Gillespie, J. H. (1968). *Cornell Vet.* **58,** 217-235.
Lancaster, J. E. (1963a). *Vet. Bull. (London)* **33,** 221-226, 279-285.
Lancaster, J. E. (1963b). *Vet. Bull. (London)* **33,** 347-360.
Lancaster, J. E. (1964). *Vet. Bull. (London)* **34,** 58-76.
Lancaster, J. E. (1966). *Can., Dep. Agric., Monog.* No. 3.
Lancaster, J. E. (1977). *World's Poult. Sci. J.* **33,** 155-165.
Lancaster, J. E., and Alexander, D. J. (1975). *Can., Dep. Agric., Monog.* No. 11.
Lang, G., Gagnon, A., and Howell, J. (1975). *Can. Vet. J.* **16,** 233-237.
Lazar, E. C., Swango, L. J., and Binn. L. N. (1970). *Proc. Soc. Exp. Biol. Med.* **135,** 173-176.
Lehmkuhl, H. D., and Cutlip, R. C. (1979). *Am. J. Vet. Res.* **40,** 512-514.
Lehmkuhl, H. D., Gough, P. M., and Reed, D. E. (1979). *Am. J. Vet. Res.* **40,** 124-126.
McCandlish, I. A. P., Thompson, H., Cornwell, H. J. C., and Wright, N. G. (1978). *Vet. Rec.* **102,** 298-301.
McFerran, J. B., Connor, T. J., Allen, G. M., Purcell, D. A., and Young, J. A. (1973). *Res. Vet. Sci.* **15,** 116-118.
McFerran, J. B., Connor, T. J., Allan, G. M., and Adair, B. (1974). *Arch. Gesamte Virusforsch.* **46,** 281-900.

Marshall, R. G. (1964). *J. Bacteriol.* **88,** 267–268.

Marshall, R. G., and Frank, G. H. (1975). *Am. J. Vet. Res.* **36,** 1085–1089.

Matumoto, M., Inaba, Y., Kurogi, H., Sato, K., Omori, T., Goto, Y., and Hirose, O. (1974). *Arch. Gesamte Virusforsch.* **44,** 280–290.

Mills, K. C., and Dochez, A. R. (1944). *Proc. Soc. Exp. Biol. Med.* **57,** 140–143.

Mills, K. C., Dochez, A. R. (1945). *Proc. Soc. Exp. Biol. Med.* **60,** 141–143.

Mims, C. A., and Murphy, F. A. (1973). *Am. J. Pathol.* **70,** 315–324.

Mohanty, S. B., Ingling, A. L., and Lillie, M. G. (1975). *Am. J. Vet. Res.* **36,** 417–419.

Mohanty, S. B., Lillie, M. G., and Ingling, A. L. (1976). *J. Infect. Dis.* **134,** 409–413.

Moore, N. F., and Burke, D. C. (1974). *J. Gen. Virol.* **25,** 275–289.

Morris, J. A., Blount, R. E., and McCown, J. M. (1956). *Cornell Vet.* **46,** 525–531.

Morris, J. A., Bozeman, F. M., Aulisio, C. G., and Shirai, A. (1963). *Proc. Soc. Exp. Biol. Med.* **113,** 296–300.

Moteane, M., Babiuk, L. A., and Schiefer, B. (1978). *Can. J. Comp. Med. Vet. Sci.* **42,** 246–248.

Mustaffa-Babjee, A., Spradbrow, P. B., and Samuel, J. L. (1974). *Avian Dis.* **18,** 226–230.

Nerome, K., Nakayama, M., Ishida, M., and Fukumi, H. (1978). *J. Gen. Virol.* **38,** 293–301.

Nguyen-Ba-Vy (1967a). *Recl. Med. Vet.* **143,** 29–48.

Nguyen-Ba-Vy (1967b). *Recl. Med. Vet.* **143,** 141–161.

Nymadava, P., Konstantinow-Siebelist, I., Schulze, P., and Starke, G. (1977). *Acta Virol.* **21,** 443.

Omar, A. R., Jennings, A. R., and Betts, A. O. (1966). *Res. Vet. Sci.* **7,** 379–388.

Paccaud, M. F., and Jacquier, C. (1970). *Arch. Gesamte Virusforsch.* **30,** 327–342.

Parker, J. C., and Reynolds, R. K. (1968). *Am. J. Epidemiol.* **88,** 112–125.

Parker, J. C., Tennant, R. W., and Ward, T. G. (1964a). *Fed. Proc., Fed. Am. Soc. Exp. Biol.* **23,** 580.

Parker, J. C., Tennant, R. W., Ward, T. G., and Rowe, W. P. (1964b). *Science* **146,** 936–938.

Parker, J. C., Tennant, R. W., and Ward, T. G. (1966). *Natl. Cancer Inst. Monogr.* **20,** 25–36.

Parker, J. C., Whiteman, M. D., and Richter, C. B. (1978). *Infect. Immun.* **19,** 123–130.

Pennington, T. H. (1978). *Arch. Virol.* **56,** 345–351.

Potgieter, L. N. D., and Aldridge, P. L. (1977). *Am. J. Vet. Res.* **38,** 1341–1343.

Proctor, S. J., Erickson, G. A., and Gustafson, G. A. (1974). *Proc. 17th Annu. Meet. Am. Assoc. Vet. Lab. Diagnosticians* pp. 115–121.

Profeta, M. L., Lief, F. S., and Plotkin, S. A. (1969). *Am. J. Epidemiol.* **89,** 316–324.

Reisinger, R. C. (1962). *Ann. N. Y. Acad. Sci.* **101,** 576–582.

Reisinger, R. C., Heddleston, K. L., and Manthei, C. A. (1959). *J. Am. Vet. Med. Assoc.* **135,** 147–152.

Richardson, L. S., Yolken, R. H., Belshe, R. B., Camargo, E., Kim, H. W., and Chanock, R. M. (1978). *Infect. Immun.* **20,** 660–664.

Robinson, T. W. E., Cureton, R. J. R., and Heath, R. B. (1968). *J. Med. Microbiol.* **1,** 89–95.

Rosenberg, F. J., Lief, F. S., Todd, J. D., and Reif, J. S. (1971). *Am. J. Epidemiol.* **94,** 147–165.

Rosenquist, B. D. (1974). *J. Infect. Dis.* **130,** 177–182.

Rossi, C. R., and Kiesel, G. K. (1978). *Arch. Virol.* **56,** 227–236.

Saona Black, L., and Lee, K. M. (1970). *Cornell Vet.* **60,** 120–134.

Sawicki, L. (1962). *Acta Virol.* **6,** 347–351.

Schiøtt, C. R., and Jensen, C. H. (1962). *Acta Pathol. Microbiol. Scand.* **56,** 479–480.

Schloer, G. (1974). *Infect. Immun.* **10,** 724–732.

Schloer, G. M., and Hanson, R. P. (1968a). *J. Virol.* **2,** 40–47.

Schloer, G., and Hanson, R. P. (1968b). *Am. J. Vet. Res.* **29,** 883–895.

Shah, K. V., Schaller, G. B., Flyger, V., and Herman, C. M. (1965). *Bull. Wildl. Dis. Assoc.* **1,** 31–32.

Shimokata, K., Nishiyama, Y., Yasuhiko, I., Kimura, Y., and Nagata, I. (1977). *Infect. Immun.* **16,** 706–708.

Sibinovic, K. H., Woods, G. T., Hardenbrook, H. J., and Marquis, G. (1965). *VM/SAC, Vet. Med. Small Anim. Clin.* **60,** 600–604.

Singh, K. V., and Baz, T. I. (1967). *Acta Virol.* **11,** 229–237.

Singh, K. V., and El Cicy, I. F. (1967). *Can. J. Comp. Med. Vet. Sci.* **31,** 70–79.

Sinha, S. K., and Abinanti, F. R. (1962). *Adv. Vet. Sci.* **7,** 225–271.

Smith, M. H., Frey, M. L., and Dierks, R. E. (1975). *Arch. Virol.* **47,** 237–247.

Spalatin, J., and Hanson, R. P. (1974). *Avian Dis.* **19,** 573–582.

Spalatin, J., Hanson, R. P., and Beard, P. D. (1970). *Avian Dis.* **14,** 542–549.

Starke, G., Alexander, D. J., Nymadawa, P., and Konstantinow-Siebelist, I. (1977). *Acta Virol.* **21,** 503–506.

Stevenson, R. G., and Hore, D. E. (1970). *J. Comp. Pathol.* **80,** 613–618.

Stone, H. D., Boney, W. A., Jr., and Coria, M. F. (1975). *Avian Dis.* **19,** 651–656.

Swift, B. L. (1973). *J. Am. Vet. Med. Assoc.* **163,** 861–862.

Takahashi, E., Inaba, Y., Kurogi, H., Sato, K., Goto, Y., and Omori, T. (1975). *Natl. Inst. Anim. Health. Q.* **15,** 179–185.

Tennant, R. W., and Ward, T. G. (1962). *Proc. Soc. Exp. Biol. Med.* **111,** 395–398.

Tennant, R. W., Parker, J. C., and Ward, T. G. (1966). *Natl. Cancer Inst. Monogr.* **20,** 93–104.

Thomas, L. H., Stott, E. J., Jebbett, J., and Hamilton, S. (1976). *Arch. Virol.* **52,** 251–258.

Thompson, H., Wright, N. G., and Cornwell, H. J. C. (1975). *Vet. Bull. (London)* **45,** 479–488.

Tikasingh, E. S., Jonkers, A. H., Spence, L., and Aitken, T. H. G. (1966). *Am. J. Trop. Med. Hyg.* **15,** 235–238.

Tsai, K. S., and Thomson, R. G. (1975a). *Infect. Immun.* **11,** 770–782.

Tsai, K. S., and Thomson, R. G. (1975b). *Infect. Immun.* **11,** 783–803.

Tyrrell, D. A. J., and Coid, C. R. (1970). *Vet. Rec.* **86,** 164–165.

Van Der Maaten, M. J. (1969). *Can. J. Comp. Med. Vet. Sci.* **33,** 141–147.

Van Der Veen, J., Poort, Y., and Birchfield, D. J. (1970). *Arch. Gesamte Virusforsch.* **31,** 237–246.

Van Der Veen, J., Poort, Y., and Birchfield, D. J. (1974). *Lab. Anim. Sci.* **24,** 48–50.

Van Nunen, M. C. J., and Van Der Veen, J. (1967). *Arch. Gesamte Virusforsch.* **22,** 388–397.

Ward, J. M. (1974). *Lab. Anim. Sci.* **24,** 938–942.

Ward, J. M., Houchins, D. P., Collins, M. J., Young, D. M., and Regan, R. L. (1976). *Vet. Pathol.* **13,** 36–46.

Wellemans, G. (1977). *Vet Sci. Commun.* **1,** 179–189.

Wellemans, G., and Leunen, J. (1975). *Ann. Med. Vet.* **119,** 359–369.

Wellemans, G., Leunen, J., and Luchsinger, E. (1970). *Ann. Med. Vet.* **114,** 89–93.

Woods, G. T. (1968). *J. Am. Vet. Med. Assoc.* **152,** Part 2, 771–777.

Woods, G. T. (1974). *Adv. Vet. Sci. Comp. Med.* **18,** 273–286.

Woods, G. T., Sibinovic, K., and Marquis, G. (1965a). *Am. J. Vet. Res.* **26,** 52–56.

Woods, G. T., Sibinovic, K., and Starkey, A. L. (1965b). *Am. J. Vet. Res.* **26,** 262–266.

Chapter 6

Morbillivirus Diseases of Animals and Man

MAX J. G. APPEL, E. PAUL J. GIBBS, SAM J. MARTIN,
VOLKER TER MEULEN, BERT K. RIMA,
JOHN R. STEPHENSON, AND WILLIAM P. TAYLOR

I.	Introduction	235
II.	Properties of the Virus	237
	A. Physicochemical Properties	237
	B. Biological Properties	242
III.	Interaction with Organisms	251
	A. Measles Infection in Man	251
	B. Canine Distemper	259
	C. Rinderpest	272
	D. Peste des Petits Ruminants	281
	References	287

I. INTRODUCTION

Measles virus (MV), canine distemper virus (CDV), and rinderpest virus (RV) have been known for a long time to be serologically related members within the Paramyxoviridae family (DeLay *et al.*, 1965). Several attempts were made to classify this group separately, e.g., "medipest viruses" (Melnick and McCombs, 1966) until the official genus *Morbillivirus* was established (Fenner, 1976). It has been proposed to add the virus of peste des petits ruminants (PPRV) to this group (Gibbs *et al.* 1979). There may be additional members in the future, as for example the Virus 107, isolated from a case of sporadic bovine meningoencephalitis (Bachman *et al.*, 1975), the cell-associated virus isolated

COMPARATIVE DIAGNOSIS
OF VIRAL DISEASES, VOL. IV

ISBN 0-12-429704-8

from a calf with experimental malignant catarrhal fever (Coulter and Storz, 1979), or the paramyxovirus causing fatal gastroenterocolitis in monkeys (Fraser *et al.*, 1978). However, sufficient evidence is not available at the present time for including additional members in the genus *Morbillivirus*. The presentation in this chapter, therefore, is restricted to MV, CDV, RV, and PPRV.

Many textbooks and review articles are available on MV and the diseases caused by it, such as those by Morgan and Rapp (1977) and Fraser and Martin (1978). Although the classical acute disease caused by MV in children has been greatly reduced after a successful vaccination program (Krugman, 1977), considerable interest has been retained in the more latent forms, as for example, subacute sclerosing panencephalitis (SSPE) (ter Meulen *et al.*, 1972a) or, as has been speculated, multiple sclerosis (MS).

Canine distemper (CD) has been recognized for many centuries as the most serious disease in dogs. Like measles in man, CD in dogs has been well controlled by efficient vaccination. Because unvaccinated dogs and CDV-susceptible wildlife species continue to spread the virus, eradication of the disease is not possible. Reviews on CDV and its associated disease cover earlier knowledge (Gorham, 1960; Bindrich, 1962; Gillespie, 1962; Appel and Gillespie, 1972). More recently, different "biotypes" of CDV have been found which cause a different but predictable course of the disease in dogs. Persistent infections with CDV can be induced *in vivo* and *in vitro*; however, the mechanism of CDV persistence, as well as MV persistence, is still poorly understood. For these reasons, research on CDV has recently received increased attention and new information is accumulating.

Rinderpest is primarily a disease of cattle, associated with high morbidity, high mortality, and a history of epi- or panzootic spread going back over many centuries. Previously more widespread, rinderpest is now largely confined to a belt of African countries immediately south of the Sahara and a group of Asian countries centered on India. It is a disease which can and has been effectively controlled by mass application of vaccine and one which has been reduced to such a low incidence that global eradication can now be contemplated. The excellent reviews of Scott (1964) and Plowright (1968) have been used as starting points for the present discussion.

Peste des petits ruminants (PPR) is a virus disease of considerable economic importance, affecting, as the name implies, small domestic ruminants. It is at present known only from West Africa, where it was first described in sheep and goats by Gargadennec and Lalanne (1942). In 1956 a relationship to bovine rinderpest was recognized (Mornet *et al.*, 1956), and for many years thereafter PPRV was considered to be a mutant of RV. This conception has only recently changed (Hamdy *et al.*, 1975), and PPRV is now proposed as an addition to the genus *Morbillivirus* of the Paramyxoviridae family (Gibbs *et al.*, 1979).

In 1967 Whitney *et al.* described a disease from the south of Nigeria which they called "Kata" and from which Johnson and Ritchie (1968) isolated a virus with a cytopathic effect identical to that now regarded as characteristic of PPRV. Rowland and Bourdin (1970) and Rowland *et al.* (1971) were unable to distinguish Kata from PPR clinically or histologically; therefore, in this chapter the two conditions will be considered synonymous.

We have attempted briefly to summarize former knowledge and to introduce more recent information on the morbilliviruses, their associated diseases, and diagnostic procedures. Because the physicochemical and biological properties of the different morbilliviruses are very similar, they have been compiled together. Data have not always been related to individual viruses, but references have been given for more detailed information. It has been assumed that similar properties would be expected in other members of the group. The interaction with the organism has been treated separately because considerable differences exist between the diseases induced in animals and man by the different morbilliviruses.

II. PROPERTIES OF THE VIRUS

A. Physicochemical Properties

1. Structure of the Virion

a. Morphology. The members of this virus group include MV, CDV, RV, and PPRV. They form a subgroup of the Paramyxoviridae family (ICNV cryptogram: R/1:4–8/1:S/E:v/o) (Kingsbury *et al.,* 1978).

All members of the group have a spherical or sometimes filamentous structure and are highly pleomorphic, with diameters that range from 100 to 700 nm (Fig. 1). The outer layer, or envelope, consists of a lipid bilayer, 5–8 nm wide, from which protrude spikes 9–15 nm in length. This heterogeneity in the size of the spikes may be due to the presence of more than one envelope protein. The lipid bilayer surrounds a coiled, hollow, rod-shaped nucleocapsid having the characteristic "herringbone" pattern of subunits (Fig. 2), similar in appearance to that found in other paramyxoviruses (Waterson *et al.,* 1961; Cruickshank *et al.,* 1962; Plowright *et al.,* 1962; Norrby *et al.,* 1963; Norrby and Magnusson, 1965; Waterson, 1965; Bourdin and Laurent-Vautier, 1967; Tajima *et al.,* 1971; Gibbs *et al.,* 1979).

b. Buoyant Density. Reported values for the buoyant density of the purified virus particle in either CsCl, sucrose, or potassium tartrate vary between 1.25 and 1.22 gm/cm^3 for infectious virus and drop to 1.20 gm/cm^3 for virus with a reduced infectivity after undiluted passages (Oddo *et al.,* 1961; Norrby, 1964; Chiarini and Norrby, 1970; Hall and Martin, 1973; Phillips and

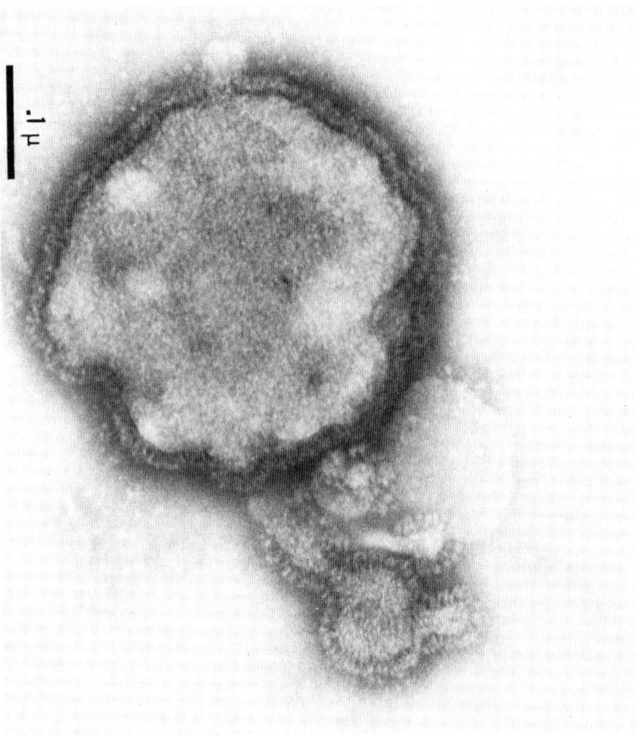

Fig. 1. Canine distemper virus (Onderstepoort strain replicated in Vero cells) negatively stained with phosphotungstate. Courtesy of Dr. H. Greisen.

Bussell, 1973; Underwood and Brown, 1974; Armitage *et al.*, 1975). It is unlikely that this change in density is directly related to the presence of subgenomic nucleocapsid material or defective interfering particles (Fraser and Martin, 1978), but it may be related to the enrichment of host membrane components which appears to occur during nonproductive infections (Rima and Martin, 1979).

c. Nucleocapsids. Nucleocapsids isolated from purified virus have a density of 1.30–1.31 gm/cm^3 and sediment between 100 and 200 S. When examined by electron microscopy these nucleocapsids are hollow, have an external diameter of 15–20 nm, an internal diameter of 7 nm, a pitch of 6.6 nm, and lengths of up to 1,000 nm, although shorter species predominate; rarely, circular and branched forms can be seen (Waterson *et al.*, 1961; Norrby and Magnusson, 1965; Waters *et al.*, 1972; Bussell *et al.*, 1974; Waters and Bussell, 1974; Thorne and McDermott, 1976; Gibbs *et al.*, 1979). Nucleocapsids from purified measles virus in the cytoplasm of infected cells have a rough appearance when examined by electron microscopy, whereas those from the nucleus have a smooth

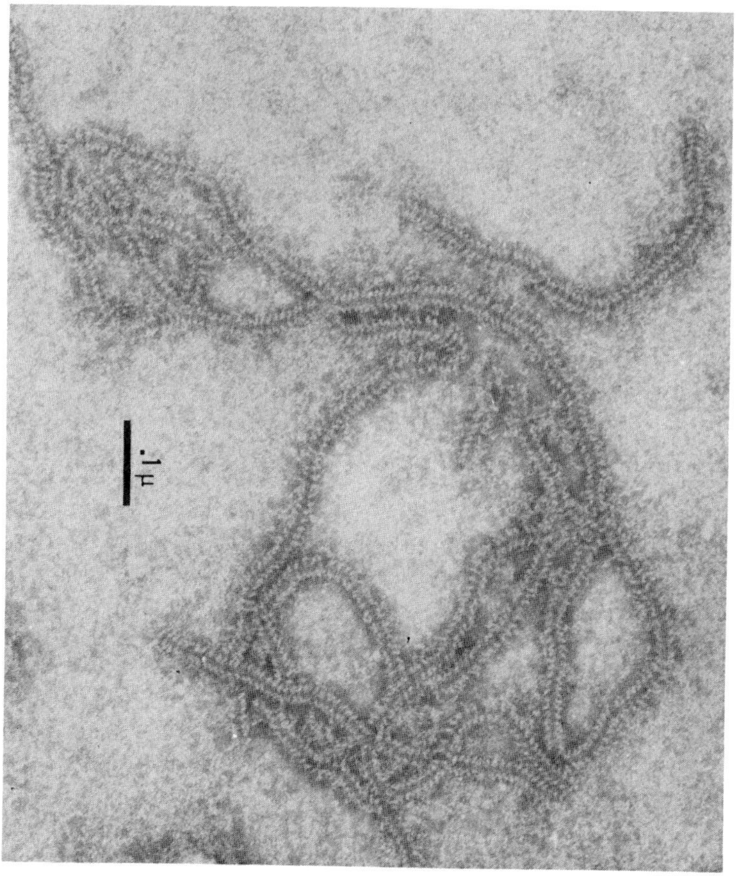

Fig. 2. Canine distemper virus nucleocapsid (Onderstepoort strain replicated in Vero cells) negatively stained with phosphotungstate. Courtesy of Dr. H. Greisen.

texture. This subcellular difference in nucleocapsid morphology is invariant whether the specimen comes from lytically infected cells, persistently infected cells, or fresh autopsy material (Oyanago *et al.*, 1971; ter Meulen *et al.*, 1973; Kratzsch *et al.*, 1977). The nucleocapsid contains about 5% by weight of RNA.

d. RNA. The RNA from purified virus is single-stranded, sediments at about 50 S, and has a molecular weight of approximately 5–6 $\times$ 10^6 on polyacrylamide gels (Schluederberg, 1971; Hall and Martin, 1973; Bussell *et al.*, 1974; Underwood and Brown, 1974; Wild *et al.*, 1974). In most preparations of virus, however, the major component is a smaller RNA fraction which probably represents defective interfering (subgenomic RNA) rather than degradation products. At present no unequivocal data are available as to whether the virion RNA is positive- or negative-stranded or if it has messenger acitivty. However, the virion

RNA is, in analogy to other paramyxoviruses, probably negative-stranded, as poly(A)-containing RNA isolated from Vero cells infected with MV, hybridized with RNA extracted from virions (Hall and ter Meulen, 1976). Furthermore, it has been reported that the viruses contain RNA-dependent RNA polymerase, but this has yet to be confirmed (Siefried *et al.*, 1978).

e. Polypeptides. Although there has been some discussion in the literature as to the precise number and molecular weight of morbillivirus polypeptides, the genome of this group of viruses appears to specify seven primary gene products. The molecular weights of these proteins bear some similarity among different members of the group but are not identical in every case. The nomenclature used here follows the scheme proposed by Graves *et al.* (1978) for measles virus. There are two glycoproteins in the viral envelope which by analogy to Sendai virus are assumed to be the hemagglutinin (H) or cell attachment factor, and the hemolysin or fusion protein (F). However, the corresponding biological activities of hemagglutination, hemadsorption, and hemolysis have been reported only for MV in spite of several attempts to detect them in CDV and RV. But cell fusion is seen in CDV- and RV-infected cells, and attachment is dependent upon the integrity of the surface antigens (Fisher and Bussell, 1977). The H glycoprotein of these viruses has a molecular weight of 76,000–80,000. The fusion protein is found as a glycosylated precursor (Fo) of molecular weight 59,000–62,000 in infected cells but is found only in the cleaved form in the virus. The cleavage products have molecular weights of 40,000 (F1) and 16,000–23,000 (F2), and their glycosylation varies. A small major virion polypeptide (M) has a molecular weight of 34,000–39,000 daltons and is assumed to be the protein lining the inner membrane of the virion. The nucleocapsid contains at least two proteins, a minor species of 66,000–73,000 daltons (P) and a major species of 58,000–62,000 daltons (N), both phosphorylated. The largest polypeptide in the virion (L = 100,000–200,000) has not been related to function or location (Waters and Bussell, 1973; Underwood and Brown, 1974; Mountcastle and Choppin, 1977; Graves *et al.*, 1978; Hall *et al.*, 1978c; Norrby, 1978; Tyrrell and Norrby, 1978; Rima *et al.*, 1979; Stallcup *et al.*, 1979; Campbell *et al.*, 1980; Hall *et al.*, 1980). In addition to the structural polypeptides, a putative nonstructural protein S (15,000–18,000) was found in CDV infected cells (Hall *et al.*, 1980; Campbell *et al.*, 1980). Other proteins, especially actin, are also found associated with virus particles and the proportions of these are increased following a nonproductive infection (Rima and Martin, 1979). Other minor bands that consistently appear following ^{35}S-labeling may be cleavage products of the N or larger viral polypeptides, as they disappear when infected cells are treated with inhibitors of proteolytic enzymes (B. K. Rima and S. J. Martin, unpublished).

Comparison of the polypeptides of several strains of MV, SSPE and CDV have been made (Hall *et al.*, 1978a; Wechsler and Fields, 1978; Rima *et al.*, 1979; Campbell *et al.*, 1980). Except for a consistent difference in the mobilities of the M-polypeptides of CDV compared to those of MV and SSPE isolates,none of the minor differences observed in the other proteins could be correlated with the type or origin of the virus.

2. Resistance to Physical and Chemical Treatment

a. Heat. Inactivation of morbilliviruses at various temperatures varies greatly, depending upon the substrate, the viral concentration, and the strain of virus used. Viruses suspended in liquid media had half-lives of 2 to 3.4 minutes at 56°C, 10 minutes at 45°C, 1 to 3 hours at 37°C, 2 hours at 21°C, and 9 to 11 days at 4°C (Bussell and Karzon, 1962; Lo *et al.*, 1965; Plowright, 1968; W. P. Taylor, unpublished). Addition of sodium or magnesium sulfate delayed heat inactivation (Rapp *et al.*, 1965; Robin and Bourdin, 1965). Lyophilized RV was stable for at least 3 years at 4°C and for at least 4.5 years at −20°C. At room temperature (20–22°C) the half-life was 14.3 weeks and at 37°C it was 3.2 weeks, but at 56°C not all of the virus was rapidly inactivated. Virus decayed at a half-life of 2.5 hours, but active virus was found up to 9 days (Plowright *et al.*, 1970). Freeze-dried vaccine virus reconstituted with distilled water to the original volume gave mean figures of 7.7 hours at 37°C, 17.8 hours at 25°C, and 23.3 hours at 4°C (Plowright *et al.*, 1971). The discrepancy before and after lyophilization remained unexplained.

b. Radiation. Morbilliviruses are slowly inactivated by visible light (Cutchins and Dayhuff, 1962; Nemo and Cutchins, 1966; Plowright *et al.*, 1971). RV was completely inactivated (Mirchamsy *et al.*, 1974) after short exposure to ultraviolet light. Palm and Black (1961) showed that CDV is inactivated by ultraviolet light at a rate of 1.0 log units/22,000ergs/cm^2. Irradiation with 10^5 of gamma quanta reduced the titer 0.4 log units at −70°C and 1 log unit at 0°C.

c. pH Stability. Reports on morbillivirus pH stability range from pH 4.5–9.0 for CDV (Bindrich, 1951) to 5.6–9.6 for RV and PPRV, with an optimal range of 7.2–8.0 (Liess and Plowright, 1963a; DeBoer and Barber, 1964; W. P. Taylor, unpublished).

c. Chemical Agents. Like other enveloped viruses, morbilliviruses are inactivated by lipid solvents like chloroform and ether (Bindrich, 1954; Palm and Black, 1961; Laurent, 1968; Gibbs *et al.*, 1979). Formalin at a concentration of 0.05% inactivated virulent CDV in 4 hours at 37°C (Bindrich, 1954). At a concentration of 0.013% and a pH of 7.1, 48 hours at 37°C were required for RV inactivation (Mirchamsy *et al.*, 1974). Stone and DeLay (1961) showed that 0.4% BPL inactivated $10^{5.0}$ of rabbit infectious doses of RV in 30 minutes at pH 8.0. Plowright (1968) found that 1 *M* hydroxylamine inactivated attenuated and virulent RV with half-lives of 26 and 74 minutes, respectively. CDV was inactivated by hydroxylamine (Kimes and Bussell, 1968) and two derivatives, *O*-methylhydroxylamine and *N*-methylhydroxylamine (Newlin and Bussell, 1972). Crystalline trypsin inactivated RV in 15 minutes at 37°C (Plowright,

1968) A 0.75% solution of phenol inactivated CDV in 10 minutes at 4°C and 0.3% Roccal, a commonly used quaternary ammonium disinfectant, had the same effect.

Metabolic inhibitors like bromodeoxyuridine did not inhibit morbillivirus replication (Gibbs *et al.*, 1979), while 6-azouridine at a concentration of 0.025 m*M* inhibited 99% of CDV replication (Phillips and Bussell, 1974). Reports on the effect of actinomycin D are contradictory (Anderson and Atherton, 1964; Matumoto, 1966; Nakai *et al.*, 1969; Follett *et al.*, 1976).

B. Biological Properties

1. Comparative Antigenicity

There have been many conflicting reports in the literature concerning antigenic relationships among the morbilliviruses (reviewed by Imagawa, 1968; Plowright, 1968; Appel and Gillespie, 1972). The studies of DeLay *et al.* (1965) using standard techniques and the same virus isolates in reciprocal tests demonstrated that many factors, such as variation in virus strain, species of animal employed, number of inoculations, relative amount of antigens, and type of test used, all influenced the results observed. Furthermore, limited allowance is made in the literature for the role of cell-mediated immunity when discussing protection by heterologous antigens. Such a cell-mediated immunity has been shown to possess a high degree of heterogeneity in its reaction to viruses which elicit distinct humoral antibody responses (Doherty *et al.*, 1977; McMichael *et al.*, 1977). Unequivocal evidence of antigenic cross reactivity has been demonstrated for MV, CDV, RV, and PPRV from studies on antigens (Waterson *et al.*, 1963; Orvell and Norrby, 1974; Gibbs *et al.*, 1979; Stephenson and ter Meulen, 1979; Hall *et al.*, 1980).

a. Antigenic Relationships between MV and CDV. It has been known for some time that human measles convalescent sera cross-react with CDV (Adams, 1953). However, sera taken during the acute phase of MV infection showed no heterologous activity (Adams and Imagawa, 1957). Experimental animals infected with MV gave similar results (Imagawa *et al.*, 1960; Karzon, 1962). Sato *et al.* (1973) reported that sera from SSPE patients cross-react to a much higher degree with CDV than do either sera from patients with acute atypical measles, convalescents, vaccines, or from hyperimmune animals. However, it is difficult to compare different studies in a quantitative fashion, as sera of widely varying titers are used. Data on the reverse process, i.e., the MV neutralization activity of CDV antisera, is equivocal. In general, dog CDV antiserum contains little (Carlström, 1958) or no neutralizing activity to MV (Karzon, 1962). However, high levels interfere with MV replication in dogs (Schultz *et al.*, 1977). The above studies on virus neutralization presumably detect only surface antigens responsible for attachment and penetration; other studies on infected cells, involving immune fluorescence (Yamanouchi *et al.*, 1970) and electron microscopy of ferritin-tagged antibody (Breese and DeBoer, 1973), show a much higher degree of cross reactivity, presumably because reactions with the internal viral antigens are detected. Similar high levels of cross reactivity have been observed using radioimmune assay directed against whole disrupted virions (Hall *et al.*, 1978b).

Studies on antigenic similarities between individual virus components reveal that the nucleocapsid of MV and CDV are identical antigenically (Orvell and Norrby, 1974). These authors also report that sera raised against CDV have a high level of antibody against MV hemolysin and low levels of activities against MV-H. More recently (Stephenson and ter Meulen, 1979; Hall *et al.*, 1980) an antigenic cross-reaction was found between all polypeptides of MV and CDV except the H proteins.

Protection against CD by immunization with MV appears to be species specific. Ferrets appear to be poorly protected by MV (Adams and Imagawa, 1957), whereas dogs are well protected from clinical disease, probably by an anamnestic immune response (Gillespie and Karzon, 1960; Moura and Warren, 1961; Slater and Murdock, 1963). However, they are not protected from CDV infection and virus replication in lymphatic cells because MV antibody in dogs, unless hyperimmunized (Roberts, 1965; Shishido *et al.*, 1967), does not neutralize CDV. In addition, MV in dogs did not stimulate immune-specific cell-mediated cytotoxicity (CMC) against CDV-infected cells, or the response was below a measurable level. An anamnestic-type response was found after challenge with CDV. Dogs after vaccination with CDV had immune lymphocytes that were cytotoxic for both CDV- and MV-infected cells (Shek, 1980). A cross-reacting CMC had been claimed earlier by skin testing (Brown and McCarthy, 1974) and by lymphocyte blastogenesis (Gerber and Marron, 1976). The reciprocal cross protection has not been convincingly demonstrated (Adams *et al.*, 1959), and children inoculated with live CDV showed no CF antibody to MV (Millian *et al.*, 1960). However, these children have not been exposed to virulent MV for testing protection against disease.

b. Antigenic Relationships between CDV and RV. Neutralization of CDV with RV antiserum was found by Goret *et al.* (1959). Mornet *et al.* (1959), Imagawa *et al.* (1960), and Orvell and Norrby (1974), but not DeLay *et al.* (1965). Neutralization of RV with CDV antiserum has been demonstrated by several groups (Goret *et al.*, 1960; DeLay *et al.*, 1965; Gibbs *et al.*, 1979). Heterologous reactions were always of a lower order than those with homologous sera. Monkeys and dogs did not elicit antibody to CDV when inoculated with RV (DeLay *et al.*, 1965), but the dogs were protected against subsequent challenge with CDV. Similar results have also been obtained by other workers with ferrets as well as dogs (Goret *et al.*, 1957; Polding *et al.*, 1959). Polding *et al.* (1959) and Goret *et al.* (1960) failed to protect animals against CDV when passively immunized with RV antiserum. In the reverse process, monkeys inoculated with CDV had low levels of RV antibody, but cattle did not (DeLay *et al.*, 1965). Goret *et al.* (1958) reported successful protection from rinderpest in cattle inoculated with ferret-adapted CDV, but these authors had less success with egg-adapted CDV, and others have failed to show such protection either in cattle (Polding *et al.*, 1959; DeLay *et al.*, 1965) or in rabbits (Plowright, 1962).

A detailed antigenic comparison has not yet been made. Yamanouchi *et al.* (1970) reported a high level of cross reactivity between the antigens on the surface of the infected cells when examined by fluorescent staining. A similar increased cross reactivity has been reported using a radioimmune assay directed against whole disrupted virus (Hall *et al.*, 1978b). RV antiserum precipitated all structural polypeptides of both, MV and CDV (Hall *et al.*, 1980).

c. Antigenic Relationship between MV and RV. Measles convalescent serum has been shown to neutralize RV (Plowright and Ferris, 1959), and RV antiserum contains neutralizing acitvity against MV (Imagawa *et al.*, 1960; Plowright, 1962; Gibbs *et al.*, 1979). With one exception, the heterologous activity was much less than for the homologous virus. Anti-RV activity in measles convalescent sera was enhanced by the addition of guinea pig complement (Plowright, 1962). Studies on antigenic relationships using electron microscopy (Breese and DeBoer, 1973) and immune fluroescence (Yamanouchi *et al.*, 1970) showed a high degree of cross reactivity, as did studies on disrupted virus using a radioimmune assay (Hall *et al.*, 1978b). Orvell and Norrby (1974) have also shown that rinderpest serum contains hemolysin inhibition activity and hemagglutinin inhibition activity against MV, with the former predominating in many cases. Heterologous response to immunization appears to be species specific. Dogs and children vaccinated with MV both produced antibodies against RV, but cattle did not. Calves inoculated with RV produced antibodies against MV (Plowright, 1962; DeLay *et al.*, 1965). Again, the heterologous response was of a lower order than the homologous response. No convincing protection against rinderpest was achieved by inoculating cattle with MV, although rabbits were protected after three or more injections (Plowright, 1962; DeLay *et al.*, 1965). Studies to protect monkeys against measles by vaccinating with RV were inconclusive (DeLay *et al.*, 1965).

d. Antigenic Relationship of PPRV to Other Morbilliviruses. Early reports showed that PPRV conferred immunity in calves against rinderpest and could be neutralized by rinderpest serum *in vitro* (Mornet *et al.*, 1956; Gilbert and Monnier, 1962). At the time of these investigations, it was concluded that PPRV was a variant of RV adapted to growth in small ruminants.

Hamdy *et al.* (1975) reinvestigated the relationship between RV and PPRV. In goats, reciprocal cross protection and reciprocal cross complement-fixing antibody was demonstrated, establishing that the viruses were related. In a further study (Hamdy *et al.*, 1976), reciprocal cross-precipitating antibody was shown, but in neither of these studies nor in a further experiment involving American white-tailed deer (Hamdy and Dardiri, 1976) could cross neutralization be shown. They concluded, therefore, that the viruses were not identical. Taylor and Abegunde (1979) confirmed that the viruses were distinct but were able to show the existence of a low level of cross neutralization. Dardiri *et al.* (1976) showed that PPRV antiserum fixed complement with MV, CDV, RV, and PPRV, while Gibbs *et al.* (1979) investigated the group relationships using neutralization tests. They showed that the viruses were all related to each other but that no two agents were identical.

2. *Growth in Cell Culture*

a. Virus Isolation. Virulent MV, CDV, RV, and PPRV can best be isolated in primary cell cultures; MV in human or monkey embryo kidney cells (Enders and Peebles, 1954; Matumoto, 1966; Lennette and Schmidt, 1969); CDV in dog or ferret macrophages (Appel and Jones, 1967; Poste, 1971); RV in renal or testicular cells from a variety of domestic ruminants (Plowright and Ferris, 1959; Bansal *et al.,* 1975; Das and Datt, 1975) in bovine thyroid cells (Plowright and Ferris, 1962), or in transformed lung macrophages (M. J. P. Lawman, personal communication); and PPRV in lamb or goat kidney cells (Gilbert and Monnier, 1962; Johnson and Ritchie, 1968; Taylor and Abegunde, 1979) or in goat lung cells. In contrast to MV, RV, and PPRV, which can be grown in primary kidney cells without adaptation, virulent CDV grows to a very limited extent in dog kidney cells unless adapted (Appel, 1978). Growth in leukocyte cultures, especially in stimulated lymphocytes, has been reported for MV (Sullivan *et al.,* 1975), CDV (Poste, 1970), and RV (Tokuda *et al.,* 1962). Isolation in continuous cell lines can best be achieved by inoculation of buffy coat suspensions from infected animals (Hamdy *et al.,* 1976). Virus-induced cell fusion may lead to initiation of infection.

Cell cultures prepared from tissues of infected animals replicate virus. If circulating viral antibody is present, it may be the only way to isolate virus, as for example, from SSPE patients. Brain cultures from these patients yield incomplete cell-bound virus only. Released virus becomes available after co-cultivation with different cells (Chen *et al.,* 1969; Horta-Barbosa *et al.,* 1969). The same effect can be achieved by overlaying minced SSPE brain tissue directly on a cell monolayer, e.g., Vero cells. Virus spreads to Vero cells presumably by cell fusion (V. ter Meulen, unpublished). In contrast, direct brain (Fig. 3), kidney, spleen, or lung cell cultures from dogs with distemper produce free CDV without co-cultivation (Vantsis, 1959; Hall *et al.,* 1979a; Imagawa *et al.,* 1979). Although not yet reported, similar results could be assumed from RV infected bovine or PPRV infected small ruminant tissues.

b. Passaged Virus. Once adapted, morbilliviruses grow in a large number of established cell lines from different species. Cell lines used for MV include

Vero, HeLa, Hep-2, AV3, BSC-1, MDCK cells (Matumoto, 1966), and CV-1 (Wechsler and Fields, 1978); for CDV, canine tumor cell lines, MDCK, Vero, BSC-1, AV3, HeLa, and Hep-2 cells (reviewed by Appel and Gillespie, 1972); for RV, MDBK (Johnson, 1962) HeLa (Liess and Plowright, 1963b), MS (Singh and El Cicy, 1966), BHK (Anonymous, 1966) and Vero cells (Shishido *et al.*, 1967; Mirchamsy *et al.*, 1974); and for PPRV, Vero cells (Hamdy *et al.*, 1976; Taylor and Abegunde, 1979).

In addition to cell lines, morbilliviruses can be adapted to a large number of primary cell cultures from different species, e.g., CDV has been grown in epithelial or fibroblast cells of canine, feline, mustelid, avian, bovine, simian, and human origin (Bussell and Karzon, 1965; reviewed by Appel and Gillespie, 1972). Bovine, simian, and human cells were used for the propagation of PPRV (Laurent, 1968).

c. Virus Multiplication. It is believed that morbillivirions attach by means of a glycoprotein (H) to specific cellular receptors. Fusion of the cellular membrane with the viral envelope probably causes release of nucleocapsids into the cytoplasm.

Morbillivirus RNA and antigens are produced in the cytoplasm of infected cells, and complete virus is formed by budding from the plasma membrane. Virus release appears to occur when nucleocapsids align themselves along the inside of the plasma membrane in areas rich in viral glycoproteins. By analogy with myxo- and rhabdoviruses, the assembly is probably regulated and mediated by the presence of M protein (reviewed by Morgan and Rapp, 1977). Morbillivirus replication is assumed to be similar to the replication of other paramyxoviruses (Kingsbury, 1972; Choppin and Compans, 1975).

The growth cycle of morbilliviruses is extremely variable, depending on the viral strain, the type of cell culture, and the multiplicity of infection. The time of maximal adsorption of virus to cells, which may be greatly influenced by the volume of suspending fluid, by temperature and pH, varied from 30 minutes for RV (Taylor and Perry, 1970) to 2 hours for CDV (Bussell and Karzon, 1962). The eclipse phase varied from 8 to 18 hours for CDV (Bussell and Karzon, 1962; Shishido *et al.*, 1967) when free virus was first detected. A plateau was reached between 18 and 24 hours postinfection (PI). Using different strains of CDV in the same cell cultures at the same multiplicity of infection, a plateau may be reached only after 7 days. Peak values of RV occurred between 4 and 10 days PI, and the first free virus was released after 14 to 24 hours (Plowright, 1964b; Plowright *et al.*, 1969). By electron microscopy, budding of PPRV was first seen 12 hours PI (Laurent, 1968).

By immunofluorescence (IF) staining, small fluorescent granules in the cytoplasm around the nucleus were seen by 12 to 24 hours post MV, CDV, and RV infection of cell cultures. The number and size of granules increased with time.

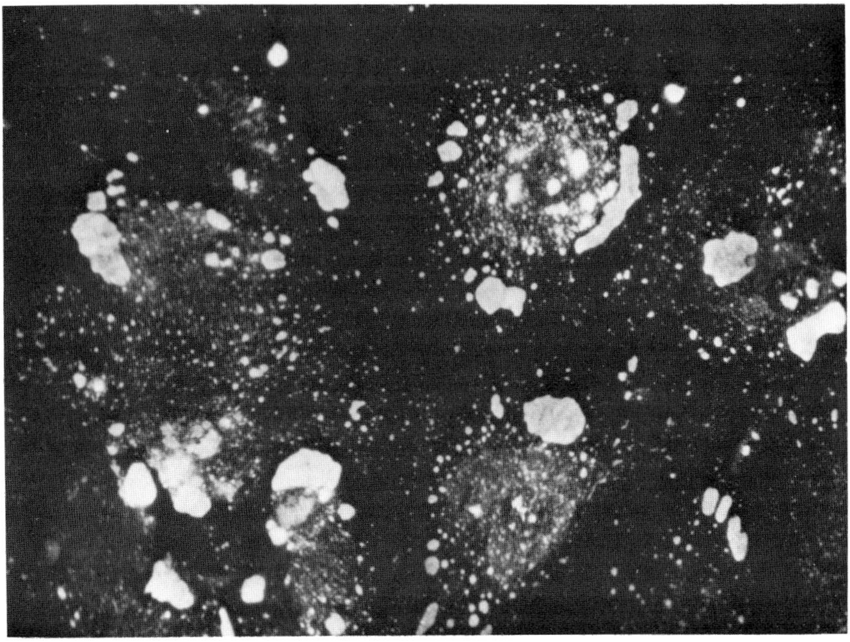

Fig. 3. Canine distemper viral antigen in brain cell culture prepared from a dog infected with CDV (Cornell A75-17 strain). Immunofluorescence. × 250.

Intranuclear fluorescence and cell fusion occurred later (Ushijima *et al.*, 1968; Yamanouchi *et al.*, 1970).

Although variable, the majority of infectious morbillivirus particles remain cell associated (Plowright, 1964b; Mannweiler, 1965; Matumoto, 1966; Shishido *et al.*, 1967; Ushijima *et al.*, 1968). Virus budding appears to be reduced when syncytial formation occurs (Laurent, 1968; Rentier *et al.*, 1978). Cell fusion inhibition during viral replication appears to increase the yield of infectious virus (Graves *et al.*, 1978).

d. Cytopathic Changes. The lytic morbillivirus infection in cell culture is characterized by intracytoplasmic and intranuclear eosinophilic inclusion bodies and multinucleated giant cells (Figs. 4, 5, 6) (Enders and Peebles, 1954; Rockborn, 1958; Plowright and Ferris, 1959; Johnson and Ritchie, 1968). These syncytia appear to occur from the action of the virus-specific fusion protein rather than from abnormal nuclear division (McCarthy, 1959). In some isolates these syncytia show a characteristic spindle or stellate form which appears to be dependent upon the nutritional state of the cells, their previous passage history and multiplicity of infection, and the isolate used (Black, 1959; Toyoshima *et al.*,

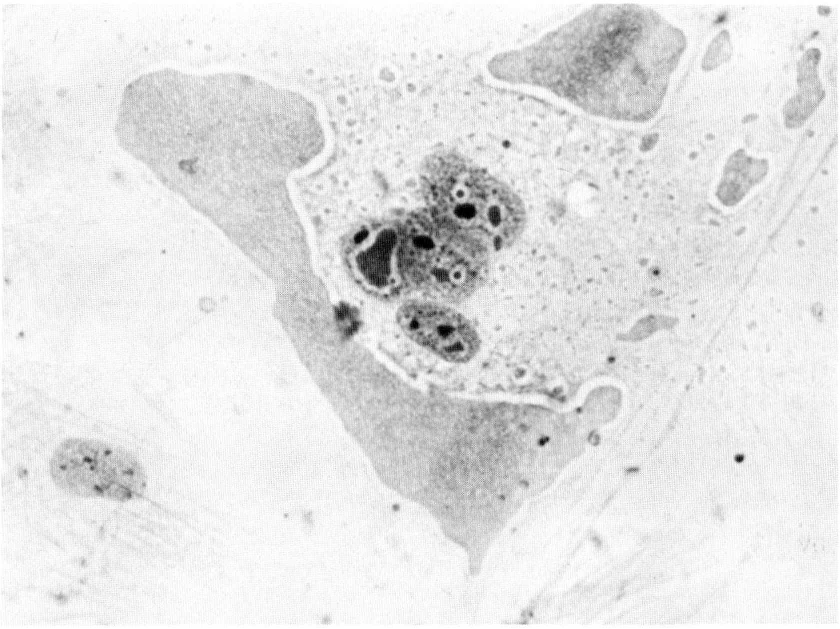

Fig. 4. Intranuclear and intracytoplasmic inclusion bodies and syncytia formation in spleen cell culture prepared from a dog infected with CDV (Snyder Hill strain). Shorr's stain. × 450.

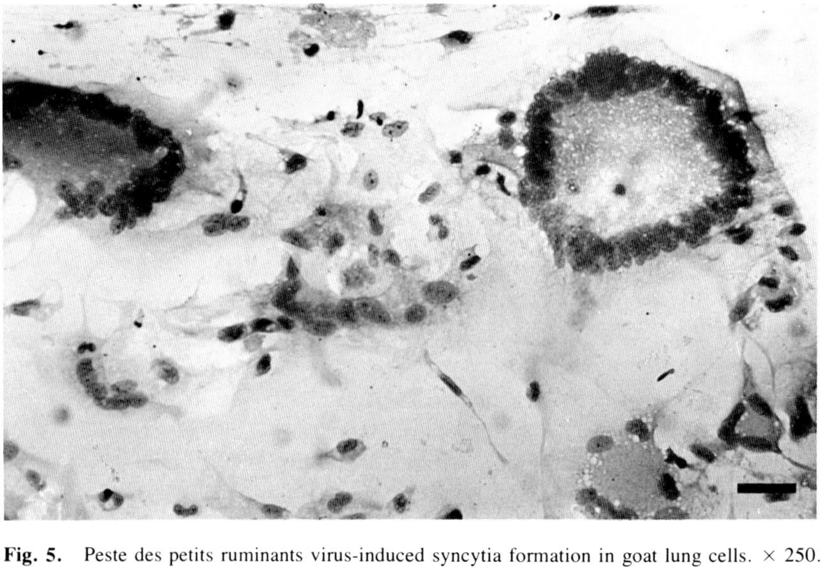

Fig. 5. Peste des petits ruminants virus-induced syncytia formation in goat lung cells. × 250.

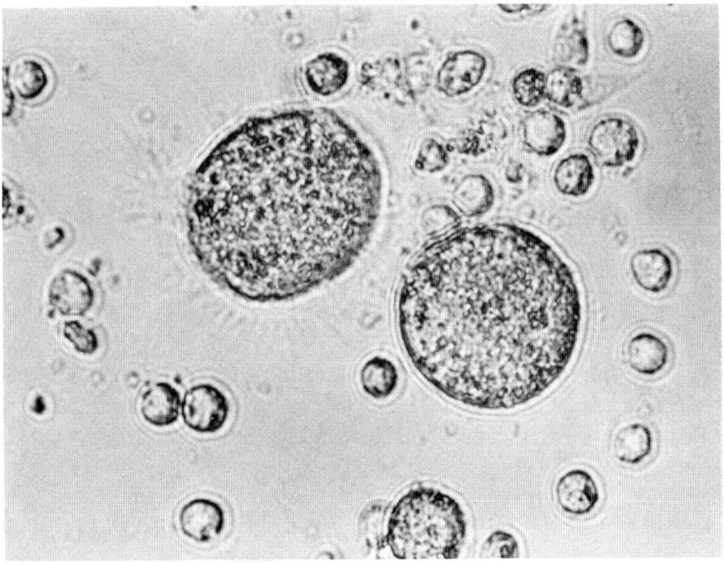

Fig. 6. Syncytia formation in dog lung macrophage cultures induced by virulent CDV (R252 strain). × 250.

1959; Hamdy *et al.*, 1976). Intracellular inclusion bodies show positive histochemical staining for RNA and contain nucleocapsid-like structures when examined by electron microscopy (Müller and ter Meulen, 1969). Chromosome breakage of MV-infected cells has been reported (Nichols and Levan, 1965).

e. Plaquing. Plaquing of morbilliviruses depends greatly on the strain of virus and the type of cell culture used. In some systems, plaquing is not possible or plaques develop slowly and remain small (Bussell and Karzon, 1962, 1965; Taylor and Perry, 1970). Vero cells are suitable for propagation and plaquing of morbilliviruses (Shishido *et al.*, 1967; Gourlay, 1970; Ozawa and Nelson, 1973). Other cell systems have been used with various results (Hsiung *et al.*, 1958; Underwood, 1959; McKercher, 1964; Harrison *et al.*, 1968; Das and Datt, 1975).

f. Persistent Infections. A persistent infection with MV was first established in HeLa cells by Rustigian (1962) and with CDV in Vero cells by ter Meulen and Martin (1976), i.e., an infection in which most cells produce virus antigen but little or no infectious virus, and their growth rates, morphology, and plating efficiency are indistinguishable from those of uninfected cells. The importance of these early experiments has been realized more recently since it has been shown that SSPE is closely associated with a persistent infection by measles virus. Using the same "cell survivor" technique, many other cell cultures persistently infected with MV have been subsequently established in HeLa cells, KB cells, primate lung cells, mammalian kidney cells, and rodent brain cells (summarized in Fraser and Martin, 1978). Although some of these cultures are of relatively short duration, they all produce little or no infectious virus and show a minimal amount of cytopathic change. Broadly speaking, these persistent cultures can be described by whether or not they produce virus, CPE, and cell surface antigens and whether these characteristics can be activated by various techniques. The HeLa cultures of Rustigian (1966) produced no virus or CPE, and the level of surface and nuclear antigen decreased with time. The culture established by Gould (1974) in Hep-2 cells again produced no CPE but had constant levels of antigen both in the nucleus and on the cell surface. The culture produced a significant amount of virus which was increased by the addition of actinomycin D. Although most persistent cultures do not show CPE, the Lu106 line (now considered to be of HeLa origin) (Norrby, 1967a,b) does contain a significant CPE and produces infectious virus. Whereas this line was unaffected by actinomycin D treatment, there was an increase in CPE and released virus on incubation at a lower temperature. Other cultures which demonstrate a CPE are those of Doi *et al.* (1972), Burnstein *et al.* (1974), and Kratzsch *et al.* (1977), all from SSPE brain isolates. None of these cultures produced infectious virus, and all attempts to rescue virus were negative; failure of viral M protein synthesis may prevent production of complete virus (Hall and Choppin, 1979); however, in all cases the neurovirulence of the agent in the cells was demonstrated by intracerebral injection into animals. Several authors have reported that subgenomic species are more prevalent in persistent infections than in lytic infections (Schluederberg *et al.*, 1972; Winston *et al.*, 1973; Hall *et al.*, 1974; Kratzsch *et al.*, 1977; Rima *et al.*, 1977; Fisher and Rapp, 1979a). Thus the situation in morbilliviruses may resemble that in the rhabdoviruses, in which subgenomic particles play a major role in the establishment of persistent infections. However, these studies did not distinguish between the production of subgenomic mRNA and nucleocapsid RNA.

Recent studies on proteins from persistently infected cells show several polypeptides with mobilities slightly different than those of the parent virus (Wechsler *et al.*, 1979a; Rozenblatt *et al.*, 1979). The latter authors also reported that the antigenicity of at least some of the proteins from acutely or chronically infected cells, made either *in vivo* or *in vitro*, appears to be unaltered in spite of changes in their electrophoretic mobilities. The relevance of such differences is not clear as similar differences occur between different isolates when grown in lytically infected cells (Wechsler and Fields, 1978).

Temperature-sensitive viral mutants may be important in viral persistence. Ju *et al.* (1978) induced persistent infection in human lymphoblastoid cell lines with MV (Edmonston strain). Little infectious virus was detected when cells were cultured at 37°C, but a temperature-sensitive population of virus was released at 31°C. Differences in the expression of virus structural proteins at permissive and nonpermissive temperatures were found in hamster embryo fibroblasts persistently infected with MV (Fisher and Rapp, 1979b).

The mechanism of establishment for persistent infections is not known. The main theoretical models implicate virus mutation, defective interfering (DI) particles, interferon, selection of a susceptible host cell, and integration of virus genetic material into that of the host cell by means of reverse transcription (reviewed by Morgan and Rapp, 1977; Rima and Martin, 1977). Evidence for all of these phenomena has been found in many persistently infected cultures, although the precise role of any of them is not clear.

III. INTERACTION WITH ORGANISMS

A. Measles Infection in Man

Measles virus infection occurs in nature only in man and monkeys which are in close contact with man. Only these species are susceptible to this virus by the natural route of infection.

1. Route of Infection

Based on clinical and experimental observations, it is almost certain that measles virus invasion takes place by the respiratory route. Primary infection occurs in the upper respiratory epithelial cells and the associated lymphoid tissue. After dissemination, measles virus is found in the respiratory tract, intestinal tract, and renal tract as well as in the nervous system. In addition, virus can be detected in lymphoid tissues in many parts of the body, as demonstrated by the presence of measles virus antigen and structures in these organs. This association indicates a certain tropism of measles virus for this tissue. During the viremic phase of measles, lymphocytes carry the virus and transport it widely within the body. In tissue culture experiments, virus can be shown to replicate in these cells when artificially stimulated (Osunkoya *et al.*, 1974a,b). Such stimulated T or B lymphocytes produce *in vitro* viral antigen and infectious virus. Moreover, in the 11–13 days after measles virus infection, a marked leukopenia occurs in man which affects mainly lymphocytes and eosinophilic leukocytes.

2. Symptomatology of the Disease

Measles is an acute disease occurring mainly in childhood. After an incubation period of approximately 10–12 days, the disease develops with a certain stereotyped appearance. The prodromal phase, which usually lasts 3–4 days, is characterized by mild fever, cough, coryza, conjunctivitis, and photophobia. These symptoms precede the Koplik's spots, the earliest pathognomonic sign of measles. These spots are grayish-white dots with a slight reddish areola located usually opposite the lower molars. They appear and disappear rapidly within 24 hours before the rash develops. The rash starts as faint macules usually on the neck and cheeks and spread successively within 3 days over the entire body. With this spread the initial lesion becomes increasingly maculopapular, developing into the typical measles exanthema. With the appearance of the rash, the temperature usually rises abruptly to 39°–40°C and subsides rapidly in uncomplicated cases when the rash has reached the lower part of the legs. The fading of the rash proceeds downward in the same sequence as its appearance. Usually lymph nodes are enlarged in the posterior cervical region and at the angle of the jaw. Sometimes a slight splenomegaly may be present. Variations in the type of rash may occur, such as petechiae and ecchymoses in severe cases or as urticarial or scarlatiniformal exanthema at the early prodromal stage. Complete absence of a rash similar to that of scarlet fever can be observed in persons who have received measles antibodies during the incubation period.

3. Measles Complications

The main complications of measles are otitis media, pneumonia, and encephalitis. Infections of the lung lead to an interstitial lesion which is pathologically recognized as giant cell pneumonia. This disease was once related to CD (Pinkerton *et al.*, 1945), but after isolation procedures became available, the relationship of this interstitial pneumonia to measles virus could be established (Enders *et al.*, 1959). However, it was noted that patients with such a complication exhibit a suppression of antibody formation, suggesting a pathogenetic relationship between impairment of the immune response and giant cell pneumonia (Mitus *et al.*, 1959).

Neurological complications are more common in measles than in any of the other viral infections associated with exanthema. The incidence of encephalomyelitis is estimated to be 0.5 to 1 per 1000 reported cases of measles. No correlation has been found between the acuteness of measles and the occurrence of encephalitis. This complication usually develops 2 to 5 days after the appearance of rash, but cases have been observed at the prodromal phase of measles or at the time of rash. The prognosis of this central nervous system (CNS) complication is not related to severity of the disease process. The pathogenesis has not yet been determined. Infectious measles virus cannot be recovered normally from brain material with standard techniques. Only in a few instances during (McLean *et al.*, 1966) and after encephalitis (ter Meulen *et al.*, 1972a,b) has isolation of measles virus been accomplished. These studies indicate that measles virus is present in diseased brain areas and show that the virus replicates in CNS cells. Whether virus infection alone or in conjunction with an immune pathological reaction leads to this encephalitis is unknown. Beside these complications, another CNS disease has been associated with measles virus. SSPE, a slow virus infection, is a chronic, inflammatory, slowly progressing disorder of the CNS which occurs mainly in children and young adults (ter Meulen *et al.*, 1972a; Agnarsdottir, 1977; ter Meulen and Hall, 1978). This disease has certain clinical and laboratory features which are pathognomonic for this neurological condition. Clinically a stereotyped pattern is observed which consists of insidious behavioral changes at onset followed weeks or months later by characteristic neurological symptoms, progressing to decerebration and death. This course can last for months or years. Laboratory investigations have revealed a characteristic electroencephalogram (EEG) pattern, along with a marked elevation of IgG in the cerebrospinal fluid (CSF), as well as high measles antibody titers in serum and CSF. The resultant reduction in the serum/CSF ratio is indicative of an intracerebral production of IgG antibodies. Moreover, oligoclonal IgG can be found in CSF which carry measles antibody activities. Virological studies revealed measles virus antigen in CNS cells and led to isolation of measles-like virus (referred to as "SSPE virus") from brain and lymph node material. Antibody to the viral M protein was not found in sera from SSPE patients (Hall *et al.*, 1979b; Stephenson and ter Meulen, 1979a; Wechsler *et al.*, 1979b). However, despite these findings, which incriminate the measles virus as the causative agent, the pathogenesis of SSPE and its connection to acute measles are unknown. Epidemiological surveys have shown that SSPE is a rare disease ($1{:}10^6$ of the childhood population with a history of an acute measles infection). On the average the acute measles infection occurs 4–6 years before the onset of SSPE, although a high percentage of cases had measles before the age of 2. In view of the wide dissemination of measles virus in the body during acute measles, it is conceivable that virus also reaches the CNS. This is supported by the frequent findings of pleocytosis in CSF (Ojala, 1974) and EEG changes (Gibbs *et al.*, 1959) during uncomplicated measles. If, under certain circumstances, measles virus can escape the defense mechanisms of the host, virus persistency could be established which, after an incubation period of several years, becomes activated by events as yet unknown. The understanding of these disease processes will depend to a great extent on the role of the immune response to this CNS infection and on our knowledge of the virus–host relationship by which the virus persists in brain tissue.

4. *Virus Strains*

A virus has been isolated from blood, throat washings, and fecal material of patients in the acute stages of disease. Virus isolates have been made from brains of patients with measles encephalitis and SSPE. Although some of these SSPE virus isolates have revealed some biochemical and biological differences from the standard measles strains (Hall and ter Meulen, 1976; Hall *et al.*, 1978a; Stephenson and ter Meulen, 1979b; Campbell *et al.*, 1980), it still remains unknown whether the viruses causing SSPE are only closely related or identical with the measles viruses causing acute measles (Rima *et al.*, 1979). All MV isolates so far made show no antigenic differences when compared by standard serological techniques. Details of the origins of these and other human morbillivirus isolates can be found in Majer (1972).

5. *Immunity*

Measles virus infection in man induces a humoral and cell-mediated immune response which provides lifelong immunity. Viral antibodies are formed against the different antigenic components of the virus during acute infection which are already recognizable at the acute stage of the disease. As with other infectious agents, measles virus stimulates the induction of specific IgM and IgG antibodies which in the case of IgM disappear shortly after the course of infection. Measles-specific IgM is normally absent from late convalescent and adult sera, whereas measles-specific IgG antibody persist throughout life. In addition, measles-specific IgA has been found, but its role in the event of measles infection and immunity has not been characterized yet (reviewed by Fraser and Martin, 1978). However, clinical experiences have revealed that the functional humoral immune response of a patient during measles is not essential in controlling this virus infection. Children with agammaglobulinemia overcome acute measles without major difficulties, in contrast to those with an insufficient cellular immune reaction. It has been observed that patients with congenital, acquired, or iatrogenic T-cell deficiency develop a severe measles infection often accompanied by fatal complications (Gatti and Good, 1970). Similar complications have been found in children with leukemia and other systemic diseases which lead to immunosuppression (Breitfeld *et al.*, 1973), indicating that an intact cell-mediated immune system is vital for the host to survive.

On the other hand, measles virus itself leads to a direct effect on lymphocytes during acute measles. In lymphoid organs, multinucleated giant cells are detectable which reveal destruction of lymphocytes by virus replication. In addition, peripheral lymphocytes from measles patients show an increased frequency of chromosome breakage, which has been linked to the virus infection (Nichols *et al.*, 1962; Tanzer *et al.*, 1963; Mori, 1969). The involvement of the lymphoid organ in measles infection can lead to a functional impairment, which von Pirquet originally observed in 1908 and which has been confirmed by others (Starr and Berkovich, 1964; Fireman *et al.*, 1969). During or shortly after acute measles, tuberculous children develop a partial or total skin anergy against tuberculin and other antigens. Moreover, an activation of a tuberculous process or a high rate of secondary infections is often seen during measles attack. In most instances, this infection of the lymphoid system is not harmful to healthy children. However, it is conceivable that occasionally such infections may lead to an impairment of the immune system against measles virus itself. The possibility that the immune system fails to control and eliminate measles virus has been suggested as a pathogenetic mechanism in subacute sclerosing panencephalitis (Burnet, 1968).

Efficient laboratory assays to test the complete spectrum of cell-mediated immune reactions are not yet available for measles virus. Unlike other viruses, measles virus does not stimulate, but depresses, the blastogenic response of lymphocytes. This effect is also observed on concurrent stimulation with measles virus and mitogens. The mechanism of this depression may be mediated by interferon (A. Neighbour and B. R. Bloom, personal communication). An effector not associated with the measles virion which suppresses host DNA synthesis *in vitro* has been described in the supernatant of measles-infected cells (Minagawa *et al.*, 1974). So far, only one aspect of cell-mediated immunity has been investigated. Lymphocytes from individuals with and without a history of measles are cytotoxic for measles-infected cells *in vitro*, provided these target cells carry an FC receptor on their cell membranes (Valdimarsson *et al.*, 1974; Kreth *et al.*, 1975). This observation indicates that normally killer lymphocytes are present to eliminate cells which harbor measles virus. In addition, measles-sensitized T lymphocytes may also play a role in the control of measles infections, as indicated by preliminary studies. It has been observed that during the acute phase of measles, measles-specific T lymphocytes can be detected in patients provided effector and target cells share histocompatibility antigens (Kreth *et al.*, 1979). Such studies are difficult to carry out, but they will provide an understanding of the immune mechanisms which are essential in overcoming the infectious disease process induced by measles virus.

6. *Epidemiology*

Measles, a worldwide childhood disease, has been known for more than 1000 years. A Syrian physician, El-Yehudi, described this disease in the year 68 B.C., but Razes, a Persian, separated it from other infections. However, our understanding of measles, especially its epidemiology, dates from 1846, when Panum (1847) observed and described a measles epidemic on the Färoe Islands. Measles had been absent for over 60 years on these remote islands, and the introduction of the virus to the inhabitants caused an epidemic effect in almost everyone below the age of 65 who did not have measles in early childhood. This natural experiment revealed several important epidemiological facts: The virus spreads from man to man and does not have an animal reservoir, since in its absence from the human community it cannot be acquired from the natural environment. Moreover, patients who have recovered from the infection do not spread the virus afterward and maintain a lifelong immunity without the requirement for frequent reexposures. Mathematical models suggest, in accordance with other infectious diseases, that measles can be maintained only in communities with over 250,000 inhabitants. Below this population level the disease will disappear, since an effective spread is inhibited. If less than 75% of a population is immune, a measles epidemic can occur. If 20% or less are susceptible, measles is maintained at an endemic level (Fraser and Martin, 1978). The incidence in the Färoe Islands also indicates that once measles virus is eradicated in a human population, for example by a vaccination program, it could never return, as there is no known animal reservoir.

7. *Prevention and Control*

Measles virus is a very contagious agent, which is reflected by the fact that the disease usually occurs predominantly in early childhood. Even in the second decennium, more than 90% of a given population have experienced this disease and are seropositive to measles. Since this virus is mainly transmitted by droplets from infected persons during the prodromal phase of the disease, when its presence is usually not suspected, little can be achieved by quarantine procedures. The only secure way to prevent measles is immunization. The occurrence of neurological complications has stimulated studies to develop vaccines after the virus was isolated and characterized. Attenuated measles strains were isolated and used as a live vaccine for active immunization. The data accumulated over the past 15 years indicate that a single successful inoculation with such attenuated virus produces an active immunization and may provide lifelong protection against the disease. Symptoms after vaccination are minimal and may occasionally consist of low fever, minor toxicity, and a faint rash 5–10 days after immunization. In general, live measles vaccine should be administered at or shortly after 12 months of age, since application before this time often does not induce a successful seroconversion. This could be due to neutralization of the vaccine by maternal antibodies and thus prevent active immunization. Therefore, vaccine given before 12 months should be followed by a second dose at the age of 12 to 16 months to insure adequate protection.

When an immunization at a later age occurs, it must be borne in mind that measles virus does block cell-mediated immune reactions to tuberculin, and immunization during an active tuberculosis could lead to an enhancement of this disease process. Moreover, live attenuated measles vaccine should not be administered to patients who exhibit leukemia, lymphoma, other generalized malignancies, and diseases with impaired cell-mediated immunity during therapy with steroids, radiation, antimetabolites, or alkylating substances, all of which depress resistance and immunity. The wide application of live measles vaccine in the United States has led to a dramatic decrease in acute measles and has almost eliminated measles encephalitis. An etiological association of this vaccine with SSPE has not been observed. On the contrary, the number of cases has declined rather than increased (Halsey *et al.*, 1978).

At the beginning of measles vaccination programs, an inactivated vaccine was produced which was proved inferior to the attenuated variety. The application of killed measles virus induced not only a short-lived immunity but also occasionally unusual local and systemic reactions when recipients were later exposed to live vaccine or encountered natural measles. Local erythema, swelling, and vesicular or hemorrhagic lesions occurred at the site of injection. Moreover, respiratory symptoms with high fever and often rashes atypical in appearance and in distribution were noted. It is assumed that these local reactions represent an "Arthus-like" phenomenon. The systemic complications certainly indicate an insufficient immunity to live measles virus, which is supported by the finding that children immunized with a tween 80 and diethyl ether-treated inactivated vaccine did not reveal antibodies to MV hemolysin. The lack of antibody to this viral glycoprotein may account for the failure to protect the vaccinees (Norrby *et al.*, 1975).

Passive immunization with an immune serum globulin containing measles antibodies can be administered under certain circumstances in an attempt to prevent acute measles. If such globulin (0.25 ml/kg/body weight/i.m.) is given within the first few days after exposure, measles can be prevented or suppressed. The person should obtain live vaccine 8 weeks after passive immunization.

8. Diagnosis

a. Differential Diagnosis. Despite the fact that acute measles in its typical appearance is clinically a unique disease, differential diagnosis is sometimes required if rubella and exanthema subitum are suspected. In rubella, Koplik's spots are not present and patients rarely develop conjunctivitis, photophobia, or cough. In addition, rubella infection is accompanied by enlargement of suboccipital, posterior cervical, postauricular, and sometimes cubital lymph nodes. The rashes are sometimes similar, but in rubella peripheral blood counts reveal a relative lymphocytosis with the presence of eosinophilic leukocytes. In exanthema subitum the rash appears as fever subsides, whereas in measles, fever and rash rise and subside together.

b. Pathology of the Skin Lesions. Histological examinations of measles skin lesions show, beneath the hypertrophic horny layer, epidermal multinucleated giant cells filled with measles nucleocapsids. Similar virus structures were found in the endothelium of dermal capillaries (Suringa *et al.*, 1970; Kimura *et al.*, 1975). These findings suggest that the viral exanthema in measles is a result of immune reactions to virus antigen present in skin tissue.

c. Rapid Viral Diagnosis of Measles by Immunofluorescence. The fact that measles virus replicates during the prodromal phase or at the onset of the disease in the epithelial cells of the nasopharynx provides access to cell preparations which can be examined for the presence of measles virus antigens. By IF techniques, measles antigen can be detected in the cytoplasm or nucleus of infected cells, which histologically consist mainly of giant cells (Fulton and Middleton, 1975). The percentage of positive cells can be enhanced when treatment at low pH is applied to dissociate antigen complexes. Nasopharyngeal cells can be obtained by suction. With this technique the diagnosis of measles can be made within a few hours. Using this approach, Fulton and Middleton (1975) detected measles virus antigen in 10 out of 11 patients.

d. Virus Isolation. Techniques and procedures to isolate measles virus from patients have been well established and described (Lennette and Schmidt, 1969). During the prodromal stage of the disease, virus is recoverable from throat washings, conjunctival secretions, blood, and urine. Just before the onset of rash the infectivity of measles virus reaches its peak, and isolation attempts at this time are the most successful. It is noteworthy that virus isolations require the use of primary human kidney and primary monkey kidney cells. In these cells a cytopathic effect occurs that is characteristic for measles virus (see section on growth of measles virus in tissue culture). This cytopathic effect can be recognized by histological staining, revealing giant cell formation and intracytoplasmic and intranuclear inclusion bodies. Similarly, direct or indirect IF staining techniques with antimeasles antisera or hemadsorption with monkey red blood cells will demonstrate the presence of virus. Only after several tissue culture passages in primary cells can virus be adapted to other cell lines from different species. The adaptational process has not been investigated yet, but it is conceivable that host-controlled factors and selection of genetic variants of measles virus are the main underlying mechanisms of adaptation. Different measles virus isolates have been successfully transmitted to embryonated eggs, mice, hamsters, ferrets, dogs, lambs and calves, but only in monkeys could a typical acute measles be induced. In the other animals, the aspect of neurotropism of measles virus was most frequently analyzed as a consequence of the association of measles virus with SSPE and measles encephalitis. It is noteworthy that the CNS infection in general could be induced only when these animals received the virus preparation by intracerebral inoculation. Virus administered by the natural route of infection does not cause a CNS disease process (reviewed by Fraser and Martin, 1978). Isolation of SSPE viruses requires cell fusion techniques or cocultivation, since infectious virus is not directly present in CNS tissue (ter Meulen *et al.*, 1972a; Agnarsdottir, 1977).

e. Serological Diagnosis. Several tests have been developed to detect measles antibodies in the course of measles infection. Depending on the antigen used in the assay system (see section on antigenicity), the hemagglutination inhibition, hemolysin inhibition, complement fixation, or neutralization tests can be applied (Lennette and Schmidt, 1969). Moreover, solid phase radioimmune assays have been developed which exhibit a high sensitivity and a low differentiation of measles-specific IgG and IgM antibodies (Halonen *et al.*, 1977; Kiessling *et al.*, 1977; Vuorimaa *et al.*, 1978). In addition, the IF techniques are used to determine quantitatively the antibodies reacting with intracellular antigens present in infected cells. Complement-dependent cytotoxic antibodies to measles-infected cells have been measured which probably play a role in the elimination of infected cells within the body (Kibler and ter Meulen, 1975). In general, all these antibody activities are low in titer at the onset of measles and rise during the course of infection. An increase in titer of fourfold or greater is regarded as being compatible with the diagnosis of measles.

B. Canine Distemper

1. Route of Infection

The natural exposure of animals to CDV appears to be by droplet and aerosol exposure (Laidlaw and Dunkin, 1926). Virus replication occurs first in macrophages and lymphocytes in the respiratory tract and local lymph nodes. Virus spreads within a few days to all lymphatic organs, including the spleen, thymus, lymph nodes, and bone marrow. If virus-neutralizing antibody is produced within 10 days PI and progressively thereafter, the infection remains almost inapparent and virus can no longer be isolated from suspended tissues or blood. If animals fail to produce protective levels of neutralizing antibody within the second week PI, virus spreads to epithelia of many organs and to the CNS, resulting in clinical disease with a duration and mortality that depend upon the host response and the CDV strain involved (Liu and Coffin, 1957; Appel, 1969; McCullough *et al.*, 1947a,b).

2. Clinical Signs

Clinical disease after CDV infection varies greatly. Ferrets and probably other members of the Mustelidae family have nearly a 100% mortality after a period of anorexia, emaciation, and dehydration that may last from several days to several weeks, depending on the strain of virus. The disease in dogs ranges from virtually no visible disease in gnotobiotics (Gibson *et al.*, 1965) to severe disease with an approximately 50% mortality (Appel, 1969; McCullough, 1974a,b). The duration and severity of the disease appear to depend on virus strains as well as on individual host resistance. The mortality in very young dogs is higher than in weaned or adult dogs. (Krakowka and Koestner, 1976).

CD is usually described as an acute contagious disease with a diphasic temperature curve, the first peak at 3 to 6 days after exposure and the second beginning several days later, then becoming intermittent until death or recovery. Leukopenia, especially lymphopenia, anorexia, nasal discharge, and conjunctivitis, are common signs as well as gastrointestinal and respiratory signs, the latter often enhanced by secondary infections. Skin pustules, often seen in natural cases, were not found in experimental dogs and were believed to be caused by bacterial complications (Carré, 1905; Laidlaw and Dunkin, 1926; Lauder *et al.*, 1954; Cornwell *et al.*, 1965).

Some dogs develop nervous signs, often combined with or following systemic disease, or the nervous form may follow a subclinical systemic infection. Disturbance in gait and posture, incoordination, convulsive seizures, and psychic changes are most often seen. Myoclonus, paresis, torticollis, nystagmus, meningeal signs of hyperesthesia and cervical rigidity, or optic nerve signs of abnormal light reflex and optic placing reactions may be seen (McGrath, 1960; Innes and Saunders, 1962). Optic neuritis and retinal lesions in dogs with CD have been reported (Fischer, 1971; Fischer and Jones, 1972). Dogs with nervous signs usually die, but some recover, sometimes with residual signs, as for example myoclonus, which may persist indefinitely.

Enamel hypoplasia of teeth in growing dogs after CD infection is a common observation (Bodingbauer, 1960; Dubielzig, 1979).

Hyperkeratosis of the foot pads and nose (hard pad disease) was frequently seen in the late 1940s (MacIntyre *et al.*, 1948) and may still be seen occasionally today. Although originally thought to be caused by different agent than classic distemper virus, it was serologically indistinguishable (Koprowski *et al.*, 1950).

Old dog encephalitis (ODE) has been postulated to be caused by CDV (Lincoln *et al.*, 1971, 1973; Adams *et al.*, 1975; Imagawa *et al.*, 1979). This disease of middle-aged or older dogs is rare today. It is characterized by progressive motor and mental deterioration and is ultimately fatal (Vandevelde *et al.*, 1980). Sometimes it is difficult to differentiate between typical distemper in older dogs and ODE (Vandevelde *et al.*, 1974).

Great variations in the duration and severity of clinical disease have been found in experimental dogs infected with different strains of CDV (Laidlaw and Dunkin, 1926; Bindrich, 1954; Rockborn, 1957; Cornwell *et al.*, 1965; Gibson *et al.*, 1965; Appel, 1969; McCullough *et al.*, 1974a,b).

3. *Pathogenesis*

Early infection of lymphatic tissues causes lymphopenia and necrosis or depletion of these tissues (McCullough *et al.*, 1974c), and immunosuppression is common (Krakowka *et al.*, 1975a; Mangi *et al.*, 1976). Dogs that recover early succeed with a potent immune response that clears virus from lymphatic tissues. Failure of the immune response leads to spread of virus to epithelium of the respiratory, alimentary, and urogenital tract, to endocrine and exocrine glands, and to the brain, causing generalized infection and disease (reviewed by Appel and Gillespie, 1972; Yamanouchi, 1980).

Although the systemic infection may subside later in some dogs, followed by recovery and proliferation of lymphatic tissue, virus may persist in various tissues in the CNS and, directly or indirectly, may be the cause of demyelinating encephalomyelitis. Because CD encephalomyelitis simulates lesions in SSPE and, to a certain degree, lesions in MS, this aspect of the disease has been of considerable interest (Koestner *et al.*, 1974; Vandevelde *et al.*, 1974; Koestner, 1975; Lampert, 1978).

Virus appeared first in the CNS in a perivascular location within infected lymphocytes 8 to 10 days PI with CDV strain Cornell A75-17 (Summers *et al.*, 1978). Meningeal and ependymal cells were infected early (Appel, 1969). Infiltration of virus into the CNS appeared to be initiated from the ventricles. Astroglia, microglia, neurons, and rarely oligodendroglia became infected. Infection may spread by cell fusion, which may result in syncytia formation (Summers *et al.*, 1979). By 15 to 20 days PI dogs were extremely lymphopenic, and lymphocytes rarely were found in the brain at that time. Experimentally, the earliest demyelination was found 24 days PI, with virus-infected glial cells always in areas of myelin loss (Wisniewski *et al.*, 1972; Summers *et al.*, 1979). Blood-derived inflammatory cells were not found in areas of preliminary demyelination (Fig. 7A) (McCullough *et al.*, 1974a,b; Raine, 1976; Summers *et al.*, 1979). A cellular immune response to myelin, therefore, could be ruled out for primary distemper demyelination. Humoral factors may be involved (Krakowka *et al.*, 1974, 1975b, 1978c). A direct viral effect on oligodendroglia appears to be unlikely because infection has rarely been seen in these cells. Fusion of virus-infected cells with myelin membranes has been suggested as an initiating step (Summers *et al.*, 1979), followed by macrophage, microglia, and possibly astrocyte phagocytosis of myelin (Raine, 1976).

Following primary demyelination, approximately 30 days PI, lymphocytes were found in a perivascular position and in areas of demyelination in the brain of recovering dogs (Fig. 7B). Recovery and sometimes proliferation of lymphatic tissues as well as normal blood lymphocyte counts were found in these dogs in spite of viral persistence in the brain (McCullough *et al.*, 1974a,b; Summers *et al.*, 1979). Perivascular infiltration of lymphocytes in the brain without or with mild demyelination is typical for ODE (Fig. 7C) (Vandevelde *et al.*, 1980).

Persistence of virus in the brain may be affected by several mechanisms (reviewed by Morgan and Rapp, 1977). Incomplete virus may be involved, as for example in SSPE. However, in contrast to SSPE, complete infectious virus can be isolated from persistently infected CD brain by direct brain cell culture (Hall *et al.*, 1979a; Imagawa *et al.*, 1979). Temperature-sensitive mutants of CDV isolated from brain have not yet been reported. Interferon, which may affect viral persistence, was found in CSF but not in serum of dogs with CD demyelinating encephalitis (Tsai, 1979), and CDV could be isolated by direct brain culture whenever interferon was found in CSF (M. Appel, unpublished).

4. *Host Range*

The natural and experimental host range of CDV has been summarized (Appel and Gillespie, 1972). Briefly, most of the terrestrial Carnivora are susceptible to

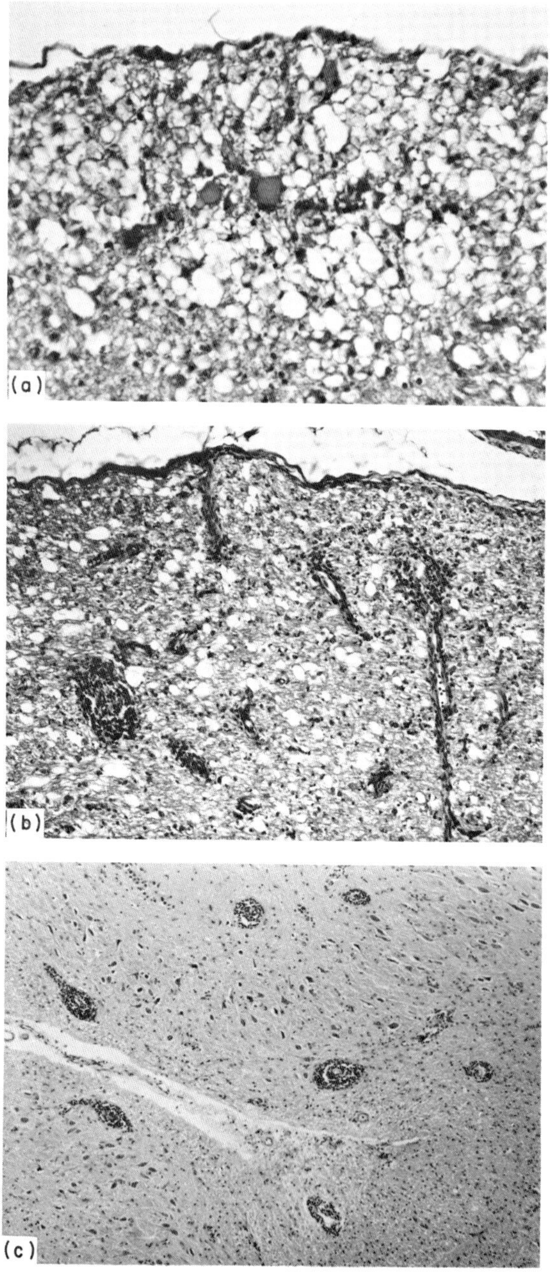

Fig. 7. (a) Syncytia formation of astroglial cells within area of early demyelination. Note the absence of inflammatory mononuclear cells. Medulla oblongata of a dog infected with CDV (Cornell A75-17 strain). H&E stain. × 250. (b) Perivascular and diffuse infiltration of mononuclear cells and gliosis in a later phase of demyelination. Optic tract of a dog infected with CDV (Cornell A75-17 strain). H&E stain. × 120. (c) Extensive perivascular cuffing and diffuse gliosis without demyelination in a dog with old dog encephalitis. Hippocampus. H&E stain. × 90. (a,b) Courtesy of Dr. B. Summers; (c) courtesy of Dr. A. de Lahunta.

natural infection. All animals in the Canidae family (e.g., dog, dingo, fox, coyote, wolf, jackal), the Mustelidae family (e.g., weasel, ferret, mink, skunk, badger, stoat, marten, otter), and the Procyonidae family (e.g., kinkajou, coati, bassariscus, raccoon, panda) were found to succumb to CDV infection or to develop clinical signs of disease. Members of the Felidae family (e.g., cats, lions, tigers) are probably susceptible to infection without developing disease, as has been found in cats (Appel *et al.*, 1974). Animals in the Hyaenidae family (hyenas), Ursidae family (bears), and Viverridae family (e.g., mongoose, meerkat, binturong) have not been adequately tested for susceptibility, but natural disease was reported only in the binturong (Goss, 1948).

Experimentally, embryonated chicken eggs, mice, hamsters, monkeys, cats, and pigs have been infected with CDV (reviewed by Appel and Gillespie, 1972). Encephalitis in mice was induced (Lyons *et al.*, 1980). Cats and pigs can readily be infected with CDV but the infection remains restricted to lymphatic tissues without virus shedding and clinical signs (Appel *et al.*, 1974). Nonhuman primates are susceptible to CDV (DeLay *et al.*, 1965), and encephalomyelitis with CDV antibody in CSF was induced after intracerebral inoculation (Yamanouchi *et al.*, 1977, 1979). One could speculate that measles free humans are susceptible to CDV as well; however, only one report on human volunteers is available (Nicolle, 1931).

It has been suggested that CDV may be involved in multiple sclerosis in man (Cook *et al.*, 1978, 1979). This has been disputed (Burridge, 1978; Nathanson *et al.*, 1978; Chandy *et al.*, 1979), and there is evidence now that CDV is not a causative agent (Stephenson *et al.*, 1980; Appel *et al.*, 1981).

5. *Virus Strains*

Immunological strain variations between CDV isolates have not yet been reported. Laidlaw and Dunkin (1928) found immunized ferrets to be protected against various strains of CDV. This concept was challenged by MacIntyre *et al.* (1948) when hard pad disease in dogs was believed to be caused by a different serotype of CDV. This postulation was not substantiated. Virus strains isolated from hard pad disease dogs were later found to be serologically identical to other CDV strains (Koprowski *et al.*, 1950; Cabasso and Cox, 1952; Gorham, 1960). However, kinetic studies for determining antigenic differences between CDV strains have not been reported.

In contrast, biological variations between CDV strains may be considerable. When ferrets were inoculated with CDV, incubation periods varied with different strains (Green and Carlson, 1945; MacIntyre *et al.*, 1948; Reculard and Guillon, 1967, 1972). Based on these findings, McCullough *et al.* (1974a,b) tested various CDV strains in dogs and found strain R252 to induce neurological sequelae with demyelinating encephalomyelitis in almost 50% of gnotobiotic dogs between 4 and 12 weeks PI. Similarly, one of six field isolates tested (Cornell

A75-17) produced fatal demyelinating encephalomyelitis in approximately 50% of specific pathogen-free (SPF) beagles between 30 and 60 days PI (M. Appel, unpublished). Although delayed distemper encephalomyelitis had commonly been observed in natural cases (Vandevelde *et al.*, 1974), this observation was in contrast to earlier tested strains, which induced a disease of shorter duration in dogs with a rare occurrence of demyelinating encephalomyelitis (Dunkin and Laidlaw, 1926; Bindrich, 1954; Rockborn, 1957; Cornwell *et al.*, 1965; Gibson *et al.*, 1965; Appel, 1969).

Intracerebral ferret inoculation with equal virus titers resulted in death within 2 weeks using CDV-SH, within 4 weeks following CDV-R252, and within 4 to 6 weeks after CDV-A75-17, while the Onderstepoort (Haig, 1956) and Rockborn (1960) strains induced subclinical infection only. In contrast, the Vero cell-adapted Onderstepoort strain produced acute encephalitis and death in adult hamsters within 2 weeks PI and in adult mice within 3 weeks PI, while infections with strains SH, R252, A75-17, and Rockborn were subclinical in these animals (M. Appel, unpublished). The Onderstepoort strain was used by Lyons *et al.* (1980) to produce encephalitis in mice and by Yamanouchi *et al.* (1977) in monkeys.

Biological CDV strain differences *in vitro* are extensive. While virulent virus strains replicate best in macrophages (Appel and Jones, 1967; Poste, 1971; Appel, 1978), attenuated strains grow better in epithelial and fibroblast cell cultures (Bussell and Karzon, 1965; reviewed by Appel and Gillespie, 1972). Onderstepoort, SH, and R252 strains have been compared after adaptation to Vero cells by light (Confer *et al.*, 1975a) and electron microscopy (Confer *et al.*, 1975b). The time of appearance and type of cytopathic effects varied greatly among strains.

6. *Immunity*

Isolated dogs surviving virulent CDV were resistant to challenge exposure 7 years later (J. H. Gillespie, unpublished). Ferrets vaccinated with attenuated CDV were immune 5.5 years later (Cabasso *et al.*, 1957). A lifelong immunity, therefore, can be assumed. Occasionally, older dogs develop CD and die several years after repeated vaccinations (unpublished). In addition, vaccinated dogs may lose neutralizing antibody as early as 1 or 2 years postvaccination (Robson *et al.*, 1959). Because immunity in dogs has frequently been based on levels of neutralizing antibody (Gillespie *et al.*, 1958), yearly vaccinattion of dogs has been recommended (Gillespie *et al.*, 1966). Although levels of maternally derived neutralizing antibody correlate well with protection, dogs that have previously been exposed to CDV may be protected from disease in spite of lacking neutralizing antibody (Gillespie, 1965). Anamnestic antibody response and cell-mediated immunity may both be important.

Dogs passively immunized with serum or by maternal antibody are protected against parenteral and aerosol challenge with virulent virus. Maternal antibody is transferred in dogs mainly by colostral feeding on the first of day of life; 2 to 18% of the maternal antibody level can be transferred *in utero* (Gillespie *et al.*, 1958; Krakowka *et al.*, 1978a). The combined placental–colostral antibody is equivalent to an average of 77% of the bitch's serum titer. The half-life of maternally transferred antibody in pups is 8.4 days. Pups with neutralizing antibody titers of 1:100 or greater were fully protected against intracerebral CDV challenge; susceptibility increased with decreasing titers. Maternal antibody interfered with active immunization. Using these data, based upon the serum antibody titer of the bitch, a nomograph was devised that predicted the earliest age at which the offspring could be vaccinated (Gillespie *et al.*, 1958).

Dogs vaccinated with inactivated CDV vaccine are not protected against infection by virulent virus but are protected against clinical disease (Gillespie, 1965). At least two sequential inoculations are necessary for inducing neutralizing and CF antibody (Karzon *et al.*, 1961). Protection is of short duration, probably not more than 6 months.

In contrast, long-lasting immunity can be induced with live attenuated virus. Virus-neutralizing, cytotoxic, and CF antibody were found in serum 6 days postvaccination. Titers increased and were optimal at 14 days. Virus-neutralizing and cytotoxic antibody levels remained high for at least 14 months (Fig. 8), while CF antibody disappeared after 3 months. Virus-specific IgM was found between 5 and 18 days postvaccination. Immune-specific, cell-mediated cytotoxicity (CMC), which was genetically restricted and not blocked by CDV antibody began at 6 days PI, was optimal at 7 to 10 days PI and was no longer measurable in most dogs after 15 days PI (Fig. 8). In contrast, a spontaneous, natural CMC of CDV-infected target cells, which could be blocked by CDV antiserum and was not genetically restricted, was found in dogs equally before infection and after the immune CMC became apparent (Shek, 1980). Lymphocyte-associated inhibition of CDV-induced syncytia in tissue culture was found in dogs between 5 and 18 days post vaccination (Krakowka *et al.*, 1978b). Antibody-dependent, cell-mediated cytotoxicity may be important in viral clearance (Ho and Babiuk, 1979). Cerebrospinal fluid remains free of measurable antibody in vaccinated dogs.

Immune responses in dogs after virulent CDV infection are greatly dependent on virus strains used and on the host response to infection. During the early phase of infection, dogs are always lymphopenic and immunosuppressed (Krakowka *et al.*, 1975a; Mangi *et al.*, 1976; Stevens and Osburn, 1976). Necrosis or depletion of lymphatic tissues is commonly seen, probably due to a direct virus effect on lymphocytes (McCullough *et al.*, 1974c). IgA and IgG levels may be reduced, while IgM levels in serum remain unchanged (R. D. Schultz and S. Krakowka, personal communication). Dogs that recover early may respond almost like vaccinated dogs, while dogs that die from acute CD may not produce

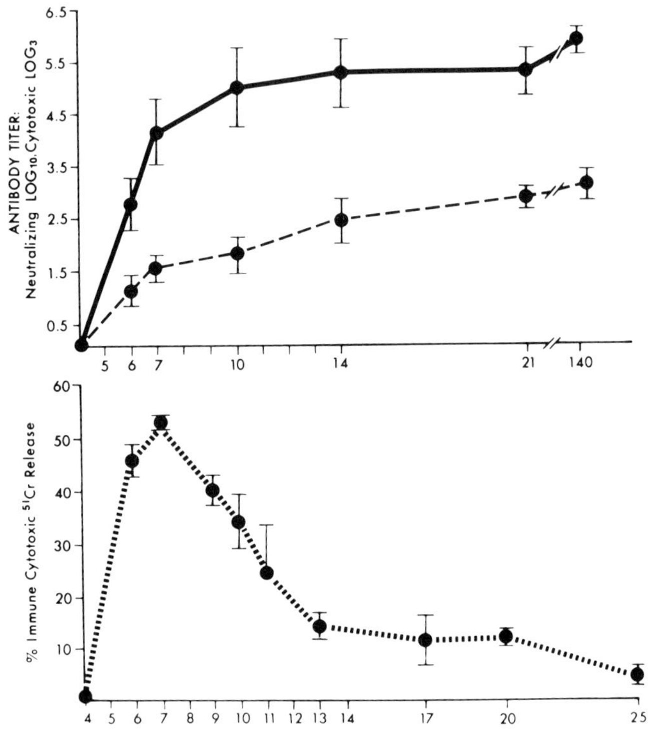

Fig. 8. Humoral and cellular immune response of dogs after vaccination with CDV (Rockborn strain). ●———● Complement-dependent cytotoxic antibody response in log 3 (days after inoculation). ●-----● Virus-neutralizing antibody response in log 10 (days after inoculation). ● ····· ● Immune-specific cytotoxic lymphocyte effect on CDV-infected autologous target cells (days after inoculation). Courtesy of Dr. W. Shek.

measurable antibody (Appel, 1969). Krakowka *et al.* (1975b) reported antibody formation against nucleocapsids and viral glycoprotein in dogs that recovered completely from CDV strain R252, reduced glycoprotein antibody in dogs with persistent brain lesions, and lack of antibody in dogs that died with demyelinating encephalomyelitis. These results were based on attenuated virus neutralization tests in Vero cells. Similar results were obtained from dogs infected with strain Cornell A75-17. However, when serum samples from dogs with demyelinating encephalomyelitis without CDV-neutralizing capacity in Vero cells were tested against virulent virus in dog lung macrophage cultures, high antibody titers were found. Antigenic differences did not account for this observation because the sera were tested against the same virus stock in the third Vero cell passage, and the sera were free of interferon. Macrophages may deal with "incomplete" virus-antibody complexes in a different manner than Vero cells.

Further studies are needed to determine the antigenic site of this antibody and its role in inducing persistent infections (M. Appel, unpublished).

Neutralizing and cytotoxic antibody in dogs after virulent virus infections persisted for at least 2 years, while complement-fixing antibody and virus-specific IgM were measurable for only 3 months or less (M. Appel, unpublished). Recovered dogs frequently have measles virus HLI antibody but not HI antibody, while both can be found in hyperimmune serum (E. Norrby, personal communication).

Lymphocyte-associated syncytia inhibition in cell culture was still present 30 days PI in dogs recovering from CDV strain R252 (Krakowka and Wallace, 1979). Immune-specific cytotoxic lymphocytes were found in dogs up to 75 days PI with strain Cornell A75-17, but only up to 21 days PI with strain SH (M. Appel, unpublished).

CF of dogs that recover early is free of antibody. Some, but not all, dogs with demyelinating encephalomyelitis have neutralizing antibody in the CSF (Appel, 1969, 1970). High concentrations of IgG and the presence of IgM in these dogs were reported by Cutler and Averill (1969).

7. *Epizootiology*

The epizootiology of CD has been reviewed by Gorham (1966). The disease is enzootic throughout the world with few exceptions. Many wildlife species are susceptible to CDV and are a constant source of infection in addition to diseased dogs. Dogs with clinical disease shed virus in all body secretions, beginning approximately 7 days after exposure. The main route of transmission appears to be by aerosol (Laidlaw and Dunkin, 1926). Animal contact is probably responsible for most infections, because CDV is not stable in the environment, except when frozen, and vectors are not involved. Transplacental transmission has been reported (Krakowka *et al.*, 1974, 1977).

CD has the typical age pattern and incidence for a widespread disease with a long-lasting immunity. Earlier studies in urban communities, where dogs were constantly exposed to CDV, showed that most pups were protected by maternal antibody until approximately 3 months of age (Rockborn, 1958). More than 70% of 4- to 6-month-old dogs were susceptible, with a constantly decreasing number thereafter due to exposure to CDV (Hoffman, 1949). In vaccinated dog populations with lower maternal antibody titers, most pups became susceptible to CDV at 6 to 8 weeks of age (J. L. Bittle, personal communication). Unvaccinated dogs in isolated communities remained susceptible to CDV for life (Gorham, 1966). Mortality in unvaccinated natural and experimental dog populations was estimated to be between 24 and 50% (Rockborn, 1958; Appel, 1969; McCullough, 1974a,b).

A seasonal prevalence of CD is usually observed in countries with temperate climates, with the highest incidence in autumn and spring. However, in kennels where young pups are constantly introduced, CD can be a continuous problem.

8. *Prevention and Control*

Eradication of pathogenic animal viruses may be desirable but is not possible for CDV. Much of the wildlife population is susceptible to this virus and a constant source of infection in addition to diseased dogs. Unvaccinated dogs are still a major source of CDV. Prevention by vaccination appears to be the best approach to controlling the disease in dogs.

Recommendations and guidelines for CD immunization were summarized in a symposium (Gillespie *et al.*, 1966). The recommendations are still valid today. A more recent review deals in more detail with CD immunization (Appel and Gillespie, 1972).

Inactivated vaccines were used extensively before live attenuated vaccines became available in the 1950s. Because inactivated CDV vaccine protects dogs against disease but not against infection with CDV and because the induced immunity is of short duration, use of these products in dogs has been discontinued. They are still indicated for animals in the Mustelidae and Procyonidae families, because CDV attenuated for dogs can still be virulent for these species (Bush *et al.*, 1976; Carpenter *et al.*, 1976).

Live attenuated virus vaccines are predominantly used today. After Haig (1948) and Cabasso and Cox (1949) attenuated CDV to embryonated hen eggs and later to chick embryo cell cultures (Cabasso *et al.*, 1959) and after Rockborn (1960) attenuated CDV to dog kidney (DK) cell cultures, live attenuated CDV vaccines replaced inactivated vaccines rapidly. Both the egg-adapted and the cell culture-adapted strains induce immunity in dogs, not only against disease but also against CDV infection. Because maternal antibody interferes with active immunization (see Section III,B,6), vaccination in dogs should be made according to the nomograph after determination of maternal antibody levels (Gillespie *et al.*, 1958) or at least twice in pups between 6 and 14 weeks of age.

Heterotypic (measles) virus vaccines are effective in dogs in the presence of maternal antibody against CDV (Gillespie and Karzon, 1960; Warren *et al.*, 1960; Slater and Murdock, 1963; Baker, 1963). Like inactivated CDV vaccines, MV vaccines protect dogs against CD disease but not against CDV infection. The original use of MV vaccines was recommended for pups 2 to 4 weeks of age and was not successful in the field, although it was tested in disease-free animals. An explanation for this discrepancy was later found by Schultz *et al.* (1977). While low titers of CDV maternal antibody in 6- to 10-week-old pups were tolerated, antibody titers of $\geqslant$ 1:300 in 2- to 4-week-old pups did interfere with MV vaccination. While the general use of MV vaccine in 2- to 4-week-old pups has been discontinued, a combination of attenuated MV and CDV is still commonly used in 6- to 10-week-old pups.

Passive immunization against CDV was used extensively before attenuated CDV vaccines became available. Hyperimmune serum with a neutralizing titer of

1:6000 protected dogs for 10 days against challenge with virulent CDV when 1 ml per 0.45 kg was administered subcutaneously (Benson, 1960). Because passive immunization enhances the effect of maternal antibody and may interfere with active immunization, the use of passive immunization has almost entirely been discontinued. Few situations, as for example, colostrum-deprived puppies or susceptible dogs known to have been exposed to virulent CDV, would warrant passive immunization. The World Health Organization Export Committee on Biological Standardization has established the international standard for anti-CDV serum (Stewart *et al.*, 1968).

9. Diagnosis

a. Differential Diagnosis. CD cannot easily be confused with other diseases in dogs if combined systemic and CNS signs are present. However, if one or the other is absent, a variety of other diseases have to be considered.

The systemic form of CD with fever, depression, anorexia, and dehydration should be differentiated from infectious canine hepatitis (ICH), canine parvovirus infection, Ehrlichiosis, and leptospirosis. Urine analysis may differentiate between leptospirosis and CD. While leptospirosis would respond to antibiotic therapy, virus infections would not. Corneal opacity may occur in dogs recovering from ICH. Dogs infected with *Ehrlichia canis* show aplastic anemia, hemorrhages, and peripheral edema (Buhles *et al.*, 1974). Isolation of the causative agent and serology may be needed for a differential diganosis.

Respiratory form: Canine adenovirus type 2 (Swango *et al.*, 1970) or canine parainfluenza virus (Binn *et al.*, 1968), in combination with mycoplasma (Rosendahl, 1972) or *Bordetella bronchiseptica* (Bemis *et al.*, 1977), can produce a respiratory disease that would simulate CD. These infections remain local in the respiratory tract and do not induce the severe immunosuppression and lymphopenia which are usually seen in CD.

Enteric form: Canine parvovirus, canine coronavirus (Appel *et al.*, 1979), and possibly canine rotavirus complicated by bacterial infections can induce enteritis in dogs as it may be seen in CD. Coccidiosis, campylobacter (Blaser *et al.*, 1978), and clostridial infections (Prescott *et al.*, 1978) may be similar. A systemic disease, besides CD, would be expected only by the parvovirus infection.

CNS form: Differential diagnosis of the CNS form is more difficult. Only myoclonus is considered to be pathognomonic for CD. Neurological disorders, including seizures, are frequent CNS signs in CD in dogs. They can also be found in idiopathic epilepsy, in malformations like hydrocephalus or portacaval shunts, in deficiency syndromes (e.g., thiamine or hypocalcemia), in lead, mercury, arsenic, strychnine, chlorinated hydrocarbons, organic phosphates, metaldehyde or sodium fluoroacetate poisoning, in primary or secondary neoplasia, in leukodystrophy, reticulosis, histiocytic or granulomatous encephalitis of unknown etiology, or in infectious diseases like rabies or toxoplasmosis (Innes and

Saunders, 1962; de Lahunta, 1977). Other CD–CNS signs should also be related to these syndromes. Cerebellar signs may be simulated by cerebellar hypoplasia, which may be hereditary in some breeds. Reticulosis, toxoplasmosis, lead poisoning, and thiamine deficiency probably induce the most common CNS disorders that may mimic CD encephalitis. CSF examination would separate infectious from noninfectious disease, and the presence of CDV-specific IgM in CSF or serum would confirm a recent CDV infection.

b. Pathology. The only consistent finding on postmortem inspection in uncomplicated CD is a considerable reduction in thymus size. Lungs often contain areas of consolidation. All other signs may be complicated by secondary infections. Pathognomonic histopathological changes are a demyelinating, nonsuppurative encephalomyelitis with intranuclear or intracytoplasmic eosinophilic inclusions in glial cells and neurons. Predominantly intracytoplasmic inclusions can be seen in epithelial cells of the respiratory, alimentary, and urogenital tracts. Intersititial pneumonia is commonly seen with bronchopneumonia when complicated by bacterial infections, e.g., *B. bronchiseptica*.

c. Identification of Specific Antigens. The IF test has been used extensively for the diagnosis of CD (summarized by Appel and Gillespie, 1972). Conjunctival or vaginal imprints as well as buffy coat cell preparations, cells from CSF, or frozen sections of foot pad biopsies are all suitable. However, these preparations are often negative after the onset of antibody production, and negative specimens do not rule out distemper. Viral antigen sometimes can be found in foot pad biopsies from dogs with subacute CD. Postmortem specimens of choice are usually imprints or frozen sections of brain, lung, stomach, urinary bladder, lymph nodes, and spleen. Paraffin-embedded tissues can also be used (Kristensen and Vandevelde, 1978) after special preservation in 95% alcohol. IF usually appears granular, from small to large deposits in the cytoplasm of epithelial cells, macrophages, and lymphocytes, and frequently intranuclear in glial cells and neurons.

Distribution of eosinophilic inclusion bodies is similar in appearance to IF. Not all cells that react by IF reveal inclusion bodies and IF, therefore, appears to be a more sensitive test. However, inclusion bodies tend to persist longer in cells than viral antigen (von Mickwitz and Schröder, 1968), and staining for inclusion bodies may be useful in addition to IF. Viral nucleocapsids can be seen by electron microscopy in inclusion bodies (Watson and Wright, 1974). However, nonspecific inclusion bodies may be found in the canine urinary tract (Dagle *et al.*, 1979).

Demonstration of viral antigen by complement fixation in a wide variety of tissues had been used earlier as a diagnostic test (Bindrich, 1954). Mansi (1958) and White and Cowan (1962) used the Ouchterlony test for demonstration of viral antigen in various tissues. This test has not been widely used for CD diagnosis.

d. Virus Isolation. Inoculation of CD-susceptible ferrets with tissue suspensions from CD suspects had been used earlier for CD diagnosis. If CDV was present, ferrets would die between 10 and 30 days postinoculation, depending on strain and dose of virus. When virus neutralization with specific antibody prevented the infection, the diagnosis was confirmed. Because dogs with CD frequently have neutralizing antibody, virus in tissue suspensions is neutralized and not infectious for ferrets.

CDV isolation in tissue culture is difficult because only macrophages appear to be highly susceptible to virulent virus (Appel and Jones, 1967; Poste, 1971). Few laboratories have susceptible dog macrophage cultures available for that purpose. In addition, tissue suspensions from CD dogs with neutralizing antibody are noninfectious for macrophages. A better approach is direct culture of lung macrophages, which may show syncytia formation as early as 18 hours after cultivation, or trypsinization and direct culture of brain, lung, and kidney or urinary bladder cells from fresh postmortem tissues of CD dogs (Wright *et al.*, 1974; Imagawa *et al.*, 1979). In contrast to SSPE virus, free infectious CDV can be transferred in culture medium from these preparations without co-cultivation.

e. Serological Diagnosis. Demonstration of neutralizing (Appel and Robson, 1973), precipitating (White and Cowan, 1962), or cytotoxic antibody (Shek, 1979) in dogs with CD is not significant for a diagnosis because most dogs are vaccinated and carry antibody for extended periods. Complement-fixing antibody in CD dogs lasts for approximately 3 months and may be present longer in dogs after reexposure to CDV (Karzon *et al.*, 1961). Only crude antigen preparations have been used for this test, and CF antibody response to defined viral antigens has not yet been reported.

The presence of CDV-specific IgM is more indicative of recent infection. By indirect IF on CDV-infected target cells with the patient's serum as the first step and anticanine heavy-chain-specific IgM labeled with FITC as the second, the following pattern was found: After vaccination with live attenuated virus, CDV-specific IgM appeared in serum of dogs 5 days PI and lasted for approximately 10 days. In dogs experimentally infected with virulent strains of CDV, the earliest onset of CDV-specific IgM was found 10 days PI. Dogs that recovered quickly responded similarly to vaccinated dogs, while dogs with clinical disease had CDV-specific serum IgM for up to 80 days PI. Reexposure to CDV did not elicit a new IgM response (M. Appel, unpublished). A quantitative approach was taken by Noon *et al.* (1980). Using the ELISA test in CDV antigen-coated microplates with the patient's serum and anticanine enzyme-linked, heavy-chain-specific IgM, results were similar. This test appears to be most promising for the diagnosis of CDV. It can be used in combination with IF demonstration of viral antigen in conjunctival imprints in dogs without antibody to CDV.

CD antibody that inhibits MV hemagglutination (Waterson *et al.*, 1963; Enders-Ruckle, 1964; Norrby, 1967a) was found in vaccinated dogs, while MV hemolysis inhibition was demonstrated in dogs after exposure to virulent virus (Orvell and Norrby, 1974). These findings may be significant for the pathogenesis of CD but less for the diagnosis.

f. Cerebrospinal Fluid. Increases in protein and mononuclear cells in the CSF of CD dogs are nonspecific indicators of encephalitis (Wright, 1978). High concentrations of IgG and the presence of IgM are typical for CD (Cutler and Averill, 1969). Specific CDV-neutralizing antibody was found in some dogs with subacute CD encephalitis, but not in dogs that were either vaccinated, recovered early from CD infection, or died from acute CD infection (Appel, 1969, 1970). However, as not all dogs with subacute CD encephalitis have neutralizing antibody in CSF, a negative result does not rule out CD. Nonspecific damage to the blood–brain barrier could allow leakage of serum-derived CDV antibody into the CSF. Comparisons should be made, therefore, with common antibody in serum of dogs, for example, to canine adenovirus.

g. Lymphocyte Blastogenesis. Total blood lymphocyte counts are usually low in dogs with acute CD; however, they often return to normal values in dogs with subacute CD encephalitis. Mitogen blastogenesis of blood lymphocytes from CD dogs is always greatly reduced, even after lymphocyte counts have returned to normal values (Krakowka *et al.*, 1975a; Schultz *et al.*, 1977). Other virus infections in dogs that have been tested (paramyxovirus SV5, canine adenoviruses 1 and 2, canine herpesvirus, reovirus, canine coronavirus) do not result in reduced mitogen blastogenesis of blood lymphocytes. The only other condition in dogs in which a reduction was found was generalized demodicosis (Scott *et al.*, 1974), and dogs with canine parvovirus infection have not been tested.

C. Rinderpest

1. Route of Infection

Plowright (1964a) considered the nasal epithelium to represent the portal of virus entry. Although primary multiplication in the nasal mucosa could not be detected, animals killed early in the incubation period before generalized infection occurred contained virus in the tonsil or in the lymph nodes draining the nasal cavity. Taylor *et al.* (1965) examined cattle infected by contact to obtain further information on the route of infection. Once more, virus was detected in lymph nodes draining the nose or pharynx, but replication in the mucous membranes was not detected; in two animals, involvement of the bronchial lymph node pointed to virus entry via the lower respiratory tract.

Following multiplication in local lymph nodes, virus entered the blood stream, presumably in infected efferent lymph, and was distributed to target organs throughout the body. At the end of the incubation period, evidence of virus replication could be found in hemolymph nodes, carcase and visceral lymph nodes, spleen, lymph follicles in the intestine, lungs, bone marrow, and mucosae of the alimentary canal, but not in the brain. In cephalic lymph nodes draining the site of entry, levels exceeding $10^{6.0}$ $TCID_{50}$/gm of tissue were found.

Throughout the prodromal period, high levels of virus were maintained in all target organs already infected, and virus began to appear in moderate amounts in the nasal mucosa of most animals. Shortly after the onset of the specific disease signs, which characterize the start of the mucosal phase, the level of virus commenced to decline, hastened perhaps by the first appearance of neutralizing antibody. Virus detection in the ensuing convalescent phase was extremely rare.

Liess and Plowright (1964) examined virus excretion by various routes in cattle infected with a virulent field strain. Nasal excretion reached a peak at the end of the prodromal period and the beginning of the mucosal phase. Nearly all infected animals shed virus by this route, with occasional titers exceeding $10^{5.0}$ $TCID_{50}$/ml. Fecal and urinary excretion rates reached peak values late in the mucosal phase with incidence values of 40 and 62.5%, respectively; titers were lower than those found in nasal secretions. Saliva also contained virus, but figures have not been given (Plowright, 1965).

Taylor and Plowright (1965) examined the multiplication of an attenuated strain in cattle and found that virus growth occurred only in lymphoreticular tissues.

2. *Clinical Signs*

The following description of the clinical disease in infected cattle is largely based on a detailed account by Plowright (1965).

The disease can be divided into four periods, of which the *incubation period* is the first, running from the time of infection to the onset of pyrexia. With a virulent East African field strain, the incubation period was from 8 to 11 days following contact infection or 3 to 5 days following parenteral administration. Mild field strains required 10 to 15 days for incubation after contact but only 2 to 5 days if inoculated subcutaneously.

The *prodromal period* follows next and lasts from the onset of pyrexia to the development of the first specific clinical sign, which usually takes 3 days. During this period a rapidly mounting fever is commonly seen; temperatures reach peak values of 40° to 42°C on the second or third day. After the first day of pyrexia, animals become depressed and increasingly anoretic. All visible mucous membranes become congested, and there is usually some degree of constipation.

The onset of mouth lesions (Fig. 9) marks the commencement of the *mucosal period,* which is characterized by the development of the classic signs of rinderpest. This period lasts between 3 and 5 days and is associated with the majority of case fatalities. Mouth lesions often start on the inner surface of the lower lip or the adjacent gum and consist of small, gray, slightly raised foci of necrosis. Further examination reveals similar lesions on the underside of the free portion of

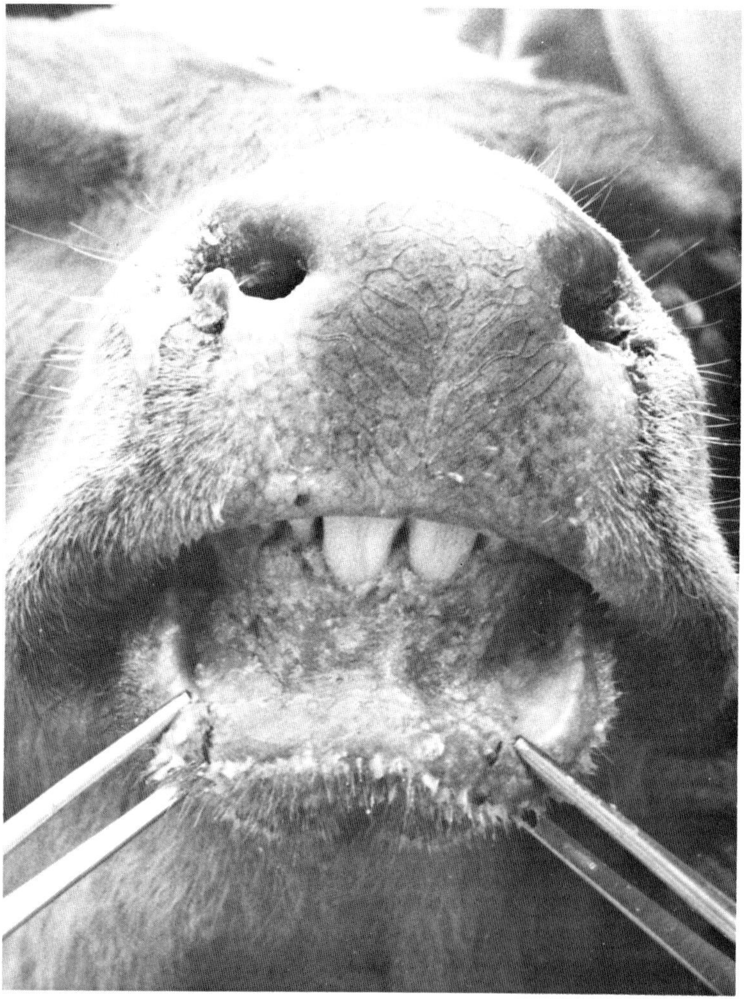

Fig. 9. Oral necrosis in the lower lip and gum of a calf infected with RV. From Gibbs *et al.* (1979), with permission of the editor.

the tongue, cheeks, cheek papillae, and palate. With time these lesions increase in size and extent and may also spread to the dorsum, base, and sides of the tongue. In a severe case, necrosis and erosion of the entire oral epithelium may be seen by the third or fourth day of the mucosal phase; a fetid odor is associated with this oral necrosis.

A mucopurulent nasal discharge also develops, frequently accompanied by necrosis in the anterior nares. Coughing is not a feature of the disease, but late in the disease course there may be dyspnea accompanied by a characteristic checked expiration. A mucopurulent ocular dischange may be seen, usually overflowing at the medial canthus and staining the cheeks. The eyelids do not mat together.

Diarrhea, the last of the characteristic signs, commences 1 or 2 days after the onset of mouth lesions. Copious watery feces are passed which may contain blood and mucus, and in severe cases there is straining and shedding of necrotic epithelium. Diarrhea is the most debilitating feature of the disease, and the dehydration that it causes accounts for the high mortality.

Genital mucosae may show necrosis and erosion. Skin lesions have also been described, varying from urticarial eruptions to poxlike lesions with scabs. Virus or virus-specific antigens have been isolated from these lesions (Kataria *et al.*, 1974; Joshi *et al.*, 1977).

Surviving animals enter the *convalescent period.* Regeneration of the oral epithelium may be dramatic and often is complete within 24 or 48 hours of commencement; at the same time diarrhea ceases and, accompanied by a return of appetite, the animal's general condition quickly improves.

Mortality is variable and depends on the virulence of the strain involved. Mild field strains with poorly developed clinical signs and no mortality were described by Plowright (1963). On the other hand, Liess and Plowright (1964) recorded a 50 to 80% mortality with a virulent field strain. To some extent, breed susceptibility also detemines the outcome of infection, the most frequently quoted example being the contrast between the resistant plains cattle and the susceptible hill cattle of India (Holmes, 1904).

3. Host Range

a. Domestic Animals. Among the domestic animals, cattle, either *Bos taurus* or *B. indicus,* water buffalo, *Bubalus* spp., yaks, sheep, goats, and pigs are all susceptible to infection with RV. The one-humped camel is also susceptible but shows only a brief pyrexia (Singh and Ata, 1967).

Rinderpest infection of sheep and goats became important in India, where it is reportedly increasing in incidence, and in Africa, where it must now be clearly differentiated from PPR.

Using virulent bovine strains, African sheep may be infected parenterally or by contact with infected bovines or infected sheep (Plowright, 1952; Ata and Singh, 1967; Zwart and Macadam, 1967a). Infected sheep did not transmit the virus to contact cattle (Plowright, 1952; Ata and Singh, 1967; Zwart and Macadam, 1976b). Sheep generally showed a transient, irregular pyrexia, but in one experiment by Zwart and Macadam (1967a) pyrexia, mouth lesions, diarrhea and death were observed.

Egyptian goats inoculated with a virulent bovine strain showed no clinical signs and did not transmit the disease to cattle or goats (Ata and Singh, 1967). On the other hand, Beaton (1930), using West African goats, produced severe reactions with pyrexia, necrosis, diarrhea, and death. The disease could be transmitted from cattle to goats, between goats, and from goats back to cattle. Zwart and Macadam (1967a,b) obtained similar results.

In spite of these results, reports of field outbreaks of rinderpest in small ruminants in Africa are rare and known only from the reports of a sheep outbreak associated with the disease in cattle (Johnson, 1958) and an outbreak involving cattle, sheep, and goats in the Sudan (El Hag Ali, 1971). The antibody evidence of Zwart and Rowe (1966) suggests that subclinical disease in sheep may occur in association with the disease in cattle. There is no evidence at this time that African sheep or goats play any part in the epizootiology of rinderpest in the absence of the disease in cattle. Taylor (1979a) found that in Nigeria, where the bovine disease does not now occur, small ruminants showed no evidence of infection.

In India, experimental rinderpest can be produced in sheep with bovine or ovine strains (Krishnan and Ranga Rao, 1972; Ramani *et al.*, 1974), the disease being characterized by pyrexia, diarrhea, and death. Mouth lesions, which were absent in experimental animals, were recorded in the field cases observed by Narayanaswamy and Ramani (1973). Outbreaks have been noted involving only sheep and goats, or sheep and goats with subsequent spread to cattle and buffaloes (Narayanaswamy and Ramani, 1973; Ramani *et al.*, 1974; Rao *et al.*, 1974), and it is possible that rinderpest is now enzootic in Indian sheep and goats.

Asiatic pigs may be severely affected (Bansal *et al.*, 1974), but pigs of European origin show only a transient pyrexia. Infection can be acquired by contact with sick cattle and can be transmitted to other cattle or to pigs (Scott *et al.*, 1959, 1962; DeLay and Barber, 1962).

b. Wild Animals. A comprehensive list of the many species in the order Artiodactyla that are susceptible to rinderpest has been given by Plowright (1968).

c. Laboratory Animals.

i. Goats. Goats have been used extensively for growth of virus for vaccine production (for historical details of the importance of this topic, see Plowright, 1968). In both India and Kenya, strains of bovine rinderpest have been serially passaged in goats, with a subsequent loss of virulence for cattle. In Kenya the virus could be used as an attenuated cattle vaccine at the 250th goat passage (Daubney, 1949), although such virus was highly pathogenic for goats.

ii. Rabbits. Rabbits have also been used for vaccine production. The Nakamura III strain of rabbit-adapted RV became "fixed" in its attenuated character after 600 rabbit passages (Cheng and Fischman, 1949) and is still used today for the production of hyperimmune rabbit serum.

iii. Embryonated eggs. The Nakamura III rabbit-adapted strain has been further passaged in fertile hens' eggs, leading to an additional attenuation for cattle (Scott, 1964, describes the history).

4. *Virus Strains*

Different virus strains vary greatly in their pathogenicity. However, antigenic variation has not been recorded for RV, and all strains are grouped in a single serotype. Minor variations may exist but have not been demonstrated convincingly.

5. *Immunity*

Young bovines acquire a passive maternal immunity to RV following the ingestion of colostrum. The half-life of these passively acquired antibodies is 37 days, and the extinction point is 10.9 months (Brown, 1958).

An animal which has recovered from infection with RV develops an immunity to reinfection that can be extremely long-lived (Bansal *et al.*, 1971) and that is usually accompanied by the production of serum-neutralizing antibodies. This has led to the development of a series of live attenuated virus vaccines culminating in the cell vaccine of Plowright and Ferris (1962). Plowright and Taylor (1967) examined the long-term results of the administration of a single field dose of this vaccine and showed that a group of East African Boran cattle, aged between 11 and 13 months were vaccinated, maintained a constant antibody level over the following 4 years. Rweyemamu *et al.* (1974) took four animals from this same group 11 years after vaccination and challenged them by the intranasal instillation of a large dose of virulent field virus. None of these animals showed any clinical reaction, viremia, or nasal excretion of virus and did not infect contact control animals. Neutralizing antibody levels did not increase in three animals and rose only slightly in the fourth.

Plowright and Taylor (1967) also examined Ankole cattle and showed that neutralizing antibody levels were stable over a 2-year period. In grade cattle, a decline in neutralizing antibody level occurred between 12 and 18 months, but thereafter it remained virtually unchanged up to 4.5 years. Rweyemamu *et al.* (1974) challenged further grade cattle 6 years after vaccination by intranasal administration of virulent virus. Clinical disease was not seen, and neither viremia nor virus excretion could be detected.

Unfortunately, the experience of Provost *et al.* (1969b) is at variance with the highly satisfactory East African results. They found that a proportion of West African cattle with high antibody levels could be infected by contact 2 years after vaccination and that, although infection was subclinical, virus could be recovered from their nasal cavity; such animals were also able to transmit infection by contact to further susceptible animals. Provost *et al.* also showed that a decline in neutralizing antibody could occur within 3 years and that when a group of these animals were challenged, clinical rinderpest was produced, together with virus transmission.

Although it is impossible to reconcile the results of these two experimental studies, it must be clearly pointed out that cell culture rinderpest vaccine has enjoyed tremendous success and has largely been responsible for the dramatic decline in the incidence of rinderpest in Africa in the last 2 decades.

Immunosuppression may be induced by RV. Tajima and Ushijima (1971) examined the histopathology of infected bovine lymph nodes and found severe depletion of lymphocytes, reticular cell hyperplasia, and syncytia formation. Similar changes occurred in the rabbit. Yamanouchi *et al.* (1974a,b) showed that rabbits infected with the Nakamura III strain underwent a marked suppression of antibody and cell-mediated immune responses, yet still produced both interferon and neutralizing antibody, together with recovery from the disease.

Kobune *et al.* (1976b) examined the recovery mechanisms in similarly infected rabbits. Animals which had been treated with antithymocyte serum (ATS) and in which delayed type hypersensitivity reactions to PPD had been abolished, or animals treated by a combination of thymectomy and antithymocyte serum and in which the stimulation ratio to phytohemagglutinin had been reduced, recovered normally. ATS treatment alone did not affect interferon or antibody titers. On the other hand, a combination of thymectomy and ATS significantly delayed antibody production. It was suggested that thymus-dependent, cell-mediated immunity might not be required for recovery from rinderpest. The development of autoimmune antibodies, either cold hemagglutinating or antinuclear, has been reported in rabbits (Fukada and Yamanouchi, 1976).

6. *Epizootiology*

Rinderpest is transmitted by contact infection, contagion coming from either infected aerosols or contaminated fomites; no vector is involved. Compared to highly infectious diseases with short incubation periods, such as foot-and-mouth disease, rinderpest is a relatively slow-spreading disease with incubation periods up to 11 days and irregular transmissibility (Taylor *et al.*, 1965). Nevertheless, it retains the potential to cause pandemics if allowed to move in an unchecked manner in a totally susceptible population. The outbreaks in Iran, Syria, Jordan, and Turkey in 1969 and 1970 are a timely reminder of the continuing threat posed by this virus.

In the recent past, game animals have undoubtedly contributed to the epizootiology of the disease (Stewart, 1964; Plowright and McCulloch, 1967), but it is doubtful if, on their own, they can ever assume a lasting role as a reservoir of infection (Taylor and Watson, 1967). Young bovines with a waning maternal immunity appear able to perpetuate the virus in enzootic areas of Africa (Plowright, 1968). Provost (1972) produced disease and death in contact cattle following the intranasal challange of calves with demonstrable maternal immunity; these calves did not show clinical signs, but they did subsequently develop a local and systemic immunity. Add to this the previous report of Provost *et al.* (1969c) concerning virus transmission by vaccinated adults following contact with clinical disease, and mechanisms begin to appear that may account for the continued inability of West African countries to rid themselves of the virus.

The role of small ruminants in the epizootiology of rinderpest has already been discussed.

7. *Prevention and Control*

In enzootic countries, control relies largely on the mass application of vaccine. In India and Africa a strain of virus attenuated by serial passage through primary calf kidney cells (Plowright and Ferris, 1962) is widely used. In East Africa the virus has been passaged over 90 times, but in Nigeria virus passaged about 70 times is preferred; neither variant causes any clinical signs in bovines.

The vaccine is generally given subcutaneously, although Provost and Borredon (1972a) showed that it could be used intranasally in animals of different ages. In calves with low residual levels of maternal antibody, a serological conversion occurred in up to 95% of animals, many of which also developed local antibody in the nasal mucus.

Various attempts have been made to improve the keeping quality of this vaccine by selecting thermostable strains, and at times these strains have been incorporated into bivalent vaccines together with an attenuated variant of contagious bovine pleuropneumonia (Provost and Borredon, 1972b, 1974). Data concerning the stability of the virus in both lyophilized and reconstituted states have been given by Plowright *et al.* (1970, 1971), while the production methods have been tabulated in an internationally recognized form (Anonymous, 1970).

Inactivated vaccines have been produced (Provost *et al.*, 1969c; Mirchamsy *et al.*, 1974), but these have not replaced the live attenuated strains used in national campaigns.

Control in countries outside the enzootic areas depends primarily on measures designed to prevent the importation of the virus in live animals or animal products. In the event of a breach of these defenses, quarantine and slaughter policies, if vigorously pursued, should ensure quick eradication. In enzootic countries in Africa, in spite of a greatly reduced disease incidence, final eradication is proving troublesome.

8. *Diagnosis*

a. Differential Diagnosis.

i. Foot-and-mouth disease. Mouth lesions differ in character and distribution between the two diseases, and lameness does not occur with rinderpest. Rinderpest spreads slowly compared with foot-and-mouth disease.

ii. Mucosal disease/virus diarrhea. There is considerable similarity between the clinical and postmortem pictures of mucosal disease and rinderpest, and in areas where both viruses are enzootic, differential diagnosis requires laboratory assistance. Soluble mucosal disease antigens can be detected by agar gel diffusion (Darbyshire, 1962), or the virus can be isolated in cell cultures.

iii. Malignant catarrhal fever. The morbidity is usually low but is associated with a very high mortality. Nervous signs may be seen, and a characteristic ocular opacity develops.

b. Pathology. RV produces a necrotic stomatitis, gastroenteritis, lymphoid necrosis, and leukopenia. Provost and Borredon (1963) considered that a diagnosis could be made only if syncytia were seen together with both intranuclear and intracytoplasmic inclusions, although Plowright (1964a) found intranuclear inclusions to be rare. Stratified squamous epithelia of the upper alimentary tract, lymph nodes, and Peyer's patches should be examined.

c. Identification of Specific Antigen. The agar gel diffusion precipitation test remains the most frequently used test for the demonstration of viral antigen. Extracts of carcase or mesenteric lymph nodes are diffused against rinderpest hyperimmune rabbit serum in a system that contains known positive and known negative lymph node preparations. The test often fails to pick up antigen in dead animals, and in an outbreak it is preferable to test a variety of tissues from several animals. Provost *et al.* (1963) developed a kit for field use in which standard reagents were impregnated on filter paper discs and applied to an agar plate at the time of use, together with a similar preparation from the test sample.

Details of the methods that have been used to detect rinderpest antigens by complement fixation are reviewed by Scott (1967). Provost and Joubert (1973) concluded that the complement fixation test for rinderpest was complex, costly, and a poor choice. These authors also described an indirect hemagglutination test based on the inhibitory effect rinderpest antisera have for measles hemagglutination. In their test the unknown sample was mixed with rinderpest antisera, which was then tested for residual inhibiting activity. Provost (1970) also attempted fluorescent antibody staining with tissues from infected animals but found that an accurate diagnosis was difficult to make.

d. Virus Isolation. This can be undertaken in live animals, usually cattle, and it is customary to inoculate suspect material into immune and susceptible animals at the same time.

Equally satisfactory is the isolation of virus in primary calf kidney cell cultures. The characteristic cytopathic effect must include syncytial development, and it must be neutralized by RV-immune serum or specific antigen must be demonstrated by the fluorescent antibody technique. Blood, spleen, and lymph nodes from animals in the prodromal phase should be used for virus isolation.

e. Serological Diagnosis. Paired sera collected in the acute and convalescent stages of the disease should be compared. A diagnosis is then made on the basis of a rising antibody titer which may be demonstrated by neutralization, complement fixation, or HAI tests.

Neutralizing antibodies are found in serum following recovery from clinical RV. They can be titrated in a tube test (Plowright and Ferris, 1961) employing calf kidney or Vero cell monolayers; a micro-adaptation of this test has been described (Rioche, 1969). Antibody levels are usually between 1:50 and 1:1000, although Provost *et al.* (1965) have described a hypogammaglobulinemia and poor antibody response in a small proportion of West African cattle.

Complement-fixing antibodies are also produced in bovine convalescent sera but are often short-lived (Nakamura, 1958), and the test remains popular only in the hands of Japanese workers.

An HAI test has been described in which the hemagglutination of Patas monkey erythrocytes by measles antigen can be inhibited by rinderpest immune sera (Bögel *et al.*, 1964). Rowe *et al.* (1967) found a 15% discrepancy between the HAI test and the tube neutralization test. Provost *et al.* (1969a) and Maurice *et al.* (1969) found that inhibiting antibodies waned rapidly but concluded that the HAI test was of value in the examination of sera from recently vaccinated or recovered animals. The test has found widespread application in India in survey work following vaccination compaigns (Ramachandran and Scott, 1972; Ramakrishna Rao *et al.*, 1976).

Ishii and Watanabe (1971) developed an indirect hemagglutination test using RV antigen coupled with tanned goat erythrocytes, but convalescent bovine sera were not always positive in this test. Kobune *et al.* (1976a) have described the detection of RV antibodies using an indirect fluorescent antibody technique.

D. Peste des Petits Ruminants

1. Route of Infection

Direct animal contact appears to be the main source of infection. A systematic study of the distribution of the virus in host animals has not been undertaken. Scrapings from the mucosa of the large intestine yield a ready source of virus and titers of up to $10^{5.0}$ $TCID_{50}$/gm have been recorded. Somewhat lower levels of virus are found in the mucosa of the small intestine and in mesenteric lymph nodes (A. Abegunde, unpublished). Virus has also been recovered from whole lung and lung macrophages from goats with PPR (M. J. P. Lawman, unpublished).

E. P. J. Gibbs and W. P. Taylor (unpublished) showed that levels of virus ranging up to $10^{3.5}$ $TCID_{50}$/gm could be recovered from saliva, nasal secretions, and feces of British goats during, but not beyond, the clinical stage. In British sheep the same secretions were infected, but much less regularly and at lower levels. Again, no persistent virus was found. Virus has not been isolated from the brain.

2. *Clinical Signs*

In West Africa most field and experimental observations have been made on goats, and French workers consider the disease to be less prevalent in sheep. Nonetheless we have observed severe outbreaks of PPR in sheep in both Ghana and Nigeria, in which the clinical picture was identical to that of caprine PPR.

The disease usually runs an acute course with an incubation period of 4 to 5 days. Thereafter, temperatures climb rapidly, reaching peak values of 40 to 41°C by the third day of pyrexia. These values are usually sustained for 1 or 2 days, after which the temperature falls back to normal, the whole febrile period lasting 5 to 7 days. Animals are depressed, listless, and anorexic, but in addition to general malaise, a further series of pathognomonic signs develops during this episode.

Most animals show a nasal discharge, serous at first but rapidly becoming profuse and catarrhal, causing sneezing. In the latter stages of a severe case, this discharge flows out of the nose or dries into a thick crust around the nostrils. Small areas of necrosis may sometimes be visible on the mucous membrane on the floor of the nasal cavity. The conjunctiva is frequently congested, and there may be a small amount of crusting at the medial canthus. Some animals develop a profuse catarrhal conjunctivitis, and the conjunctival sac fills with a thin yellowish fluid which eventually thickens and causes the eyelids to mat together. Tear staining of the cheeks is not a feature.

In the oral cavity a necrotic stomatitis (Fig. 10) is often observed, commencing as small roughened areas of superficial necrosis on the mucosal surface of the lower lip, the lower gum, and around the insertions of the incisor teeth. In severe cases the extent of necrosis rapidly increases to involve the dental pad, palate, cheeks and cheek papillae, and both surfaces of the tongue. There is obviously considerable discomfort, and the tongue works constantly. At this stage the breath has an unpleasant fetid smell highly reminiscent of acute rinderpest in the bovine. There may also be an excess of saliva, but this never hangs from the mouth; rather, animals are found to be wet under the chin.

Nearly all animals develop diarrhea, often profuse but not usually bloody. A cough is also common in the later stages of disease, and bronchopneumonia can often be diagnosed. Females may show engorgement and necrosis of the genital mucosa, and pregnant animals may abort.

Whitney *et al.* (1967) found in the south of Nigeria that a serous nasal discharge actually slightly preceded the onset of pyrexia, and that oral necrosis appeared coincident with the end of incubation. In northern Nigeria, goats infected with strain 75/1 usually had a 2- to 3-day prodromal period before the development of specific clinical signs.

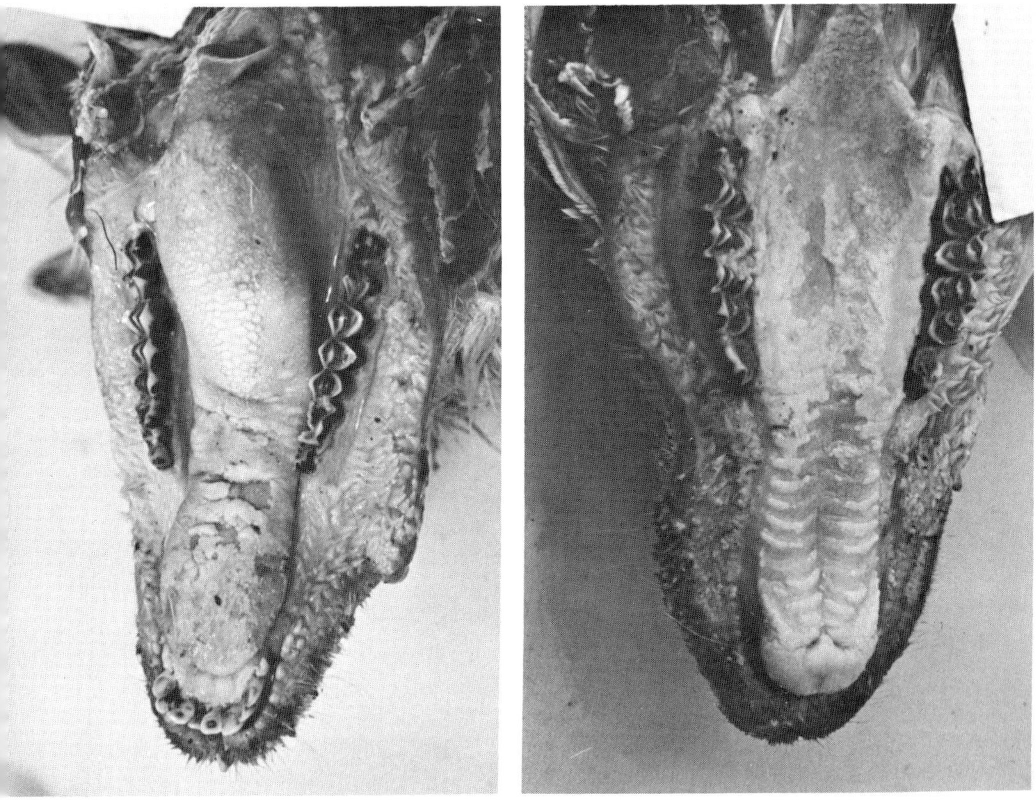

Fig. 10. Necrotic stomatitis in the tongue and palate of a sheep infected with PPRV. From Gibbs *et al.* (1979), with permission of the editor.

During an outbreak of PPR in goats at Vom in northern Nigeria, all animals developed a fever, but only 50% showed oral necrosis. Diarrhea and ocular signs were seen in 80% of cases and nasal discharge in 97%. The morbidity was 77% and the case mortality was 82%; the average survival time from the end of the incubation period was 5 days. Similar figures were recorded by Nduaka and Ihemelandu (1973), but in addition, 90% of their animals developed a bronchopneumonia. This latter condition undoubtedly contributes to the high mortality, and it is not uncommon to isolate species of *Pasteurella* postmortem. It is probable that a primary virus pneumonia precedes a secondary bacterial invasion.

The development of proliferative labial scabs in sheep and goats convalescing from PPR has been described in the south of Nigeria (Whitney *et al.*, 1967; Johnson and Ritchie, 1968; Isoun and Mann, 1972). These lesions are usually described as "Orf-like," and the recent identification of the virus of contagious pustular dermatitis (CPDV) in the same study area (Obi and Gibbs, 1979) leads us to speculate that these lesions are in fact due to CPDV. Significantly, perhaps, such lesions have not been encountered in cases of PPR in the north of Nigeria.

3. Host Range

a. Sheep and Goats. Sheep and goats throughout West Africa are susceptible to PPR, although animals from the Sahel region in Senegal are said to be more resistant (Mornet *et al.*, 1956). North American sheep and goats are fully susceptible (Hamdy *et al.*, 1975). British goats are also fully susceptible, but Dorset horn sheep are only mildly affected (E. P. J. Gibbs and W. P. Taylor, unpublished).

b. Cattle. Mild pyrexia or oral erosions have been seen in calves (Mornet *et al.*, 1956; Gilbert and Monnier, 1962). Conversely, Dardiri *et al.* (1976) failed to produce clinical disease in 18-month-old Hereford steers, as did Gibbs *et al.* (1979) in yearling Jersey heifers or Taylor and Abegunde (1979) in humped Nigerian cattle. In each instance, cattle previously given PPRV were resistant to challenge with virulent RV. Taylor and Abegunde (1979) demonstrated low levels of RV-neutralizing antibody in PPR-immune steers.

Dardiri *et al.* (1976) showed that sick goats could transmit subclinical PPR to cattle, but Gibbs *et al.* (1979) were unable to transmit the virus by contact between cattle, which can therefore be viewed as "dead-end" hosts. Gargadennec and Lalanne (1942) observed during naturally occurring outbreaks of disease that cattle exposed to sick sheep and goats did not show clinical signs; it is not known if subclinical infection of such animals occurs.

c. Pigs. Pigs roam freely in many West African villages but do not appear to be involved in the epizootiology of PPRV. Nawathe and Taylor (1979) found that pigs, too, are dead-end hosts. Goats could transmit virus to pigs, but pigs did not transmit it to goats or other pigs. Clinical signs did not appear, but inoculated pigs developed neutralizing antibodies to PPRV.

d. American White-Tailed Deer. Hamdy and Dardiri (1976) showed that *Odocoileus virginianus* was fully susceptible to PPR infection, developing pyrexia, nasal discharge, conjunctivitis, necrotic stomatitis, and diarrhea. Infection spread by contact between deer and recovered animals resisted challenge with virulent RV.

e. Small Animals. PPRV has not yet been adapted to small laboratory animals.

4. Virus Strains

There appears to be only one serological type of PPRV. Using neutralization tests, Taylor and Abegunde (1979) showed that four Nigerian field strains were identical and that one of them was indistinguishable from a Senegalese strain. Hamdy *et al.* (1975) showed that an earlier Nigerian isolate was identical to the same Senegalese virus, as was a Ghanian isolate (E. P. J. Gibbs, M. J. P.

Lawman, and M. A. Bonniwell, unpublished). It is not known if field strains vary in virulence.

5. *Immunity*

Animals recovering from PPR develop neutralizing antibodies and resist reinfection, but a homologous vaccine has yet to be developed. Instead, use was made of the cross relationship between PPRV and RV, and trials with RV as a heterologous vaccine were instituted but not fully reported (Bourdin *et al.*, 1970; Bourdin, 1973). In a recently completed laboratory trial, Taylor (1979a) showed that a single inoculation of RV gave 12 months of immunity to goats against parenteral challenge with PPRV. These animals developed a secondary antibody response when challenged with PPRV. Gibbs *et al.* (1979) found that although sheep and goats vaccinated with RV did not develop clinical disease when challenged parenterally with PPRV, the challenge virus was subsequently detected in nasal secretions.

6. *Epizootiology*

PPR has been reported from Senegal (Mornet *et al.*, 1956), the Ivory Coast (Gargadennec and Lalanne, 1942), Ghana (M. A. Bonniwell, unpublished), Togo (Bourdin, 1973), Benin (formerly Dahomey) (Mornet *et al.*, 1956), and Nigeria (Whitney *et al.*, 1967), and probably also occurs in the republics of Tchad and Cameroun (A. Provost, personal communication). Further African countries may be added to this list as familiarity with the disease increases.

Neither the detailed epizootiology nor the full extent of losses due to PPR are known. Outbreaks involving small numbers of animals probably fail to be reported, and it is highly probable that a low level of infection exists continuously in enzootic areas. With periodic increases in the susceptible population, epizootics which would receive veterinary attention might then be anticipated. Such epizootics were reported from Senegal in 1956 and 1962 (Bourdin *et al.*, 1970), and a similar episode may have occurred in Nigeria in 1975 and 1976. During epizootics the virus makes contact with totally susceptible populations, and at the village level, decimation of entire sheep and goat populations has been recorded.

Transmission is by contact, and although the portal of entry is not defined, a respiratory route may be suspected. The secretions of sick animals provide a ready source of infective virus.

Taylor (1979b) has described a cell culture screening test for PPRV and a neutralization test that differentiates between a PPRV and an RV antibody response. Using these tests, it was possible to detect large numbers of previously infected sheep and goats entering abbatoirs in the north of Nigeria. At the same time, it was shown that RV was not present in this ruminant population.

7. *Prevention and Control*

Control measures against PPR in enzootic areas have not yet been formulated. In the first instance it would probably be necessary to reduce the extent of the disease by vaccination, but further measures such as quarantine, movement control, and compulsory reporting would, at present, be difficult to implement.

8. Diagnosis

a. Differential Diagnosis.

i. Rinderpest. In Africa, rinderpest infection of sheep and goats remains a rare occurrence (Johnson, 1958; El Hag Ali, 1971). Although the clinical lesions are indistinguishable, rinderpest may be suspected if the disease simultaneously involves cattle. Confirmation requires virus isolation and cross neutralization.

ii. Bluetongue. The virus of bluetongue is widely distributed in West Africa but, as far as is known, infections are mostly subclinical except in imported stock. Here the swelling of the lips, muzzle, and oral mucosa, together with edema of the head region, should serve to differentiate bluetongue from PPR. Coronitis is not a feature of PPR.

iii. Contagious pustular dermatitis. The virus of CPD is present in West Africa, but the lesions of CPD are proliferative rather than necrotic and involve the lips rather than the entire oral cavity. Nasal discharges and diarrhea are not typical of the disease.

iv. Sheep/goat pox. Sheep and goat pox infections have been seen in Ghana (M. A. Bonniwell, personal communication) and may be common elsewhere in West Africa. Pox lesions occur all over the body, including the mouth, but there should be no difficulty in differentiating these from PPR. Postmortem pox lesions are found on internal organs.

v. Foot-and-mouth disease. Foot-and-mouth disease occurs in West Africa. The severity of the disease in sheep and goats is variable but usually mild. Mouth lesions are often transient, and the most characteristic sign is lameness; the latter is not a feature of PPR.

b. Pathology. The gross and histopathological changes associated with PPR have been dealt with in a series of studies by Whitney *et al.* (1967), Rowland *et al.* (1969, 1971), and Rowland and Bourdin (1970).

Lesions visible only postmortem consist of necrosis of the pharyngeal mucous membrane and multiple linear erosions extending down the esophagus. Areas of congestion may be found in the abomasum, but the mucosa of the small intestine is normal. The Peyer's patches are prominent and edematous. Highly characteristic changes are seen in the large intestine, where engorgement of the capillaries of the mucosa, particularly along the linear folds of the cecum, colon, and rectum, produce the so-called "Zebra striping" effect, a name which aptly describes the black discoloration of these folds which is seen in advanced cases.

c. Identification of Antigens. Laboratory confirmation can be given on the basis of a positive reaction in agar gel against a rinderpest hyperimmune serum. Mesenteric lymph node extracts and scrapings of the mucosa of the large intestine provide a source of antigen for gel diffusion tests.

In the event of spread to a country hitherto uninfected, it would be as well to classify all outbreaks as rinderpest until a differentiation can be made. A positive gel diffusion test against rinderpest hyperimmune serum would indicate that one or the other virus was involved, but a differential diagnosis requires virus isolation and cross neutralization. In such circumstances it would be advisable to attempt virus isolation.

d. Virus Isolation. Virus isolation should be attempted in primary LK and BK cells. PPRV would grow better in LK cells and RV in BK cells. Lung tissue and scrapings of the mucosa of the large intestine are suitable for virus isolation.

e. Serological Diagnosis. The immunoelectron microscope test described by Hamdy *et al.* (1976) may possibly assist in the rapid differential diagnosis of PPR from rinderpest. When PPRV-infected cells were reacted with PPRV or RV-immune goat sera, followed by ferritin-conjugated rabbit antigoat immunoglobulin, tagging was intense only in the homologous system. However, it must be emphasized that this test has not been assessed in field trials.

ACKNOWLEDGMENTS

Contributions and the critical review of parts of this chapter by Dr. Brian Summers and the excellent secretarial work by Mrs. Ann Signore are greatly appreciated.

REFERENCES

Adams, J. M. (1953). *Pediatrics* **11,** 15-27.
Adams, J. M., and Imagawa, D. T. (1957). *Proc. Soc. Exp. Biol. Med.* **96,** 240-244.
Adams, J. M., Imagawa, D. T., Wright, S. W., and Tarjan, G. (1959). *Virology* **7,** 351-353.
Adams, J. M., Brown, W. J., Snow, H. D., Lincoln, S. D., Sears, A. W., Jr., Barenfus, M., Holliday, T. A., Cremer, N. E., and Lennette, E. H. (1975). *Vet. Pathol.* **12,** 220-226.
Agnarsdottir, G. (1977). *Recent Adv. Clin. Virol.* **1,** 21-48.
Anderson, C. D., and Atherton, J. G. (1964). *Nature (London),* **203,** 670-671.
Anonymous (1966). *Rev. Elev. Med. Vet. Pays Trop.* **19,** 365-413.
Anonymous (1970). *W.H.O. Tech. Rep. Ser.* **444,** 23-57.
Appel, M. (1969). *Am. J. Vet. Res.* **30,** 1167-1182.
Appel, M., Glickman, L. T., Raine, C. S., and Tourtellotte, M. W. (1981). *Neurology.* In press.
Appel, M. (1970). *J. Am. Vet. Med. Assoc.* **156,** 1681-1684.
Appel, M. J. G. (1978). *J. Gen. Virol.* **41,** 385-393.
Appel, M., and Jones, O. R. (1967). *Proc. Soc. Exp. Biol. Med.* **126,** 571-574.
Appel, M., and Robson, D. S. (1973). *Am. J. Vet. Res.* **34,** 1459-1463.
Appel, M., Sheffy, B. E., Percy, D. H., and Gaskin, J. M. (1974). *Am. J. Vet. Res.* **35,** 803-806.

Appel, M., and Gillespie, J. H. (1972). *Virol. Monogr.* **11,** 1-96.
Appel, M., Cooper, B. J., Greisen, H., Scott, F., and Carmichael, L. E. (1979). *Cornell Vet.* **69,** 123-144.
Armitage, A. M. T., Cornwell, H. J. C., Wright, N. G., and Weir, A. R. (1975). *Arch. Virol.* **47,** 319-329.
Ata, F. A., and Singh, K. V. (1967). *Bull. Epizoot. Dis. Afr.* **15,** 213-220.
Bachmann, P. A., ter Meulen, V., Jentsch, G., Appel, M., Iwasaki, Y., Meyermann, R., Koprowski, H., and Mayr, A. (1975). *Arch. Virol.* **48,** 107-120.
Baker, J. A. (1963). *Proc. Gaines Vet. Symp., 12th, 1963,* pp. 7-8.
Bansal, R. P., Chawla, S. K., Sharma, S. D., and Menon, M. S. (1971). *Indian J. Anim. Sci.* **41,** 18-26.
Bansal, R. P., Joshi, R. C., and Kumar, S. (1974). *Bull. Off. Int. Epizoot.* **81,** 305-312.
Bansal, R. P., Joshi, R. C., and Kumar, S. (1975). *Indian J. Anim. Sci.* **11,** 913-914.
Beaton, W. G. (1930). *J. Comp. Pathol.* **43,** 301-307.
Bemis, D. A., Greisen, H. A., and Appel, M. J. G. (1977). *J. Infect. Dis.* **135,** 753-762.
Benson, T. F. (1960). M. S. Thesis, Cornell University, Ithaca, New York.
Bindrich, H. (1951). *Arch. Exp. Veterinaer med.* **4,** 120-126.
Bindrich, H. (1954). *Arch. Exp. Veterinaer med.* **8,** 131-162, 263-315.
Bindrich, H. (1962). *Kleintier-Prax.* **7,** 161-171, 171-187.
Binn, L. N., Lazar, E. C., Rogul, M., Shepler, V. M., Swango, L. J., Claypoole, T., Hubbard, D. W., Asbill, S. G., and Alexander, A. D. (1968). *Am. J. Vet. Res.* **29,** 1809-1815.
Black, F. L. (1959). *Virology* **7,** 184-192.
Blaser, M., Powers, B. W., Cravens, J., and Wang, W. L. (1978). *Lancet* **2,** 979-981.
Bodingbauer, J. (1960). *Vet. Rec.* **72,** 636-637.
Bögel, K., Enders-Ruckle, G., and Provost, A. (1964). *C. R. Hebd. Seances Acad. Sci.* **259,** 482-484.
Bourdin, P. (1973). *Rev. Elev. Med. Vet. Pays Trop.* **26,** 71a-74a.
Bourdin, P., and Laurent-Vautier, A. (1967). *Rev. Elev. Med. Vet. Pays Trop.* **20,** 383-386.
Bourdin, P., Rioche, M.. and Laurent, A. (1970). *Rev. Elev. Med. Vet. Pays Trop.* **23,** 295-300.
Breese, S. S., Jr., and DeBoer, C. J. (1973). *J. Gen Virol.* **20,** 121-125.
Breitfeld, V., Hashida, Y., Sherman, F. E., Odagiri, K., and Yunis, E. J. (1973). *Lab. Invest.* **28,** 279-291.
Brown, A. L., and McCarthy, R. E. (1974). *Nature (London)* **248,** 344-345.
Brown, R. D. (1958). *J. Hyg.* **56,** 427-434.
Buhles, W. C., Jr., Huxsoll, D. L., and Ristic, M. (1974). *J. Infect. Dis.* **130,** 357-367.
Burnet, F. M. (1968). *Lancet* **2,** 610-613.
Burnstein, T., Jacobsen, L. B., Zeman, W., and Chen, T. T. (1974). *Infect. Immun.* **10,** 1378-1382.
Burridge, M. J. (1978). *J. Am. Vet. Med. Assoc.* **173,** 1439-1444.
Bush, M., Montali, R. J., Brownstein, D., James, A. E., Jr., and Appel, M. J. G. (1976). *J. Am. Vet. Med. Assoc.* **169,** 959-960.
Bussell, R. H., and Karzon, D. T. (1962). *Virology* **18,** 589-600.
Bussell, R. H., and Karzon, D. T. (1965). *Arch. Gesamte Virusforsch.* **17,** 163-182, 183-202.
Bussell, R. H., Waters, D. J., Seals, M. K., and Robinson, W. S. (1974). *Med. Microbiol. Immunol.* **160,** 105-124.
Cabasso, V. J., and Cox, H. R. (1949). *Proc. Soc. Exp. Biol. Med.* **71,** 246-250.
Cabasso, V. J., and Cox, H. R. (1952). *Cornell Vet.* **42,** 96-107.
Cabasso, V. J., Johnson, D. W., Stebbins, M. R., and Cox, H. R. (1957). *Am. J. Vet. Res.* **18,** 414-418.
Cabasso, V. J., Kiser, K., and Stebbins, M. R. (1959). *Proc. Soc. Exp. Biol. Med.* **100,** 551-554.

Campbell, J. J., Cosby, S. L., Scott, J. K., Rima, B. K., Martin, S. J., and Appel, M. (1980). *J. Gen. Virol.* **48,** 149–159.
Carlström, G. (1958). *Arch. Gesamte Virusforsch.* **8,** 527–538.
Carpenter, J. W., Appel, M. J. G., Erickson, R. C., and Novilla, M. N. (1976). *J. Am. Vet. Med. Assoc.* **169,** 961–964.
Carré, H. (1905). *C. R. Hebd. Seances Acad. Sci.* **140,** 689–690, 1489–1491.
Chandy, K. G., John, T. J., Mukundan, P., and Cherian, G. (1979). *Lancet* **1,** 381.
Chen, T. T., Watanabe, I., Zeman, W., and Mealey, J., Jr. (1969). *Science* **163,** 1193–1194.
Cheng, S. C., and Fischman, H. R. (1949). *FAO Agric. Stud.* **8,** 47–63.
Chiarini, A., and Norrby, E. (1970). *Arch. Gesamte Virusforsch.* **29,** 205–214.
Choppin, P. W., and Compans, R. W. (1975). *In* "Comprehensive Virology" (H. Fraenkel-Conrat and R. R. Wagner, eds.), Vol. 4, pp. 95–178. Plenum, New York.
Confer, A. W., Kahn, D. E., Koestner, A., and Krakowka, S. (1975a). *Am. J. Vet. Res.* **36,** 741–748.
Confer, A. W., Kahn, D. E., Koestner, A., and Krakowka, S. (1975b). *Infect. Immun.* **11,** 835–844.
Cook, S. D., Dowling, P. C., and Russell, W. C. (1978). *Lancet* **1,** 605–606.
Cook, S. D., Dowling, P. C., Norman, J., and Jablon, S. (1979). *Lancet* **1,** 380–381.
Cornwell, H. J. C., Campbell, R. S. F., Vantsis, J. T., and Penny, W. (1965). *J. Comp. Pathol.* **75,** 3–17.
Coulter, G. R., and Storz, J. (1979). *Am. J. Vet. Res.* **40,** 1671–1677.
Cruickshank, J. G., Waterson, A. P., Kanarek, A. D., and Berry, D. M. (1962). *Res. Vet. Sci.* **3,** 485–486.
Cutchins, E. C., and Dayhuff, T. R. (1962). *Virology* **17,** 420–425.
Cutler, R. W. P., and Averill, D. R. (1969). *Neurology* **19,** 1111–1114.
Dagle, G. E., Zwicker, G. M., Adee, R. R., and Park, J. F. (1979). *Vet. Pathol.* **16,** 258–259.
Darbyshire, J. H. (1962). *Res. Vet. Sci.* **3,** 118–124.
Dardiri, A. H., DeBoer, C. J., and Hamdy, F. M. (1976). *Proc. 19th Annu. Meet. Am. Assoc. Vet. Lab. Diagnosticians* pp. 337–344.
Das, S. K., and Datt, N. S. (1975). *Indian J. Anim. Health* **14,** 109–112.
Daubney, R. (1949). *FAO Agric. Stud.* **8,** 6–18.
DeBoer, C. J., and Barber, T. L. (1964). *Arch. Gesamte Virusforsch.* **15,** 98–108.
deLahunta, A. (1977). *In* "Veterinary Neuroanatomy and Clinical Neurology," pp. 234–307. Saunders, Philadelphia, Pennsylvania.
DeLay, P. D., and Barber, T. L. (1962). *Proc. U.S. Livestock Sanit. Assoc., 66th Annu. Meet.* pp. 132–136.
DeLay, P. D., Stone, S. S., Karzon, D. T., Katz, S., and Enders, J. (1965). *Am. J. Vet. Res.* **26,** 1359–1373.
Doherty, P. C., Effros, R. B., and Bennik, J. (1977). *Proc. Natl. Acad. Sci. U.S.A.* **74,** 1209–1213.
Doi, Y., Sanpe, T., Nakajima, M., Okawa, S., Katoh, T., Itoh, H., Sato, T., Oguchi, K., Kumanshi, T., and Tsubaki, T. (1972). *Jpn. J. Med. Sci. Biol.* **25,** 321–333.
Dubielzig, R. R. (1979). *Vet. Pathol.* **16,** 268–270.
Dunkin, G. W., and Laidlaw, P. P. (1926). *J. Comp. Pathol.* **39,** 213–221.
El Hag Ali, B. (1971). *Bull. Epizoot. Dis. Afr.* **21,** 421–428.
Enders, J. F., and Peebles, T. C. (1954). *Proc. Soc. Exp. Biol. Med.* **86,** 277–286.
Enders, J. F., McCarthy, K., Mitus, A., and Cheatham, W. J. (1959). *N. Engl. J. Med.* **261,** 875–881.
Enders-Ruckle, G. (1964). *Proc. Int. Symp. Stand. Vaccines Against Measles Rubella, 1964,* pp. 192–400.

Fenner, F. (1976). *Intervirology* **7,** 59-60.
Fireman, P., Friday, G., and Kumate, J. (1969). *Pediatrics* **43,** 264-272.
Fischer, C. A. (1971). *J. Am. Vet. Med. Assoc.* **158,** 740-752.
Fischer, C. A., and Jones, G. T. (1972). *J. Am. Vet. Med. Assoc.* **160,** 68-79.
Fisher, L. E., and Bussell, R. H. (1977). *Intervirology* **8,** 218-225.
Fisher, L. E., and Rapp, (1979a). *J. Virol.* **30,** 64-68.
Fisher, L. E., and Rapp, F. (1979b). *Virology* **94,** 55-60.
Follett, E. A. C., Pringle, C. R., Pennington, T. H., and Shirodaria, P. (1976). *J. Gen. Virol.* **32,** 163-175.
Fraser, C. E. O., Chalifoux, L., Shegal, P., Hunt, R. D., and King, N. W. (1978). *Primates Med.* **10,** 261-270.
Fraser, K. B., and Martin, S. J. (1978). "Measles Virus and Its Biology." Academic Press, New York.
Fukuda, A., and Yamanouchi, K. (1976). *Infect. Immun.* **13,** 1449-1453.
Fulton, R. E., and Middleton, P. J. (1975). *J. Pediatr.* **86,** 17-22.
Gargadennec, L., and Lalanne, A. (1942). *Bull. Serv. Zootech. Epizoot. A. O. F.* **5,** 16-21.
Gatti, J. M., and Good, R. A. (1970). *Med. Clin. North Am.* **54,** 281-307.
Gerber, J. D., and Marron, A. E. (1976). *Am. J. Vet. Res.* **37,** 133-138.
Gibbs, E. P. J., Taylor, W. P., Lawman, M. J. P., and Bryant, J. (1979). *Intervirology* **11,** 268-274.
Gibbs, F. A., Gibbs, E. L., Carpenter, P. R., and Spies, H. W. (1959). *J. Am. Med. Assoc.* **171,** 1050-1055.
Gibson, J. P., Griesemer, R. A., and Koestner, A. (1965). *J. Pathol. Vet.* **2,** 1-19.
Gilbert, Y., and Monnier, J. (1962). *Rev. Elev. Med. Vet. Pays Trop.* **15,** 321-335.
Gillespie, J. H. (1962). *Ann. N. Y. Acad. Sci.* **101,** 540-547.
Gillespie, J. H. (1965). *Cornell Vet.* **55,** 3-8.
Gillespie, J. H., and Karzon, D. T. (1960). *Proc. Soc. Exp. Biol. Med.* **105,** 547-551.
Gillespie, J. H., Baker, J. A., Burgher, J., Robson, D., and Gilman, B. (1958). *Cornell Vet.* **48,** 103-126.
Gillespie, J. H., Brueckner, A. L., Hejl, J., Jackson, R., McClelland, R. B., and Ott, R. L. (1966). *J. Am. Vet. Med. Assoc.* **149,** 714-717.
Goret, P., Mornet, P., Gilbert, Y., and Pilet, C. (1957). *C. R. Hebd. Seances Acad. Sci.* **245,** 2564-2566.
Goret, P., Mornet, P., Gilbert, Y., and Pilet, C. (1958). *Bull. Off. Int. Epizoot.* **49,** 501-506.
Goret, P., Fontaine, M., Mackowiak, C., and Pilet, C. (1959). *C. R. Hebd. Seances Acad. Sci.* **248,** 2143-2144.
Goret, P., Brion, A., and Fountaine, M. (1960). *Bull. Acad. Vet. Fr.* **33,** 343-347.
Gorham, J. R. (1960). *Adv. Vet. Sci.* **6,** 287-351.
Gorham, J. R. (1966). *J. Am. Vet. Med. Assoc.* **149,** 610-622.
Goss, L. J. (1948). *Am. J. Vet. Res.* **9,** 65-68.
Gould, E. (1974). *Med. Microbiol. Immunol.* **160,** 211-219.
Gourlay, J. A. (1970). *Cornell Vet.* **60,** 613.
Graves, M. C., Silver, S. M., and Choppin, P. W. (1978). *Virology* **86,** 254-263.
Green, R. G., and Carlson, W. E. (1945). *J. Am. Vet. Med. Assoc.* **107,** 131-142.
Haig, D. A. (1948). *J. Vet. Sci. Anim. Ind.* **23,** 149-155.
Haig, D. A. (1956). *J. Vet. Res.* **27,** 19-53.
Hall, W. W., and Martin, S. J. (1973). *J. Gen. Virol.* **19,** 175-188.
Hall, W. W., and ter Meulen, V. (1976). *Nature (London)* **264,** 474-477.
Hall, W. W., and Choppin, P. W. (1979). *Virology* **99,** 443-447.
Hall, W. W., Martin, S. J., and Gould, E. (1974). *Med. Microbiol. Immunol.* **160,** 155-164.
Hall, W. W., Kiessling, W. R., and ter Meulen, V. (1978a). *Nature (London)* **272,** 460-462.

Hall, W. W., Kiessling, W. R., and ter Meulen, V. (1978b). *In* "Negative Strand Viruses and the Host Cell" (R. D. Barry and B. W. J. Mahy, eds.), pp. 143–156. Academic Press, New York.

Hall, W. W., Nagashima, K., Kiessling, W. R., and ter Meulen, V. (1978c). *In* "Negative Strand Viruses and the Host Cell" (R. D. Barry and B. W. T. Mahy, eds.), pp. 765–770. Academic Press, New York.

Hall, W. W., Imagawa, D. T., and Choppin, P. W. (1979a). *Virology* **98,** 283–287.

Hall, W. W., Lamb, R. A., and Choppin, P. W. (1979b). *Proc. Natl. Acad. Sci. U.S.A.* **76,** 2047–2051.

Hall, W. W., Lamb, R. A., and Choppin, P. W. (1980). *Virology* **100,** 433–449.

Halonen, P., Matikainen, M. T., Salmi, A., Vuorimaa, T., and Ziola, B. R. (1977). *Lancet* **1,** 1201.

Halsey, N. A., Modlin, J. F., and Jabbour, J. T. (1978). *In* "Persistent Viruses." (J. G. Stevens, G. J. Todaro, and C. F. Fox, eds.), pp. 101–114. Academic Press, New York.

Hamdy, F. M., and Dardiri, A. H. (1976). *J. Wildl. Dis.* **12,** 516–522.

Hamdy, F. M., Dardiri, A. H., Breese, S. S., Jr., and DeBoer, C. J. (1975). *Proc. 79th Annu. Meet. U.S. Anim. Health Assoc.* pp. 168–179.

Hamdy, F. M., Dardiri, A. H., Nduaka, O., Breese, S. S., Jr., and Ihemelandu, E. C. (1976). *Can. J. Comp. Med.* **40,** 276–284.

Harrison, M. J., Oxer, D. T., and Smith, F. A. (1968). *J. Comp. Pathol.* **78,** 133–139.

Ho, C. K., and Babiuk, L. A. (1979). *Immunology* **37,** 231–239.

Hoffman, F. (1949). *Acta Vet. Hung.* **1,** 89–92.

Holmes, J. D. E. (1904). *J. Comp. Pathol.* **17,** 317–326.

Horta-Barbosa, L., Fuccillo, D. A., Sever, J. L., and Zeman, W. (1969). *Nature (London)* **221,** 974.

Hsiung, G. D., Mannini, A., and Melnick, J. L. (1958). *Proc. Soc. Exp. Biol. Med.* **98,** 68–70.

Imagawa, D. T. (1968). *Prog. Med. Virol.* **10,** 160–193.

Imagawa, D. T., Goret, P., and Adams, J. M. (1960). *Proc. Natl. Acad. Sci. U.S.A.* **46,** 1119–1123.

Imagawa, D. T., Howard, E. B., Van Pelt, L. F., and Ryan, C. P. (1979). *Proc. 79th Annu. Meet. Am. Soc. Microbiol.* p. 306.

Innes, J. R. M., and Saunders, L. Z. (1962). "Comprehensive Neuropathology." Academic Press, New York.

Ishii, S., and Watanabe, M. (1971). *Natl. Inst. Anim. Health Q.* **11,** 55–63.

Isoun, T. T., and Mann, E. D. (1972). *Bull. Epizoot. Dis. Afr.* **20,** 167–174.

Johnson, R. H. (1958). *Vet. Rec.* **70,** 457–461.

Johnson, R. H. (1962). *Br. Vet. J.* **118,** 107–116.

Johnson, R. H., and Ritchie, J. S. D. (1968). *Bull. Epizoot. Dis. Afr.* **16,** 411–417.

Joshi, R. C., Chaudhary, P. G., and Bansal, R.P. (1977). *Indian Vet. J.* **54,** 871–873.

Ju, G., Udem, S., Rager-Zisman, B., and Bloom, B. R. (1978). *J. Exp. Med.* **147,** 1637–1652.

Karzon, D. T. (1962). *Ann. N. Y. Acad. Sci.* **101,** 527–539.

Karzon, D. T., Gillespie, J. H., and Bussell, R. H. (1961). *Am. J. Vet. Res.* **22,** 1069–1073.

Kataria, R. S., Majumdar, S. S., and Srivastava, S. N. (1974). *Indian Vet. J.* **51,** 736–737.

Kibler, R., and ter Meulen, V. (1975). *J. Immunol.* **114,** 93–98.

Kiessling, W. R., Hall, W. W., Yung, L. L., and ter Meulen, V. (1977). *Lancet* **1,** 324–327.

Kimes, R. C., and Bussell, R. H. (1968). *Arch. Gesamte Virusforsch.* **27,** 387–395.

Kimura, A., Tosaka, K., and Nakao, T. (1975). *Arch. Virol.* **47,** 295–307.

Kingsbury, D. W. (1972). *Curr. Top. Microbiol. Immunol.* **59,** 1–33.

Kingsbury, D. W., Bratt, M. A., Choppin, P. W., Hanson, R. P., Hosaka, Y., ter Meulen, V., Norrby, E., Plowright, W., Rott, R., and Wunner, W. H. (1978). *Intervirology* **10,** 137–152.

Kobune, F., Ito, M., and Yamanouchi, K. (1976a). *Jpn. J. Med. Sci. Biol.* **29,** 171–176.

Kobune, F., Chino, F., and Yamanouchi, K. (1976b). *Jpn. J. Med. Sci. Biol.* **29,** 265–275.

Koestner, A. (1975). *Am. J. Pathol.* **78,** 361–364.
Koestner, A., McCullough, B., Krakowka, G. S., Long, J. F., and Olsen, R. G. (1974). *In* "Slow Virus Diseases" (W. Zeeman and E. H. Lennette, eds.), pp. 86–101. Williams & Wilkins, Baltimore, Maryland.
Koprowski, H., Jervis, G. A., James, T. R., Burkhart, R. L., and Poppensiek, G. C. (1950). *Am. J. Hyg.* **51,** 63–75.
Krakowka, S., and Koestner, A. (1976). *J. Infect. Dis.* **134,** 629–632.
Krakowka, S., and Wallace, A. L. (1979). *Am. J. Vet. Res.* **40,** 669–672.
Krakowka, S., McCullough, B., Koestner, A., and Olsen, R. (1973). *Infect. Immun.* **8,** 819–827.
Krakowka, S., Confer, A., and Koestner, A. (1974). *Am. J. Vet. Res.* **35,** 1251–1253.
Krakowka, S., Cockerell, G., and Koestner, A. (1975a). *Infect. Immun.* **11,** 1069–1078.
Krakowka, S., Olsen, R., Confer, A., Koestner, A., and McCullough, B. (1975b). *J. Infect. Dis.* **132,** 384–392.
Krakowka, S., Hoover, E. A., Koestner, A., and Ketring, K. (1977). *Am. J. Vet. Res.* **38,** 919–922.
Krakowka, S., Long, D., and Koestner, A. (1978a). *J. Infect. Dis.* **137,** 605–608.
Krakowka, S., Wallace, A. L., and Koestner, A. (1978b). *J. Clin. Microbiol.* **7,** 292–297.
Krakowka, S., Mandor, R. A., and Koestner, A. (1978c). *Acta Neuropathol.* **43,** 235–241.
Kratzsch, V., Hall, W. W., Nagashima, K., and ter Meulen, V. (1977). *J. Med. Virol.* **1,** 139–154.
Kreth, H. W., Käckell, Y. M., and ter Meulen, V. (1975). *J. Immunol.* **114,** 1042–1046.
Kreth, H. W., ter Meulen, V., and Eckert, G. (1979). *Med. Microbiol. Immunol.* **165,** 203–214.
Kirshnan, R., and Ranga Rao, D. V. (1972). *Cheiron* **1,** 1–7.
Kristensen, B., and Vandevelde, M. (1978). *Am. J. Vet. Res.* **39,** 1017–1021.
Krugman, S. (1977). *J. Pediatr.* **90,** 1–12.
Laidlaw, P. P., and Dunkin, G. W. (1926). *J. Comp. Pathol.* **39,** 213–221.
Laidlaw, P. P., and Dunkin, G. W. (1928). *J. Comp. Pathol.* **41,** 1–17.
Lampert, P. W. (1978). *Am. J. Pathol.* **91,** 176–198.
Lauder, I. M., Martin, W. B., Gordon, E. D., Lawson, D. D., Campbell, R. S. F., and Watrach, A. M. (1954). *Vet. Rec.* **66,** 607–611, 623–631.
Laurent, A. (1968). *Rev. Elev. Med. Vet. Pays Trop.* **21,** 297–308.
Lennette, E. H., and Schmidt, N. J. (1969). "Diagnostic Procedures for Viral and Rickettsial Infections," 4th ed. Am. Public Health Assoc., New York.
Liess, B., and Plowright, W. (1963a). *J. Hyg.* **61,** 205–211.
Liess, B., and Plowright, W. (1963b). *Arch. Gesamte Virusforsch.* **14,** 27–38.
Liess, B., and Plowright, W. (1964). *J. Hyg.* **62,** 81–100.
Lincoln, S. D., Gorham, J. R., Ott, R. L., and Hegreberg, G. A. (1971). *Vet. Pathol.* **8,** 1–8.
Lincoln, S. D., Gorham, J. R., Davis, W. C., and Ott, R. L. (1973). *Vet. Pathol.* **10,** 124–129.
Liu, C., and Coffin, D. L. (1957). *Virology* **3,** 115–131.
Lo, J. P., Dickson, W. M., and Gorham, J. R. (1965). *Arch. Gesamte Virusforsch.* **15,** 74–90.
Lyons, M. J., Hall, W. W., Petito, C., Cam, V., and Zabriskie, J. B. (1980). *Neurology* **30,** 92–98.
McCarthy, K. (1959). *Br. Med. J.* **15,** 201–204.
McCullough, B., Krakowka, S., and Koestner, A. (1974a). *Lab. Invest.* **31,** 216–222.
McCullough, B., Krakowka, S., Koestner, A., and Shadduck, J. (1974b). *J. Infect. Dis.* **130,** 343–350.
McCullough, B., Krakowka, S., and Koestner, A. (1974c). *Am. J. Pathol.* **74,** 155–165.
McGrath, J. T. (1960). "Neurologic Examination of the Dog," 2nd ed. Lea & Febiger, Philadelphia, Pennsylvania.
MacIntyre, A. B., Trevan, D. J., and Montgomerie, R. (1948). *Vet. Rec.* **60,** 635–642.
McKercher, P. D. (1964). *Can. J. Comp. Med. Vet. Sci.* **28,** 113–120.
McLean, D. M., Best, J. M., Smith, P. A., Larke, R. P. B., and McNaughton, G. A. (1966). *Can. Med. Assoc. J.* **94,** 905–910.

McMichael, A. J., Ting, A., Zweerink, H. J., and Askonas, B. A. (1977). *Nature (London)* **270,** 524-526.
Majer, M. (1972). *In* "Strains of Human Viruses" (M. Majer and S. A. Plotkin, eds.), pp. 131-141. Karger, Basel.
Mangi, R. J., Munyer, T. P., Krakowka, S., Jacoby, R. O., and Kantor, F. S. (1976). *J. Infect. Dis.* **133,** 556-563.
Mannweiler, K. (1965). *Arch. Gesamte Virusforsch.* **16,** 89-96.
Mansi, W. (1958). *Nature (London)* **181,** 1289-1290.
Matumoto, M. (1966). *Bacteriol. Rev.* **30,** 152-176.
Maurice, Y., Provost, A., and Borredon, C. (1969). *Rev. Elev. Med. Vet. Pays Trop.* **20,** 1-8.
Melnick, J. L., and McCombs, R. M. (1966). *Prog. Med. Virol.* **8,** 400-409.
Millian, S., Maisel, J., Kempe, C. H., Plotkin, S., Pagano, J., and Warren, J. (1960). *J. Bacteriol.* **79,** 616-618.
Minagawa, T., Nakaya, C., and Iida, H. (1974). *J. Virol.* **13,** 1118-1125.
Mirchamsy, H., Shafyi, A., Bahrami, S., Nazari, P., and Akbarzadeh, J. (1974). *Res. Vet. Sci.* **17,** 242-247.
Mitus, A., Enders, J. F., Craig, J. M., and Holloway, A. (1959). *N. Engl. J. Med.* **261,** 882-889.
Morgan, E. M., and Rapp, F. (1977). *Bacteriol. Rev.* **41,** 636-666.
Mori, M. (1969). *Kromosomo* **77-78,** 2510-2516.
Mornet, P., Orue, J., Gilbert, Y., Thiery, G., and Sow Mamadou (1956). *Rev. Elev. Med. Vet. Pays Trop.* **9,** 313-342.
Mornet, P., Goret, P., and Gilbert, Y. (1959). *Bull. Epizoot. Dis. Afr.* **7,** 255-263.
Mountcastle, W. E., and Choppin, P. W. (1977). *Virology* **78,** 463-474.
Moura, R. A., and Warren, J. (1961). *J. Bacteriol.* **82,** 702-705.
Müller, D., and ter Meulen, V. (1969). *Acta Neuropathol.* **12,** 227-243.
Nakai, T., Shand, F. L., and Howatson, A. F. (1969). *Virology* **38,** 50-67.
Nakamura, J. (1958). "Complement Fixation Reaction in Rinderpest Study. Guide for Technique and Application." Int. Off. Epizoot., Paris.
Narayanaswamy, M., and Ramani, K. (1973). *Indian Vet. J.* **50,** 829-832.
Nathanson, N., Palsson, P. A., and Gudmundsson, G. (1978). *Lancet* **2,** 1127-1129.
Nawathe, D. R., and Taylor, W. P. (1979). *Trop. Anim. Health Prod.* **11,** 120-122.
Nduaka, O., and Ihemelandu, E. C. (1973). *Bull. Epizoot. Dis. Afr.* **21,** 87-98.
Nemo, G. T., and Cutchins, E. C. (1966). *J. Bacteriol.* **91,** 798-802.
Newlin, G. E., and Bussell, R. H. (1972). *Arch. Gesamte Virusforsch.* **39,** 303-306.
Nichols, W. W., and Levan, A. (1965). *Arch. Gesamte Virusforsch.* **16,** 168-174.
Nichols, W. W., Levan, A., Hall, B., and Östergen, G. (1962). *Hereditas* **48,** 367-370.
Nicolle, C. (1931). *Arch. Inst. Pasteur Tunis* **20,** 321-323.
Noon, K. G., Rogul, M., Binn, L. N., Keefe, T. J., Marchwicki, R. H., and Appel, M. J. (1980). *Am. J. Vet. Res.* **41,** 605-609.
Norrby, E. (1964). *Arch. Gesamte Virusforsch.* **14,** 306-318.
Norrby, E. (1967a). *In* "Recent Advances in Medical Microbiology" (A. P. Waterson, ed.), pp. 1-53. Churchill, London.
Norrby, E. (1967b). *Arch. Gesamte Virusforsch.* **20,** 215-224.
Norrby, E., and Magnusson, P. (1965). *Arch. Gesamte Virusforsch.* **17,** 443-447.
Norrby, E., Friding, B., Rockborn, G., and Gard, S. (1963). *Arch. Gesamte Virusforsch.* **13,** 335-344.
Norrby, E., Enders-Ruckle, G., and ter Meulen, V. (1975). *J. Infect. Dis.* **132,** 262-269.
Obi, T. U., and Gibbs, E. P. J. (1979). *Trop. Anim. Health Prod.* **10,** 233-235.
Oddo, F. G., Flaccomio, R., and Sinatra, A. (1961). *Virology* **13,** 550-553.
Ojala, A. (1974). *Ann. Med. Intern. Fenn.* **36,** 321-331.

Orvell, C., and Norrby, E. (1974). *J. Immunol.* **113,** 1850–1858.
Osunkoya, B. O., Cooke, A. R., Ayeni, O., and Adejumo, T. A. (1974a). *Arch. Gesamte Virusforsch.* **44,** 313–322.
Osunkoya, B. O., Adeleye, G. I., Adejumo, T. A., and Salimonu, L. S. (1974b). *Arch. Gesamte Virusforsch.* **44,** 323–329.
Oyanagi, S., ter Meulen, V., Katz, M., and Koprowski, H. (1971). *J. Virol.* **7,** 176–187.
Ozawa, Y., and Nelson, R. T. (1973). *Bull. Epizoot. Dis. Afr.* **21,** 287–289.
Palm, C. R., and Black, F. L. (1961). *Proc. Soc. Exp. Biol. Med.* **107,** 588–590.
Panum, P. L. (1847). *Arch. Pathol. Anat. Physiol. Klin. Med.* **1,** 492–512.
Phillips, L. A., and Bussell, R. H. (1973). *Arch. Gesamte Virusforsch.* **41,** 310–318.
Phillips, L. A., and Bussell, R. H. (1974). *Am. J. Vet. Res.* **35,** 821–824.
Pinkerton, H., Smiley, W. L., and Anderson, W. A. D. (1945). *Am. J. Pathol.* **21,** 1–23.
Plowright, W. (1952). *Br. Vet. J.* **108,** 450–457.
Plowright, W. (1962). *Ann. N. Y. Acad. Sci.* **101,** 548–563.
Plowright, W. (1963). *Res. Vet. Sci.* **4,** 96–108.
Plowright, W. (1964a). *J. Hyg.* **62,** 257–281.
Plowright, W. (1964b). *Arch. Gesmate Virusforsch.* **14,** 431–448.
Plowright, W. (1965). *Vet. Rec.* **77,** 1431–1438.
Plowright, W. (1968). *Virol. Monogr.* **3,** 25–110.
Plowright, W., and Ferris, R. D. (1959). *J. Comp. Pathol.* **69,** 152–172.
Plowright, W., and Ferris, R. D. (1961). *Arch. Gesamte Virusforsch.* **11,** 516–533.
Plowright, W., and Ferris, R. E. (1962). *Res. Vet. Sci.* **3,** 172–182.
Plowright, W., and McCulloch, B. (1967). *J. Hyg.* **65,** 343–358.
Plowright, W., and Taylor, W. P. (1967). *Res. Vet. Sci.* **8,** 118–128.
Plowright, W., Cruickshank, J. G., and Waterson, A. P. (1962). *Virology* **17,** 118–122.
Plowright, W., Herniman, K. A. J., and Rampton, C. S. (1969). *Res. Vet. Sci.* **10,** 373–381.
Plowright, W., Rampton, C. S., Taylor, W. P., and Herniman, K. A. J. (1970). *Res. Vet. Sci.* **11,** 71–81.
Plowright, W., Herniman, K. A. J., and Rampton, C. S. (1971). *Res. Vet. Sci.* **12,** 40–46.
Polding, J. B., Simpson, R. M., and Scott, G. R. (1959). *Vet. Rec.* **71,** 643–645.
Poste, G. (1970). *Transplantation* **10,** 106.
Poste, G. (1971). *J. Comp. Pathol.* **81,** 49.
Prescott, J. F., Johnson, J. A., and Patterson, J. M. (1978). *Vet. Rec.* **103,** 116–117.
Provost, A. (1970). *Bull. Off. Int. Epizoot.* **73,** 923–928.
Provost, A. (1972). *Rev. Elev. Med. Vet. Pays Trop.* **25,** 155–159.
Provost, A., and Borredon, C. (1963). *Rev. Elev. Med. Vet. Pays Trop.* **16,** 445–526.
Provost, A., and Borredon, C. (1972a). *Rev. Elev. Med. Vet. Pays Trop.* **25,** 141–153.
Provost, A., and Borredon, C. (1972b). *Rev. Elev. Med. Vet. Pays Trop.* **25,** 507–520.
Provost, A., and Borredon, C. (1974). *Rev. Elev. Med. Vet. Pays Trop.* **27,** 251–263.
Provost, A., and Joubert, L. (1973). *Rev. Elev. Med. Vet. Pays Trop.* **26,** 383–396.
Provost, A., Queval, R., Borredon, C., and Maurice, Y. (1963). *Rev. Elev. Med. Vet. Pays Trop.* **16,** 287–297.
Provost, A., Borredon, C., and Queval, R. (1965). *Rev. Elev. Med. Vet. Pays Trop.* **18,** 385–393.
Provost, A., Maurice, Y., and Borredon, C. (1969a). *Rev. Elev. Med. Vet. Pays Trop.* **22,** 9–15.
Provost, A., Maurice, Y., and Borredon, C. (1969b). *Rev. Elev. Med. Vet. Pays Trop.* **22,** 453–464.
Provost, A., Borredon, C., and Maurice, Y. (1969c). *Rev. Elec. Med. Vet. Pays Trop.* **22,** 473–479.
Raine, C. S. (1976). *J. Neurol. Sci.* **30,** 13–28.
Ramachandran, S., and Scott, G. R. (1972). *Indian Vet. J.* **49,** 1060–1062.
Ramakrishna Rao, M., Sambamurti, B., and Sarma, B. J. R. (1976). *Indian Vet. J.* **53,** 653–658.

Ramani, K., Samuel Charles, Y., and Ramachandran, S. (1974). *Indian Vet. J.* **51,** 129–138.
Rao, M., Indra Devi, T., Ramachandran, S., and Scott, G. R. (1974). *Indian Vet. J.* **51,** 439–450.
Rapp, F., Butel, J. S., and Wallis, C. (1965). *J. Bacteriol.* **90,** 132–135.
Reculard, P., and Guillon, J. C. (1967). *Bull. Acad. Vet. Fr.* **40,** 507.
Reculard, P., and Guillon, J. C. (1972). *Ann. Inst. Pasteur, Paris* **123,** 477–487.
Rentier, B., Hooghe-Peters, E. L., and Dubois-Dalcq, M. (1978). *J. Virol.* **28,** 567–577.
Rima, B. K., and Martin, S. J. (1977). *Med. Microbiol. Immunol.* **162,** 89–118.
Rima, B. K., and Martin, S. J. (1979). *J. Gen. Virol.* **42,** 603–608.
Rima, B. K., Davidson, W. B., and Martin, S. J. (1977). *J. Gen Virol.* **35,** 89–97.
Rima, B. K., Martin, S. J., and Gould, E. A. (1979). *J. Gen. Virol.* **43,** 603.
Rioche, M. (1969). *Rev. Elev. Med. Vet. Pays Trop.* **22,** 465–471.
Roberts, J. A. (1965). *J. Immunol.* **94,** 622–628.
Robin, P., and Bourdin, P. (1965). *Rev. Elev. Med. Vet. Pays Trop.* **19,** 451–456.
Robson, D. S., Kenneson, R., Gillespie, J. H., and Benson, T. F. (1959). *Proc. Gaines Vet. Symp., 9th, 1959,* pp. 10–14.
Rockborn, G. (1957). *Arch. Gesamte Virusforsch.* **7,** 168–182.
Rockborn, G. (1958). *Arch. Gesamte Virusforsch.* **8,** 485–492.
Rockborn, G. (1960). *J. Small Anim. Pract.* **1,** 53.
Rosendahl, S. (1972). *Acta Vet. Scand.* **13,** 137–139.
Rowe, L. W., Zwart, D., and Kouwenhoven, B. (1967). *Bull. Epizoot. Dis. Afr.* **15,** 301–306.
Rowland, A. C., and Bourdin, P. (1970). *Rev. Elev. Med. Vet. Pays Trop.* **23,** 301–307.
Rowland, A. C., Scott, G. R., and Hill, D. H. (1969). *J. Pathol.* **98,** 83–87.
Rowland, A. C., Scott, G. R., Ramachandran, S., and Hill, D. H. (1971). *Trop. Anim. Health Prod.* **3,** 241–245.
Rozenblatt, S., Gorecki, M., Shure, H., and Prives, C. L. (1979). *J. Virol.* **29,** 1099–1106.
Rustigian, R. (1962). *Virology* **16,** 101–104.
Rustigian, R. (1966). *J. Bacteriol.* **92,** 1805–1811.
Rweyemamu, M. M., Reid, H. W., and Okuna, N. (1974). *Bull. Epizoot. Dis. Afr.* **22,** 1–9.
Sato, T. A., Yamanouchi, K., and Shishido, A. (1973). *Arch. Gesamte Virusforsch.* **42,** 36–41.
Schluederberg, A. (1971). *Biochem. Biophys. Res. Commun.* **42,** 1012–1015.
Schluederberg, A., Williams, C. A., and Black, F. L. (1972). *Biochem. Biophys. Res. Commun.* **48,** 657–661.
Schultz, R. D., Appel, M., Carmichael, L. E., and Farrow, B. (1977). *In* "Current Veterinary Therapy: Small Animal Practice" (R. W. Kirk, ed.), 6th ed., pp. 1271–1275. Saunders, Philadelphia, Pennsylvania.
Scott, D., Farrow, B., and Schultz, R. (1974). *J. Am. Anim. Hosp. Assoc.* **10,** 233–244.
Scott, G. R. (1964). *Adv. Vet. Sci.* **9,** 113–224.
Scott, G. R. (1967). *FAO Argic. Stud.* **71.**
Scott, G. R., DeTray, D. E., and White, G. (1959). *Bull. Off. Int. Epizoot.* **51,** 694–698.
Scott, G. R., DeTray, D. E., and White, G. (1962). *Am. J. Vet. Res.* **23,** 452–456.
Shek, W. R. (1979). M. S. Thesis, Cornell University, Ithaca, New York.
Shek, W. R., Schultz, R. D., and Appel, M. J. G. (1980). *Infect. Immun.* **28,** 724–734.
Shishido, A., Yamanouchi, K., Hikita, M., Sata, T., Fukada, A., and Kobune, F. (1967). *Arch. Gesamte Virusforsch.* **22,** 364–380.
Siefried, A. S., Albreacht, P., and Milstein, J. B. (1978). *J. Virol.* **25,** 781–787.
Singh, K. V., and Ata, F. (1967). *Bull. Epizoot. Dis. Afr.* **15,** 19–23.
Singh, K. V., and El Cicy, I. F. (1966). *Nature (London)* **211,** 314–315.
Slater, E. A., and Murdock, F. M. (1963). *Vet. Med. (Kansas City, Mo.)* **58,** 717–723.
Stallcup, K. C., Wechsler, S. L., and Fields, B. N. (1979). *J. Virol.* **30,** 166–176.
Starr, S., and Berkovich, S. (1964). *N. Engl. J. Med.* **270,** 386–391.

Stephenson, J. R., and ter Meulen, V. (1979). *Proc. Natl. Acad. Sci. U.S.A.* **76,** 6601-6605.
Stephenson, J. R., and ter Meulen, V. (1979). *In* "Humoral Immunity in Neurological Diseases." D. Karcher, A. Lowenthal, A. D. Strosberg (eds.), Plenum Press, New York, London, pp. 105-113.
Stephenson, J. R., ter Meulen, V., and Kiessling, W. (1980). *Lancet* **2,** 772-775.
Stevens, D. R., and Osburn, B. I. (1976). *J. Am. Vet. Med. Assoc.* **168,** 493-498.
Stewart, D. L., Hebert, C. N., and Davidson, I. (1968). *Bull. W.H.O.* **39,** 917-924.
Stewart, D. R. M. (1964). *Bull. Epizoot. Dis. Afr.* **12,** 39-42.
Stone, S. S., and DeLay, P. D. (1961). *J. Immunol.* **87,** 464-467.
Sullivan, J. L., Barry, D. W., Lucas, S. J., and Albrecht, P. (1975). *J. Exp. Med.* **142,** 773-784.
Summers, B. A., Greisen, H. A., and Appel, M. J. G. (1978). *Lancet* **2,** 187-189.
Summers, B. A., Greisen, H. A., and Appel, M. J. G. (1979). *Acta Neuropathol.* **46,** 1-10.
Suringa, D. W. R., Bank, L. J., and Ackerman, A. B. (1970). *N. Engl. J. Med.* **283,** 1139-1142.
Swango, L. J., Wooding, W. L., Jr., and Binn, L. N. (1970). *J. Am. Vet. Med. Assoc.* **156,** 1687-1696.
Tajima, M., and Ushijima, T. (1971). *Am. J. Pathol.* **62,** 221-228.
Tajima, M., Motohashi, T., Kishi, S., and Nakamura, J. (1971). *Jpn. J. Vet. Sci.* **33,** 1-10.
Tanzer, J., Stoitchkov, Y., Harel, P., and Boiron, M. (1963). *Lancet* **2,** 1070-1071.
Taylor, W. P. (1979a). *Res. Vet. Sci.* **26,** 236-242.
Taylor, W. P. (1979b). *Res. Vet. Sci.* **27,** 321-324.
Taylor, W. P., and Abegunde, A. (1979). *Res. Vet. Sci.* **26,** 94-96.
Taylor, W. P., and Perry, C. T. (1970). *Arch Gesamte Virusforsch.* **32,** 269-282.
Taylor, W. P., and Plowright, W. (1965). *J. Hyg.* **63,** 263-275.
Taylor, W. P., and Watson, R. M. (1967). *J. Hyg.* **65,** 537-545.
Taylor, W. P., Plowright, W., Pillinger, R., Rampton, C. S., and Staple, R. F. (1965). *J. Hyg.* **63,** 497-506.
ter Meulen, V. (1974). *Med. Microbiol. Immunol.* **160,** 165-172.
ter Meulen, V., and Hall, W. W. (1978). *J. Gen. Virol.* **41,** 1-25.
ter Meulen, V., and Martin, S. J. (1976). *J. Gen. Virol.* **32,** 431-440.
ter Meulen, V., Katz, M., and Müller, D. (1972a). *Curr. Top. Microbiol. Immunol.* **57,** 1-38.
ter Meulen, V., Müller, D., Käckell, Y. M., Katz, M., and Meyermann, R. (1972b). *Lancet* **1,** 1172-1175.
ter Meulen, V., Katz, M., and Käckell, Y. M. (1973). *Ann. Clin. Res.* **5,** 293-297.
Thorne, H. V., and McDermott, E. (1976). *Nature (London)* **264,** 473-474.
Tokuda, G., Fukusho, K., Morimoto, T., and Watanabe, M. (1962). *Natl. Inst. Anim. Health Q.* **2,** 189-200.
Toyoshima, K., Takahashi, M., Hata, S., Kunita, N., and Okuno, Y. (1959). *Biken J.* **2,** 305-312.
Tsai, S. C. (1979). Ph.D. Thesis, Cornell University, Ithaca, New York.
Tyrrell, D. L. J., and Norrby, E. (1978). *J. Gen. Virol.* **39,** 219-230.
Underwood, G. E. (1959). *J. Immunol.* **83,** 198-205.
Underwood, B., and Brown, F. (1974). *Med. Microbiol. Immunol.* **160,** 125-132.
Ushijima, T., Tajima, M., and Kishi, S. (1968). *Jpn. J. Vet. Sci.* **31,** 43-49.
Valdimarsson, H., Agnarsdottir, G., and Lachmann, P. J. (1974). *Proc. R. Soc. Med.* **67,** 1125-1129.
Vandevelde, M., Fatzer, R., and Fankhauser, R. (1974). *Schweiz. Arch. Tierheilkd.* **116,** 391-404.
Vandevelde, M., Kristensen, B., Braund, K. G., Greene, C. E., Swango, L. J., and Hoerlein, B. J. (1980). *Vet. Pathol.* **17,** 17-29.
Vantsis, J. T. (1959). *Vet. Rec.* **71,** 99-100.
von Mickwitz, C. U., and Schröder, H. D. (1968). *Zentralbl. Veterinaer med., Reihe B* **15,** 453.

von Pirquet, C. (1908). *Dtsch. Med. Wochenschr.* **34,** 1297-1300.

Vuorimaa, T. O., Pertii, P. A., Ziola, B. A., Salmi, A. A., Hänninen, P. T., and Halonen, P. E. (1978). *J. Med. Virol.* **2,** 271-278.

Warren, J., Nadel, K., Slater, E., and Millian, S. J. (1960). *Am. J. Vet. Res.* **21,** 111-119.

Waters, D. J., and Bussell, R. H. (1973). *Virology* **55,** 554-557.

Waters, D. J., and Bussell, R. H. (1974). *Virology* **61,** 64-79.

Waters, D. J., Hersh, R. T., and Bussell, R. H. (1972). *Virology* **48,** 278-281.

Waterson, A. P. (1965). *Arch. Gesamte Virusforsch.* **16,** 57-80.

Waterson, A. P., Cruickshank, J. G., Lawrence, G. D., and Kanarek, A. D. (1961). *Virology* **15,** 379-382.

Waterson, A. P., Rott, R., and Enders-Ruckle, G. (1963). *Z. Naturforsch., B: Anorg. Chem., Org. Chem., Biochem., Biophys., Biol.* **18B,** 377-384.

Watson, A. D. J., and Wright, R. G. (1974). *J. Comp. Pathol.* **84,** 417-427.

Wechsler, S., and Fields, B. N. (1978). *Nature (London)* **272,** 458-459.

Wechsler, S. L., Rustigian, R., Stallcup, K. C., Byers, K. B., Winston, S. H., and Fields, B. N. (1979a). *J. Virol.* **31,** 677-684.

Wechsler, S. L., Weiner, H. L., and Fields, B. N. (1979b). *J. Immunol.* **12,** 884-889.

White, G., and Cowan, K. M. (1962). *Virology* **16,** 209-211.

Whitney, J. C., Scott, G. R., and Hill, D. H. (1967). *Bull. Epizoot. Dis. Afr.* **15,** 31-41.

Wild, F., Underwood, B., and Brown, F. (1974). *Med. Microbiol. Immunol.* **160,** 133-141.

Winston, S. H., Rustigian, R., and Bratt, M. A. (1973). *J. Virol.* **11,** 926-932.

Wisniewski, H., Raine, C. S., and Kay, W. J. (1972). *Lab. Invest.* **26,** 589-599.

Wright, J. A. (1978). *Vet. Rec.* **103,** 48-51.

Wright, N. G., Cornwell, H. J. C., Thompson, H., and Lauder, I. M. (1974). *Vet. Rec.* **94,** 86-92.

Yamanouchi, K. (1980). *Jpn. J. Med. Sci. Biol.* **33,** 41-66.

Yamanouchi, K., Kobune, F., Fukuda, A., Hayami, M., and Shishido, A. (1970). *Arch. Gesamte Virusforsch.* **29,** 90-100.

Yamanouchi, K., Chino, F., Kobune, F., Fukuda, A., and Yoshikawa, Y. (1974a). *Infect. Immun.* **9,** 199-205.

Yamanouchi, K., Fukuda, A., Kobune, F., Yoshikawa, Y., and Chino, F. (1974b). *Infect. Immun.* **9,** 206-211.

Yamanouchi, K., Yoshikawa, Y., Sato, T. A., Katow, S., Kobune, F., Kobune, K., Uchida, N., and Shishido, A. (1977). *Jpn. J. Med. Sci. Biol.* **30,** 241-257.

Yamanouchi, K., Sato, T. A., Kobune, F., and Shishido, A. (1979). *Infect. Immun.* **23,** 185-191.

Zwart, D., and Macadam, I. (1967a). *Res. Vet. Sci.* **8,** 37-47.

Zwart, D., and Macadam, I. (1967b). *Res. Vet. Sci.* **8,** 53-57.

Zwart, D., and Rowe, L. W. (1966). *Res. Vet. Sci.* **7,** 504-511.

Part V

CORONAVIRIDAE

Chapter 7

Coronaviruses: Diagnosis of Infections

EDWARD H. BOHL

I.	Introduction	301
II.	Characteristics of Coronaviruses	302
III.	Antigenic Composition	304
IV.	Clinical and Pathological Features of Infections	305
	A. Infections of the Intestinal Tract	305
	B. Infections of the Respiratory Tract and Other Organs	308
V.	Laboratory Diagnosis	311
	A. Detection of Virus by Electron Microscopy	311
	B. Detection of Viral Antigens	313
	C. Isolation and Identification of Virus	315
	D. Detection of Antibodies	318
VI.	Prevention	322
	References	323

I. INTRODUCTION*

This chapter will briefly comment on the diseases or clinical syndromes associated with coronaviral infections in animals and birds, with emphasis on methods of diagnosis.

*List of Abbreviations: AGP, agar gel precipitin; CCV, canine coronavirus; CDCV, calf diarrhea coronavirus; CF, compelment fixation (or complement fixing); CIE, counterimmunoelectrophoresis; CPE, Cytopathic effect; DEAE, diethylaminoethyl; ELISA, enzyme-linked immunosorbent assay; EM, electron microscopy; FA, fluorescent antibody; FIPV, feline infectious peritonitis virus; HA, hemagglutination; HAd, hemadsorption; HCV, human coronavirus; HECV, human enteric

COMPARATIVE DIAGNOSIS
OF VIRAL DISEASES, VOL. IV

ISBN 0-12-429704-8

In the mid-1960s, several viruses which had been isolated from man, mice, and swine were found to have a common morphology with the avian infectious bronchitis virus (IBV) (Berry *et al.*, 1964). As a result, "coronavirus" was the name proposed for a group of viruses having distinctive and similar characteristics (Tyrrell *et al.*, 1968). The morphological characteristic from which the name is derived is a fringe of club-shaped projections, 12–24 nm long, around a pleomorphic 60–220-nm viral particle, having a resemblance to a solar corona. Within the Coronaviridae family of viruses, the *Coronavirus* genus is the only one presently recognized, and the avian IBV has been designated the type species (Tyrrell *et al.*, 1978).

Cororaviruses cause infections in man, animals, and birds. Table I gives a list of coronaviruses, their hosts, and their associated diseases or clinical syndromes. As indicated, intestinal diseases occur in swine, cattle, dogs, mice, turkeys, and possibly man; respiratory diseases in man, rats, and chickens; hepatitis in mice and cats; encephalomyelitis in mice and swine; and sialodacryoadenitis in rats.

The coronaviruses cause diseases which are of considerable economic importance, even with our present limited knowledge. In man, they are one of the myriad causes of upper respiratory tract illness, often resulting in losses due to sick leave. Intestinal infections in swine, cattle, dogs, mice, and turkeys and the resulting diarrhea can cause high mortality and great economic loss. In chickens, the economic loss from infectious bronchitis is appreciable from the mortality due to bronchitis in chicks, decreased egg production, or vaccination costs. Additional efforts and costs are necessary for rearing mice and rats free of coronaviral infections so as to provide high-quality research animals.

The ability to recognize or diagnose the infections caused by coronaviruses is often a prerequisite for developing a control or preventive program. Thus, the development and use of appropriate diagnostic procedures are important.

II. CHARACTERISTICS OF CORONAVIRUSES

The basic characteristics of coronaviruses have been adequately reviewed (McIntosh, 1974; Kapikian, 1975; Pensaert and Callebaut, 1978; Tyrrell *et al.*, 1978) and will not be repeated in this chapter. Some of the basic characteristics

---

coronavirus; HEHA, hemadsorption-elution-hemagglutination; HEV, hemagglutinating encephalomyelitis virus; HI, hemagglutination inhibition; IB, infectious bronchitis; IBV, infectious bronchitis virus; IEM, immune electron microscopy, IFA, indirect fluorescent antibody; IHA, indirect hemagglutination; IgA, immunoglobulin A; LIVIM, lethal intestinal virus of infant mice; MHV, murine hepatitis virus; RCV, rat coronavirus; SDAV, sialodacryoadenitis virus; SRH, single radial hemolysis; TECV, turkey enteritis coronavirus; TGE, transmissible gastroenteritis; TGEV, transmissible gastroenteritis virus; VN, virus neutralization.

TABLE I
The Coronaviruses, Their Hosts, and Associated Diseases

Virus Species[a]	Host	Disease or syndrome
Avian infectious bronchitis virus (IBV)	Chickens	Avian infectious bronchitis, nephrosis, salpingitis
Calf diarrhea coronavirus (CDCV)	Cattle	Diarrhea
Canine coronavirus (CCV)	Dogs	Diarrhea
Feline infectious peritonitis virus (FIPV)	Cats	Feline infectious peritonitis, peritonitis, hepatitis
Human coronavirus (HCV)	Man	Upper respiratory tract disease
Human enteric coronavirus (HECV)	Man	Possible diarrhea
Lethal intestinal virus of infant mice (LIVIM)[b]	Mice	Diarrhea
Murine hepatitis virus (MHV)	Mice	Hepatitis, encephalitis, diarrhea
Porcine transmissible gastroenteritis virus (TGEV)	Swine	Porcine transmissible gastroenteritis
Porcine hemagglutinating encephalomyelitis virus (HEV)	Swine	Encephalomyelitis, vomiting, and wasting disease
Rat coronavirus (RCV)	Rats	Pneumonia in newborns
Sialodocryoadenitis virus of rats (SDAV)[c]	Rats	Sialodocryoadenitis
Turkey enteritis coronavirus (TECV)	Turkeys	Turkey coronaviral enteritis, bluecomb disease

[a]The terminology is that which is in common use, but should be considered only tentative. More suitable terminology is indicated for some species. In addition, unnamed or unconfirmed coronavirus-like agents have been detected in specimens from pigs, rhesus and cynomologous monkeys, foals, parrots, and rabbits. See text.

[b]Also considered a murine hepatitis virus.

[c]Also considered a rat coronavirus.

TABLE II
Some Basic Characteristics of the Coronaviruses[a]

Diameter	
Negative staining	60–220 nm, ave. about 100 nm
Thin section in tissue	50–70 nm
Morphology (neg. staining)	Round or elongated, pleomorphic
Surface projections	12–24 nm long, widely spaced
Buoyant density	1.16–1.23 gm/cm^3 in sucrose 1.23–1.24 gm/cm^3 in CsCl
Nucleic acid	RNA, single-stranded
Morphogenesis	Budding into cytoplasmic vesicles
Stability of infectivity	Ether and chloroform labile Rapidly inactivated at 56°C Variable at pH 3

[a]Abstracted from Tyrrell *et al.* (1978).

of the *Coronavirus* genus are summarized in Table II, and others which are relevant for diagnostic purposes will be indicated later.

III. ANTIGENIC COMPOSITION

Knowledge of antigenic relationships among coronaviruses is limited because some coronaviruses have been detected only recently and because suitable serological tests have not been available that can be uniformly applied to different viral strains. Antigenic similarities among coronaviruses occur (McIntosh, 1974; Kapikian, 1975; Pedersen *et al.*, 1978), but no common or group antigens, as are found with the influenza viruses, have been reported (Bradburne, 1970). Only one serotype is known for each of the coronavirual species which cause enteric infections (Table I). In contrast, several serotypes have been reported for some of the coronaviral species which primarily infect the respiratory tract. For example, at least 13 serotypes have been reported for IBV which have been isolated from chickens in the United States (Johnson and Marquardt, 1976; Hofstad, 1978) and 3 serotypes for human coronavirus (HCV) (Kapikian 1975). Serotypes for murine hepatitis virus (MHV) have not been delineated, although all known strains are antigenically related, either by complement fixation (CF) or neutralization tests, but no two strains are identical (McIntosh, 1974). The MHV-S/CDC strain causing lethal enteritis in newborn mice was reported closely related to MHV-S, unilaterally related to MHV-JHM, and more distantly related to MHV-1, MHV-3, MHV-A59, and human coronavirus OC43 (Hierholzer *et al.*, 1979).

A study of eight coronaviruses revealed that they could be segregated into two groups on the basis of antigenic cross reactivity by immunofluorescence (Pederson *et al.*, 1978). One group was composed of MHV-3, hemagglutinating encephalomyelitis virus (HEV)-67N, calf diarrhea coronavirus (CDCV), and HCV-OC43; and the other group of transmissible gastroenteritis virus (TGEV), feline infectious peritonitis virus (FIPV), HCV-229E, and canine coronavirus (CCV). Although viruses in each group had varying antigenic similarities, no antigenic similarities were detected between viruses of one group and those of the other group. Although no antigenic relationship was detected between the two porcine coronaviruses, TGEV and HEV, a previous report (Phillip *et al.*, 1971) had indicated cross reactivity by an immunodiffusion test.

Other reports have also shown an antigenic relationship for TGEV and FIPV (Osterhaus *et al.*, 1977; Reynolds *et al.*, 1977; Witte, *et al.*, 1977), TGEV and CCV (Binn *et al.*, 1974), and HEV and HCV-043 (Kaye *et al.*, 1977). Based on the one-way antigenic relationship between TGEV and FIPV, fluorescein-labeled anti-FIPV was used for detecting TGEV antigen in pigs, but an anti-TGEV conjugate did not detect FIPV antigen in cats (Wirahadiredja *et al.*, 1978). In contrast, IBV does not appear to be antigenically related to any other coronavirus (Bradburne, 1970).

IV. CLINICAL AND PATHOLOGICAL FEATURES OF INFECTIONS

In most cases, the diagnosis of a disease begins by the association of clinical signs or lesions with one or more specific etiological factors. Thus, some of the clinical signs or lesions will be indicated which are suggestive of coronaviral infections and which would lead to the use of appropriate laboratory methods for a specific diagnosis.

Coronaviruses tend to have a predilection for cells of the intestinal or respiratory tracts. Currently, intestinal coronaviral infections are receiving attention, partly because of their economic importance and the limited information that is available on this topic.

A. Infections of the Intestinal Tract

Enteric infections caused by coronaviruses which result in diarrhea have been described for swine, cattle, mice, dogs, and turkeys. Available information suggests that the clinical signs, pathogenesis, and lesions associated with these infections are similar. These features will be described for swine infected with TGEV, since this infection has been more extensively studied than the others (Haelterman and Hooper, 1967; Haelterman, 1972). Following ingestion, the virus infects the epithelial cells lining the villi of the small intestine. The infected cells are sloughed from the villi, resulting in a shortening and blunting of the villi. A marked decrease in digestive enzymatic activity occurs, resulting in a malabsorption syndrome and diarrhea. Diarrhea is severe and mortality is often high in neonates because of their milk diet and their dependence on readily available nutrients and because the replacement of sloughed epithelial cells on the villi is less rapid than in more mature animals (Moon *et al.*, 1973). The lesions are primarily confined to the small intestines and are characterized by blunting, shortening, or atrophy of the villi, which are helpful in diagnosis but are not pathognomonic (Fig. 1).

For differential diagnosis of coronaviral diarrhea it is necessary to consider the role of other infectious agents, either alone or as concurrent infections with coronaviruses. Rotaviruses, as well as coronaviruses, can cause diarrhea in swine (Woode *et al.*, 1976), cattle (Mebus *et al.*, 1969), and mice (Much and Zajac, 1972). The pathogenesis and lesions—mainly villous atrophy of the small intestine—occurring in rotaviral and coronaviral diarrheas are similar (Moon, 1978). Enterotoxigenic *Escherichia coli* is a common cause of diarrhea in the

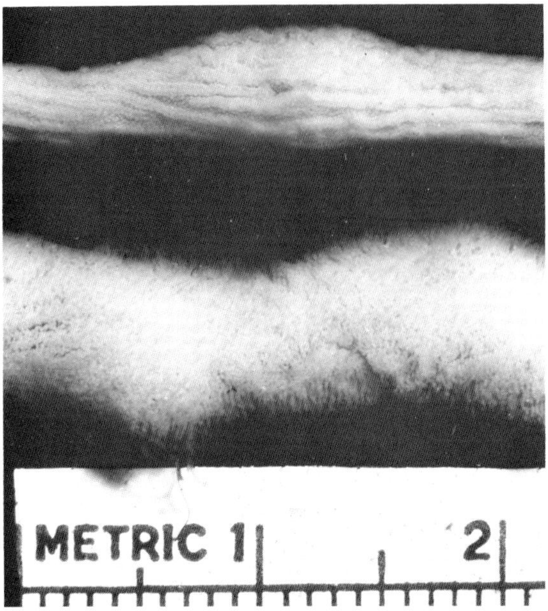

Fig. 1. Mucosal surface of mid-jejunal segment of gnotobiotic pig infected with TGEV (top) and with *Escherichia coli* (bottom). Villous atrophy is evident in the top segment; normal-appearing villi are in the bottom segment. Reproduced with permission from the New York Academy of Sciences from Bohl and Cross (1971).

young of many animal species, especially pigs and calves, but villous atrophy is infrequently observed, at least in the early stage of diarrhea. Cryptosporidia have been reported recently as a probable factor in diarrhea in calves (Morin *et al.*, 1976; Pohlenz *et al.*, 1978). A parvovirus-like agent has been associated with a diarrheal syndrome in dogs which resembles coronaviral diarrhea (Appel *et al.*, 1978).

1. Swine

TGEV causes a highly contagious disease characterized by diarrhea in swine of all ages and a high mortality in pigs under 3 weeks of age. In the newborn, vomiting is usually first observed, followed by a profuse, watery, yellowish diarrhea, dehydration, and mortality of about 100%. Mortality in swine over 3 weeks of age is low, and subclinical or mild disease can occur in mature animals. When an acute, highly contagious type of diarrhea occurs in swine of all ages, the cause is most probably TGEV. When transmissible gastroenteritis (TGE) is enzootic in a herd and the mothers are immune, diarrhea is usually observed only in 1- to 7-week-old pigs and the mortality is low (Bohl, 1975). Enzootic TGE can be confused with rotaviral diarrhea (Bohl *et al.*, 1978).

Recently, a coronavirus-like agent, distinct from TGEV, has been associated with diarrhea in swine but has not been cultivated in cell culture (Chasey and Cartwright, 1978; Pensaert and de Bouck, 1978).

2. Cattle

CDCV has been recognized only recently as a cause of diarrhea, occurring most commonly in 5- to 21-day-old calves (Stair *et al.*, 1972; Mebus *et al.*, 1973b). After oral administration of the virus to newborn calves, diarrhea begins in about 20 hours as liquid, yellowish feces, and mortality can be high. In addition to shortened and fused small intestinal villi, the spiral colon has atrophic ridges, and the epithelial cells vary from cuboidal to low columnar (Mebus *et al.*, 1973b; Doughri and Storz, 1977). Infection is probably common and widespread, but only limited information is available on its clinical, pathological, and epidemiological features under field conditions (Zygraich *et al.*, 1975; Morin *et al.*, 1976; Wellemans *et al.*, 1977).

3. Dogs

In 1978 widespread outbreaks of a vomiting and diarrheal syndrome in dogs were reported to the Cornell Research Laboratory for Disease of Dogs (Appel *et al.*, 1978). From some of these outbreaks, coronavirus-like particles were detected in stool samples by electron microscopy (EM) and, subsequently, a coronavirus-like agent was isolated. Deaths occurred in some young pups. Previously, Binn *et al.* (1974) had isolated a coronavirus, CCV, from U.S. military dogs in Germany during an epizootic of a diarrheal disease. Experimental infection of pups with this isolate resulted in a diarrheal syndrome which was similar to that seen in coronaviral intestinal infections in other animals (Keenan *et al.*, 1976; Takeuchi *et al.*, 1976).

4. Mice

A devastating diarrheal disease of infant mice was first described by Kraft (1962), and the etiological agent was called "lethal intestinal virus of infant mice" (LIVIM). The cause of this disease was identified as an MHV (Broderson *et al.*, 1976). Further studies revealed that the virus was a new strain, and was designated "MHV-S/CDC" (Hierholzer *et al.*, 1979). Previously, Rowe *et al.* (1963) had incriminated MHV as a cause of a lethal enteric disease in infant mice. Infection of mice under 7 days of age usually results in death but diarrhea is not a consistent clinical sign, while 7- to 21-day-old mice usually display diarrhea with a lowered mortality (Kraft, 1966). Both intestinal and hepatic lesions resulted from experimental infection with a Japanese untyped MHV strain (Ishida *et al.*, 1978), while only intestinal lesions resulted from infection with the MHV-S/CDC strain (Hierholzer *et al.*, 1979).

5. *Turkeys*

Turkey enteritis coronavirus (TECV) causes an acute, severe enteric disease in turkeys of all ages, characterized by anorexia, depression, loss in weight, and frothy or watery droppings. The disease is now referred to as "coronaviral enteritis of turkeys" but formerly as "bluecomb," "mud fever," or "transmissible enteritis." Mortality may vary from 5 to 50%, being highest in young poults maintained under adverse environmental conditions. Lesions are confined to the intestinal tract. For a review, see Pomeroy (1978).

6. *Man*

Coronavirus-like particles have been detected by EM in feces of subjects with tropical sprue (Mathan *et al.*, 1975) and with diarrhea (Caul *et al.*, 1975; Maass *et al.*, 1978; Peigue *et al.*, 1978), and in Australian aboriginals with or without diarrhea (Schnagl *et al.*, 1978). A coronavirus was isolated from feces of one diarrheal patient, using human embryonic intestinal organ culture (Caul and Clarke, 1975; Caul and Egglestone, 1977). The role of coronavirus as a cause of intestinal disease or diarrhea in humans has not been delineated.

B. Infections of the Respiratory Tract and Other Organs

1. *Chickens*

Avian infectious bronchitis (IB) is a highly contagious respiratory disease characterized by tracheal rales, coughing, and sneezing, especially in young chicks. For a review, see Hofstad (1978). In young chicks the respiratory signs can be severe, and mortality can approach 25%. Mortality is increased by concurrent infections with *E. coli* or mycoplasma. In layers, the respiratory signs can be minimal, with the major loss due to reduced egg production and poor-quality eggs. Nephrosis has also been observed. Infectious bronchitis virus (IBV) infects the respiratory tract (especially the trachea and lungs), kidneys, bursa, and oviduct, producing lesions at these sites. Chickens are the only natural hosts. Modified live virus vaccines are commonly used in large commercial flocks in the United States.

2. *Swine*

Present information suggests that the upper respiratory tract and pharyngeal region are the sites where infection with HEV is initiated, and that secondary localization of the virus occurs in the central nervous system of very young pigs. For a review, see Greig (1975). Two clinical syndromes have been described: (a) encephalomyelitis (Greig *et al.*, 1962) and (b) vomiting and wasting disease (Roe and Alexander, 1958). The epidemiological characteristics and clinical signs are similar except that neurological signs are absent or minimal in the latter syndrome. Differences in the neurotropic tendencies of viral strains may account for some of the variations in clinical signs. Infection of swine is considered common and widespread, and only rarely are clinical signs observed (Girard *et al.*, 1964; Mengeling, 1975). Only one serotype is known.

Clinical signs are usually limited to pigs under 2 weeks of age and include vomiting, depression, anorexia, hyperesthesia, muscle tremors, paddling, constipation, and emaciation. Morbidity in affected litters has ranged from 20 to 100%, with the majority of very young pigs dying. Illness in older pigs is mild or inapparent. In field cases, there are usually no significant gross lesions, and the microscopic lesions are those of a nonsuppurative encephalomyelitis, which probably account for most, if not all, of the clinical signs (Greig, 1975; Werdin *et al.*, 1976).

3. *Rats*

Two distinct but antigenically related coronaviruses have been isolated from rats. Parker *et al.* (1970) isolated a virus, named "rat coronavirus (RCV)," which is prevalent in colony-reared and wild rats and which causes a fatal pneumonitis in newborn rats. Resistance to disease occurs with increasing age, so that rats over 21 days of age have no clinical signs, although rhinitis, tracheitis, and pneumonitis can be detected histologically (Parker *et al.*, 1970; Bhatt and Jacoby, 1977).

The second coronavirus, referred to as "sialod*a*cryoadenitis (SDAV)," infects the ductal epithelial lining cells of the submaxillary and parotid salivary glands, the harderian gland, trachea, and lungs (Jonas *et al.*, 1969; Bhatt *et al.*, 1972; Jacoby *et al.*, 1975). Clinical signs include cervical edema, rhinitis, and ocular discharge. In contrast to SDAV, infection of the salivary and lacrimal glands with RCV is minimal. Bhatt and Jacoby (1977) propose that SDAV and RCV be considered different strains of rat coronavirus.

4. *Cats*

A coronavirus was identified as the cause of feline infectious peritonitis and is tentatively referred to as "FIPV" (Osterhaus *et al.*, 1976; Pedersen, 1976a; Horzinek *et al.*, 1977). Clinically, the disease in both domestic and wild Felidae characterized by fever, depression, anorexia, weight loss, and ascites. Lesions inlcude diffuse fibrinous peritonitis, pleuritis, and focal necrosis in parenchymal organs (Ward *et al.*, 1974). Infection is probably common and widespread but relatively few infected cats develop disease, with mortality high in those that do (Pedersen, 1976b). For a review, see Horzinek and Osterhaus (1979).

5. *Mice*

A number of antigenically related strains of MHV (MHV-1, MHV-2, MHV-3, A59, JHM, S) have been isolated (Calisher and Rowe, 1966; Piazza, 1969).

However, they have not been adequately compared as to antigenic relatedness or pathogenic potential to determine which can be associated with distinct infectious processes or clinical syndromes. As previously indicated, LIVIM has only recently been recognized as an enteric strain of MHV which causes fatal diarrhea. The newly isolated strain was designated "MHV-S/CDC" (Hierholzer *et al.*, 1979).

Another distinct, naturally occurring, infectious condition is that caused by strain MHV-1, which is considered a hepatotropic strain of low virulence (Gledhill and Andrewes, 1951; Carthew, 1978a). This strain appears to be widespread in mouse colonies and generally causes no apparent disease. However, a fatal hepatitis and/or encephalitis can occur when mice have a concurrent infection with *Eperythrozoon coccoides* (Gledhill and Niven, 1955) or a thymic deficiency, as seen in homozygous nude mice (Sebesteny and Hill, 1974; Carthew, 1978a).

Most of the other strains of MHV can produce hepatitis or encephalitis, but usually only under certain prescribed conditions (Piazza, 1969). Without reference to specific strains, Calisher and Rowe (1966) reported that enteric infection of conventional laboratory mice with MHV is common and that disease does not usually result unless nonimmune newborns are exposed to MHV, in which case mortality can approach 50 to 75%, with death resulting primarily from encephalitis.

6. *Man*

HCV has been shown to cause upper respiratory tract infections. Symptoms resemble those of the common cold and are characterized by pharyngitis, cough, nasal congestion, and fever. For reviews, see Kapikian (1975) and McIntosh (1978). Three serotypes (B814, 229E, and OC43) have been described (Kapikian, 1975). Infections with the 229E and OC43 serotypes are common and widespread, as judged by serological surveys (Kaye and Dowdle, 1975; Kaye *et al.*, 1977).

7. *Parrots*

A coronavirus-like agent was isolated in chick embryos from parrots. *Amazona* sp. (Hirai *et al.*, 1979). The virus was antigenically distinct from IBV, TGEV, and CCV. It was isolated from parrots with anorexia, diarrhea, and lethargy; otherwise, the disease has not been characterized. The virus was pathogenic for young chickens and budgerigars, resulting in lesions characterized by hemorrhagic necrosis in the liver and spleen.

8. *Rabbits*

A new disease of rabbits was described which was thought to be caused by a coronavirus having a two-way cross reactivity with HCV-229E and a one-way cross reactivity with HVC-OC43. The disease was characterized by multifocal myocardial degeneration and necrosis; thus, the name "rabbit infectious cardiomyopathy" was suggested. "It is uncertain whether the agent is a natural pathogen of rabbits or a coronavirus contaminant from another species, possibly human" (Small *et al.*, 1979, p. 709).

V. LABORATORY DIAGNOSIS

Successful laboratory diagnosis begins with the proper selection and collection of appropriate clinical specimens. This point cannot be overly stressed. Knowledge of the viral replicating sites in the body, for the disease in question, will help in selecting suitable specimens at the appropriate stage of infection. Laboratory diagnosis is usually accomplished by microscopic detection of virus, detection of viral antigen, isolation and identification of virus, or detection of a significant antibody response.

A. Detection of Virus by Electron Microscopy

With the electron microscope becoming more readily available and with simplification of techniques such as negative contrast staining, EM is being increasingly used by diagnostic laboratories for detecting viruses in clinical specimens and cell culture harvests. For reviews, see Doane and Anderson (1977) and Flewett (1978). The characteristic morphology of coronaviruses can aid in their detection and presumptive identification when examining clinical or cell culture specimens. However, particles rather similar to coronaviruses are often seen in fecal specimens, and are characterized by projections which are shorter (less than 15 nm in length) and more closely spaced than those of known coronaviruses (Ritchie *et al.*, 1973; Pensaert and de Bouck, 1978; Woode *et al.*, 1978). They are believed to be fragments of intestinal brush border membranes (Schnagl *et al.*, 1978).

By negative contrast EM, coronaviruses or coronavirus-like particles have been demonstrated in intestinal contents or feces from calves (Stair *et al.*, 1972; England *et al.*, 1976), pigs (Saif *et al.*, 1977; Pensaert and de Bouck, 1978), foals (Bass and Sharpee, 1975), mice (Broderson *et al.*, 1976), dogs (Appel *et al.*, 1978; Schnagl and Holmes, 1978), turkeys (Ritchie *et al.*, 1973), chickens (McFerran *et al.*, 1978), man (Caul *et al.*, 1975; Mathan *et al.*, 1975), and monkeys (rhesus and cynomologous) (Caul and Egglestone, 1979). Illustrating the productiveness of using negative contrast EM for detecting coronaviruses in diarrheic fecal specimens are the following reports: (a) from calves, Marsolais *et al.* (1978) reported 69% of 134 specimens, England *el al.* (1976) reported 20% of 91 specimens, and Langpap *et al.* (1979) reported 13.4% of 629 specimens as positive; and (b) from infants and young children in Germany, Maass *et al.* (1977) reported 23% of 195 cases as positive.

Immune electron microscopy (IEM) has advantages over the conventional EM technique in that it is more sensitive for detecting viruses and can provide serological identification of viruses, either from clinical specimens or from cell culture, organ culture, or chick embryo harvests (Almeida and Waterson, 1969; Kapikian *et al.*, 1975). By this procedures, known specific antiserum, appropriately diluted, is added to viral specimens, incubated, and then examined by negative contrast EM for evidence of antibody-clumped viral particles (Fig. 2). IEM has been used for detecting and/or identifying coronaviruses from man (Kapikian *et al.*, 1973), calves (Hajer and Storz, 1978), pigs (Saif *et al.*, 1977), chickens (McFerran *et al.*, 1978), and turkeys (Ritchie *et al.*, 1973). McFerran *et al.* (1978) have reported that IEM is of special value for identifying IBV in allantoic fluid of chick embryos which have been inoculated with clinical specimens, thus shortening the time for laboratory diagnosis.

EM examination of thin sections of intestinal mucosa has been used to detect coronaviruses in pigs (Thake, 1968; Pensaert *et al.*, 1970b; Wagner *et al.*, 1973), calves (Mebus *et al.*, 1973b; Chasey and Lucas, 1977), dogs (Takeuchi *et al.*, 1976), and turkeys (Adams *et al.*, 1972a; Pomeroy *et al.*, 1978). However, this technique demands more time, equipment, and expertise than does negative contrast EM and, thus, is more relevant for research than diagnosis.

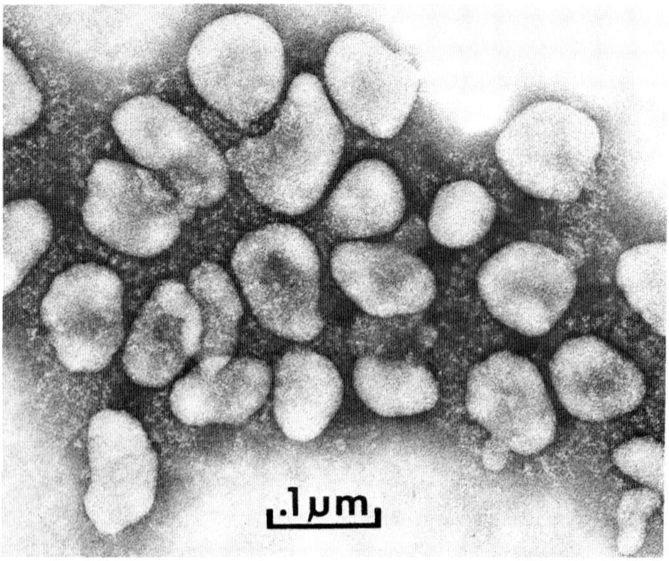

Fig. 2. Immune electron micrograph. Aggregates observed after incubating cell-cultured TGEV with anti-TGEV serum. Reproduced with permission from the American Veterinary Medical Association. See Saif *et al.* (1977).

B. Detection of Viral Antigens

The detection of coronaviral antigens in cells, either from clinical specimens or from infected cell or organ cultures, is of great diagnostic value. Either the immunofluorescent or the immunoperoxidase technique may be used, but the former is more commonly used and will be described in more detail. Detection of coronaviral antigens in tissue extracts has had limited use for diagnostic purposes.

1. Fluorescent Antibody Test for Viral Antigens in Cells

The detection of coronaviral antigen in epithelial cells of the small intestines is probably the simplest and most rapid (2 hours) method for diagnosing coronaviral diarrhea, if suitable specimens are available. Either the direct or indirect fluorescent antibody (FA) test may be used, but the former is more commonly used. The accuracy of these procedures will largely depend on the specificity and quality of the fluorescein-conjugated immunoglobulin or antiserum used. Preferably, these reagents should not contain antibodies against extraneous enteric viruses, especially rotaviruses.

For best results, intestinal specimens should be collected as soon after onset of diarrhea as possible since the highest number of infected epithelial cells are on the villi at that time, but they rapidly become sloughed so that few, if any, can be detected after a few days. Pensaert *et al.* (1970a) detected infected cells from 5 hours to 7 days, but not later, after inoculating pigs with TGEV. In contrast, Patel *et al.* (1976) detected infected cells from 1 to 28 days after inoculating turkeys with TECV. Mucosal specimens should be collected from the jejunum or, preferably, from several sites of the small intestines. However, with calves, specimens from the spiral colon (Mebus *et al.*, 1973b) or rectum (Wellemans *et al.*, 1977) are preferred, probably because infected cells persist longer at these sites even after death of the calf. If possible, specimens should be obtained from euthanized animals in the early stage of diarrhea. Decomposition of small intestinal mucosa occurs rapidly after death, and such samples will give less satisfactory results. Intestinal specimens have been prepared for FA staining by frozen sections (Konishi and Bankowski, 1967) or paraffin-embedded sections (Pospisil *et al.*, 1969), but the mucosal scraping and compression techniques described by Black (1971) is simple, rapid, and probably best suited for diagnostic purposes (Fig. 3).

Immunofluorescence of intestinal epithelial cells has been used for diagnosing coronaviral diarrhea in pigs (Konishi and Bankowski, 1967; Pospisil *et al.*, 1969; Pensaert *et al.*, 1970a), calves (Mebus *et al.*, 1973b; England *et al.*, 1976; Marsolais *et al.*, 1978), dogs (Keenan *et al.*, 1976), mice (Ishida *et al.*, 1978), and turkeys (Patel *et al.*, 1975). Immunofluorescence of tracheal smears has been used as a rapid method for diagnosing avian infectious bronchitis (Braune and Gentry, 1965; Lucio and Hitchner, 1970). A conjugate prepared from poly-

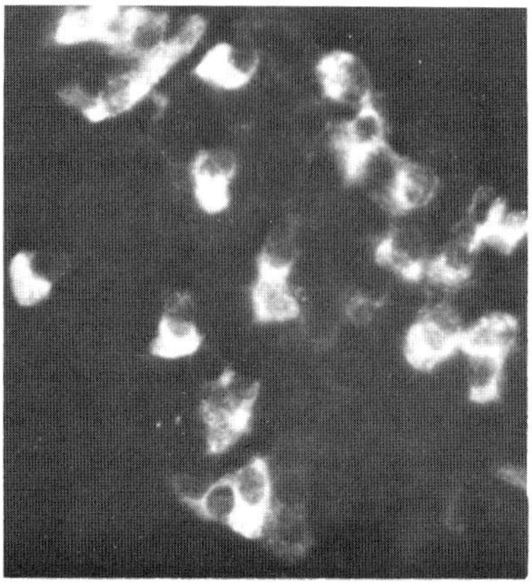

Fig. 3. Immunofluorescing cells from a TGEV-infected pig. A compression smear was made from a mucosal scraping of the jejunum and stained by the direct FA test. × 300.

valent antisera is preferred because immunofluorescent cross reactivity may not occur between different IBV serotypes.

The indirect FA test was used for demonstrating viral antigen in epithelial cells shed from the nasopharynx of symptomatic human volunteers who had received HCV (McIntosh *et al.*, 1978). This procedure offers a new and promising approach for the rapid diagnosis of the respiratory-pharyngeal form of human coronaviral infections. The FA test can be used for demonstrating viral antigen in nasal mucosa of rats infected with RCV (Bhatt and Jacoby, 1977). The FA test is also of great value for identifying coronaviruses in cell or organ cultures which have been inoculated with clinical specimens, and will be further discussed in Section V,C.

2. *Immunoperoxidase Techniques for Viral Antigens in Cells*

This technique has applications similar to those of the FA test, and may have some advantages in that it is highly sensitive and a light microscope is used for observing the staining reaction (Kurstak *et al.*, 1977). However, it has had only limited use for diagnosis of coronaviral infections. It has been used for detecting (i) TGE viral antigen in frozen sections of small intestines and in infected cell cultures (Becker *et al.*, 1974; Yen *et al.*, 1976). and (ii) mouse hepatitis viral antigen in infected cell cultures (Carthew, 1978a,b).

3. *Viral Antigens in Tissue Extracts*

Coronaviral antigens have been detected in alkaline intestinal extracts (i) by the agar gel precipitation test from infected pigs (Bohac *et al.*, 1975) and calves (Hajer and Storz, 1978); (ii) by the hemagglutination test from infected calves (Hajer and Storz, 1978); and (iii) by immunoelectrophoresis and counterimmunoelectrophoresis (CIE) from pigs (Bohac and Derbyshire, 1975) and by CIE from calves (Dea *et al.*, 1979). More positive calves were detected by CIE than EM and, unexpectedly, 30% (5/17) of healthy calves were positive by the combined tests (Dea *et al.*, 1979). However, these methods have had limited use for routine diagnosis. A hemadsorption-elution-hemagglutination assay (HEHA) was developed for detecting CDCV antigen in feces (Van Balken *et al.*, 1979). It is based on the observation that antigen or virus is adsorbed onto mouse erythrocytes at 4°C, followed by elution at 37°C. The eluate can then be tested for hemagglutinating activity, which also can be inhibited by specific antibody. A fluorescent virus precipitin test has been described whereby fecal extracts are mixed with fluorescein isothiocyonate conjugated antibody and the antibody-aggregated viral particles are viewed under an epifluorescent microscope at a magnification of about 1250× (Foster *et al.*, 1975). By this procedure, coronavirus and rotavirus have been detected in the feces of infected calves (Petersen *et al.*, 1976). An enzyme-linked immunosorbent assay (ELISA) was used for detecting CDCV antigen in calf feces (Ellens *et al.*, 1978). It gave fast and reproducible results and was found more sensitive for detecting infected calves than EM or HEHA.

C. Isolation and Identification of Virus

1. Selection of Laboratory Hosts

Suitable laboratory hosts for the isolation or detection of viable virus from clinical specimens may include susceptible animals, embryonated eggs, organ cultures, or cell cultures (primary, secondary, or continuous). For most diagnostic laboratories, because of economics and time, cell cultures will be preferred, whenever possible, for isolating mammalian coronaviruses, while embryonated eggs may be preferred for avian coronaviruses. However, in some cases, susceptible animals may be the only currently available host, as is the pig for isolating and detecting a newly recognized coronavirus which is associated with piglet diarrhea (Pensaert and de Bouck, 1978). Also, an animal may be the most sensitive host for demonstrating coronavirus; for example, pigs were more sensitive than cell cultures for detecting TGEV (Dulac *et al.*, 1977). The "antibody production" test may well be the most sensitive method for detecting viruses from clinical specimens, as has been described for detecting murine viruses, including MHV (Rowe *et al.*, 1963). By this procedure the specimen is inoculated into serologically negative susceptible animal hosts, and after about 21 days their sera are tested for antibodies.

Some species of coronaviruses grow poorly, if at all, in cell cultures, especially on primary isolation. Consequently, embryonated eggs or organ cultures are often more suitable for primary isolation, as is generally true for IBV, CDCV, FIPV, HCV, and TECV. Table III indicates the host systems that are

TABLE III
Primary Isolation of Coronaviruses Using Embryonated Eggs, Cell Cultures, or Organ Cultures

Virus species	Isolation systems and references
IBV	Embryonated chicken eggs (Hofstad, 1978); chick embryo tracheal organ cultures (Cook *et al.*, 1976)
CDCV	Primary and continuous bovine fetal kidney cell cultures (Mebus *et al.*, 1973a); bovine fetal tracheal organ cultures (Stott *et al.*, 1976); Vero, Madin-Darby bovine kidney, and porcine kidney-15 continuous cells (Dea *et al.*, 1980)
CCV	Primary canine kidney cell cultures (Binn *et al.*, 1974)
FIPV	Feline intestinal organ cultures (Hoshino and Scott, 1978); feline embryonic lung cells (O'Reilly *et al.*, 1979)
HCV	Human embryonic tracheal organ cultures (Tyrrell and Bynoe, 1965); secondary human kidney cell cultures (Hamre and Procknow, 1966); semicontinuous human embryonic intestinal cell cultures, MA-177 (Kapikian *et al.*, 1969); continuous human embryo lung cell cultures, L-132 (Bradburne and Tyrrell, 1969); continuous human rhabdomyosarcoma cell cultures (Schmidt *et al.*, 1979)
HECV	Human embryonic intestinal organ cultures (Caul and Clarke, 1975)
MHV	Continuous mouse liver cell cultures, NCTC 1469 (Hartley and Rowe, 1963); continuous mouse fibroblast cell cultures, L-929 (Carthew, 1978a,b); continuous mouse brain tumor cell cultures, DBT (Hirano *et al.*, 1974)
TGEV	Primary porcine cell cultures from kidneys (Harada *et al.*, 1963); thyroid, salivary glands (Witte and Easterday, 1967); continuous porcine testis cell cultures (McClurkin and Norman, 1966); porcine intestinal organ cultures (Rubenstein *et al.*, 1970)
HEV	Primary porcine cell cultures from kidneys (Greig *et al.*, 1962); thyroid and fetal kidneys (Mengeling, 1973)
RCV	Primary rat kidney cell cultures (Parker *et al.*, 1970)
SDAV	Primary rat kidney cell cultures (Bhatt *et al.*, 1972)
TECV	Embryonated turkey eggs (Adams and Hofstad, 1971); turkey embryo intestinal organ cultures (Adams *et al.*, 1972b)

currently used for primary isolation of coronaviruses. Undoubtedly, cell culture methods will be increasingly used for isolation and propagation purposes when more attention is given to selecting appropriate cells and milieu.

The addition of diethylaminoethyl (DEAE) dextran (Pharmacia), at levels of 25 to 100 μg per ml, to cell culture media improved viral titers and/or expression of cytopathic effect (CPE) for CCV (Binn *et al.*, 1974), MHV (Takayama and Kirn, 1978), and RCV (Parker *et al.*, 1970). Cell culture media buffered with HEPES resulted in increased viral yields for TGEV (Pocock and Garwes, 1975) at pH 6.5, IBV (Alexander and Collins, 1975) at pH 6.5, and MHV (Takayama and Kirn, 1978) at pH 7.2. The author has found that the addition of pancreatin to cell culture media will improve CPE when TGEV is exposed to continuous porcine testis cell cultures.

Bovine tracheal organ cultures were found more sensitive for isolating CDCV than were primary calf kidney cell cultures (Bridger *et al.*, 1978b). With a previously isolated CDCV, hemagglutinin titers were higher in harvests from infected bovine fetal intestinal organ cultures than from tracheal organ cultures, suggesting that the former may be better for primary isolation purposes, although this has not yet been determined (Bridger *et al.*, 1978a).

A cell culture system may be used for the propagation and study of coronaviruses even though it may not be applicable for primary viral isolation. Chick embryo-passaged IBV will grow and produce CPE in primary chicken kidney cell cultures (Kawamura *et al.*, 1961). Cell culture-adapted CDCV will grow and produce marked CPE in continuous bovine embryonic kidney (BEK-1) cell cultures (Inaba *et al.*, 1976).

Some viruses, especially in early cell culture passages, will produce discernible plaques when the cellular monolayer is overlapped with agar but will not produce a readily evident CPE when liquid medium is used, as has been reported for TGEV (Bohl and Kumagai, 1965). The reason may be that under liquid medium, infected cells may be replaced with new cells almost as rapidly as they are destroyed, thus leading to a transient and indefinite expression of CPE. In contrast, under agar medium, infected cells are not replaced and are not stained with a vital dye, such as neutral red. Plaque assays of human coronaviruses 229E, B814, LP, OC43, and OC38 were 15- to 30-fold more efficient than by detecting CPE, using a human rhabdomyosarcoma or a fetal tonsil cell line (Schmidt *et al.*, 1979).

2. *Detection of Viral Replication and Identification of Isolates*

Following inoculation of the host system, appropriate techniques must be used for detecting and/or identifying a replicating virus. Since viral replication in embryonated eggs, organ cultures, or cell cultures may not be evident by such conventional indicators as embryo death, ciliostasis, or CPE, respectively, other methods must also be used which indicate not only whether viral replication has occurred but also the probable identity of the replicating virus. The methods most commonly used include immunofluorescence, the immunoperoxidase test, EM, IEM, hemagglutination (HA), or hemadsorption (HAd).

A viral isolate can be identified as a coronavirus by EM or by one of the following serological tests employing appropriate antisera: virus neutralization, immunofluorescent, immunoperoxidase, immunodiffusion, hemagglutination inhibition (HI), CF, or IEM. For identification of a coronaviral species, the virus neutralization test is generally considered most reliable, but it can be time-consuming and expensive; thus, more rapid and inexpensive tests are preferred whenever possible and will be given emphasis in the remainder of this section.

The detection of virus-specific antigens in cells of the laboratory host system by immunofluorescent or immunoperoxidase staining is being increasingly used as a simple and rapid method for determining viral replication and identification (see Section V, B, 1 and 2). Some examples are as follows: (i) embryonated eggs: IBV (Clarke *et al.*, 1972), TECV (Patel *et al.*, 1975; Pomeroy *et al.*, 1978); (ii) organ cultures: IBV (Jones, 1973), CDCV (Stott *et al.*, 1976; Bridger *et al.*, 1978a,b), HCV (McIntosh *et al.*, 1969), HECV (Caul and Clarke, 1975); (iii) cell cultures: TGEV (McClurkin, 1965; Solorzano *et al.*, 1978), CDCV (Mebus *et al.*, 1973a), HEV (Mengeling *et al.*, 1972), MHV (Ishida *et al.*, 1978), HCV (Hamre *et al.*, 1967). The combined cell culture-immunostaining technique is probably the most rapid and simple method whereby viable coronaviruses can be isolated or detected from clinical specimens. It is admirably suited for most diagnostic laboratories. The test in its entirety—cell culture inoculation, staining, and microscopy—can be conducted in wells of microtiter plates, making it a simple procedure. By this procedure, coronaviral-infected cells can be detected in 18 to 48 hours after inoculation. The requirements of this procedure involve the use of (i) susceptible cells for the virus being isolated, (ii) an appropriate milieu for viral replication, and (iii) highly specific immune staining reagents. A concerted effort should be made to satisfy these requirements for each coronaviral species.

EM or IEM has been used for the detection of coronaviruses in laboratory host systems and in some situations is being increasingly used for diagnostic purposes. Some examples are IBV (McFerran *et al.*, 1978) and TECV (Pomeroy *et al.*, 1978) in embryonated eggs; CDCV (Stott *et al.*, 1976), FIPV (Hoshino and Scott, 1978), and HCV (McIntosh *et al.*, 1967; Kapikian *et al.*, 1973) in organ cultures.

Three coronaviral species are known to agglutinate erythrocytes spontaneously and to cause hemadsorption: the OC38/43 strain of HCV (Kaye and Dowdle, 1969), HEV (Greig and Girard, 1963), and CDCV (Sharpee *et al.*, 1976). The demonstration of HA or HAd can be used for detecting viral replication in organ and cell cultures. Viral identification can be made by inhibition of HA or HAd by specific antisera.

D. Detection of Antibodies

The serological tests most commonly used for detecting coronaviral antibodies are virus neutralization (VN), HI, CF, indirect fluorescent antibody (IFA), and agar gel precipitin (AGP).

There are many different ways by which information from serological tests can be used for attempting to diagnose or control infectious diseases. Serological diagnosis of a current disease problem is most reliable when there is a fourfold or greater rise in antibody titers between acute and convalescent serum samples. However, paired serum samples are often not available for testing. An alternative procedure that may be useful for diagnosing a current herd or flock disease problem is the simultaneous collection of sera from several animals in different stages (acute and convalescent) of the disease. Also, in the absence of paired serum samples, high antibody titers or the presence of CF or precipitin antibodies are suggestive of recent infections. The history of the herd or flock in respect to disease or serological status can be used in helping to interpret serological findings.

Serological tests are useful when attempting to diagnose a continuing herd or flock disease problem. It is often more suitable to test sera against one or several antigens than to collect and test specimens for viral isolation. Serological tests are increasingly used to monitor periodically the infection status of specific pathogen-free animal colonies, herds, or flocks. To help avoid the transmission of certain infectious diseases, only serologically negative animals should be introduced into herds or flocks.

The VN test, in one form or another, has been reported for each of the coronaviral species except FIPV and HECV. It is probably the most specific and versatile of the serological tests but is usually expensive and time-consuming to conduct. Laboratory hosts for determining viral neutralization may include animals, embryonated eggs, organ cultures, or cell cultures. Many of the VN tests are now being conducted in microtiter plates, utilizing cell cultures as the host system. There has been a continuing effort to devise and utilize serological tests that are more rapid, safe, and inexpensive than the VN test.

Three coronaviruses (HCV-OC43, HEV, CDCV) spontaneously agglutinate erythrocytes, and some strains of IBV will do so after treatment with the enzyme phospholipase C. Thus, the simple and rapid HI test is often preferred for detecting antibodies against HCV-OC43 (Kaye and Dowdle, 1969), HEV (Girard *et al.*, 1964), CDCV (Sharpee *et al.*, 1976), and IBV (Bingham *et al.*, 1975).

The CF test has been used frequently for detecting coronaviral antibodies from man, mice, and rats, but infrequently or not at all from swine, cattle, dogs, cats, chickens, or turkeys. The CF test has been reported to be less sensitive than the HI test for detecting HCV-OC43 infections of man (Kaye *et al.*, 1971). CF antibodies do not persist as long after infections as do neutralizing antibodies and, thus, their presence suggests a recent infection or reinfection (Gerna *et al.*, 1978). Likewise, CF tests are generally unsuitable for conducting seroepidemiological surveys. CF antibodies could not be detected in sera from pigs convalescing from TGEV infection but were detected in hyperimmune sera (Stone *et al.*, 1976; Dulac *et al.*, 1977).

The IFA test offers the advantage of a rapid (about 2 hours) and versatile test that is being increasingly used. In this test, infected cell cultures or sections from infected tissues are used an antigen substrates, and fluorescent labeled antispecies globulins serve as indicators for antigen–antibody reactions. It has been used for detecting antibodies against HCV strains OC43 and 229E (Monto and Rhodes, 1977), TGEV (Benfield *et al.*, 1978), HEV (Black, 1975), IBV (Jones, 1975), FIPV (Pedersen, 1976b), and TECV (Patel *et al.*, 1976).

The AGP test has been used for detecting antibodies against IBV (Witter, 1962), HEV (Mengeling, 1975), TGEV (Bohac and Derbyshire, 1975), and CDCV (Hajer and Storz, 1978). Although this test is simple to conduct, it is insensitive compared to other serological tests, and the duration of precipitating antibodies may be only brief after infection (Witter, 1962). The latter characterisitcs may be helpful in detecting recent infections but detracts from its usefulness for seroepidemiological surveys. The availability of an adequate amount of antigen for large-scale routine testing is a problem.

An indirect hemagglutination (IHA) test, also referred to as a ''passive hemagglutination test,'' has been used for detecting antibodies against IBV (Brown *et al.*, 1962), HCV-229E (Kaye *et al.*, 1972), and TGEV (Labadie *et al.*, 1977; Shimizu and Shimizu, 1977). In this test, tannic acid-treated erythrocytes are sensitized with concentrated virus, which in the presence of viral antibody will result in hemagglutination. In tests with HCV-229E, the IHA test was found to be as sensitive and specific as the CF and VN tests but more simple to conduct. As reported by Shimizu and Shimizu (1977), a marked limitation of the test for detecting TGEV antibodies is the need for highly purified and concentrated virus for sensitizing erythrocytes.

A single radial hemolysis (SRH) test, also referred to as a ''radial hemolysis-in-gel test,'' has been developed for detecting antibodies against the 229E and OC43 strains of HCV (Hierholzer and Tannock, 1977; Riski *et al.*, 1977). In this test, viral-sensitized erythrocytes and complement are incorporated in agar. Serum samples are added to wells in the agar, and if specific antibody is present a zone of hemolysis occurs, the diameter of which is proportional to antibody concentration. It is reported to be sensitive, reproducible, and specific for quantitating antibody, and serial serum dilutions are unnecessary. Furthermore, it is applicable with nonhemagglutinating viruses, such as the 229E strain of HCV.

ELISA for detecting coronaviral antibody will, undoubtedly, be more extensively used, but at present it has had very limited use. This procedure was found more sensitive for detecting MHV antibody than were the CF or VN tests (Peters *et al.*, 1979).

A summary of the serological tests that have been used for detecting antibodies against each species of coronaviruses is as follows:

1. *Avian infectious bronchitis virus*. The VN test (Page and Cunningham, 1962) has been the standard serological test and the one most commonly used, although it is time-consuming and expensive. The test is conducted in chick embryonated eggs using embryo-adapted IBV. A specific and reproducible HI test has been developed (Bingham *et al.*, 1975; Alexander and Chettle, 1977). It is simple and rapid and can serve as an alternative to the VN test (Bahl *et al.*,

1977; Gough and Alexander, 1978; MacPherson and Feest, 1978). Other tests, such as CF (Marquart, 1974), AGP (Witter, 1962), IFA (Jones, 1975), IHA (Brown *et al.*, 1962), and plaque reduction (Lukett, 1966), have not had wide acceptance for diagnositc serology. The VN test performed in chick embryo tracheal organ cultures has been suggested as being especially suitable for determining antigenic relationships between IBV isolates (Darbyshire *et al.*, 1979).

2. *Calf diarrhea coronavirus*. The following serological tests have been reported: VN (Zygraich *et al.*, 1975; Sharpee *et al.*, 1976), HI (Sharpee *et al.*, 1976; Sato *et al.*, 1977), and AGP (Hajer and Storz, 1978).

3. *Canine coronavirus*. Neutralizing antibodies have been detected using primary canine kidney cell cultures, with unneutralized virus detected by CPE (Binn *et al.*, 1974).

4. *Feline infectious peritonitis virus*. A heterologous VN test has been utilized for detecting FIPV antibodies (Reynolds *et al.*, 1977; Witte *et al.*, 1977), making use of the one-way antigenic relationship between FIPV and TGEV. In this test, sera or peritoneal fluids from cats affected with FIP will neutralize TGEV, as conducted in porcine cell culture systems. The use of a homologous VN test now becomes a reality from a finding (Osterhaus *et al.*, 1978) that FIPV can replicate in the brains of 1-day-old laboratory mice. A heterologous IFA test (Osterhaus *et al.*, 1977) has been described wherein TGEV-infected porcine thyroid cell cultures were used as a substrate antigen. Also, a homologous IFA (Pederson, 1976a) has been used wherein cryostat sections of livers from infected kittens were employed as a substrate antigen.

5. *Human coronavirus*. Reviews by McIntosh (1974) and Monto (1974) briefly described the VN, CF, HI, and IHA tests. Later, additional tests were reported. An SRH test was used to detect antibodies against the OC43 and 229E strains (Hierholzer and Tannock, 1977). An IFA test has been developed for use with 229E and OC43 strains, with the results correlating reasonably well with the VN (Monto and Rhodes, 1977). An immune-adherence hemagglutinating test was reported to be superior to the VN, CF, and IHA tests for detecting remote 229E infections (Gerna *et al.*, 1978).

6. *Human enteric coronavirus*. No reference was found for detecting antibodies against HECV.

7. *Mouse hepatitis virus*. The CF test (Rowe *et al.*, 1963; Parker *et al.*, 1975), utilizing a polyvalent antigen, the VN test (Hierholzer *et al.*, 1979), ELISA (Peters *et al.*, 1979), and the SRH test (Hierholzer *et al.*, 1979) have been reported.

8. *Transmissible gastroenteritis virus*. The VN test has been most commonly employed, utilizing cell culture-adapted viruses in cell culture systems by a variety of procedures, as follows: inhibition of CPE in tubes (Harada *et al.*, 1963) or in microtiter plates (Toma and Benet, 1976), stained monolayer test (Witte and Easterday, 1968), micro-color test (Witte, 1971), and plaque reduc-

tion (Bohl and Kumagai, 1965; Thomas and Dulac, 1976). The IFA test was shown to be rapid (3 hours), reliable, and sensitive (Benfield *et al.*, 1978). Although the AGP test (Bohac and Derbyshire, 1976; Stone *et al.*, 1976) could detect antibodies, it was rather insensitive. CF antibodies could not be demonstrated in convalescing swine (Stone *et al.*, 1976; Dulac *et al.*, 1977). An immunooperoxidase antibody test was described as a rapid and sensitive test, yielding titers which were about seven times higher than those obtained by the VN test (Kodoma *et al.*, 1980).

An *in vitro* leukocyte aggregation assay was positive for swine which had been infected at least 3 days previously with TGEV (Wood, 1976). In this test, leukocytes in heparinized blood samples aggregated when mixed with TGEV antigen. This test does not demonstrate antibodies but, rather, a response that appears related to cell-mediated immunity.

9. *Hemagglutinating encephalomyelitis virus.* The HI test (Girard *et al.*, 1964) has been most commonly used but other tests have been reported, as follows: VN and AGD (Mengeling, 1975), and IFA (Black, 1975). The VN test was found most sensitive, followed by the HI and AGP tests (Mengeling, 1975).

10. *Rat coronavirus and sialacryoadenitis virus.* CF (Parker *et al.*, 1970; Bhatt *et al.*, 1972) and neutralizing (Bhatt *et al.*, 1972) antibodies have been detected for these two viruses, which are considered different strains of RCV.

11. *Turkey enteritis coronavirus.* An IFA test was developed in which cryostat sections of intestine from experimentally infected turkey embryos served as substrate antigen (Patel *et al.*, 1976). This test was more rapid and specific than the VN test (Pomeroy *et al.*, 1975) conducted in 1- to 4-day-old poults.

VI. PREVENTION

The use of appropriate disease control measures, such as introduction of only serologically negative animals, can be used to help maintain stocks of animals free of coronaviral infections. Licensed, commercial vaccines are available for three coronaviral infections: IBV, TGEV, and CDCV. Vaccination of chicks and pullets with live modified IBV vaccines is commonly practiced. IBV vaccines can be administered by a variety of routes; nasal, conjunctival sac, or in drinking water. Because of multiple serotypes, there has been controversy as to the value of monovalent versus multivalent vaccines, with the deciding factor usually depending on the serotype(s) prevalent in a given locality (Hofstad, 1978). A live modified TGEV vaccine is available for intramuscular injection of pregnant swine so as to provide passive immunity to suckling pigs. However, limited immunity is provided by this procedure because the ensuing antibody in colostrum or milk is predominantly of the immunoglobulin (Ig)G class rather than of the IgA class, which is more protective but occurrs in mammary secretions only as a result of an intestinal infection (Bohl *et al.*, 1975). A combined live modified CDCV and bovine rotavirus vaccine is available for oral administration to newborn calves, preferably as soon after birth as possible. This vaccine was shown to be beneficial for preventing or reducing diarrheal problems under both experimental (Mebus, 1977) and field (Thurber *et al.*, 1976) conditions.

REFERENCES

Adams, N. R., and Hofstad, H. S. (1971). *Avian Dis.* **15,** 426–433.

Adams, N. R., Ball, R. A., Annis, C. L., and Hofstad, M. S. (1972a). *J. Comp. Pathol.* **82,** 187–196.

Adams, N. R., Hofstad, M. S., and Fry, M. L. (1972b). *Arch Gesamte Virusforsch.* **38,** 97–99.

Alexander, D. J., and Chettle, N. J. (1977). *Avian Pathol.* **6,** 9–17.

Alexander, D. J., and Collins, M. S. (1975). *Arch. Virol.* **49,** 339–348.

Almeida, J. D., and Waterson, A. P. (1969). *Adv. Virus Res.* **15,** 307–338.

Appel, M. J. G., Cooper, B. J., Greisen, H., and Carmichael, L. E. (1978). *J. Am. Vet. Med. Assoc.* **173,** 1516–1518.

Bahl, A. K., Newman, J. A., and Pomeroy, B. S. (1977). *Proc. Am. Assoc. Vet. Lab. Diagn.* **20,** 225–236.

Bass, E. P., and Sharpee, R. L. (1975). *Lancet* **1,** 822.

Becker, W., Teufel, P., and Mields, W. (1974). *Zentralbl. Veterinaer med., Reihe B* **21,** 59–65.

Benfield, D. A., Haelterman, E. O., and Burnstein, T. (1978). *Can. J. Comp. Med.* **42,** 478–482.

Berry, D. M., Cruickshank, J. G., Chu, H. P., and Wells, R. J. H. (1964). *Virology* **23,** 403–407.

Bhatt, P. N., and Jacoby, R. O. (1977). *Arch. Virol.* **54,** 345–352.

Bhatt, P. N., Percy, D. H., and Jonas, A. M. (1972). *J. Infect. Dis.* **126,** 123–130.

Bingham, R. W., Madge, M. H., and Tyrrell, D. A. J. (1975). *J. Gen Virol.* **28,** 381–390.

Binn, L. N., Lazar, E. C., Keenan, K. P., Huxsoll, D. L., Marchwicki, R. H., and Strano, A. J. (1974). *Proc. U.S. Anim. Health Assoc.* **78,** 359–366.

Black, J. W. (1971). *Proc. U.S. Anim. Health Assoc.* **75,** 492–498.

Black, J. W. (1975). *Proc. Am. Assoc. Vet. Lab. Diagn.* **18,** 39–47.

Bohac, J., and Derbyshire, J. B. (1975). *Can. J. Microbiol.* **21,** 750–753.

Bohac, J., and Derbyshire, J. B. (1976). *Can. J. Comp. Med.* **40,** 161–165.

Bohac, J., Derbyshire, J. B., and Thorsen, J. (1975). *Can. J. Comp. Med.* **39,** 67–75.

Bohl, E. H. (1975). *In* "Diseases of Swine" (H. W. Dunne and A. D. Leman, eds.), 4th ed., pp. 168–188. Iowa State Univ. Press, Ames.

Bohl, E. H., and Cross, R. (1971). *Ann. N. Y. Acad. Sci.* **176,** 150–161.

Bohl, E. H., and Kumagai, T. (1965). *Proc. U.S. Livestock Sanit. Assoc., 69th Annu. Meet.* pp. 343–350.

Bohl, E. H., Frederick, G. T., and Saif, L. J. (1975). Am. J. Vet. Res. **36,** 267–271.

Bohl, E. H., Kohler, E. M., Saif, L. J., Cross, R. F., Agnes, A. G., and Theil, K. W. (1978). *Am. J. Vet. Res.* **172,** 458–463.

Bradburne, A. F. (1970). *Arch. Gesamte Virusforsch.* **31,** 352–364.

Bradburne, A. F., and Tyrrell, D. A. J. (1969) *Arch. Gesmate Virusforsch.* **28,** 133–150.

Braune, M. O., and Gentry, R. F. (1965). *Avian Dis.* **9,** 535–545.

Bridger, J. C., Caul, E. O., and Egglestone, S. I. (1978a). *Arch. Virol.* **57,** 43–51.

Bridger, J. C., Woode, G. N., and Meyling, A. (1978b). *Vet Microbiol.* **3,** 101–113.

Broderson, J. R., Murphy, F. A., and Hierholzer, J. C. (1976). *Lab. Anim. Sci.* **26,** 824.

Brown, W. E., Schmittle, S. C., and Foster, J. W. (1962). *Avian Dis.* **6,** 99–106.

Calisher, C. H., and Rowe, W. P. (1966). *Natl. Cancer Inst. Monogr.* **20,** 67-75.
Carthew, P. (1978a). *J. Infect. Dis.* **138,** 410-413.
Carthew, P. (1978b). *Vet. Rec.* **103,** 188.
Caul, E. O., and Clarke, S. K. R. (1975). *Lancet* **2,** 953-954.
Caul, E. O., and Egglestone, S. I. (1977). *Arch. Virol.* **54,** 107-117.
Caul, E. O., and Egglestone, S. I. (1979). *Vet. Rec.* **104,** 168-169.
Caul, E. O., Paver, W. K., and Clarke, S. K. R. (1975). *Lancet* **1,** 1192.
Chasey, D., and Cartwright, S. F. (1978). *Res. Vet. Sci.* **25,** 255-256.
Chasey, D., and Lucas, M. (1977). *Vet. Rec.* **100,** 530-531.
Clarke, J. K., McFerran, J. B., and Gay, F. W. (1972). *Arch Gesamte Virusforsch.* **36,** 62-70.
Cook, J. K. A., Darbyshire, J. H., and Peters, R. W. (1976). *Arch Virol.* **50,** 109-18.
Darbyshire, J. H., Rowell, J. G., Cook, J. K. A., and Peters, R. W. (1979). Arch. Virol. **61,** 227-238.
Dea, S., Roy, R. S., and Bégin, M. E. (1979). *J. Clin. Microbiol.* **10,** 240-244.
Dea, S., Roy, R. S., and Bégin, M. E. (1980). *Am. J. Vet. Res.* **41,** 30-38.
Doane, F. W., and Anderson, N. (1977). *In* "Comparative Diagnosis of Viral Diseases" (E. Kurstak and C. Kurstak, eds.), Vol. 2, pp. 505-539. Academic Press, New York.
Doughri, A. M., and Storz, J. (1977). *Zentralbl. Veterinaer med. B* **24,** 367-385.
Dulac, G. C., Ruckerbauer, G. M., and Boulanger, P. (1977). *Can. J. Comp. Med.* **41,** 357-363.
Ellens, D. J., Van Balken, J. A. M., and de Leeuw, P. W. (1979). *In* "Proceedings of the Second International Symposium on Neonatal Diarrhea. Veterinary Infectious Disease Organization" (S. D. Acres, ed.) pp. 321-329. University of Saskatchewan, Canada.
England, J. J., Frye, C. S., and Enright, E. A. (1976). *Cornell Vet.* **66,** 172-181.
Flewett, T. H. (1978). *J. Am. Vet. Med. Assoc.* **173,** 538-543.
Foster, L. G., Petersen, M. W., and Spendlove, R. S. (1975). *Proc. Soc. Exp. Biol. Med.* **150,** 155-160.
Gerna, G., Achilli, G., Cattaneo, E., and Cereda, P. (1978). *J. Med. Virol.* **2,** 215-223.
Girard, A., Greig, A. S., and Mitchell, D. (1964). *Res. Vet. Sci.* **5,** 294-302.
Gledhill, A. W., and Andrewes, C. H. (1951). *Br. J. Exp. Pathol.* **32,** 559-568.
Gledhill, A. W., and Niven, J. S. F. (1955). *Vet. Rev. Annot.* **1,** 82-90.
Gough, R. E., and Alexander, D. J. (1978). *Vet. Microbiol.* **2,** 289-301.
Greig, A. S. (1975). *In* "Diseases of Swine" (H. W. Dunne and A. D. Leman, eds.), 4th ed., pp. 385-390. Iowa State Univ. Press, Ames.
Greig, A. S., and Girard, A. (1963). *Res. Vet. Sci.* **4,** 511-517.
Greig, A. S., Mirchell, D., Corner, A. H., Bannister, G. L., Meads, E. B., and Julian, R. J. (1962). *Can. J. Comp. Med.* **26,** 49-56.
Haelterman, E. O. (1972). *J. Am. Vet. Med. Assoc.* **160,** 534-538.
Haelterman, E. O., and Hooper, B. E. (1967). *Gastroenterology,* **53,** 109-116.
Hajer, I., and Storz, J. (1978). *Am. J. Vet. Res.* **39,** 441-444.
Hamre, D., and Procknow, J. J. (1966). *Proc. Soc. Exp. Biol. Med,* **121,** 190-193.
Hamre, D., Kindig, D. A., and Mann, J. (1967). *J. Virol.* **1,** 810-816.
Harada, K., Kumagai, T., and Sasahara, J. (1963). *Natl. Inst. Anim. Health Q.* **3,** 166-167.
Hartley, J. W., and Rowe, W. P. (1963). *Proc. Soc. Exp. Biol. Med.* **113,** 403-406.
Hierholzer, J. C., and Tannock, G. A. (1977). *J. Clin. Microbiol.* **5,** 613-620.
Hierholzer, J. C., Broderson, J. R., and Murphy, F. A. (1979). *Infect. Immun.* **24,** 508-522.
Hirai, K., Hitchner, S. B., and Calnek, B. W. (1979). *Avian Dis.* **23,** 515-525.
Hirano, N., Fujiwara, K., Hino, S., and Matumoto, M. (1974). *Arch. Gesamte Virusforsch.* **44,** 298-302.
Hofstad, M. S. (1978). *In* "Diseases of Poultry" (M. S. Hofstad *et al.*, eds.), 7th ed., pp. 487-503. Iowa State Univ. Press, Ames.

Horzinek, M. C., and Osterhaus, A. D. M. E. (1979). *Arch. Virol.* **59,** 1–15.
Horzinek, M. C., Osterhaus, A. D. M. E., and Ellens, D. J. (1977). *Zentralbl. Veterinaer med., Reihe B* **24,** 398–405.
Hoshino, Y., and Scott, F. W. (1978). *Cornell Vet.* **68,** 411–417.
Inaba, Y., Sato, K., Kuragi, H., Takahashi, E., Ito, Y., Omori, T., Goto, Y., and Matumoto, M. (1976). *Arch. Virol.* **50,** 339–342.
Ishida, T., Taguchi, F., Lee, Y. S., Yamada, A., Tamura, T., and Fujiwara, K. (1978). *Lab. Anim. Sci.* **28,** 269–276.
Jacoby, R. O., Bhatt, P. N., and Jones, A. M. (1975). *Vet. Pathol.* **12,** 196–209.
Johnson, R. B., and Marquardt, W. W. (1976). *Avian Dis.* **20,** 382–386.
Jonas, A. M., Craft, J., Black, C. L., Bhatt, P. N., and Hilding, D. (1969). *Arch. Pathol.* **88,** 613–622.
Jones, R. C. (1973). *Proc. World Vet. Poult. Assoc.* **2,** 780–786.
Jones, R. C. (1975). *J. Comp. Pathol.* **85,** 473–479.
Kapikian, A. Z. (1975). *Dev. Biol. Stand.* **28,** 42–64.
Kapikian, A. Z., James, H. D., Jr., Kelly, S. J., Dees, J. H., Turner, H. C., McIntosh, K., Kim, H. W., Parrott, R. H., Vincent, M. M., and Chanock, R. M. (1969). *J. Infect. Dis.* **119,** 282–290.
Kapikian, A. Z., James, H. D., Jr., and Vaughn, A. L. (1973). *Infect. Immun.* **7,** 111–116.
Kapikian, A. Z., Feinstone, S. M., Purcell, R. H., Wyatt, R. G., Thornhill, T. S., Kalica, A. R., and Chanock, R. M. (1975). *Prespec. Virol.* **9,** 9–47.
Kawamura, H., Isogai, S., and Isubahara, H. (1961). *Natl. Inst. Anim. Health Q.* **1,** 190–198.
Kaye, H. S., and Dowdle, W. R. (1969). *J. Infect. Dis.* **120,** 576–581.
Kaye, H. S., and Dowdle, W. R. (1975). *Am. J. Epidemiol.* **101,** 238–244.
Kaye, H. S., Marsh, H. B., and Dowdle, W. R. (1971). *Am. J. Epidemiol.* **94,** 43–49.
Kaye, H. S., Ong, S. B., and Dowdle, W. R. (1972). *Appl. Microbiol.* **24,** 703–707.
Kaye, H. S., Yarbrough, W. B., Reed, C. J., and Harrison, A. K. (1977). *J. Infect. Dis.* **135,** 201–209.
Keenan, K. P., Jervis, H. R., Marchwicki, R. H., and Binn, L. N. (1976). *Am. J. Vet. Res.* **37,** 247–256.
Kodoma, Y., Ogata, M., and Shimizu, Y. (1980). *Am. J. Vet. Res.* **41,** 133–135.
Konishi, S., and Bankowski, R. A. (1967). *Am. J. Vet. Res.* **28,** 937–942.
Kraft, L. M. (1962). *Science* **137,** 282–283.
Kraft, L. M. (1966). *Natl. Cancer Inst. Monogr.* **20,** 55–61.
Kurstak, E., Tijssen, P., and Kurstak, C. (1977). *In* "*Comparative Diagnosis of* Viral Diseases" (E. Kurstak and C. Kurstak, eds.), Vol. 2, pp. 403–457. Academic Press, New York.
Labadie, J. P., Aynaud, J. M., Vaissaire, J., and Renault, L. (1977). *Recl. Med. Vet.* **153,** 931–936.
Langpap, T. V., Bergland, M. E., and Reed, D. C. (1979). *Am. J. Vet. Res.* **40,** 1476–1478.
Lucio, B., and Hitchner, S. B. (1970). *Avian Dis.* **14,** 9–24.
Lukert, P. D. (1966). *Avian Dis.* **10,** 305–313.
Maass, G., Baumeister, H. G., and Freitag, N. (1977). *Muench. Med. Wochenschr.* **119,** 1029–1034.
McClurkin, A. W. (1965). *Can. J. Comp. Med.* **29,** 46–53.
McClurkin, A. W., and Norman, J. O. (1966). *Can. J. Comp. Med.* **30,** 190–198.
McFerran, J. B., McNulty, M. S., and Curran, W. L. (1978). *Am. J. Vet. Res.* **39,** 505–508.
McIntosh, K. (1974). *Curr. Top. Microbiol. Immunol.* **63,** 85–129.
McIntosh, K. (1978). *In* "Comparative Diagnosis of Viral Diseases" (E. Kurstak and C. Kurstak, eds.), Vol. 1, pp. 609–619. Academic Press, New York.
McIntosh, K., Dees, J. H., Becker, W. B., Kapikian, A. Z., and Chanock, R. M. (1967). *Proc. Natl. Acad. Sci. U.S.A.* **57,** 933–940.

McIntosh, K., Kapikian, A. Z., Hardison, K. A., Hartley, J. W., and Chanock, R. M. (1969). *J. Immunol.* **102,** 1109–1118.
McIntosh, K., McQuillin, J., Reed, S. E., and Gardner, P. S. (1978). *J. Med. Virol.* **2,** 341–346.
Macpherson, I., and Feest, A. (1978). *Avian Pathol.* **7,** 337–347.
Marquardt, W. W. (1974). *Avian Dis.* **18,** 105–110.
Marsolais, G., Assaf, R., Montpetit, C., and Marois, P. (1978). *Can. J. Comp. Med.* **42,** 168–171.
Mathan, M., Mathan, V. L., Swaminathan, S. P., Yesudoss, S., and Baker, S. J. (1975). *Lancet* **1,** 1068–1069.
Mebus, C. A. (1977). *Am. J. Clin. Nutr.* **30,** 1851–1856.
Mebus, C. A., Underdahl, M. R., Rhodes, M. B., and Twiehaus, M. J. (1969). *Stn. Bull.—Nebr., Agrig. Exp. Stn.* **233,** 1–16.
Mebus, C. A., Stair, E. L., Rhodes, M. B., and Twiehaus, M. J. (1973a). *Am. J. Vet. Res.* **34,** 145–150.
Mebus, C. A., Stair, E. L., Rhodes, M. R., and Twiehaus, M. J. (1973b). *Vet. Pathol.* **10,** 45–64.
Mengeling, W. L. (1973). *Am. J. Vet. Res.* **34,** 779–783.
Mengeling, W. L. (1975). *Am. J. Vet. Res.* **36,** 821–823.
Mengeling, W. L., Boothe, A. D., and Ritchie, A. E. (1972). *Am. J. Vet. Res.* **33,** 297–308.
Monto, A. S. (1974). *Yale J. Biol. Med.* **47,** 234–251.
Monto, A. S., and Rhodes, L. M. (1977). *Proc. Soc. Exp. Biol. Med.* **155,** 143–148.
Moon, H. W. (1978). *J. Am. Vet. Med. Assoc.* **172,** 443–448.
Moon, H. W., Norman, J. O., and Lambert, G. (1973). *Can. J. Comp. Med.* **37,** 157–166.
Morin, M., Lariviere, S., and Lallier, R. (1976). *Can. J. Comp. Med.* **40,** 228–240.
Much, D. H., and Zajac, I. (1972). *Infect. Immun.* **6,** 1019–1024.
O'Reilly, K. J., Fishman, B., and Hitchcock, L. M. (1979). *Vet. Rec.* **104,** 348.
Osterhaus, A. D. M. E., Horzinek, M. C., and Ellens, D. J. (1976). *Berl. Muench. Tieraerztl. Wochenschr.* **89,** 135–137.
Osterhaus, A. D. M. E., Horzinek, M. C., and Reynolds, D. J. (1977). *Zentralbl. Veterinaer med., Reihe B* **24,** 835–841.
Osterhaus, A. D. M. E., Horzinek, M. C., and Wirahadiredja, R. M. S. (1978). *Zentralbl. Veterinaer med., Reihe B* **25,** 301–307.
Page, C. A., and Cunningham, C. H. (1962). *Am. J. Vet. Res.* **23,** 1065–1071.
Parker, J. C., Tennant, R. W., Ward, T. G., and Rowe, W. P. (1965). *JNCI J. Natl. Cancer Inst.* **34,** 371–380.
Parker, J. C., Cross, S. S., and Rowe, W. P. (1970). *Arch. Gesamte Virusforsch.* **31,** 293–302.
Patel, B. L., Deshmukh, D. R., and Pomeroy, B. S. (1975). *Am. J. Vet. Res.* **36,** 1265–1267.
Patel, B. L., Pomeroy, B. S., Gonder, E., and Cronkite, C. E. (1976). *Am. J. Vet. Res.* **37,** 1111–1112.
Pedersen. N. C. (1976a). *Am. J. Vet. Res.* **37,** 567–572.
Pedersen, N. C. (1976b). *Am. J. Vet. Res.* **37,** 1449–1453.
Pedersen, N. C., Ward, J., and Mengeling, W. L. (1978). *Arch. Virol.* **58,** 45–53.
Peigue, H., Beytout-Monghal, M., Laveran, H., and Bourges, M. (1978). *Ann. Microbiol. (Paris) 129 B,* 101–106.
Pensaert, M. B., and Callebaut, P. (1978). *Ann. Med. Vet.* **122,** 301–322.
Pensaert, M. B., and de Bouck, P. (1978). *Arch. Virol.* **58,** 243–247.
Pensaert, M. B., Haelterman, E. O., and Burnstein, T. (1968). *Can. J. Comp. Med.* **32,** 555–561.
Pensaert, M. B., Haelterman, E. O., and Burnstein, T. (1970a). *Arch. Gesamte Virusforsch.* **31,** 321–334.
Pensaert, M. B., Haelterman, E. O., and Hinsman, E. J. (1970b). *Arch. Gesmate Virusforsch.* **31,** 335–351.

Peters, R. L., Collins, M. J., O'Beirne, A. J., Howton, P. A., Hourihan, S. L., and Thomas, S. F. (1979). *J. Clin. Microbiol.* **10,** 595-597.
Petersen, M. W., Spendlove, R. S., and Smart, R. A. (1976). *J. Clin. Microbiol.* **3,** 376-377.
Phillip, J. I. H., Cartwright, S. F., and Scott, A. C. (1971). *Vet. Rec.* **88,** 311-312.
Piazza, M. (1969). "Experimental Viral Hepatitis." Thomas, Springfield, Illinois.
Pocock, D. H., and Garwes, D. J. (1975). *Arch. Virol.* **49,** 239-247.
Pohlenz, J., Moon, H. W., Cheville, N. F., and Bembrick, W. J. (1978). *J. Am. Vet. Med. Assoc.* **172,** 452-457.
Pomeroy, B. S. (1978). *In* "Diseases of Poultry" (M. S. Hofstad, ed.), 7th ed., pp. 633-640. Iowa State Univ. Press, Ames.
Pomeroy, B. S., Larsen, C. T., Deshmukh, D. R., and Patel, B. L. (1975). *Am. J. Vet. Res.* **36,** Part 2, 553-555.
Pomeroy, K. A., Patel, B. L., and Larsen, C. T. (1978). *Am. J. Vet. Res.* **39,** 1348-1354.
Pospisil, Z., Mesaros, E., and Stepanek, J. (1969). *Zentralbl. Veterinaer med., Reihe B* **16,** 840-846.
Reynolds, D. J., Garwes, D. J., and Gaskells, C. J. (1977). *Arch. Virol.* **55,** 77-86.
Riski, R., Hovi, T., Väänänen, P., and Penttinen, K. (1977). *Scand. Infect. Dis.* **9,** 75-77.
Ritchie, A. E., Deshmukh, D. R., Larsen, C. T., and Pomeroy, B. S. (1973). *Avian Dis.* **17,** 546-558.
Roe, C. K., and Alexander, T. J. L. (1958). *Vet. Rec.* **22,** 305-307.
Rowe, W. P., Hartley, J. W., and Capps, W. I. (1963). *Proc. Soc. Exp. Biol. Med.* **112,** 161-165.
Rubenstein, D., Tyrrell, D. A. J., Derbyshire, J. B., and Collins' A. P. (1970). *Nature (London)* **227,** 1348.
Saif, L. J., Bohl, E. H., Kohler, E. M., and Hughes, J. H. (1977). *Am. J. Vet. Res.* **38,** 13-20.
Sato, K., Inaba, Y., Takahashi, E., Satoda, K., Omori, T., and Matumoto, M. (1977). *Vet. Microbiol.* **2,** 83-87.
Schmidt, O. W., Cooney, M. K., and Kenny, G. E. (1979). *J. Clin. Microbiol.* **9,** 722-728.
Schnagl, R. D., and Holmes, I. H. (1978). *Vet. Rec.* **102,** 528-529.
Schnagl, R. D., Holmes, I. H., and Mackay-Scollay, E. M. (1978). *Med. J. Aust.* **1,** 307-309.
Sebesteny, A., and Hill, A. C. (1974). *Lab. Anim.* **8,** 317-326.
Sharpee, R. L., Mebus, C. A., and Bass, E. P. (1976). *Am. J. Vet. Res.* **37,** 1031-1041.
Shimizu, M., and Shimizu, Y. (1977). *J. Clin. Microbiol.* **60,** 91-95.
Small, J. D., Aurelian, L., Squire, R. A., Strandberg, J. D., Melby, E. C., Jr., Turner, T. B., and Newman, B. (1979). *Am. J. Pathol.* **95,** 709-730.
Solorzano, R. F., Morin, M., and Morehouse, L. G. (1978). *Can. J. Comp. Med.* **42,** 385-391.
Stair, E. L., Rhodes, M. B., White, R. G., and Mebus, C. A. (1972). *Am. J. Vet. Res.* **33,** 1147-1156.
Stone, S. S., Kemeny, L. J., and Jensen, M. T. (1976). *Infect. Immun.* **13,** 521-525.
Stott, E. J., Thomas, L. H., Bridger, J. C., and Jebbett, N. J. (1976). *Vet. Microbiol.* **1,** 65-69.
Takayama, N., and Kirn, A. (1978). *Arch. Virol.* **58,** 29-34.
Takeuchi, A., Binn, L. N., Jervis, H. R., Keenan, K. P., Hildebrandt, P. K., Valas, R. B., and Bland, F. F. (1976). *Lab. Invest.* **34,** 539-549.
Thake, D. C. (1968). *Am. J. Pathol.* **53,** 149-166.
Thomas, F. C., and Dulac, G. C. (1976). *Can. J. Comp. Med.* **40,** 171-174.
Thurber, E. T., Bass, E. P., and Beckenhauer, W. H. (1976). *Can. J. Comp. Med.* **41,** 131-136.
Toma, B., and Benet, J. J. (1976). *Recl. Med. Vet.* **152,** 565-568.
Tyrrell, D. A. J., and Bynoe, M. L. (1965). *Br. Med. J.* **1,** 1467-1470.
Tyrrell, D. A. J., Almeida, J. D., Berry, D. M., Cunningham, C. H., Hamre, D., Hofstad, M. S., Mallucci, L., and McIntosh, K. (1968). *Nature (London)* **220,** 650.

Tyrrell, D. A. J., Alexander, D. J., Almeida, J. D., Cunningham, C. H., Easterday, B. C., Garwes, D. J., Hierholzer, J. C., Kapikian, A., MacNaughton, M. R., and McIntosh, K. (1978). *Intervirology* **10,** 321-328.

Van Balken, J. A. M., de Leeuw, P. W., Ellens, D. J., and Straver, P. J. (1979). *Vet. Microbiol.* **3,** 205-211.

Wagner, J. E., Beamer, P. D., and Ristic, M. (1973). *Can. J. Comp. Med.* **37,** 177-188.

Ward, J. M., Gribble, D. H., and Dungworth, D. L. (1974). *Am. J. Vet. Res.* **35,** 1271-1275.

Wellemans, G., Antoine, H., Botton, Y., and Van Opdenbosch, E. (1977). *Ann. Med. Vet.* **121,** 411-420.

Werdin, R. E., Sorensen, D. K., and Stewart, W. C. (1976). *J. Am. Vet. Med. Assoc.* **168,** 240-246.

Wirahadiredja, R. M. S., Anakotta, J., and Osterhaus, A. D. M. E. (1978). *Zentralbl. Veterinaermed., Reihe B* **25,** 775-778.

Witte, K. H. (1971). *Arch. Gesamte Virusforsh.* **33,** 171-176.

Witte, K. H., Easterday, B. C. (1967). *Arch. Gesamte Virusforsh.* **20,** 327-350.

Witte, K. H., Easterday, C. (1968). *Am. J. Vet. Res.* **29,** 1409-1417.

Witte, K. H., Tuch, K., Dubenkropp, H., and Walther, C. (1977). *Berl. Muench. Tieraerztl. Wochenschr.* **90,** 396-401.

Witter, R. L. (1962). *Avian Dis.* **6,** 478-492.

Wood, R. D. (1976). *Am. J. Vet. Res.* **37,** 1405-1408.

Woode, G. N., Bridger, J., Hall, G. A., Jones, J. M., and Jackson, G. (1976). *J. Med. Microbiol.* **9,** 203-209.

Woode, G. N., Bridger, J. C., and Meyling, A. (1978). *Vet. Rec.* **102,** 15-16.

Yen, S., Solorzano, R. F., and Morehouse, L. G. (1976). *Proc. Am. Assoc. Vet. Lab. Diag.* **19,** 21-46.

Zygraich, N., Georges, A. M., and Vascoboinic, E. (1975). *Ann. Med. Vet.* **119,** 105-113.

Part VI

TOGAVIRIDAE

Chapter 8

Togaviral Diseases of Domestic Animals

THOMAS P. MONATH AND DENNIS W. TRENT

I. Introduction 332
II. Morphology, Physicochemistry, and Antigenic Composition of Togaviruses 335
  A. Alphavirus Particles 335
  B. Flavivirus Particles 356
III. Comparative Biology and Pathogenesis 367
  A. Viruses Causing Encephalomyelitis 367
  B. Viruses Causing Abortion and Stillbirth 391
  C. Other Clinicopathological Entities 394
IV. Epizootiology 395
  A. Eastern Equine Encephalitis 395
  B. Western Equine Encephalitis 400
  C. Venezuelan Equine Encephalitis 403
  D. Japanese Encephalitis 408
  E. Louping Ill Virus 410
  F. Israel Turkey Meningoencephalitis 410
  G. Wesselsbron 410
  H. Other Arboviruses 411
V. Comparative Diagnosis 413
  A. Hemagglutination Inhibition Test 415
  B. Complement Fixation Test 416
  C. Neutralization Test 416
  D. Other Tests 418
  E. Choice and Sequence of Tests 419
  F. Interpretation 419
VI. Prevention and Control 421
  A. Eastern and Western Equine Encephalitis Vaccine 421
  B. Venezuelan Equine Encephalitis Vaccine 422
  C. Japanese Encephalitis Vaccines 424
  D. Vector Control 425
  References 426

COMPARATIVE DIAGNOSIS
OF VIRAL DISEASES, VOL. IV

ISBN 0-12-429704-8

I. INTRODUCTION

Sixteen arthropod-borne viruses in the family Togaviridae are known or suspected to cause disease in domestic animals (Table I). Three alphaviruses (group A arboviruses) which produce encephalomyelitis in horses, mules, and donkeys—Venezuelan, eastern, and western equine encephalitis viruses—are pathogens of major veterinary public health importance responsible for intermittent epizootics and high rates of morbidity and mortality. Of the arthropod-borne flaviviruses (group B arboviruses), Japanese encephalitis (JE), Wesselsbron, louping ill, and Israel turkey meningoencephalitis viruses are also implicated in animal disease outbreaks. With the exception of JE, which is a major disease problem in many parts of Asia, the other flaviviral infections are highly localized or infrequently encountered. Seven arboviruses listed in Table I are suspected (but not proven) to cause naturally acquired disease in animals; illness associated with these agents has been sporadic and infrequently reported.

The association between clinical expression of infection and viral etiology is shown in Table II. The most frequent syndrome is encephalomyelitis, and equines are the most susceptible host. Of 11 viruses which cause equine encephalomyelitis, 10 are transmitted by mosquitoes. Louping ill virus produces encephalomyelitis primarily in sheep, but also in cattle, horses, and swine, and is tick-borne. Israel turkey meningoencephalitis is a mosquito-borne infection of turkeys; eastern equine encephalitis virus is also implicated in epornitics among domestic and penned exotic birds (pheasants, chukar partridges) in North America. Abortion and stillbirth are associated with infection by JE virus of swine, Wesselsbron virus of sheep, and possibly Middelburg virus of equines. Getah virus causes a febrile disease with rash in equines in Japan. The clinical spectrum of infection with these agents is undoubtedly much broader than indicated in the table. Experimentally inoculated animals generally exhibit undifferentiated febrile syndromes as well as the full expression of disease (encephalomyelitis, abortion, etc.). Human illness is reported for many of the viruses pathogenic for livestock (Table II). Overt infection of man takes the form of either nonspecific influenza-like illness or central nervous system (CNS) disease.

The known geographic distribution of each virus and the epidemiological pattern of animal disease are summarized in Table I. Epizootic equine encephalomyelitis principally affects the Western Hemisphere. JE virus is an important cause of abortion and stillbirth of swine in Asia, where it is also associated with epidemic encephalomyelitis in horses. Louping ill is restricted to the British Isles, and Wesselsbron disease is of veterinary importance in southern Africa. The distribution of some viruses, e.g., Wesselsbron and West Nile viruses, is wider than that of the recognized disease.

TABLE I
Togaviruses Known or Suspected to Produce Disease in Domesticated Animals

Genus	Etiology proven	Etiology suspected	Geographic distribution
Alphavirus	Venezuelan equine encephalitis (VEE)		South and Central America
	Eastern equine encephalitis (EEE)		North America, Caribbean,
	Western equine encephalitis (WEE)		North and South America.
	Getah (GET)		Southeast Asia, Japan
		Middelburg (MID)	Africa
		Aura	South America
		Una	South America
		Semliki Forest (SF)	Africa
		Ross River (RR)	Australia, South Pacific
		Ndumu (NDU)	South and Central Africa
Flavivirus	Japanese encephalitis (JE)		Asia
	West Nile (WN)		Africa, Middle East, Europe
	Wesselsbron (WSL)		Africa
	Louping ill (LI)		British Isles
	Israel turkey meningoencephalitis (IT)		Israel
		Murray Valley encephalitis (MVE)	Australia, New Guinea

TABLE II
Association between Clinical Expression of Infection in Domestic Animals and Viral Etiology

	Meningoencephalitis					Abortion/stillbirth				Other (see text)			Human illness reported (natural and laboratory infection)
	Equines	Cattle	Sheep	Swine	Fowl	Equines	Cattle	Sheep	Swine	Equines	Cattle	Swine	
Alphaviruses													
VEE	+					+							+
EEE	+			+	+								+
WEE	+				+								+
MID						+							
AURA	+												
UNA	+												
SF	+												+
RR	+									+			+
NDU											+		
GET										+			
Flaviviruses													
JE	+			+					+			+	+
WN	+												+
WSL								+			+		+
LI	+	+	+	+									+
IT					+								
MVE	+												+

[a] VEE, Venezuelan equine encephalitis; EEE, eastern equine encephaltis; WEE, western equine encephalitis; MID, Middelburg; AURA, aura; UNA, una; SF, Semliki Forest; RR, Ross River; NDU, Ndumu; GET, getah; JE, Japanese encephalitis; WN, West Nile; WSL, Wesselsbron; LI, louping ill; IT, Israel turkey meningoencephalitis; MVE, Murray Valley encephalitis.

II. MORPHOLOGY, PHYSICOCHEMISTRY, AND ANTIGENIC COMPOSITION OF TOGAVIRUSES

A. Alphavirus Particles

1. Chemical and Physical Properties

The chemical composition of purified alphavirus particles has been determined for Sindbis (SIN) and Semliki Forest (SF) viruses (Pfefferkorn and Hunter, 1963a,b; Kaariainen and Gomatos, 1969; Simmons and Strauss, 1972; Laine *et al.*, 1973; Garoff and Simons, 1974; Burke and Keegstra, 1976). Analyses have not been reported for other alphaviruses, although it would be expected that their composition is similar to that reported for SIN and SF viruses. Purified alphaviruses contain RNA, 5.5-6.3%; protein, 57-61%; carbohydrate, 61.4%; and lipid, 27-31%. The morphogenesis of the alphavirus is so specific that host cell membrane proteins are not found in the virus particles (Garoff and Simons, 1974); however, as much as 0.2% of the protein in the virions may be of host cell origin (Strauss, 1978).

Alphaviruses are reported to have sedimentation coefficients of 240-300 S in sucrose gradients. This wide disparity in values is undoubtedly due to variation in experimental conditions (reviewed by Horzinek, 1973b). Eastern equine encephalitis (EEE) virus purified by DEAE-column chromatography and sedimented in sucrose and CsCl gradients was experimentally determined to have an S value of 240, a mathematically calculated S value of 227, and a particle molecular weight of 58×10^6 (Fuscaldo *et al.*, 1971). This is somewhat smaller than the value of 260 S reported by Aaslestad *et al.* (1968) for EEE virus and of 280 S for SIN virus with a particle molecular weight of 70×10^6 (Horzinek and Mussgay, 1969). The buoyant density of the alphavirus particle in Tris-buffered sucrose containing EDTA ranges from 1.21 to 1.20 gm/cm^3 (Mussgay and Rott, 1964; Faulkner and McGee-Russell, 1968; Horzinek and Mussgay, 1969; Igarashi, 1969; Compans, 1971; Fuscaldo *et al.*, 1971; Stinski and Gruber, 1971; Uryrayen *et al.*, 1971). In CsCl isopycnic gradients, the alphavirus has an apparent density of 1.18 to 1.20 gm/cm^3 (Aaslested *et al.*, 1968; Igrashi, 1969; Uryrayen *et al.*, 1971). The buoyant density of western equine encephalitis (WEE), Venezuelan equine encephalitis (VEE), and EEE viruses in glycerol-potassium tartrate equilibrium:viscosity gradients is 1.20 (D. W. Trent, unpublished data, 1979).

2. Morphological Structure and Chemical Organization

Alphaviruses have a similar morphology as defined by negative contrast electron microscopy of purified particles (reviewed by Murphy, 1980a). The virus particles are quite uniform in their appearance: spherical particles with a diameter of 50 to 65 nm (Pfefferkorn and Shapiro, 1974). The particles contain a nuc-

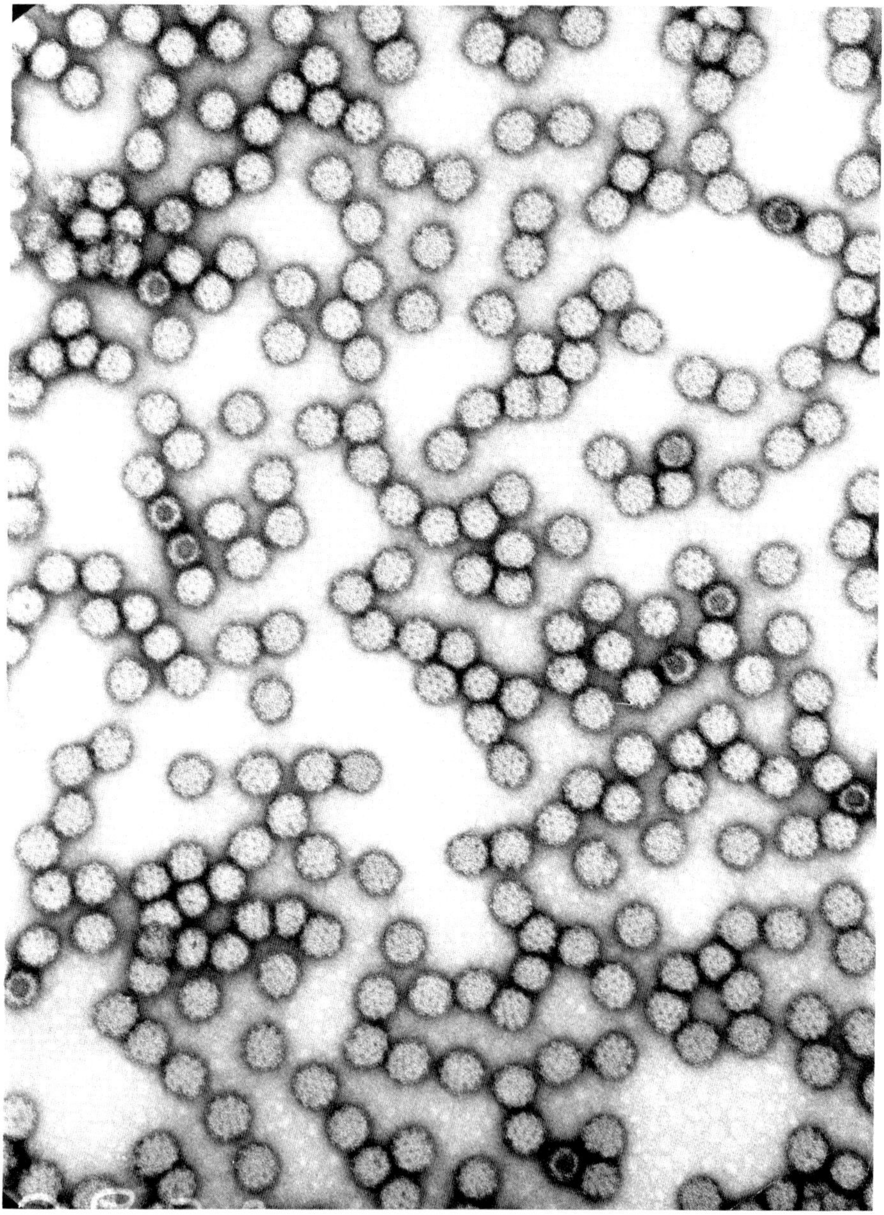

Fig. 1. Semliki Forest virus, negative contrast with uranyl acetate. The typical $T = 4$ icosahedral structure of the alphavirus surface is evident in the preparation of highly purified virus. × 200,000. Micrograph courtesy of Dr. C. H. von Bonsdorff.

leocapsid of 28 to 49 nm surrounded by a halo and a finely textured fuzzy or ragged surface layer (Fig. 1) (Simpson and Hauser, 1968a; Igarashi *et al.*, 1970; Compans, 1971; Murphy and Harrison, 1971; von Bonsdorff, 1973).

The organization of the spikes or surface projections on the particle surface remains somewhat unresolved; however, there is good evidence that they are viral specific and their arrangement is symmetrical (Simpson and Hauser, 1968a,c). Examination of highly purified virions of (SF) virus revealed that the individual surface projections are 3 to 3.5 nm thick, 6 to 10 nm long, and spaced at intervals of 4 to 6 nm over the virion surface (Osterrieth, 1968; von Bonsdorff, 1973). The spikes appear to be inserted radially into the lipid bilayer or "looped" as a continuous filament (Horzinek, 1973a). Polyacrylamide gel electrophoresis (PAGE) of purified virions indicates that the particles are composed of two large glycoproteins, E_1 and E_2, of approximately 50,000 to 60,000 daltons and a nonglycosylated core protein; a third small glycoprotein (E_3) of 10,000 molecular weight may also be present, and a nonglycosylated core protein (Strauss and Strauss, 1976).

Von Bonsdorff and Harrison (1975, 1978) reported that the glycoproteins which make up the projections on SIN and SF virus surfaces are organized in a trimer cluster to form a $T = 4$ (42-subunit) icosahedral lattice. The envelope glycoproteins (E_1, E_2, and E_3) are present in equimolar quantities of 240 subunits each in the completed virion. Utermann and Simons (1974) observed that E_1 and E_2 glycoproteins of SF virus have hydrophobic tails which penetrate the lipid bilayer. Using cross-linking techniques, Garoff *et al.* (1974) demonstrated that the tails of at least one of the glycoproteins extend across the lipid bilayer and are in contact with the nucleocapsid. It is suggested that the nucleocapsid functions as a molecular contact point for budding of the virion. When SIN-infected cells are treated with nonionic detergent, at a concentration sufficient to separate the cores from the membranes, the viral glycoproteins in the plasma membrane fuse to form crystalline arrays similar to those observed on the virion surface (Garoff *et al.*, 1974; von Bonsdorff and Harrison, 1975, 1978). This suggests that the local geometry of the glycoprotein–glycoprotein interaction does not depend upon the presence of the viral nucleocapsid.

The molecular weights of the structural proteins of various alphaviruses have been determined by PAGE in several laboratories. All alphaviruses contain two large envelope glycoproteins and a nucleocapsid protein (Strauss *et al.*, 1969; Schlesinger and Schlesinger, 1972; Schlesinger *et al.*, 1972; von Bonsdorff, 1973; Garoff *et al.*, 1974; Pedersen *et al.*, 1974; Clegg, 1975; Burke and Keegstra, 1976; Strauss and Strauss, 1976; Wiebe and Scherer, 1979; Trent and Grant, 1980). An electropherogram showing the structural proteins of EEE, VEE, and WEE viruses is shown in Fig. 2. Alphavirus structural proteins are designated "E_1," "E_2," "E_3," and "C," with the designation of "E_1" indicating that it is the first envelope protein synthesized in its final form (Baltimore *et al.*, 1976). The

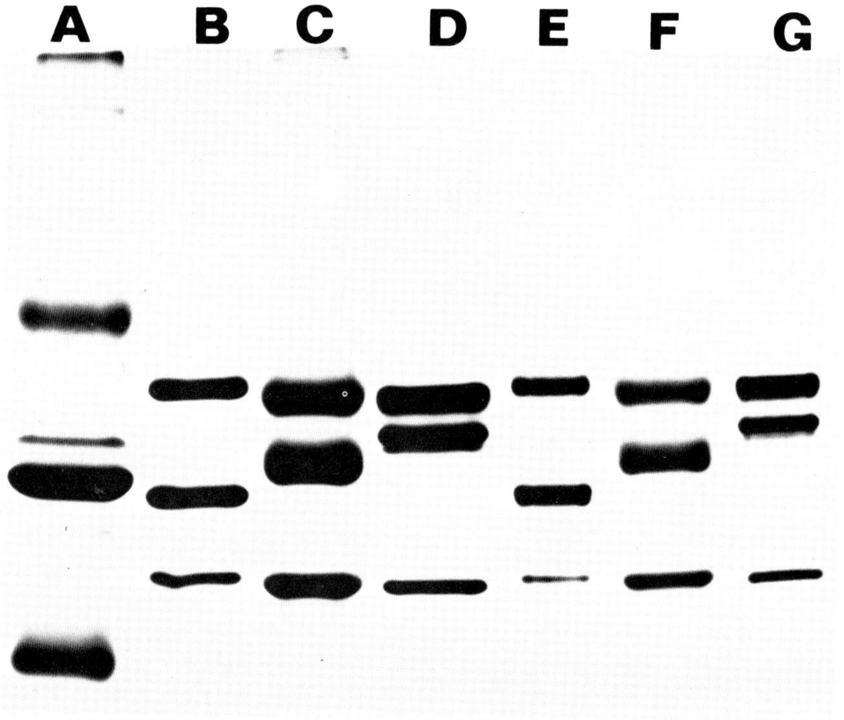

Fig. 2. Polyacrylamide gel electrophoresis of the structural proteins of (A) vesicular stomatitis, (B) eastern equine encephalitis, North American subtype, (C) western equine encephalitis, (D) eastern equine encephalitis, South American subtype, (E) Highlands J, and (F) Venezuelan equine encephalitis, IAB subtype.

largest glycoproteins of SIN and WEE viruses are designated E_1 and the smaller E_2 (Schlesinger and Schlesinger, 1973; Simons *et al.*, 1973; Morser and Burke, 1974; Baltimore *et al.*, 1976; Lachmi and Kaariainen, 1976; Wirth *et al.*, 1977; France *et al.*, 1979; Trent and Grant, 1980). The largest VEE and SF glycoproteins are formed after the smaller envelope protein E_1 and therefore are designated "E_2" (Simons *et al.*, 1973; Lachmi and Kaariainen, 1976). In addition to these three proteins, SF virus particles contain a low-molecular-weight envelope glycoprotein (Garoff *et al.*, 1974). A similar protein is synthesized in SIN (Welch and Sefton, 1979) and VEE (D. W. Trent, unpublished data) virus-infected cells but is not incorporated into the virus particle.

The molecular weights of the alphavirus structural proteins are similar, although variation between those of different serotypes is observed (Table III). Comparative analysis of structural polypeptides from 29 strains of VEE virus revealed considerable heterogeneity in the electrophoretic mobility and apparent number of polypeptides (Pedersen and Eddy, 1975; France *et al.*, 1979; Wiebe

TABLE III
Molecular Weights of Structural Proteins and Representative Alphaviruses

	Molecular weights of proteins ($\times 10^{-3}$)				
Virus	E_1	E_2	E_3	C	References
Eastern equine encephalitis	53–56	45–48	—	29–31	Pedersen *et al.* (1974)
Western equine encephalitis	56	47	—	32	Trent and Grant (1980)
Venezuelan equine	49	59	—	36	France *et al.* (1979)
encephalitis (IA)	50–53	56–59	—	29–31	Pedersen and Eddy (1975)
Venezuelan equine encephalitis (II, Everglades)	49	58	—	36	France *et al.* (1979) Wiebe and Scherer (1979) Pedersen and Eddy (1975)
Venezuelan equine encephalitis (III, Mucambo)	50	55	—	36	France *et al.* (1979) Wiebe and Scherer (1979) Pedersen and Eddy (1975)
Venezuelan equine encephalitis (IV, Pixuna)	48	53	—	36	France *et al.* (1979) Wiebe and Scherer (1979) Pedersen and Eddy (1975)
Semliki Forest	49	52	10	34	Helenius *et al.* (1974) Garoff *et al.* (1974)
Sindbis	56–59	40–50	—	29–31	Strauss and Strauss (1976) Pedersen *et al.* (1974)

and Scherer, 1979). All strains had two common structural proteins: the nucleoprotein of 36,000 daltons and a glycoprotein of approximately 50,000 daltons. Virions of all strains contained a second glycoprotein whose size varied from 51,000 to 58,000, while some of the virus isolates contained a third glycoprotein of approximately 56,000 to 58,000 (Wiebe and Scherer, 1979; France *et al.*, 1979). Pedersen and Eddy (1975) reported that the PAGE patterns of VEE virus isolates could be used to classify strains according to serotype and epidemiological characteristics. Although more recent studies have not confirmed these observations, it is apparent that considerable variation exists in the electrophoretic mobilities of the structural proteins of the different serological subtypes.

Glycoproteins of WEE virus McMillan strain, labeled with glucosamine, have molecular weights of 56,000 (E_1) and 47,000 (E_2). The nucleocapsid of WEE virus has a molecular weight of 32,000 (Trent and Grant, 1980).

The amino acid composition of E_1 and E_2 glycoproteins and nucleocapsid protein has been determined for SIN virus grown in chicken embryo and BHK cells (Burke and Keegstra, 1976; Dalrymple *et al.*, 1976; Bell *et al.*, 1979). The structural proteins are rich in lysine, aspartate, threonine, serine, glutamate, proline, glycine, alanine, valine, leucine, and isoleucine. The amino acid compositions of the individual viral structural proteins are quite similar, with the exception that the nucleocapsid protein contains relatively more proline and the basic amino acids lysine and arginine. No cysteine could be detected in the capsid protein. NH_2-terminal sequences of about 50 residues have been determined for each of the two membrane glycoproteins (Bell *et al.*, 1978).

Both E_1 and E_2 NH_2-terminal regions contain more nonbasic than basic polar amino acids and more aromatic than aliphatic nonpolar residues. Proline appears to be somewhat enriched in the NH_2-terminal portion of E_1. The amino terminal residues are tyrosine for E_1 and serine for E_2 (Bell *et al.*, 1978). The capsid protein contains a blocked NH_2-terminal amino group, which is consistent with other studies indicating that viral proteins derived from the NH_2 terminus of precursor molecules are blocked (Bell *et al.*, 1978).

It has been reported that there is a selective binding of lipid to the SIN virus envelope glycoproteins (Schmidt *et al.*, 1979). The lipid is covalently bound to E_1 and E_2 at some late stage in glycoprotein maturation, possibly at the time the glycoproteins migrate to the plasma membrane. The role which these fatty acids plays in viral morphogenesis and glycoprotein structure is unknown.

Purified preparations of SF and SIN virus hemolyse red blood cells from several species of animals (Karabatsos, 1963; Vaananen and Kaariainen, 1979). The hemolytic activity resides in the viral glycoproteins but can be demonstrated only in virus particles which have been damaged during purification (Vaananen and Kaariainen, 1979). E_2 is probably the envelope protein involved in the hemolysis.

The capsid and two membrane proteins of SIN virus grown in chicken cells have been reported to contain 0.03 to 0.1 mole of phosphate per mole of virus protein (Waite *et al.*, 1974). Bell *et al.* (1979) reported less than 0.1 molecule of phosphate per molecule of protein. The role of the phosphate is unknown. Perhaps it is involved in the maintenance and integrity of the virion structure. Various phosphorylated states and their relative occurrence seem to be characteristic for a specific virus (Sokol and Clark, 1973; Pal *et al.*, 1975).

The carbohydrate compositions of the SIN virus E_1 and E_2 glycoproteins are shown in Table IV. Polysaccharides account for 8 to 12% of the mass of E_1 and E_2 and 45% of E_3 (Burge and Strauss, 1970; Sefton and Keegstra, 1974; Burke and Keegstra, 1976). The carbohydrates are found as polysaccharides of A-type, which contain glucosamine, mannose, galastose, and sometimes sialic acid and fucose, or B-type, which contain only glucosamine and mannose (Sefton and Keegstra, 1974). Each glycoprotein contains three different A-type oligosaccharide species which differ primarily in the degree to which they are sialylated (Keegstra *et al.*, 1975). The type B oligosaccharides of SIN virus, which are rich in mannose, are quite heterologous by column chromatography (Sefton and Keegstra, 1974). The carbohydrate content of each envelope protein is sufficient for three or four carbohydrate chains per each molecule of protein inserted in the alphavirus envelope.

When SIN virus that is grown in chicken embryo cells and the purified E_1

TABLE IV
Carbohydrate Composition of Sindbis Virus Glycoproteins[a]

Sugar	μg of carbohydrate/mg of protein	
	Glycoprotein E_1	Glycoprotein E_2
N-acetyle glucosamine	24	19
Mannose	19	26
Galactose	12	9
Fucose	3	1
Sialic acid	5	7
Total carbohydrate	63	62

[a] Data from Burke and Keegstra (1976).

protein are digested with pronase and chromatographed, four glycopeptides—S_1, S_2, S_3, and S_4—are separated (Burge and Strauss, 1970; Sefton and Keegstra, 1974; Keegstra *et al.*, 1975; Burke and Keegstra, 1976, 1979). The E_1 glycoprotein from virus grown in BHK cells contains less of the mannose-rich small glycopeptide, S_4. The E_1 glycoprotein from virus grown in BHK cells contains more galactose than E_1 from chick embryo cell-grown virus. Apparently glycopeptides S_1, S_2, and S_3 contain progressively less galactose and sialic acid. The carbohydrate structure of the SIN virus glycoprotein E_2 has been defined (Burke and Keegstra, 1979). They found that the three largest glycopeptides (S_1, S_2, and S_3) have similar carbohydrate structures which differ only in the extent of their sialylation.

The function of the B-type oligosaccharide structure on the E_1 glycoprotein is obscure, since virus grown in BHK cells which have less E_1 have a particle/PFU ratio equal to that of virus grown in chicken embryo cells, which have equal molar amounts of the B-type oligosaccharides. The biological and/biochemical role(s) of the carbohydrate moiety of the envelope proteins is unknown. Treatment of purified virus with neuraminidase-released sialic acid raises the isoelectric point of the virions but does not alter the infectivity, hemagglutinin (HA) activity, or antigenic properties (Clegg and Kennedy, 1974). Treatment of virus with mixed sugar hydrolases lowers infectivity and HA activity but does change the antigenic properties of the virus. Viral proteins produced in cells in the presence of tunicamycin (which inhibits glycoprotein synthesis) are antigenically indistinguishable from the glycosilated forms (Levitt *et al.*, 1976). However, treatment of virus with bromelain or phospholipase C destroyed the virus particle and all virion surface antigenic activity (Friedman and Pastan, 1969). Limited digestion of virions with pronase (Osterrieth, 1968), bromelain (Compans, 1971), or thermolysin (Ghamberg *et al.*, 1972) leaves a naked particle which is not infectious or antigenically reactive. These observations suggest that it is the protein portion of the envelope glycoproteins which is responsible for infectivity and antigenic reactivity of the virion.

Fine detail of the organization of the glycoproteins on the virion surface is unknown; however, studies have permitted the construction of a theoretical model for the surface organization (McCarthy and Harrison, 1977). Most of the carbohydrate is not accessible to either endo- or exoglycosidases. The carbohydrate side chains therefore appear to be exposed at their nonreducing termini or not at all. The only completely accessible carbohydrate structures on the SIN envelope proteins are type A oligosaccharides on E_2 which can be digested by mixed glycosidases. These residues must be near the carbohydrate linkage on E_2 as well as external sugars on this protein which are exposed. External regions of some of the E_1 type A oligosaccharides are exposed to permit removal of sialic acid with loss of some galactose and glucosamine. The carbohydrate–peptide linkage of E_1 type A oligosaccharides is not accessible to enzymatic digestion by endoglycosidase. The type B oligosaccharides on the E_1 and E_2 membrane proteins are inaccessible to glycosodiase digestion on the intact SIN virus particle (McCarthy and Harrison, 1977).

When intact SIN virus is enzymatically iodinated, E_2 glycoprotein is labeled more than E_1; however, both proteins are labeled equally well following sodium dodechyl sulfate (SDS) disruption of the intact virus (Sefton *et al.*, 1973). These results indicate a differential availability of tyrosine residues in E_1 and E_2 glycoprotein which parallels the carbohydrate side chain exposure on E_1 and E_2 to enzyme digestion. It appears, therefore, that the portion of the SIN E_1 protein which is exposed on the virion surface contains those determinants which are responsible for hemagglutination (Dalrymple *et al.*, 1976).

Harrison *et al.* (1971) and McCarthy and Harrison (1977) have summarized the chemical and physical descriptions of the alphavirus surface and proposed a model for the arrangement of the membrane glycoproteins. The outer polar groups of the lipid bilayer lie approximately 25 nm from the center of the virus particle, and the glycoprotein subunits project 8 to 9 nm from the outer surface of the bilayer. A hydrophobic "tail" of each subunit penetrates the lipid bilayer (Gahmberg *et al.*, 1972; Utermann and Simons, 1974). X-ray scattering experiments have shown that the highest subunit contacts occur between 26 and 30 nm from the particle center, indicating that the distal ends of the glycoproteins must taper. Electron micrographs show that the glycoproteins are tightly clustered as trimers which must have some intersubunit association to maintain their structural configuration (von Bonsdorff and Harrison, 1975, 1978). The intertrimer contacts appear to be less extensive than those within one trimer cluster, revealing deep grooves between trimers which extend to almost the lipid bilayer. Triton X-100 nonionic detergent disrupts the surface lattice, dissociates the subunits at the surface, and makes the carbohydrate side chain susceptible to glycosidase digestion but does not denature or unfold the subunit (McCarthy and Harrison,

1977). It is postulated that the type A carbohydrate side chains on E_2 are linked near the tip of the protein or to a lateral surface facing outward from the clustered trimer (McCarthy and Harrison, 1977). The other oligosaccharides are buried: Those of type B are completely so, and those of type A on E_1 are accessible only at the outermost limits.

The envelope layer of the alphavirus is a modified unit membrane derived from the host cell membranes during the process of virion morphogenesis (Achesen and Tamm, 1967; Harrison *et al.*, 1971; Brown *et al.*, 1972; Birdwell and Strauss, 1974; Hirschberg and Robbins, 1974). The overall phospholipid composition of the alphavirus envelope is similar to that of the plasma membrane of the host cell (Pfefferkorn and Hunter, 1963b; David, 1971; Heydrick *et al.*, 1971; Renkonen *et al.*, 1971; Laine *et al.*, 1973). The major phospholipids are phosphatidylocholine, sphingomyelin, phosphatidylserine, and phosphatidylethanolamine. SIN virus grown in chick embryo fibroblasts has a cholesterol–phospholipid molar ratio of 0.68–0.80, which is similar to that of the plasma membrane (Renkonen *et al.*, 1971). SF virus grown in hamster kidney cells has a slightly higher cholesterol–phospholipid molar ratio of 0.94, characteristic of the hamster cell plasma membrane (Ciarcini *et al.*, 1968; Renkonen *et al.*, 1971).

The microviscosity of the lipid membrane of SIN and SF virus is higher than that of the cells in which the virus was grown (Moore *et al.*, 1976). SIN virus grown in BHK cells at 37°C has a higher microviscosity than virus grown in *Aedes albopictus* cells at 22°C. Presumably, the increased microviscosity is due to the hydrophobic tails of the envelope glycoproteins which are embedded in the alphavirus membrane (Utermann and Simons, 1974; Hirschberg and Robbins, 1974). Other factors such as the ratio of phospholipid species or the degree of saturation of the phospholipid and sphingomyelin side chains may also influence the fluidity of the viral membrane (Moore *et al.*, 1976).

The alphavirus nucleocapsid has been studied extensively by several investigators. Nucleocapsids generally have a spherical shape that is often seen in a deformed configuration, suggesting some flexibility in its organization (Igarashi *et al.*, 1966; Horiznek and Mussgay, 1969; Acheson and Tamm, 1970; Murphy, 1980a). Formaldehyde-fixed SIN and SF virus nucleocapsids have a radial symmetry which is best described as a geometric $T = 3$ (32-capsomer) or $T = 4$ (42-capsomer) configuration (Simpson and Hauser, 1968a,b; Osterrieth, 1968). SIN virus nucleocapsids released from virions by sodium deoxycholate (DOC) have 12–14 nm ring-like subunits arranged in a $T = 3$ icosahedral lattice (Horzinek and Mussgay, 1969). On the basis of capsomer arrangement, von Bonsdorff (1973) suggested that the SF nucleocapsid has a $T = 4$ configuration the same as that proposed for the symmetry of the glycoproteins in the membrane of the virion (von Bonsdorff and Harrison, 1978). The structure of the nucleocapsid has also been studied by freeze-etch methods (Brown *et al.*, 1972). On the basis of these physical observations, it was proposed that the nucleocapsid was composed of substructures 7 nm in diameter which are organized in an icosahedron with $T = 9$ symmetry (Brown *et al.*, 1972). The nucleocapsid is not a tightly closed structure; the viral RNA is digested within the structure by ribonuclease (Kaariainen and Soderlund, 1971). Kaariainen and Soderlund (1971) suggested that viral RNA forms an integral structural element of the nucleocapsid in the form of loops which interact with and stabilize the capsid proteins. Much remains to be learned about the structure and organization of the alphavirus nucleocapsid. Resolution of the ultrastructural chemistry of this viral subunit will depend upon the development of new methods for resolving the substructure of "native" particles.

3. Alphavirus Genome RNA

The genomes of alphaviruses such as WEE, VEE, EEE, SIN, and SF are composed of single-stranded RNA of 4.1×10^6 daltons (Levin and Friedman, 1971; Simmons and Strauss, 1972; Horzinek, 1973a; Martin and Burke, 1974; Strauss and Strauss, 1976). The RNA has a sedimentation coefficient of 42 to 49 S depending upon conditions of analysis (Levin and Friedman, 1971; Simmons and Strauss, 1972; Horzinek, 1973a; Martin and Burke, 1974). The naked RNA is infectious and therefore must code for synthesis of the viral structural and nonstructural proteins (Friedman *et al.*, 1966; Pfefferkorn and Shapiro, 1974; Strauss and Strauss, 1976).

The virion RNA contains a polyadenylic[poly(A)] sequence of 30 to 90 nucleotides at the 3′ terminus (Eaton and Faulkner, 1972; Johnston and Bose, 1972; Clegg and Kennedy, 1974; Sawicki and Gomatos, 1976; Wengler and Wengler, 1976). The existence of poly(U) in double-stranded RNA of both SIN and SF viral RNAs strongly suggests that the poly(A) synthesis in the alphaviruses is genetically coded (Bruton and Kennedy, 1975; Sawicki and Gomatos, 1976).

Sequence analysis of the 5′ end of the alphaviral RNA indicates that it has a cap O-type terminal sequence of 7mG(5′)pppApUpGp (Hefti *et al.*, 1976). Electron microscopic studies of native SIN genome RNA have revealed that the 49 S molecules form circles, suggesting that the terminal 3′ and 5′ sequences may contain terminal inverted sequences which are repeated (Hsu *et al.*, 1973).

4. Antigens

Alphaviruses have been divided by viral neutralization into groups or complexes: VEE, WEE, SF, EEE, Ndumu, and Middelburg (Karabatsos, 1975; Chanas *et al.*, 1976). Within each complex, individual viruses are recognized which are antigenically closely related but serologically distinguishable from the prototype strain (Casals, 1963, 1964; Karabatsos, 1963; Henderson, 1964; Chanas *et al.*, 1977; Woodroofe *et al.*, 1977).

As seen in Table V, by neutralization testing the WEE complex contains the viruses WEE, SIN, Whataroa, Aura, Highlands J (HJ), and Y62-33 (Karabatsos, 1975). The VEE complex contains VEE, Mucambo (MUC), and Pixuna (PIX); the SF complex contains Mayaro (MAY), Chikungunya (CHIK), Una, O'nyong 'nyong (ONN), Bebaru (BEB), Getah (GET), and Ross River (RR). Three viruses, EEE, Ndumu (NDU), and Middelburg (MID), do not relate to each other or to any of the other known alpha togaviruses. On the basis of these and other serological studies (Casals, 1957; Porterfield, 1961), the alphaviruses may be divided into at least six complexes (Calisher *et al.*, 1981). Each virus complex may contain a single virus with no known antigenic relatives or several closely related viruses. Some virus types have been further divided on the basis of their antigenic relationships into subtypes and varieties (Table VI).

The molecular basis for antigenic specificity among the alphaviruses has been determined by immunological characterization of the individual structural proteins. Treatment of purified virus with nonionic detergents, DOC, Tween-80-ether, or 8 *M* urea destroys viral infectivity but preserves complement-fixing, hemagglutinating, and neutralizing antibody blocking activities (Mussgay and Rott, 1964; Mussgay and Horzinek, 1966; Mussgay, 1967; Appleyard *et al.*, 1970; Bose and Sagik, 1970; Dalrymple *et al.*, 1973). The envelope glycoproteins and nucleocapsid may be separated by rate zonal centrifugation in sucrose gradients (Dalrymple *et al.*, 1973) or isopycnic centrifugation in CsCl (Mussgay and Rott, 1964). The E_1 and E_2 glycoproteins have also been separated by physical and chemical methods, including isoelectric focusing (Dalrymple *et al.*, 1976; France *et al.*, 1979; Hashimoto and Simizu, 1979; Trent and Grant, 1980), centrifugation in sucrose gradients containing DOC (Helenius and Simons, 1975; Helenius *et al.*, 1976), calcium phosphate chromatography (Garoff *et al.*, 1974), differential solubility in nonionic detergents at low ionic strength (Burke and Keegstra, 1976), electrophoresis in polyacrylamide gels (Pedersen and Eddy, 1974), and chromatography on glass wool (Bell *et al.*, 1979). The structural proteins separated by these techniques have been used to immunize animals for the production of antibodies to specific viral subunits (Dalrymple *et al.*, 1976; Symington *et al.*, 1977; France *et al.*, 1979; Trent and Grant, 1980).

Immunological studies with antisera to individual alphavirus structural proteins of SIN virus indicate that the nucleocapsid is antigenically unrelated to the envelope glycoproteins and that the E_1 and E_2 glycoproteins are antigenically related but distant from each other (Dalrymple *et al.*, 1976). In the radioimmune precipitation test (RIP), antisera to SIN E_1 (pI 6) reacted with SIN E_1 (pI 6) and E_2 (pI 9) proteins; however, the E_2 (pI 9) antiserum reacted only with the E_2 (pI 9) protein. Antiserum to E_1 reacted in the hemagglutination inhibition (HI) test

TABLE V
Relationships of Alphaviruses by Serum Dilution Neutralization Test

	Immune ascitic fluid or antiserum[a]									
Virus[a]	WEE	HJ	(Y62-33)	SIN	WHA	Aura	VEE	MUC	PIX	SF
WEE	700[b]	70	80	0[c]	0	0	0	0	0	0
HJ	440	2,400	230	140	10	0	0	0	0	0
(Y62-33)	50	0	49,000	0	0	0	0	0	0	0
SIN	150	0	0	770	65	0	0	0	0	0
WHA	120	0	0	50	100	0	0	0	0	0
Aura	0	0	0	0	0	30	0	0	0	0
VEE	0	0	0	0	0	0	1,400	40	20	0
MUC	0	0	0	0	0	0	190	380	55	0
Pixuna	0	0	0	0	0	0	25	20	19,200	0
SF	0	0	0	0	0	0	0	0	0	2,300
MAY	0	0	0	0	0	0	0	0	0	90
CHIK	0	0	0	0	0	0	0	0	0	10
Una	0	0	0	0	0	0	0	0	0	100
ONN	0	0	0	0	0	0	0	0	10	15
BEB	0	0	0	0	0	0	0	0	0	40
GET	0	0	0	0	0	0	0	0	0	40
RR	0	0	0	0	0	0	0	0	0	30
EEE	0	0	0	0	0	0	0	0	0	0
NDU	0	0	0	0	15	0	0	0	0	0
MID	0	0	0	0	0	0	0	0	0	0

Virus[a]	Immune ascitic fluid or antiserum[a]									
	MAY	CHIK	Una	ONN	BEB	GET	RR	EEE	NDU	MID
WEE	0	0	0	0	0	0	0	0	0	0
HJ	0	0	0	0	0	0	0	0	0	0
(Y62-33)	0	0	0	0	0	0	0	0	0	0
SIN	0	0	0	0	0	0	0	0	0	0
WHA	0	0	0	0	0	0	0	0	0	0
Aura	0	0	0	0	0	0	0	0	0	0
VEE	0	0	0	0	0	0	0	0	0	0
MUC	0	0	0	0	0	0	0	0	0	0
Pixuna	0	0	0	0	0	0	0	0	0	0
SF	230	0	0	0	0	120	20	0	0	0
MAY	2,400	0	0	0	0	200	10	0	0	0
CHIK	25	11,000	0	0	20	30	0	0	0	0
Una	100	0	200	0	25	80	0	0	0	0
ONN	25	7,800	0	8,000	0	100	0	0	0	0
BEB	0	0	0	0	2,800	55	0	0	0	0
GET	200	0	0	0	0	9,000	100	0	0	0
RR	10	0	0	0	10	80	1,000	0	0	0
EEE	0	0	0	0	0	0	0	6,500	0	0
NDU	0	0	0	0	0	0	0	0	1,700	0
MID	0	0	0	0	0	0	10	0	10	6,200

[a] Abbreviations: WEE, western equine encephalitis; HJ, Highlands J; SIN, Sindbis; WHA, Whataroa; VEE, Venezuelan equine encephalitis; MUC, Mucambo; PIX, Pixuna; SF, Semliki Forest; MAY, Mayaro; CHIK, Chikungunya; ONN, O'nyong'nyong; BEB, Bebaru; GET, Getah; RR, Ross River; EEE, eastern equine encephalitis; DNU, Ndumu; MID, Middelburg. [b] Reciprocal of dilution giving 50% plaque reduction. The homologous PRNT titers of Highlands J, Y62-33, Sindbis, Whataroa, Aura, Mucambo, Pixuna, Semliki Forest, Una, Ross River, EEE, and Middelburg as well as WEE antibody with Highlands J and Whataroa, Y62-33 antibody with Highlands J, Sindbis antibody with Highlands J, Whataroa antibody with Ndumu, VEE antibody with Mucambo, Semliki antibody with Una and Bebaru, Mayaro antibody with Chikungunya and Getah, Chikungunya antibody with ONN, Getah antibody with Semliki Forest and Ross River antibody with Getah were geometric mean titers of two or more determinations.

[c] 0 = <10. Source: From Karabatsos (1975), with permission.

TABLE VI
Classification of the Alphaviruses Based on Their Serological Relationships[a]

Complex	Virus	Subtype	Variety
EEE[b]	EEE		
MID	MID		
NDU	NDU		
SF	SF		
	CHIK	(1) CHIK	Several
		(2) ONN	
	GET	(1) GET	
		(2) SAG	
		(3) BEB	
		(4) RR	
		(1) MAY	
		(2) UNA	
VEE	VEE	(1) VEE	At least
		(2) EVE	
		(3) MUC	MUC
			TON
		(4) PIX	
WEE	WEE		
	Y62-33		
	HJ		
	FM		
	SIN	(1) SIN	
		(2) WHA	
		(3) KZL	
	AURA		

[a] After Calisher *et al.* (1981), with permission.
[b] Abbreviations: EEE, eastern equine encephalitis; MID, Middelburg; NDU, Ndumu; SF, Semliki Forest; CHIK, Chikungunya; ONN, O'nyong nyong; GET, getah; SAG, Sagiyama; BEB, Bebaru; RR, Ross River; MAY, Mayaro; UNA, Una; VEE, Venezuelan equine encephalitis; EVE, Everglades; MUC, Mucambo; TON, Tonate; PIX, Pixuna; WEE, western equine encephalitis; HJ, Highlands J; FM, Fort Morgan; SIN, Sindbis; WHA, Whataroa; KZL, Kyzylagach; AURA, Aura.

with homologous SIN virus and to a much lower titer with WEE virus. Antiserum to the SIN E_2 (pI 9) glycoprotein neutralized only homologous virus (Dalrymple *et al.*, 1976). Trent and Grant (1980) observed that the E_1 glycoprotein of WEE virus had a pI of 6.4 in urea isoelectric focusing gradients and induced HI antibodies which were cross-reactive with HJ virus, a serologically related virus in the WEE complex. The E_2 glycoprotein of WEE virus had a pI of 8.4 and induced antibody which was virus specific by neutralization but cross-reacted with HJ virus in the RIP test. The E_1 glycoprotein of HJ virus induced antibody which was specific for the homologous virus in the HI, neutralization, and RIP tests. Isolated E_2 of HJ virus had a pI of 9.1 and induced antibodies which were cross-reactive at equal titer with both WEE and HJ viruses in the RIP test (Trent and Grant, 1980).

Pedersen and Eddy (1974) reported that the largest VEE glycoprotein E_2 was responsible for the induction of antibodies reactive in the HI and neutralization tests. Studies in our laboratory have confirmed this observation and determined that this envelope glycoprotein contains antigenic determinants which are virus type specific (France *et al.*, 1979). The E_1 glycoprotein of VEE virus contains antigenic determinants which are subgroup specific but do not distinguish virus variants IAB, C, D, or E within the complex (Trent *et al.*, 1979).

Data from these studies are summarized in Table VII. It is obvious that although all alphaviruses contain two large glycoproteins, the biological functions and antigenic specificities vary from virus to virus even within a specific complex. It is clear that within the alphavirus genus no single model for the biological and immunological functions of these envelope glycoproteins can be assumed.

5. *Variation*

Alphavirus isolates differ in their plaque size (Marshall *et al.*, 1962; Ushijima *et al.*, 1962; Hannoun *et al.*, 1964; Takemoto, 1966; Simizu and Takayama, 1969; Stim, 1969; Zarate and Scherer, 1969; Bose *et al.*, 1970a; Pedersen *et al.*, 1972a; Jahrling, 1976; Chanas *et al.*, 1977; Gresikova and Batikova, 1978), virulence (Dunayevich *et al.*, 1961; Casals, 1964; Young and Johnson, 1969; Zarate and Scherer, 1969; Johnson and Martin, 1974; Scherer and Pancake, 1977; Woodroofe *et al.*, 1977; Hayes, 1978; Chanas *et al.*, 1979; France *et al.*, 1979; Trent *et al.*, 1979; Trent and Grant, 1980), antigenic character (Dunayevich

TABLE VII
Antigenic Characteristics and Biological Functions of Alphavirus Envelope Glycoproteins

Virus	Glycoprotein E_1	Glycoprotein E_2	Reference
Sindbis	Group-reactive hemagglutination	Type-specific neutralization	Dalrymple *et al.* (1976)
Western equine encephalitis	Group-reactive hemagglutination	Type-specific neutralization	Trent and Grant (1980)
Highlands J	Type-specific neutralization hemagglutination	Group reactive	Trent and Grant (1980)
Venezuelan equine encephalitis	Complex reactive	Type-specific hemagglutination, neutralization	Pedersen and Eddy (1974) France *et al.* (1979)
Semliki Forest		Hemagglutination	Helenius *et al.* (1976)

et al., 1961; Zarate and Scherer, 1969; Bradish *et al.*, 1971, 1972; Jahrling and Scherer, 1973a; Brown and Officer, 1975; Jahrling, 1976; Scherer and Chin, 1977), host range in cell culture (Marshall *et al.*, 1962; Symington and Schlesinger, 1978), heat stability (Simpson and Hauser, 1968b), and charge differences (Bose *et al.*, 1970b; Pedersen *et al.*, 1972b; Jahrling and Eddy, 1977). Although variation in a wide variety of biological and chemical features is a common phenomenon, the genome structure of each of these viruses is distinct enough to allow identification by their oligonucleotide fingerprint of its genome RNA (Wengler *et al.*, 1977; Trent *et al.*, 1979; Trent and Grant, 1980). The molecular basis for variation has not been determined, although biochemical and immunological studies have compared variants of SIN, VEE, and SF viruses with their prototypes (Bose *et al.*, 1970b; Pedersen *et al.*, 1973; Trent *et al.*, 1979). Large and small plaque mutants of SIN (Pedersen *et al.*, 1973), WEE virus (Jahrling, 1976), and virulent and avirulent clones of VEE virus (Jahrling and Eddy, 1977) elute differently from calcium phosphate columns, suggesting that the surface charge on the virions differs. Isoelectric focusing of mixtures of large plaque and small plaque SIN variants with prototype virus failed to separate the two plaque type viruses on the basis of this isoelectric point (Pedersen *et al.*, 1973). Symington and Schlesinger (1978) found the E_1 and E_2 envelope proteins of a SIN virus variant to be slightly more acidic than that of the prototype. In order to determine if the difference in net charge of the individual glycoproteins was due to changes in the primary sequence of the viral protein, tryptic peptides of the whole virus (Bose *et al.*, 1970b) or separated glycoproteins and nucleocapsids were analyzed (Symington and Schlesinger, 1978). No differences in the number of tryptic peptides in the prototype and variant viruses were found.

The TC-83 attenuated vaccine strain and the Trinidad donkey strain of VEE virus have been compared by calcium phosphate chromatography (Jahrling and Eddy, 1977) and oligonucleotide mapping of their viral RNAs (Trent *et al.*, 1979). The avirulent TC-83 strain elutes from calcium phosphate at a molarity of 0.23, in contrast to the parent, which elutes at 0.18 (Jahrling and Eddy, 1977). A six-oligonucleotide difference was shown in the RNA oligonucleotide fingerprints of the two viruses. Three of the oligonucleotide differences are found in the 3′ one-third of the VEE genome which codes for the virion structural proteins (D. W. Trent, unpublished data). This suggests that the virion mRNA of these two viruses may code for proteins which do have differences in their amino acid sequences. Those differences may be detected by tryptic map analyses of their structural proteins.

Oligonucleotide maps of SF virus strains of high and low virulence have been examined and compared (Wengler and Wengler, 1976). Differences were detected only in the sequences of the 26 S RNA, the oligonucleotides present exclusively in the 42 S RNA being identical. Thus, like the TC-83 strain of VEE, genetic differences in virulent and avirulent viruses seem to be related to the nucleotide sequence coding for the structural proteins. If this is the case, then perhaps immunological differences in the virus strains might be expected. These strains are closely related serologically and cannot be distinguished by cross neutralization and double adsorption tests (Bradish *et al.*, 1971, 1972). These authors conclude that the expression of virulence is not related to a dominant antigenic mechanism but is related to minor antigenic determinants or regulated by host-specified mechanisms (Bradish *et al.*, 1971, 1972; Walder and Bradish, 1979). These authors suggest that virulence is genetically determined and that naturally occurring or spontaneously derived avirulent strains differ from those in which chemical mutation is imposed upon a single genome population.

The small plaque variant of SIN virus is more extensively neutralized than is the large plaque variant by antiserum raised against homologous virus (Pedersen *et al.*, 1973). The SIN virus variant which is more infectious for mouse cells than standard virus is neutralized just as well as the parent virus, suggesting that there are no immunological differences between these strains (Symington and Schlesinger, 1978).

In summary, variation among alphaviruses appears to be genetically controlled and the variants, whether antigenic, plaque type, or virulent, are genetically stable. On the basis of differences in oligonucleotide fingerprints of virulent and avirulent SF and VEE virus in 26 S RNA, virion structural proteins may be involved as virulence determinants. The molecular organization of the alphavirus genome determining virus variation must await further definition.

The importance of biological variation in alphavirus disease can be best exemplified in the case of VEE infections of man and animals (Young and Johnson, 1969; Johnson and Martin, 1974). VEE virus is widely distributed in the Western Hemisphere and has been serologically divided into at least four subtypes (I–IV); subtype I is further separable into four antigenic variants designated IAB, IC, ID, and IE (Young and Johnson, 1969; Johnson and Martin, 1974; France *et al.*, 1979). The subtypes are distinguishable by standard serological tests using hyperimmune antisera, but specialized techniques and reagents are required for the identification of the variants of subtype I (short-incubation HI tests with antisera prepared in spiny rats or antisera to the E_2 glycoprotein). Specific identification is epidemiologically important because only the IAB and IC variants produce epizootic disease in equines and associated human epidemics. The ID and IE variants and subtypes II, III, and IV are enzootic infections of low virulence for equines. The enzootic VEE viruses are distinct antigenically, biologically, and genetically from the equine virulent subtypes (Young and Johnson, 1969; Zarate and Scherer, 1969; Young, 1972; Johnson and Martin, 1974; Scherer and Chin, 1977; Jahrling and Eddy, 1977; Trent *et al.*, 1979). The vector–host relationships of epizootic and enzootic strains are also distinct (see Section IV,C,2). Fig. 3 shows the known geographic distribution of the serological subtypes and variants. VEE, IAB, and IC viruses are entrenched in northern South America (Peru, Ecuador, Colombia, Venezuela, Guyana, Trinidad), but on at least one occasion spread through Central America,

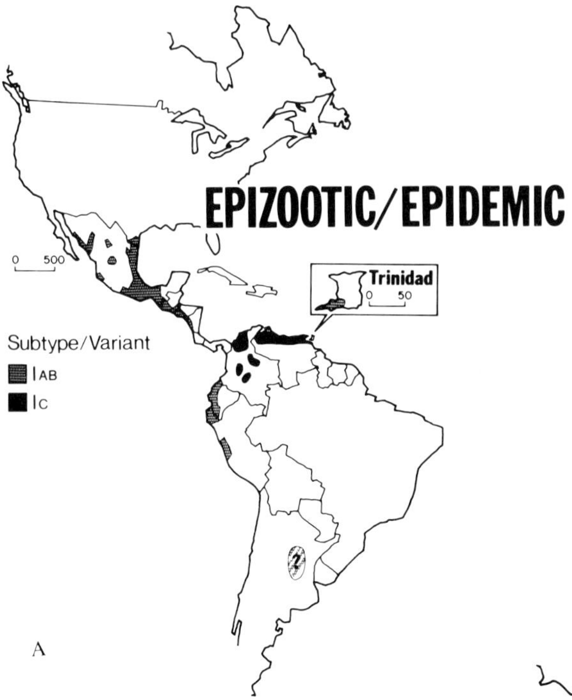

A

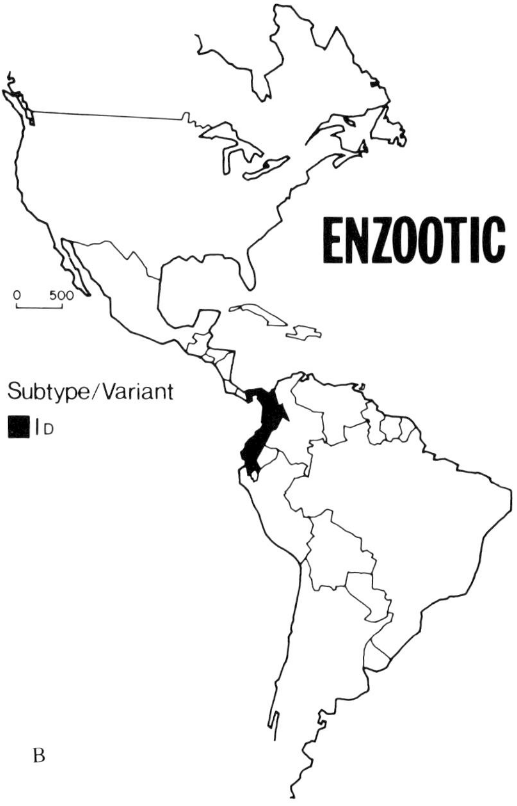

B

Fig. 3. Geographic distribution of antigenic subtypes and variants of Venezuelan equine encephalitis virus. (A) Subtypes IAB and IC; (B) subtype ID.

(*continued*)

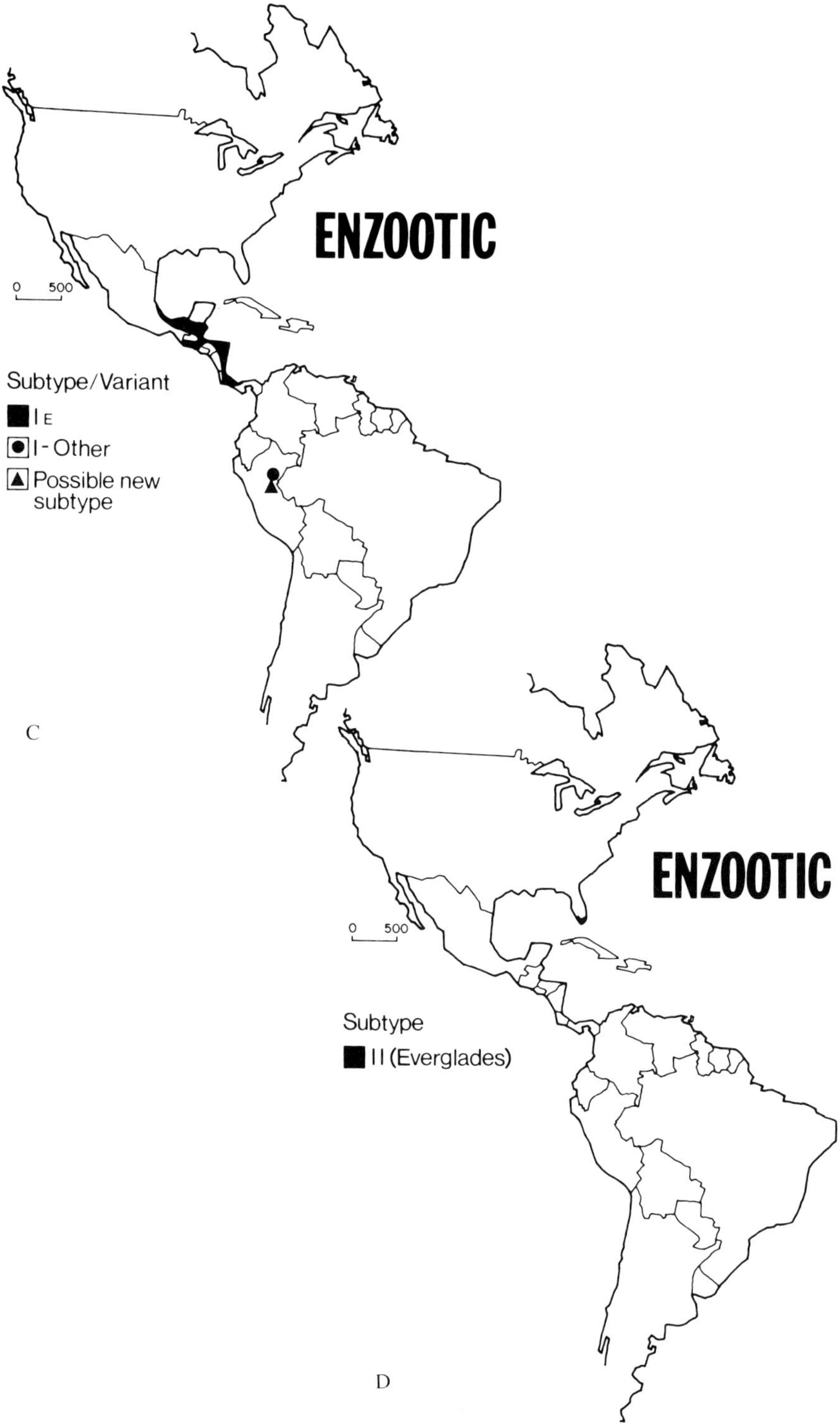

Fig. 3. (C) Subtype IE and possible new subtypes; (D) subtype II (Everglades virus).

(continued)

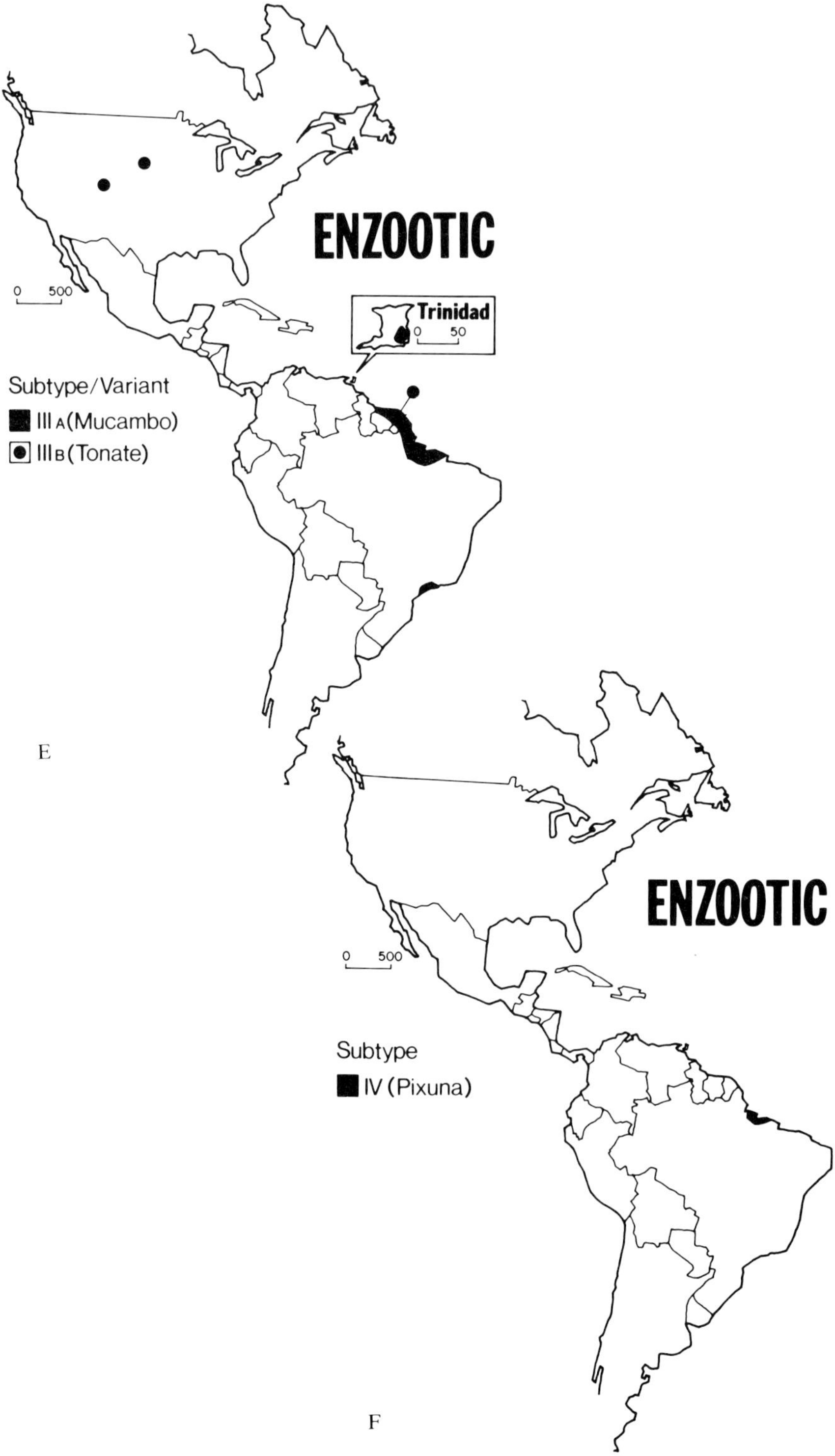

Fig. 3. (E) Subtype III; (F) subtype IV.

Mexico, and the United States (Young and Johnson, 1969; Groot, 1972; Johnson and Martin, 1974). The enzootic VEE virus variants, ID, and IE have been found in Colombia, Panama, Nicaragua, Honduras, Guatemala, Mexico, and Belize (Young and Johnson, 1969; Franck, 1972). The other enzootic subtypes are found in Florida (subtype II), Brazil, French Guiana, Surinam, Guyana, Trinidad (subtype III), and Brazil (subtype IV) (Young and Johnson, 1969; Franck, 1972).

Antigenic variants of EEE virus were initially described by Casals (1964), who separated the viruses by kinetic HI tests into North American and South American subtypes. The North American subtype has been responsible for outbreaks occurring in the northern Caribbean islands (Hispaniola), possibly having been introduced by birds migrating south from the United States. The South American subtype has been recovered from northward migrants entering the southern United States (Calisher *et al.*, 1971), but not from enzootic sites on the continent. The evidence suggests that the two subtypes are permanently maintained in their respective regions, without significant exchange.

6. Host Range

The laboratory host range of alphaviruses is shown in Table VIII.

TABLE VIII
Host Range of Togaviruses Known to Cause Disease in Domestic Animals[a]

		Virus							
Host (age)	Route	EEE	WEE	VEE	JE	WN	WESS	LI	IT
Mice (1–4 day)	ic	D	D	D	D	D	D	D	D
	ip	D	D	D	D	D	D	—	D
Mice (3–4 weeks)	ic	D	D	D	D	D	D	D	D
	ip	D	D	D	D	D	D	occ D	—
Chick (1 day)	sc	D	D	D	occ D	occ D	—	—	D
Hamster (3–4 weeks)	ic	D	D	D	D	D	—	D	N
	ip	D	IA	D	IA	D	IA	occ D	—
Guinea pig (3–4 weeks)	ic	D	D	D	IA	IA	—	D	N
	sc	—	IA	D	IA	IA	—	—	—
Rabbit	sc	IA	IA	D	IA	IA	—	IA	—
Embryonated eggs	ca	D	D	D	D	D	—	D	—
	ys	D	D	D	D	D	—	D	—
Cell cultures									
Chick or duck embryo		P, CPE	P, CPE	P, CPE	P, CPE	P, CPE	—	P	P
BHK-21		P, CPE	P, CPE	P, CPE	P, CPE	P, CPE	CPE	—	—
Vero, LLC-MK2		P, CPE	P, CPE	P, CPE	P, CPE	P, CPE	P	P	P

[a] EEE, eastern equine encephalitis; WEE, western equine encephalitis; VEE, Venezuelan equine encephalitis (epizootic subtypes); JE, Japanese encephalitis; WN, West Nile; WSL, Wesselsbron; LI, louping ill; IT, Israel turkey meningoencephalitis; D, death; occ D, occasional death; ic, intracerebral; ip, intraperitoneal; sc, subcutaneous; IA, inapparent infection, antibody; ac, chorioallantoic; ys, yolk sac; N, no evidence of infection; P, plaques; CPE, cytopathic effect; —, no information.

B. Flavivirus Particles

1. Chemical and Physical Properties

The overall chemical composition of the flavivirus St. Louis encephalitis (SLE) is presented in Table IX. The flaviviruses contain RNA, protein, phospholipid, and carbohydrate. The flavivirus virion is constructed from three protein species: a glycoprotein of 5 to 59 × 10^3 daltons, a nucleoprotein of 13.5 × 10^3 daltons, and a small membrane protein of 7.7 to 8.7 × 10^3 daltons. The structural proteins of all of the flaviviruses which have been examined are similar in size and number (Shapiro *et al.*, 1971). The molecular weights of the structural proteins of some of the more thoroughly studied flaviviruses of human and veterinary importance are shown in Table X. The largest structural protein is glycosylated and exposed on the outer surface of the virus particle, where it forms an integral part of the viral envelope and major viral antigen (Kitano *et al.*, 1974; Trent, 1977). Treatment of the virus particles with nonionic detergents or sodium deoxycholate solubilizes the envelope and releases the nucleocapsid, which is composed of a nonglycosylated protein of 13.5 × 10^3 daltons (Stollar, 1969; Westaway and Reedman, 1969; Shapiro *et al.*, 1971; Trent and Qureshi, 1971). The solubilized material contains both the large glycoprotein and a small nonglycosylated protein of 7 to 9 × 10^3 daltons (Westaway and Reedman (1969).

The physical and chemical properties of the flaviviruses have been reviewed by Horzinek (1973b) and recently by Karabatsos (1980).

The density of flavivirus particles has been determined by isopycnic sedimen-

TABLE IX
Composition of Flavivirus St. Louis Encephalitis[a]

Protein	
% total mass	66.0
Structural protein (mol. wt. × 10^{-3})	
Envelope	53.0
Membrane	8.5
Nucleocapsid	13.5
RNA	
% total mass	6.0
Molecular weight × 10^{-6}	3.8
GC content (%)	53.3
Carbohydrate	
% total mass	9.0
Lipid	
% total mass	17.0
% phospholipid	12.0

[a] From Trent and Naeve (1980), with permission.

TABLE X
Flavivirus Structural Proteins

Virus	Envelope glycoprotein ($\times 10^{-3}$)	Nucleo-capsid protein ($\times 10^{-3}$)	Membrane protein ($\times 10^{-3}$)	Reference
Dengue-2	59	13.5	7.7	Stollar (1969)
Japanese encephalitis	53	13.5	8.7	Shapiro *et al.* (1971)
Kunjin	51	13.5	8.5	Westaway and Reedman (1969)
Murray Valley encephalitis	56	16.0	8.5	Westaway (1975)
St. Louis encephalitis	53	13.5	8.5	Trent and Qureshi (1971)

tation in gradients of sucrose, potassium tartrate, and cesium chloride. The density of dengue and tick-borne encephalitis virus virions in cesium chloride is 1.24 gm/cm^3 (Stevens and Schlesinger, 1965; Slavik *et al.*, 1970). SLE virus has a density of 1.228 gm/cm^3 in potassium tartrate (Trent and Qureshi, 1971). In sucrose gradients, JE and SLE viruses sediment to a density of 1.20 gm/cm^3 (Cardiff *et al.*, 1971; D. W. Trent, unpublished data).

The flavivirus virion has a sedimentation rate of approximately 208 S in sucrose, compared to 280 S for the alphavirus, SIN (Slavin *et al.*, 1970; Smith *et al.*, 1970; Shapiro *et al.*, 1971; Stinski and Gruber, 1971; Boulton and Westaway, 1972; Della-Porta and Westaway, 1972). Rate zonal sedimentation of flavivirus preparations facilitates the separation of three different virus antigens, each having distinct biophysical and antigenic characteristics (Stevens and Schlesinger, 1965; Brandt *et al.*, 1970; Smith *et al.*, 1970; Cardiff *et al.*, 1971; Stinski and Gruber, 1971; Della-Porta and Westaway, 1972). The fastest sedimenting particle is the infectious virion, composed of three structural proteins. A second particle is donut shaped, 13–14 nm in diameter, is not infectious, sediments at 65 to 80 S, and hemagglutinates. This particle is composed of the envelope glycoprotein, a membrane protein, and usually a viral nonstructural protein. The third antigen present in sucrose gradients is designated the "soluble complement-fixing antigen (SCF)" (Brandt *et al.*, 1970; Smith *et al.*, 1970; Cardiff *et al.*, 1971; Eckles *et al.*, 1975). The dengue SCF antigen is a single protein of 30,000 daltons which is resistant to inactivation by urea or sodium deodecylsulfate (SDS) and 2-mercaptoethanol (2-ME) (Brandt *et al.*, 1970). JE virus SCF is larger than the dengue antigen, 53,000 versus 30,000 daltons. The JE virus antigen is also resistant to inactivation by SDS, urea, and 2-ME.

The envelope of the flavivirus particle is composed of viral protein(s) embedded in a bilayer of lipid and carbohydrate which are of host origin (Trent and Naeve, 1980). Flaviviruses, like alphaviruses, are rapidly inactivated by lipid solvents such as ether, chloroform, or DOC (Sulkin and Zarafonetis, 1947; Andrewes and Horstman, 1947). Ada and Anderson (1959) observed that treatment of Murray Valley encephalitis (MVE) virus released infectious RNA, indicating that the removal of the viral lipid destroyed the integrity of both the viral envelope and the nucleocapsid (Franklin, 1962). It has also been demonstrated that treatment of purified flaviviruses SLE (Trent and Qureshi, 1971), Kunjin (Boulton and Westaway, 1972), dengue (Stollar, 1969), and JE (Kitano *et al.*, 1974; Eckles *et al.*, 1975) with nonionic detergents and DOC releases fragments of the envelope and envelope glycoproteins.

When SLE virus was grown in host cells of different animal origin, the phospholipids in the virion were different (Trent and Naeve, 1980). The quantitative differences in the phospholipids in the different cell types and SLE virus grown in those cells were similar. This strongly suggests that during the process of flavivirus morphogenesis the virus-modified cellular membrane contains virus-specified proteins, but the lipids in the membranes are left unmodified (Shapiro *et al.*, 1972a; Kos *et al.*, 1975; Stohlman *et al.*, 1975; Boulton and Westaway, 1976). The relatively high phosphatidylcholine and phosphatidylethanolamine content of the SLE virion and low cholesterol–phospholipid ratio suggests that flaviviruses mature in association with the internal membranes (Yasuzumi *et al.*, 1964; Murphy *et al.*, 1968; Renkonen *et al.*, 1971).

The flavivirus genome is composed of single-stranded RNA of molecular weight of approximately 3.78 to 4.2 $\times$ 10^6 (Harley *et al.*, 1975; Boulton and Westaway, 1976; Naeve and Trent, 1978). The sedimentation coefficient of the genomic RNA from several different flaviviruses is reported to be in the range 38 to 45 S (Igarashi *et al.*, 1964; Stollar *et al.*, 1966; Trent *et al.*, 1969; Boulton and Westaway, 1972; *Vezza et al.*, 1980). The range in molecular weights reported for the flaviviral genome probably reflects variation in sedimentation conditions or molecular weight markers rather than differences in the size of the RNA.

The molar base ratios of viral RNAs for several of the flaviviruses have been reported (Table XI). The overall base composition of the RNAs of these serologically diverse viruses is very similar. The RNA species are rich in A and G, with an average A/G ratio of 1:10.

Relatively little is known about the molecular structure of the flaviviral genome. Unlike alphavirus 42 and 26 S RNAs which contain poly(A) tracts of 30 to 90 residues at their 3′ terminus (Eaton and Faulkner, 1972; Johnston and Bose, 1972; Clegg and Kennedy, 1974; Wengler *et al.*, 1977), the flavivirus 42 S genome does not have a similar structure (Wengler *et al.*, 1978; Trent and Naeve, 1980; Vezza *et al.*, 1980). It is possible that a short poly(A) tract would elude detection in the oligonucleotide fingerprints of T_1 RNAse-resistant material; however, SLE genome RNA does not bind to oligo(dT) columns under conditions in which 60–70% of alphavirus 42 S binds (Naeve and Trent, 1978). These data suggest that either the virion RNA does not contain a poly(A) tract or

TABLE XI
Base-Ratio Analysis of the Genome RNA of Selected Flaviviruses

	Nucleotide				
Virus	CMP	AMP	GMP	UMP	Reference
Dengue 1	20.7	31.6	26.2	21.5	Vezza *et al.* (1980)
Dengue 2	23.0	31.3	22.7	21.6	Stollar *et al.* (1976)
	20.6	32.5	25.7	21.2	Vezza *et al.* (1980)
Dengue 3	20.7	31.6	26.4	27.2	Vezza *et al.* (1980)
Dengue 4	19.6	30.1	27.2	23.1	Vezza *et al.* (1980)
Murray Valley encephalitis	21.5	25.5	27.5	25.5	Ada *et al.* (1962)
Japanese encephalitis	24.9	23.3	31.0	20.7	Blair (1977)
St. Louis encephalitis	21.7	30.7	26.2	21.4	Trent *et al.* (1969)

contains a short tract in the molecule whose secondary structure is such that it cannot bind to the oligo(dT) column. Obviously, this point needs further clarification.

Recent data from several laboratories indicate that the 5′ terminus of the flaviviral RNA possesses a methylated cap of the form 7mG(5′)ppp(5′) Amp . . . (Wengler *et al.*, 1978; Cleaves and Dubin, 1979; Vezza *et al.*, 1980). This type I cap is different from the type O sequence of 7mGpppApUp present on the 5′ terminus of the alphavirus genome (Hefti *et al.*, 1976). The significance of this difference is not known; however, it is unusual for positive-stranded cytoplasmic RNA viruses. All RNA viruses previously shown to exhibit penultimate ribose methylation are negative-strand viruses (Rothman, 1978). The absence of internal m^6Ade on the dengue viral RNA may correlate with the subcellular location of dengue RNA replication. Viral mRNAs synthesized in the nucleus contain internal m^6Ade, while mRNAs formed in the cytoplasm generally lack them (Boone and Moss, 1976; Beemon and Keith, 1977).

The secondary structure of SLE RNA appears to be different from that of SIN virus. The 42 S of SLE virus eluted from CF_{11} cellulose at a lower alcohol concentration than did the SIN viral RNA, suggesting that the SLE genome has less secondary structure (Engelhardt, 1972; Brawner *et al.*, 1977).

2. *Morphological Structure and Chemical Organization*

Excellent reviews of the morphology and morphogenesis of flaviviruses have been published (Pfefferkorn and Shapiro, 1974; Schlesinger, 1977; Murphy, 1980a,b). In all studies, the particles of antigenically diverse flaviviruses have been observed to be morphologically very similar. Virions of dengue (Brandt *et al.*, 1970; Smith *et al.*, 1970; Matsumura *et al.*, 1971), JE (Nishimura *et al.*, 1968; Yoshinaka and Hotta, 1971; Kitano *et al.*, 1974), tick-borne encephalitis virus (Weekstrom and Nyholm, 1965; Slavik *et al.*, 1967, 1970), Langat (Boulton *et al.*, 1971), and West Nile (WN) (Chippaux-Hyppolite *et al.*, 1970) have been characterized by negative contrast electron microscopy. The particles have been described by all investigators as being spherical with a mean diameter of 43 nm. Morphologically the viral envelope has been shown to be a unit membrane with surface projections (Nishimura *et al.*, 1968; Smith *et al.*, 1970; Matsumura *et al.*, 1971; Kitano *et al.*, 1974; Hayashi *et al.*, 1978; Murphy, 1980a,b). Surface projections of JE virus particles could be removed by pronase and bromelain, leaving a smooth intact particle (Kitano *et al.*, 1974). Smith *et al.*, (1970) reported that the dengue virus particle surface was covered by "doughnuts" 7 nm in diameter with a 2- to 3-nm hole. Other workers have not observed this type of surface detail and usually described the surface layer as being quite thin (< 5 nm) and with very little surface detail (Nishimura *et al.*, 1968; Matsumura *et al.*, 1971; Kitano *et al.*, 1974; Murphy, 1980a,b).

In an extensive study of the structure of SLE virus particles, Murphy (1980b) used various staining techniques. Studies in our laboratory with uranyl acetate-stained SLE virus have facilitated a physical description of the virions (Fig. 4). Particles were spherical in shape, averaging 49 nm in diameter, with a core of approximately 30 nm. The surface of the virion was covered by a thin layer of

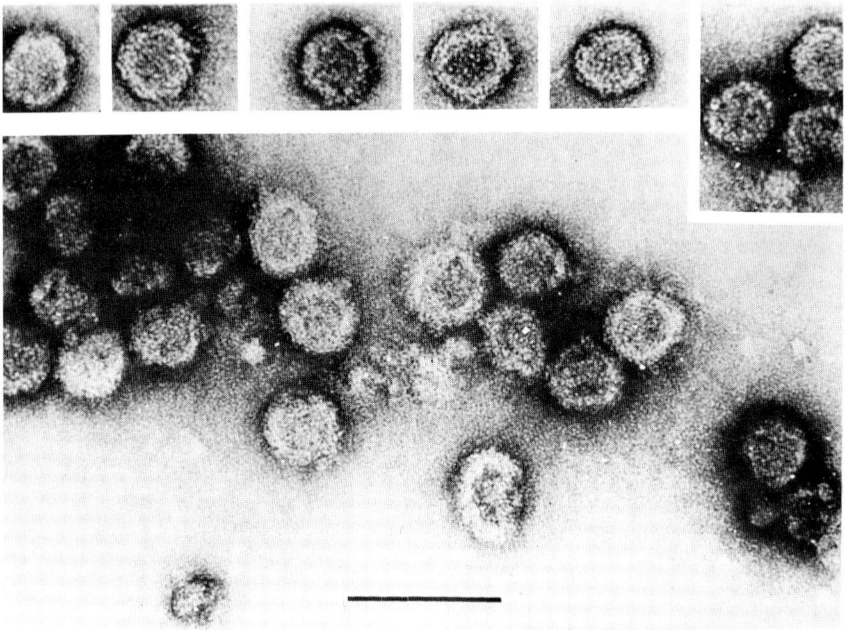

Fig. 4. St. Louis encephalitis virus stained with uranyl acetate (0.5%, pH 4.5). Virus particles are penetrated, revealing a membranous fringed envelope and an inner core particle. Composite; × 420,000; bar is 100 nm.

projections (< 5 nm) which were not well resolved. Murphy (1980a,b) found that virions stained with phosphotungstate and silicotungstate were similar in appearance to those stained with uranyl acetate, without resolving any symmetry in the viral envelope or surface projections. Ammonium molybdate-stained virions had a mean diameter of 48.8 nm; the projection layer was very thin (closely apposed to the envelope), and the individual projections were not resolved.

The core or nucleocapsid structure of the flavivirus particle does not appear to have a well-defined geometric structure or regularly arranged subunits (Murphy 1980a,b). Freeze-fracture techniques have revealed the dengue nucleocapsid without any symmetrical subunits to be surrounded by a unit membrane envelope which has 7-nm surface projections (Demsey *et al.*, 1974).

3. *Antigenic Composition*

The flavivirus genus of the Togaviridae contains 57 viruses which share certain physical and biochemical properties and are differentiated from each other by their serological relationships (Westaway, 1966; Theiler and Downs, 1973; DeMadrid and Porterfield, 1974). DeMadrid and Porterfield (1974) used a serum dilution neutralization test to study the serological relationships of 42 flaviviruses. They defined seven subgroups (Table XII) or complexes within the flavivirus genus. Six tick-borne viruses fell into one subgroup; 7 flaviviruses associated with small rodents and bats and 1 tick-borne virus were placed in a second group; 22 mosquito-borne viruses fell into a third major group. Four mosquito-borne, one tick-borne, and one bat virus were antigenically distinct from each other and from all other flaviviruses tested.

Many of the biological and immunological properties of the flavivirus particle depend upon the properties of the proteins on the virion surface. It is known that the glycoprotein of 51,000 to 59,000 molecular weight is located on the surface of the particle and is reactive in the neutralization and hemagglutination tests (Kitano *et al.*, 1974). This surface glycoprotein has been isolated from purified virions by solubilization of the virus with detergents followed by isopycnic centrifugation in CsCl gradients (Kitano *et al.*, 1974), rate zonal centrifugation in sucrose (Della-Porta and Westaway, 1977), molecular sieving chromatography (Eckles *et al.*, 1975), or isoelectric focusing (Trent, 1977). JE viral envelope glycoprotein purified by solubilization of virus with Nonidet P-40 and isopycnic centrifugation had a density of 1.256 gm/cm^3, hemagglutinated goose red blood cells, and induced neutralizing antibody in mice (Kitano *et al.*, 1974). Purified Kunjin envelope glycoprotein also hemagglutinates and induces neutralizing and HI antibodies in rabbits (Della-Porta and Westaway, 1977). Treatment of JE virus with SDS destroyed all hemagglutinating activity of the glycoprotein; however, the isolated protein retained its complement-fixing activity (Eckles *et al.*, 1975). The envelope glycoprotein of SLE virus which was sepa-

TABLE XII
Results of Cross-Neutralization Test with 42 Flaviviruses and Their Respective Antisera

		Antiserum and immunization course																							
		NEG 1	LGT 3	KFD 3	LI 5	OHF Ad	TBE Ad	KDM 3	APO 3	DB 1	EB 3	BB 1	RB 3	MOD 1	CR 1	JBE 3	MVE 1	WN 3	SLE 5	KUN 1	USU 1	KOK 1	STR 5	ALF 3	SPO Ad
Negishi	NEG	**4**				2	4																		
Langat	LGT		**7**				6																		
Kyasanur Forest	KFD		**27**																						
Louping ill	LI				**3**	2	3																		
Omsk HF	OHF					**5**	3																		
Tick-borne E	TBE		2			2	**5**																		
Kadam	KDM							**7**																	
Apoi	APO	2	1					2	**7**	2			2												
Dakar Bat	DB									**4**		3													
Entebbe Bat	EB							2		2	**7**	3	2	1		4		3							
Bukalasa Bat	BB									1		**5**		1		2									
Rio Bravo	RB									2		2	**5**	2		2	2								
Modoc	MOD													**7**	3										
Cowbone Ridge	CR													3	**7**										
Japanese BE	JBE															**5**	6								
Murray Valley E	MVE															2	**3**	5	3		5				
West Nile	WN															2	2	**7**	2	2	2				
St. Louis E	SLE															3	1	7	**7**						2
Kunjin	KUN															2		7		**5**					
Usutu	USU															4	3	3			**2**				
Kokobera	KOK																3					**7**	4		
Stratford	STR																						**2**		
Alfuy	ALF																3				3			**2**	

		SPO Ad	ZIK 1	CHU 1	IT 3	NTA 3	TMU 1	BAN 1	UGS Ad	EH 3	D 3	D2 5	D3 Ad	D4 Ad	YF 3	BSQ 3	WSL Ad	ILH 3	MML 1	POW Ad
Spondweni	SPO	**4**		6																
Zika	ZIK	3	**7**																	
Chuku	CHU	5		**7**																
Israel turkey	IT				**4**		2													
Ntaya	NTA				6	**6**	3													
Tembusu	TMU				1		**7**													
Banzi	BAN							**5**	5											
Uganda S	UGS							5	**7**											
Edge Hill	EH								2	**3**										
Dengue 1	D 1										**6**									
Dengue 2	D 2										3	**5**	5							
Dengue 3	D 3												**4**	2						
Dengue 4	D 4												4	**3**						
Yellow fever	YF														**7**					
Bussuquara	BSQ															**5**				
Wesselsbron	WSL																**6**			
Ilheus	ILH																	**5**		
Montana ML	MML																		**5**	
Powassan	POW																			**6**

The number following each antiserum represents the immunization schedule used for that particular antiserum: 1, 1 dose of antigen intravenously; 3, 3 doses of antigen intravenously on days 1, 8, and 15; 5, 5 doses of antigen intravenously on days 1, 2, 3, 4, and 5; Ad., 1 dose of antigen intramuscularly with Freund's adjuvant. The numbers represent activities in neutralization as the reciprocals of antibody titers. Thus final serum dilutions of 1:20, 1:40, . . . , 1:1280 are shown as 1, 2, . . . , 7. Homologous reactions are in bold type. All cross reactions not numbered showed no neutralization at 1:20 serum dilution.

Source: Reproduced with permission from A. T. DeMadrid, and J. S. Porterfield. The flaviviruses (group B arboviruses): A cross-neutralization study. *J. Gen. Virol.* **23,** 91 (1974). Cambridge University Press, New York.

rated by isoelectric focusing has a pI of 7.8 and retained its antigenicity as measured by its reactivity in solid-phase radioimmunoassay (Trent *et al.*, 1976; Trent, 1977). The envelope glycoproteins of several flaviviruses have been isolated from infected cells after detergent solubilization and ion-exchange chromatography (Qureshi and Trent, 1973a,b,c). Concanavalin A affinity chromatography has also been used to isolate virus-specific dengue virus glycoproteins from infected cells solubilized with nonionic detergent (Stohlman *et al.*, 1976).

Early antigenic analysis of the virion antigens involved in the HI reaction indicated that there were three different types of antigenic determinants on the flavivirus particle (Clarke, 1960). These were described as (1) type-specific antigenic determinants which are unique for each virus, (2) complex-reactive antigenic determinants which are shared by closely related viruses, and (3) group-reactive antigenic determinants which are shared by all serologically related flaviviruses. The extensive cross reactions which occur in the HI test, due to common antigenic determinants in the envelope glycoprotein, are also seen in the neutralization test, although the latter is usually more virus specific (Westaway, 1965a,b, 1966; DeMadrid and Porterfield, 1969, 1974). On the basis of complement fixation (CF), immunodiffusion, and competition radioimmunoassay analysis, Trent and his associates determined that the envelope glycoprotein contains mostly type-specific antigenic determinants, fewer complex antigens, and only a few flavivirus group-reactive determinants (Qureshi and Trent, 1973a,b,c; Trent *et al.*, 1976; Trent, 1977). In another study, Eckles *et al.* (1975) showed by CF test that the JE virion glycoprotein was cross-reactive with other closely related flaviviruses, implying that it possessed both type-specific and group-reactive determinants. These observations are consistent with the antigenic-mosaic model of the flavivirus surface proposed by Della-Porta and Westaway (1977). On the basis of viral neutralization studies, these workers proposed that the flavivirus particle is neutralized by a multi-hit process which results in viral neutralization only after the "critical area" has reacted with antibody specific for that antigenic determinant (Westaway, 1965b; Della-Porta and Westaway, 1977).

The nucleocapsid proteins of SLE, JE, and dengue-2 viruses have been shown by competition radioimmunoassay to contain only group-reactive determinants (Trent, 1977). These findings are consistent with the alpha togaviruses, whose nucleocapsids also contain group-reactive determinants (Dalrymple *et al.*, 1973).

Two nonstructural flavivirus antigens have been isolated and characterized from infected cells and tissues (Russell *et al.*, 1970, Brandt *et al.*, 1970). The dengue virus nonstructural antigen has a molecular weight of 39,000 daltons and fixes complement in the presence of antibodies in the serum of experimentally infected animals or humans which have acquired natural infections (Brandt *et al.*, 1970; Smith *et al.*, 1970; Cardiff *et al.*, 1971). This antigen is not inactivated by SDS, urea, or 2-ME, which clearly differentiates it on the basis of its physical and chemical properties from the virion glycoprotein (Brandt *et al.*, 1970). This antigen, designated "SCF" (soluble complement fixing), is more type specific than the envelope protein, although the major immunological response in naturally infected humans is directed to the envelope protein (Falkner *et al.*, 1977). Eckles *et al.* (1975) reported on antigen with similar physical properties in mouse brain infected with JE virus. This protein had a molecular weight similar to that of the JE envelope protein of 53,000 daltons; however, its stability to inactivation with reducing agents and isoelectric point were very different. The significance of this protein in the replication of the virus and the role of antibodies to it are yet unknown.

Trent and his associates isolated and characterized the largest flaviviral nonstructural protein of approximately 100,000 molecular weight from cells infected with a variety of flaviviruses (Qureshi and Trent, 1973a,b,c). In contrast to the envelope glycoprotein, this viral-specific antigen was type specific. Antibodies to this nonstructural protein were found by CF and radioimmunoassay in sera from experimentally infected mice and humans after natural infection (Trent *et al.*, 1976). As with the SCF antigens of dengue and JE, the role of this protein in the biology of the flaviviruses is unknown, although Westaway (1977) has suggested that it may function in the viral RNA polymerase.

4. Biological and Antigenic Variation

Biological and antigenic variation among isolates of various flaviviruses is well known and has been documented for JE (Huang and Wong, 1963; T. Okuno *et al.*, 1968; Makino *et al.*, 1971), WN (Hammam *et al.*, 1965; Price and O'Leary, 1967; Gaidamovich *et al.*, 1973; Umrigar and Pavri, 1977); tick-borne encephalitis (Clarke, 1964), dengue (Cole and Wisseman, 1969), and SLE (Bowen *et al.*, 1980; Monath *et al.*, 1980; Trent *et al.*, 1980).

When strains of WN virus were examined in the kinetic HI test, they could be grouped into two antigenically related sets (Hammam *et al.*, 1965). One group contained viruses isolated from African–Middle Eastern areas, and the other isolates from India. In a previous study, Clarke (1960) used antibody adsorption techniques to show that yellow fever isolates from the Americas were distinct from African isolates. Although some antigenic variation among strains of SLE virus were suggested by Cassals (1963) and Sather and Hammon (1971), a more detailed study by the kinetic HI test failed to establish significant differences among the isolates (J. Casals, personal communication).

A detailed comparison of SLE isolates from the Americas has been reported which has facilitated separation of the strains according to virulence and biochemical markers. Although SLE is not an animal pathogen, studies on variation of this virus are revealing and suggest the need to define further differences

TABLE XIII
Comparison of St. Louis Encephalitis Virus Isolates by Animal Virulence, Oligonucleotide Fingerprint Similarity, and Geographical Distribution[a]

Geographical area/ epidemiology	Percent of strains with oligonucleotides similar to geographical prototype	Virulence		
		High	Intermediate	Low
Central and Atlantic states: epidemic	92	+		
Florida: epidemic	80	+		
Florida: enzootic	100			+
Western United States: endemic-epidemic	67		+	
Central and South America: enzootic	88	+	+	+
South America: enzootic				+

[a] Modified from Trent *et al.* (1981), with permission.

between geographic strains of other flaviviruses. Monath *et al.* (1980) found that SLE isolates from the Americas could be divided on the basis of their peripheral virulence for 3-week-old mice into three sets or groups—those of high virulence, intermediate virulence, and relatively low virulence for weanling mice. The biological virulence marker was also true for rhesus monkey neurovirulence and viremia in nestling house sparrows (Bowen *et al.*, 1980). These biological markers correlate with epidemiological characteristics of human case-fatality rates, attack rates, virus amplification, and immunity rates in wild avian hosts (Monath *et al.*, 1980). Isolates of SLE virus from the Mississippi–Ohio river basin are highly virulent, whereas most strains from the western United States, the Caribbean, and South America are of low or intermediate virulence. An analysis of the genome RNA of isolates with different virulence properties and from different geographic areas by the oligonucleotide map technique revealed a wide genetic difference among strains (Trent *et al.*, 1980). The genomes of SLE virus isolates from the Central and Atlantic states are quite similar to each other and distinct from those of strains from the far western United States and South America. A comparison of the virulence and oligonucleotide maps of isolates throughout the Americas has facilitated their classification into six sets (Trent *et al.*, 1981) (Table XIII). SLE strains within a given geographic area or ecosystem are more similar to each other than they are to viruses in other areas. The importance of these findings in the evolution of flavivirus disease potential among strains is not yet known.

5. *Host Range*

The laboratory host range of flaviviruses known to produce disease in domestic livestock is shown in Table VIII.

III. COMPARATIVE BIOLOGY AND PATHOGENESIS

A. Viruses Causing Encephalomyelitis

Inflammatory disease of the brain and spinal cord, generally also involving the meninges, is a final common expression of domestic animal infection with 12 arthropod-borne viruses (Table II). The clinical syndrome is characterized by varied and diffuse signs of CNS involvement, including altered states of consciousness (diminished alertness, lethargy, stupor, coma), convulsions, motor disturbances (paresis, paralysis), autonomic and extrapyramidal abnormalities, sensory deficits, abnormal reflexes, and cerebellar disturbances. Because the clinical features of many togaviral CNS infections are similar, they are never as important in the differential etiological diagnosis as the epidemiological features (geographic location, season) and specific laboratory tests.

The neuropathology of infection with the arthropod-borne togaviruses also demonstrates a commonality of features. All these viruses predominantly affect the gray matter, but brain damage involves not only nerve cells but also interstitial tissues and blood vessels. Neuronal degeneration and death, neuronophagia, patechiael hemorrhages, focal tissue necrosis, microglial proliferation and nodule formation, and round cell infiltration of the Virchow–Robin perivascular spaces are lesions found in all these diseases. Certain viruses also produce lesions in organs other than the neuraxis.

1. *Clinical and Laboratory Features*

a. Eastern, Western, and Venezuelan Equine Encephalitis Virus Infection in Equidae. Descriptions of the clinical disease acquired in nature are similar for all three infections. Fever, decreased locomotor activity, depression, and anorexia are common features which may be present without progression to encephalitic illness. Signs of CNS involvement include drooping of the head, chewing movements, incoordination, circling, a tendency to veer to the left or right, and excessive salivation. As the disease progresses over several days to a week, signs of increasingly severe brain dysfunction appear which include head pressing, inability to stand, flaccidity of the lips, apparent blindness, partially closed eyelids, drooping ears, movements of the legs, convulsions, and hyperexcitability. In the terminal stages, the animals lie on their sides and exhibit paddling movements (running movements while lying on the ground), nystagmus, difficult breathing, and coma (see Figs. 5 and 6). This clinical description was

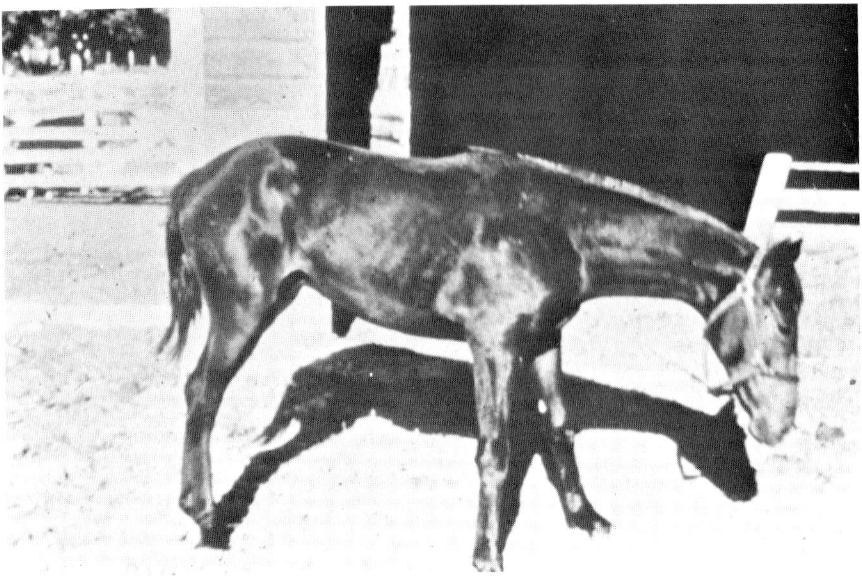

Fig. 5. Horse with western equine encephalomyelitis. Signs apparent in the photograph include drooping of the head, ataxia (wide-based stance), and partially closed eyelids.

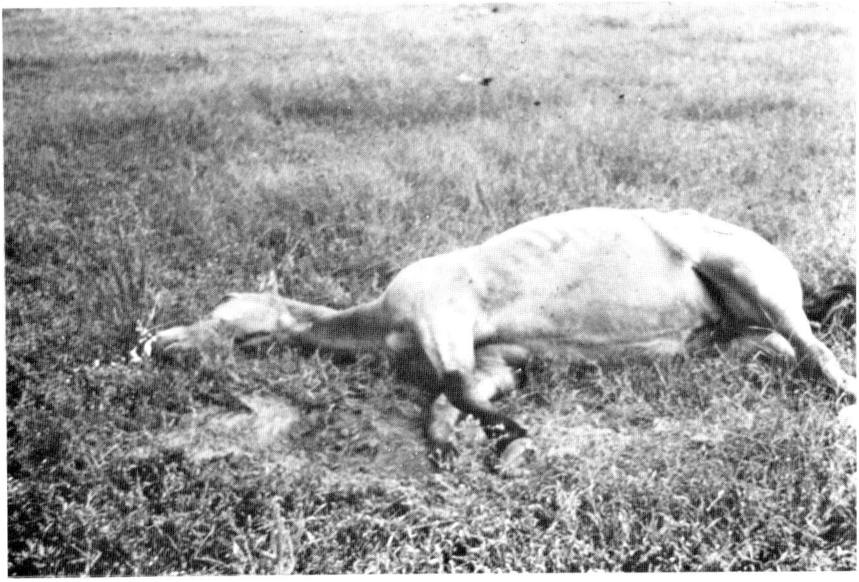

Fig. 6. Horse in terminal stages of Venezuelan equine encephalomyelitis. Note the ground worn bare by paddling movements.

noted as long ago as 1831 during an epizootic (presumably of EEE) in Massachusetts (Hanson, 1957) and has been confirmed in numerous subsequent reports (Giltner and Shahan, 1933; Doby *et al.*, 1966; Blood and Henderson, 1974).

The severity and lethality of encephalitis caused by EEE virus are greater than for VEE, and WEE virus is the least pathogenic of the three agents. In various epizootics, case fatality rates in Equidae have been reported as 75–98% for EEE, 32–86% for VEE, and 10–50% for WEE. These rates should be considered estimations only, because of the difficulty in obtaining accurate numerator and denominator data during outbreaks. Available information suggests that young horses (less than 1 year of age) may be somewhat more susceptible to lethal infection, but in general, age appears to be a relatively minor determinant of susceptibility in equines. In contrast, EEE, WEE, and VEE infections in man are characterized by lower inapparent:apparent infection ratios, higher mortality, and higher incidence of neurological sequelae in infants and children.

Experimental infection studies have led to a further definition of the clinical, viremic, and antibody responses of equines to EEE, WEE, and VEE viruses. Comparative descriptions suffer, however, from the lack of uniformity of experimental animals and methods used for the assay of virus.

In 1935, TenBroeck *et al.* showed transmission of EEE virus from an experimentally infected horse to a normal horse by *Aedes sollicitans* mosquitoes. Three days after exposure to the infected mosquitoes, the horse developed fever and peak viremia (approximately 3.5 log weanling mouse (WM) LD_{50}/ml). The horse was febrile and viremic for 2 days but failed to develop CNS signs. Giltner and Shahan (1936) also demonstrated brief EEE viremias in experimentally infected horses. In a more definitive study, Kissling *et al.* (1954) infected horses with a low-passage strain of EEE virus (originally isolated from *Culiseta melanura*) by both inoculation and mosquito bite. The lowest dose which produced illness and viremia detectable by intracerebral (ic) inoculation of weanling mice was approximately 2.5 log LD_{50}. All four animals which received $>$ 3.5 log LD_{50} died with encephalitis. Of two animals given 2.5 log LD_{50}, one developed encephalitis but recovered and the other had a brief, self-limited febrile illness without CNS signs. Animals which received 2.2 log LD_{50} were neither ill nor viremic but developed serum antibodies.

In horses given the larger virus doses, fever developed 40 hours postinoculation, reached a peak of 105°F, and returned to normal by 64 hours. Virus was present in the blood during this first febrile phase, after which neutralizing antibodies were detected. After an afebrile and nonviremic period of 3–7 days, there was a second rise in temperature without reappearance of circulating virus but accompanied by signs of CNS dysfunction. The duration of clinical encephalitis was 1–4 days before death. Virus was recoverable from the brains of horses dying 1–2 days after the appearance of CNS signs, but not from a horse dead after 4 days of encephalitis.

Viremia appeared at 24–40 hours and persisted for 1–3 days (mean 28 hours). Titers ranged from 2.0 to 6.7 log LD_{50}/ml, with a mean peak of 4.0. From these and other results, it was concluded that, under natural conditions, the horse would rarely serve as a source of infection for vector mosquitoes (Kissling *et al.*, 1954; Schaeffer and Arnold, 1954; Kissling, 1958a). Successful horse-mosquito-horse transmission has been achieved (TenBroeck and Merrill, 1935; Sudia *et al.*, 1956) but probably requires an unusually high viremia, present only in occasional animals.

These studies with EEE virus demonstrated several features of infection which may be considered generally applicable to the disease in horses: (1) The onset, intensity, and duration of viremia and the severity of disease are correlated with virus dose. (2) Three patterns of infection occur: (a) a biphasic course in which fever and viremia are followed by a period of remission and then return of fever and appearance of CNS signs; (b) abortive infection characterized a single temperature rise accompanied by viremia but without development of subsequent CNS signs; (c) inapparent infection without fever or detectable viremia. (3) Viremia is brief and generally of low titer. (4) Antibodies appear at about the time of disappearance of viremia, at a variable interval (often several days) before the second rise in temperature and onset of CNS signs. (5) Virus is not recoverable from brain tissue in some fatal cases (especially those with a duration of encephalitic illness of 3 days or more before death).

Byrne *et al.* (1964) experimentally infected burros with EEE virus. In these studies, an EEE viral strain recovered from an affected horse in Maryland was used. Of 19 animals inoculated, 18 developed a febrile response, 17 developed viremia, but only 2 had clinical encephalitis. Viremia titers and duration were similar to those reported by Kissling *et al.* (1954) in horses, but there was a less striking dose dependency.

Experimental studies on the response of horses to WEE virus have been reported by Kissling (1958a) and Hetrick (1960). The studies reported by Kissling were conducted by Stamm, Sudia, Chamberlain, and Kissling at the Center for Disease Control, Montgomery, Alabama, in 1958 but were not formally published. Two WEE virus strains, both from the eastern United States, were inoculated into horses; one of these (L2-34a) is now recognized to be a strain of HJ virus, antigenically related to but distinct from WEE virus. The other strain (936) was isolated from the brain of a sick horse in Tennessee; its antigenic relationships are not known. These virus strains produced no clinical signs of illness and only trace levels of detectable viremia.

In contrast to these results, Byrne *et al.* (1964) observed fever accompanied by anorexia and depression in all eight burros without preexisting heterologous EEE or VEE antibodies given varying doses of the BFS-1703 strain of WEE virus (from *Culex tarsalis,* California). Viremia was detected (by inoculation of wet chicks) in all burros but was of low titer (generally < 3 log LD_{50}/ml) and brief duration (mean: 2.7 days). No encephalitis developed, and the temporal relationships among fever, viremia, antibody, and CNS signs in WEE infection of equines thus remain unknown. The discrepancies between Byrne's study and the earlier reports of WEE infection of horses probably relate to the different viral strains used.

The clinical responses of horses and burros to VEE virus have been described by various investigators (Kissling *et al.,* 1956; Gleiser *et al.,* 1962; Gochenour *et al.,* 1962; Byrne *et al.,* 1964; Garman *et al.,* 1968; Henderson *et al.,* 1971; Walton *et al.,* 1973; Dietz *et al.,* 1978). In published studies of horses and burros inoculated with epizootic strains (subtypes IAB and IC), the frequency of encephalitic illness has varied from 25 to 100% and mortality from 25 to 50% (Table XIV). Infection is characterized by (1) fever without clinical signs of illness, (2) fever with anorexia and mild depression, or (3) fever with subsequent development of signs of CNS dysfunction. Fever, accompanied by depression and anorexia, generally appears earlier (by 24 hours) than in horses infected with EEE virus and is followed at 3–5 days postinoculation by CNS signs. A clear biphasic febrile course (as described for EEE-infected horses) has only rarely been a feature of VEE. The appearance of CNS signs coincides with the disappearance of circulating virus and the detection of serum antibodies. Deaths generally occur during the second week after inocualtion. Diarrhea has been described (Kissling *et al.,* 1956; Walton *et al.,* 1973) but is not a consistent feature of the disease. Loss of hair may occur in survivors (Walton *et al.,* 1973), probably as a result of high sustained fever.

In contrast to horses and burros infected experimentally with EEE and WEE viruses, those inoculated with epizootic VEE virus circulate virus in their blood for up to 6 days, generally at high titers (Table XIV). In 16 horses naturally infected with VEE IAB virus, viremia levels during the 48 hours after onset of illness ranged from 2.2 to 8.3 LD_{50}/ml, with a geometric mean of 6.1 (Calisher and Maness, 1975). Equines are thus effective viremic hosts for vector mosquitoes, and no difficulty has been experienced in infecting various mosquito species in laboratory experiments (Sudia *et al.,* 1971). Viremia levels of > 5.0 log LD_{50}/ml have been considered to represent a high potential for vector infection (Sudia and Newhouse, 1975); titers in equine sera often exceed this threshold. Epizootic but not enzootic strains of VEE virus have also been recovered from nasal, oral, and conjunctival washings as well as from milk, but not from urine or feces of experimentally infected horses.

Hematological abnormalities have also been reported in VEE-infected equines, including leukopenia and changes in the hematocrit and platelet count (Kissling *et al.,* 1956; Gleiser *et al.,* 1962; Walton *et al.,* 1973). Leukopenia consists of a fall in both granulocytic and lymphocytic elements, but the

TABLE XIV
Review of Results of Experimental Studies in Equines Infected with Venezuelan Equine Encephalomyelitis Virus Subtypes and Variants

				No. ill/no. inoc				Viremia			
Virus subtype (origin)	Epidemiologic type	Species	Route inoc.	Fever only	CNS signs	Death	No. viremic/ no. inoc.	Onset	Duration (days)	Range of max. titer $\log_{10} LD_{50}$/ml	Reference
IAB (Trinidad, 1943)	epizootic	horse	sc	2/5	3/5	2/5	5/5	10–40 hr	4–6	3.9–8.1	Kissling *et al.* (1956)
IAB (Trinidad, 1943)	epizootic	horse	sc	9/12	3/12	3/12	11/12	1–3 d	2–4	2.4–6.7	Dietz *et al.* (1978)
IAB (Honduras, 1969 and Colombia, 1967) IC (Venezuela, 1964)	epizootic epizootic epizootic	horse horse horse	sc	3/16	13/16	8/16	16/16	NS[a]	4–5.5	>8.2	Walton *et al.* (1973)
IAB (Guatemala, 1969)	epizootic	horse	sc	0/4	4/4	¼ + 1 sacrif	4/4	1 d	4–5	5.2–7.2	Henderson *et al.* (1971)
IAB (Trinidad, 1943)	epizootic	burro	im	1/3	2/3	1/3	3/3	1 d	3–5	2.5–7.6	Gleiser *et al.* (1962)
IAB (Trinidad, 1943)	epizootic	burro	im	1/3	2/3	1/3	3/3	1 d	3–5	2–8.1	Gochenour *et al.* (1962)
I (Colombia, 1957)	epizootic	burro	im	1/3	2/3	1/3	3/3	1 d	2	0–6.9	*Ibid.*
IE (various)	enzootic	horse	sc	N.S.	0/4	0/4	3/4	1–2 d	1–3.5	2.4–7.4	Dietz *et al.* (1978)
IE (Mexico, 1965)	enzootic	horse	sc	1/1	—	0/1	1/1	2 d	4	1.6	Garman *et al.* (1968)
IE (Nicaragua, 1967)	enzootic	horse	sc	8/8	—	0/8	8/8	N.S.	N.S.[b]	tr –2.5	Walton *et al.* (1973)
ID (Panama, 1963)	enzootic	horse	sc	±4/4	—	0/4	1/4	N.S.	2	2.6	*Ibid.*
II (Florida, 1963)	enzootic	horse	sc	1/6	0/6	0/6	2/6	2–4 d	1–3	2.2–3.9	Henderson *et al.* (1971)
III (Brazil, 1954)	enzootic	horse	im	1/1	—	0/1	1/1	1 d	3	3.2	Shope *et al.* (1964)
IV (Brazil, 1961)	enzootic	horse	im	0/1	—	0/1	0/1	—	—	—	*Ibid.*

[a] N.S., not specified. [b] Viremia persisted > 12 hr. in 6/8 horses.

granulocytopenia is more marked. The leukocyte count begins to fall within the first postinoculation day and reaches a nadir toward the end of the period of viremia. The appearance of immature polymorphonuclear leukocytes has been noted during the terminal stages of fatal cases and during the recovery phase in survivors. No consistent changes in monocytes, eosinophils, or basophils are reported. On days 2–5, marked decreases in the packed red cell volume occurred in horses which developed encephalitis after inoculation with VEE, IAB, and IC viruses. This was followed by hemoconcentration, which was especially marked before death. Abnormal clot retraction, decreased platelet counts, and prolonged prothrombin time have also been described.

In contrast to these findings, enzootic strains of VEE virus produce inapparent or abortive, nonfatal infections in equines (Table XIV). Fever, with or without anorexia and depression, is a feature of VEE subtypes IE and III in experimental infection, but it less frequently follows inoculation with subtypes ID and II. Viremia is generally of low titer and brief duration, and the potential for infection of vector mosquitoes is low. Hematological changes are absent or minor.

b. Eastern and Western Equine Encephalitis Virus Infection in Domestic Birds. The occurrence of EEE as an epornitic disease of pheasants and other penned exotic birds was first described by Tyzzer *et al.* (1938) during an extensive outbreak in New England. Many outbreaks have occurred since in the eastern United States, and the disease in game farms of ring-necked pheasants (*Phasianus colchicus torquatus*) and chukar partriges (*Alectoris graeca*) has been recognized as a major economic problem (Beaudette *et al.*, 1954; Wallis and Main, 1974). Natural infection and disease have also been reported in other species of veterinary importance, including domestic pigeons (Fothergill and Dingle, 1938), white Pekin ducklings (Dougherty and Price, 1960), and bronze turkeys (Spalatin *et al.*, 1961). Although a high prevalence of infection has been demonstrated in chickens by antibody surveys, they rarely show signs of illness. An encephalomyelitic disease of young chickens in New England, described as "epidemic tremor" by Jones (1934), was never etiologically defined but may have been due to EEE virus (Tyzzer and Sellards, 1941). In experimental studies, young chickens (Tyzzer and Sellards, 1941) and bobwhite quail (Tyzzer *et al.*, 1938; Tyzzer and Sellards, 1941) have been shown to develop illness and lethal infection with EEE virus. A variety of wild avian species are also susceptible to infection but rarely demonstrate signs of illness.

Natural infections with WEE virus of many avian species, including domestic chickens, are common, but overt disease is rare. WEE virus was suspected on serological grounds as the cause of an encephalomyelitis outbreak in turkeys in Nebraska (Woodring, 1957). An epornitic of encephalitis in chukar partridges in Florida was attributed to WEE virus, and experimental subcutaneous inoculation of young chukars with WEE virus produced a disease characterized bz somnolence, ruffled feathers, prostration, and death (Ranck *et al.*, 1965).

The clinical features of EEE in pheasants include signs of systemic illness (droopiness, diarrhea, inactivity, ruffled feathers) and signs of neurological dysfunction (ataxia, paralysis, etc.) (Tyzzer *et al.*, 1938; Beaudette *et al.*, 1952). The incubation period (after experimental infection) is between 1 and 7 days, depending upon the dose and route of inoculation (Jungherr *et al.*, 1958). Illness has been noted in 40–100% of experimentally infected birds, with a mortality of 25–100% (Kissling, 1958b; Satriano *et al.*, 1958; Hanson *et al.*, 1968). Mortalities of up to 80% are recorded in natural epornitics (Sussman *et al.*, 1957–1958). Differences in mortality in experimental studies are explained by variations in the viral strains, dosages, routes of inoculation, and age of the host. Susceptibility to EEE decreases rapidly with age. In one study (Hanson *et al.*, 1968) infection resulted in 100% mortality in chicks 1–3 days of age, 50% mortality in chicks 21 days of age, and < 10% mortality in chicks 28 days old. In pheasants exposed by contact (the usual mode of infection; see Section IV,A,3) the mean incubation period is 4.5 days (Jungherr *et al.*, 1958). The relative incidence of clinical signs in naturally and experimentally infected pheasants is presented in Table XV. Death occurs within 2–4 days of onset; occasional recovery from paralytic illness is described.

Experimental studies have defined the course of EEE viremia and virus excretion in pheasants (Holden, 1955; Satriano *et al.*, 1958; Hanson *et al.*, 1968).

TABLE XV
Relative Incidence of Clinical Signs in Ring-Necked Pheasants Naturally or Experimentally Infected with EEE Virus

Sign	Percent with sign	
	Natural infection	Experimental infection
Systemic		
Droopiness	38	56
Ruffled feathers	0	32
Soiled vent region	4	7
Gasping	2	0
Neurological		
Incordination	25	26
Paresis		
Unilateral	2	6
Bilateral	6	11
Paralysis	25	12
Coarse tremor of head	4	0
Torticollis	8	1
Circling	2	0
Retropulsion	4	1

When the virus is inoculated by the subcutaneous, intramuscular, or intravenous routes, viremia is detected within 24 hours, lasting up to 72 hours, but titers are generally quite low, and pheasants are probably not an important source of vector infection. Oral infection also produces viremic infection and clinical signs. Virus can be obtained from oral secretions, fecal material, and feather quills. Direct pheasant–pheasant transmission of EEE by feather picking and cannibalism has been repeatedly shown in experimental studies and represents the most important mode of spread in penned birds.

The disease in chukar partridge chicks was characterized by the appearance of anorexia, listlessness, and ataxia, after which the chicks became prostrate. The mortality in an affected flock was 40%. Infection of Pekin ducklings 1–2 weeks of age resulted in sudden onset of illness; posterior paresis followed by paralysis were the most consistent signs. Mortality was estimated at 2–60% in naturally affected flocks, and the virus was apparently spread by contact (Dougherty and Price, 1960).

Outbreaks of EEE in Wisconsin turkey flocks have resulted in low mortality (approximately 5%) despite high rates of infection as indicated by serological tests. The clinical illness is characterized by muscular incoordination, tremors, and paralysis of legs or wings (Spalatin *et al.*, 1961). Experimentally inoculated turkeys were shown to be quite resistant. Clinical encephalitis developed in only 3 of 28 birds, and viremia titers on days 1–4 postinoculation were low ($\leq$ 1.5 log chicken embryo LD_{50}/ml). Young chicks (1–2 days of age) were more susceptible than poults 6–8 weeks old.

Newly hatched chickens are highly susceptible to infection with EEE and WEE viruses and were used formerly for the primary isolation and assay of these agents. By 4 days of age, some resistance to EEE virus lethal infection is evident, and susceptibility declines thereafter. Chicks 12 days of age or more are refractory (Byrne and Robbins, 1961). The clinical illness in young chicks (Fig. 7) is characterized by inactivity, weakness, ruffled feathers, stupor, coma, and death, generally without obvious paralysis (Tyzzer and Sellards, 1941). The incubation period is 2–7 days depending on the age of the chick. Two- to 4-week-old chickens inoculated with 30–1000 $TCID_{50}$ developed viremias of 3.7–8.1 $TCID_{50}$/ml lasting 1–2 days (Hayes *et al.*, 1964). Young (Hammon and Reeves, 1946) and adult chickens (LaMotte *et al.*, 1967) also develop brief viremias following inoculation with WEE virus. On the basis of the viremia data for EEE and WEE and the low prevalence of EEE antibodies in domestic poultry, chickens are probably not especially important in the natural transmission cycle of these viruses (Kissling, 1958a).

c. Infection of Other Domestic Animals (Cows, Pigs, Dogs) with EEE, WEE, and VEE Viruses. Antibodies to EEE virus in 18–26% of feral and domestic swine have been reported from Georgia (Karstad and Hanson, 1958), Massachusetts (Feemster *et al.*, 1958), and Wisconsin (Karstad and Hanson,

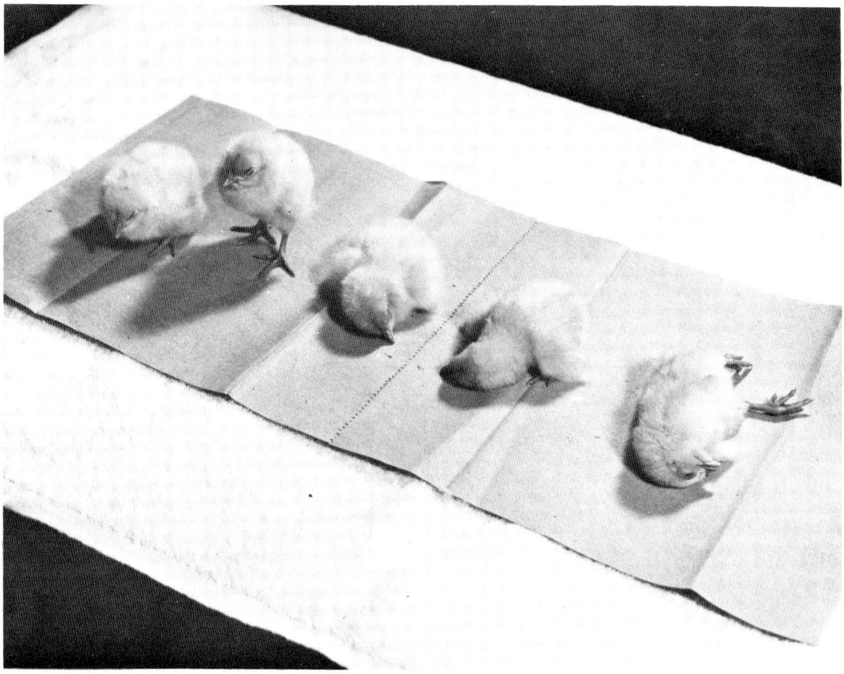

Fig. 7. Newly hatched chicks inoculated with eastern equine encephalomyelitis virus, demonstrating the progression of illness from inactivity (left) to coma to death (right).

1959), but clinically apparent infection is unusual. A single report of EEE virus isolation from the brain of a naturally infected 3-week-old pig with signs of CNS disturbance (Pursell *et al.*, 1972) indicates the potential pathogenicity of EEE virus in swine. Histopathological evidence suggested that other pigs in the same herd were similarly affected.

In an early study (Karstad and Hanson, 1959) with two 3-month-old pigs experimentally infected with varying doses by the intranasal, intradermal, intravenous, and intracerebral routes, no clinical illness or viremia was observed, but all animals developed antibody. Minor histopatholigical changes of encephalitis were found in one of three peripherally inoculated animals examined. In a later investigation (Pursell *et al.*, 1972) using nursing pigs and original or first-passage virus from a pig with naturally acquired, clinical EEE, four of four animals inoculated ic but none of five peripherally inoculated pigs developed overt encephalitis and evidence for virus replication in brain tissues. No viremia was detected in any animal. It is concluded that pigs are not a link in the natural transmission cycle of EEE virus and that natural infections are nearly always subclinical. When the blood–brain barrier is breached (by experimental ic infection or in rare instances after natural infection of young animals), pigs are susceptible to clinical encephalitis.

WEE virus has been isolated from the liver and spleen of a hog also infected with hog cholera virus (McNutt and Packer, 1943). Clinical illness in experimentally infected pigs occurred only in 3-day-old animals; older swine developed antibodies.

Antibodies to VEE virus have been found in a high proportion of pigs surveyed in tropical America (Sudia and Newhouse, 1975). Intramuscular infection of 3.5- and 8-month-old Duroc pigs with VEE IAB (Trinidad) virus (Hogge, 1964) resulted in a brief febrile response in some animals without signs of illness, low-titer viremias (maximal titer 2.9 log LD_{50}/ml) lasting 1–4 days, and minimal or absent virus in brain tissues. Inoculation of pregnant sows did not result in abortion or transplacental infection of the fetal brain. Virus was recoverable from the milk of lactating swine and resulted in subclinical infection of nursing pigs. In a later study of pigs less than 6 months old subcutaneously inoculated with two strains of VEE IAB virus (isolated in Guatemala, 1969), there were no signs of illness, but all pigs developed a brief (2–3 day) viremia (mean maximal titer of 2.4 $\log_{10}$ chick embryo cell plaque-forming units per milliliter) (Dickerman *et al.*, 1973). The viremia titers observed are considered too low to serve as a source of infection for most important vectors of VEE virus.

Relatively little is known about the course of EEE and WEE infection in bovines. Young calves inoculated with EEE virus by the ic route developed fatal encephalitis (Helmboldt *et al.*, 1955). High prevalences of virus antibodies to VEE have been found in cows in parts of tropical America, but there is a question about the specificity of the viral inhibitors found in bovine sera (Scherer *et al.*, 1972). About 50% of cows inoculated with various strains of VEE virus develop brief viremias accompanied by neutropenia and a transient febrile response but no symptoms (Walton and Johnson, 1972a; Dickerman *et al.*, 1973). Pharyngeal excretion of VEE IE virus in inoculated cows was also reported. Viremia levels were generally too low to be effective for mosquito infection, but occasional animals had titers of 5.5 log LD_{50}/ml. Experimentally inoculated cows developed specific antibodies.

The possibility that dogs are involved in the transmission cycle of VEE has been investigated by Taber *et al.* (1965) and Davis *et al.* (1966). In various published reports, the prevalence of natural VEE immunity in dogs in tropical America is 7–70% (Sudia and Newhouse, 1975). Susceptible purebred beagles were infected with VEE IAB (Trinidad) virus by Taber *et al.* (1965). Fever and viremia were found within 24 hours. Three dogs exhibited atypical clinical signs (increased aggressiveness); none had encephalitic signs, but two died (58 and 70 hours postinoculation). Results obtained by Davis *et al.* (1966) were generally comparable; all beagles inoculated with VEE IAB (Trinidad) virus became febrile and 60% died. The high mortality may have been due to use of a general anesthetic in these experiments. Viremias of 1–4 days' duration ranged from 1.5–6.2 $\log_{10}$WM $IPLD_{50}$/ml, with a mean maximal titer of 4.3. Infection of *Aedes triseriatus* (a highly susceptible vector) was documented after feeding upon viremic dogs.

d. Japanese Encephalitis Virus Infection of Equidae. The infection in equines is often subclinical (Kii *et al.*, 1939). Abortive infection with fever, lethargy, and mild systemic symptoms lasting 5–9 days also has been described (Witherington, 1953). The frequency with which infection results in encephalitis and death of horses is not well defined, but localized encephalitis outbreaks and widespread epizootics have been well-documented events. During epizootics, approximately 30–40% of clinically affected horses have died. The clinical disease is characterized by onset of fever and anorexia, followed in 2–4 days by depression, photophobia, muscular tremor, incoordination and ataxia, falling, circling, paresis, paralysis, increased respiratory rate, hyperexcitability, congestion of mucous membranes, stupor, and death.

Experimental studies have confirmed the susceptibility of equines to infection (Kii *et al.*, 1937). Successful transmission from chick to horse, from horse to horse, and from horse to chick was accomplished by *Culex tritaeniorhynchus* mosquitoes (Gould *et al.*, 1964). Of three horses exposed to infected mosquitoes, all became infected (as determined by viremia and antibody), but only one developed fever and CNS signs (10 days postexposure and 4 days after cessation of viremia). Viremia determined by suckling mouse inoculation was of low magnitude (ranging from less than 2.0 to 2.5 $\log_{10}$ LD_{50}/ml with a peak 5 days after exposure.

e. Japanese Encephalitis Virus Infection of Swine. Transplacental JE virus infection of swine fetuses resulting in invasion of the brain and fatal encephalitis is an important cause of stillbirth of pigs in parts of Asia. The disease is considered in detail later in this chapter (Section III, B).

f. Louping Ill of Sheep, Cattle, Horses, and Swine. Louping ill is an important cause of encephalitis in sheep on hill farms of Scotland and Northern Ireland; the incidence of fatal encephalitis in sheep known to have naturally acquired viremic infection may exceed 50% (Brotherson *et al.*, 1971). The incubation period under natural conditions varies from 6 to 18 days. Early features of the acute disease are high fever and listlessness lasting 1–2 days, during which virus may be recovered from the blood. This stage is often followed by a period of remission. In abortive infections the animal then makes a complete recovery, but in others fever returns on the fourth or fifth day, accompanying signs of a CNS disorder characterized by incoordination, tremors, cerebellar ataxia (causing the staggering gait responsible for the name of the disease), paresis, and paralysis. The clinical encephalitic process may be rapidly fatal (within several days) or persist for several weeks in more chronic cases (Gordon *et al.*, 1962). The disease may also affect cows, pigs, horses (Brownlee and Wilson, 1932; Smith, 1962; Timoney *et al.*, 1976), and captive (and wild) red grouse (*Lagopus lagopus scoticus*) (Timoney, 1972). A high proportion (approximately 67%) of experimentally infected lambs and sheep develop viremia and fatal neurological illness 7–12 days after inoculation (Reid and Doherty, 1971a,b).

g. Israel Turkey Meningoencephalitis. Outbreaks on turkey farms in Israel in the late 1950s affected birds of both sexes and all breeds. Clinical manifestations included apathy, staggering gait, drooping wings, and diarrhea, with progression to paralysis of wings and legs and death (Komarov and Kalmar, 1960). Experimentally infected 5- and 6-week-old turkeys manifested a similar syndrome 7–11 days after inoculation. No natural disease has been noted in chickens and ducks, and chickens over 3 weeks of age, ducks, and pigeons are resistant to intracerebral infection. Wet chicks are only partially susceptible by the ic and peripheral routes.

h. Other Arboviruses. WN virus has been isolated occasionally from horses with encephalitis in Africa (Schmidt and El Mansoury, 1963) and Europe (Panthier *et al.*, 1966). The prevalence of antibodies among equines is high in the Nile Delta (and other enzootic areas), but overt disease has been very rarely recognized, probably not only as a result of the high level of natural immunity but also because the virus has a low virulence for equines.

Schmidt and El Mansoury (1963) described the disease in a 12-year-old horse in Egypt. The illness was characterized by colic, hematuria, and urinary retention, followed by ataxia, progressive paresis of the hind quarters, inability to stand, and death 60 hours after onset. WN virus was recovered from brain tissue.

Experimentally infected donkeys and horses showed a high degree of resistance to WN virus (Schmidt and El Mansoury, 1963). Viremia, detected in two of six donkeys and none of three horses, was minimal (less than 1.5 $\log_{10} LD_{50}$/ml), with onset on days 4 and 6 and a duration of one day. Low-grade fevers were present in viremic donkeys, but there were no other signs of illness.

MVE and Ross River (RR) viruses have been suspected to cause nervous system disease in equines in Australia. Only serological evidence exists, and neither virus has been isolated from sick equines (Gard *et al.*, 1977). Antibodies to both viruses are found in sera taken during surveys of equine populations, and most infections are apparently subclinical. The etiological relationship to equine CNS disease thus remains uncertain and is further complicated by the existence of flaviviruses other than MVE and alphaviruses other than RR in Australia. In addition to the encephalitic syndromes, serological evidence suggests that RR virus might be responsible for a muscle disease in horses characterized by stiffness, muscle and joint swelling, and muscle tenderness. This presentation is a logical one, since RR virus produces joint and muscle disease in mice and humans.

RR virus has been inoculated subcutaneously (sc) into sheep and pigs. No disease was noted, and high viremias were present only in young lambs (Spradbrow, 1973).

MVE virus produces equine encephalitis following ic inoculation, but peripherally challenged horses remained clinically normal in one series of experiments (Gard *et al.*, 1977).

SF virus has been implicated on serological grounds alone in an outbreak of equine encephalomyelitis in Senegal (Robin *et al.*, 1974). The etiology of the disease in horses was not definitely established, and further observations will be required to assess the equine virulence of this alphavirus.

There are isolated, unpublished reports of the recovery of Una and Aura viruses from horses with possible encephalitis in Argentina (M. Sabattini, personal communication).

2. *Pathology*

a. Eastern, Western, and Venezuelan Equine Encephalitides. Invasion of the CNS by these viruses is by way of the bloodstream; this accounts for the widespread distribution of histopathological lesions and the appearance of vascular damage and inflammation. In general, the CNS reactions produced by EEE, WEE, and VEE viruses are similar. However, differences in the intensity and character of lesions have been described which may relate to differences in the virulence of the infecting virus strain or in the duration of neurological illness prior to death or sacrifice. No comprehensive comparative pathological study has been made of the three diseases; consequently, it is difficult to present a clear picture of the neuropathological features which distinguish each disease.

The basic gross and microscopic components of the encephalitis produced by EEE, WEE, and VEE viruses are (1) gross congestion and edema of the cerebrum; (2) perivascular reaction composed of lymphocytes and neutrophils in varying proportions; (3) petechial and ring hemorrhages; (4) neuronal degeneration, necrosis, neuronophagia, and inflammatory infiltration of perineuronal spaces; (5) nodular formation and diffuse infiltration of microglia; (6) meningeal infiltration with inflammatory cells. The gray matter is primarily affected, and damage is generally most severe in the cerebral cortex. These features are described in varying degrees in all natural and experimentally induced infections.

EEE is often characterized by a prominent polymorphonuclear reaction, but this is probably a reflection of the fulminating course of the disease and the brevity of the neurological inflammatory process. In horses surviving 1 day after the onset of CNS symptoms, there was a diffuse infiltration of the gray matter with polymorphonuclear leukocytes and perivascular infiltration of polymorphonuclear and lymphocytic cells in approximately equal numbers (Hurst, 1934; Kissling and Rubin, 1951). In these hyperacute cases, endothelial cell swelling, thrombus formation, and perivascular hemorrhages were prominent changes. In the brains of animals surviving for longer periods of time, the perivascular reaction was predominantly lymphocytic, and the diffuse polymorphonuclear infiltration was replaced by diffuse and nodular gliosis. Necrosis and fragmentation of nerve cells, neuronal vacuolation and degeneration, and chomatolysis have been described (Miller *et al.*, 1973), but these changes may be minimal, especially in the hyperacute cases. Damage involves all areas of the brain but is most pronounced in the central cortex, hippocampus, hypothalamus, and dorsal nuclei of the medulla. No consistent changes are noted in extraneural organs.

The neuropathological alterations in WEE (Fig. 8) are generally less severe; the inflammatory reaction is primarily lymphocytic and more focal in nature (Miller *et al.*, 1973). Perivascular cuffing and hemorrhages, gliosis, and neuronal changes are present (Haring *et al.*, 1931; Larsell *et al.*, 1934) but are usually less prominant than in EEE. No consistent pathological changes in extraneural organs are described.

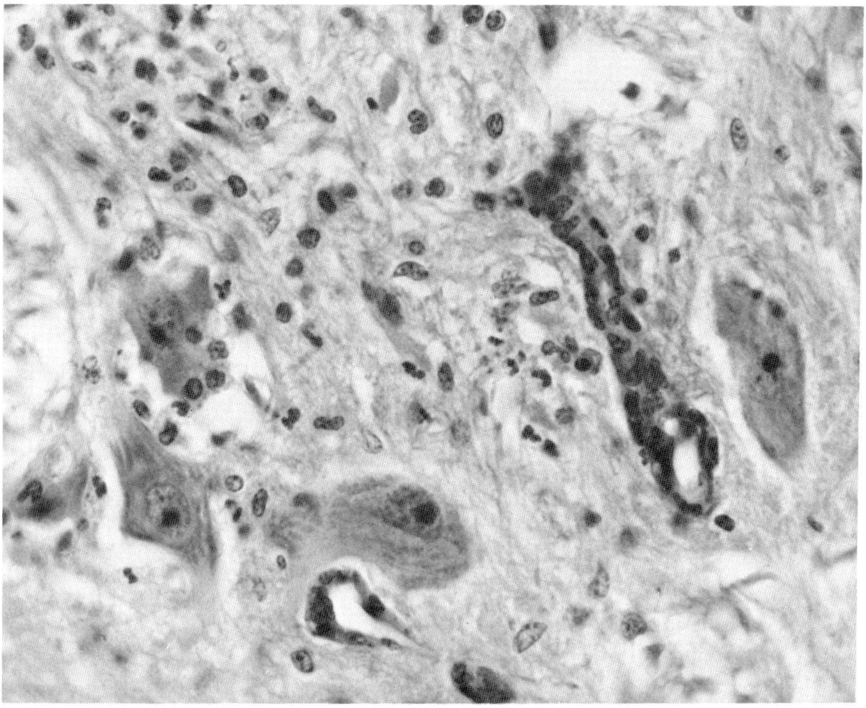

Fig. 8. Cerebellar cortex of a pony inoculated intracerebrally with western equine encephalitis virus and killed with terminal encephalitis, showing necrosis, neuronal degeneration, and perivascular and perineuronal inflammatory reaction; hematoxylin and eosin. × 400. Courtesy of Dr. N. F. Cheville.

Descriptions of the histopathology of VEE have been more variable. In a study of 15 horses which died during epizootics in Colombia, the inflammatory changes were generally similar to those reported for EEE (Roberts *et al.*, 1970). Perivascular infiltrates were most intense in the cerebral cortex and were composed of lymphocytes or mixtures of neutrophils and lymphocytes. Endothelial hyperplasia and thrombosis of small arteries and vascular necrosis were prominent features, and lysis of perivascular inflammatory cells was seen in the more severe cases. Diffuse perineuronal infiltration of neutrophils and astrocytes, diffuse gliosis, satellitosis, and neuronal necrosis were present. Similar lesions, but less intense and characterized by a purer lymphocytic inflammatory reaction, were noted in the caudate, nuclear-thalamic, and pontine-medullary zones and in the cerebellum. Extensive necrosis of Purkinje cells was described. Hemorrhages were not massive and did not obliterate normal histological structures. Similar pathological features have been described in burros experimentally infected with VEE IAB (Trinidad, 1943) virus (Gleiser *et al.*, 1962) and horses inoculated with an epizootic strain from Honduras (Dietz *et al.*, 1978). However, Monlux and Luedke (1973) found extensive liquefactive necrosis, hemorrhage, and demyelination of the cerebral cortex, the claustrum, and especially the lobus orbitalis of horses experimentally infected with VEE IAB strains (Trinidad, 1943, and Texas, 1971), and proposed that these changes distinguished VEE from the other encephalitides. Vasculitis and vascular necrosis were not prominent in their study.

Unlike EEE and WEE viruses, VEE virus causes lesions in extraneural tissues of donkeys and horses with fatal infections. The principal damage is to hematopoietic and lymphoid structures, with depletion of myeloid, erythroid, and megakaryocytic elements of the bone marrow, and necrosis, hemorrhage, and neutrophilic infiltration of lymphatic tissue to the extent that the normal architecture is destroyed (Gleiser *et al.*, 1962; Gochenour *et al.*, 1962). Other lesions include focal degeneration and necrosis of the hepatic parenchyma, myocardial necrosis, and necrosis of the proximal convoluted tubules of the kidney. In Kissling and colleagues' study of horses (1956), necrosis of the acinar cells of the pancreas was described, but Gleiser *et al.* (1962) and Gochenour *et al.* (1962) found no pancreatic changes. In contrast to these studies, which employed the 1943 Trinidad strain of VEE virus, a later study utilizing several isolated epizootic strains showed no bone marrow changes and only minimal lymphoid depletion of germinal centers (Johnson and Martin, 1974). Pancreatic changes were noted in only one of seven horses.

The alterations of lymphatic, hematopoietic, and pancreatic tissues and of the myocardium probably reflect the true tropisms and pathogenic potential of VEE virus, since experimental studies of small laboratory animals have shown similar results (Section III,A,3). The variable extent of damage to extraneural tissues in the equine studies may be due to differences in host susceptibility, infecting viral strain, or duration of illness.

Recovery of virus from tissues of equines dying of EEE, WEE, and VEE has been inconsistent. In general, the longer the interval between cessation of viremia or onset of CNS signs and death, the less likely that virus will be recovered from brain or other tissues. EEE virus was isolated from four horses dying 1-2 days after onset of CNS signs, but not from a horse dying 4 days after onset (Kissling *et al.*, 1954). The highest titers of virus were found in the basal ganglia and thalamus. VEE virus has been recovered from the brains of approximately half of the animals studied, and tissue titers have been low. In addition to the brain, visceral organs have occasionally yielded virus.

b. Eastern and Western Equine Encephalitis of Domestic Birds. The pathology of EEE in pheasants has been described by Tyzzer *et al.* (1938) and Jungherr *et al.* (1958). The microscopic changes consist of vasculitis, patchy necrosis, gliosis, neuronal degeneration, and meningeal inflammation. Endothelial hyperplasia and hypertrophy of capillaries and arterioles were the most consistent findings. Perivascular infiltrates were not prominent and, when present, the cells were invariably mononuclear; these findings were dissimilar to the prominent perivascular cuffing and polymorphonuclear responses in equines. Widespread microgliosis was present in the form of focal nodules and diffuse infiltration, primarily in the vicinity of affected vessels. Patchy necrosis of ground substance in association with vascular lesions indicated damage due to microinfarction. Neuronal degeneration and necrosis, neuronophagia, and satellitosis, though present, were not prominent. The rostral portions of the brain were most severely affected, and the spinal cord showed few changes.

On the basis of their analysis, Jungherr *et al.* (1958) pointed out that the neuropathological changes of EEE in pheasants more closely resembled those of WEE in horses and man than of EEE in these mammalian hosts.

The microscopic lesions noted in ducklings with EEE infection consisted of edema of the white matter of the spinal cord, patchy lymphocytic meningitis, and mild microgliosis (Dougherty and Price, 1960). Perivascular infiltration, endothelial changes, and necrosis were not described.

The brains of turkeys naturally and experimentally infected with EEE virus showed calcification of blood vessel walls and of cells and cell processes, chiefly of the outer layers of cerebral cortex, extremities of the cerebellar folia, and the basal part of the medulla (Spalatin *et al.*, 1961). Inflammatory changes were observed in birds inoculated ic with EEE virus and consisted of perivascular infiltration of lymphocytes, gliosis, neuronal degeneration, and endothelial cell swelling; birds dying before the sixth day postinoculation failed to show mineralization changes.

Young chickens were found to present still another histopathological picture

(Tyzzer and Sellards, 1941). The principal disturbance, and the presumptive cause of death, was marked dilation of the heart with severe diffuse degeneration and fragmentation of myocardial cells, frank necrosis, and infiltration of primitive mesenchymal cells, lymphocytes, and neutrophils. Other extraneural tissues—liver, spleen, and lymph nodes—showed inconsistent changes. Neuropathological lesions were found in less than half the chicks and were generally unimpressive.

It appears that the histopathology produced by EEE virus in chukar partridges has features in common with that in chickens. Although classic brain lesions (gliosis, sattelitosis, and lymphocytic perivascular infiltration) are present, myocardial necrosis with lymphocytic and monocytic inflammation is a prominent feature (Ranck *et al.*, 1965).

c. Japanese Encephalitis in Equidae. The pathological features of JE in equines are reviewed by Miyake (1964). The results of studies on experimentally infected horses are confused somewhat by the administration of vaccine prior to viral challenge, a procedure which may enhance exudative responses. Lesions in the brains of control (unvaccinated) horses consisted of meningeal and perivascular lymphocytic inflammation, nerve cell degeneration, neuronophagia, hemorrhages, and glial proliferation, especially around blood vessels (Sugawa *et al.*, 1949; Hale and Witherington, 1953; Miyake, 1964). In naturally infected equines, the distribution of lesions was prominent in the corpus striatum and thalamus rather than being widely dispersed throughout the cerebrum and spinal cord.

A variety of pathological changes in extraneural tissues have been described. These consist of hyperplasia of the germinal centers of lymph nodes, enlargement of splenic Malpighian bodies, interstitial myocarditis, swelling and karyorrhexis of hepatic Kupffer cells, and pulmonary interalveolitis.

d. Japanese Encephalitis Virus Infection of Swine. See Section III, B, below.

e. Louping Ill of Sheep, Cattle, Horses, and Swine. The pathology in sheep has been described by Brownlee and Wilson (1932) and reviewed by Innes and Saunders (1962). The disease consists of an acute encephalomyelitis and leptomeningitis, with a prominence of lesions in the cerebellar cortex (Doherty and Reid, 1971). The pathological features include chromatolysis and destruction of the Purkinje and Golgi cells, reactive gliosis and astrocytosis, and the typical appearance of perivascular cuffing by lymphocytes and plasmacytes, with occasional phagocytes and polymorphonuclear cells (Doherty *et al.*, 1971). Lesions may also be found in the brain stem, spinal cord, and cerebral cortex but are never as frequent and severe as in the cerebellum, explaining the clinical predominance of ataxic gait and disturbed balance.

The neuropathological features of louping ill in the red grouse (Buxton and Reid, 1975) differ in the more rostral distribution of lesions, the less prominent neuronal degeneration, and the nature of the inflammatory cell infiltrate, and resemble more closely the changes produced by EEE virus in pheasants.

f. Israel Turkey Meningoencephalitis. The histopathology in turkeys and experimentally infected chicks is characterized by lymphocytic meningoencephalitis (Komarov and Kalman, 1960). Detailed descriptions are not available.

g. Other Arboviruses. Histopathological changes noted in the cerebellum of a single horse with a natural WN virus infection consisted of petechial hemorrhages in the granular and molecular layers, Purkinje cell degeneration, neuronophapia, and glial proliferation without perivascular cuffing (Schmidt and El Mansoury, 1963). Other tissues were not examined.

No descriptions are extant of the pathological features of the other exotic arboviruses causing occasional encephalitis in domestic animals.

3. Pathogenesis

Studies of the pathogenesis of the neurotropic alpha- and flaviviruses have relied extensively on the use of laboratory animal models, especially mice and hamsters. The outcome of infection in laboratory rodents, as well as in larger experimental or natural hosts, is influenced by a variety of important factors, including dose and route of infection, species and age of host, and the inherent virulence of the infecting viral strain. Other host factors that affect the pathogenetic process, but are less well studied, include body temperature, reproductive and nutritional status, stress (plasma corticosteroid level), heavy metal poisoning, and genetic influences. The last are most clearly understood in mouse models and will be discussed in more detail below. Host defenses which determine the outcome of primary infection in an individual host include the humoral and cellular immune responses, interferon, and reticuloendothelial clearance mechanisms. Concurrent infections with an unrelated microbial agent may also determine pathogenic events. Infections with herpes simplex virus and with *Toxocara canis* enhance JE viral infection in mice, presumably by compromising the blood–brain barrier.

High dose and intracranial route of infection both predispose to lethal encephalitis in the host. These artificial laboratory maneuvers have little relevance for the pathogenesis of natural infections, which result from relatively low-dose inoculation by arthropod vectors (in the range of 10^3 LD_{50}) by the peripheral (intradermal or intravascular) route.

The immature brain is generally most susceptible to infection and evolution of a lethal encephalitic process. The mechanisms underlying age-dependent resistance are not entirely understood. Rates of viral replication and brain virus titers are higher in neonatal than in older animals, in which a smaller proportion of the neuronal cell population becomes infected (El Dadah and Nathanson, 1967). Immune responses have not been implicated as primary factors underlying age differences in togaviral replication. No age-specific differences were noted in antibody responses of mice to EEE virus (Morgan, 1941) or in humoral and cellular immune responses to SIN virus infection (Griffin, 1976). In mice infected with SIN virus, antibodies were not detected in weanlings until 3 days after infection, but viral brain titers in newborn animals exceeded those in weanlings as early as 24 hours (Hackbarth *et al.*, 1973). Moreover, cyclophosphamide-induced immunosuppression does not alter the slow rate of viral replication in West Nile virus-infected adult rats (Nathanson and Cole, 1971). Similarly, interferon has generally been discounted as a factor since many studies have shown that togavirus and interferon levels are directly correlated, with the highest interferon production in newborn animals having peak brain viral titers. *In vitro* experiments (Pattyn *et al.*, 1975; Fleming, 1977) have shown lower alphavirus susceptibility of tissues from adult than from neonatal animals. The available data thus suggest that cellular permissiveness to viral replication decreases with age and that the slower progression of replication allows intervention of the immune response before clinical signs become apparent.

Differences in virulence between viral strains are extremely important pathogenetic variables. Classic examples are the enzootic strains of subtype I VEE virus (variants D and E), which infect but do not produce disease in equines, whereas variants IAB and IC are equine virulent. Correlates to virulence of VEE viruses in natural hosts exist in laboratory rodent models, including mice (Calisher and Maness, 1974), hamsters (Austin and Scherer, 1971), and guinea pigs (Scherer and Chin, 1977), the last being most closely associated with equine pathogenicity. Differences between strains of the other viruses under consideration have been less intensively investigated. Virus strains (HJ) closely related to WEE virus isolated from *Culiseta melanura* and other sources in the eastern United States have reduced equine and human virulence compared to WEE viral strains from the West; a number of the strains from the eastern United States are also nonpathogenic for suckling mice. In addition, certain wintertime WEE virus isolates from *Culex tarsalis* had characteristics of attenuation for laboratory mice (Reeves *et al.*, 1958). No studies have been directed toward elucidating differences between strains of EEE virus or between the North and South American subtypes of the virus. Hanson *et al.* (1968) suggested that EEE viral strains from the eastern seaboard were more virulent than a Wisconsin isolate for pheasants.

Virulence differences between naturally occurring JE viral strains have been investigated. In mice, the Pekin strain produced high viremia, caused endothelial infection, and had high neuroinvasiveness and peripheral lethality, whereas the Nakayama strain was noninvasive and nonlethal (Huang and Wong, 1963). Adaptation of JE virus by repeated intravenous passage resulted in a strain with a higher capacity to produce viremia in mice, and this change was correlated with increased neuroinvasiveness (Rokutanda, 1969). In general, virulent viral strains have been found to be more efficient in terms of their ability to replicate in brain and extraneural tissues than less virulent strains, but the virus-specified and host-specified factors underlying these phenomena are still poorly understood. The evidence is conflicting that interferon plays a role in differential rates of viral replication in togavirus virulence models (Cole and Wisseman, 1969; Rokutanda, 1969; Jahrling *et al.*, 1976). Studies comparing mouse-virulent and avirulent strains of an alphavirus (SF) by Bradish and co-workers (1975) have stressed the importance of immune host responses in the expression of virus virulence. In their model, regulation of viral replication by immune stimulation of B and T cell function appears to underlie virulence heterogeneity of strains. However, nonimmune mechanisms also clearly determine differences in alphavirus virulence. When benign and virulent clones of VEE virus were injected intravenously into hamsters, clearance of the virulent clone by hepatic reticuloendothelial cells was much slower than for the benign clone (Jahrling and Gorelkin, 1975). Reduced clearance, which was shown to correlate with low virion surface charge, predisposed to high viremia and, thus, neuroinvasiveness. In studies with guinea pigs, however, some benign strains of VEE virus were cleared inefficiently, and it was concluded that restriction of virus replication was responsible for low virulence (Jahrling *et al.*, 1977). Strain differences in vascular clearance rates have also been shown for SIN virus (Postic *et al.*, 1969).

A general schema for the course of togaviral encephalitis is shown in Fig. 9.

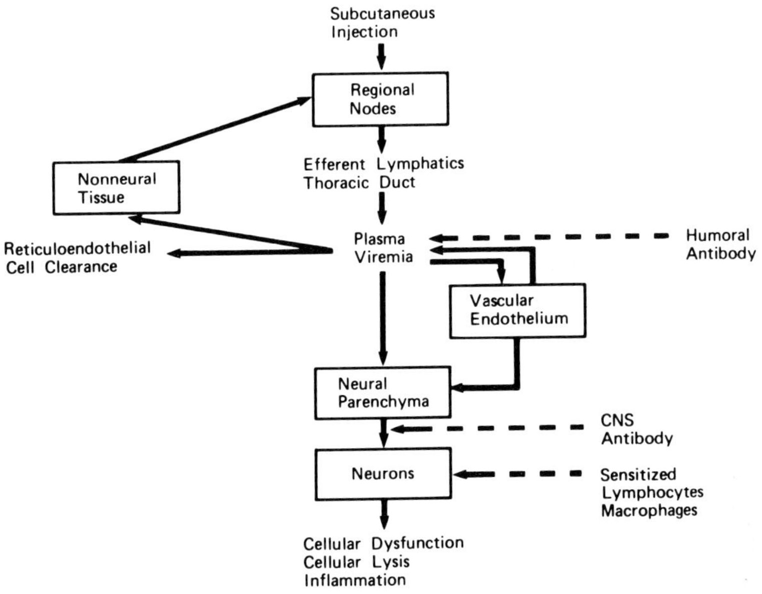

Fig. 9. General schema of pathogenesis of togaviral encephalitis. Modified from Nathanson (1980), with permission.

After inoculation into the skin, the virus may replicate at the site of entry or spread, probably by way of afferent lymphatics, to regional lymph nodes, where primary replication occurs. Virus is then carried by efferent lymphatics to the thoracic duct and into the bloodstream. This primary viremia disseminates the infection and seeds extraneural tissues and possibly vascular endothelium, which in turn support further cycles of replication and serve as a source for continued release of virus into the circulation. Viremia levels are first modulated by the turnover rate of replication in extraneural sites and by the rate of viral clearance by macrophages, and are later terminated by the appearance of humoral antibodies. If viremia is intense or prolonged enough for the neural parenchyma to be invaded, infection of the CNS follows, with or without ensuing clinical disease.

The sites of extraneural infection have been investigated by various workers. In the case of most alphaviruses, striated muscle is a major site of replication and serves as an important source for viremic invasion of the CNS (Grimley and Friedman, 1970; Murphy *et al.*, 1970; Liu *et al.*, 1970). In some alphaviral models, e.g., RR virus in the mouse, however, myotropism predominates, and, despite high viremias, there is little evidence for CNS invasion. Alphaviral myositis and myocarditis may be clinically significant in the mouse (Murphy *et al.*, 1973; Monath *et al.*, 1978). There is some suggestion that a muscular disorder may similarly occur in RR virus-infected equines, and myalgic disease without encephalitis is a feature of chikungunya, RR, and SIN infections of man.

Another pattern has been described in the case of VEE. This virus is lympho- and myelotropic and replicates also in myocardium and skeletal muscle. In equines and man, replication in and release from these extraneural sites contribute to CNS invasion and development of encephalitis. However, in two laboratory models, the hamster and guinea pig, severe lymphoreticular and myeloid necrosis (Fig. 10) are the predominant findings and cause death without evidence of neuroparenchymal disease. The mechanism of death has been reported to be bacteremic/endotoxic shock induced by viral destruction of Peyer's patches and ileal ulceration (Gorelkin and Jahrling, 1975).

Flaviviral replication in extraneural tissues has been less extensively investigated. Replication occurs in a wide variety of sites, including smooth muscle, heart, lymphoreticular tissues, and vascular endothelium (Huang and Wong, 1963).

Penetration of viruses across the blood–brain junction remains poorly understood. Some togaviruses which replicate in cerebrovascular endothelium may thereby gain access to the CNS (Johnson, 1965). It is more likely that passive capillary permeability permits circulating virus to enter the brain (Pathak and Webb, 1974). Increased vascular permeability by use of CO_2 or vasoactive amines (Sellers, 1969) has been shown to promote viral neuroinvasion.

The location and characteristics of pathological lesions produced in the CNS

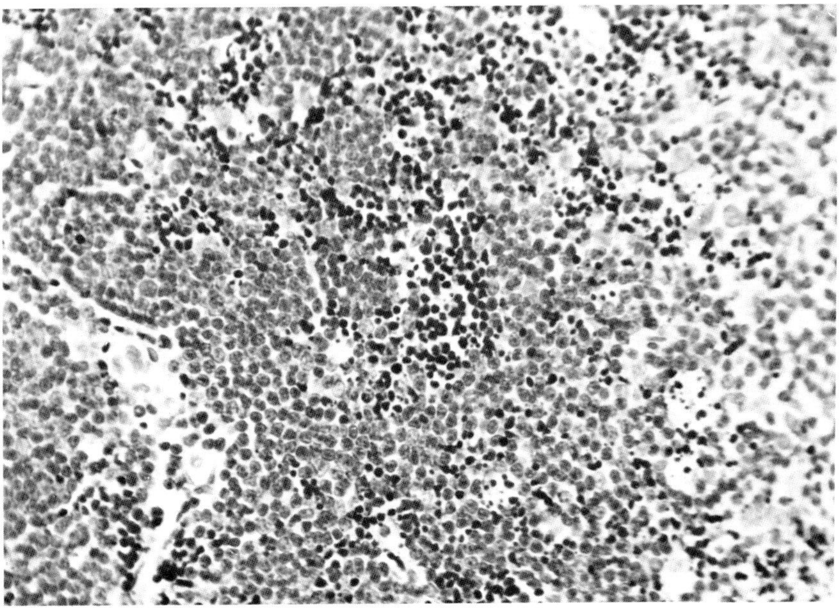

Fig. 10. Necrosis of thymic lymphocytes in hamsters inoculated two days previously with Venezuelan equine encephalomyelitis virus. Focal lesions consist of irregular pycnotic cells. (Reproduced with permission from Walker *et al.* (1976).

vary somewhat with the virus-host system. Two basic processes are common to all togaviral encephalitides: (1) neuronal and glial damage, caused directly by intracellular viral infection, and (2) an inflammatory response involving the migration of viral antigen-specific immunologically active cells into the perivascular space and brain parenchyma. The degree to which both elements of the pathological process will be evident on histopathological examination is a function of the time course and severity of infection; thus the rapidly fatal course of EEE or WEE in infant mice is characterized by severe neuronal damage without much inflammation (Murphy and Whitfield, 1970). The localization of neural injury also varies from virus to virus and reflects differences in specific tropisms. An example is the special predilection of louping ill virus for the Purkinje cells of the cerebellum of mice and sheep (Doherty and Reid, 1971).

The immune responses are instrumental in clearance of and recovery from togaviral infection, but they may also play a pathogenetic role. Nathanson and his colleagues developed the concept of a race in time between neuroparenchymal virus-induced cytopathological processes and the expression of immune clearance mechanisms. Experiments in which immunosuppression of mice by cyclophosphamide was shown to potentiate WN encephalitis demonstrated the role of the humoral immune response in recovery from sublethal flaviviral infection (Camenga *et al.*, 1974). Passive administration of antibodies to immunosuppressed mice as late as 6 days after ip infection reduced mortality from > 90 to 15%, presumably by suppression and delay of viremia and neuroinvasion. In the SIN-mouse model, passive transfer of immune serum to animals with established CNS infection was protective (Griffin and Johnson, 1977). A protective role for antibody within the CNS has not been thoroughly investigated, however, although it is clear from studies of louping ill virus in sheep that antibody may be locally produced in the brain (Reid *et al.*, 1971). A role for cell-mediated immunity in viral clearance has been demonstrated in a SIN-mouse model (McFarland *et al.*, 1972); the mononuclear cell inflammation observed 4–8 days after infection is virus specific, since sensitized lymphocytes (but not serum) transferred to immunosuppressed infected mice can transfer this response. Moreover, *in vitro* studies have demonstrated the antiviral activity of immune spleen cells in experimental VEE infection of mice (Rabinowitz and Proctor, 1974). Murine macrophages have been shown to mediate antibody-dependent celluar cytotoxicity against cells infected with SF virus (McFarland *et al.*, 1977).

A deleterious effect of the immune mechanism has been suggested by various investigators (Gleiser *et al.*, 1962; Doherty and Vantsis, 1973; Camenga and Nathanson, 1975; Nathanson *et al.*, 1975; Semenov *et al.*, 1975; Woodman *et al.*, 1975). An immunopathological component is well documented for the tick-borne flaviviruses, e.g., louping ill, both in laboratory rodents and in the natural host (Doherty and Vantsis, 1973). The immune mechanism is responsible for the inflammatory histopathological response seen in all togaviral infections. In severe or lethal infections characterized by rapid, high-titer viral growth, the immune response may enhance the extent of the lesions and accelerate death.

In one flaviviral infection of man (dengue), enhancement of viral replication in peripheral blood monocytes from immune (but not nonimmune) individuals has been observed (Marchette *et al.*, 1976). This phenomenon may occur in the case of superinfections with heterologous flaviviruses, but there is no evidence that it is of pathogenetic significance, except possibly in dengue hemorrhagic fever/shock syndrome.

The possibility that togaviral infection may produce alterations in the host's immune response has received relatively little attention. VEE virus has been reported to prevent tolerance induction (by injection of aggregate-free gamma globulin) (Howard *et al.*, 1969) and to impair antibody formation in experimental animals (Hruskova *et al.*, 1972). This aspect deserves further attention because of the pathological evidence for lymphoreticular damage in VEE and some flaviviral infections.

Genetic determinants play an undisputed role in pathogenesis of togaviral infections. Classic studies by Webster (1933) and Sabin (1952) showed that resistance of nonimmune mice to lethal flavivirus infection was determined by a single autosomal dominant allele. The resistance allele has been incorporated into the susceptible C3H inbred mouse strain to yield two histocompatible lines (C3H/He and C3H/RV) which differ in susceptibility to flaviviruses (Groschel and Koprowski, 1965). Studies by Koprowski and his colleagues showed that WN virus yields in brains of resistant (C3H/RV) mice were significantly lower than in susceptible mice, but failed to demonstrate differences in interferon or humoral antibody responsiveness between the two strains (Vainio *et al.*, 1961; Goodman and Koprowski, 1962a). *In vitro* studies with cells derived from the two mouse strains indicated that (1) macrophages from resistant mice do not support flavivirus replication as well as macrophages from the susceptible strain (Goodman and Koprowski, 1962b) and (2) the lower WN virus yields observed in cells from RV mice may be due to greater production of defective interfering virus particles (Darnell and Koprowski, 1974).

Using the same mouse model, Jacoby and Bhatt (1976) investigated susceptibility to the flavivirus Banzi. When inoculated by the ic route, this virus produced identical infections in He and RV mice in terms of mortality, virus yields, and immunofluorescent staining. After intraperitoneal (ip) inoculation, viral yields in lymphatic tissues of both mouse strains were similar, but RV mice developed lower brain virus titers and less histopathology and had lower mortality. Thus, in this model, genetic resistance did not appear to depend upon permissiveness/resistance of tissues to virus replication. Further studies employing immunosuppression have indicated that immunological factors appear to underlie the genetic differences in Banzi viral pathogenesis and that T cells play a role in determining resistance (Bhatt and Jacoby, 1976).

Heritable factors which determine cellular susceptibility and immunological responsiveness to togaviral infection of domestic livestock and humans have not yet been elucidated. In the case of domestic animal stock which has and will continue to undergo genetic manipulations, future investigations of genetic resistance-susceptibility will be especially interesting.

B. Viruses Causing Abortion and Stillbirth

Abortion and stillbirth is a feature of JE in swine, Wesselsbron virus disease of ovines, and possibly Middelburg and VEE virus infections of Equidae (Table II). Congenital anomalies in newborn lambs infected *in utero* with Wesselsbron virus are also described. These syndromes are the result of active viral infection of the developing fetal CNS and/or visceral organs.

1. Clinical and Pathological Features

a. Japanese Encephalitis in Swine. The association of JE virus infection with stillbirths in swine was made during summer outbreaks of human and equine encephalitis in Japan in 1935 and again in 1947–1948, when the virus was also isolated from the brains of stillborn piglets (Shimizu and Kawakami, 1949; Burns, 1950). The incidence of stillbirth in the 1948 outbreak was estimated as 56.6% (Itosoya *et al.*, 1950). Subsequently, many isolations of JE virus were made from stillborn piglets, and histopathological examinations confirmed the presence of nonpurulent encephalitis. Stillborn piglets were often mummified or hydranencephalic. Pregnant gilts were usually not clinically affected, although some were noted to develop anorexia and brief fevers at various stages of pregnancy (Burns, 1950).

JE viremias in naturally infected adult swine have been shown to last 3–5 days (Ueba *et al.*, 1972). Experimentally infected pigs develop mild fever and circulate viremias of sufficient titers to infect vector mosquitoes (Scherer *et al.*, 1959d; Nakamura *et al.*, 1964), and swine are considered to be the most important amplifying host for JE virus in nature (Scherer *et al.*, 1959b; Ueba *et al.*, 1971). The high and relatively prolonged viremias and high incidence of natural infections in swine are relevant factors in the epidemiological importance of stillbirth caused by JE virus. The potential dampening effect of congenital infection on pig populations and hence viral transmission is offset by the growth of swine farming in Asia and by husbandry techniques which schedule breeding to avoid intrauterine infections.

Pregnant gilts experimentally infected with two strains of JE virus at varying times of gestation developed viremias lasting 1 to 4 days, beginning 24 hours postinoculation (Shimizu *et al.*, 1954). Some pigs given one virus strain delivered mummified, stillborn, or live fetuses with subcutaneous edema and hydranencephaly. Fetuses which were examined 7–22 days after maternal inoculation contained JE virus in lung, spleen, kidney, blood, and nervous tissues. Fetuses tested after longer intervals were virus negative. It is probable that infection early in pregnancy is more likely to produce fetal damage than later infection.

b. Wesselsbron Disease. Abortions and deaths of newborn lambs and pregnant ewes in southern Africa were described in the 1940s and early 1950s. Some of these cases were attributed to Rift Valley (RV) fever virus, but in the late summer of 1954–1955 Wesselsbron virus was isolated from a dead 8-day-old lamb during an outbreak in the Wesselsbron District, Orange Free State (Weiss *et al.*, 1956). A second report of naturally acquired Wesselsbron disease in newborn sheep with acute, subacute, and chronic infections and in older animals with mild or abortive infections was made by Belonje (1958).

Weiss *et al.* (1956) experimentally infected 32 ewes in advanced stages of pregnancy. The ewes developed febrile reactions 24–72 hours after inoculation, and six died; pathological examination revealed changes consistent with Wesselsbron disease. One fatal case of Wesselsbron disease (defined by pathological study) was produced in a newborn lamb, nine ewes aborted full-term fetuses, and four weak lambs were born which either died or were killed *in extremis*. Wesselsbron virus was recovered from several fetuses aborted 6–14 days after maternal infection. It was concluded that abortion during the primary febrile reaction of the ewe was nonspecific, since virus could not be recovered from expelled fetuses; however, if early abortion did not occur, the fetus became infected *in utero,* resulting in fetal death, stillbirth, or fatal illness in newborn lambs.

The clinical features of the disease in newborn lambs are ill defined. After an incubation period of 24–72 hours, young lambs become febrile and die within 1–2 days. The mortality approaches 100%.

A syndrome of severe abdominal distension (*Hydrops amnii*) in pregnant ewes with prolonged gestation, maternal deaths, and fetal malformation has been described in South Africa (Coetzer and Barnard, 1977). The fetuses show arthrogyphosis, hydranencephaly, hypoplasia, or segmental aplasia of the spinal cord, neurogenic muscular atrophy, and inferior brachygnathia. The syndrome was associated epidemiologically with use of a live, attenuated Wesselsbron viral vaccine during the first trimester of pregnancy; experimental studies confirmed that both the vaccine and wild-type virus were responsible. There is some suggestion that the attenuated vaccine may have increased neurotropic properties (LeRoux, 1959).

The most striking pathological changes in newborn lambs with Wesselsbron disease are in the liver (Weiss *et al.,* 1956; LeRoux, 1959; Coetzer *et al.,* 1978). Gross autopsy findings include icterus, hepatomegaly, and widely disseminated petechial hemorrhages. On microscopic examination, mild to extensive necrosis of isolated liver cells scattered through the hepatic parenchyma is present, without any definite pattern. Councilman-like bodies are frequent. Other features of the hepatic pathology include mild lymphocytic and neutrophilic infiltration around central and portal veins and in the sinusoidal spaces, fatty infiltration, Kupffer cell proliferation, cholestasis, and bile duct proliferation. Changes in other organs are not striking; congestion and hemorrhages may be seen in the myocardium, kidney, spleen, and gall bladder.

Wesselsbron virus has been suspected to cause a disease in cattle similar to that described in sheep. The virus produces mild febrile reactions in experimentally infected cows. In a preliminary experimental study, cows appeared to be quite refractory to intrauterine infection, and seroepidemiological evidence further suggested that Weselsbron virus was not a significant cause of abortion or disease in cattle (Blackburn, 1977).

The naturally or laboratory-acquired disease in man is characterized by influenza- or dengue-like symptoms. Jaundice is not described, but in one case, hepatosplenomegaly and hepatic tenderness were present. Congenital infections are not reported, and the pathogenic potential for the human fetus is unknown.

c. Other Viruses. Venezuelan equine encephalitis virus is known to cause severe encephalitis, with massive cerebral necrosis in human fetuses infected transplacentally during the first trimester of pregnancy (Wenger, 1963, 1967). Similar observations have not been reported in equines or other domestic animals, although Jochim *et al.* (1973) described abortion in two horses shortly after infection with VEE IAB virus in Texas. Walton *et al.* (1973), however, reported birth of a normal foal to a mare infected 1 month previously with epizootic virus.

There is some suggestion that Middelburg virus may cause abortion in horses (B. M. McIntosh, personal communication). The virus has been isolated from sick horses in South Africa, and antibody is very prevalent in mares which have aborted.

WEE virus has been reported to cause transplacental infection and neonatal disease in man (Shinefield and Townsend, 1953).

C. Other Clinicopathological Entities

1. Disorder of Spermatogenesis in Boars Infected with Japanese Encephalitis Virus

JE virus has been suspected on epidemiological grounds to be a cause of summer-autumn reduced libido, hypo- and aspermia in boars in Japan (Habu *et al.*, 1977). Experimentally infected boars developed fever and viremia, and virus was recovered from semen for up to 17 days after inoculation (Habu *et al.*, 1977; Ogasa *et al.*, 1977). After recovery from fever, defective spermatogenesis (reduced concentration and motility, and abnormal sperm morphology) was found in a high proportion of animals. Histopathological changes included deposition of fibrinous material and inflammatory and proliferative changes of the surface of the tunica vaginalis communis and the tunica testis. The lamina propria and interstices of the epididymis were infiltrated with lymphocytes, and spermatogenic arrest was demonstrated in the testis. Artificial insemination with infected sperm resulted in viremia and lack of fertilization of recipient sows.

These studies suggest that JE virus may reduce fertility of swine both by transplacental infection and stillbirth (Section III, B) and by reduced fertility of boars. Venereal virus transmission in swine has also been demonstrated, but its epidemiological significance remains speculative.

2. Febrile Exanthem of Horses Caused by Getah Virus

In 1978, a new clinical entity was recognized during an epizootic among race horses in Ibaragi Prefecture, Japan (Kamada *et al.*, 1980). Clinical manifestations included a temperature increase to 40°C, a vesicular exanthem (Fig. 11), and edema of the hind legs, lasting up to 1 week. There were no cases with CNS symptoms. Getah virus was isolated from sera of 62 of 209 horses tested. Horses experimentally inoculated with the virus developed a disease similar to the natural infection.

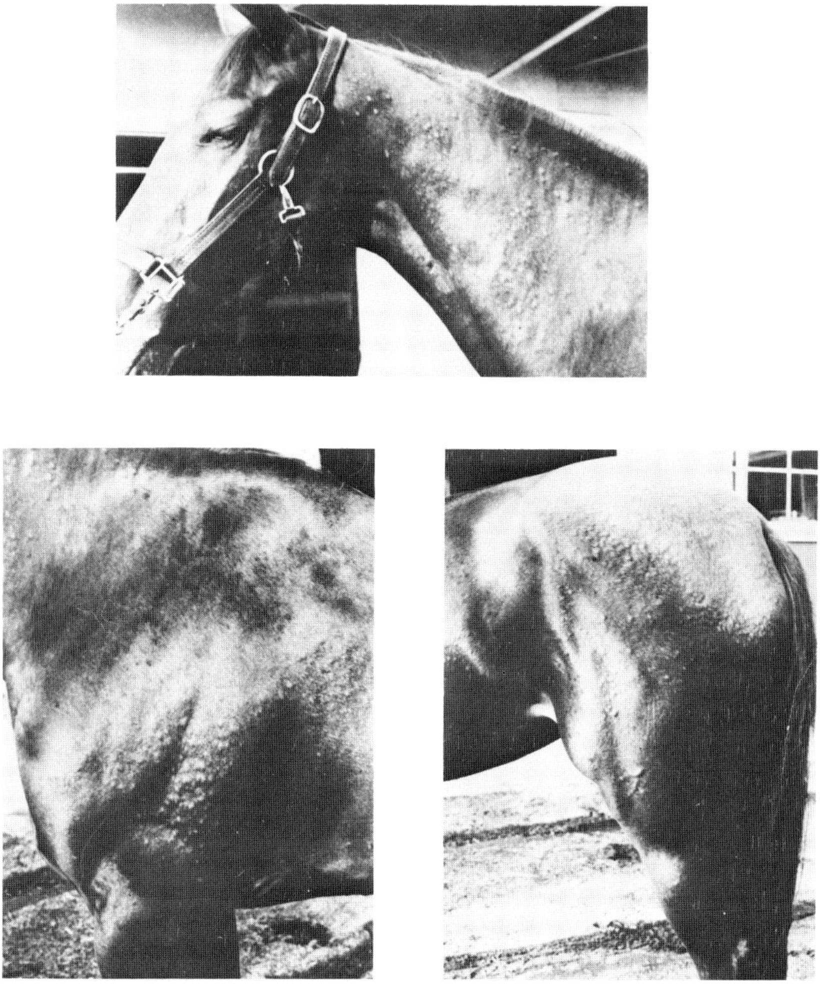

Fig. 11. Exanthem in equines caused by getah virus. The infection which occurred in epizootic form in racehorses in Japan in 1978 was characterized by fever, rash, and hindlimb edema. (Photo courtesy of Dr. T. Kumanomido.)

IV. EPIZOOTIOLOGY

A. Eastern Equine Encephalitis

1. Occurrence and Geographic Distribution

The early history of this disease is reviewed by Beadle (1952). Equine outbreaks occurring in the eastern United States as early as 1847 and designated "blind staggers," "forage poisoning," or "cerebrospinal meningitis" were in retrospect probably due to EEE. In 1933, the etiological agent was first isolated during an outbreak in the tidewater areas of Virginia which involved at least 1000 horses with a mortality of about 90%. An extensive equine epizootic affected Massachusetts, Rhode Island, and Connecticut in 1938, with 381 reported cases and a mortality of 93%. During this outbreak it was first recognized that EEE was also a disease of humans and of encephalitis in ring-necked pheasants and pigeons. In 1941 a focal epizootic occurred in Brownsville, Texas; this was the first indication of EEE viral transmission along the Gulf Coast. In 1942 an outbreak in horses was recognized in an area of southwestern Michigan, characterized by freshwater marshes near a large bird sanctuary. This outbreak signaled the presence of EEE virus in inland areas. The largest recorded equine outbreak in the United States occurred in 1947 in the bayous of coastal Louisiana and adjacent Texas. As many as 14,334 equine cases with 11,722 deaths and 10 human cases were attributed to this epizootic. In 1955 and 1956 small outbreaks of the disease affected horses and humans in Massachusetts, and in 1959 the disease appeared in epizootic-epidemic form in New Jersey; 66 fatal equine cases and 32 human cases (22 fatal) were recorded (Goldfield and Sussman, 1968). Between 1960 and 1978 small outbreaks occurred in eastern Maryland (1960, 1968); Oswego County, New York (1971, 1976); eastern Canada (1972); Massachusetts-New Hampshire (1973); and central Florida (1978). Epornitics in penned pheasants and chukar partridges have been described in New Jersey, Connecticut, Pennsylvania, and Florida, often in association with outbreaks of disease in horses and humans. Between 1939 and 1959, 56 outbreaks in commercial flocks were reported in New Jersey alone, with morbidity of up to 29.2 per 1000 flocks in years of high viral activity (Beaudette *et al.*, 1952; Goldfield and Sussman, 1968).

Accurate information on the incidence of equine disease during epizootics is generally lacking. In the 1959 New Jersey outbreak, the overall attack rate was 8.1 per 1000, but in the most severely affected county, the incidence reached 60 per 1000.

Human morbidity has been low; the largest epidemics occurred in 1938 in Massachusetts and in 1959 in New Jersey (27 and 32 cases, respectively). The mortality has, however, been high (65–70%), and neurological sequelae are frequent in survivors. Abortive or subclinical infections occur, but the inapparent:apparent infection ratio is quite low (estimated at 8:1 in children 0–4 years old, approximately 50:1 in young adults, and 15–20:1 in the elderly (Goldfield *et al.*, 1968).

In tropical America, equine epizootics have been described in Jamaica (1962) (Hart *et al.*, 1964); the Dominican Republic (1948–1949, 1978) (Eklund *et al.*, 1951); Panama (1973) (Vigilancia Epidemiologica, 1973); and Venezuela (1976) (Vigilancia Epidemiologica, 1976). In Guyana, Brazil, and Argentina, outbreaks of mixed etiology (due to WEE, VEE, and EEE) have occurred. In addition to the localities affected by epizootics, countries in which the virus has been isolated include Mexico, Colombia, Ecuador, and Trinidad.

In the United States the habitats associated with viral activity are characterized by freshwater swamps and marshes, which support the principal mosquito vector, *Culiseta melanura,* and large populations of wild birds, the enzootic hosts. Various studies have shown that the densities of *C. melanura,* the rate at which this vector is infected, and the frequency of seroconversion in sentinel birds are higher within than in localities immediately outside freshwater swamps (Hayes *et al.,* 1960; Williams *et al.,* 1974). Equine and pheasant outbreaks and sporadic cases of human disease occur in proximity to freshwater swamp habitats. In coastal regions, however, equine outbreaks and, especially, human epidemics have also been associated with the transitional zone between freshwater swamps and salt marshes, mangrove swamps, and saline lagoons, where vector species such as *Aedes sollicitans* have been implicated in transmission.

2. *Seasonal Distribution*

Outbreaks of EEE occur during late summer and early fall. The first clinical warning of viral activity is often the appearance in late June or July of affected pheasant flocks or sporadic horse infections, which may precede recognition of an epizootic and the first human case by 4–6 weeks. Fig. 12 shows the chronological relationships among pheasant, equine, and human cases during the 1959 epidemic in New Jersey. In some areas the sentinel value of equine and pheasant cases has been reduced by effective immunization programs. In Mas-

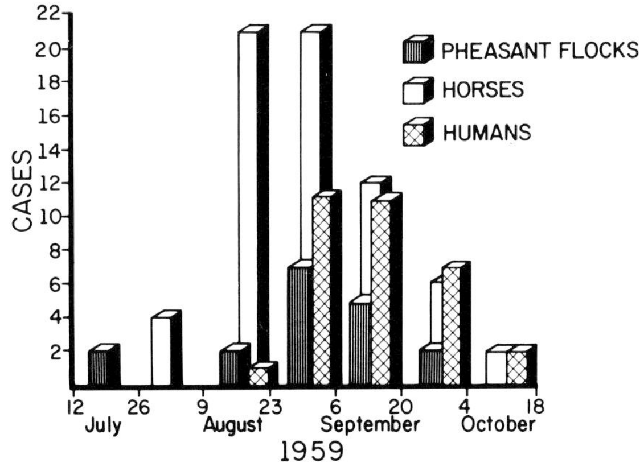

Fig. 12. Chrolologic relationships between cases of eastern equine encephalomyelitis in pheasants, horses, and humans during the 1959 outbreak in New Jersey. (Modified from Goldfield and Sussman (1968).

sachusetts, widespread equine cases in 1973 presaged human infections, whereas in 1974, after completion of an equine vaccination campaign, three human deaths occurred in the absence of equine cases (Grady *et al.*, 1978).

Viral activity can be detected every year in areas of the eastern United States, with or without associated equine or human cases. In nonepidemic years, studies employing sentinel birds have detected EEE viral transmission as early as June in some areas of the eastern United States; peak viral transmission as evidenced by viral isolations from mosquitoes and birds, however, occurs in August–September, in concert with the peak abundance of the enzootic vector, *C. melanura* (Main *et al.*, 1968; Saugstad *et al.*, 1972). Interestingly, the appearance and peak rate of transmission of the WEE complex virus (HJ) precede those of EEE virus by several weeks or more; these agents share identical habitat, vectors, and hosts.

Climatological factors appear to influence EEE viral activity (Hayes and Hess, 1964). Excessive rainfall during the autumn preceding and the summer of the outbreak was associated with epizootics. Rainfall correlated also with the occurrence of high population densities of *C. melanura*. Increased and early peak abundance of vector populations have been shown to coincide with the appearance of epizootics (Wallis *et al.*, 1974).

In the Caribbean, outbreaks caused by the North American serotype of EEE virus have generally begun in October–December, lending some support for the possibility that the virus was introduced from the United States by southward-migrating birds.

3. *Cycle of Transmission*

Many field studies have established the role of *C. melanura* as the principal enzootic vector of EEE virus in North America. The high rate of success in obtaining virus isolates from field-collected *C. melanura* compared to other species (Chamberlain *et al.*, 1958), the association between high population densities of *C. melanura* and equine and pheasant epizootics, and the high vector efficiency demonstrated in experimental studies on colonized mosquitoes (Howard and Wallis, 1974) support this view. Minimal infection rates in *C. melanura* (often 1:1000 or greater) have been consistently much higher than in other mosquito species. *C. melanura* is responsible for viral transmission between wild birds in freshwater swamp habitats (Fig. 13). It is a highly ornithophilic species, and mammalian feedings are so infrequent that this species cannot account for transmission to equines or humans. Other species, in particular *Aedes vexans*, *A. sollicitans*, and *Coquillettidia perturbans* have been implicated as epizootic vectors (Kandle, 1960; Wallis *et al.*, 1960). Occasional recoveries of EEE virus from other mosquito species (*Culex restuans*, *C. pipiens*, *C. salinarius*, *C. nigripalpus*, *C.* (*Melanoconion*) sp., *A. mitchellae*, *A. atlanticus*, *A. triseriatus*, and *Anopheles* spp.) as well as from chicken mites, chicken lice,

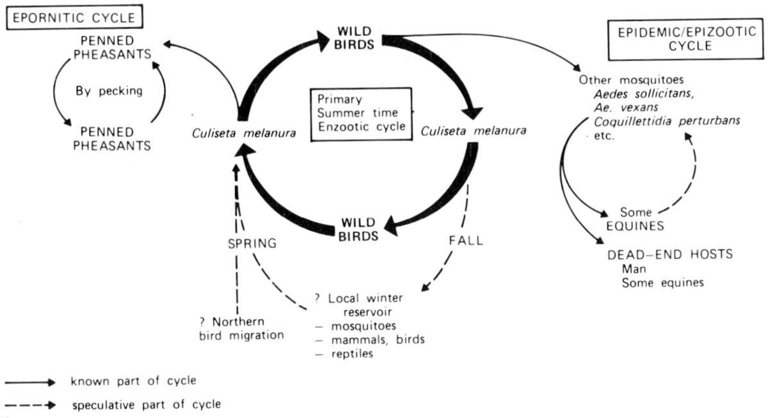

Fig. 13. Transmission cycle of eastern equine encephalomyelitis. The primary enzootic cycle involves transmission between wild birds in a freshwater swamp habitat by the agency of *Culiseta melanura* mosquitoes.

simuliid flies, and *Culicoides* midges have been reported but are not epidemiologically significant.

Both field and laboratory studies have established the importance of wild birds, primarily the smaller species of *Passeriformes*, as maintenance and amplifying hosts in the transmission cycle. Virus isolations and antibodies have been found in a wide variety of species. Antibody prevalences of 10–20% in wild birds are reported during outbreaks (Kissling, 1958a; Hayes *et al.*, 1964), and in focal areas of intense viral activity, prevalences of 50% of juvenile birds have been described (Williams *et al.*, 1974). In experimental studies, all avian species that have been investigated (more than 20) develop viremias, usually lasting 4 days. Small passeriform birds are especially susceptible and have maximum viremia titers in excess of 10^6/ml. Conversely, wild mammals are infrequently infected in nature. Antibody prevalences of less than 1 to 1.5% have been reported in studies of rodents, rabbits, and bats in Massachusetts (Daniels *et al.*, 1960; Hayes *et al.*, 1964). In experimental studies, a variety of rodent species were found susceptible to infection but failed to develop detectable viremias (Karstad *et al.*, 1961). Occasional viral isolations have been reported from wild mammals but are probably incidental to the natural history of EEE. As discussed above, horses develop brief and usually low-titered viremias and may occasionally serve as a source of biological or mechanical infection of arthropods. Reptiles have been suggested as alternate hosts in the cycle, mainly on the basis of experimental studies (Hayes *et al.*, 1964), but little evidence supports an important role in nature.

The spread of EEE virus within pheasant flocks is apparently not by the agency of arthropod vectors but rather by contact transmission, through pecking and cannibalism (Holden, 1955). Epornitics are probably initiated by mosquito-borne infection of one or more birds in a flock. A study of the possible spread of EEE virus to pheasant farmers showed no evidence for contact transmission (Liao, 1955).

An overall view of the transmission cycle (Fig. 13) may be summarized as follows: In the early summer, the virus begins to circulate between wild birds and *C. melanura* mosquitoes in freshwater swamps in the eastern United States. Both adult and nestling birds and mosquito populations are increasing at this time, and viral activity amplifies as more hosts and vectors become infected. In areas where equines or humans come into close contact with the restricted habitats supporting the enzootic cycle, *C. melanura* may account for sporadic infections. However, the presence of abundant species, such as *Aedes* or *Coquillettidia,* which feed on birds but which also have a proclivity to bite mammals, may also become infected and then serve as epizootic vectors. The abundance of these species and their high biting rates account for their role in viral transmission to horses, despite a low frequency of feeding on birds and low infection rates. Equine morbidity is inversely proportional to natural and vaccine-induced immunity rates. Transmission declines and ceases in the fall as the proportion of susceptible avian hosts decreases and cold weather reduces vector populations. The overwintering mechanism is not known.

B. Western Equine Encephalitis

1. Occurrence and Distribution

Just as for EEE, this disease was observed in epizootic form long before isolation and identification of the etiological agent in 1930 in California (Meyer *et al.,* 1931). In the next decade, notable epizootics occurred in the western United States, Minnesota, and southern Saskatchewan, and the virus was shown to be transmissible by mosquitoes in the laboratory and was isolated from wild *Culex tarsalis* in Yakima Valley, Washington (Hammon *et al.,* 1941). The most extensive outbreak and the first to involve large numbers of humans occurred in 1941 in the north-central United States and bordering areas of Canada; hundreds of thousands of equines and at least 3000 humans were affected. A number of smaller outbreaks have subsequently occurred in Manitoba (1947), the Central Valley of California (1952), Kansas (1958), Utah (1958), western Texas (1963–1964), Colorado (1965), and Minnesota–North Dakota–Manitoba (1975). These outbreaks have involved hundreds (occasionally more) of equine cases; the number of reported human cases is approximately 10–20% the recognized equine morbidity.

WEE virus is widely distributed from Argentina to northern Saskatchewan. A closely related virus (HJ) is present in the eastern United States; HJ has only rarely been associated with disease in equines. Its ecology has been extensively reviewed by Hayes and Wallis (1977).

As for EEE, the disease in equines precedes the appearance of human cases, and surveillance of horses (in areas where naturally acquired and vaccine-induced immunity is low) may provide a warning of risk of human infection. Attack rates in equines have not been accurately estimated during outbreaks of the disease. In man, incidence rates of 5 to 15/100,000 have been documented in many outbreaks.

2. *Transmission Cycle*

In most of North America, the primary transmission cycle involves *C. tarsalis* mosquitoes and wild birds of various species (Fig. 14). In the United States, *C. tarsalis* is relatively common west of the Mississippi River, where its distribution and abundance are closely correlated with the presence of irrigated farmlands. Irrigation wastewater, rice fields, vegetated margins of lakes and ponds, and areas flooded by overflow or excess snowmelt are favored habitats. Peak activity occurs within 2 hours after sunset. The vector is strongly attracted to avian hosts (Tempelis, 1975) but also feeds on man, horses, and other vertebrates. Studies in several of the western states (Tempelis *et al.*, 1965) have shown a seasonal shift in the feeding habits of *C. tarsalis*. During the spring and early summer the vector is primarily ornithophilic, but in midsummer it feeds increasingly on mammalian hosts. Mammalian feedings peak at the time of the occurrence of equine and human cases. Population peaks in most areas occur between June and September, but in the Rio Grande Valley, the species is most abundant between November and February. Climatological factors greatly influence vector activity.

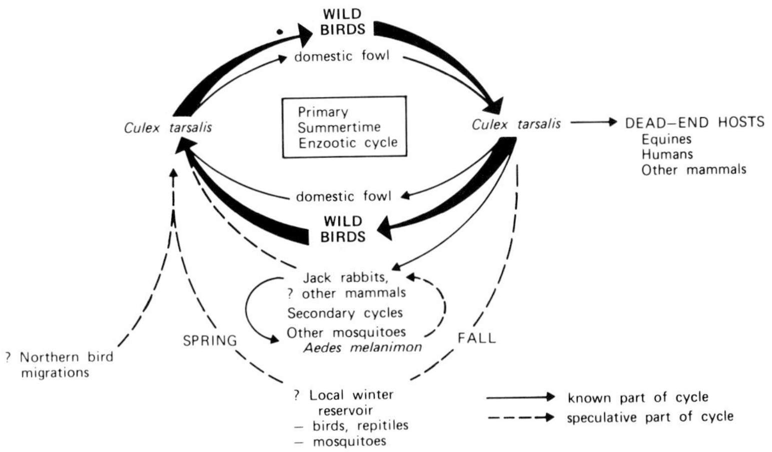

Fig. 14. Transmission cycle of western equine encephalomyelitis in North America.

Emergence of hibernating *C. tarsalis* correlates with the time of spring inversion of soil temperatures (Bennington *et al.*, 1958). Maximum activity of WEE virus has been noted in years when spring temperatures were unusually low and emergence of the vector occurred later in the spring (Hess *et al.*, 1963).

Studies in the Central Valley of California have helped to define the relationship between *C. tarsalis* population density and incidence of WEE. Human disease did not occur when the seasonal light trap index (number of female *C. tarsalis* per New Jersey light trap per night) fell below 10. Transmission of the virus in the silent enzootic cycle continued at light trap indices of 2–9 but ceased at indices below 1 (Reeves, 1971a).

On the basis of serological evidence and virus isolation, wild birds have been incriminated as the primary vertebrate hosts in the transmission cycle. In the course of an epizootic, it is not unusual to find 15–60% antibody prevalences in wild bird populations. WEE virus recovery rates from the blood of wild house sparrow nestlings were shown to correlate closely with disease incidence in west Texas (Holden *et al.*, 1973).

Many species of wild birds are susceptible to experimental infection, including house sparrows, house finches, tricolor blackbirds, white-crowned sparrows, white-throated sparrows, blue jays, purple finches, and pigeons (Hammon *et al.*, 1951; Kaplan *et al.*, 1955; Kissling *et al.*, 1957). Viremia appears on the first day after inoculation and lasts for 2 to 6 days, with peak titers (4–7 $\log_{10} LD_{50}$/ml) on the first or second day.

WEE virus has also been associated with mosquito species other than *C. tarsalis* and with mammalian, reptilian, and amphibian vertebrates, but they play a secondary or incidental role in the cycle. In California, the black-tailed jackrabbit (*Lepus californicus*) and *A. melanimon* are believed to constitute a secondary cycle of transmission (Hardy *et al.*, 1977). In Canada WEE virus has been isolated from various *Aedes* species (*campestris, spencerii, flavescens*), and from *Culiseta inornata*, and there is reason to believe that some of these mosquitoes may play a role in early springtime viral amplification, before *C. tarsalis* is active (McLintock *et al.*, 1970). In this regard, it is important to mention that there is some evidence that certain small mammals (the principal source of blood feeding for the implicated aedine species) are effective viremic hosts and become infected in nature (Leung *et al.*, 1975). Absence of antibody in natural populations of some rodent species (e.g., kangaroo rats) may reflect a high mortality of infected animals (Hardy *et al.*, 1974).

Considerable debate revolves around the role of poikilothermic vertebrates. Both neutralizing substances (Spalatin *et al.*, 1964) and virus isolations (Bowen, 1977) have been found in natural populations; experimentally inoculated reptiles have been shown to become viremic, sometimes for prolonged periods spanning wintertime hibernation (Gebhardt *et al.*, 1964). At present, the consensus is that reptiles play a potentially important but unproven role in the overwinter maintenance of the virus.

The overwintering mechanism is not presently understood (see the review by Reeves, 1974). The evidence favors one or more local winter reservoirs; renewed interest in the survival of virus in hibernating adult *C. tarsalis* has been stimulated by the findings of Bailey and co-workers (1978) that this mechanism accounts, at least in part, for overwintering of SLE virus in *C. pipiens*.

C. Venezuelan Equine Encephalitis

Excellent reviews of the epidemiology of VEE have been published by Johnson and Martin (1974) and the Pan American Health Organization (1972).

1. Occurrence and Distribution

The first published record of an epizootic of VEE occurred in Colombia in 1935, but the virus was not isolated until 1938. Since that time, there have been numerous outbreaks in northern South America, the areas most severely affected being Colombia, Venezuela (especially the Guajira Peninsula), Ecuador, northern Peru, Trinidad, and Guyana. Epizootics have appeared at intervals of 5–10 years, a pattern reflecting the time necessary for accumulation of numbers of nonimmune equines adequate to support a high rate of viral transmission. The capacity of the virus to invade new territory was exhibited in 1969, when the epizootic (subtype IAB) virus invaded the Pacific coast of Guatemala and spread in waves southward to Costa Rica and northward through Mexico to southern Texas in 1971 (Spertzl and McKinney, 1972; Fig. 15). An epizootic–epidemic in Ecuador (Gutierrez *et al.*, 1975) and northern Peru preceded appearance of the disease in Central America.

The magnitude of these outbreaks has assured a place for VEE as a major veterinary public health problem in tropical America. In Colombia in 1967, for example, over 100,000 equine deaths probably occurred (Pan American Health Organization, 1972); in this outbreak the case fatality was estimated at 20–36%. In the wake of the outbreak, approximately 70% of surviving equines sampled had neutralizing antibodies, amply demonstrating the exhaustion of immunologically susceptible animals.

Associated human morbidity has also been great; recognition of the extent of this problem was not appreciated until an outbreak in Venezuela between 1962 and 1964, when it is estimated that over 23,000 human cases occurred (Avilán, 1966). The predominant syndrome produced is a self-limited influenza-like illness, and only about 4% of the cases exhibit CNS disease. The latter is more frequent in children than adults. During the 1969 epidemic in Ecuador, 80% of the hospitalized patients with neurological manifestations were 12 years

Fig. 15. Introduction of Venezuelan equine encephalitis into the Pacific lowlands of Guatemala in 1969 initiated epizootic waves that spread southward to Costa Rica and northward to Mexico and Texas between 1969 and 1972. At least 20,000 equines died of the disease.

of age or younger. The case fatality in children 0–5 years with encephalitis was 35%, but it was less than 10% in older age groups (Gutierrez *et al.*, 1975).

2. *Transmission Cycles*

The vector relationships of the epizootic virus subtypes (IAB and IC) have been reviewed by Sudia and Newhouse (1975). These virus subtypes have been isolated from at least 34 different mosquito species belonging to 8 genera, reflecting in part the wide variety of ecological habitats invaded during outbreaks and the very high virus titers in the blood of infected equine hosts. The most complete studies conducted during the outbreaks in Guatemala, Costa Rica, Mexico, and Texas between 1969 and 1972 established the importance of *Psorophora confinnis, P. discolor, A. sollicitans, A. taeniorhynchus, A. thelcter, Mansonia titillans,* and *Deinocerites pseudes*. Extraordinarily high field infection rates (up to 2–4% of some species of mosquitoes tested) were found during these outbreaks, explaining the explosive nature and high attack rates observed in VEE epizootics.

Experimental studies have confirmed the capacity for biological transmission by a number of mosquito species. Those species for which both laboratory proof of vector competence and evidence for infection in nature have been accumulated are *P. confinnis, A. aegypti, A. sollicitans, M. tittilans, M. indubitans, C. tarsalis,* and *A. taeniorhynchus*. The threshold for infection of *Psorophora* and *Aedes* appears to be approximately 5 log/ml viremic blood. Since equines may circulate virus in their blood at titers above this level for up to 5 days, they serve as a source of infection for large numbers of vectors (Fig. 16).

Peak equine viremia titers of over 10 log/ml have occasionally been documented and probably explain the occurrence of mechanical transmission of virus by a variety of other biting flies which are incapable of the biological virus transmission. Thus, VEE virus has been isolated from *Culicoides* and *Simulium*, and *Simulium* blackflies were the responsible vectors during an outbreak in 1967 in Colombia (Sanmartin *et al.*, 1973).

Although equines are the undisputed primary vertebrate hosts for epizootic VEE virus subtypes, the roles of man, wild mammals, and birds in the transmission cycles are still in dispute. High-titer viremias occur in infected persons, who also may shed the virus from the pharynx. In general, it is conceded that man has so far played a minor role in viral transmission. However, person-to-person spread by droplet or aerosol infection has been suspected on occasion, and the potential for a future outbreak sustained by interhuman transmission by vector mosquitoes (e.g., *A. aegypti*) must be recognized. Dogs and pigs have been

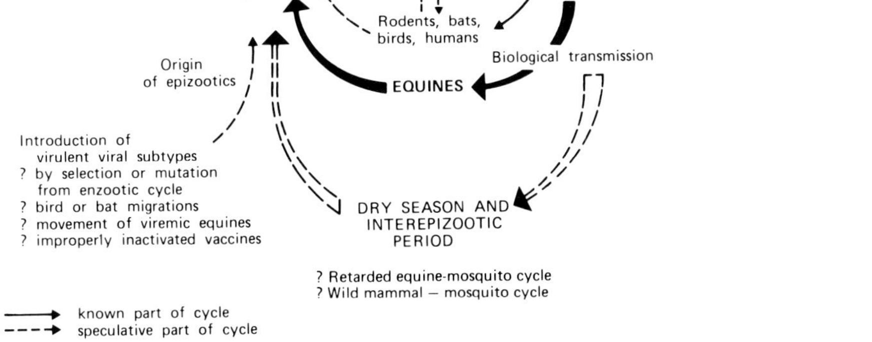

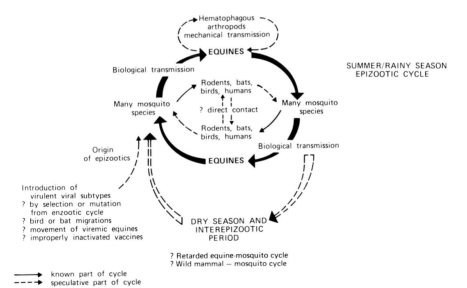

Fig. 16. Transmission cycle of epizootic subtypes of Venezuelan equine encephalomyelitis virus. Equidae are the principal viremic hosts, and many vector species are implicated in transmission. Because of the high viremias in equines, mechanical transmission (e.g., by Simulidae) is possible.

considered to be potentially important hosts on the basis of their abundance, high antibody prevalence in some areas, and experimental data showing moderate viremias (see Section III, C). A number of wild mammals have been investigated as hosts for epizootic virus subtypes (Johnson and Martin, 1974; Sudia and Newhouse, 1975; Bowen, 1976), but there is little evidence that they play an epidemiologically important role. VEE virus has been isolated from bats, and antibodies have been demonstrated in wild bat populations. Experimental infection of five Neotropical bat species with epizootic (and enzootic) virus strains have shown viremia levels of sufficient magnitude to infect vector mosquitoes (Seymour *et al.*, 1978). It is possible that bats play a role in VEE virus maintenance or dispersal.

A variety of birds develop brief viremias of sufficient magnitude to infect vector mosquitoes (Chamberlain *et al.*, 1956; Dickerman *et al.*, 1976; Bowen and McLean, 1977). Antibodies (at low prevalence) have been found in wild birds, but it has not been clearly established whether such infections were caused by enzootic or epizootic virus subtypes. Most of the major vector species are not highly ornithophilic. Birds are considered to play at best a minor role in virus amplification and maintenance, but the possibility cannot be excluded that they may occasionally function in dispersal of the virus.

The interepidemic maintenance cycle of the epizootic virus subtypes and the mechanism whereby the virus can emerge in epizootic form are still unknown. The possibilities, as stated by Johnson and Martin (1974), p. 98–99 are:

1. The virus is maintained in silent mosquito-vertebrate cycles until conditions are conducive for emergence of an epizootic.
2. Epizootic viruses represent mutations of avirulent enzootic strains.
3. The virus is maintained through indolent but continuous horse-to-horse passage in areas where most horses are immune or where vector populations are very low.
4. The virus is maintained as a latent infection in wild vertebrates without need of arthropods. . . .
5. Epizootics result from the . . . use of improperly prepared killed virus vaccines.

The first two of these possibilities have been the subject of extensive investigations by Scherer and his colleagues (1976a,b; Scherer, 1977). In several areas of Central America, field studies conducted after the 1969 epizootic failed to demonstrate the persistence of epizootic VEE virus or the presence of an enzootic cycle of transmission of subtype IAB virus. Investigation of the possibility that the virulent subtype might arise by mutation or selection from enzootic virus strains was, however, limited by the lack of sensitive methods for detecting small amounts of epizootic virus possibly present in enzootic strains. Jahrling and Eddy (1977) have described the separation of subtypes of hydroxylapatite chromatography, a technique that allows detection of minute virion subpopulations. This method is now being applied to field isolates, but results of the studies are pending.

Little or no evidence exists to support or refute hypotheses 3 or 4.

With regard to the introduction of epizootic virus by use of improperly inactivated vaccines, there is little doubt that this has occurred in some areas [e.g., Argentina (de Diego *et al.*, 1975) and Nicaragua (Spertzl and McKinney, 1972)]. It is impossible to determine how many other outbreaks have had an "iatrogenic" origin. Oligonucleotide mapping of the viral genome of vaccine and field strains is a technique which may provide clues to the origin of future outbreaks.

In addition to the hypotheses presented, transovarial transmission of VEE virus in mosquito vectors deserves serious consideration as a means of virus survival.

Only brief mention will be made of the epidemiology of the enzootic subtypes (ID, IE, II, III, and IV). These virus subtypes are not important animal pathogens, although some (e.g., ID) may be responsible for heretofore underrecognized endemic-epidemic human illness (Franck and Johnson, 1970; Dietz *et al.*, 1979). These viruses are primarily transmitted in forest-swamp habitats in cycles involving wild rodents (*Sigmodon, Proechimys, Zygodontomys, Oryzomys*) and *Culex* (*Melanoconion*) mosquitoes (Fig. 17). Other mammals and some birds (especially the Ciconiidae) may be secondary hosts. Equines play no role in transmission. One virus (Tonate, subtype IIIB) from French Guiana (Digoutte and Girault, 1976) is transmitted by *C.* (*Mel.*) *portesi* and may be exceptional in

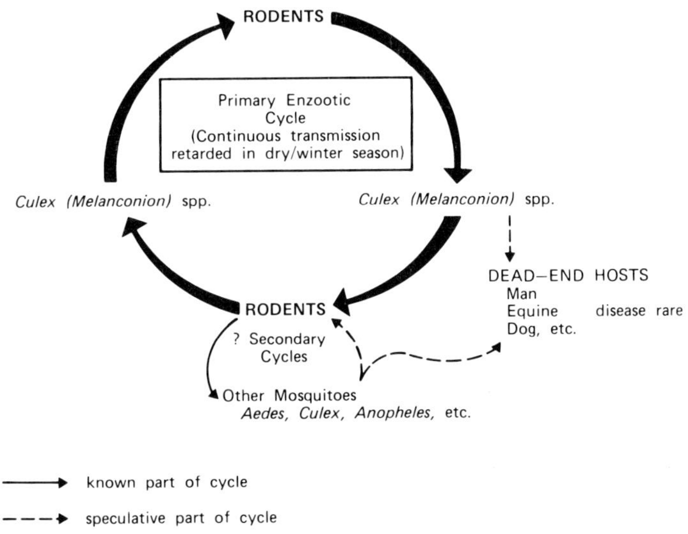

Fig. 17. Transmission of enzootic strains of Venezuelan equine encephalomyelitis virus; small rodents (cotton rats, etc.) and *Culex* (*melanoconion*) *aikenii*, *opisthopus*, and other species constitute the cycle.

that birds appear to constitute the principal enzootic hosts. Tonate virus has also been isolated in North America (Monath *et al.*, 1980).

D. Japanese Encephalitis

The disease in man was recognized in 1871 in Japan and the etiological agent first isolated in 1934 from a fatal human case. JE is principally a human health problem in Asian countries. The virus is present in Siberia, Japan, mainland China, Taiwan, Korea, the Philippines, Vietnam, the Lao People's Democratic Republic, Malaysia, Singapore, Thailand, Indonesia, Sri Lanka, Burma, and India, and has occasionally invaded the Pacific islands. Repeated major human outbreaks with up to 5000 cases occurred in Japan after World War II, but the incidence has declined dramatically since 1967. Endemic and epidemic JE has also been a major public health problem in Korea and Taiwan. In recent years, important outbreaks have occurred in northern Thailand (Grossman *et al.*, 1973) and the West Bengal area of India. Major outbreaks occur in the summer season in countries of the temperate zones, whereas in tropical areas epidemics correspond to the rainy season, with endemic cases being reported throughout the year. The disease primarily affects children under 10 years of age and old persons, with a high case-fatality (2–70% in various outbreaks) and incidence of neurological sequelae. Reviews of the human epidemiology are given by Okuno (1978) and Kono and Ho Kim (1969).

A disease of horses resembling JE was described in nineteenth-century Japan, and an accurate report of the clinical and pathological features was made by Oguni in 1926. The virus was isolated from the brains of sick equines in 1937. During the epidemics of 1948 and 1949, a total of 4895 equine deaths were reported, with a morbidity of 4.4 per 1000 unvaccinated equines (Nakumura, 1970). The disease has not been a veterinary public health problem in Japan since, due to vaccination practices, a declining horse population, and diminishing viral activity. However, it is a major disease in China.

The ecology of JE was first clearly established by Scherer and colleagues in Japan (Scherer *et al.*, 1959a,b,c; Buescher and Scherer, 1959), and their findings have been confirmed and extended by many other workers in other areas of Asia (Fig. 18). Both mammals and birds play a role in transmission, but swine are the principal amplifying hosts. In temperate areas the virus first appears in July as mosquitoes, and within 1–3 weeks infections are found in pigs and birds, principally ardeid species (black-crowned night herons and egrets). Human infections are noted only several weeks thereafter. Pigs are the most common domestic animals in the Far East, develop effective viremic infections, and have high reproductive rates and population turnover due to slaughtering. Pig breeding in some areas is scheduled to avoid maternal infection during gestation (because of the problem of stillbirth), and consequently, at the time of summer viral trans-

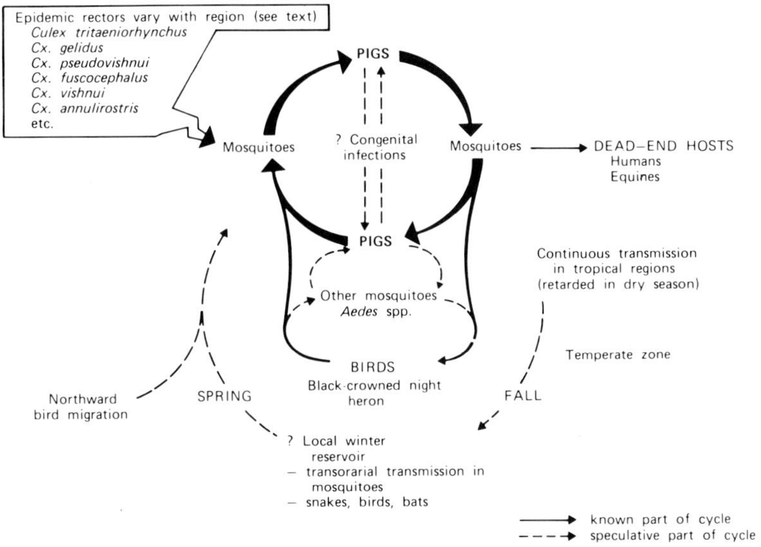

Fig. 18. Transmission cycle of Japanese encephalitis. Swine are the principal amplifier host, and ardeid birds probably represent important hosts in the maintenance of transmission.

mission, large numbers of susceptible piglets reach the age at which maternal antibody has waned (Scherer *et al.*, 1959d).

The overwinter maintenance of JE in temperate areas and over the dry season in tropical areas remains to be clearly established. In tropical areas such as Thailand, Sarawak, and southern India, a simple pattern of year-round transmission of the virus between mosquitoes, birds, and pigs seems plausible (Grossman *et al.*, 1974; Simpson *et al.*, 1974). Among the many possible mechanisms for viral persistence and reemergence in temperate (as well as tropical) regions, transovarial transmission in mosquitoes, especially *Aedes* spp., demonstrated in the laboratory (Rosen *et al.*, 1978), is receiving current emphasis, but survival of virus in hibernating *C. tritaeniorhynchus*, hibernating bats and reptiles, and reintroduction by migratory birds remain possible alternatives.

The principal epidemic vectors are *Culex* species. *C. tritaeniorhynchus*, the most important vector in Japan and many other areas, is a rice paddy breeder, and populations reach high densities during the hot summer months. The decline in JE in Japan and Taiwan in recent years is in part attributable to improved rice farming techniques and use of agricultural insecticides. Other implicated species include *C. annulus* (Taiwan), *C. gelidus* and *fuscocephalus* (Malaysia, Thailand), *C. vishnui* and *pseudovishnui* (India), and *C. annulorostris* (Guam). The *Culex* vector species are primarily attracted to large mammals (including swine) but also occasionally feed on avians. Experimental transmission of JE virus has been documented for *C. tritaeniorhynchus* (Gresser *et al.*, 1958) and *C. fuscophalus* (Muangman *et al.*, 1972).

E. Louping Ill Virus

Louping ill virus is transmitted by *Ixodes ricinus* ticks during the spring and summer in Scotland and northern England. The tick larvae and nymphs parasitize and may transmit the virus between small mammals (hares, field mice, shrews) and ground-dwelling birds. The virus has been isolated from the brains of sick red grouse (*Lagopus l. scoticus*) in Scotland, and this species has been shown experimentally to develop fatal infection and viremias sufficient to infect ixodid tick vectors. Sheep and other large mammals are hosts for the adult stage of *I. ricinus*, but sheep may also serve as amplifying hosts (Smith, 1962). Transovarial transmission of the virus in *I. ricinus* has not been demonstrated. Man is rarely infected in nature, but laboratory infections and contact spread from affected sheep to butchers and veterinarians have been frequent. Tick-borne fever, a concurrent rickettsial (*Rickettsia phagocytophilia*) infection of sheep, appears to aggravate louping ill, presumably by favoring neuroinvasion (Gordon *et al.*, 1962).

F. Israel Turkey Meningoencephalitis

Two summer–fall outbreaks, in 1958 and 1959, involving turkey farms in the Shomron area of Israel have been described. Spread within affected flocks and to other flocks was slow and irregular. The virus is presumed to be mosquito-borne on the basis of its serological relationships and susceptibility of *A. aegypti* to experimental infection (Komarov and Kelmar, 1960), but the vector species and wild vertebrate hosts are unknown.

G. Wesselsbron

The disease in sheep was first recognized and the virus isolated in South Africa in 1955 (Weiss *et al.*, 1956). Subsequently, there have been numerous instances of sporadic or epizootic disease in sheep characterized by abortion of pregnant ewes and severe and fatal hepatitis in newborn lambs. The disease may be confused with (or occur concurrently with) Rift Valley (RV) fever. Although cattle are frequently infected and may serve as viremic hosts, they are not believed to be clinically affected (Blackburn, 1977).

The virus is mosquito borne. It has been repeatedly isolated from *A. caballus* and *A. circumtuleolus* in South Africa and from *A. lineatopennis* in Rhodesia (Kokernot *et al.*, 1960; Worth *et al.*, 1961); experimental transmission has been documented for the first two species. Sheep and cattle appear to be viremic, amplifying hosts. The antibody prevalence in domestic livestock approaches 50% in some areas of South Africa (Kokernot *et al.*, 1961). Antibody rates in man are lower; human infections result not only from mosquito bite but also (in butchers, veterinarians, ranchers) by contact spread.

The maintenance cycle is poorly known. An isolate was made from a species of *Desmodillus,* which was also shown to become viremic after experimental inoculation (McIntosh and Gear, 1975). However, other rodents belonging to the genera *Mystromys, Mastomys, Aethomys, Tatera,* and *Saccostomys* were resistent (McIntosh, 1961). Antibodies have been found in wild birds (Paterson *et al.*, 1957), and three species of *Anatidae* have been shown to circulate the virus after infection (McIntosh and Gear, 1975).

Wesselsbron virus has been isolated in Cameroun, Nigeria, and Thailand. A closely elated agent (Sepik virus) has been recovered in New Guinea. No disease has been reported from these areas.

H. Other Arboviruses

WN virus has a wide distribution in Africa, the Middle East, India, Europe, and the Soviet Union. In Egypt, where an isolated case of equine encephalitis was reported (Schmidt and El Mansoury, 1963), antibodies have been found in up to 81% of equines, with a north-to-south increase in prevalence. The primary transmission cycle involves wild birds and *Culex* mosquitoes (Taylor *et al.*, 1956; Work, 1971; McIntosh *et al.*, 1976), the species varying with geographic area. In Africa, *C. univittatus,* an ornithophilic mosquito, is the principal vector. During a recent human outbreak in South Africa, infection rates in this species were very high (39/1000 mosquitoes) (McIntosh *et al.*, 1969; 1976). In Europe, *C. modestus* has been implicated, and in India, mosquitoes of the *C. vishnui* complex. *C. pipiens quinquefasciatus* (= *fatigans*) is susceptible to experimental inoculation. The virus has also been isolated from and shown to replicate in bird-feeding ticks, which are suspected to play a role in overwintering. Because WN virus infection is so prevalent, disease is so rare, and viremia levels are minimal, equines are considered to be incidental to the transmission cycle and to be quite resistant to clinical infection. Bovines do not develop viremia after experimental inoculation, but antibodies in cattle are prevalent. Most wild rodents investigated are quite resistant to infection (McIntosh, 1961).

SF virus has been isolated in East, Central, and West Africa from mosquitoes (mainly *Aedes* spp.), sentinel mice, wild birds, and a hedgehog. Serosurveys indicate that the virus (or a close relative) is present in southeast Asia and southern Africa. Antibodies in Africa are prevalent in man and nonhuman primates and have also been found in rodents and domestic animals. The maintenance cycle is unknown, and the association with equine disease based solely on serology is problematic.

Aura virus has been recovered from mosquitoes in Brazil and Argentina. The virus has also been isolated from a sick equine in Argentina. Antibodies have been found in rodents, marsupials, and (at low prevalence) horses (Causey *et al.*, 1963). The maintenance cycle has not been elucidated. Una virus has been isolated from two horses in Argentina and from a variety of mosquito species in Colombia, Brazil, Trinidad, and Panama, many of which are primarily large mammal feeders.

RR virus was first isolated in 1963 and suggested as the etiological agent of epidemic polyarthritis, a dengue-like human disease described in Australia as early as 1927 (Doherty *et al.*, 1964). The disease occurs in relatively small outbreaks in eastern Australia; in 1979 an explosive epidemic was recognized in Fiji involving tens of thousands of human cases. The virus has been repeatedly isolated from *A. vigilax* and *C. annulirostris,* which are considered the major vector species (Gard *et al.*, 1973); experimental studies with these species support this concept (Kay *et al.*, 1975). Antibodies have been found mainly in man, domestic livestock, and wild mammals (wallabies, kangaroos, rats, bats). It has been suggested that the large marsupials may be important reservoir hosts. The demonstration of high viremias in experimentally infected young lambs (Kay *et al.*, 1975) indicates that livestock may also play a role in amplification. The association of RR with muscle and joint disease in horses has been discussed above.

MVE has been the cause of small outbreaks of CNS infections in man in Australia, first recognized in 1917; the virus was isolated from fatal cases in 1951 (French, 1952). The virus is enzootic in tropical areas of northern Australia and New Guinea, with *C. annulirostris* (and possibly other species) and wild birds constituting the basic transmission cycle (Anderson, 1954; Doherty *et al.*, 1976). Outbreaks follow invasion of the temperate zone during periods favorable to vector activity. Antibodies are prevalent in a variety of birds and mammals, including livestock. As discussed previously, on serological grounds, the virus is suspected to cause CNS disease in equines.

Middelburg virus has been isolated from mosquitoes (mainly *Aedes* spp.) in South Africa, Senegal, Kenya, Cameroun, and the Central African Empire. Antibodies have been found in man in Mozambique, and in sheep, goats, and cattle in South Africa (Kokernot *et al.*, 1961). Antibodies have been found to be present in a high proportion of mares that have aborted, and the virus has been isolated from sick horses (B. M. McIntosh, personal communication). Experimentally inoculated lambs develop fever and viremia and are suspected to be involved in virus amplification in nature, with *A. caballus* as the probable vector. Rodents of five genera develop viremias after experimental infection (McIntosh, 1961). The maintenance cycle is not known.

Getah virus caused an outbreak of fever and rash among 722 (38%) of 1903 thoroughbred and Anglo-Arabian horses at a race horse training center in central

Japan in 1978 (Kamada *et al.*, 1980). This was the first indication that the virus was of medical importance. Getah virus was first isolated in 1955 from *C. gelidus* in Malaysia and subsequently has been recovered from other mosquitoes (*C. tritaeniorhynchus, C. bitaeniorhynchus, Anopheles amictus,* and *A. vexans*) and from pigs in Malaysia, Cambodia, northern Australia, the Soviet Union, and Japan. Studies in Japan were conducted during and after the epizootic. The virus was repeatedly isolated from *A. vexans,* and pigs showed high antibody rates (T. Kumanomido, personal communication).

V. COMPARATIVE DIAGNOSIS

Other viral and nonviral CNS diseases of equines may be confused on clinical grounds with EEE, WEE, VEE, JE, etc. Plant poisonings (with *Senechio, Crotalaria,* yellow star thistle, etc.), botulism, listerial meningoencephalitis, and heavy metal poisoning are among the nonviral and rabies virus infections among the viral diseases to be considered. The diagnosis of arboviral equine encephalitis is suggested by the epizootic nature of the disease, the seasonal and geographic distribution of cases, and the presence of consistent histopathological lesions in the CNS. Borna disease in Europe is distinguished on epidemiological grounds, since (with the possible exception of WN virus) arboviral equine encephalitis is not recognized in this region.

The differential diagnosis of abortion in swine, sheep, and equines includes a wide array of considerations, including *Salmonella, Streptococcus, Brucella, Vibrio,* and *Listeria* infections, toxoplasmosis, equine viral rhinopneumonitis, Q fever, equine arteritis, and nitrate poisoning. In southern Africa, abortion in sheep due to Wesselsbron virus infection must be differentiated from Rift Valley fever and *Tribulosis* poisoning.

Specific diagnosis depends upon isolation of the virus and/or demonstration of a serological response. Virus isolation may be attempted by inoculation of serum or suspension of tissues (usually the major target organ, e.g., brain, placenta, liver) into a susceptible host system. Serum or tissue suspensions collected in the field should generally be preserved in dry ice, liquid nitrogen, or a mechanical freezer (−70°C) until tested for virus. The alphaviruses (e.g., EEE), however, are reportedly quite stable at ambient temperatures for several days in blood dried on paper discs (Karstad, 1964), but preservation at ultra-low temperatures is to be preferred.

Infant mice (2–4 days old) inoculated by the ic route have been most widely used for virus isolation and, for the viruses under consideration, are at least as sensitive, or 1–2 logs more sensitive than cell cultures (Kissling, 1957; Karabatsos and Buckley, 1967; Gorman *et al.,* 1975). Suckling hamsters and chicken embryos have also been employed. For the diagnosis of EEE and WEE, brain tissue of equines is the best source of virus, whereas serum less frequently yields virus (Rowan *et al.*, 1968). In cases of VEE, isolation attempts from brain tissue are sometimes unsuccessful. A useful technique in the field is to examine and sample apparently well horses in the same vicinity as an encephalitic case; virus isolations can often be made from the sera of febrile horses which are sustaining early or inapparent infections.

If possible, it is worthwhile to inoculate material for virus isolation attempts into both mice and one or more cell culture substrates known to be susceptible to the suspected agent(s). The average survival time in mice and the time of appearance and morphology of plaques in cell cultures may allow an early presumptive identification. Infected mouse brain tissue may be used for specific serological identification by the CF and neutralization tests. For viruses that readily hemagglutinate (these include most, if not all, of the agents under consideration), HA antigens for use in preliminary serological identification of viral isolates can also be prepared from mouse brain or infected cell cultures (Della-Porta and Westaway, 1972), including a harvest of monolayer cultures used for primary isolation by plaque assay (Kirk and Holden, 1974). Rapid identification of virus in cell culture can also be achieved by immunofluorescence (Metzger *et al.*, 1961) or solid-phase radioimmunoassay (Levitt *et al.*, 1976). In general, the success of identification of the agent is enhanced by at least one passage to increase titer in the most appropriate substrate.

A variety of cell types are suitable for primary isolation attempts. Primary Pekin duck embryo (DE) or chick embryo fibroblast cultures are widely used for the isolation of EEE, WEE, and VEE viruses, but continuous cell lines (e.g., BHK, Vero) are also useful. Many flaviviruses produce CPE or plaques in DE, BHK-21, CER, Vero, LLC-MK2, or porcine kidney cells. In general, if cell cultures are to be used for primary isolation attempts, their susceptibility to unpassaged viruses under consideration should be assessed and compared to suckling mice.

In recent years, continuous cell lines derived from mosquitoes have received attention, especially for the isolation of viruses (such as dengue) which are difficult to recover in conventional test systems. With other togaviruses, both CPE (usually syncytia formation) and plaques have been reported, depending upon the virus and mosquito cell line tested (Singh and Paul, 1968; Race *et al.*, 1980; Hsu *et al.*, 1978). Potential advantages of the use of arthropod cell lines are the ease with which they can be grown and maintained, their optima for growth (and viral replication) at relatively low temperatures (~ 28°C), and their use under field conditions whereby clinical specimens can be inoculated, without the need for ultra-low temperature preservation. No comparative assessment has yet been made of the usefulness of mosquito cells for primary isolation of the togaviruses under consideration in this chapter.

Inoculation of live mosquitoes is also a useful technique for the isolation and identification of some fastidious arboviruses (e.g., dengue). Mosquitoes are injected intrathoracically with putative infected material, held for an appropriate extrinsic incubation period, and then examined by immunofluorescent techniques or CF for antigen. Details of the procedure are given by Rosen and Gubler (1974) and Kuberski and Rosen (1977).

A special problem is presented by the isolation of VEE virus from equines during the course of an epizootic in which live, attenuated TC-83 vaccine is being used. In this instance, it is not unusual to isolate TC-83 virus from equines bled up to 14 days after vaccination (Calisher and Maness, 1975). TC-83 virus has also occasionally been recovered from mosquitoes presumably infected by feeding on a viremic vaccinated horse (Pedersen *et al.*, 1972c). Differentiation in the laboratory of TC-83 from epizootic viruses may be epidemiologically important. Epizootic subtypes IAB and IC are readily distinguished from TC-83 by their virulence for adult guinea pigs, hamsters, or mice inoculated by the peripheral route (Calisher and Maness, 1974, 1975).

Definitive serological subtyping of EEE, WEE, and VEE virus isolates may be epidemiologically important and may provide information about the origin of an outbreak, but this requires specialized test procedures and/or reagents (see Section II, A). For example, in the event of an EEE epizootic in Central America or the Caribbean, it would be important to establish whether the North or South American serotype was responsible. Determination of the infecting viral subtype is most readily made by serological (or physicochemical) examination of a virus isolate; however, VEE subtype and variant-specific neutralizing antibodies are produced by equines (Walton *et al.*, 1973) and serological tests may be employed for the retrospective determination of the antigenic type responsible for infection.

Serological diagnosis is achieved by demonstrating the presence of specific antibodies in appropriately timed serum samples. The presence of antibodies in a single convalescent-phase serum provides a presumption of infection, but the relationship to the current illness is uncertain. Demonstration of a rising (or falling) antibody titer between paired serum samples confirms the diagnosis, but it is less often achieved in the case of domestic animals than humans. A variety of tests have been employed for serodiagnosis, but the conventional HI, CF, and neutralization techniques are most widely employed.

A. Hemagglutination Inhibition Test

HA by specific antibody forms the basis for the HI test (Clarke and Casals, 1958; Hammon and Sather, 1969). The test is relatively simple, and results are rapidly obtained. Serological cross reactivities between members of some virus groups are great; thus, inclusion of an antigen in the HI test may provide a useful screening procedure for diagnosis of infection with heterologous members of the same virus group. The marked cross reactivity of HI antibodies to many togaviruses is, however, also a limitation of the technique, since virus-specific diagnosis may be impossible. The alphaviruses elicit a more specific primary HI antibody response than the flaviviruses. IgG antibodies show marked cross reactivity within antigenic subgroups of the flaviviruses, whereas IgM HI antibodies are more specific (Westaway *et al.*, 1974); measurement of IgM HI antibodies may provide a specific etiological diagnosis in certain flaviviral infections.

Another limitation of the technique is the presence of nonspecific inhibitors of HA in serum phospholipid fractions which must be removed by acetone extraction or kaolin adsorption. In the case of avian serum, acetone-insoluble, nonspecific inhibitors are present and must be inactivated with protamine sulfate (Holden *et al.*, 1966).

HI antibodies appear during the first week after primary infection (Hetrick, 1960). Peak antibody titers are attained 14–30 days after infection and often decline over the next month. Thereafter, HI titers decline very slowly, and antibody may persist for many years.

B. Complement Fixation Test

CF antibodies to togaviruses generally appear during the second or third week after primary infection and reach peak titer at 1 or 2 months. Antibody may decline to undetectable levels 1 to 2 years after infection; therefore, the test is most useful for the diagnosis of infection within a definite period prior to collection of the serum sample. The CF test is more specific than the HI test for the diagnosis of togavirus infections and can be carried out with viral antigens which fail to hemagglutinate. The technique is not useful for most avian sera, since with the exception of sera from pigeons and parrots, guinea pig complement is not bound. An indirect (inhibition) CF test may, however, be employed for avian sera (Tesh and McCammon, 1979). There have been few reports of the CF antibody responses of equines to natural infections with encephalitis viruses; from the available data it appears that the CF test is rather insensitive in equines. Bivalent EEE/WEE vaccine stimulates production of detectable CF antibodies in only about 10% of equines, whereas the live attenuated TC-83 VEE vaccine induces CF antibodies in the majority of equines (see Section VI, B).

C. Neutralization Test

The basis of the neutralization (N) test is the ability of an immune serum to render a suspension of virus nonpathogenic for a susceptible host, either a laboratory animal or cell culture. The test may be performed by mixing undiluted test serum (and normal serum as a control) with varying tenfold dilutions of virus and inoculating the mixtures into a susceptible host system (usually litters of suckling mice by the ic route); the reduction in virus titer over control is expressed as the neutralization index. A more widely employed test performed with monolayer cell cultures measures the reduction in viral plaque infectivity of a constant amount of virus afforded by serial dilutions of the test serum. In order to eliminate the possibility that differences in accessory factor concentrations between paired sera could affect the interpretation of N test results, it is advisable to heat-inactivate the samples prior to testing. Addition of a standardized source of accessory factor (Chappell *et al.*, 1971) back to the test system is optional.

Nonspecific neutralizing substances may occasionally present problems. These substances may be present in the sera of birds that have been shot (Scherer *et al.*, 1964). Nonspecific viral inhibitors (in plaque reduction N and HI tests) of EEE, WEE, and VEE viruses have been described in cattle sera (Scherer *et al.*, 1972).

The choice of viral strain for use in N (or other tests with high specificity) may be important. Walton *et al.* (1973) have shown that horses infected with epizootic VEE virus subtypes (IAB and IC) develop significantly higher N antibody titers to homologous test virus than to subtypes IE or II (Everglades virus). For biohazard reasons, it may be advisable to use TC-83 virus (the attenuated vaccine derivative of subtype IAB) as a test virus in N tests for the detection of antibodies to epizootic VEE virus.

Neutralization of an alpha togavirus (SIN) grown in mosquito cells (but not in mammalian cells) by antisera to antigens of purportedly uninfected vector mosquitoes has been reported (Feinsod *et al.*, 1975). It was proposed that virion envelopes produced in mosquito cells contain mosquito surface antigenic sites. Confirmation of these results and elucidation of the importance of the phenomenon in nature are needed.

As opposed to the N test in mice, the plaque-reduction test offers many advantages including sensitivity, quantitative precision, specificity, economy of serum, and simplicity of the titration of antibody. Because the N test is often essential to specific diagnosis, its advantages far outweigh its disadvantages, which include the expense of maintaining cells and the relatively long time required to obtain results. A modification of the N test technique used in some laboratories is the focus reduction method with fluorescent antibody or peroxidase-antiperoxidase staining (Okuno *et al.*, 1978). This method reduces the time required for incubation of test virus and is useful for viruses which do not plaque in cell cultures or are not pathogenic for mice.

The plaque-forming characteristics of arboviruses and use of plaque N tests for arboviruses have been studied by a number of investigators (Earley *et al.*, 1967; Bergold and Mazzali, 1968; Stim, 1969; Gorman *et al.*, 1975). Plaque formation and size are determined by a wide variety of factors, including the viral strain, cell substrate, pH, and composition of overlay.

The mechanism of neutralization and the variables involved are beyond the scope of this chapter and have been recently reviewed by Mandel (1978). Conditional neutralization, dependent upon the cell substrate used, has been described for some togaviruses, may be related to intracellular processing of virus–antibody complexes, and is an important consideration in selecting a system for diagnostic or survey serological tests (Fig. 19). The presence of a thermolabile neutralizing accessory factor in serum from patients convalescent from togaviral infections is well known. However, the enhancement of neutralization by complement (guinea pig serum) seems to depend upon the host–virus pairing, the developmental stage of the immune response, and the substrate used to grow challenge virus.

N antibody may appear within the first few days after infection. Peak titers are reached at 1 to 2 months; antibody persists for a prolonged period, probably for life.

D. Other Tests

The radioimmunoassay is rapid, sensitive, precise, and quite specific for the detection of arboviral antibodies (Trent *et al.*, 1976; Jahrling *et al.*, 1978), and it is readily adaptable for the measurement of antibody in the individual immunoglobulin classes. Use of purified viral structural and nonstructural proteins as antigens in the radioimmunoassay may provide highly sensitive and specific results. The technique can be adapted for rapid testing of large numbers of sera. Drawbacks are the expense of radioisotopes and the scintillation counter, the specialized nature of some reagents, and the need to control radioactive sub-

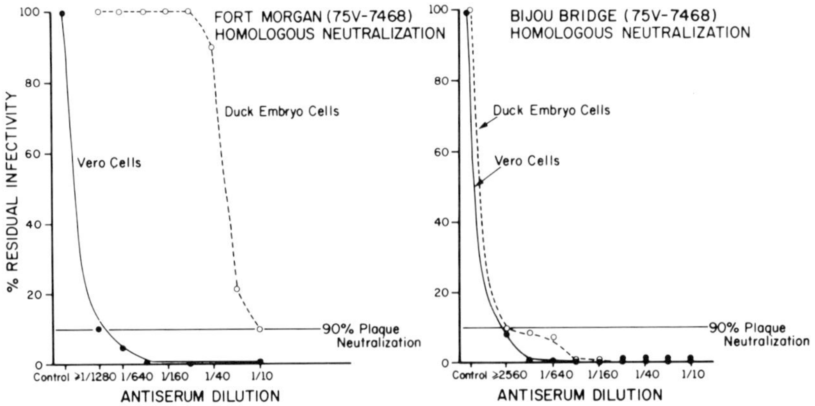

Fig. 19. Neutralization of an alphavirus, conditional upon the cell culture substrate used. Fort Morgan virus neutralization is demonstrated in Vero but not in primary duck embryo cells, whereas Bijou Bridge virus (a strain of Tonate virus) neutralization can be assayed in both substrates. (From Monath *et al.* (1980), with permission.)

stances. The most applicable method employs staphylococcal protein A as a solid-phase immunoabsorbent for [^{3}H]labeled viruses complexed with IgG antibody (Jahrling *et al.*, 1978).

The ELISA technique has also been applied to the measurement of arboviral antibodies (Hofmann *et al.*, 1979). Like the radioimmunoassay, ELISA lends itself to the determination of IgG and IgM antibodies and is more sensitive than the HI test. The specificity of the test for cross-reactive viruses (e.g., some flaviviruses) has not been investigated. ELISA has advantages over the radioimmunoassay and N tests, respectively, in that work with radioactive substances and infectious viruses is avoided.

The indirect fluorescent antibody test is applicable to the diagnosis of togaviral infections, but we are not aware of studies on its use in detecting antibodies in domestic or wild animals.

E. Choice and Sequence of Tests

For serological diagnosis of arbovirus infections, an acute serum collected 0 to 7 days after onset and an early convalescent serum collected 10 to 21 days after the first serum specimen are needed. A late convalescent serum collected 6 to 8 weeks after onset may also be useful for detection of a decline in HI or CF antibody titers. The CF test may be used as the primary diagnostic procedure; however, because of the earlier appearance of HI antibody and the greater sensitivity of the test, it is preferable initially to screen sera by HI against an appropriate battery of antigens. If a serum is positive to one or more antigens used in the HI test, the acute and convalescent sera should then be tested by the CF and N tests for extension and confirmation of results. To assure comparable serological results, paired acute and convalescent sera must be examined in the same test.

Depending upon the expertise and experience of the laboratory, one of the less widely used tests (radioimmunoassay, ELISA, etc.) may be selected. The sensitivity and specificity of these newer tests should be examined with reference to conventional serological procedures before they are used for routine diagnosis or surveys.

F. Interpretation

A positive result is indicated by an HI antibody titer of $\geq$ 10 or a CF titer of $\geq$ 8. In the serum dilution N test in cell culture, $\geq$ 90% plaque reduction by an undiluted serum is often considered positive and 70 to 89% plaque reduction equivocal; however, less rigorous limits of interpretation are sometimes used. Lack of correlation between HI or CF and N test results may indicate a technical error or infection with an antigenically related virus which induces antibody that is cross-reactive by HI or CF but not by the more specific viral neutralization.

Interpretation of the newer tests such as radioimmunoassay and ELISA should be based upon comparative titrations of standard sera using conventional methods (HI, CF, N). The radioimmunoassay procedure of Jahrling *et al.* (1978) was about three times more sensitive than and at least as specific as the plaque reduction N test for the detection of alphavirus antibodies. A positive indirect fluorescent antibody test at low dilutions of serum (1:4–1:8) may be significant, but convalescent phase titers are frequently $\geqslant$ 128.

When only a single serum is available, evidence for infection at some time in the past is provided by a positive result in any of the tests described, but a relationship with the current illness remains problematic. Presumptive evidence for such a relationship may be assumed if the HI antibody titer is high and/or if CF antibodies, which reflect recent infection, are present.

For serological confirmation, it is necessary to interpret comparative titrations of paired sera collected at appropriate intervals in relation to the onset of clinical symptoms. A fourfold or greater rise or fall in antibody titer confirms recent infection with the virus used in the test and indicates that the illness was etiologically related to it. Twofold changes in antibody titers are regarded as inconsequential and may be found if the paired serum samples are taken within too short an interval. Stable HI and/or N antibody titers in both an early acute and a convalescent serum and absence of CF antibody in the latter suggest remote infection unrelated to the current illness. The absence of detectable HI antibody in a serum obtained 2 weeks or more after onset of clinical symptoms is presumptive evidence against infection, but absence of neutralization antibody is more conclusive.

Primary infection is usually followed by the development of antibodies which are immunologically monotypic. With time, antibodies develop which are more cross-reactive with heterologous antigens. Advantage can be taken of the cross reactivity of togavirus HI antibodies, since infection with a member of a serological group may be detected despite omission from the test of the specific antigen responsible for infection. Because of the antigenic sharing between members of the alphavirus and flavivirus groups and different specificities of the available tests, serological interpretation is most reliable when HI and CF antibody titers are compared with a variety of antigens within the serogroup and when confirmation by the N test has been achieved. A fourfold or higher titer to one virus antigen than to other members of a serogroup generally indicates infection with that agent, although heterotypic reactions may occur. Results should be reviewed with regard to their compatibility with the known seasonal activity of arthropod-borne virus infection and with the geographical distribution of the viruses in question.

Definitive diagnosis is most difficult when infection occurs in an individual previously naturally infected or vaccinated with an antigenically related virus. This problem is particularly evident in tropical regions where two or more related viruses are endemic, although in the United States vaccination with EEE/WEE vaccine may complicate interpretation of serological results. Superinfection usually results in the development of a rapid and broad anamnestic response that may preclude the demonstration of a rise in titer or prevent specific diagnosis. In some instances, the antibody response may be heterotypic with a higher titer to the original antigen than to the agent responsible for the current illness.

VI. PREVENTION AND CONTROL

A. Eastern and Western Equine Encephalitis Vaccine

Inactivated EEE and WEE vaccines (commercially available in the United States as a bivalent preparation or as a trivalent vaccine incorporating inactivated attenuated VEE virus) are used for the prevention of disease in individual equines. Since equines are not important viremic hosts for EEE and WEE viruses, vaccination does not preclude enzootic/epizootic transmission and spread of the virus to unvaccinated equines or to man. EEE/WEE vaccine is derived from the fluids of infected chicken embryo fibroblast cell cultures, which are inactivated with formalin. The vaccine is administered in two divided doses at an interval of 3 or 4 weeks, and revaccination is recommended at yearly intervals. Immune responses to bivalent EEE/WEE and trivalent killed EEE/WEE/VEE vaccines and challenges of immunized horses with virulent viruses have been described (Gutekunst, 1969; Barber *et al.*, 1978). One week after administration of the first vaccine dose to horses, 72% had detectable N antibodies to VEE and 52% had antibodies to WEE and EEE; 93 to 97% were seropositive at 2 weeks after the first vaccine dose. A second dose of vaccine was given at 3 weeks, after which solid immunity was present to all three viruses. Highest titers of VEE N antibodies (geometric mean titer (GMT), 512) and of WEE N antibodies (GMT, 188) appeared 1 week after the second vaccine dose; highest titers of EEE N antibodies (GMT, 148) were present 3 weeks after the second dose. The lower EEE and WEE antibody responses as compared to VEE antibodies were also reflected in briefer persistence. All horses remained seropositive to VEE and EEE for 5 months and to WEE for 3 months after completion of the two-dose vaccination. However, at 1 year, only 6% of the horses had EEE and WEE antibodies, whereas 94% remained positive to VEE. Despite the absence of detectable N antibodies, vaccinated horses resisted challenge with virulent EEE virus 12 months after primary immunization; WEE challenge experiments were inconclusive. A single dose of trivalent vaccine given at 12 months provided maximal antigenic stimulation.

Bivalent EEE/WEE vaccine is also widely used for the protection of pheasant flocks against EEE epornitics.

B. Venezuelan Equine Encephalitis Vaccine

The development and evaluation of the live attenuated VEE (TC-83) vaccine and its use in man and equines have been reviewed by the Pan American Health Organization (1972), Spertzl (1973), and Johnson and Martin (1974).

The reactions of equines to the live vaccine are characterized by a mild febrile response (often biphasic and generally between the second and fifth days after vaccination), variably detectable viremia, leukopenia, minor clinical illness with depression and anorexia in up to 50% of equines (if carefully examined), and rapid development of HI, CF, and N antibodies and resistance to challenge with virulent VEE virus (Spertzl and Kahn, 1971; Walton *et al.*, 1972).

Viremias in vaccinated equines have been of low magnitude (generally < 3.0 $\log_{10}$ units/ml, occasionally up to 3.7 $\log_{10}$ units/ml) and last from 1 to 4 days between the first and sixth days after vaccination. The titer of viremia is thus considered to be below the threshold for vector mosquito infection, and the risk of reintroduction of vaccine virus into nature is considered minimal. The vaccine virus, nevertheless, has been recovered (on at least two occasions) from wild-caught mosquitoes, suggesting that the occasional equine circulates higher virus titers or that some mosquito species may be more susceptible than presently credited. There is some basis for the possibility of reversion to virulence in the equine host (or, rather, selection and replication of a subpopulation of virions capable of inducing higher viremia). P. Jahrling (personal communication), by use of hydroxylapatite chromatography, has separated clones from TC-83 vaccine which have elution profiles of parent (Trinidad donkey) virus and are more virulent for hamsters than uncloned TC-83 virus. Virus strains recovered from some human vaccinees with clinical reactions also have similar characteristics. Oligonucleotide fingerprinting and tryptic peptide analyses of these clones, however, have not shown significant differences between hamster-virulent clones and standard TC-83 virus (D. W. Trent, unpublished data).

Passage of TC-83 virus in mice and hamsters may restore virulence (McKinney *et al.*, 1963; Jahrling and Scherer, 1973a). Back passage of TC-83 in equines resulted in increased frequency and severity of clinical reactions and high viremia, but did not alter the virus' lack of virulence for guinea pigs (Luedke *et al.*, 1972).

Live TC-83 vaccination is followed by the appearance of N, HI, and CF antibodies as early as 5, 6.5 and 8 days, respectively (Walton *et al.*, 1972). By 8 days, all vaccinated animals have demonstrable N antibodies; there is ample evidence to suggest that protection against lethal infection with virulent VEE virus is present within 3–4 days after administration of the vaccine (Eddy *et al.*, 1972). The vaccine induces highest N antibody titers to itself (TC-83); long-lasting, high-titer antibodies are also formed to the epizootic subtypes IAB and IC but not to enzootic ID and IE viruses (Walton *et al.*, 1972). Horses retain heterologous antibodies to TC-83 and to the epizootic subtypes as long as 30 months after TC-83 vaccination (Walton and Johnson, 1972b), and have been shown to resist challenge with virulent virus as long as 14 months after vaccination (Walton *et al.*, 1972). Horses with plaque-neutralizing serum titers (to vaccine virus) of $\geqslant 40$ have been shown to be refractory to challenge (Jochim *et al.*, 1973); this level of immunity is apparently maintained for 30 months or longer (Walton and Johnson, 1972b).

In a study by Ferguson *et al.* (1979), foals of VEE-immune mares were passively immunized by way of colostrum. The half-life of maternal antibodies was 20 days, and the persistence of immunity in the foal was related to colostrum antibody titer. Nonimmune foals vaccinated at 1.7–5.4 months of age became immunized, apparently without untoward reactions. Passively acquired maternal antibodies prevented active immunization.

A number of studies have shown that the presence of preexisting heterologous EEE and/or WEE antibodies in equines interferes with VEE immunization with live TC-83 vaccine (Calisher *et al.*, 1973; Jochim and Barber, 1974; Ferguson *et al.*, 1978) (a result which was not unexpected on the basis of the well-established cross protection between these alphaviruses in experimental studies with a variety of species). Fig. 20 shows the suppressed antibody responses to TC-83 vaccine of horses with preexisting WEE antibodies. In another study of 60 horses which had multiple bivalent EEE/WEE vaccinations prior to receiving TC-83, only 57% had detectable N antibodies 18 months later compared to 100% of 30 horses with no record of EEE/WEE immunization (Vanderwagen *et al.*, 1975). In contrast to the interference by heterologous antibodies with replication and antigenic stimulation of live TC-83 vaccine, preexisting EEE or WEE immunity appears to enhance the response to administration of killed TC-83 vaccine.

The safety and efficacy of live TC-83 vaccine in equines are attested to by the absence of recognized severe reactions among several million animals which have received the vaccine in the field. Only minor, reversible histopathological changes occur in the brains of some vaccinated adult equines 5–9 days after inoculation (Monlux *et al.*, 1972). The vaccine has been shown in several field trials to be highly efficacious, in terms of both induction of antibodies in over 90% of equines and interruption of virus transmission and disease (Eddy *et al.*, 1972). Mass vaccination undoubtedly precluded further spread of the virus in Texas in 1971. The live vaccine has the obvious advantage of inducing long-lasting immunity, a feature of critical importance to the protection of equines and man in tropical, developing areas where annual revaccination with killed vaccine is impractical. One potential complication, still problematic, is the teratogenicity and abortogenicity of the vaccine, although available evidence is generally negative.

At the present time, most arbovirologists favor use of an inactivated vaccine prepared from the attenuated TC-83 strain (to avoid potential problems with

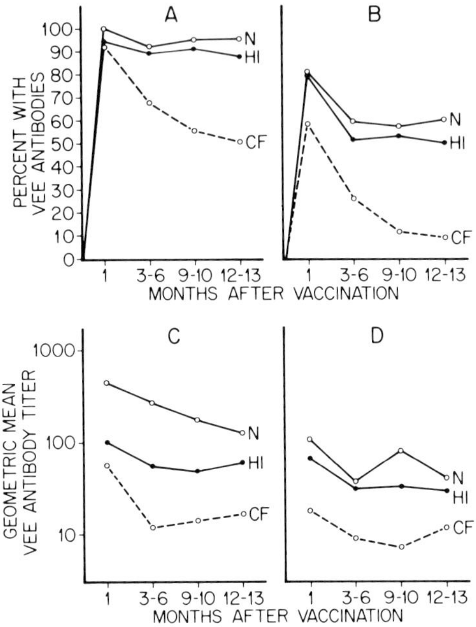

Fig. 20. Percentage of horses with VEE antibodies and geometric mean antibody titers after live attenuated VEE TC-83 vaccination. (A, C) Horses without prevaccination WEE antibodies; (B, D) horses seropositive to WEE prior to vaccination. (Reproduced with modification from Ferguson *et al.* (1978), with permission.)

escape of virulent virus from improperly formalinized preparations). As mentioned previously, an efficacious inactivated trivalent vaccine is licensed and commercially available in the United States. Use of killed TC-83 vaccine avoids the potential complications of virulence reversion, infection of and transmission by vectors, abortogenesis, and suppression of immunogenicity by heterologous antibodies. In the event of an epizootic, however, recourse must be made to the live vaccine, since rapid immunization is mandatory.

C. Japanese Encephalitis Vaccines

A 1% formalinized mouse brain vaccine was used on a wide scale for equine vaccination in Japan after World War II (Kitaoka, 1971) and is credited with a marked decline in the incidence in equine encephalitis (Ito, 1975). Use of killed mouse brain vaccine in humans was authorized in 1954 in Japan; at present the vaccine for human use is partially purified by alcohol–protamine sulfate treatment or ultracentrifugation, and it has been used for mass prophylaxis mainly in school children in Japan and elsewhere in Asia.

Scherer *et al.* (1959a) first suggested immunization of the principal amplifying hosts of JE (swine) as a means of controlling transmission of the virus and reducing human morbidity. The benefits for veterinary public health were also recognized, since immunization of swine prior to the epidemic season would reduce losses due to fetal infection and stillbirths. In the mid-1960s attempts to demonstrate efficacy of administration of inactivated vaccine to swine were inconclusive, and it is apparent that the ordinary 1% tissue concentration formalinized vaccine (used in equines and man) is not of sufficient antigenic potency for use in swine. A review of field trials in swine with killed vaccines at various dose levels is presented by Nakamura (1971). Field trials in Japan have shown a significant reduction of stillbirth in sows immunized prior to the epidemic season. Trials in pigs with inactivated vaccine and adjuvant have shown enhanced immune responses (Ogata *et al.*, 1970).

A live attenuated JE vaccine (M strain) was developed by passage through mouse fibroblast cells. The vaccine failed to produce viremia in pigs and lost its ability to replicate in mosquitoes (Kodama *et al.*, 1968; Tsuchiya, 1970; Inoue, 1971). Colostrum-deprived newborn piglets less than 48 hours of age were shown to be a suitable host for demonstrating attenuation of the vaccine. During 1967–1968 a field trial was conducted on Iki Island (Takahashi *et al.*, 1971). The results indicated significant antibody responses, a lower rate of mosquito infection, reduced incidence of natural swine infection, and an apparent reduction in human cases. Immunization of piglets was proposed as an effective means of reducing viral amplification and the risk of human disease.

Live attenuated vaccines have been commercially available in Japan since 1970. Four different strains available from six producers are employed for vaccination of swine. A single dose of 2 ml containing 10^5 to 10^6 suckling mouse IC LD_{50} has been shown to induce N antibodies in 100% of susceptible pigs within 1 week of vaccination (T. Okuno, unpublished data).

Because swine are such efficient amplifying hosts, a high prevalence of immunity is required to prevent human infections. One of the difficulties in applying this measure is timing of administration of vaccination. Since pigs are slaughtered at 6–8 months of age, vaccination must be targeted during a specific and narrow period—a practical difficulty in swine husbandry.

Live attenuated vaccines are widely used in equines in China without clinical reactions and with a seroconversion rate of approximately 90% (Wang *et al.*, 1976). Foals born to immunized mares acquired maternal antibodies which persisted for 4–5 months and interfered with active immunization.

D. Vector Control

In areas affected by repeated occurrences of togaviral disease, preventive measures aimed at reducing vector populations may be warranted. Such programs depend upon (1) a basic understanding of the natural history of the arbovirus in question and the biology of the vector(s) and (2) establishment of surveillance systems to define vector population density and distribution and, if possible, the level of viral activity in vectors or enzootic reservoir hosts. Preventive measures include reduction of vector breeding by environmental modification and use of biological and chemical techniques. In some instances, residual insecticides to kill adult vectors in proximity to vertebrate hosts have been employed.

In the event of an established outbreak or surveillance data indicating intense viral amplification, emergency measures must be employed to reduce the population of adult infected vectors. In the case of the mosquito-borne togaviruses, space sprays with ultra-low volume applications of organophosphorus compounds, are delivered by ground-operated or aerial equipment.

For reviews of the subjects of surveillance of vectors and viral activity and of vector-reduction methods utilized for prevention and control to togaviral infections, see publications by Mitchell *et al.* (1969), Lofgren (1970), Reeves (1970, 1971b), and the Pan American Health Organization (1972).

REFERENCES

Aaslested, H. F., Hoffman, E. J., and Brown, A. (1968). *J. Virol.* **2,** 972–978.

Acheson, N. H., and Tamm, I. (1967). *Virology* **32,** 128–143.

Acheson, N. H., and Tamm, I. (1970). *Virology* **41,** 306–320.

Ada, G. L., and Anderson, S. G. (1959). *Virology* **8,** 270–271.

Ada, G. L., Abbot, A., and Anderson, S. G. (1962). *J. Gen. Microbiol.* **29,** 165–170.

Anderson, S. G. (1954). *J. Hyg.* **52,** 447–468.

Andrewes, C. H., and Horstman, D. M. (1947). *J. Gen. Microbiol.* **3,** 290–306.

Appleyard, G., Oram, J. D., and Stanley, J. L. (1970). *J. Gen. Virol.* **9,** 179–189.

Austin, F. J., and Scherer, W. F. (1971). *Am. J. Pathol.* **62,** 195–219.

Avilán, J. (1966). *Rev. Venez. Sanid. Asist. Soc.* **31,** Suppl. 3, 787–805.

Bailey, C. L., Eldridge, B. F., Hayes, D. E., Watts, D. M., Tammariello, R. F., and Dalrymple, J. M. (1978). *Science* **199,** 1346–1349.

Baltimore, D., Burke, D. C., Horzinek, M. C., Huang, A. S., Kaariainen, L., Pfefferkorn, E. R., Schlesinger, M. J., Schlesinger, S., Schlesinger, W. R., and Scholtisser, C. (1976). *J. Gen. Virol.* **30,** 273.

Barber, T. L., Walton, T. E., and Lewis, K. J. (1978). *Am. J. Vet. Res.* **39,** 621–625.

Beadle, L. D. (1952). *Mosq. News* **12,** 102–107.

Beaudette, F. R., Black, J. J., Hudson, C. B., and Bivens, J. A. (1952). *J. Am. Vet. Med. Assoc.* **121,** 478–483.

Beaudette, F. R., Holden, P., Black, J. J., Bivens, J. A., Hudson, C. B., and Tudor, B. S. (1954). *Proc. U. S. Livestock Sanit. Assoc., 58th Annu. Meet.* pp. 309–321.

Beeman, K., and Keith, J. (1977). 4a2J. Mol. Biol. **113,** 165–179.

Bell, J. R., Hunkapiller, M. W., Hood, L. E., and Strauss, J. H. (1978). *Proc. Natl. Acad. Sci. U.S.A.* **75,** 2722–2726.

Bell, J. R., Strauss, E. G., and Strauss, J. H. (1979). *Virology* **97,** 287–294.

Belonje, C. W. A. (1958). *J. S. Afr. Vet. Med. Assoc.* **29,** 1-12.
Bennington, E. E., Sooter, C. A., and Baer, H. (1958). *Mosq. News* **18,** 299-304.
Bergold, G. H., and Mazzali, R. (1968). *J. Gen. Virol.* **2,** 273-284.
Bhatt, P. N., and Jacoby, R. O. (1976). *J. Infect. Dis.* **134,** 166-173.
Birdwell, C. R., and Strauss, J. H. (1974). *J. Virol.* **14,** 366-374.
Blackburn, N. K. (1977). Masters of Philosophy Thesis, Faculty of Medicine, University of Rhodesia, Salisbury.
Blair, C. (1977). *Am. Soc. Microbiol., Abstr. Annu. Meet.* p. 287.
Blood, D. C., and Henderson, J. A. (1974). "Veterinary Medicine," 4th ed., pp. 527-530. Williams & Wilkins, Baltimore.
Boone, R., and Moss, B. (1976). *Am. Soc. Microbiol., Abstr. Annu. Meet.* pp. 227-242.
Bose, H. R., and Sagik, B. (1970). *J. Virol.* **5,** 410-412.
Bose, H. R., Carl, Z., and Sagik, B. P. (1970a). *Arch. Gesamte Virusforsch.* **29,** 83-87.
Bose, H. R., Brundige, M. A., Carl, G. Z., and Sagik, B. P. (1970b). *Arch. Gesamte Virusforsch.* **31,** 207-213.
Boulton, P. S., Webb, H. E., Fairbairn, G. E., and Illavia, S. J. (1971). *Brain* **94,** 403-410.
Boulton, R. W., and Westaway, E. G. (1972). *Virology* **49,** 283-289.
Boulton, R. W., and Westaway, E. G. (1976). *Virology* **69,** 416-430.
Bowen, G. S. (1976). *Am. J. Trop. Med. Hyg.* **25,** 891-899.
Bowen, G. S. (1977). *Am. J. Trop. Med. Hyg.* **26,** 171-175.
Bowen, G. S., and McLean, R. G. (1977). *Am. J. Trop. Med. Hyg.* **26,** 808-814.
Bowen, G. S., Monath, T. P., Kemp, G. E., Kerschner, J. H., and Kirk, L. J. (1980). *Am. J. Trop. Med. Hyg.* **29,** 1411-1419.
Bradish, C. J., Allner, K., and Maber, H. B. (1971). *J. Gen. Virol.* **12,** 141-160.
Bradish, C. J., Allner, K., and Maber, H. B. (1972). *J. Gen. Virol.* **16,** 359-372.
Bradish, C. J., Allner, K., and Fitzgeorge, R. (1975). *J. Gen. Virol.* **28,** 225-237.
Brandt, W. E., Cardiff, R. D., and Russell, P. K. (1970). *J. Virol.* **6,** 500-506.
Brawner, T. A., Lee, J. C., and Trent, D. W. (1977). *Arch. Virol.* **54,** 147-151.
Brotherson, J. G., Bannatyne, C. C., Mathieson, A. O., and Nicolson, T. B. (1971). *J. Hyg.* **69,** 479-489.
Brown, A., and Officer, J. E. (1975). *Arch. Virol.* **47,** 123-138.
Brown, D. T., Waite, M. R. F., and Pfefferkorn, E. R. (1972). *J. Virol.* **10,** 524-536.
Brownlee, A., and Wilson, D. R. (1932). *J. Comp. Pathol.* **45,** 67-92.
Bruton, C. J. and Kennedy, S. I. T. (1975). *J. Gen. Virol.* **28,** 111-127.
Buescher, E. L., and Scherer, W. F. (1959). *Am. J. Trop. Med. Hyg.* **8,** 719-722.
Burge, B. W., and Strauss, J. H. (1970). *J. Mol. Biol.* **47,** 449-466.
Burke, D. J., and Keegstra, K. (1976). *J. Virol.* **20,** 676-686.
Burke, D. J., and Keegstra, K. (1979). *J. Virol.* **29,** 546-554.
Burns, K. F. (1950). *Proc. Soc. Exp. Biol. Med.* **75,** 621-625.
Buxton, D., and Reid, H. W. (1975). *J. Comp. Pathol.* **85,** 231-235.
Byrne, R. J., and Robbins, M. L. (1961). *J. Immunol.* **86,** 13-16.
Byrne, R. J., French, G. R., Yancey, F. S., Gochenour, W. S., Jr., Russell, P. K., Ramsburg, H. H., Brand, O. A., Schneider, F. G., and Buescher, E. L. (1964). *Am. J. Vet. Res.* **25,** 24-31.
Calisher, C. H., and Maness, K. S. C. (1974). *Appl. Microbiol.* **28,** 881-884.
Calisher, C. H., and Maness, K. S. C. (1975). *J. Clin. Microbiol.* **2,** 198-205.
Calisher, C. H., Maness, K. S. C., Lord, R., and Coleman, P. H. (1971). *Am. J. Epidemiol.* **94,** 172-178.
Calisher, C. H., Sasso, D. R., and Sather, G. E. (1973). *Appl. Microbiol.* **26,** 485-488.
Calisher, C. H., Brandt, W. E., Casals, J., Karabatsos, N., Murphy, F. A., Tesh, R. B., and Wiebe, M. E. (1981). *Intervirology* (in press).

Camenga, D. L., and Nathanson, N. (1975). *J. Neuropathol. Exp. Neurol.* **34,** 492-500.

Camenga, D. L., Nathanson, N., and Cole, G. A. (1974). *J. Infect. Dis.* **130,** 634-641.

Cardiff, R. D., Brandt, W. E., McCloud, T. G., Shapiro, D., and Russell, P. K. (1971). *J. Virol.* **7,** 15-23.

Casals, J. (1957). *Trans. N.Y. Acad. Sci.* [2] **19,** 219-235.

Casals, J. (1963). *An. Microbiol.* **11,** 13-34.

Casals, J. (1964). *J. Exp. Med.* **119,** 547-565.

Causey, O. R., Casals, J., Shope, R. E., and Udomsakdi, S. (1963). *Am. J. Trop. Med. Hyg.* **12,** 777-781.

Chamberlain, R. W., Kissling, R. E., Stamm, D. D., Nelson, D. B., and Sikes, R. K. (1956). *Am. J. Hyg.* **63,** 261-273.

Chamberlain, R. W., Sudia, W. D., Burbutis, P. P., and Bozne, M. D. (1958). *Mosq. News* **18,** 305-308.

Chanas, A. C., Johnson, B. K., and Simpson, D. I. H. (1977). *J. Gen. Virol.* **35,** 455-462.

Chanas, A. C., Johnson, B. K., and Simpson, D. I. H. (1976). *J. Gen. Virol.* **32,** 295-300.

Chanas, A. C., Hubalek, D., Johnson, B. K., and Simpson, D. I. H. (1979). *Arch. Virol.* **59,** 231-238.

Chappell, W. A., Sasso, D. R., Toole, R. F., and Monath, T. P. (1971). *Appl. Microbiol.* **21,** 79-83.

Chippaux-Hyppolite, C., Choux, R., Olmer, H., and Tamalet, J. (1970). *C. R. Hebd. Seances Acad. Sci.* **270,** 3162-3164.

Ciarcini, E., Fadda, G., Salvadori, C., Serventi, G., and Turano, A. (1968). *G. Microbiol.* **16,** 155-162.

Clarke, D. H. (1960). *J. Exp. Med.* **111,** 1-23.

Clarke, D. H. (1964). *Bull. W.H.O.* **31,** 45-56.

Clarke, D. H., and Casals, J. (1958). *Am. J. Trop. Med. Hyg.* **7,** 561-573.

Cleaves, G. R., and Dubin, D. T. (1979). *Virology* **96,** 159-165.

Clegg, J. C. S. (1975). *Nature (London)* **254,** 454-455.

Clegg, J. C. S., and Kennedy, S. I. T. (1974). *J. Gen. Virol.* **22,** 331-345.

Coetzer, J. A. W., and Barnard, B. J. H. (1977). *Onderstepoort J. Vet. Res.* **44,** 119-126.

Coetzer, J. A. W., Theodoridis, A., and Van Heerden, A. (1978). *Onderstepoort J. Vet. Res.* **45,** 93-106.

Cole, G. A., and Wisseman, C. L. (1969). *Am. J. Epidemiol.* **89,** 669-680.

Compans, R. W. (1971). *Nature (London) New Biol.* **229,** 114-116.

Dalrymple, J. M., Vogel, S. N., Teramoto, A. Y., and Russell, P. K. (1973). *J. Virol.* **12,** 1034-1042.

Dalrymple, J. M., Schlesinger, S., and Russell, P. K. (1976). *Virology* **69,** 93-103.

Daniels, J. B., Stuart, G., Wheeler, R. E., Gifford, C., Ahearn, J. P., Hayes, R. O., and MacCready, R. A. (1960). *N. Engl. J. Med.* **263,** 516-520.

Darnell, N. B., and Koprowski, H. (1974). *J. Infect. Dis.* **129,** 248-256.

David, A. (1971). *Virology* **46,** 711-720.

Davis, M. H., Hoegge, A. L., Jr., Corristan, E. C., and Ferrell, J. F. (1966). *Am. J. Trop. Med. Hyg.* **15,** 227-230.

de Diego, I. A., Grela, M. E., and Barrera Oro, J. G. (1975). *Gac. Vet.* **37,** 404-418.

Della-Porta, A. J., and Westaway, E. G. (1972). *Appl. Microbiol.* **23,** 158-160.

Della-Porta, A. J., and Westaway, E. G. (1977). *Infect. Immun.* **15,** 874-882.

DeMadrid, A. T., and Porterfield, J. S. (1969). *Bull. W.H.O.* **40,** 113-123.

DeMadrid, A. T., and Porterfield, J. S. (1974). *J. Gen. Virol.* **23,** 91-96.

Demsey, A., Steeve, R. L., Brandt, W. E., and Veltri, B. J. (1974). *J. Ultrastruct. Res.* **46,** 103-116.

Dickerman, R. W., Baker, G. J., Ordonez, J. V., and Scherer, W. F. (1973). *Am. J. Vet. Res.* **34,** 357-361.
Dickerman, R. W., Bonacossa, C. M., and Scherer, W. F. (1976). *Am. J. Epidemiol.* **104,** 678-683.
Dietz, W. H., Jr., Alvarez, O., Jr., Martin, D. H., Walton, T. E., Ackerman, L. J., and Johnson, K. M. (1978). *J. Infect. Dis.* **137,** 227-237.
Dietz, W. H., Jr., Peralta, P. H., and Johnson, K. M. (1979). *Am. J. Trop. Med. Hyg.* **28,** 329-334.
Digoutte, J. P., and Girault, G. (1976). *Ann. Microbiol. (Paris)* **127B,** 429-437.
Doby, P. B., Schnurrenberger, P. R., Martin, R. J., Hanson, L. E., Sherrick, G. W., and Schoenholz, W. K. (1966). *J. Am. Vet. Med. Assoc.* **148,** 422-427.
Doherty, P. C., and Reid, H. W. (1971). *J. Comp. Pathol.* **81,** 531-536.
Doherty, P. C. and Vantsis, J. T. (1973). *J. Comp. Pathol.* **83,** 481-491.
Doherty, P. C., Reid, H. W., and Smith, W. (1971). *J. Comp. Pathol.* **81,** 545-549.
Doherty, R. L., Gorman, B. M., Whitehead, R. H., and Carley, J. G. (1964). *Australas. Ann. Med.* **13,** 322-327.
Doherty, R. L., Carley, J. G., Kay, R. H., Filippich, C., and Marks, E. N. (1976). *Aust. J. Exp. Biol. Med. Sci.* **54,** 237-243.
Dougherty, E., III, and Price, J. I. (1960). *Avian Dis.* **4,** 247-258.
Dubin, D. T., and Stollar, V. (1975). *Biochem. Biophys. Res. Commun.* **66,** 1373-1379.
Dunayevich, M., Johnson, H. N., and Burleson, W. (1961). *Virology* **15,** 295-298.
Earley, E., Peralta, P. H., and Johnson, K. M. (1967). *Proc. Soc. Exp. Biol. Med.* **125,** 741-747.
Eaton, B. T., and Faulkner, P. (1972). *Virology* **50,** 865-873.
Eckles, K. H., Hetrick, F. M., and Russell, P. K. (1975). *Infect. Immun.* **11,** 1053-1060.
Eddy, G. A., Martin, D. H., Reeves, W. C., and Johnson, K. M. (1972). *Infect. Immun.* **5,** 160-163.
Eklund, C. M., Bell, J. F., and Brennan, J. M. (1951). *Am. J. Trop. Med.* **31,** 312-320.
El Dadah, A. H., and Nathanson, N. (1967). *Am. J. Epidemiol.* **86,** 776-790.
Engelhardt, D. L. (1972). *J. Virol.* **9,** 903-908.
Falkner, W. A., Jr., Diwan, A. R., and Halstead, S. B. (1977). *J. Immunol.* **111,** 1804-1809.
Faulkner, P., and McGee-Russell, S. M. (1968). *Can. J. Microbiol.* **14,** 153-159.
Feemster, R. F., Wheeler, R. E., Daniels, J. B., Rose, H. D., Schaeffer, M., Kissling, R. E., Hayes, R. O., Alexander, E. R., and Murray, W. A. (1958). *N. Engl. J. Med.* **259,** 107-113.
Feinsod, F. M., Spielman, A., and Warner, J. L. (1975). *Am. J. Trop. Med. Hyg.* **24,** 533-536.
Ferguson, J. A., Reeves, W. C., Milby, M. M., and Hardy, J. L. (1978). *Am. J. Vet. Res.* **39,** 371-376.
Ferguson, J. A., Reeves, W. C., and Hardy, J. L. (1979). *Am. J. Vet. Res.* **40,** 5-10.
Fleming, P. (1977). *J. Gen. Virol.* **37,** 93-105.
Fothergill, L. D., and Dingle, J. H. (1938). *Science* **88,** 549-560.
France, J. K., Wyrick, B. C., and Trent, D. W. (1979). *J. Gen. Virol.* **44,** 725-740.
Franck, P. T. (1972). *Sci. Publ.—Pan Am. Health Organ.* **243,** 322-328.
Franck, P. T., and Johnson, K. M. (1970). *Am. J. Trop. Med. Hyg.* **19,** 860-865.
Franklin, R. M. (1962). *Prog. Med. Virol.* **4,** 1-53.
French, E. L. (1952). *Med. J. Aust.* **1,** 100-103.
Friedman, R. M., Levy, H., and Carter, W. B. (1966). *Proc. Natl. Acad. Sci. U.S.A.* **56,** 440-447.
Friedman, R. M., and Pastan, I. (1969). *J. Mol. Biol.* **40,** 107-
Fuscaldo, A. A., Aaslestad, H. G., and Hoffman, E. J. (1971). *J. Virol.* **7,** 233-240.
Gaidamovich, S. Ya., and Sokey, J. (1973). *Acta Virol.* **17,** 343-350.
Gard, G. P., Marshall, I. D., and Woodroofe, G. M. (1973). *Am. J. Trop. Med. Hyg.* **22,** 551-560.
Gard, G. P., Marshall, I. D., Walker, K. H., Acland, H. M., and De Sarem, W. G. (1977). *Aust. Vet. J.* **53,** 61-66.
Garman, J. L., Scherer, W. F., and Dickerman, R. W. (1968). *Bol. Of. Sanit. Panam.* **69,** 238-251.
Garoff, H., and Simons, K. (1974). *Proc. Natl. Acad. Sci. U.S.A.* **71,** 3988-3992.

Garoff, H., Simons, K., and Renkonen, O. (1974). *Virology* **61,** 493-504.
Gebhardt, L. P., Stanton, G. J., Hill, P. W., and Collett, G. C. (1964). *N. Engl. J. Med.* **271,** 172-177.
Ghamberg, C. G., Utermann, G., and Simons, K. (1972). *FEBS Lett.* **28,** 179-182.
Giltner, L. T., and Shahan, M. S. (1933). *North Am. Vet.* **14,** 25-27.
Giltner, L. T., and Shahan, M. S. (1936). *J. Am. Vet. Med. Assoc.* **88,** 363-374.
Gleiser, C. A., Gochenour, W. S., Jr., Berge, T. O., and Tigertt, W. D. (1962). *J. Infect. Dis.* **110,** 80-97.
Gochenour, W. S., Jr., Berge, T. O., Gleiser, C. A., and Tigertt, W. D. (1962). *Am. J. Hyg.* **75,** 351-362.
Goldfield, M., and Sussman, O. (1968). *Am. J. Epidemiol.* **87,** 1-10.
Goldfield, M., Welsh, J. N., and Taylor, B. F. (1968). *Am. J. Epidemiol.* **87,** 32-38.
Goodman, G. T., and Koprowski, H. (1962a). *J. Cell. Physiol.* **59,** 333-373.
Goodman, G. T., and Koprowski, H. (1962b). *Proc. Natl. Acad. Sci. U.S.A.* **48,** 160-165.
Gordon, W. S., Brownlee, A., Wilson, D. R., and MacLeod, J. (1962). *Symp. Zool. Soc. London*, 1-27.
Gorelkin, L., and Jahrling, P. B. (1975). *Lab. Invest.* **32,** 78-85.
Gorman, B. M., Leer, J. R., Filippich, C., Goss, P. D., and Doherty, R. L. (1975). *Aust. J. Med. Technol.* **6,** 65-71.
Gould, D. H., Byrne, R. J., and Mayes, D. E. (1964). *Am. J. Trop. Med. Hyg.* **13,** 742-746.
Grady, G. F., Maxfield, H. K., Hildreth, S. W., Timperi, R. J., Jr., Gilfillan, R. F., Roseman, B. J., Francy, D. B., Calisher, C. H., Marcus, L. C., and Madoff, M. A. (1978). *Am. J. Eipdemiol.* **107,** 170-178.
Gresikova, M., and Batikova, M. (1978). *Acta Virol.* **22,** 162-166.
Gresser, I., Hardy, J. L., Hu, S. M. K., and Scherer, W. F. (1958). *Am. J. Trop. Med. Hyg.* **7,** 365-373.
Griffin, D. E. (1976). *J. Infect. Dis.* **133,** 456-464.
Griffin, D. E., and Johnson, R. T. (1977). *J. Immunol.* **118,** 1070-1075.
Grimley, P. M., and Friedman, R. M. (1970). *J. Infect. Dis.* **122,** 45-52.
Groot, H. (1972). *Sci. Publ.—Pan Am. Health Organ.* **243,** 7-16.
Groschel, D., and Koprowski, H. (1965). *Arch. Gesamte Virusforsch.* **17,** 379-391.
Grossman, R. A., Edelman, R., and Chiewanich, P. (1973). *Am. J. Epidemiol.* **98,** 121-132.
Grossman, R. A., Edelman, R., and Gould, D. J. (1974). *Am. J. Epidemiol.* **100,** 69-76.
Gutekunst, D. E. (1969). *J. Am. Vet. Med. Assoc.* **155,** 368-374.
Gutierrez, V. E., Monath, T. P., Alava, A. A., Uriguen, B. D., Arzube, R. M., and Chamberlain, R. W. (1975). *Am. J. Epidemiol.* **102,** 400-413.
Habu, A., Murakami, Y., Ogasa, A., and Fujisaki, Y. (1977). *Virus* **27,** 21-26. (In Japanese).
Hackbarth, S. A., Reinarz, A. B. G., and Sagik, B. P. (1973). *Res, J. Reticuloendothel. Soc.* **14,** 405-425.
Hale, J. H., and Witherington, D. H. (1953). *J. Comp. Pathol.* **63,** 195-198.
Hammam, M. H., Clarke, D. H., and Price, W. H. (1965). *Am. J. Epidemiol.* **82,** 40-55.
Hammon, W. McD., and Reeves, W. C. (1946). *J. Exp. Med.* **83,** 163-173.
Hammon, W. McD., and Sather, G. E. (1969). *In* "Diagnostic Procedures for Viral and Rickettsial Infections" (E. H. Lennette and N. J. Schmidt, eds.), pp. 227-280. Am. Public Health Assoc., New York.
Hammon, W. McD., Reeves, W. C., Brookman, B., and Izumi, E. M. (1941). *Science* **94,** 328-330.
Hammon, W. McD., Reeves, W. C., and Sather, G. E. (1951). *J. Immunol.* **67,** 357-367.
Hannoun, C., Cisso, J., and Ardoin, P. (1964). *Ann. Inst. Pasteur, Paris* **107,** 598-603.
Hanson, R. P. (1957). *Am. J. Trop. Med. Hyg.* **6,** 858-862.

Hanson, R. P., Vadlamudi, S., Trainer, D. O., and Anslow, R. (1968). *Am. J. Vet. Res.* **29,** 723-727.

Hardy, J. L., Reeves, W. C., Rush, W. A., and Nir, Y. D. (1974). *Infect. Immun.* **10,** 553-564.

Hardy, J. L., Milby, M. M., Wright, M. E., Beck, A. J., Presser, S. B., and Bruen, J. P. (1977). *J. Wildl. Dis.* **13,** 383-392.

Haring, C. M., Howarth, J. A., and Meyer, K. F. (1931). *North Am. Vet.* **12,** 29-36.

Harley, E. H., Losman, M. J., Hall, E., and Naudie, W. DuT. (1975). *S. Afr. J. Sci.* **71,** 305-308.

Harrison, S. C., David, A., Jumblatt, J., and Darnell, J. E. (1971). *J. Mol. Biol.* **60,** 523-528.

Hart, K. L., Keen, D., and Belle, E. A. (1964). *Am. J. Trop. Med. Hyg.* **13,** 331-334.

Hashimoto, K., and Simizu, B. (1979). *Arch. Virol.* **60,** 299-309.

Hayashi, K., Akashi, M., and Ueda, Y. (1978). *Trop. Med.* **20,** 1-14.

Hayes, C. G. (1978). *Acta Virol.* **22,** 401-409.

Hayes, C. G., and Wallis, R. C. (1977). *Adv. Virus Res.* **21,** 37-83.

Hayes, R. O., and Hess, A. D. (1964). *Am. J. Trop. Med. Hyg.* **13,** 851-858.

Hayes, R. O., LaMotte, L. C., and Hess, A. D. (1960). *Mosq. News* **20,** 85-87.

Hayes, R. O., Daniels, J. B., Maxfield, H. K., and Wheeler, R. E. (1964). *Am. J. Trop. Med. Hyg.* **13,** 595-606.

Hefti, E., Bishop, D. H. L., Dubin, D. T., and Stollar, V. (1976). *J. Virol.* **17,** 149-159.

Helenius, A., and Simons, K. (1975). *Biochim. Biophys. Acta* **415,** 29-79.

Helenius, A., Fries, E., Garoff, H., and Simons, K. (1976). *Biochim. Biophys. Acta* **436,** 319-334.

Helmboldt, C. F., Luginbuhl, R. E., Hammar, A. H., Satriano, S. F., and Jungherr, E. L. (1955). *Am. J. Vet. Res.* **16,** 57-63.

Henderson, B. E., Chappell, W. A., Johnston, J. G., Jr., and Sudia, W. D. (1971). *Am. J. Epidemiol.* **93,** 194-205.

Henderson, J. R. (1964). *J. Immunol.* **93,** 452-461.

Hess, A. D., Cherubin, C. E., and LaMotte, L. C. (1963). *Am. J. Trop. Med. Hyg.* **12,** 657-667.

Hetrick, F. M. (1960). Master of Science Thesis, University of Maryland, College Park.

Heydrick, F. P., Comer, J. F., and Wachter, R. F. (1971). *J. Virol.* **7,** 642-645.

Hirschberg, C. B., and Robbins, P. W. (1974). *Virology* **61,** 602-608.

Hofmann, H., Frisch-Niggemeyer, W., and Heinz, F. (1979). *J. Gen. Virol.* **42,** 505-511.

Hogge, A. L. (1964). *Annu. Prog. Rep. U.S. Army Med. Unit, Ft. Detrick, Md.* pp. 99-112.

Holden, P. (1955). *Proc. Soc. Exp. Biol. Med.* **88,** 607-610.

Holden, P., Muth, D., and Shriner, R. B. (1966). *Am. J. Epidemiol.* **84,** 67-73.

Holden, P., Hayes, R. O., Mitchell, C. J., Francy, D. B., Lazuick, J. S., and Hughes, T. B. (1973). *Am. J. Trop. Med. Hyg.* **22,** 244-253.

Horzinek, M. C. (1973a). *J. Gen. Virol.* **20,** 87-103.

Horzinek, M. C. (1973b). *Prog. Med. Virol.* **16,** 109-156.

Horzinek, M., and Mussgay, M. (1969). *J. Virol.* **4,** 514-520.

Howard, J. S., and Wallis, R. C. (1974). *Am. J. Trop. Med. Hyg.* **23,** 522-525.

Howard, R. J., Craig, C. P., Trevino, G. S., Dougherty, S. F., and Mergenhagen, S. E. (1969). *J. Immunol.* **103,** 699-707.

Hruskova, J., Rychterova, V., and Kliment, V. (1972). *Acta Virol.* **16,** 125.

Hsu, M. T., Kung, H. J., and Davidson, N. (1973). *Cold Spring Harbor Symp. Quant. Biol.* **38,** 943-950.

Hsu, S. H., Huang, M. H., Wang, B. T., Wong, W. J., and Lin, S. N. (1978). *J. Med. Entomol.* **14,** 581-584.

Huang, C. H., and Wong, C. (1963). *Acta Virol.* **7,** 322-334.

Hurst, E. W. (1934). *J. Exp. Med.* **59,** 529-542.

Igarashi, A. (1969). *Biken J.* **12,** 161-168.

Igarashi, A., Fukunga, T., and Fukai, K. (1964). *Biken J.* **7,** 111-119.

Igarashi, A., Fukuoka, T., and Fukai, K. (1966). *Biken J.* **12,** 245-257.
Igarashi, A., Fukuoka, T., Nishiuthat, P., Hsu, L. C., and Fukai, K. (1970). *Biken J.* **13,** 93-107.
Innes, J. R. M., and Saunders, L. Z. (1962). "Comparative Neuropathology." Academic Press, New York.
Inoue, Y. K. (1971). *In* "Immunization for Japanese Encephalitis" (W. McD. Hammon, M. Kitaoka, and W. G. Downs, eds.), p. 204. Williams & Wilkins, Baltimore, Maryland.
Ito, Z. (1975). *J. Public Health* **18,** 17-26 (in Japanese).
Itosoya, H., Matumoto, M., and Iwasa, S. (1950). *Jpn. J. Exp. Med.* **20,** 587-595.
Jacoby, R. O., and Bhatt, P. N. (1976). *J. Infect. Dis.* **134,** 158-165.
Jahrling, P. B. (1976). *J. Gen. Virol.* **32,** 121-128.
Jahrling, P. B., and Eddy, G. A. (1977). *Am. J. Epidemiol.* **106,** 408-417.
Jahrling, P. B., and Gorelkin, L. (1975). *J. Infect. Dis.* **132,** 667-676.
Jahrling, P. B., and Scherer, W. F. (1973a). *Infect. Immun.* **7,** 905-910.
Jahrling, P. B., and Scherer, W. F. (1973b). *Infect. Immun.* **8,** 456-462.
Jahrling, P. B., Navarro, E., and Scherer, W. F. (1976). *Arch. Virol.* **51,** 23-35.
Jahrling, P. B., Heisey, G. B., and Hesse, R. A. (1977). *Infect. Immun.* **17,** 356-360.
Jahrling, P. B., Hesse, R. A., and Metzger, J. F. (1978). *J. Clin. Microbiol.* **8,** 54-60.
Jochim, M. M., and Barber, T. L. (1974). *J. Am. Vet. Med. Assoc.* **165,** 621-625.
Jochim, M. M., Barber, T. L., and Luedke, A. J. (1973). *J. Am. Vet. Med. Assoc.* **162,** 280-283.
Johnson, K. M., and Martin, D. H. (1974). *Adv. Vet. Sci. Comp. Med.* **18,** 79-116.
Johnson, R. T. (1965). *Am. J. Pathol.* **46,** 929-937.
Johnston, R. E., and Bose, H. R. (1972). *Biochem. Biophys. Res. Commun.* **46,** 712-718.
Jones, E. J. (1934). *J. Exp. Med.* **59,** 781-798.
Jungherr, E. L., Helmboldt, C. F., Satriano, S. F., and Luginbuhl, R. E. (1958). *Am. J. Hyg.* **67,** 10-20.
Kaariainen, L., and Gomatos, P. J. (1969). *J. Gen. Virol.* **5,** 251.
Kaariainen, L., and Soderlund, H. (1971). *Virology* **43,** 291-299.
Kamada, M., Ando, Y., Fukunaga, Y., Kumanomido, T., Imagawa, H., Wada, R., and Akiyama, Y. (1980). *Am. J. Trop. Med. Hyg.* **29,** 984-988.
Kandle, R. P. (1960). *Proc. Annu. Meet. N. J. Mosq. Exterm. Assoc.* **47,** 11-15.
Kaplan, W., Winn, J. F., and Palmer, D. F. (1955). *J. Immunol.* **75,** 225-226.
Karabatsos, N. (1963). *J. Immunol.* **91,** 76-82.
Karabatsos, N. (1975). *Am. J. Trop. Med. Hyg.* **24,** 527-532.
Karabatsos, N. (1980). *In* "St. Louis Encephalitis" (T. P. Monath, ed.), pp. 105-158. Am. Public Health Assoc., Washington, D. C.
Karabatsos, N., and Buckley, S. M. (1967). *Am. J. Trop. Med. Hyg.* **16,** 99-105.
Karstad, L. (1964). *Zoonoses Res.* **3,** 59-64.
Karstad, L. H., and Hanson, R. P. (1958). *Trans. North Am. Wildl. Conf.* **23,** 185-186.
Karstad, L. H., and Hanson, R. P. (1959). *J. Infect. Dis.* **105,** 293-296.
Karstad, L. H., Spalatin, J., and Hanson, R. P. (1961). *Zoonoses Res.* **1,** 87-96.
Kay, B. H., Carley, J. G., and Filippich, C. (1975). *J. Med. Entomol.* **12,** 279-283.
Keegstra, K., Sefton, B., and Burke, D. (1975). *J. Virol.* **16,** 613-620.
Kii, N., Sato, K., Okubo, K., Ando, K., Nakayama, T., Ichikawa, S., and Yamada, M. (1937). *Zikken Igaku Zasshi* **21,** 1849-1857 (in Japanese).
Kii, N., Sato, K., Ando, K., Yamada, M., and Nakayama, T. (1939). *Jpn. J. Exp. Med.* **23,** 595-560 (in Japanese).
Kirk, L. J., and Holden, P. (1974). *Can. J. Microbiol.* **20,** 215-217.
Kissling, R. E. (1957). *Proc. Soc. Exp. Biol. Med.* **96,** 290-294.
Kissling, R. E. (1958a). *Ann. N. Y. Acad. Sci.* **70,** 320-327.
Kissling, R. E. (1958b). *J. Am. Vet. Med. Assoc.* **132,** 466-468.

Kissling, R. E., and Rubin, H. (1951). *Am. J. Vet. Res.* **12,** 100-105.
Kissling, R. E., Chamberlain, R. W., Eidson, M. E., Sikes, R. K., and Bucca, M. A. (1954). *Am. J. Hyg.* **60,** 237-250.
Kissling, R. E., Chamberlain, R. W., Nelson, D. B., and Stamm, D. D. (1956). *Am. J. Hyg.* **63,** 274-287.
Kissling, R. E., Chamberlain, R. W., Sudia, W. D., and Stamm, D. D. (1957). *Am. J. Hyg.* **66,** 48-55.
Kitano, T., Suzuki, K., and Yamaguchi, T. (1974). *J. Virol.* **14,** 631-639.
Kitaoka, M. (1971). *In* "Immunization for Japanese Encephalitis" (W. McD. Hammon, M. Kitaoka, and W. G. Downs, eds.), pp. 5-11. Williams & Wilkins, Baltimore, Maryland.
Kodama, K., Sasahi, N., and Inoue, Y. K. (1968). *J. Immunol.* **100,** 194-200.
Kokernot, R. H., Smithburn, K. C., Paterson, H. E., and de Meillon, B. (1960). *S. Afr. J. Med. Res.* **34,** 871-874.
Kokernot, R. H., Smithburn, K. C., and Kluge, E. (1961). *Ann. Trop. Med. Parasitol.* **55,** 73-85.
Komarov, A., and Kalmar, E. (1960). *Vet. Res.* **72,** 257-261.
Kono, R., and Ho Kim, K. (1969). *Bull. W.H.O.* **40,** 263-277.
Kos, K., Shapiro, D., Vaituzis, Z., and Russell, P. K. (1975). *Arch. Virol.* **47,** 217-224.
Kuberski, T. T., and Rosen, L. (1977). *Am. J. Trop. Med. Hyg.* **26,** 538-543.
Lachmi, B. C., and Kaariainen, L. (1976). *Proc. Natl. Acad. Sci. U.S.A.* **73,** 1936-1940.
Laine, R., Soderlund, H., and Renkonen, O. (1973). *Intervirology* **1,** 110-118.
LaMotte, L. C., Jr., Crane, G. T., Shriner, R. B., and Kirk, L. J. (1967). *Am. J. Trop. Med. Hyg.* **16,** 348-356.
Larsell, O., Haring, C. M., and Meyer, K. F. (1934). *Am. J. Pathol.* **10,** 361-373.
LeRoux, J. M. W. (1959). *Onderstepoort J. Vet. Res.* **28,** 237-243.
Leung, M-K., Burton, A., Iversen, J., and McLintock, J. (1975). *Can. J. Microbiol.* **21,** 954-958.
Levin, J. G. and Friedman, R. M. (1971). *J. Virol.* **7,** 504-513.
Levitt, N. H., Miller, H. V., and Eddy, G. A. (1976). *J. Clin. Microbiol.* **4,** 382-384.
Liao, S. J. (1955). *Yale J. Biol. Med.* **27,** 287-296.
Liu, C., Voth, D. W., Rodina, P., Shauf, L. R., and Gonzalez, G. (1970). *J. Infect. Dis.* **122,** 53-63.
Lofgren, C. S. (1970). *Annu. Rev. Entomol.* **15,** 321-342.
Luedke, A. J., Barber, T. L., Foster, N. M., Batalia, D., and Mercado, S. (1972). *J. Am. Vet. Med. Assoc.* **161,** 824-831.
McCarthy, M., and Harrison, S. C. (1977). *J. Virol.* **23,** 61-73.
McFarland, H. F., Griffin, D. E., and Johnson, R. T. (1972). *J. Exp. Med.* **136,** 216-226.
McFarland, R. I., Burns, W. H., and White, D. O. (1977). *J. Immunol.* **119,** 1569-1574.
McIntosh, B. M. (1961). *Trans. R. Soc. Trop. Med. Hyg.* **55,** 63-68.
McIntosh, B. M., and Gear, J. H. S. (1975). *In* "Diseases Transmitted from Animals to Man" (W. T. Hubbert, W. F. McColloch, and P. R. Schnurrenberger, eds.), 6th ed., pp. 939-967. Thomas, Springfield, Illinois.
McIntosh, B. M., Dickinson, D. B., and McGillivray, G. M. (1969). *S. Afr. J. Med. Sci.* **34,** 77-82.
McIntosh, B. M., Jupp, P. G., Dos Santos, T., and Meenehan, G. M. (1976). *S. Afr. J. Sci.* **72,** 295-300.
McKinney, R. W., Berge, T. O., Sawyer, W. D., Tigertt, W. D., and Crozier, D. (1963). *Am. J. Trop. Med. Hyg.* **12,** 597-603.
McLintock, J., Burton, A. N., McKiel, J. A., Hall, R. R., and Rempel, J. G. (1970). *J. Med. Entomol.* **7,** 446-454.
McNutt, S. H., and Packer, A. (1943). *Vet. Med. (Kansas City, Mo.)* **38,** 22-25.
Main, A. J., Jr., Hayes, R. O., and Tonn, R. J. (1968). *Mosq. News* **28,** 619-626.

Makino, S., Fujita, N., Aoki, H., Takehara, M., and Hotta, S. (1971). *Kobe J. Med. Sci.* **17,** 75-84.
Mandel, B. (1978). *Adv. Virus Res.* **23,** 205-268.
Marchette, N. J., Halstead, S. B., and Chow, J. S. (1976). *J. Infect. Dis.* **133,** 274-282.
Marshall, L. D., Scrivani, R. P., and Reeves, W. C. (1962). *Am. J. Hyg.* **76,** 216-224.
Martin, B. A. B. and Burke, D. C. (1974). *J. Gen. Virol.* **24,** 45-66.
Matsumura, T., Stollar, V., and Schlesinger, R. W. (1971). *Virology* **46,** 344.
Metzger, J. F., Banks, I. S., Smith, C. W., and Hoggan, M. D. (1961). *Proc. Soc. Exp. Biol. Med.* **106,** 212-214.
Meyer, K. F., Haring, C. M., and Howitt, B. (1931). *Science* **74,** 227-228.
Miller, L. D., Pearson, J. E., and Muhm, R. L. (1973). *Proc. 67th Annu. Meet., U. S. Anim. Health Assoc.* pp. 629-631.
Mitchell, C. J., Hayes, R. O., Holden, P., Hill, H. R., and Hughes, T. B., Jr. (1969). *J. Med. Entomol.* **6,** 155-162.
Miyake, M. (1964). *Bull. W.H.O.* **30,** 153-160.
Monath, T. P., Kemp, G. E., Cropp, C. B., and Chandler, F. W. (1978). *J. Infect. Dis.* **138,** 59-66.
Monath, T. P., Lazuick, J. S., Cropp, C. B., Calisher, C. H., Kinney, R. M., Trent, D. W., Kemp, G. E., Bowen, G. S., and Francy, D. B. (1980). *Am. J. Trop. Med. Hyg.* **29,** 948-962.
Monlux, W. S., and Luedke, A. J. (1973). *Am. J. Vet. Res.* **34,** 465-473.
Monlux, W. S., Luedke, A. J., and Bowne, J. (1972). *J. Am. Vet. Med. Assoc.* **161,** 265-269.
Moore, N. F., Barenholz, Y., and Wagner, R. R. (1976). *J. Virol.* **19,** 126-136.
Morgan, I. M. (1941). *J. Exp. Med.* **74,** 115-124.
Morser, M. J., and Burke, D. C. (1974). *J. Gen. Virol.* **22,** 395-409.
Muangman, D., Edelman, R., Sullivan, M. J., and Gould, D. J. (1972). *Am. J. Trop. Med. Hyg.* **21,** 482-486.
Murphy, F. A. (1980a). *In* "The Togaviruses" (R. W. Schlesinger, ed.). Academic Press, New York (in press).
Murphy, F. A. (1980b). *In* "St. Louis Encephalitis" (T. P. Monath, ed.), pp. 65-104. Am. Public Health Assoc., Washington, D. C.
Murphy, F. A., and Whitfield, S. G. (1970). *Exp. Mol. Pathol.* **13,** 131-146.
Murphy, F. A. and Harrison, A. K. (1971). *In* "Venezuelan Encephalitis," Sci. Pub. No. 243, pp. 28-29. Pan Am. Health Organ., Washington, D.C.
Murphy, F. A., Harrison, A. K., Gary, G. W., Whitfield, S. G., and Forrester, F. T. (1968). *Lab. Invest.* **19,** 652-662.
Murphy, F. A., Harrison, A. K., and Collin, W. K. (1970). *Lab. Invest.* **22,** 318-328.
Murphy, F. A., Taylor, W. P., Mims, C. A., and Marshall, I. D. (1973). *J. Infect. Dis.* **127,** 129-138.
Mussgay, M. (1967). *Arch. Gesamte Virusforsch.* **21,** 144-154.
Mussgay, M., and Horzinek, M. (1966). *Virology* **29,** 199-204.
Mussgay, M., and Rott, R. (1964). *Virology* **23,** 573-581.
Naeve, C. W. and Trent, D. W. (1978). *J. Virol.* **25,** 535-545.
Nakamura, H. (1970). *Equine Vet. J.* **4,** 1-2.
Nakamura, H. (1971). *In* "Immunization for Japanese Encephalitis" (W. McD. Hammon, M. Kitaoka, and W. G. Downs, eds.), pp. 305-312. Williams & Wilkins, Baltimore, Maryland.
Nakamura, J., Nakamura, H., and Nozaki, I. (1964). *Virus* **14,** 252-253 (in Japanese).
Nathanson, N. (1980). *In* "St. Louis Encephalitis" (T. P. Monath, ed.), pp. 201-237. Am. Public Health Assoc., Washington, D. C.
Nathanson, N., and Cole, G. A. (1971). *Fed. Proc. Fed. Am. Soc. Exp. Biol.* **30,** 1822-1824.
Nathanson, N., Monjan, A. A., Panitch, H. S., Johnson, E. D., Petursson, G., and Cole, G. A. (1975). *In* "Viral Immunology and Immunopathology" (A. L. Notkins, ed.), pp. 357-391. Academic Press, New York.

Nishimura, C., Nomura, M., and Kitaoka, M. (1968). *Jpn. J. Med. Sci. Biol.* **21,** 1-12.
Ogasa, A., Yokoki, Y., Fujisaki, Y., and Habu, A. (1977). *Jpn. J. Anim. Reprod.* **23,** 171-175.
Ogata, M., Nagao, Y., Jitsunari, F., Kikui, R., and Kitamura, N. (1970). *Acta Med. Okayama* **24,** 579-587.
Oguni, H. (1926). *Rep. Res. Stn. Epiz. Koyukai* **5,** 27-31 (in Japanese).
Okuno, T. (1978). *World Health Stn. Q.* **31,** 120-131.
Okuno, T., Okada, T., Suzuki, M., Kobayashi, M., and Oya, A. (1968). *Bull. W.H.O.* **38,** 547-563.
Okuno, Y., Igarashi, A., and Fukai, K. (1978). *Biken J.* **21,** 137-147.
Osterrieth, P. M. (1968). *Mem. Soc. R. Sci. Liege, Collect. 8* pp. 11-27.
Pal, B. K., McAllister, R. M., Gardner, M. B., and Roy-Burman, P. (1975). *J. Virol.* **16,** 123-131.
Pan American Health Organization (1972). "Venezuelan Encephalitis," Sci. Publ. 243. PAHO, Washington, D.C.
Panthier, R., Hannoun, C., Oudar, J., Beytout, D., Corniou, B., Jouberi, L., Guillon, J.-Cl., and Mouchet, J. (1966). *Presse Med.* **74,** 2495.
Paterson, H. E., Kokernot, R. H., and Davis, D. H. S. (1957). *S. Afr. J. Med. Sci.* **22,** 63-69.
Pathak, S., and Webb, H. E. (1974). *J. Neurol. Sci.* **23,** 175-183.
Pattyn, S. R., de Vleesschauwer, L., and Van Der Groen, G. (1975). *Arch. Virol.* **49,** 33-37.
Pedersen, C. E., Jr., and Eddy, G. A. (1974). *J. Virol.* **14,** 740-744.
Pedersen, C. E., Jr., and Eddy, G. A. (1975). *Am. J. Epidemiol.* **101,** 245-252.
Pedersen, C. E., Jr., Slocum, D. R., and Robinson, D. M. (1972a). *Infect. Immun.* **6,** 799-784.
Pedersen, C. E., Jr., Slocum, D. R., and Levitt, N. H. (1972b). *Appl. Microbiol.* **24,** 91-96.
Pedersen, C. E., Jr., Robinson, D. M., and Cole, F. E., Jr. (1972c). *Am. J. Epidemiol.* **95,** 490-496.
Pedersen, C. E., Jr., Barrera, C. R., and Sagik, B. P. (1973). *Arch. Gesamte Virusforsch.* **41,** 28-39.
Pedersen, C. E., Jr., Marker, S. C., and Eddy, G. A. (1974). *Virology* **60,** 312-314.
Pfefferkorn, E. R., and Clifford, R. L. (1964). *Virology* **23,** 217-223.
Pfefferkorn, E. R., and Hunter, H. S. (1963a). *Virology* **20,** 433-445.
Pfefferkorn, E. R., and Hunter, H. S. (1963b). *Virology* **20,** 446-456.
Pfefferkorn, E. R., and Shapiro, D. (1974). *In* "Comprehensive Virology" (H. Frankel-Conrat and R. R. Wagner, eds.), Vol. 2, Chapter 4, pp. 171-230. Plenum, New York.
Porterfield, J. S. (1961). *Bull. W.H.O.* **24,** 735-741.
Postic, B., Schleupner, C. J., Armstrong, J. A., and Ito, M. (1969). *J. Infect. Dis.* **120,** 339-347.
Price, W. H., and O'Leary, W. (1967). *Am. J. Epidemiol.* **85,** 84-87.
Pursell, A. R., Peckham, J. C., Cole, J. R., Stewart, W. C., and Mitchell, F. E. (1972). *J. Am. Vet. Med. Assoc.* **161,** 1143-1147.
Qureshi, A. A., and Trent, D. W. (1973a). *Infect. Immun.* **7,** 242-248.
Qureshi, A. A., and Trent, D. W. (1973b). *Infect. Immun.* **8,** 985-992.
Qureshi, A. A., and Trent, D. W. (1973c). *Infect. Immun.* **8,** 993-999.
Rabinowitz, S. G., and Proctor, R. A. (1974). *J. Immunol.* **112,** 1070-1076.
Race, M. W., Williams, M. C., and Agostini, C. F. M. (1980). *Trans. R. Soc. Trop. Med. Hyg.* **73,** 18-22.
Ranck, F. M., Gainer, J. H., Hanley, J. E., and Nelson, S. L. (1965). *Avian Dis.* **9,** 8-20.
Reeves, W. C. (1970). *Proc. Pap. Annu. Conf. Calif. Mosq. Control Assoc.* **37,** 3-6.
Reeves, W. C. (1971a). *In* "Ecology and Physiology of Parasites" (A. M. Fallis, ed.), pp. 223-230. Univ. of Toronto Press, Toronto.
Reeves, W. C. (1971b). *Mosq. News* **31,** 319-325.
Reeves, W. C. (1974). *Prog. Med. Virol.* **17,** 193-220.
Reeves, W. C., Bellamy, R. E., and Scrivani, R. P. (1958). *Am. J. Hyg.* **67,** 78-89.
Reid, H. W., and Doherty, P. C. (1971a). *J. Comp. Pathol.* **81,** 291-298.

Reid, H. W., and Doherty, P. C. (1971b). *J. Comp. Pathol.* **81,** 521-529.
Reid, H. W., Doherty, P. C., and Dawson, A. M. (1971). *J. Comp. Pathol.* **81,** 537-546.
Renkonen, O., Kaariainen, L., Simons, K., and Ghamberg, C. (1971). *Virology* **46,** 318-326.
Roberts, E. D., Sanmartin, C., Payan, J., and Mackenzie, R. B. (1970). *Am. J. Vet. Res.* **31,** 1223-1229.
Robin, Y., Bourdin, P., Le Gonidec, G., and Heme, G. (1974). *Ann. Microbiol. (Paris)* **125A,** 235-241.
Rokutanda, H. K. (1969). *J. Immunol.* **102,** 662-667.
Rosen, L., and Gubler, D. (1974). *Am. J. Trop. Med. Hyg.* **23,** 1153-1160.
Rosen, L., Tesh, R. B., Lien, J. C., and Cross, J. H. (1978). *Science* **199,** 199-200.
Rothman, F. M. (1978). *Int. Rev. Biochem.* **17,** 45-73.
Rowan, D. F., Goldfield, M., Taylor, B. F., and Sussman, O. (1968). *Am. J. Epidemiol.* **87,** 11-17.
Russell, P. K., Chiewsilp, D., and Brandt, W. E. (1970). *J. Immunol.* **105,** 838-845.
Sabin, A. B. (1952). *Proc. Natl. Acad. Sci. U.S.A.* **38,** 540-546.
Sanmartin, C., Mackenzie, R. B., Trapido, H., Barreto, P., Mullenax, C. H., Gutierrez, E., and Lesmes, C. (1973). *Bol. Of. Sanit. Panam.* **74,** 108-137.
Sather, G. E., and Hammon, W. McD. (1971). *In* "Immunization for Japanese Encephalitis" (W. McD. Hammon, M. Kitaoka, and W. G. Downs, eds.), pp. 23-25. Williams & Wilkins, Baltimore, Maryland.
Satriano, S. F., Luginbuhl, R. E., Wallis, R. C., Jungherr, E. L., and Williamson, L. A. (1958). *Am. J. Hyg.* **67,** 21-34.
Saugstad, E. S., Dalrymple, J. M., and Eldridge, B. F. (1972). *Am. J. Epidemiol.* **96,** 114-122.
Sawicki, D. L., and Gomatos, P. J. (1976). *J. Virol.* **4,** 117-122.
Schaeffer, M., and Arnold, E. H. (1954). *Am. J. Hyg.* **60,** 231-236.
Scherer, W. F. (1977). *Am. J. Trop. Med. Hyg.* **26,** 167-170.
Scherer, W. F., and Chin, J. (1977). *Am. J. Trop. Med. Hyg.* **26,** 307-312.
Scherer, W. F., and Pancake, B. A. (1977). *J. Clin. Microbiol.* **6,** 578-585.
Scherer, W. F., Buescher, E. L., and McClure, H. E. (1959a). *Am. J. Trop. Med. Hyg.* **8,** 689-697.
Scherer, W. F., Moyer, J. T., Izumi, T., Gresser, I., and McCown, J. (1959b). *Am. J. Trop. Med. Hyg.* **8,** 698-706.
Scherer, W. F., Kitaoka, S. E., Okuno, T., and Ogata, T. (1959c). *Am. J. Trop. Med. Hyg.* **8,** 707-715.
Scherer, W. F., Moyer, J. T., and Izumi, T. (1959d). *J. Immunol.* **83,** 620-626.
Scherer, W. F., Hardy, J. L., Gresser, I., and McClure, H. E. (1964). *Am. J. Trop. Med. Hyg.* **13,** 859-866.
Scherer, W. F., Reeves, W. C., Hardy, J. L., and Miura, T. (1972). *Am. J. Trop. Med. Hyg.* **21,** 182-193.
Scherer, W. F., Anderson, K., Pancake, B. A., Dickerman, R. W., and Ordonez, J. V. (1976a). *Am. J. Epidemiol.* **103,** 576-588.
Scherer, W. F., Ordonez, J. V., Dickerman, R. W., and Navarro, J. E. (1976b). *Am. J. Epidemiol.* **104,** 60-73.
Schlesinger, M. J., Schlesinger, S., and Burge, B. W. (1972). *Virology* **47,** 539-541.
Schlesinger, R. W. (1977). "Virology Monograph 16." Springer-Verlag, Berlin and New York.
Schlesinger, S., and Schlesinger, M. J. (1972). *J. Virol.* **10,** 925-932.
Schlesinger, S., and Schlesinger, M. (1973). *J. Virol.* **10,** 925-932.
Schmidt, J. R., and El Mansoury, H. K. (1963). *Ann. Trop. Med. Parasitol.* **57,** 415-427.
Schmidt, M. F., Bracha, M., and Schlesinger, M. J. (1979). *Proc. Natl. Acad. Sci. U.S.A.* **76,** 1687-1691.
Sefton, B. M., and Keegstra, K. (1974). *J. Virol.* **14,** 522-530.
Sefton, B. M., Wickus, G., and Burge, B. W. (1973). *J. Virol.* **11,** 730-735.

Sellers, M. (1969). *J. Exp. Med.* **129,** 719-725.
Semenov, B. F., Khozinsky, V. V., and Vargin, V. V. (1975). *Med. Biol.* **53,** 531.
Seymour, C., Dickerman, R. W., and Martin, M. S. (1978). *Am. J. Trop. Med. Hyg.* **27,** 297-306.
Shapiro, D., Brandt, W. E., Cardiff, R. D., and Russell, P. K. (1971). *Virology* **44,** 108-124.
Shapiro, D., Kos, K., Brandt, W. E., and Russell, P. K. (1972a). *Virology* **48,** 360-372.
Shapiro, D., Trent, D., Brandt, W. E., and Russell, P. K. (1972b). *Infect. Immun.* **6,** 206-209.
Shimizu, T., and Kawakami, Y. (1949). *Bull. Natl. Inst. Anim. Health* **22,** 117-128 (in Japanese).
Shimizu, T., Kawakami, Y., Fukuhara, S., and Matumoto, M. (1954). *Jpn. J. Exp. Med.* **24,** 363-375.
Shinefield, H. R., and Townsend, T. E. (1953). *J. Pediatr.* **43,** 21-25.
Shope, R. E., Causey, O. R., de Andrade, A. H. P., and Theiler, M. (1964). *Am. J. Trop. Med. Hyg.* **13,** 723-727.
Simizu, B., and Takayama, N. (1969). *J. Virol.* **4,** 799-800.
Simmons, D. T., and Strauss, J. (1972). *J. Mol. Biol.* **71,** 599.
Simons, K., Keranen, S., and Kaariainen, L. (1973). *FEBS Lett.* **29,** 87-91.
Simpson, D. I. H., Bowen, E. T. W., Way, H. J., Platt, G. S., Hill, M. N., Kamath, S., Teong Wah, L., Bendell, P. J. E., and Heathcote, O. H. U. (1974). *Ann. Trop. Med. Parsitol.* **68,** 394-404.
Simpson, R. W., and Hauser, R. E. (1968a). *Virology* **34,** 358-361.
Simpson, R. W., and Hauser, R. E. (1968b). *Virology* **34,** 361-364.
Simpson, R. W., and Hauser, R. E. (1968c). *Virology* **34,** 568-570.
Singh, K. R. P., and Paul, S. D. (1968). *Curr. Sci.* **37,** 65-67.
Slavik, I., Mayer, V., and Mrena, E. (1967). *Acta Virol.* **11,** 66.
Slavik, I., Mrena, E., and Mayer, V. (1970). *Acta Virol.* **14,** 8-16.
Smith, C. E. G. (1962). *In* "Aspects of Disease Transmission by Ticks", pp. 199-221, Zool. Soc. London, London.
Smith, T. J., Brandt, W. E., Swanson, J. C., McCown, J. M., and Beuscher, E. C. (1970). *J. Virol.* **5,** 524-532.
Sokol, F., and Clark, H. F. (1973). *Virology* **52,** 246-263.
Spalatin, J., Karstad, L., Anderson, J. R., Lauerman, L., and Hanson, R. P. (1961). *Zoonoses Res.* **1,** 29-48.
Spalatin, J., Connell, R., Burton, A. N., and Gollop, B. J. (1964). *J. Comp. Med. Vet. Sci.* **28,** 131-142.
Spertzl, R. O. (1973). *Proc. Int. Conf. Equine Infect. Dis. 3rd, 1972* pp. 146-156.
Spertzl, R. O., and Kahn, D. E. (1971). *J. Am. Vet. Med. Assoc.* **159,** 731-738.
Spertzl, R. O., and McKinney, R. W. (1972). *Mil. Med.* **137,** 441-445.
Spradbrow, P. B. (1973). *Aust. Vet. J.* **49,** 403-404.
Stevens, T. M., and Snhlesinger, R. W. (1965). *Virology* **27,** 103-112.
Stim, T. B. (1969). *J. Gen. Virol.* **5,** 329-338.
Stinski, M. F., and Gruber, J. (1971). *Proc. Soc. Exp. Biol. Med.* **136,** 1340-1346.
Stohlman, S. A., Weisseman, C. L., Jr., Eylar, O. R., and Silverman, D. L. (1975). *J. Virol.* **16,** 1017-1026.
Stohlman, S. A., Eylar, O. R., and Wisseman, C. L., Jr. (1976). *J. Virol.* **18,** 132-140.
Stollar, V. (1969). *Virology* **39,** 426-438.
Stollar, V., Stevens, T. M., and Schlesinger, R. W. (1966). *Virology* **30,** 303-312.
Stollar, V., Stollar, B. D., Koo, R., Harrap, K. A., and Schlesinger, R. W. (1976). *Virology* **69,** 104-115.
Strauss, E. G. (1978). *J. Virol.* **28,** 466-474.
Strauss, J. H., and Strauss, E. G. (1976). *In* "The Molecular Biology of Animal Viruses" (D. P. Nayak, ed.), pp. 111-116. Dekker, New York.

Strauss, J. H., Jr., Burge, B. W., and Darnell, J. E. (1969). *Virology* **37,** 367-376.
Sudia, W. D., and Newhouse, V. F. (1975). *Am. J. Epidemiol.* **101,** 1-13.
Sudia, W. D., Stamm, D. D., Chamberlain, R. W., and Kissling, R. E. (1956). *Am. J. Trop. Med. Hyg.* **5,** 802-808.
Sudia, W. D., Newhouse, V. F., and Henderson, B. E. (1971). *Am. J. Epidemiol.* **93,** 206-211.
Sugawa, Y., Mochizuki, H., and Yamamoto, S. (1949). *Bull. Natl. Inst. Anim. Health* **22,** 9-25 (in Japanese).
Sulkin, S. E., and Zarafonetis, C. (1947). *J. Exp. Med.* **85,** 559-564.
Sussman, O., Cohen, D., Gerende, J. E., and Kissling, R. E. (1957-1958). *Ann. N.Y. Acad. Sci.* **70,** 328-341.
Symington, J., and Schlesinger, M. J. (1975). *J. Virol.* **15,** 1037-1041.
Symington, J., and Schlesinger, M. J. (1978). *Arch. Virol.* **58,** 127-136.
Symington, J., McCann, A. K., and Schlesinger, M. J. (1977). *Infect. Immun.* **15,** 720-725.
Taber, L. E., Hogge, A. L., Jr., and McKinney, R. W. (1965). *Am. J. Trop. Med. Hyg.* **14,** 647-651.
Takahashi, K., Matsuo, R., Kuma, M., Baba, S., Noguchi, H., Inoue, Y. K., Sasaki, N., and Kodama, K. (1971). *In* "Immunization for Japanese Encephalitis" (W. McD. Hammon, M. Kitaoka, and W. G. Downs, eds.), pp. 292-303. Williams & Wilkins, Baltimore, Maryland.
Takemoto, K. K. (1966). *Prog. Med. Virol.* **8,** 314-348.
Taylor, R. M., Work, T. H., Hulbut, H. S., and Rizk, F. (1956). *Am. J. Trop. Med. Hyg.* **5,** 579-620.
Tempelis, C. H. (1975). *J. Med. Entomol.* **11,** 635-653.
Tempelis, C. H., Reeves, W. C., Bellamy, R. E., and Lofy, M. F. (1965). *Am. J. Trop. Med. Hyg.* **14,** 170-177.
TenBroeck, C., and Merrill, M. H. (1935). *Am. J. Pathol.* **11,** 847-851.
TenBroeck, C., Hurst, E. W., and Traub, E. (1935). *J. Exp. Med.* **62,** 677-685.
Tesh, M. J., and McCammon, J. R. (1979). *Am. J. Vet. Res.* **40,** 299-301.
Theiler, M., and Downs, W. G. (1973). *In* "The Arthropod-Borne Viruses of Vertebrates" (M. Theiler and W. G. Downs, eds.), pp. 147-208. Yale Univ. Press, New Haven, Connecticut.
Timoney, P. J. (1972). *Br. Vet. J.* **128,** 19-23.
Timoney, P. J., Donnelly, W. J. C., Clements, L. O., and Fenlon, M. (1976). *Equine Vet. J.* **8,** 113-117.
Trent, D. W. (1977). *J. Virol.* **22,** 608-618.
Trent, D. W., and Grant, J. A. (1980). *J. Gen. Virol.* **47,** 261-282.
Trent, D. W., and Naeve, C. W. (1980). *In* "St. Louis Encephalitis" (T. P. Monath, ed.), pp. 159-200. Am. Public Health Assoc., Washington, D. C.
Trent, D. W., and Qureshi, A. A. (1971). *J. Virol.* **7,** 379-388.
Trent, D. W., Swenson, C. C., and Qureshi, A. A. (1969). *J. Virol.* **3,** 385-394.
Trent, D. W., Harvey, C. L., Qureshi, A., and LeStourgeon, D. (1976). *Infect. Immun.* **13,** 1325-1333.
Trent, D. W., Clewley, J. P., France, J. K., and Bishop, D. H. L. (1979). *J. Gen. Virol.* **43,** 365-381.
Trent, D. W., Grant, J. A., Vorndam, A. V., and Monath, T. P. (1981). *Virology* (in press).
Tsuchiya, N. (1970). *Virus* **20,** 290-300 (in Japanese).
Tyzzer, E. E., and Sellards, A. W. (1941). *Am. J. Hyg.* **33,** 69-81.
Tyzzer, E. E., Sellards, A. W., and Bennett, B. L. (1938). *Science* **88,** 505-506.
Ueba, N., Maeda, A., Buei, K., Mitsuda, B., Otsu, K., Kimoto, T., Kunita, N., and Arai, H. (1971). *Jpn. J. Public Health* **18,** 267-275 (in Japanese).
Ueba, N., Maeda, A., Otsu, K., Mitsuda, B., Kimoto, T., Fujito, S., and Kunita, N. (1972). *Biken J.* **15,** 67-79.

Umrigar, M. D., and Pavri, K. M. (1977). *Indian J. Med. Res.* **65,** 603-612.
Uryrayen, L. V., Zhdanov, V. M., Yershov, F. I., and Bykovsky, A. F. (1971). *Arch. Gesamte Virusforsch.* **33,** 281-287.
Ushijima, R. N., Hill, D. W., Dolona, G. H., and Gebhardt, L. D. (1962). *Virology* **17,** 356-357.
Utermann, G., and Simons, K. (1974). *J. Mol. Biol.* **85,** 569-587.
Vaananen, P., and Kaariainen, L. (1979). *J. Gen. Virol.* **43,** 593-601.
Vainio, R., Awatkin, R., and Koprowski, H. (1961). *Virology* **14,** 385-390.
Vanderwagen, L. C., Pearson, J. L., Franti, C. E., Tamm, E. L., Riemann, H. P., and Behymer, D. E. (1975). *Am. J. Vet. Res.* **36,** 1567-1571.
Vezza, A. C., Rosen, L., Replik, P., Dalrymple, J., and Bishop, D. H. L. (1980). *Am. J. Trop. Med. Hyg.* **29,** 643-652.
Vigilancia Epidemiologica (1973). *Cent. Panam. Zoonosis, Buenos Aires* **2,** 1.
Vigilancia Epidemiologica (1976). *Cent. Panam. Zoonosis, Buenos Aires* **5,** 1.
von Bonsdorff, C. H. (1973). *Commentat. Biol.* **74,** 1-53.
von Bonsdorff, C. H., and Harrison, S. C. (1975). *J. Virol.* **16,** 141-145.
von Bonsdorff, C. H., and Harrison, S. C. (1978). *J. Virol.* **28,** 578-583.
Waite, M. R. F., Lubin, M., Jones, K. J., and Bose, H. R. (1974). *J. Virol.* **13,** 244-246.
Walder, R. and Bradish, C. J. (1979). *J. Gen. Virol.* **44,** 373-382.
Walker, D. H., Harrison, A., Murphy, K., Femister, M., and Murphy, F. A. (1976). *Am. J. Pathol.* **4,** 351-370.
Wallis, R. C., and Main, A. J., Jr. (1974). *Mem. Conn. Entomol. Soc.* pp. 117-144.
Wallis, R. C., Taylor, R. M., and Henderson, J. R. (1960). *Proc. Soc. Exp. Biol. Med.* **103,** 442-444.
Wallis, R. C., Howard, J. J., Main, A. J., Jr., Frazier, C., and Hayes, C. (1974). *Mosq. News* **34,** 63-65.
Walton, T. E., and Johnson, K. M. (1972a). *Infect. Immun.* **5,** 155-159.
Walton, T. E., and Johnson, K. M. (1972b). *J. Am. Vet. Med. Assoc.* **161,** 916-918.
Walton, T. E., Alvarez, O., Jr., Buckwalter, R. M., and Johnson, K. M. (1972). *Infect. Immun.* **5,** 750-756.
Walton, T. E., Alvarez, O., Jr., Buckwalter, R. M., and Johnson, K. M. (1973). *J. Infect. Dis.* **128,** 271-282.
Wang, Y.-J., Chen, J., and Zhang, Z. (1976). *Acta Microbiol. Sin.* **16,** 17-20 (in Chinese with English summary).
Webster, L. T. (1933). *J. Exp. Med.* **57,** 793-817.
Weekstrom, P., and Nyholm, M. (1965). *Nature (London)* **205,** 211.
Weiss, K. E., Haig, D. A., and Alexander, R. A. (1956). *Onderstepoort J. Vet. Res.* **27,** 183-195.
Welch, W. J., and Sefton, B. M. (1979). *J. Virol.* **29,** 1186-1195.
Wenger, F. (1963). *Invest. Clin.* **7,** 21-45.
Wenger, F. (1967). *Invest. Clin.* **21,** 13-31.
Wengler, G., and Wengler, G. (1976). *Virology* **73,** 190-199.
Wengler, G., Wengler, G., and Filipe, A. R. (1977). *Virology* **78,** 124-134.
Wengler, G., Wengler, G., and Gross, H. J. (1978). *Virology* **89,** 423-437.
Westaway, E. G. (1965a). *Virology* **26,** 517-527.
Westaway, E. G. (1965b). *Virology* **26,** 528-537.
Westaway, E. G. (1966). *Am. J. Epidemiol.* **84,** 439-456.
Westaway, E. G. (1975). *J. Gen. Virol.* **27,** 283-293.
Westaway, E. G. (1977). *Virology* **80,** 320-355.
Westaway, E. G., and Reedman, B. M. (1969). *J. Virol.* **4,** 688-693.
Westaway, E. G., Della-Porta, A. J., and Reedman, B. M. (1974). *J. Immunol.* **112,** 656-663.
Wiebe, M. E., and Scherer, W. F. (1979). *Virology* **94,** 474-478.

Williams, J. E., Young, O. P., and Watts, D. M. (1974). *J. Med. Entomol.* **11,** 352-354.
Wirth, D. F., Katz, F., Small, B., and Lodish, H. F. (1977). *Cell* **10,** 253-263.
Witherington, D. H. (1953). *J. Comp. Pathol.* **63,** 195-201.
Woodman, D. R., McManus, A. T., and Eddy, G. A. (1975). *Infect. Immun.* **12,** 1006-1011.
Woodring, F. R. (1957). *J. Am. Vet. Med. Assoc.* **130,** 151-152.
Woodroofe, G., Marshall, I. D., and Taylor, W. D. (1977). *Aust. J. Exp. Biol. Med. Sci.* **55,** 79-88.
Work, T. H. (1971). *Am. J. Trop. Med. Hyg.* **20,** 169-186.
Worth, C. B., Paterson, H. E., and de Meillon, B. (1961). *Am. J. Trop. Med. Hyg.* **10,** 583-592.
Yasuzumi, G., Tusuba, I., Sugihara, R., and Nakai, Y. (1964). *J. Ultrastruct. Res.* **11,** 213-229.
Yoshinaka, Y., and Hotta, S. (1971). *Proc. Soc. Exp. Biol. Med.* **131,** 1047-1052.
Young, N. A. (1972). *Sci. Publ.—Pan Am. Health Organ.* **243,** 84-89.
Young, N. A., and Johnson, K. M. (1969). *Am. J. Epidemiol.* **89,** 286-307.
Zarate, M. L., and Scherer, W. F. (1969). *Am. J. Epidemiol.* **89,** 489-502.

Chapter 9

Nonarbo Togavirus Infections of Animals: Comparative Aspects and Diagnosis

MARIAN C. HORZINEK

I. Introduction and Classification 442
II. Comparative Characteristics 443
  A. Virion Morphology and Substructure 443
  B. Physicochemical Properties 446
  C. Antigenic Structure and Relationship 449
III. Pathology and Pathogenesis 451
  A. Hog Cholera 451
  B. Bovine Viral Diarrhea 452
  C. Equine Arteritis 453
  D. Infections of Mice with LDV 454
  E. Simian Hemorrhagic Fever 455
IV. Epidemiological Aspects 455
  A. Hog Cholera 455
  B. Bovine Viral Diarrhea 456
  C. Equine Arteritis 457
  D. Infections of Mice with LDV 457
  E. Simian Hemorrhagic Fever 458
V. Comparative Diagnosis 459
  A. Identification of Virus and Viral Antigen 459
  B. Detection of Antibodies 465
VI. Conclusions 467
  References 468

COMPARATIVE DIAGNOSIS
OF VIRAL DISEASES, VOL. IV

ISBN 0-12-429704-8

I. INTRODUCTION AND CLASSIFICATION

A definition of the Togaviridae family, as compiled from the more recent taxonomic publications, may be given as follows: Togavirions are spherical particles measuring 40–70 nm in diameter and consist of an isometric, probably icosahedral nucleocapsid, closely surrounded by a lipoprotein envelope. The viral membrane contains host cell lipid and one to three virus-specified polypeptides, one or more of which are glycosylated. The nucleocapsid is constructed from one nonglycosylated polypeptide and contains a single colinear molecule of single-stranded RNA. Its molecular weight is about 4×10^6. The RNA is infectious when extracted and assayed under appropriate conditions. Togaviruses multiply in the cytoplasm and mature by budding.

Initially, the Togaviridae family contained the two well-known arthropod-borne genera only, the alphaviruses (formerly group A arboviruses) and the flaviviruses (from the type species of the former group B arboviruses, yellow fever virus). These agents and the infections they cause have been reviewed in Chapter 15, Volume I, of this series. The collective term "nonarthropod-borne" or "nonarbo" togaviruses has been coined by the author (Horzinek, 1973a) as a convenient label for all those viruses which resemble the classic arthropod-borne togaviruses more or less closely but lack any serological relationship with them. The most prominent human pathogen, rubella virus, has been reviewed in Chapter 16, Volume I. It is the sole member of the genus *Rubivirus*. In 1973, the term "pestivirus" was introduced as a generic name for the only other antigenically related togaviruses, hog cholera virus (HCV) or swine fever virus and bovine diarrhea virus (BDV) or mucosal disease virus (Horzinek, 1973b). In abandoning the ecologically based scheme of arbovirus classification, the genera *Rubivirus* and *Pestivirus* were incorporated into the Togaviridae family during the third International Congress for Virology in Madrid (Fenner, 1975–1976). In the opinion of the Togavirus Study Group of the International Committee on the Taxonomy of Viruses, equine arteritis virus (EAV) and lactic dehydrogenase virus of mice (LDV) are additional members outside the existing genera (Porterfield *et al.*, 1978), and simian hemorrhagic fever virus (SHFV) is another possible member. For the sake of completeness, it must be emphasized that togaviruses have also been identified in lower animals and plants.

In this chapter, the epidemiologically and economically important infections of vertebrates with nonarbo togaviruses are surveyed. The literature on this subject is scattered over many (veterinary and other) journals. Some comparative aspects and distinguishing characteristics have been discussed by Horzinek (1973a,b, 1975, 1981). Treatises reviewing individual viruses and the respective diseases will be given for a more detailed reference.

After the classic description by Dunne (1964), the properties of HCV were surveyed by a group of workers sponsored by the Commission of the European

Communities (van Aert *et al.*, 1969). Diagnostic and prophylactic aspects of the disease have been abstracted by Sasahara (1970). A comprehensive monograph (Mahnel and Mayr, 1974) was followed by a bibliography edited by the U.S. Department of Agriculture (1976) and a Food and Agricultural Organization paper (1976) on hog cholera eradication. The epidemiology of the disease has been discussed by Ellis *et al.* (1977). The bovine viral diarrhea/mucosal disease syndrome and its causative virus have been reviewed by Pritchard (1963) and by Saurat *et al.* (1972). The virology of equine arteritis has been reviewed by Bürki (1970). After the original survey by Notkins (1965) on LDV of mice, a later monograph appeared (Rowson and Mahy, 1975). The thesis by Wood (1972) is a useful source of information on SHFV.

II. COMPARATIVE CHARACTERISTICS

A. Virion Morphology and Substructure

Undegraded virions of HCV measure 40–60 nm in diameter in negatively stained preparations (Horzinek, 1973b). More recent estimates are 47 ± 5 nm (from urografin gradients), 49 ± 18 nm (from glycerol or sucrose gradients; Frost *et al.*, 1977), and 42 ± 8 nm (Enzmann and Härtner, 1977; Enzmann and Weiland, 1978), supporting the validity of the lower figures. Surface projections are apparently lost during the process of purification. They were first reported by Ritchie and Fernelius (1967).[1] After formaldehyde fixation of infectious cell culture fluid and subsequent purification, Enzmann and Härtner (1977) were able to demonstrate virions with a fringe of 6–8 nm projections on their surface. The membrane surrounds an isometric core which may be liberated from the virion by e.g., detergent treatment; diameter values of 27 ± 3 nm (Horzinek *et al.*, 1971) and 29 ± 3 nm (Frost *et al.*, 1977) have been reported.

As expected from its antigenic relationship (Darbyshire, 1960), BDV is morphologically very similar to HCV (e.g., Hafez *et al.*, 1968). Spontaneously released core particles (28 ± 3 nm in diameter) occasionally show ringlike morphological subunits measuring about 12 nm across (Horzinek *et al.*, 1971). The results of thin section electron microscopy confirm the data obtained on negatively stained, purified pestivirus particles (Hermodsson and Dinter, 1962; Cheville and Mengeling, 1969; Schulze, 1971). The classic study of Scherrer *et al.* (1970) of HCV-infected porcine kidney cells *in vitro* has been supplemented by micrographs obtained by Rutili and Titoli (1977) and Rutili *et al.* (1977) of thin sections through epithelial cells of tonsillar crypts from experimentally infected pigs (Fig. 1). Accumulations of round, sometimes polygonal particles are visible, which measure 40–45 nm in diameter and contain an electron-dense nucleoid (30–35 nm).

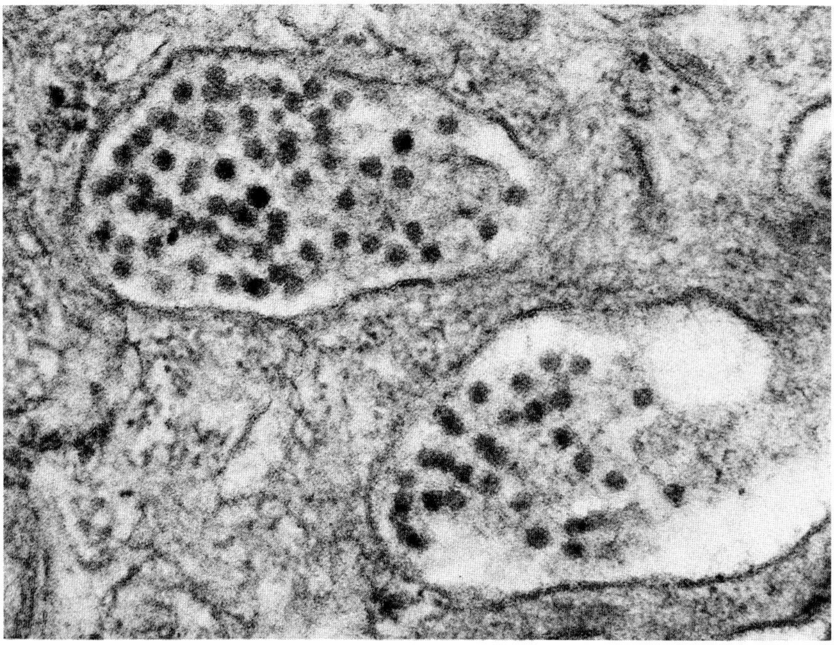

Fig. 1. Thin section through an epithelial cell in a tonsillar crypt of a pig experimentally infected with hog cholera virus; 40-nm particles are accumulated within cisternae. × 99,000. (Courtesy of Dr. D. Rutili.)

With diameter values between 55 and 70 nm, EAV particles are slightly larger than pestivirions; core structures within stain-penetrated particles measure about 35 nm in diameter (for a review, see Horzinek, 1973b). Nucleocapsids prepared by NP 40 treatment and rate zonal centrifugation had a diameter of 36 ± 3 nm (Horzinek *et al.*, 1971). On the viral envelope, annular substructures 12–15 nm in diameter were first detected by Hyllseth (1973a); although the virion surface appears smooth in most cases, the substructures can be discerned in one-sided images (Fig. 2). Diameter values from thin sections are 43 ± 2 nm for the virion and 35 ± 2 nm for the core, as determined in infected equine dermis cells as well as in tissue specimens from experimentally infected horses (Breese and McCollum, 1970, 1973). The identity of the structures was verified using the immunoferritin technique (Breese and McCollum, 1971).

Also LDV approaches alphaviruses in size, although the diameter may vary by about 60% depending on the method of fixation and staining (Horzinek *et al.*, 1975). The particles are spherical (Fig. 3) and show a smooth surface in most cases; annular substructures (8–14 nm across) have been detected (Brinton-Darnell and Plagemann, 1975). The membrane envelops a smooth core particle

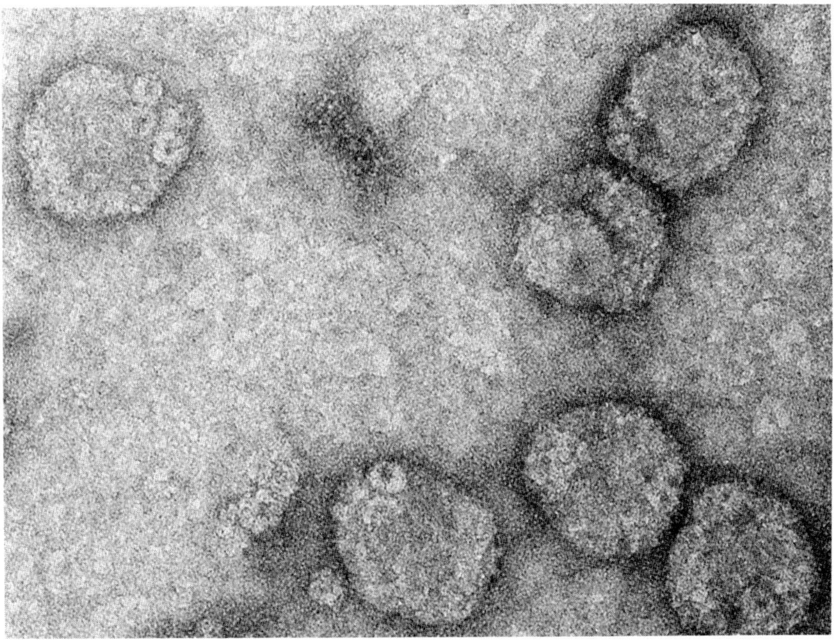

Fig. 2. Equine arteritis virus particles after gradient purification. Negative contrast preparation. Note ring-like surface substructure (e.g., upper left corner). × 350,000.

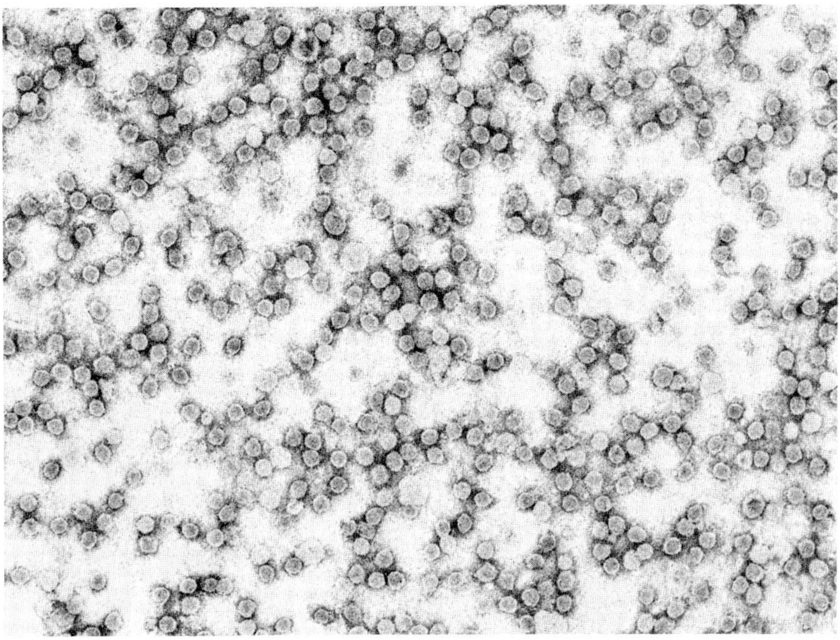

Fig. 3. Highly purified preparation of lactic dehydrogenase virus particles. Negative contrast. × 50,000.

measuring 34 ± 4 nm in diameter which has been released from the virion by detergent treatment (Brinton-Darnell and Plagemann, 1975; Horzinek *et al.*, 1975). From thin sections through infected cells, values of 50–55 nm were obtained for the virion and 25–30 nm for the core.

SHFV propagated in the MA-104 line of rhesus monkey kidney cells was partially purified by isodensity centrifugation and examined in negatively stained preparations. The virions showed a wide variation in size (40–100 nm), although 93% measured between 45 and 50 nm; they consisted of a core (25 nm in diameter) and a tight-fitting envelope (Trousdale *et al.*, 1975). These measurements confirmed earlier studies, in which 40–45-nm particles had been detected in thin sections, containing 22–25-nm electron-dense cores (Wood *et al.*, 1970).

From this short summary of virion morphology, it is obvious that virion size and shape and the presence of a membrane are important taxonomic criteria. Any enveloped spherical virus particle measuring 70 nm or less in diameter should first be considered a Togaviridae family candidate. While identification in thin sections *in situ* may be comparatively easy and straightforward, nonarbo togaviruses are easily distorted and disrupted upon purification. This has resulted in artifactual structures and in misinterpretations. On the other hand, virion disruption—either spontaneous or detergent induced—may reveal details of the particle anatomy which are relevant for classification. The cubic symmetry type of the nucleocapsid is essential to the definition of the Togaviridae family. However, the symmetry criterion has not been applied so far in togavirus classification; resolution of any substructure of the viral core has been elusive for all nonarbo togaviruses, flaviviruses, and most alphaviruses. A detailed analysis of the alphavirus Sindbis has led to controversial data on the triangulation number of the nucleocapsid. The results of Enzmann and Weiland (1979) have confirmed our earlier finding of a $T = 3$ capsomer arrangement (Horzinek and Mussgay, 1969). In this situation, in which cubic capsid symmetry is postulated but unproven for the majority of togaviruses, the morphological evidence on core architecture obtainable from negatively stained preparations comes down to "isometric, not helical" in most cases. Physicochemical and antigenic properties must be used to corroborate the preliminary classification by virion morphology.

B. Physicochemical Properties

Since the purpose of this series is to contribute to the identification of disease agents and the etiological diagnosis of a disease, only those physicochemical properties of the virion will be discussed which are relevant. This would exclude the buoyant density, which may vary according to the gradient material, the host cell used to propagate the virus, etc. A comprehensive listing of togavirus densities has been published by Horzinek (1973b).

The sedimentation coefficients of the viruses reviewed here are lower than those of alphaviruses. For pestiviruses, values between 140 and 180 S have been found (Zeegers and Horzinek, 1977; Laude, 1978a); LDV sediments at about 200 S (Horzinek *et al.*, 1975) and EAV at 227 S (Horzinek *et al.*, 1971; van der Zeijst *et al.*, 1975).

The physical properties of the ribonucleoprotein, the nucleic acid, and the structural polypeptides are of taxonomic importance and will be discussed below.

1. Ribonucleoprotein

The capsids of EAV and LDV are rather stable (in contrast to the core structure of e.g., rubella virus) and can be isolated on a preparative scale for further analysis. After treatment of labeled EAV with Triton X100 and subsequent sedimentation analysis in a sucrose gradient, all [^{3}H]uridine label and about half of the [^{35}S]methionine activity was associated with a 158 S particle. Electrophoretic analysis revealed that only the smallest of the three structural polypeptides is present in this structure (Zeegers *et al.*, 1976). The subviral RNA protein core is identical with the 36 ± 3-nm structure (Horzinek *et al.*, 1971) and represents the EAV nucleocapsid of assumed icosahedral symmetry.

LDV nucleocapsids liberated by Nonidet P40 treatment of virus pellets sedimented at 176 S in isokinetic gradients and appeared as spherical particles 34 ± 4 nm in diameter (Horzinek *et al.*, 1975). Using this isolation technique, Michaelides and Schlesinger (1973) had previously shown that the subviral structure contained all uridine label and the smallest of the three polypeptides.

2. RNA

Early evidence for the existence of RNA in nonarbo togaviruses has been obtained from studies using halogenated deoxyuridines. No inhibition of virus multiplication in their presence has been reported for the pestiviruses (Gillespie *et al.*, 1962; Hermodsson and Dinter, 1962; Dinter, 1963; Loan, 1964; Castrucci *et al.*, 1968b), for EAV (Bürki, 1965b), and for SHFV (Tauraso *et al.*, 1968). A property of the RNA essential to the definition of the Togaviridae family is its infectiousness. Diderholm and Dinter (1966) were the first to show an infectious principle in phenol extracts of BDV preparations. Release of the RNA could also be achieved by treatment with sodium deoxycholate in the presence of polyvinylsulfate (Moennig, 1971). In transfection experiments employing the hypertonic saline method, RNA plaque titers were 5–6 log 10 units lower than the respective virus titers. Also, HCV gave infectious RNA preparations (Zeegers and Horzinek, 1977). Infectious RNA was extracted from LDV using the phenol method (Notkins and Scheele, 1963) but also by treatment with diethyl ether, butanol, and chloroform after removal of RNAse (Notkins, 1964). Brinton-Darnell and Plagemann (1975) estimated a maximum RNA:virion infectivity ratio of about

10^{-5}. Figures between 10^{-6} and 10^{-7} were given for infectious RNA extracted from purified EAV (van der Zeijst *et al.*, 1975).

Sedimentation analysis of infectious or labeled RNA has resulted in values between 38 and 45 S for the pestiviruses (Moennig, 1971; Pritchett *et al.*, 1975; Felmingham and Brown, 1977; Enzmann and Rehberg, 1977). BDV RNA is single-stranded, as suggested by its sensitivity to RNase. Its hydrodynamic properties indicate secondary structure. Lowering the NaCl molarity decreased the sedimentation velocity; Mg^{2+} and Ca^{2+} ions had a contrary effect. The integrity of the molecules sedimenting differently under different conditions was demonstrated by transfection experiments (Horzinek, 1976). Somewhat higher sedimentation values were found for the RNA of EAV (48 S; van der Zeijst *et al.*, 1975) and LDV (48 S; Darnell and Plagemann, 1972). In these studies, alphaviruses were included for comparison, whose RNAs behaved almost indistinguishably from those of the nonarbo togaviral RNAs.

From the sedimentation studies and from polyacrylamide gel electrophoretic analyses the molecular weights of togaviral RNAs have been calculated. Most values are in the range between 3.0 and 4.1×10^6. Actual molecular weight estimates of alphavirus RNA are 4.5×10^6 (Strauss and Strauss, 1977).

3. Structural Polypeptides

Attempts have been made to elucidate the protein composition of pestiviruses. HCV which had been grown in the presence of actinomycin D and labeled with [^{35}S]methionine was resolved into three main polypeptides in SDS polyacrylamide gels. By labeling of concentrated and purified HCV with [^{3}H]sodium borohydride, it was shown that the polypeptides with molecular weights of 55 and 46 kd are superficially located glycoproteins, and the 36 kd is the core protein. The viral specificity of these polypeptides was verified using radioimmune precipitation (Enzmann and Rehberg, 1977; Enzmann and Härtner, 1977; Enzmann and Weiland, 1978). Three major polypeptides with molecular weights of 57, 44, and 34 kd were also found in BDV, the two larger ones being glycosylated (Matthaeus, 1979). From these results, it is evident that pestiviruses have a very similar polypeptide composition—a generic trait already noted for the alpha- and flaviviruses, respectively.

Because of the close resemblance of the polypeptide patterns and molecular weights of EAV (21, 14, and 12 kd; Zeegers *et al.*, 1976) and LDV (24–44, 17–18, and 15–17 kd; Michaelides and Schlesinger, 1973; Darnell *et al.*, 1974; Brinton-Darnell and Plagemann, 1975), we attempted to demonstrate an antigenic relationship between the two viruses. Heterologous neutralization, immunofluorescence, and radioimmune precipitation tests, however, did not reveal any cross reaction. The small differences in the molecular weights were found not to be due to technical variations but to represent a distinctive feature, as shown in parallel electrophoretic runs in polyacrylamide slab gels (van Berlo *et al.*, 1981).

Electrophoresis of purified SHFV showed four peaks of radioactivity (Trousdale *et al.*, 1975). No estimates of molecular weight, specificity criteria, or an indication for the localization of the polypeptides in the virion were given. In its glucosamine labeling of the biggest polypeptide, the relative migration distances, and the molar proportions, the general picture resembles that of EAV/LDV rather than that of the pestiviruses.

From these data, some generalizations can be drawn which are of taxonomic significance. Togaviruses possess only 3–4 structural polypeptides, the smallest of which is participating in nucleocapsid construction. Two envelope proteins are consistently present, one or both of which are glycosylated. These figures constitute actual minimum estimates. Using more refined methods of analysis, additional proteins may be found. A special problem encountered by authors attempting polypeptide analyses of nonarbo togaviruses is that of assessing the purity of their preparation and the specificity of the proteins detected. Radioimmune precipitation using defined sera would be the method of choice for obtaining unequivocal results.

C. Antigenic Structure and Relationship

Elucidation of the antigenic structure has not been achieved for any of the nonarbo togaviruses. Antigens detectable by various serological tests have been described, but their identity and localization within the virion remain to be analyzed. Obviously, the determinants responsible for the induction of neutralizing antibody are situated on the togavirion surface, which is also true for the hemagglutinating determinants. With the exception of rubella virus, however, nonarbo togaviruses are devoid of hemagglutinating activity. A whole range of erythrocyte species has been tested with HCV (Kubin, 1967), BDV (Dinter *et al.*, 1962; Borgen, 1963; Taylor *et al.*, 1963; Castrucci *et al.*, 1968a; Hafez and Liess, 1972), EAV (Doll *et al.*, 1957a), and LDV (Crispens, 1965) preparations serving as antigens. Not in all instances did the authors assay their preparations over a range of pH (< 6 to > 7) and temperatures (+4 to +37°C), which are conditions known to influence agglutination in the arthropod-borne togaviruses.

The taxonomic dilemma felt by many authors studying the viruses reviewed here induced them to perform comparative serological studies. No cross reactions were found between BDV, the flaviviruses, and lymphocytic choriomeningitis virus (Gutekunst and Malmquist, 1965), between SHFV and the alpha- and flaviviruses (Wood, 1972), and between EAV and African horse sickness virus (McCollum *et al.*, 1970) or epizootic hemorrhagic disease of deer virus (Shope *et al.*, 1960). Saurat *et al.* (1972) listed some 20 animal viruses belonging to different families which lacked an antigenic relationship with BDV. Antiserum prepared against bee chronic paralysis virus did not react with LDV in an Ouchterlony gel diffusion test (Gibbs, 1969). Also, within the nonarbo togaviruses there was no cross reaction between rubella virus, EAV, LDV, and the pestiviruses (Horzinek *et al.*, unpublished observations). The only serologically related viruses are HCV and BDV (and the closely related virus causing border disease of sheep), which constitute the genus *Pestivirus* within the Togaviridae family.

A large literature has accumulated on the *in vivo* and *in vitro* cross reactions between pestiviruses. In summarizing the results, there can be no doubt that cross neutralization between HCV and BDV does occur. Due to different degrees of antigenic relatedness, strains may show a pronounced or weak reaction—or no reaction at all—with a serum raised against a strain of the heterologous virus. Animals immunized against one pestivirus will react by a secondary (anamnestic) response when inoculated with another virus of the genus.

In an attempt to identify the assumed surface glycoprotein responsible for the pestivirus cross reaction, Dalsgaard (1976) and Dalsgaard and Overby (1977) have isolated a polypeptide by tandem-crossed immunoelectrophoresis which binds to concanavalin A. Injection into pigs resulted in their protection against an otherwise fatal challenge infection with virulent HCV. The group of animals receiving the common glycopeptide isolated from BDV-infected cultures showed some delay in the onset of clinical symptoms after challenge, but all animals eventually died. These experiments have been performed with material from detergent-treated infected cells and not with purified virions. In a study using virion antigenic material, Volenec *et al.* (1972) found no protection, whereas cell lysates did protect the animals. The observed differences may be due to the concentration of antigenic material present in a preparation and to its immunogenicity, which may depend upon the size of the membrane fragments produced by various treatments (Mussgay *et al.*, 1973).

Another pestivirus, isolated from sheep showing signs of "border disease" or "hairy shaker disease," has been found to produce neutralizing antibody against BDV (Acland *et al.*, 1972; Hamilton and Timoney, 1972). Osburn *et al.* (1973), Plant *et al.* (1973), and Lamont and Stuart (1974) included HCV in their studies and confirmed the pestiviral character of border disease virus. Ovine infectious material inoculated into pregnant heifers produced stillbirth and abortions; the cows developed antibody to BDV (Gibbons *et al.*, 1974). Similarly, BDV induced abortions and congenital defects when inoculated into pregnant ewes (Ward, 1971; French *et al.*, 1974; Snowdon *et al.*, 1975; Plant *et al.*, 1976a,b). Harkness *et al.* (1977) reported the isolation, in cell culture, of a noncytopathic virus from sheep which showed immunofluorescence and neutralization reactions with HCV antiserum.

One may question whether it is justified to consider HCV, BDV, and border disease virus as separate viral species, since the antigenic differences between some strains may be minute. Actually, they are so small that even the most exclusive tests (e.g., neutralization) may not result in unequivocal identification. The use of sera produced after a single injection of a given virus strain and comparative testing of the isolate using different reference strains are mandatory.

III. PATHOLOGY AND PATHOGENESIS

A. Hog Cholera

Hog cholera in its classic form is an acute or peracute febrile disease of pigs characterized by severe generalized vasculitis and necrosis of lymphocytes. Most affected animals die within 1 to 2 weeks, with widespread petechiation, thrombosis, splenic infarction, intestinal ulceration, and encephalitis. Occasionally a pig will survive and improve clinically, only to succumb later to an exacerbation of an acute disease. These animals may continue to excrete virus and therefore play an important role in the dissemination of the disease. Strains of widely varying virulence have been described; low-virulence strains may lead to inapparent (subclinical) infections but also to persistent infections in adult animals. In pregnant sows, the virus may pass the placental barrier and cause fetal mummification, edema, stillbirth or neonatal death, and congenital tremors. HCV can persist in piglets after an *in utero* infection and has been recovered for many months after birth. Detailed reviews of the pathology of hog cholera have been presented by Fuchs (1968) and Mahnel and Mayr (1974).

Also, the histopathological changes of the acute infection have been adequately described (for literature, see Cheville and Mengeling, 1969). The virus affects cells of the reticuloendothelial system, producing hydropic degeneration and necrosis of endothelial cells which result in widespread edema and hemorrhage. Lymphoid tissues are depleted of lymphocytes. Other changes include interlobular mononuclear cell infiltration in the liver, degenerative changes in intestinal epithelial cells and renal tubular epithelium, and adrenal hyperplasia. The pathological manifestations have been interpreted at the cellular level as a disorder of the enzyme system caused by a virus-induced production of a chymotrypsin-like protease (Korn and Matthaeus, 1977).

Although conjunctival and transdermal exposure have been found to induce hog cholera in experimental animals, the oronasal route must be considered of primary importance under natural conditions. The tonsils are the primary sites of virus multiplication (Dunne *et al.*, 1959); they are also the tissue of choice for the early detection of HCV after natural infection with virulent strains (Aiken *et al.*, 1964; Robertson *et al.*, 1965; Meyling and Schjerning-Thiesen, 1968; Ressang and de Boer, 1968; Peckham *et al.*, 1970). Virus titers in the tonsils were found to remain high from days 3–7 after experimental oral exposure. Virus excretion in the feces and urine starts about 1 week after infection (Ressang, 1973).

In its chronic form, hog cholera may be caused by strains of low or moderate virulence, but also by minimal doses of virulent virus (Dunne *et al.*, 1955) or by simultaneous exposure to virulent virus and homologous antiserum (Biester and Schwarte, 1952). Malnutrition is a predisposing factor to chronic illness caused by highly virulent virus (Schmidt *et al.*, 1965). Based on the symptomatology, three phases were distinguished in chronic hog cholera. In the acute phase, large amounts of virus and viral antigen were present in the blood and different (reticuloendothelial, lymphoid, and epithelial) tissues. Upon clinical improvement, during the second phase, virus could no longer be detected in the blood—or only at very low titers. The increase in the serum gamma globulin levels was accompanied by exaggerated plasma cell formation. Viral antigen was limited to the epithelial cells of the tonsils, ileum, salivary glands, and kidney. During the terminal phase, the virus again became disseminated throughout the body. A glomerulonephritis developed as an expression of the deposition of antigen–antibody complexes in the kidney and subsequent attraction of neutrophils (Cheville *et al.*, 1970).

Antibodies to HCV and its antigens have been demonstrated using virus neutralization, complement fixation, gel precipitation, indirect immunofluorescence, enzyme-linked immunosorbent assay, and passive hemagglutination tests. However, nonimmunogenic strains of HCV have been isolated (Korn, 1964; Korn and Hecke, 1964). Corthier (1978) noted that such strains did not induce neutralizing antibody and could only be traced serologically using passive hemagglutination (Corthier *et al.*, 1977).

B. Bovine Viral Diarrhea

The majority of infections of cattle with BDV are inapparent (Malmquist, 1968). In case a disease develops, the first clinical signs are fever, leukopenia, anorexia, nasal discharge, and general depression (Pritchard, 1963; Castrucci *et al.*, 1968a), followed by erosions of the oral mucosa and diarrhea (Dinter *et al.*, 1962; French and Snowdon, 1964). A large proportion of clinically diagnosed cases terminate fatally (Mills *et al.*, 1965; Mossman and Hanly, 1976), but some acutely infected animals may recover or may develop a chronic debilitating form of the disease. This stage is characterized by weight loss, intermittent diarrhea, erosions of the mucosal membranes of the alimentary track, and sometimes lameness due to a necrotic interdigital dermatitis or ulcerative coronitis (Dow *et al.*, 1956; French and Snowdon, 1964; Castrucci *et al.*, 1968a; Johnson and Muscoplat, 1973; Mossman and Hanly, 1976). In some epizootics, the symptomatology is dominated by respiratory signs (McKercher, 1968; Wellemans, 1969; Kahrs, 1971; Wellemans and Leunen, 1974). BDV has been found to induce teratological and also lethal effects on the bovine fetus, including cerebellar hypoplasia. ocular defects, mummification, and abortion. It has been shown to cross the placenta, causing concurrent maternal and fetal infection (Kahrs *et al.*, 1970a,b,c; Kahrs, 1971).

The chief pathological changes of a clinical BDV infection are hyperemia, hemorrhage, edema, erosion, and ulceration of the alimentary track mucosae ("mucosal disease") accompanied by inflammatory and atrophic processes in lymphatic tissues (for reviews, see Pritchard, 1963; Saurat *et al.*, 1972). Necrotic foci can be observed in the lymph nodes and spleen of fatally infected cattle; gross lesions include destruction of Peyer's patches in the small intestine and loss of differentiation between cortical and medullary areas of the lymph nodes (Muscoplat *et al.*, 1973a).

Under field conditions the virus is transmitted via the oral (Dinter, 1968) and respiratory routes (Mills and Luginbuhl, 1968). After viremic spread, viral antigen has been demonstrated by immunofluorescence in epithelial and reticuloendothelial cells, in histiocytes, in cells of the glomeruli and salivary glands, and in neurons of the cortex, hippocampus, and cerebellum (Meyling, 1970). The affinity of BDV for cells of the immune system is certainly of pathogenetic significance. Usually, BDV-infected cattle produce antibody which confer lifelong immunity (Robson *et al.*, 1960; Kahrs *et al.*, 1966). However, there are several reports in which clinically infected animals did not mount an antibody response (Dinter *et al.*, 1962; Thomson and Savan, 1963; Malmquist, 1968). A report by Muscoplat *et al.* (1973b) indicates that failure to produce antibody—which is a common denominator in fatally infected cattle (Shope, 1964)—is due to a virus-induced suppression of the secretion of immunoglobulins. Shope *et al.* (1976) concluded that antibodies play a crucial role in controlling primary infection with BDV. Immunosuppressed calves raised without maternal antibody developed a fatal viremia, whereas animals with high levels of humoral antibody were protected.

Infection of pregnant ewes with a virus closely related to BDV has been shown to cause abortion or border disease in the offspring. Affected lambs often have a history of poor growth and viability; their fleece is abnormally pigmented and hairy rather than woolly. A significant proportion of the lambs is born with characteristic tonic-clonic spasms (hairy shaker disease) that may be severe enough to cause incoordination. Histological examination of the central nervous system often reveals hypomyelinogenesis and abnormal glial formation (Barlow and Dickinson, 1965; Manktelow *et al.*, 1969).

C. Equine Arteritis

The clinical features of infection of equines with EAV are acute anorexia and fever, usually accompanied by palpebral edema, conjunctivitis, mucopurulent nasal catarrh, and edema of the legs, genitals, and abdomen. In pregnant mares, abortion occurs in about half of the exposed animals. To a lesser extent, photophobia, corneal opacity, and respiratory symptoms are observed. Edema and petechiae are most conspicuous gross pathological lesions. Histological examination of the intestine and the mesenteric lymph nodes shows marked edema, fibrin exudation, and lymphocyte depletion. Endothelial cell destruction is present in blood vessels of all sizes. The media of larger arteries are often degenerate or necrotic, resulting in edema and hemorrhage with infarction in the intestine and spleen. Kidney changes include hyperemic and hypercellular glomeruli (Jones, 1969; Estes and Cheville, 1970). Vascular necrosis and glomerulonephritis did not appear until the fourteenth day, at which time surviving animals have clinically recovered and possess antiviral antibody. It has been suggested that the renal lesions may be the result of immune complex depositions (Prickett *et al.*, 1973), since EAV could be isolated from the kidneys long after it had disappeared from other tissues (McCollum *et al.*, 1971). Thus virus excretion would occur via the urine in addition to the upper respiratory mucosae (McCollum *et al.*, 1961b). Serological evidence suggests that most infections take an inapparent course.

D. Infections of Mice with LDV

LDV has been identified as a benign passenger virus in many biological materials passaged or transplanted serially in mice. Although ordinarily nonpathogenic, it can combine synergistically with *Eperythrozoon coccoides* (Riley, 1964; Riley *et al.*, 1964; Fitzmaurice *et al.*, 1972), another frequent murine contaminant, and with most malignancies to produce modified responses. These include either suppression (Bailey *et al.*, 1965; Notkins, 1965; Riley, 1966, 1968; Brinton-Darnell and Brand, 1977) or enhancement of tumor growth (Plagemann and Swim, 1966; Michaelides and Schlesinger, 1974) depending on the timing of LDV infection and the inoculation of tumor cells (Michaelides and Schlesinger, 1974). Specific influences of an LDV infection on the host concern mainly its immune system, e.g., a stimulation of humoral antibody (Notkins *et al.*, 1966b; Mergenhagen *et al.*, 1967) and a depression of cell-mediated immunity (Howard *et al.*, 1969; Snodgrass *et al.*, 1972), partial destruction of the thymus (Santisteban *et al.*, 1972), lysis of lymphocytes (Riley, 1968), and other effects. A lifelong viremia is noted in infected mice, in which the virus is present as an infectious complex with antibody (Notkins *et al.*, 1966a, 1968). Immune complex deposits have been demonstrated in the renal glomeruli, but only mild lesions are observed (Porter and Porter, 1971).

An age-dependent polioencephalomyelitis was observed during studies of the immune response of C 58 mice to syngenic leukemia cells (Murphy *et al.*, 1970). Histopathologically the disease is characterized by a mononuclear cell infiltration of the gray matter of the spinal cord and brain stem (Homburger *et al.*, 1973; Lawton and Murphy, 1973). The original hypothesis that the condition is autoimmune mediated had to be discarded after the demonstration of its viral etiology (Martinez, 1979). It has been demonstrated that the causative agent is LDV (Brinton, personal communication).

E. Simian Hemorrhagic Fever

SHFV affects rhesus monkeys (*Macaca mulatta*) and other macaques clinically, while patas monkeys (*Erythrocebus patas*) develop a chronic inapparent infection (London, 1977). The disease is characterized by high fever, loss of appetite, dehydration, and lethargy, progressing to facial erythema and edema. Neurological symptoms may ensue, such as muscular weakness, tremors, ataxia, and epileptoid seizures. Petechial hemorrhages appear, followed by gross bleeding from the gums, the nose, and the intestines. All animals showing disease symptoms die, usually in 5 to 14 days (Lapin *et al.*, 1967; Palmer *et al.*, 1968). At necropsy, multiple hemorrhages were seen throughout the body, characteristically in the duodenum and the spleen. Pathology of the central nervous system did not resemble that of a viral encephalitis. There is also a generalized destruction of lymphoid tissue (Allen *et al.*, 1968; Abildgaard *et al.*, 1975).

IV. EPIDEMIOLOGICAL ASPECTS

A. Hog Cholera

Hog cholera affects predominantly domestic pigs, which also serve as the virus reservoir between epizootics. Wild hogs of the genus *Sus*, although susceptible, are of minimal importance in the epidemiology, and the role of other species is only a matter of conjecture. The distribution of the virus is worldwide, although eradication has been possible in some countries (Table I). During epizootics, the economic losses caused by hog cholera are higher than those due to any other porcine infectious disease. From morbidity statistics compiled in European Economic Community countries, there are indications for a 3–4 year periodicity of epizootics. Cycles in disease incidence are thought to be due to changes in the pig population density, which in turn depend upon the evolution of pork prices and the seasonal variations and movements in sow but predominantly in piglet populations (Ellis *et al.*, 1977). Airborne spread and direct contact between infected and susceptible pigs are the principal means of virus spread. Mechanical vector transmission is of little significance. Indirect spread by contaminated materials, swill, and other waste foods containing pig meat involves a lower degree of probability. Widely divergent morbidity and mortality data have been reported,

TABLE I
Countries Free from Hog Cholera Prior to 1970

Country	HCV eradicated
Finland	1917
Iceland	1933
Denmark	1933
Sweden	1945
Ireland	1958
Norway	1961
Australia	1963[a]
Czechoslovakia	1964
Canada	1964
United Kingdom	1966[b]

[a]Sporadic cases in 1967.
[b]Reintroduced in 1971 but immediately eradicated.
Source: From Mahnel and Mayr (1974) and Ellis *et al.* (1977).

depending on the virulence of the strain involved in an epizootic and the immunity level of the pig population. Variations in the antigenic character of HCV are very limited; changes in virulence, however, appear to occur even during the course of an infection.

Different policies for reducing the incidence of disease have been followed which are based on essentially two principles. The first is systematic, widespread vaccination using modified live HCV strains; the second is eradication of infection. To achieve the latter goal, all infected pigs must be slaughtered and their contacts traced. Serological testing is used to detect inapparent infection. Obviously this can be practiced only in the absence of vaccination.

B. Bovine Viral Diarrhea

The species which can be infected by BDV are cattle, sheep, goats, pigs, and several species of exotic wild ruminants (for literature, see Saurat *et al.*, 1972). The most common form of infection is the inapparent case. For example, Malmquist (1968) reported that in a closed herd of 160 head, only one clinical case had been recorded, whereas about 60% of the cattle had neutralizing antibody to the virus. This would exclude the necessity of an extrabovine reservoir for the perpetuation of BDV between epizootics. Although the virus appears to be of cosmopolitan distribution, there are differences in the percentages of seropositive animals in different countries, regions, and even herds within the same region. BDV is the most frequently found viral contaminant in fetal bovine sera; 25% of the batches carrying statements on the bottle label that the sera had been prescreened by the supplier for bovine virus contamination were found BDV positive (Kniazeff *et al.*, 1975). In addition, noncytopathic strains have been found to contaminate the organ tissue used to establish cell cultures (Nuttall, 1978).

With respect to their morbidity and mortality, BDV infections present two different epidemiological patterns. The bovine viral diarrhea symptomatology is characterized by a rapid spread within a herd and a high (50–90%) morbidity, the mortality not exceeding 5% of the diseased animals. The mucosal disease syndrome is noted only in about 1–5% of the animals in a herd with successive cases evolving during a usually long and variable period of time. Most cases (80–95%) are fatal. This is a somewhat schematic differentiation, and intermediate forms of course occur. The experimental transmission of the disease is not consistent; bovine diarrhea can be reproduced, but the mucosal disease syndrome develops only incidentally. Vaccines containing attenuated BDV alone or in combination with other bovine viruses causing respiratory infections (Gillespie, 1968) are widely used. They sometimes cause clinical reactions and are contraindicated for pregnant cattle due to the hazard of abortion. Inactivated vaccines have proved to confer seroconversion and protection (for a review, see Lambert, 1973).

C. Equine Arteritis

The knowledge of the epidemiology of EAV is based on only a few disease outbreaks. The first enzootic occurred in Bucyrus, Ohio, in 1953 and led to the isolation of the reference virus strain (Doll *et al.*, 1957a,b). In Europe the virus has been isolated in Switzerland and Austria (Bürki, 1965b, 1970). A disease similar in its symptomatology was known to occur in Europe previously, but its causative agent had not been isolated (Brion *et al.*, 1967). Serological evidence suggests that the virus has infected horses in Africa (Moraillon and Moraillon, 1978), India, Italy, Scandinavia, France, Germany (for a review, see Akashi *et al.*, 1976), and the Netherlands (de Boer *et al.*, 1978). From the limited experience, it appears that equines are the only susceptible species; inapparent infections seem to be the rule (McCollum and Bryans, 1973). During outbreaks, morbidity between 50% (Bürki and Gerber, 1966) and 75% (Doll *et al.*, 1957b) has been reported. A modified live virus vaccine has been developed which proved safe for use in horses of any age; mares late in gestation should be excluded from vaccination (Doll *et al.*, 1968).

D. Infections of Mice with LDV

LDV does not infect any species other than the mouse; all strains of laboratory mice are susceptible, and isolations from wild mice have been made in Europe,

Australia, and North America. Under laboratory colony conditions, transmission between mice in adjacent cages may occur, although infrequently. A transcutaneous mode of transmission of the virus via infected saliva during fighting seems to be relevant (Notkins *et al.*, 1964); arthropod transmission has been a matter of conjecture. If it does play a role in LDV epidemiology at all, the mode of transmission is purely mechanical. Du Buy (personal communication) found that LDV did not multiply in ticks which had been fed on infected mice. Cannibalism is also of importance. Vertical transmission occurs when the mothers are infected during pregnancy; the virus probably crosses the placenta during the acute phase of the infection, when the plasma virus titer is high. Other sources of infectious virus are the milk and feces. A detailed review of LDV ecology has been presented by Rowson and Mahy (1975).

E. Simian Hemorrhagic Fever

As in the case of EAV, there have been only a few outbreaks of SHFV on which all epidemiological experience is based. The virus was first isolated in 1964 during an outbreak of febrile hemorrhagic disease among newly arrived rhesus monkeys (*Macaca mulatta*) held in the quarantine colony of the National Institutes of Health (NIH), Bethesda, Maryland (Palmer *et al.*, 1968). A similar enzootic had occurred 3 months earlier in the monkey breeding colony of the Sukhumi Institute of Experimental Pathology and Therapy in the Soviet Union (Lapin *et al.*, 1967). Subsequent isolation and identification of the Sukhumi agent in the NIH laboratories established the Sukhumi outbreak as the first recognized appearance of SHFV (Tauraso *et al.*, 1968). Circumstantial evidence for a single agent responsible for both outbreaks was provided by the discovery that both institutions had received animals within 3 months from the same Indian supplier. Additional enzootics have been reported from Davis, California, and Sussex, England. In the latter colony, 220 out of 240 cynomolgus monkeys (*M. fascicularis*) contracted the disease in 1966–1967 and died, in addition to 7 out of 44 rhesus monkeys of the infected shipment. Further outbreaks were recorded in 1968 and 1969 in this colony; the single surviving rhesus monkey of 116 animals had antibody to the NIH strain of SHFV (Wood, 1972). Among many species housed in the Sukhumi colony, only the *Macaca* species present (*M. mulatta, M. nemestrina, M. speciosa,* and *M. assamensus*) appeared susceptible. Patas monkeys (*Erythrocebus patas*), baboons (*Papio papio*), and African green monkeys (*Cercopithecus aethiops*) were found to support persistently inapparent infections. Transmission of infection from chronically infected to susceptible monkeys does not occur by contact or aerosol. The NIH epizootic was inadvertently initiated by mechanical transfer of blood from a chronically infected patas monkey to a rhesus monkey. In the reservoir species, infectivity titers as high as

10^{12} ID_{50}/ml blood were observed (London, 1977). Only during the acute phase of the LDV infection of mice have viremia titers of similar magnitude been found so far.

V. COMPARATIVE DIAGNOSIS

The ecological diagnosis of nonarbo togavirus diseases is essential because of its epidemiological implications. The clinical pictures of, e.g., hog cholera (European swine fever) and African swine fever may be very similar, as the names suggest, but the veterinary measures to be taken when one of the causative agents is identified are quite different.

A. Identification of Virus and Viral Antigen

In this chapter, the methods of virus isolation will be reviewed, focusing on the use of cell cultures. Animal experiments are included only where they present a technical advantage, as in the case of LDV. In addition, serological methods designed to identify viral antigen either in autopsy material or *in vitro* are discussed.

1. Virus Isolation

The cell species susceptible to infection with HCV are listed in Tables II and III. The PK15 line has become most popular in diagnostic laboratories. It should be emphasized that HCV replication has to be demonstrated indirectly by e.g., immunofluorescence, since cytopathology is very exceptional. Only the HCV strains PAV-1 (Gillespie *et al.*, 1961) and CAP (Laude, 1978b) are cytopathogenic when propagated under appropriate conditions.

In Tables IV and V the spectrum of cells susceptible to BDV is given. Cytopathogenicity has been encountered after infection of bovine cells with many, but not all, BDV strains. Variations are due to intrinsic properties of a strain and to the conditions of culture. Van Bekkum and Straver (1964) underline the advantage of using bovine testicle cells, and Malmquist (1968), Scott *et al.* (1972), and Schiff and Storz (1972) favor bovine embryonic spleen cell cultures as compared to the conventional bovine kidney cell culture.

The third pestivirus, border disease virus (Acland *et al.*, 1972; Hamilton and Timoney, 1972, 1973; Plant *et al.*, 1973), can be expected to behave variably, too, with respect to cytopathic effect (CPE). Indeed, pertinent observations have been made (CPE: Vantsis *et al.*, 1976; no CPE: Hadjisavvas *et al.*, 1975; Harkness *et al.*, 1977; Terpstra, 1978).

The cell species listed in Tables VI and VII show cytopathological alterations

TABLE II
Susceptibility of Cultured Cells (Primary and Low-Passage Lines) to Hog Cholera Virus

Artiodactyla	
Porcine	Hecke (1932)
Kidney	Karasszon and Bodon (1963)
Spleen	Dale and Songer (1957)
	Frenkel *et al.* (1955)
Testis	Tsubaki (1938)
	Kumagai *et al.* (1958)
	Kubin (1964)
Trachea	Mengeling and Pirtle (1972)
Liver	Kresse *et al.* (1967)
Bone marrow	Boynton (1946)
Leukocytes	Dunne *et al.* (1957)
	Loan and Gustavson (1961)
Macrophages	Korn and Zoeth (1971)
	Korn and Lorenz (1976)
Bovine kidney	Sato *et al.* (1964)
Bovine skin, spleen, trachea (embryonal)	Pirtle and Kniazeff (1968)
Ovine kidney, esophagus (embryonal)	Pirtle and Kniazeff (1968)
	Leftheriotis *et al.* (1971)
Goat kidney, esophagus (embryonal)	Pirtle and Kniazeff (1968)
White-tailed deer, collared pekari kidney	Pirtle and Kniazeff (1968)
Carnivora	
Striped skunk, american badger kidney	Pirtle and Kniazeff (1968)
Mink lung (embryonal)	Pirtle and Kniazeff (1968)
Rodentia	
Guinea pig kidney	Pirtle and Kniazeff (1968)
	Sasahara and Kumagai (1965)
Fox squirrel kidney	Pirtle and Kniazeff (1968)
Lagomorpha	
Domestic rabbit kidney, skin	Pirtle and Kniazeff (1968)
	Leftheriotis *et al.* (1971)
Eastern cotton-tail rabbit kidney, skin	Pirtle and Kniazeff (1968)
Cetacea	
Spotted dolphin, bottle-nosed dolphin kidney	Pirtle and Kniazeff (1968)
Common dolphin testes	Pirtle and Kniazeff (1968)
Pacific pilot whale kidney	Pirtle and Kniazeff (1968)

subsequent to EAV infection. In those cases in which CPE was absent (primary calf and pig kidney cells, chick embryo cells—Bürki, 1965b), virus multiplication was not further studied.

LDV is routinely diagnosed by inoculation of suspected material into susceptible mice and subsequent determination of plasma LDH activity. The susceptibility of the test animals is confirmed by demonstrating that their LDH levels are

TABLE III
Susceptibility of Cell Lines to Hog Cholera Virus

Primates	
HeLa human cervix carcinoma	Pirtle and Kniazeff (1968)
LLC-MK2 rhesus monkey kidney	
Carnivora	
MDCK canine kidney	Pirtle and Kniazeff (1968)
Artiodactyla	
MDBK bovine kidney	Pirtle and Kniazeff (1968)
SV40-transformed bovine, porcine kidney	Diderholm and Dinter (1965)
IB-RS2 porcine kidney	Laude (1978b)
	de Castro (1973)
PK2a, PK15, SK6 porcine kidney	Pirtle and Kniazeff (1968)
	Kasza *et al.* (1972)
ST porcine testicle	Pirtle and Kniazeff (1968)

below 500 I.U./ml. Three to 4 days after intraperitoneal inoculation the mice are bled and the LDH levels are determined using, e.g., commercially available test kits. A five- to tenfold increase in enzyme activity is indicative for successful infection, provided that other LDH-elevating agents (e.g., *Eperythrozoon coccoides*) have been excluded. Infectivity titers of $\geq 10^{10}$ ID_{50}/ml of plasma have been reported (Notkins, 1965; Rowson and Mahy, 1975). The permissive mouse

TABLE IV
Susceptibility of Cultured Cells (Primary and Low-Passage Lines) to Bovine Diarrhea Virus

Artiodactyla	
Bovine kidney	Hansen *et al.* (1962)
	Hermodsson and Dinter (1962)
	Mills and Luginbuhl (1965)
Testis	Underdahl *et al.* (1957)
Spleen (embryonal)	Scott *et al.* (1972)
	Schiff and Storz (1972)
	Cancellotti *et al.* (1972)
Skin, intestine (embryonal)	Kniazeff, cited in Pritchard (1963)
Lung (embryonal)	Singh (1969b)
	Goldsmith and Barzilai (1975)
Endometrium (fetal)	Soto-Belloso (1976)
Bone marrow	Kendrick (1971)
Leukocytes	Truitt and Shechmeister (1973)
Sheep kidney	Singh (1969a)
Testis	Gillespie *et al.* (1961)
	Seibold and Dougherty (1967)
Goat kidney	Kniazeff, cited in Pritchard (1963)
Pig kidney	Malmquist *et al.* (1965)
Testis	Wellemans and Leunen (1966)

TABLE V
Susceptibility of Cell Lines to Bovine Diarrhea Virus

Primates	
HeLa human cervix carcinoma	Kniazeff, cited in Pritchard (1963)
ERK-1 (possibly HeLa)	Fernelius *et al.* (1969)
Artiodactyla	
MDBK bovine kidney	Marcus and Moll (1968)
AV-BEK fetal bovine kidney	Cancellotti and Turilli (1975)
BT bovine turbinate	Stewart *et al.* (1971)
	Carbrey (1971)
EBTr bovine trachea (embryonal)	Carbrey (1971)
Spleen	Malmquist (1968)
Skin	Coria (1969)
Cornea	Ludwig *et al.* (1969)
PK15 porcine kidney	Malmquist *et al.* (1965)
	Carbrey and Lee (1965)
Rodentia	
HaK hamster kidney	Fernelius *et al.* (1969)

cell species is the phagocytic cell of the reticuloendothelial system. Growth *in vitro* in macrophage cultures is not accompanied by gross cytopathology. Schlesinger *et al.* (1976) have demonstrated that LDV will also grow in somatic cell hybrids between mouse peritoneal macrophages and SV40-transformed human Lesch-Nyhan fibroblasts.

Of the many primary cell cultures and established cell lines tested for their susceptibility for SHFV during the Sukhumi and Bethesda epizootics, only the

TABLE VI
Susceptibility of Cultured Cells (Primary and Low-Passage Lines) to Equine Arteritis Virus

Perissodactyla	
Horse kidney	McCollum *et al.* (1961a,b)
	Bürki (1965a,b)
	Foster (1969)
Testicle	Hyllseth (1969)
Artiodactyla	
Pig kidney	Konishi *et al.* (1973)
Carnivora	
Cat kidney	Konishi *et al.* (1973)
Rodentia	
Hamster kidney	Wilson *et al.* (1962)
Lagomorpha	
Rabbit kidney	McCollum *et al.* (1962)
	Bürki (1965b)
	Hyllseth (1969)

TABLE VII
Susceptibility of Cell Lines to Equine Arteritis Virus

Primates	
B-SC-1 African green monkey kidney	Crawford and Davis (1970)
JINET Cynomolgus monkey kidney	Konishi *et al.* (1973)
LLC-MK2 rhesus monkey kidney	McCollum *et al.* (1971)
	Hyllseth, 1973b
Vero African green monkey kidney	Konishi *et al.* (1973)
	Hyllseth (1973b)
Perissodactyla	
E. derm. (NBL6) equine dermis	Breese and McCollum (1971)
	Radwan and Burger (1973a)
Rodentia	
BHK-21 baby hamster kidney	Hyllseth (1969)
	Maess *et al.* (1970a,b)
HT7; HS canine hepatitis virus-transformed hamster tumor	Shinagawa *et al.* (1976)
HmLu hamster lung	Konishi *et al.* (1973)
Lagomorpha	
LLC-RK1 rabbit kidney	Radwan and Burger (1973a)
RK 13 rabbit kidney	McCollum (1970)

MA 104 line of monkey origin showed CPE; by serial passages, SHFV could be adapted to growth in BSC-1 cells (Wood, 1972).

2. *Immunofluorescence*

Togaviral antigen can be demonstrated by immunofluorescence either directly in organ sections after biopsy or autopsy or after infection of susceptible cells (cover slip cultures). In fact, the only other *in vitro* methods of tracing the notoriously noncytopathogenic pestiviruses are based on heterologous interference (Kumagai *et al.*, 1958; Nishimura *et al.*, 1962; Kubin, 1962).

A straightforward and rapid laboratory diagnosis of HCV employs cryostat sections of tonsils, lymph nodes, spleen, pancreas, kidney, ileum, and other porcine tissues for direct immunofluorescence (Aiken *et al.*, 1964; Peckham *et al.*, 1970). The tonsils are by far the most important organ, showing positive reactions in 98% of the positive animals, followed by the ileum (71%), spleen (60%), and kidney (39%). In subacute and chronic cases, the distal part of the ileum was frequently found positive (Terpstra, 1976). HCV strains of low virulence frequently escape detection by this method. Inoculation of primary pig kidney or PK15 cells and examination by immunofluorescence 1–3 days after infection are recommended in these cases. The conditions for isolation of low-virulence strains have been outlined by Aynaud (1976), who has also reviewed the HCV immunofluorescence literature (Aynaud and Bibard, 1971).

Also, BDV infections may be diagnosed by direct immunofluorescence applied to tissue sections (Peter *et al.*, 1968; Meyling, 1970). Usually, however, coverslip cultures are inoculated with clinical specimens and examined for CPE—if present—and cytoplasmic fluorescence after 1-4 days (Fernelius, 1969; Rohde and Liess, 1970; Smithies and Robertson, 1970; Ruckerbauer *et al.*, 1971; Carbrey, 1971). It has been claimed that four bovine viruses (BDV, rhinotracheitis, parainfluenza-3, and syncytial virus) can be diagnosed in one operation, using a tetravalent serum; the time of appearance, intensity, texture, and localization of fluorescence should serve as discriminating features (Mengeling and van der Maaten, 1971). The possibility of diagnosing border disease by immunofluorescence using homologous and anti-HCV conjugates has been investigated by Terpstra (1977). Antigen could be visualized in cell cultures derived from the brain, kidneys, and testicles of affected lambs.

In studies of the pathogenesis of the EAV infection in the horse, Crawford and Henson (1973) have employed immunofluorescence. Others have demonstrated viral antigen in the cytoplasm of infected BSCl, primary equine kidney (Crawford and Davis, 1970), Vero, E.derm. (Inoue *et al.*, 1975), and transformed hamster cells (Shinagawa *et al.*, 1976). We were able to show that the capsid protein of EAV binds to protein A of *Staphylococcus aureus;* FITC conjugates of protein A have been used to identify infected cells in culture (Zeegers and Horzinek, unpublished observations).

Porter *et al.* (1969) and Rowson and Michaels (1973) employed immunofluorescence for the demonstration of LDV antigen-containing cells *in vivo*. In sections through the spleen and liver of mice infected 18 hours previously, antigen was detected. The technique was also employed with infected primary mouse embryo cells (Oldstone *et al.*, 1974) and mouse macrophage/human fibroblast hybrid cells (Schlesinger *et al.*, 1976).

3. *Agar Gel Precipitation*

Since the discovery by Darbyshire (1960) of an antigenic relationship between HCV and BDV—historically the basis for the genus *Pestivirus*—this technique has been used preferentially for the analysis of pestiviral antigens. Its diagnostic application for HCV dates back to 1954, when Molnár detected precipitating antigen in lymph node material and defibrinated blood of infected pigs. Following a suggestion by Mansi (1957), pancreas tissue has been mostly used as an antigen source. The technique must be considered obsolete since the introduction of immunofluorescence and other methods. Its popularity in diagnostic laboratories is largely due to its simplicity. Erroneous results are not infrequent; for a discussion, the reader is referred to Pirtle (1963) and Matthaeus and Korn (1975). Also, the "soluble" antigen of BDV is encountered in organ tissue of diseased animals, notably in the intestinal mucosae and mesenteric lymph nodes. In his study on the morphogenesis of SHFV, Wood (1972) employed the Ouchterlony technique; its diagnostic use has not been recommended. Because of the poor sensitivity of the radial diffusion method, the immunoelectro-osmophoresis modification has been employed, mainly for serodiagnostic purposes. Hantschel (1967) qualifies the technique as quick and sensitive for the detection of HCV antigen in the blood of infected pigs. In fact, the modification is up to 16 times more sensitive than the conventional double diffusion method (Terpstra, 1977, 1978).

B. Detection of Antibodies

Serology has been used for the retrospective diagnosis of an infection based on differences in titer between paired serum samples and for seroepidemiological screening. Under certain circumstances, demonstration of antibody in a single sample may also be of diagnostic value. Thus, the presence of specific antibody to BDV in the serum of aborted or newborn calves indicates that they were infected *in utero*. The presence of such antibody, along with fetal lesions and/or teratological defects, will aid in determining the role of BDV in cases of abortion.

1. Virus Neutralization

The neutralization of togaviruses is complicated by the fact that the virus–antibody interaction may lead to complexes which are still infectious. Only after the addition of a second antibody directed against the gamma globulin species present in the original neutralization mixture is a reduction in infectivity noted. Details of the reaction have been studied with LDV (Notkins, 1968; Notkins *et al.*, 1968) and EAV (Radwan and Burger, 1973a,b; Radwan *et al.*, 1973). On the other hand, the neutralizing event may be triggered by the addition of complement to the virus–antibody mixture, with ensuing virolysis. A thorough study using EAV has elucidated this mechanism (Radwan and Burger, 1973a,b; Radwan *et al.*, 1973; Radwan and Crawford, 1974), explaining the neutralization-enhancing effect of fresh serum observed by others (Foster, 1969; Hyllseth and Pettersson, 1970; Maess, 1971; Crawford and Henson, 1973). No information on potentiation of neutralization was found in the pestivirus literature; pronounced effects have not been observed (Terpstra, V. Bakkum, Laude, and Carthier, personal communications). Also, added mouse complement did not improve the LDV-neutralizing activity of heated antisera (Rowson, cited in Rowson and Mahy, 1975). In plaque reduction tests for the demonstration of neutralizing antibody against SHFV, however, the addition of fresh monkey serum was mandatory (Wood, 1972).

HCV-neutralizing antibodies have been quantitated using either the cytopathogenic PAV-1 strain (Coggins and Sheffy, 1961; Robson *et al.*, 1961; Coggins and Baker, 1964) or immunofluorescence. In the latter technique, serial serum dilutions are incubated with an HCV preparation containing a predetermined number of fluorescent plaque-forming units. Major antigenic types—repeatedly a matter of conjecture—do not exist in HCV. However, specific antisera against the Ames and the 331 strains neutralized the homologous viruses to a greater extent than the heterologous ones (Pirtle and Mengeling, 1971). Aynaud *et al.* (1974) defined two antigenic subgroups of HCV, the first containing virulent (Alfort) and attenuated strains (Chinese vaccine, cold mutant Thiverval vaccine strain) and the second group comprising the American 331 strain and several isolates from cases of chronic hog cholera. It is recommended that representative strains from both subgroups be used in serological tests (Aynaud, 1976). Natural infection of pigs with BDV resulting from contact with infected cattle caused problems in hog cholera diagnosis. Low cross-reacting titers against HCV induced by BDV infection were found in healthy pigs and animals tested because of clinical signs of low-virulent hog cholera. It is therefore mandatory to perform neutralization tests on one serum in parallel against HCV and BDV (Carbrey *et al.*, 1976a,b). The kinetics of appearance of the heterologous antibody have been investigated by Liess *et al.* (1976).

Neutralization tests for antibodies against BDV are performed with cytopathogenic strains using either quantal assay (preferably a micro adaptation of the method: Frey and Liess, 1971; Rossi and Kiesel, 1971) or plaque reduction (Kniazeff and Pritchard, 1961; Malmquist, 1968). Plaques are also formed by noncytopathogenic BDV strains (Straver, 1971), an observation which has been confirmed for HCV (van Bekkum and Barteling, 1970; Laude, 1978c). Testing fetal bovine sera for antiviral antibodies is of value in determining the etiology of congenital anomalies and abortion. Unlike the human infant, which transplacentally acquires the immune status of its mother, the bovine neonate lacks maternally bestowed antibodies until acquisition by nursing. If colostrum ingestion has been excluded, detection of antibody directed against BDV in aborted fetuses or presuckle calves indicates an active fetal response to infection acquired *in utero* (Kahrs *et al.*, 1971).

2. *Complement Fixation*

This common serological reaction has not been used extensively for diagnostic purposes in hog cholera (Boulanger *et al.*, 1965). Species-inherent procomplementary activity of porcine sera has precluded its widespread application. Pretreatment with formaldehyde and addition of unheated calf serum were necessary to obtain reproducible results (Gutekunst and Malmquist, 1964; Jakubik, 1969). Eskildsen (1975) and Eskildsen and Overby (1976) developed a direct test procedure for use with porcine sera in which isolated C_{1q} had to be added to the complement diluent. After treatment of the test sera with beta mercaptoethanol at 56°C for 20 minutes, their hemolytic prozone effect was eliminated. Also, for serological testing of bovine diarrhea and equine arteritis (see, e.g., Fukunaga and McCollum, 1977) complement fixation has found little application.

3. *Other Assay Methods*

The indirect immunofluorescence technique has been used for the rapid demonstration of antibodies against pestiviruses. HCV-infected PK15 cells in suspension are dried onto microscope slides and acetone fixed. After incubation with serial dilutions of the pig serum to be tested, an anti-swine FITC conjugate is used to visualize bound porcine IgG (Anonymous, 1976).

An enzyme-linked immunosorbent assay has been developed for the serodiagnosis of hog cholera, using cellulose acetate filter discs as an antigen vehicle (Saunders and Wilder, 1974). The technique has been improved by the introduction of disposable microtiter trays as antigen carriers (Saunders and Clinard, 1976). Comparative testing showed a correlation of $> 99\%$ between this assay and serum neutralization; cross reactions between HCV and BDV were observed in both systems (Saunders, 1977).

A solid phase radioimmunoassay using methanol-fixed HCV-infected PK15 cells in flat-bottom microtiter trays and iodinated antiporcine IgG has the advantage of rapidly providing quantitative estimates of antigen-bound antibody (Horzinek, unpublished observations).

Passive hemagglutination has been introduced in HCV serology by Segre (1962). After modifications concerning mainly the method of antigen coupling, the technique was suitable to detect about 85% of naturally and experimentally infected pigs with 0.5% false-positive reactions (Abdallah *et al.*, 1966). Antibodies are detected earlier than by seroneutralization and also in cases of chronic hog cholera (Labadie *et al.*, 1977).

Immunoelectro-osmophoresis is a rapid method for screening large numbers of serum samples. Although less sensitive than the fluorescent plaque reduction assay, the results of both tests are in close agreement (Terpstra, 1976).

VI. CONCLUSIONS

Nonarthropod-borne togaviruses of animals may cause the whole spectrum of symptoms from acute infections associated with high mortality (e.g., HCV, SHFV) to persistently inapparent infections (e.g., SHFV in several monkey species, LDV in the mouse). Persistent infections are sometimes characterized by extraordinarily high concentrations of infectious particles present in the blood during viremia, which, however, contribute little to the efficiency of epidemic spread. It must remain a matter of speculation whether the exorbitant viremia titers are the evolutionary rudiment of a former arthropod transmission.

There is an obvious discrepancy between the very few nonarbo togaviruses discovered so far and the vast number (> 80) of arthropod-borne representatives in the *Alphavirus* and *Flavivirus* genera. One reason for this can be given: the inconspicuous behavior of some nonarbo togaviruses in the host (inapparent infection) and in tissue culture (no apparent cytopathic effect). The virus causing hemorrhagic fever in macaques is harbored inapparently by other monkey species in which it probably would not have been detected, or only by chance. Recently, togavirus-like particles were visualized by thin section electron microscopy in the urinary bladder of a capuchin monkey (*Cebus apella*) and in the placenta of a baboon (*Papio papio*; Smith *et al.*, 1978); the animals were in other studies and have not been examined for specific disease symptoms. By using inapparently infected animals in biomedical research, the experimental data may be modified and compromised. Riley (1974) has commented on this risk with respect to LDV infections in mice. Cells may be contaminated with BDV from (fetal) calf serum. Some sublines of the Madin-Darby bovine kidney line have been found infected by noncytopathogenic strains (Carbrey, personal communication). Apart from the virus, high levels of antibody directed against BVD may be present in calf serum. In spite of the abundance of infection in cattle, isolations are rarely reported from some diagnostic laboratories. It can be suspected that homologous interference or/and specific antibody are responsible for these failures.

These two examples illustrate the importance of diagnostic procedures intended to trace adventitious nonarbo togaviruses. On the other hand, new agents will probably be identified or new syndromes caused by familiar viruses. The "rediscovery" of LDV as the causative agent of an age-dependent polioencephalomyelitis in mice (Martinez, 1979) is an example. The notorious embryopathogenicity of nonarbo togaviruses—of which rubella virus is the best-known representative—is another challenge for comparative diagnostic virology.

ACKNOWLEDGEMENTS

The author wishes to express his gratitude to Miss M. Maas Geesteranus and Mrs. J. H. M. Royackers for their assistance during preparation of the chapter.

REFERENCES

Abdallah, I. S., Böhm, K. H., and Reus, U. (1966). *Zentralbl. Veterinaermed., Reihe B* **13,** 459–472.

Abildgaard, C., Harrison, J., Espana, C., Spangler, W., and Gribble, D. (1975). *Am. J. Trop. Med. Hyg.* **24,** 537–544.

Acland, H. M., Gard, G. P., and Plant, J. W. (1972). *Aust. Vet. J.* **48,** 70.

Aiken, J. M., Hoopes, K. H., Stair, E. L., and Rhodes, M. B. (1964). *J. Am. Vet. Med. Assoc.* **144,** 1395–1397.

Akashi, H., Konishi, S.-i., and Ogata, M. (1976). *Jpn. J. Vet. Sci.* **38,** 71–73.

Allen, A. M., Palmer, A. E., Tauraso, N. M., and Shelokov, A. (1968). *Am. J. Trop. Med. Hyg.* **17,** 413–421.

Anonymous (1976). *In* "Laboratory Manual for Research on Classical and African Swine Fever," Publ. EUR 5487, Appendix 4, pp. 140–141. Comm. Eur. Communities, Brussels.
Aynaud, J. M. (1976). *In* "Studies on Virus Replication," Publ. EUR 5451, pp. 30–38. Comm. Eur. Communities, Brussels.
Aynaud, J. M., and Bibard, C. (1971). *Cah. Med. Vet.* **5,** 1–11.
Aynaud, J. M., Rigaud, C., Le Turdu, Y., Galicher, C., Lombard, J., Corthier, G., and Laude, H. (1974). *Ann. Rech. Vet.* **5,** 57–85.
Bailey, J. M., Clough, J., and Lohaus, A. (1965). *Proc. Soc. Exp. Biol. Med.* **119,** 1200–1204.
Barlow, R. W., and Dickinson, A. G. (1965). *Res. Vet. Sci.* **6,** 230–237.
Biester, H. E., and Schwarte, L. H. (1952). *Proc. Am. Vet. Med. Assoc.* pp. 65–67.
Borgen, H. C. (1963). *Nord. Veterinaermed.* **15,** 409–417.
Boulanger, P., Appel, M., Bannister, G. L., Ruckerbauer, G. M., Mori, K., and Gray, D. P. (1965). *Can. J. Comp. Med.* **29,** 201–208.
Boynton, W. H. (1946). *Vet. Med.* **41,** 346.
Breese, S. S., and McCollum, W. H. (1970). *Proc. Int. Conf. Equine Infect. Dis., 2nd, Paris, 1969* pp. 133–139.
Breese, S. S., and McCollum, W. H. (1971). *Arch. Gesamte Virusforsch.* **35,** 290–295.
Breese, S. S., and McCollum, W. H. (1973). *Proc. Int. Conf. Equine Infect. Dis., 3rd, Paris, 1972* pp. 273–281.
Brinton-Darnell, M., and Brand, I. (1977). *J. Natl. Cancer Inst.* **59,** 1027–1029.
Brinton-Darnell, M., and Plagemann, P. G. W. (1975). *J. Virol.* **16,** 420–433.
Brion, A., Fontaine, M., and Moraillon, R. (1967). *Rec. Med. Vet.* **143,** 17–27.
Bürki, F. (1965a). *Pathol. Microbiol. Suisse* **28,** 156–166.
Bürki, F. (1965b). *Pathol. Microbiol. Suisse* **28,** 939–949.
Bürki, F. (1970). *Proc. Int. Conf. Equine Infect. Dis., 2nd, Paris, 1969* pp. 125–129.
Bürki, F., and Gerber, H. (1966). *Berl. Muench. Tieraerztl. Wochenschr.* **20,** 391–395.
Cancellotti, F., and Turilli, C. (1975). *Vet. Ital.* **26,** 3–12.
Cancellotti, F., Zanin, E., Irsara, A., and Zoletto, R. (1972). *Arch. Vet. Ital.* **23,** 435–441.
Carbrey, E. A. (1971). *U.S. Anim. Health Assoc.* **75,** 629–648.
Carbrey, E. A., and Lee, L. R. (1965). *Proc. Annu. Meet. U.S. Livest. Sanit. Assoc.* **69,** 501.
Carbrey, E. A., Stewart, W. C., Kresse, J. I., and Snyder, M. L. (1976a). *In* "Diagnosis and Epizootiology of Classical Swine Fever," Publ. EUR 5486, pp. 126–158. Comm. Eur. Communities, Brussels.
Carbrey, E. A., Stewart, W. C., Kresse, J. I., and Snyder, M. L. (1976b). *J. Am. Vet. Med. Assoc.* **169,** 1217–1219.
Castrucci, G., Avelini, G., Cilli, V., Morettini, B., and Bellachioma, F. (1968a). *Boll. Ist. Sieroter. Milan* **47,** 341–351.
Castrucci, G., Cilli, V., and Gagliardi, G. (1968b). *Arch. Gesamte Virusforsch.* **24,** 48–61.
Cheville, N. F., and Mengeling, W. L. (1969). *Lab. Invest.* **20,** 261–274.
Cheville, N. F., Mengeling, W. L., and Zinober, M. R. (1970). *Lab. Invest.* **22,** 458–467.
Coggins, L., and Baker, J. (1964). *Am. J. Vet. Res.* **25,** 408–412.
Coggins, L., and Sheffy, B. E. (1961). *Proc. Annu. Meet. U.S. Livest. Sanit. Assoc.* **65,** 333–337.
Coria, M. F. (1969). *Am. J. Vet. Res.* **30,** 369–375.
Corthier, G. (1978). *Am. J. Vet. Res.* **39,** 1841–1844.
Corthier, G., Labadie, J. P., and Petit, E. (1977). *Bull. Acad. Vet. Fr.* **50,** 425–433.
Crawford, T. B., and Davis, W. C. (1970). *Fed. Proc., Fed. Am. Soc. Exp. Biol.* **29,** 286. (Abstr. No. 225.)
Crawford, T. B., and Henson, J. B. (1973). *Proc. Int. Conf. Equine Infect. Dis., 3rd, Paris, 1972* pp. 282–302.

Crispens, C. G. (1965). *J. Natl. Cancer Inst.* **35,** 975–979.

Dale, C. N., and Songer, J. R. (1957). *Am. J. Vet. Res.* **18,** 362.

Dalsgaard, K. (1976). *In* "Hog Cholera/Classical Swine Fever and African Swine Fever," Publ. EUR 5904 EN, pp. 320–325. Comm. Eur. Communities, Brussels.

Dalsgaard, K., and Overby, E. (1977). *In* "Hog Cholera/Classical Swine Fever and African Swine Fever," Publ. EUR 5904 EN, pp. 70–74. Comm. Eur. Communities, Brussels.

Darbyshire, J. H. (1960). *Vet. Rec.* **72,** 331.

Darnell, M. B., and Plagemann, P. G. W. (1972). *J. Virol.* **10,** 1082–1085.

Darnell, M. B., Collins, J. K., and Plagemann, P. G. W. (1974). *Abstr., Annu. Meet. Am. Soc. Microbiol.* **121,** V121.

de Boer, G. F., Osterhaus, A. D. M. E., and Wemmenhove, R. (1978). *Proc. Int. Conf. Equine Infect. Dis., 4th, Lyon 1976* **4,** 487–492.

de Castro, M. P. (1973). *In Vitro* **9,** 8–16.

Diderholm, H., and Dinter, Z. (1965). *Zentralbl. Veterinaermed., Reihe B* **12,** 469–475.

Diderholm, H., and Dinter, Z. (1966). *Zentralbl. Bakteriol., Parasitenkd., Infektionskr. Hyg., Abt. 1: Orig.* **201,** 270–272.

Dinter, Z. (1963). *Zentralbl. Bakteriol., Parasitenkd., Infektionskr. Hyg., Abt. 1: Orig.* **188,** 475–486.

Dinter, Z. (1968). *In* "Handbuch der Virusinfektionen bei Tieren, Band III/2, Mucosal Disease" (H. Röhrer, ed.), pp. 721–740. Fischer, Jena.

Dinter, Z., Hansen, H. J., and Ronéus, O. (1962). *Zentralbl. Veterinaermed.* **9,** 739–747.

Doll, E. R., Bryans, J. T., McCollum, W. H., and Crowe, M. E. W. (1957a). *Cornell Vet.* **47,** 3–41.

Doll, E. R., Knappenberger, R. E., and Bryans, J. T. (1957b). *Cornell Vet.* **47,** 69–75.

Doll, E. R., Bryans, J. T., Wilson, J. C., and McCollum, W. H. (1968). *Cornell Vet.* **58,** 497–524.

Dow, C., Jarret, W. F. H., and McIntyre, W. I. M. (1956). *Vet. Rec.* **68,** 620–623.

Dunne, H. W. (1964). "Diseases of Swine," 2nd ed. Iowa State Univ. Press, Ames.

Dunne, H. W., Reich, C. V., Hokanson, J. F., and Lindstrom, E. S. (1955). *Proc. AM. Vet. Med. Assoc.* pp. 148–153.

Dunne, H. W., Luedke, A., Reich, C. V., and Hokanson, J. F. (1957). *Am. J. Vet. Res.* **18,** 502.

Dunne, H. W., Hokanson, J. F., and Luedke, A. J. (1959). *Am. J. Vet. Res.* **20,** 615–618.

Ellis, P. R., James, A. D., and Shaw, A. P. (1977). *In* "Studies on the Epidemiology and Economics of Swine Fever Eradication in the EEG," Publ. EUR 5738 e. Comm. Eur. Communities, Brussels.

Enzmann, P. J., and Härtner, D. (1977). *In* "Hog Cholera/Classical Swine Fever and African Swine Fever," Publ. EUR 5904 EN, pp. 75–84. Comm. Eur. Communities, Brussels.

Enzmann, P. J., and Rehberg, H. (1977). *Z. Naturforsch., Teil C* **32,** 456–458.

Enzmann, P. J., and Weiland, F. (1978). *Arch. Virol.* **57,** 339–348.

Enzmann, P. J., and Weiland, F. (1979). *Virology* **95,** 501–510.

Eskildsen, M. (1975). *Acta Pathol. Microbiol. Scand., Sect. C* **83,** 315–324.

Eskildsen, M., and Overby, E. (1976). *Acta Vet. Scand.* **17,** 131–141.

Estes, P. C., and Cheville, N. F. (1970). *Am. J. Pathol.* **58,** 235–253.

FAO (1976). *FAO Anim. Prod. Health Pap.* No. 2, AGA-816.

Felmingham, D., and Brown, F. (1977). *In* "Hog Cholera/Classical Swine Fever and African Swine Fever," Publ. EUR 5904 EN, pp. 58–69. Comm. Eur. Communities, Brussels.

Fenner, F. (1975–1976). *Intervirology* **6,** 1–12.

Fernelius, A. L. (1969). *Arch. Gesamte Virusforsch.* **27,** 1–12.

Fernelius, A. L., Lambert, G., and Hemness, G. J. (1969). *Am. J. Vet. Res.* **30,** 1541–1550.

Fitzmaurice, M. A., Riley, V., and Santisteban, G. A. (1972). *Pathol. Biol.* **20,** 743–750.

Foster, A. G. (1969). Ph.D. Thesis, Coll. Vet. Med., Washington State Univ., Pullman.

French, E. L., and Snowdon, W. A. (1964). *Aust. Vet. J.* **40,** 99–105.
French, E. L., Hore, D. E., Snowdon, W. A., Parsonson, I. M., and Uren, J. (1974). *Aust. Vet. J.* **50,** 45–54.
Frenkel, S., Van Bekkum, J. G., and Frenkel, H. S. (1955). *Bull. Off. Int. Epizoot.* **43,** 327.
Frey, H.-R., and Liess, B. (1971). *Zentralbl. Veterinaermed., Reihe B* **18,** 61–71.
Frost, J. W., Liess, B., and Prager, D. (1977). *In* "Hog Cholera/Classical Swine Fever and African Swine Fever," Publ. EUR 5904 EN, pp. 23–30. Comm. Eur. Communities, Brussels.
Fuchs, F. (1968). *In* "Handbuch der Virusinfektionen bei Tieren" (H. Röhrer, ed.), Band III/1. pp. 15–250. Fischer, Jena.
Fukunaga, Y., and McCollum, W. H. (1977). *Am. J. Vet. Res.* **38,** 2043–2046.
Gibbons, D. F., Sinkler, C. E., Shaw, I. G., Terlecki, S., Richardson, C., and Done, J. T. (1974). *Br. Vet. J.* **130,** 357–360.
Gibbs, A. J. (1969). *J. Gen. Virol.* **5,** 447–449.
Gillespie, J. H., co-chm. (1968). *J. Am. Vet. Med. Assoc.* **152,** 713–719.
Gillespie, J. H., Coggins, L., Thompson, J., and Baker, J. A. (1961). *Cornell Vet.* **51,** 155–159.
Gillespie, J. H., Madin, S. H., and Darby, N. B. (1962). *Proc. Soc. Exp. Biol. Med.* **110,** 248–250.
Goldsmith, L., and Barzilai, E. (1975). *Am. J. Vet. Res.* **36,** 407–412.
Gutekunst, D. E., and Malmquist, W. A. (1964). *Can. J. Comp. Med.* **28,** 19–23.
Gutekunst, D. E., and Malmquist, W. A. (1965). *Arch. Gesamte Virusforsch.* **15,** 159–168.
Hadjisavvas, T. H., Harkness, J. W., Huck, R. A., and Stuart, P. (1975). *Res. Vet. Sci.* **18,** 237–243.
Hafez, S. M., and Liess, B. (1972). *Acta Virol. (Engl. Ed.)* **16,** 388–398.
Hafez, S. M., Petzoldt, K., and Reczko, E. (1968). *Acta Virol. (Engl. Ed.)* **12,** 471–473.
Hamilton, A., and Timoney, P. J. (1972). *Vet. Rec.* **91,** 468.
Hamilton, A., and Timoney, P. J. (1973). *Res. Vet. Sci.* **15,** 265–267.
Hansen, H.-J., Ronéus, O., and Dinter, Z. (1962). *Zentralbl. Veterinaermed.* **9,** 854–864.
Hantschel, H. (1967). *Arch. Exp. Vet. Med.* **20,** 1131–1145.
Harkness, J. W., King, A. A., Terlecki, S., and Sands, J. J. (1977). *Vet. Rec.* **100,** 71–72.
Hecke, F. (1932). *Zentralbl. Bakteriol., Parasitenkd., Infektionskr. Hyg., Abt. 1: Orig.* **126,** 517.
Hermodsson, S., and Dinter, Z. (1962). *Nature (London)* **194,** 893–894.
Homburger, H., Abrams, G. D., Lawton, J. W. M., and Murphy, W. H. (1973). *Proc. Soc. Exp. Biol. Med.* **144,** 979–982.
Horzinek, M. C. (1973a). *J. Gen. Virol.* **20,** 87–103.
Horzinek, M. C. (1973b). *Prog. Med. Virol.* **16,** 109–156.
Horzinek, M. C. (1975). *Med. Biol.* **53,** 406–411.
Horzinek, M. C. (1976). *In* "Studies on Virus Replication," Publ. EUR 5451, pp. 9–28. Comm. Eur. Communities, Brussels.
Horzinek, M. C. (1981). "Non-Arthropod-Borne Togaviruses," Experimental Virology. Academic Press, New York. In press.
Horzinek, M. C., and Mussgay, M. (1969). *J. Virol.* **4,** 514–520.
Horzinek, M. C., Maess, J., and Laufs, R. (1971). *Arch. Gesamte Virusforsch.* **33,** 306–318.
Horzinek, M. C., van Wielink, P. S., and Ellens, D. J. (1975). *J. Gen. Virol.* **26,** 217–226.
Howard, R. J., Notkins, A. L., and Mergenhagen, S. E. (1969). *Nature (London)* **221,** 873–874.
Hyllseth, B. (1969). *Arch. Gesamte Virusforsch.* **30,** 26–33.
Hyllseth, B. (1973a). *Arch. Gesamte Virusforsch.* **40,** 97–104.
Hyllseth, B. (1973b). Ph.D. Thesis, Univ. Stockholm.
Hyllseth, B., and Pettersson, V. (1970). *Arch. Gesamte Virusforsch.* **32,** 337–347.
Inoue, T., Yanagawa, R., Shinagawa, M., and Akiyama, Y. (1975). *Jpn. J. Vet. Sci.* **37,** 569–575.
Jakubik, J. (1969). *Dtsch. Tieraerztl. Wochenschr.* **76,** 111–115.
Johnson, D. W., and Muscoplat, C. C. (1973). *Am. J. Vet. Res.* **34,** 1139–1141.

Jones, T. C. (1969). *J. Am. Vet. Med. Assoc.* **155,** 315–317.

Kahrs, R. F. (1971). *J. Am. Vet. Med. Assoc.* **159,** 1383–1386.

Kahrs, R. F., Robson, D. S., and Baker, J. A. (1966). *Proc. U.S. Livest. Sanit. Assoc.* **70,** 145–153.

Kahrs, R. F., Scott, F. W., and de Lahunta, A. (1970a). *J. Am. Vet. Med. Assoc.* **156,** 851–857.

Kahrs, R. F., Scott, F. W., and de Lahunta, A. (1970b). *J. Am. Vet. Med. Assoc.* **156,** 1443–1450.

Kahrs, R. F., Scott, F. W., and de Lahunta, A. (1970c). *Teratology* **3,** 181–184.

Kahrs, R. F., Scott, F. W., and Hillman, R. B. (1971). *Proc. Annu. Meet. U.S. Anim. Health Assoc.* **75,** 588–594.

Karasszon, D., and Bodon, L. (1963). *Acta Microbiol. Hung.* **10,** 287–291.

Kasza, L., Shadduck, J. A., and Christifinis, G. J. (1972). *Res. Vet. Sci.* **13,** 46–51.

Kendrick, J. W. (1971). *Am. J. Vet. Res.* **32,** 533–544.

Kniazeff, A. J., and Pritchard, W. R. (1961). *Proc. U.S. Livest. Sanit. Assoc.* **64,** 344–350.

Kniazeff, A. J., Wopschall, L. J., Hopps, H. E., and Morris, C. S. (1975). *In Vitro* **11,** 400–403.

Konishi, S.-i., Akashi, H., Sentsui, H., and Ogata, M. (1973). *Proc. Meet. Jpn. Soc. Vet. Sci.* **76,** 57.

Korn, G. (1964). *Zentralbl. Veterinaermed., Reihe B* **11,** 379–392.

Korn, G., and Hecke, G. (1964). *Zentralbl. Veterinaermed., Reihe B* **11,** 40–50.

Korn, G., and Lorenz, R. J. (1976). *Zentralbl. Bakteriol., Parasitenkd., Infektionskr. Hyg. Abt. 1: Orig., Reihe A* **236,** 150–162.

Korn, G., and Matthaeus, W. (1977). *Zentralbl. Bakteriol., Parasitenkd., Infektionskr. Hyg., Abt. 1: Orig., Reihe A* **218,** 407–416.

Kresse, J. F., Lee, L. R., and Wheelock, T. L. (1967). IA Rep. 91-63. ARS, Natl. Anim. Dis. Cent., Ames, Iowa.

Kubin, G. (1962). *Bull. Off. Int. Epizoot.* **57,** 1395–1406.

Kubin, G. (1964). *Zentralbl. Veterinaermed., Reihe B* **11,** 373–378.

Kubin, G. (1967). *Zentralbl. Veterinaermed.* **14,** 543–552.

Kumagai, T., Shimizu, T., and Matumoto, M. (1958). *Science* **128,** 366.

Labadie, J. P., Corthier, G., Aynaud, J. M., Renault, L., and Vaissaire, J. (1977). *Bull. Acad. Vet. Fr.* **50,** 533–542.

Lamberg, G. (1973). *J. Am. Vet. Med. Assoc.* **163,** 874–876.

Lamont, P. H., and Stuart, P. (1974). *Proc. Int. Pig Vet. Soc. Congr., 3rd, Lyon* Sect. HC9, pp. 1–5.

Lapin, B. A., Pekerman, S. M., Yakovleva, L. A., Dzhikidze, E. K., Shevtosova, Z. V., Kuksova, M. I., Danko, L. V., Krilova, R. L., Akbroit, E. Y., and Agraba, V. Z. (1967). *Vopr. Virusol.* **12,** 168.

Laude, H. (1978a). *Int. Virol. IV, The Hague* p. 440.

Laude, H. (1978b). *Ann. Microbiol. (Paris)* **129A,** 553–561.

Laude, H. (1978c). *J. Gen. Virol.* **40,** 225–228.

Lawton, J. W. M., and Murphy, W. H. (1973). *Arch. Neurol.* **28,** 367–370.

Leftheriotis, E., Precausta, P., and Caillere, F. (1971). *Rev. Med. Vet.* **122,** 33.

Liess, B., Frey, H. R., Prager, D., Hafez, S. M., and Roeder, B. (1976). *In* "Diagnosis and Epizootiology of Classical Swine Fever," Publ. EUR 5486, pp. 99–113. Comm. Eur. Communities, Brussels.

Loan, R. W. (1964). *Am. J. Vet. Res.* **25,** 1366–1370.

Loan, R. W., and Gustavson, D. P. (1961). *Am. J. Vet. Res.* **22,** 741–745.

London, W. T. (1977). *Nature (London)* **268,** 344–345.

Ludwig, H., Paulsen, J., and Kaminjolo, J. S. (1969). *Z. Med. Mikrobiol. Immunol.* **155,** 133–149.

McCollum, W. H. (1970). *Proc. Int. Conf. Equine Infect. Dis., 2nd, Paris, 1969* pp. 143–151.

McCollum, W. H., and Bryans, J. T. (1973). *Proc. Int. Conf. Equine Infect. Dis., 3rd, Paris, 1972* pp. 256–263.

McCollum, W. H., Doll, E. R., Wilson, J. C., and Johnson, C. B. (1961a). *Am. J. Vet. Res.* **22,** 731–734.
McCollum, W. H., Doll, E. R., and Wilson, J. C. (1961b). *Am. J. Vet. Res.* **23,** 465–469.
McCollum, W. H., Wilson, J. C., and Cheatham, J. (1962). *Cornell Vet.* **52,** 452–458.
McCollum, W. H., Ozuwa, Y., and Dardiri, A. H. (1970). *Am. J. Vet. Res.* **31,** 1963–1972.
McCollum, W. H., Prickett, M. E., and Bryans, J. T. (1971). *Res. Vet. Sci.* **12,** 459–464.
McKercher, D. G. (1968). *J. Am. Vet. Med. Assoc.* **152,** 729–737.
Maess, J. (1971). *Arch. Gesamte Virusforsch.* **33,** 194–196.
Maess, J., Reczko, E., and Böhm, H. O. (1970a). *Proc. Int. Conf. Equine Infect. Dis., 2nd, Paris, 1969* pp. 130–132.
Maess, J., Reczko, E., and Böhm, H. O. (1970b). *Arch. Gesamte Virusforsch.* **30,** 47–58.
Mahnel, H., and Mayr, A. (1974). "Schweinepest." Fischer, Jena.
Malmquist, W. A. (1968). *J. Am. Vet. Med. Assoc.* **152,** 763–768.
Malmquist, W. A., Fernelius, A. L., and Gutekunst, D. E. (1965). *Am. J. Vet. Res.* **26,** 1316–1327.
Manktelow, B. W., Porter, W. L., and Lewis, K. H. C. (1969). *N. Z. Vet. J.* **17,** 245–248.
Mansi, W. (1957). *J. Comp. Pathol.* **67,** 297–303.
Marcus, S. J., and Moll, T. (1968). *Am. J. Vet. Res.* **29,** 817–819.
Martinez, D. (1979). *Infect. Immun.* **23,** 45–48.
Matthaeus, W. (1979). *Arch. Virol.* **59,** 299–305.
Matthaeus, W., and Korn, G. (1975). *Zentralbl. Veterinaermed., Reihe B* **22,** 239–253.
Mengeling, W. L., and Pirtle, E. C. (1972). *Arch. Gesamte Virusforsch.* **32,** 359–364.
Mengeling, W. L., and van der Maaten, M. J. (1971). *Am. J. Vet. Res.* **32,** 1825–1833.
Mergenhagen, S. E., Notkins, A. L., and Dougherty, S. F. (1967). *J. Immunol.* **99,** 576–581.
Meyling, A. (1970). *Acta Vet. Scand.* **11,** 59–72.
Meyling, A., and Schjerning-Thiesen, K. (1968). *Acta Vet. Scand.* **9,** 50–64.
Michaelides, M. C., and Schlesinger, S. (1973). *Virology* **55,** 211–217.
Michaelides, M. C., and Schlesinger, S. (1974). *J. Immunol.* **112,** 1560–1564.
Mills, J. H. L., and Luginbuhl, R. E. (1965). *Cornell Vet.* **55,** 344–354.
Mills, J. H. L., and Luginbuhl, R. E. (1968). *Am. J. Vet. Res.* **29,** 1367–1375.
Mills, J. H. L., Nielsen, S. W., and Luginbuhl, R. E. (1965). *J. Am. Vet. Med. Assoc.* **146,** 691–696.
Moennig, V. (1971). Diss., Vet. School, Hannover.
Molnár, I. (1954). *Acta Vet. Acad. Sci. Hung.* **4,** 247–251.
Moraillon, A., and Moraillon, R. (1978). *Proc. Int. Conf. Equine Infect. Dis., 4th, Lyon, 1976* **4,** pp. 467–473.
Mossman, D. H., and Hanly, G. J. (1976). *N. Z. Vet. J.* **24,** 108–110.
Murphy, W. H., Tan, M. R., Lanzi, R. L., Abell, M. R., and Kauffman, C. (1970). *Cancer Res.* **30,** 1612–1622.
Muscoplat, C. C., Johnson, D. W., and Stevens, J. B. (1973a). *Am. J. Vet. Res.* **34,** 753–755.
Muscoplat, C. C., Johnson, D. W., and Teuscher, E. (1973b). *Am. J. Vet. Res.* **34,** 1101–1104.
Mussgay, M., Weiland, E., Strohmaier, K., Ueberschär, S., and Enzmann, P. J. (1973). *J. Gen. Virol.* **19,** 89–101.
Nishimura, Y., Sato, U., Hanaki, T., and Kawashima, H. (1962). *Jpn. J. Vet. Sci.* **24,** 29–37.
Notkins, A. L. (1964). *Virology* **22,** 563–567.
Notkins, A. L. (1965). *Bacteriol. Rev.* **29,** 143–160.
Notkins, A. L. (1968). *Perspect. Virol.* **6,** 189–192.
Notkins, A. L., and Scheele, C. (1963). *Virology* **20,** 640–642.
Notkins, A. L., Scheele, C., and Scherp, H. W. (1964). *Nature (London)* **202,** 418–419.
Notkins, A. L., Mahar, S., Scheele, C., and Goffman, J. (1966a). *J. Exp. Med.* **124,** 81–97.

Notkins, A. L., Mergenhagen, S. E., Rizzo, A. A., Scheele, C., and Waldmann, T. A. (1966b). *J. Exp. Med.* **123,** 347–364.
Notkins, A. L., Mage, M., Ashe, W. K., and Mahar, S. (1968). *J. Immunol.* **100,** 314–320.
Nuttall, P. A. (1978). Ph.D. Thesis, Univ. of Reading.
Oldstone, M. B. A., Yamazaki, S., Niwa, A., and Notkins, A. L. (1974). *Intervirology* **2,** 261–265.
Osburn, B. J., Clarke, G. L., Stewart, W. C., and Sawyer, M. (1973). *J. Am. Vet. Med. Assoc.* **163,** 1165–1167.
Palmer, A. E., Allen, A. M., Tauraso, N. M., and Shelokov, A. (1968). *Am. J. Trop. Med. Hyg.* **17,** 404–412.
Peckham, J. C., Cole, J. R., and Pursell, A. R. (1970). *J. Am. Vet. Med. Assoc.* **157,** 1204–1207.
Peter, C. P., Duncan, J. R., Tyler, D. E., and Ramsey, F. K. (1968). *Am. J. Vet. Res.* **29,** 939–948.
Pirtle, E. C. (1963). *Can. J. Comp. Med.* **27,** 241–248.
Pirtle, E. C., and Kniazeff, A. J. (1968). *Am. J. Vet. Res.* **29,** 1033–1040.
Pirtle, E. C., and Mengeling, W. L. (1971). *Am. J. Vet. Res.* **32,** 1473–1477.
Plagemann, P. G. W., and Swim, H. E. (1966). *Proc. Soc. Exp. Biol. Med.* **121,** 1142–1146.
Plant, J. W., Littlejohns, I. R., Gardiner, A. C., Vantsis, J. T., and Huck, R. A. (1973). *Vet. Rec.* **92,** 455.
Plant, J. W., Acland, H. M., and Gard, G. P. (1976a). *Aust. Vet. J.* **52,** 57–63.
Plant, J. W., Gard, G. P., and Acland, H. M. (1976b). *Aust. Vet. J.* **52,** 247–249.
Porter, D. D., and Porter, H. G. (1971). *J. Immunol.* **106,** 1264–1266.
Porter, D. D., Porter, H. G., and Deerhake, B. B. (1969). *J. Immunol.* **102,** 431–436.
Porterfield, J. S., Casals, J., Chumakov, M. P., Gaidomovick, S. Y., Hannoun, C., Holmes, I. H., Horzinek, M. C., Mussgay, M., Okerblom, N., Russell, P. K., and Trent, D. W. (1978). *Intervirology* **9,** 129–148.
Prickett, M. E., McCollum, W. H., and Bryans, J. T. (1973). *Proc. Int. Conf. Equine Infect. Dis., 3rd, Paris, 1972* pp. 265–272.
Pritchard, W. R. (1963). *Adv. Vet. Sci.* **8,** 1–47.
Pritchett, R. F., Manning, J. S., and Zee, Y. C. (1975). *J. Virol.* **15,** 1342–1347.
Radwan, A. I., and Burger, D. (1973a). *Virology* **51,** 71–77.
Radwan, A. I., and Burger, D. (1973b). *Virology* **53,** 306–371.
Radwan, A. I., and Crawford, T. B. (1974). *J. Gen. Virol.* **25,** 229–237.
Radwan, A. I., Burger, D., and Davis, W. C. (1973). *Virology* **53,** 372–378.
Ressang, A. A. (1973). *Zentralbl. Veterinaermed., Reihe B* **20,** 256–271.
Ressang, A. A., and de Boer, J. L. (1968). *Neth. J. Vet. Sci.* **1,** 71–89.
Riley, V. (1964). *Science* **146,** 921–923.
Riley, V. (1966). *Science* **153,** 1657–1658.
Riley, V. (1968). *Methods Cancer Res.* **4,** 493–618.
Riley, V. (1974). *Cancer Res.* **34,** 1752–1754.
Riley, V., Loveless, J. D., and Fitzmaurice, M. A. (1964). *Proc. Soc. Exp. Biol. Med.* **116,** 486–490.
Ritchie, A. E., and Fernelius, A. L. (1967). *Vet. Rec.* **81,** 417–418.
Robertson, A., Bannister, G. L., Boulanger, P., Appel, M., and Gray, D. P. (1965). *Can. J. Comp. Med. Vet. Sci.* **29,** 299–305.
Robson, D. S., Gillespie, J. A., and Baker, J. A. (1960). *Cornell Vet.* **50,** 503–509.
Robson, D. S., Coggins, L., Sheffy, B. E., and Baker, J. A. (1961). *Proc. Annu. Meet. U.S. Livest. Sanit. Assoc.* **65,** 338–342.
Rohde, G., and Liess, B. (1970). *Zentralbl. Veterinaermed., Reihe B* **17,** 686–700.
Rossi, C. R., and Kiesel, G. K. (1971). *Appl. Microbiol.* **22,** 32–36.
Rowson, K. E. K., and Mahy, B. W. J. (1975). *Virol. Monogr.* **13,** 1–121.
Rowson, K. E. K., and Michaels, L. (1973). *J. Med. Microbiol.* **6,** xi.

Ruckerbauer, G. M., Girard, A., Bannister, L., and Boulanger, P. (1971). *Can. J. Comp. Med.* **35,** 230–238.
Rutili, D., and Titoli, F. (1977). *In* "Hog Cholera/Classical Swine Fever and African Swine Fever," Publ. EUR 5904 EN, pp. 40–51. Comm. Eur. Communities, Brussels.
Rutili, D., Titoli, F., and Morozzi, A. (1977). *Fol. Vet. Lat.* **7,** 165–173.
Santisteban, G. A., Riley, V., and Fitzmaurice, M. A. (1972). *Proc. Soc. Exp. Biol. Med.* **139,** 202–206.
Sasahara, J. (1970). *Natl. Inst. Anim. Health Q.* **10,** 57–81.
Sasahara, J., and Kumagai, I. (1965). Cited in Ayaud and Bibard (1971).
Sato, U., Nishimura, Y., Hanaki, T., and Nobuto, K. (1964). *Arch. Gesamte Virusforsch.* **14,** 394–403.
Saunders, G. C. (1977). *Am. J. Vet. Res.* **38,** 21–25.
Saunders, G. C., and Clinard, E. H. (1976). *J. Clin. Microbiol.* **3,** 604–608.
Saunders, G. C., and Wilder, M. E. (1974). *J. Infect. Dis.* **129,** 362–364.
Saurat, P., Gilbert, Y., and Chantal, J. (1972). "La Maladie des Muqueuses." Expansion Sci., Paris.
Scherrer, R., Aynaud, J. M., Cohen, J., and Bic, E. (1970). *C. R. Acad. Sci., Ser. D* **271,** 620–623.
Schiff, L. J., and Storz, J. (1972). *Arch. Gesamte Virusforsch.* **36,** 218–225.
Schlesinger, S., Lagwinska, E., Stewart, C. C., and Croce, C. M. (1976). *Virology* **74,** 535–539.
Schmidt, D., Bergmann, H., and Wittmann, E. (1965). *Arch. Exp. Vet. Med.* **19,** 149–156.
Schulze, P. (1971). *Arch. Exp. Vet. Med.* **25,** 413–425.
Scott, F. W., Kahrs, R. F., and Parsonson, I. M. (1972). *Cornell Vet.* **62,** 74–84.
Segre, D. (1962). *Am. J. Vet. Res.* **23,** 748–751.
Seibold, H. R., and Dougherty, E. (1967). *Am. J. Vet. Res.* **28,** 137–140.
Shinagawa, M., Yanagawa, R., Ione, T., and Akiyama, Y. (1976). *Jpn. J. Vet. Sci.* **38,** 25–32.
Shope, R. E. (1964). Ph.D. Thesis, Univ. of Minnesota, Minneapolis.
Shope, R. E., McNamara, L. G., and Mangold, R. (1960). *J. Exp. Med.* **111,** 155–170.
Shope, R. E., Muscoplat, C. C., Chen, A. W., and Johnson, D. W. (1976). *Can. J. Comp. Med.* **40,** 355–359.
Singh, K. V. (1969a). *Vet. Rec.* **84,** 230–232.
Singh, K. V. (1969b). *Appl. Microbiol.* **17,** 323–234.
Smith, G. C., Kalter, S. S., and Heberling, R. L. (1978). *Intervirology* **10,** 44–50.
Smithies, L. K., and Robertson, S. B. (1970). *J. Am. Vet. Med. Assoc.* **156,** 107–108.
Snodgrass, M. J., Lowrey, D. S., and Hanna, M. G. (1972). *J. Immunol.* **108,** 877–892.
Snowdon, W. A., Parsonson, I. M., and Brown, M. L. (1975). *J. Comp. Pathol.* **85,** 241–251.
Soto-Belloso, E. R. (1976). *Am. J. Vet. Res.* **37,** 1103–1105.
Stewart, W. C., Carbrey, E. A., Jenney, E. W., Brown, C. L., and Kresse, J. I. (1971). *J. Am. Vet. Med. Assoc.* **158,** 1891.
Strauss, J. H., and Strauss, E. G. (1977). *In* "The Molecular Biology of Animal Viruses" (D. P. Nayak, ed.), Vol. 1, pp. 111–166. Dekker, New York.
Straver, P. J. (1971). *Arch. Gesamte Virusforsch.* **34,** 131–135.
Tauraso, N. M., Shelokov, A., Palmer, A. E., and Allen, A. M. (1968). *Am. J. Trop. Med. Hyg.* **17,** 422–431.
Taylor, D. O. N., Gustafson, D. P., and Chaflin, R. M. (1963). *Am. J. Vet. Res.* **24,** 143–149.
Terpstra, C. (1976). *In* "Diagnosis and Epizootiology of Classical Swine Fever," Publ. EUR 5486, pp. 58–63. Comm. Eur. Communities, Brussels.
Terpstra, C. (1977). *In* "Hog Cholera/Classical Swine Fever and African Swine Fever," Publ. EUR 5904 EN, pp. 283–290. Comm. Eur. Communities, Brussels.
Terpstra, C. (1978). *Zentralbl. Veterinaermed., Reihe B* **25,** 350–355.
Thomson, R. G., and Savan, M. (1963). *Can. J. Comp. Med.* **27,** 207–214.

Trousdale, M. D., Trent, D. W., and Shelekov, A. (1975). *Proc. Soc. Exp. Biol. Med.* **150,** 707–711.
Truitt, R. L., and Shechmeister, J. L. (1973). *Arch. Gesamte Virusforsch.* **42,** 78–87.
Tsubaki, S. (1938). *Saikingaku Zasshi* p. 513.
Underdahl, N. R., Grace, O. D., and Hoerlein, A. B. (1957). *Proc. Soc. Exp. Biol. Med.* **94,** 795–797.
USDA, Veterinary Services (1976). "Hog Cholera Bibliography." Washington, D.C.
van Aert, A., Aynaud, J. M., Bachmann, P. A., Horzinek, M. C., Maess, J., Mussgay, M., and Torlone, V. (1969). *Bull. Off. Int. Epizoot.* **72,** 671–693.
van Bekkum, J. G., and Barteling, S. J. (1970). *Arch. Gesamte Virusforsch.* **32,** 185–200.
van Bekkum, J. G., and Straver, P. J. (1964). *Bull. Off. Int. Epizoot.* **62,** 843–849.
van Berlo, M., Zeegers, J. J. W., Horzinek, M. C., and van der Zeijst, B. A. M. (1981). In preparation.
van der Zeijst, B. A. M., Horzinek, M. C., and Moennig, V. (1975). *Virology* **68,** 418–425.
Vantsis, J. T., Barlow, R. M., Fraser, J., Rennic, J. C., and Mould, D. L. (1976). *J. Comp. Pathol.* **86,** 111–120.
Volenec, F. J., Sheffy, B. E., and Baker, J. A. (1972). *Arch. Gesamte Virusforsch.* **36,** 275–283.
Ward, G. M. (1971). *Cornell Vet.* **61,** 179–191.
Wellemans, G. (1969). *Ann. Med. Vet.* **113,** 47–60.
Wellemans, G., and Leunen, J. (1974). *Ann. Med. Vet.* **118,** 95–103.
Wilson, J. C., Doll, E. R., McCollum, W. H., and Cheatham, J. (1962). *Cornell Vet.* **52,** 200–205.
Wood, K. R., Gould, E. A., and Smith, H. (1970). *J. Gen. Virol.* **6,** 257–265.
Wood, O. L. (1972). Ph.D. Thesis, Yale Univ. Press, New Haven, Connecticut.
Zeegers, J. J. W., and Horzinek, M. C. (1977). *In* "Hog Cholera/Classical Swine Fever and African Swine Fever," Publ. EUR 5904 EN, pp. 52–57. Comm. Eur. Communities, Brussels.
Zeegers, J. J. W., van der Zeijst, B. A. M., and Horzinek, M. C. (1976). *Virology* **73,** 280–205.

Part VII

BUNYAVIRIDAE

Chapter 10

Bunyaviridae: Infections and Diagnosis

J. S. PORTERFIELD and A. J. DELLA-PORTA

I. Introduction 479
II. Akabane Disease 482
A. Virus Structure and Composition 482
B. Epidemiology 483
C. Natural History 486
D. The Disease 487
E. Serology 492
F. Laboratory Features and Diagnosis 495
III. Aino and Other Simbu Group Viruses 497
IV. Rift Valley Fever 499
A. History 499
B. Epidemiology 500
C. Control 501
D. Diagnosis 501
V. Bhanja Virus Infections 501
VI. Nairobi Sheep Disease 502
VII. Ganjam Virus Infections 503
VIII. Dugbe Virus Infections 503
IX. Crimean-Congo Viruses 504
References 504

I. INTRODUCTION

The large and recently established family Bunyaviridae contains a number of viruses that are important pathogens of man and of domestic animals as well as others, such as Rift Valley fever (RVF) virus, that can cause serious disease in

COMPARATIVE DIAGNOSIS
OF VIRAL DISEASES, VOL. IV

ISBN 0-12-429704-8

both animals and man. The name of the family is derived from that of the prototype virus, Bunyamwera virus, which was isolated in Uganda from a pool of mosquitoes collected in a forest region bearing that name (Smithburn *et al.*, 1946). Structurally, these viruses contain three separate segments of RNA of negative polarity enclosed within a roughly spherical envelope which gives the whole virion a diameter of about 100 nm; the envelope contains one or two glycoproteins in different viruses. Taxonomically, the family originally contained a single genus, *Bunyavirus,* made up of about 80 different viruses, and about another 80 viruses were classified as "possible Bunyaviruses" (Porterfield *et al.*, 1975-1976). Serologically, the family can be divided into some 18 groups, each named after a prominent member of that particular serogroup (Table I). The viruses within a serogroup are quite closely related by a number of serological tests but are clearly separable by neutralization tests. Individual viruses in some serogroups show minor cross reactions with viruses in other serogroups, and the viruses thus linked are said to fall within a "supergroup" of

TABLE I
The Family Bunyaviridae Arranged to Show Taxonomic and Serological Subdivisions and Their Relation to Important Diseases of Animals.

Genus	Serogroup	Animal disease
Bunyavirus	Bunyamwera	None
	Bwamba	None
	C. Group	None
	California	None
	Capim	None
	Guama	None
	Koogol	None
	Mirim	None
	Olifantsvlei	None
	Patois	None
	Simbu	Aino, Akabane
	Tete	None
Nairovirus	Crimean-Congo	Crimean-Congo H.F., Hazara
	Nairobi sheep disease	Nairobi sheep disease, Ganjam, Dugbe
Phlebovirus	Phlebotomus fever	Rift valley fever
Uukuvirus	Uukuniemi	None
Unclassified Bunyaviridae	Anopheles	None
	Bakau	None
	Kaisodi	None
	Mapputta	None
	Turlock	None
	Unassigned	Bhanja

10 interrelated groups, the remaining viruses in the other 8 groups falling outside the supergroup (Bishop and Shope, 1979). Although the precise chemistry of the viruses is still rather poorly characterized, these serological cross reactions are probably reflections of similarities between the surface glycoproteins in different viruses. Recently, three new genera have been proposed within the family (Bishop, D. H. L., personal communication); these are Nairovirus, named after Nairobi sheep disease virus, Phlebovirus, named after the Phlebotomus fever viruses, and Uukuvirus, named after Uukuniemi virus, isolated in Finland from tick species. Akabane and Aino viruses, in the Simbu serogroup, remain within the Bunyavirus genus. Rift valley fever virus, previously a "possible Bunyavirus," is now classified in the Phlebovirus genus. Crimean-Congo, Dugbe, Ganjam and Nairobi sheep disease viruses are all classified in the Nairovirus genus. Bhanja virus remains as an unclassified member of the family Bunyaviridae.

Biologically, the Bunyaviridae are arthropod-borne animal viruses, or arboviruses, which are capable of infecting both vertebrate and invertebrate hosts in nature, and they are normally transmitted between vertebrate hosts by the bites of blood-sucking arthropods, which may be mosquitoes, ticks, phlebotomines, culicoides, or other arthropods. The natural history of these viruses is, in most cases, very poorly understood and may be very complex. There is commonly a pattern of alternate cycles of replication in vertebrate and invertebrate hosts, but some viruses, such as members of the California group, may undergo vertical transovarial transmission in their mosquito hosts (Watts *et al.*, 1973), and others, such as Akabane virus, may spread vertically in their vertebrate hosts (cattle, sheep, and goats). As a further complication, it seems very likely that some viruses, notably RVF virus, can spread horizontally between vertebrate hosts as well as vertically in both mosquitoes and vertebrates.

There is no evidence that Bunyaviridae produce any detrimental effects in their invertebrate hosts. In vertebrates, the effects range from totally inapparent infections to infections which result in death of the host. The same virus may produce quite different disease patterns in different hosts; thus, Nairobi sheep disease virus not infrequently kills sheep, whereas the same virus in the African field rat, which may be its true natural host, produces a completely silent infection. Two viruses, Akabane and Aino, are important because they are known to cause congenital deformities in their vertebrate hosts. The Crimean-Congo viruses are described, in spite of the fact that no overt animal disease is known, because these viruses can cause severe or fatal infections in man. Rift Valley fever is remarkable for several reasons, in addition to those already mentioned. For many years, it has no close serological relatives, with the possible exception of Lunyo virus (Weinbren *et al.*, 1957), and it has recently been shown to be related to the Phlebotomus fever viruses (Shope *et al.*, 1980).

II. AKABANE DISEASE

Akabane virus was first isolated from mosquitoes caught in Japan in 1959 while the transmission of Japanese encephalitis was being investigated (Matsuyama *et al.*, 1960; Oya *et al.*, 1961). Subsequently, Akabane virus was isolated from culicoides caught in Australia in 1968 while searching for the vector of bovine ephemeral fever virus (Doherty *et al.*, 1972). Although no antibodies to Akabane virus were found in man (Oya *et al.*, 1961; Doherty *et al.*, 1972), it was found that cattle and sheep possessed neutralizing antibodies (Doherty *et al.*, 1972). However, when Akabane virus was inoculated into cattle there were no signs of a clinical response (Doherty *et al.*, 1972), and until 1974 it was thought that this virus was not associated with disease.

Major epizootics of a congenital disease of cattle, characterized by arthrogryposis (AG) and/or hydranencephaly (HE) and associated with infections with Akabane virus, have occurred in three separate areas within a 5-year span. They were in Israel in 1969–1970 (Markusfeld and Mayer, 1971; Nobel *et al.*, 1971), Japan in 1972–1974 (Omori *et al.*, 1974; Miura *et al.*, 1974), and Australia in 1974 (Della-Porta *et al.*, 1976; Shepherd *et al.*, 1978). Akabane virus infections of the fetus were associated with these epizootics when serum-neutralizing antibodies to Akabane virus were found in presuckling serum samples from the affected calves (Miura *et al.*, 1974; Omori *et al.*, 1974; Hartley *et al.*, 1975, 1977). Since then, Akabane disease has been described for cattle, sheep, and goats and confirmed by experimental infection studies (Inaba *et al.*, 1975; Parsonson *et al.*, 1975, 1977a; Kurogi *et al.*, 1977a,b; Narita *et al.*, 1979).

A. Virus Structure and Composition

Akabane virus, a member of the Simbu serological group in the family Bunyaviridae (Doherty *et al.*, 1972, Porterfield *et al.*, 1975–1976; Berge, 1975; Bishop and Shope, 1979) resembles other members in being sensitive to deoxycholate, ether, chloroform, and 5-iodo-2′-deoxyuridine and in having a buoyant density in CsCl of 1.22 gm/ml (Takahashi *et al.*, 1978). The virus will pass through 200- and 100- but not 50-nm-pore-size membrane filters, and negatively stained preparations appear roughly spherical, with a diameter of 70 to 130 nm (Takahashi *et al.*, 1978). Akabane virus is very acid labile, being rapidly inactivated at pH 3 and losing about 2 logs of infectivity at pH 5 after 1 hour at room temperature; it is also very heat labile, losing about 0.3 log of infectivity per hour at 37°C (Takahashi *et al.*, 1978).

Only very preliminary biochemical studies of Akabane virus have been made (McPhee and Della-Porta, unpublished observations). The virus contains four proteins, two of which are glycoproteins (Gl, molecular weight 104×10^3; G2, molecular weight 33×10^3), the nucleoprotein (N, molecular weight 22×10^3), and probably a large protein (L, molecular weight 150×10^3). There appear to be five proteins in virus-infected cells: L, G1, G2, N, and one smaller nonstructural protein (molecular weight 15×10^3).

B. Epidemiology

Akabane virus was first isolated in Japan by Oya *et al.* (1961), in Australia by Doherty *et al.* (1972), in Kenya by Metselaar and Robin (1976), and in South Africa by Theodoridis *et al.* (1979). Serological studies have further extended the distribution of the virus to Israel (Kalmar *et al.*, 1975), Cyprus (Sellers and Herniman, 1981), Thailand, and possibly Taiwan, Vietnam, and Bandung (Oya, 1971), Indonesia, Malaysia, and the Philippines (Inaba, personal communication). Serological evidence suggests that Akabane virus is not present in Papua-New Guinea (Doherty *et al.*, 1973; Della-Porta *et al.*, 1976; Cybinski *et al.*, 1978). Fig. 1 shows the known world distribution of Akabane virus. Lack of information of whether the virus is present in other areas of the world (Fig. 1, blank area) prevents us from completing the distribution map. It would seem likely that the distribution of the virus stretches from Australia, through Southeast Asia, to Japan and possibly also to India, and across the Middle East to Israel and Africa. However, it appears that congenital disease associated with Akabane virus is found principally at the extremities of its normal distribution.

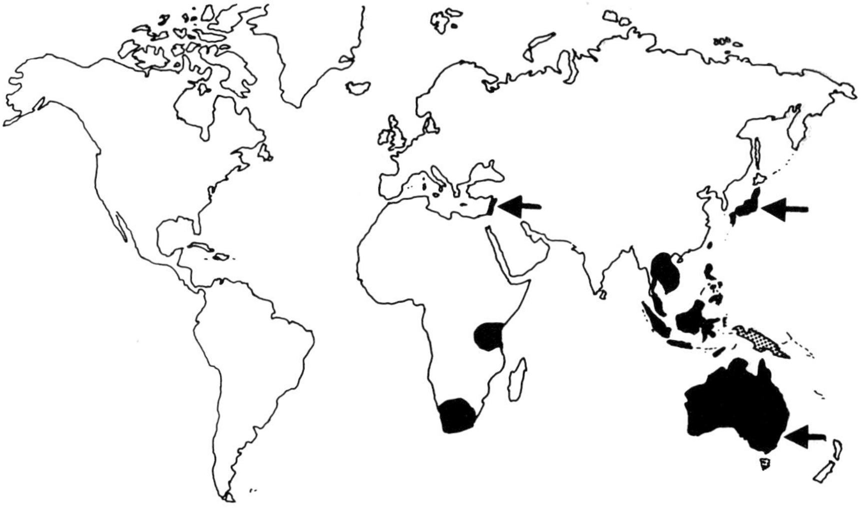

Fig. 1. Worldwide distribution of Akabane virus. Black areas: countries for which there is serological evidence for the presence of the virus or the virus has been isolated. Dotted areas: countries for which there is serological evidence which indicates that Akabane virus is absent. White areas: countries for which there are no data. Arrowed are the areas where Akabane disease has been observed.

1. Australia

Congenital bovine AG and HE have been observed as epizootics in southeastern New South Wales at least since the mid-1940s (Blood, 1956). Epizootics were observed in 1951, 1955, 1960, 1964, 1968, and 1974 (Blood, 1956; Bonnor *et al.*, 1961; Hartley and Wanner, 1974; Shepherd *et al.*, 1978). Similar, but smaller, epizootics have been observed in the Hunter Valley (Hindmarsh, 1937; Young, 1969) and the New England Tablelands (Coverdale *et al.*, 1978).

Australia-wide serological surveys for the presence of neutralizing antibodies against Akabane in cattle blood (Della-Porta *et al.*, 1976; Cybinski *et al.*, 1978) have shown that most (80%) of the animals in northern Australia become infected early in life. The distribution of seropositive animals has been linked with the distribution of *Culicoides brevitarsis* (Della-Porta *et al.*, 1976) (Fig. 2) and the disease associated with the extension of virus-infected insects beyond their nor-

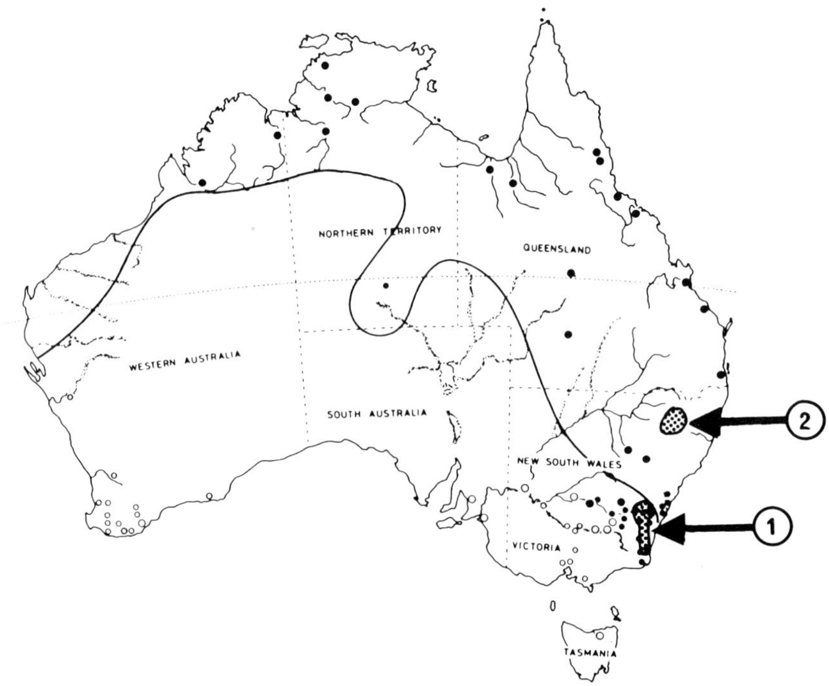

Fig. 2. The Australian distribution of serum-neutralizing antibodies against Akabane virus in cattle after the 1974 epizootic of congenital bovine arthrogryposis and hydranencephaly (Della-Porta *et al.*, 1976). ●, seropositive herds. ○, seronegative herds. The line drawn across Australia represents the known normal southward distribution of *Culicoides brevitarsis* (Murray, 1975; Standfast, personal communication), the probable insect vector of Akabane virus in Australia. Also marked are the areas where the principal epizootics of Akabane disease occur: (1) the south coast of New South Wales and (2) the New England Tablelands.

mal distribution, due to extended warm, humid summers, to areas where susceptible pregnant animals are present. These areas are either at a lower distribution limits for *C. brevitarsis*, particularly the southeastern coast of New South Wales (Fig. 2), or in elevated areas where the insects are not usually present, for example, the New England Tablelands (Fig. 2).

The outbreaks of Akabane disease are usually seen in southeastern New South Wales between July and November (Hartley and Wanner, 1974; Shepherd *et al.*, 1978). In the 1974 epizootic AG was the first syndrome observed, and the peak number of cases was seen in August (Shepherd *et al.*, 1978). The peak incidence of HE in the same epizootic occurred in September (Shepherd *et al.*, 1978). It is thought that the susceptible pregnant animals are probably infected with Akabane virus during the late summer–early autumn period. This is suggested both on serological evidence and the isolation of virus from animals which were naturally infected between February and April (Della-Porta *et al.*, 1977).

2. *Japan*

In Japan, the major outbreaks of Akabane disease have been seen in the southwestern areas (Omori *et al.*, 1974), where they have been observed since the late 1940s, with epizootics recorded in 1949–1950, 1959–1960, 1965–1966, and 1972–1974. The area is similar in geographic distribution to that where bovine ephemeral fever and Ibaraki virus infections are seen, and so it was considered likely that vectors played a role in transmission of the disease.

Akabane disease is seen from about August until March (Kurogi *et al.*, 1975). The peak of abortions is from August until December. This is followed by a peak of AG, cerebral defects (principally HE), and stillbirths from December until March. Although the virus was isolated in Japan from the mosquitoes *Aedes vexans* and *Culex tritaeniorhynchus* (Oya *et al.*, 1961), at present there is no link between the seasonal distribution of those insects and Akabane disease.

3. *Israel*

The only outbreak of Akabane disease reported in Israel occurred during 1969–1970 (Markusfeld and Mayer, 1971). It was estimated that 2 to 4% of all dairy calves born in Israel during the epizootic suffered from congenital AG/HE. In badly affected areas the syndrome was encountered in almost all herds, but its incidence within the herd was relatively low. In badly affected herds, the number of abnormal calves amounted to 30% of all calvings during the peak months of the epizootic. Congenitally deformed calves were encountered far more frequently in the north of Israel. Herds in the south, in the Negev, and in the hilly regions of the country remained practically unaffected. In the western Galilee area, AG occurred between November and February with the peak incidence in December, whereas HE occurred between November and June with the peak incidence in April.

C. Natural History

1. Host Range

Akabane disease has been reported in cattle (Blood, 1956; Whittem, 1957; Markusfeld and Mayer, 1971; Omori *et al.*, 1974), sheep (Nobel *et al.*, 1971; Hartley and Haughey, 1974), and goats (Nobel *et al.*, 1971). The virus has been isolated from naturally infected bovine (Kurogi *et al.*, 1976) and ovine (Della-Porta *et al.*, 1977) fetuses. Serological evidence indicates that Akabane virus also infects horses, buffaloes, and camels (Doherty *et al.*, 1972; Della-Porta *et al.*, 1976; Cybinski *et al.*, 1978), but not wallabies, kangaroos, rats, domestic chickens and ducks, man, or pigs. It would appear that the distribution of antibody in animal species is determined by the feeding preference of the vector. In Australia, the suspect vector *C. brevitarsis* has a feeding preference, determined by blood meal analyses, of cattle and probably horses, followed by sheep (Muller and Murray, 1977).

Akabane virus will produce experimental infections in species other than the natural hosts, and in some of these species it can cross the placenta to infect the developing fetus. Virus has been found to cross the placenta and infect the fetus of the hamster, with the young dying at birth, but appears not to cross the placenta in the mouse (Anderson and Campbell, 1978). Spradbrow and Miah (1978) found that rats and mice produced normal litters; rabbits aborted, produced deformed litters, or readsorbed dead fetuses; guinea pigs aborted or readsorbed the fetuses or produced stillborn young, but no developmental abnormalities were seen; and 4-day-old chick embryos inoculated by the yolk sac either died or developed abnormally (including the development of AG).

2. Insect Transmission

Akabane virus was first isolated in Japan from two species of mosquito, *Aedes vexans* and *Culex tritaeniorhynchus* (Oya *et al.*, 1961). These mosquitoes were collected at livestock pens (Matsuyama *et al.*, 1960), and it is possible that blood-fed insects were analyzed. In Australia, there were two isolations in 1968 from the biting midge *C. brevitarsis* (Doherty *et al.*, 1972), and subsequently there have been four further isolations from *C. brevitarsis* (St. George *et al.*, 1978; Standfast, personal communication). Further, the distribution of seroposi tive cattle to Akabane virus in Australia would suggest that it is spread by *C. brevitarsis* (Della-Porta *et al.*, 1976; Cybinski *et al.*, 1978).

In Africa, Akabane virus has also been isolated from mosquitoes and from culicoides. Metselaar and Robin (1976) isolated the virus from the mosquito *Anopheles funestus* collected in Kenya. In 1970, Akabane virus was isolated from culicoides caught in South Africa during an outbreak of bovine ephemeral fever (Theodoridis *et al.*, 1979).

Only very limited growth studies in insects have been carried out. Because of the difficulty of feeding virus to *C. brevitarsis*, it has not been possible to determine whether the virus grows in this insect (Muller, 1977). Growth of Akabane virus has been shown in the mosquito *A. vigilax* but not in *C. annulirostris* after feeding infected blood (Kay *et al.*, 1975) or intrathoracic inoculation (Standfast and Whelan, 1976). It has been reported (Standfast and Whelan, 1976) that Japanese workers have recorded that they were unable to infect *C. tritaeniorhynchus*. As yet it is not clear whether mosquitoes and culicoides can both transmit Akabane virus, and it is important that studies be undertaken to find which insect(s) are the vector(s) of the virus.

D. The Disease

1. Clinical Signs

There is no evidence that Akabane virus produces ill effects in cattle, sheep, and goats infected after birth, although virus replication, as indicated by viremia, occurs (Kurogi *et al.*, 1977a,b; Parsonson *et al.*, 1977a). Semen collected from infected bulls contained spermatozoa of normal morphology, and virus was not

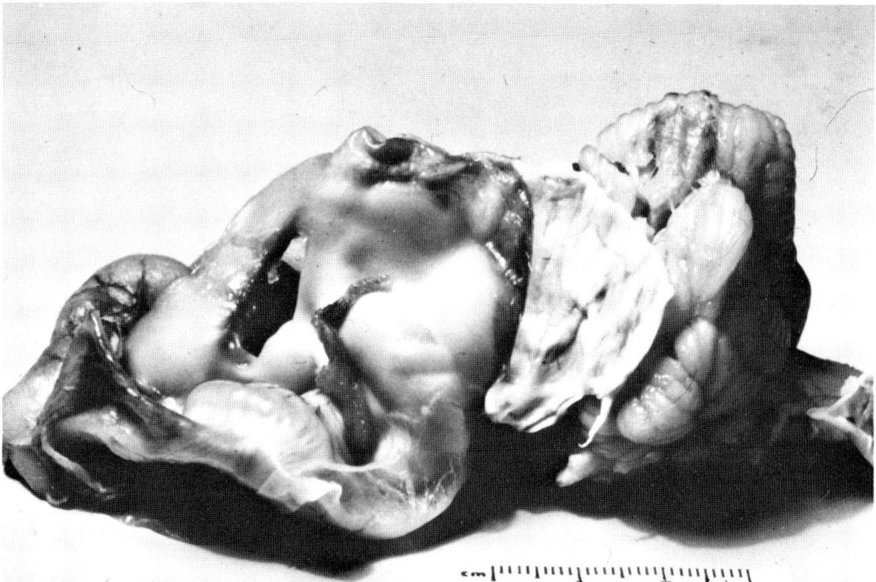

Fig. 3. The brain of a naturally infected calf from an outbreak of Akabane disease. Most of the cerebrum was replaced by fluid to show a typical case of hydranencephaly. Note the severe cavitation in the remaining cerebrum. The cerebellum showed some dysgenesis. (Parsonson, unpublished observations.)

recovered from any of the semen samples (Parsonson *et al.*, 1977b). Virgin heifers developed viremias following intrauterine inoculation at estrus with semen, followed immediately by virus (Parsonson *et al.*, 1977b), but there was no effect on fertility and four of the heifers, allowed to proceed to parturition, gave birth to normal calves.

Dystocia is usually associated with AG-affected calves, usually resulting in their death and delivery by embryotomy or cesarian section. In AG the affected joints cannot be flexed or extended through the normal range of movement, and the appearance in calves resembles that seen in lambs (Fig. 4). Calves born with HE may survive for many months if hand reared, but they never thrive. They may show signs of ataxia and lack of coordination, blindness and dysphagia, and regurgitation of food (Blood, 1956; Markusfeld and Mayer, 1971).

2. *Pathology*

The gross pathology of Akabane disease in cattle, sheep, and goats is similar, AG and HE being the main lesions seen in all species. However, these lesions can differ in their severity, the time at which they occur, and their detailed histopathology.

An epizootic of Akabane disease in cattle may be first noticed by an increased

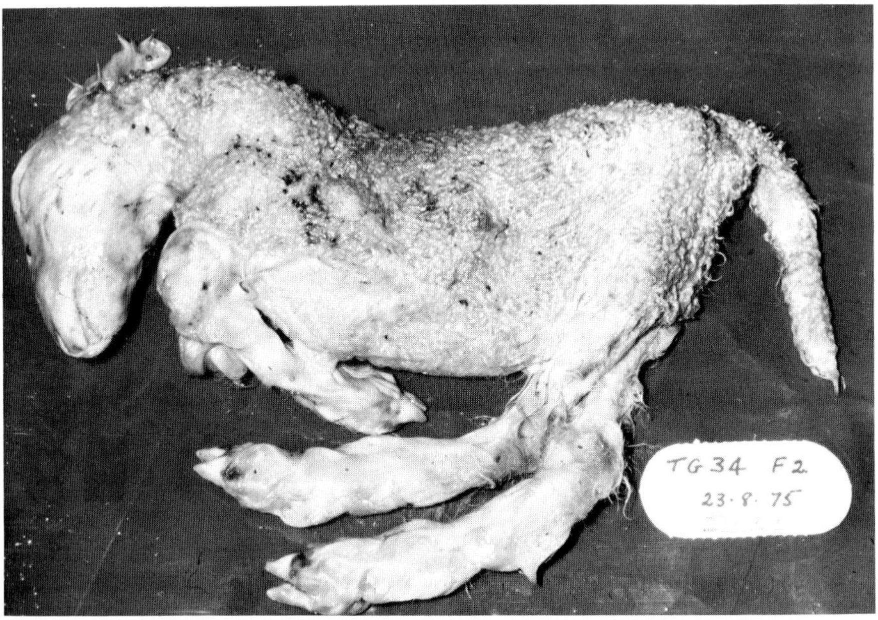

Fig. 4. Lamb showing arthrogryposis (the limb joints locked in a fixed position). This lamb also has branchygnathia (an undershot jaw) and kyphosis (a humped back). The dam was infected with Akabane virus at 34 days of gestation in an experimental study. [Parsonson *et al*. (1975, 1977a).]

incidence of abortions (Omori *et al.*, 1974; Kurogi *et al.*, 1976), followed some months later by the birth of incoordinated calves (Hartley *et al.*, 1977), calves with mild to severe AG (Whittem, 1957; Markusfeld and Mayer, 1971; Konno *et al.*, 1975; Hartley *et al.*, 1977; Shepherd *et al.*, 1978) and sometimes cervical scoliosis, torticollis, and kyphosis (Whittem, 1957; Markusfeld and Mayer, 1971; Hartley *et al.*, 1977), and finally, calves with HE sometimes associated with AG and other minor forms of brain damage (Whittem, 1957; Markusfeld and Mayer, 1971; Konno *et al.*, 1975; Hartley *et al.*, 1977). The progression in which these syndromes are seen in the field may represent the developmental stages of the fetus at which Akabane virus infection occurs (Hartley *et al.*, 1977) or greater awareness and better reporting as the epizootic progresses. Often the various pathological syndromes seen in Akabane disease overlap, with many occurring concurrently in the same animal. Experimental studies in pregnant cows indicate that infection with Akabane virus between 76 and 96 days of gestation can produce congenital malformations in calves (Kurogi *et al.*, 1977b).

AG is seen in approximately 30 to 50% of affected offspring (Whittem, 1957; Markusfeld and Mayer, 1971; Nobel *et al.*, 1971; Shepherd *et al.*, 1978). In affected animals the joint surfaces appear normal, whereas the bulk of the muscle is usually smaller and paler in color than normal (Markusfeld and Mayer, 1971; Whittem, 1957). Microscopically, there is a severe loss of myelinated fibers in the lateral and ventral funiculi of affected areas of the spinal cord, together with the loss of ventral horn neurons and marked loss of nerve fibers in ventral spinal nerves (Whittem, 1957; Konno *et al.*, 1975; Hartley *et al.*, 1977). The lesions are associated with a moderate to severe atrophy and/or adipose tissue replacement of the skeletal musculature supplied by the ventral spinal nerves (Whittem, 1957; Hartley *et al.*, 1977).

HE is seen in approximately 30 to 60% of affected animals (Nobel *et al.*, 1971; Hartley and Wanner, 1974; Shepherd *et al.*, 1978). It can be severe when the cerebrum is almost totally replaced by a fluid-filled cavity (Figs. 3 and 5) (Hartley *et al.*, 1977), or less severe and consist of extensive cavitation or porencephaly of the white matter of the dorsal and ventrolateral cerebral hemispheres with dilation of the lateral ventricles (Blood, 1956; Konno *et al.*, 1975; Hartley *et al.*, 1977).

Other syndromes have been reported for calves with Akabane disease. Early in the outbreaks, calves have been found which were incoordinate or unable to stand at birth, but no gross pathological lesions were seen in these animals (Konno *et al.*, 1975; Hartley *et al.*, 1977). Microscopically, there was a mild to moderate nonsuppurative acute encephalomyelitis, most evident in gray matter of the mid and posterior brain stem. There is no evidence to link this syndrome with Akabane virus, other than the association with epizootics of Akabane disease (Hartley *et al.*, 1977). Toward the end of the 1974 epizootic of Akabane disease in Australia, there were a number of calves with additional congenital abnor-

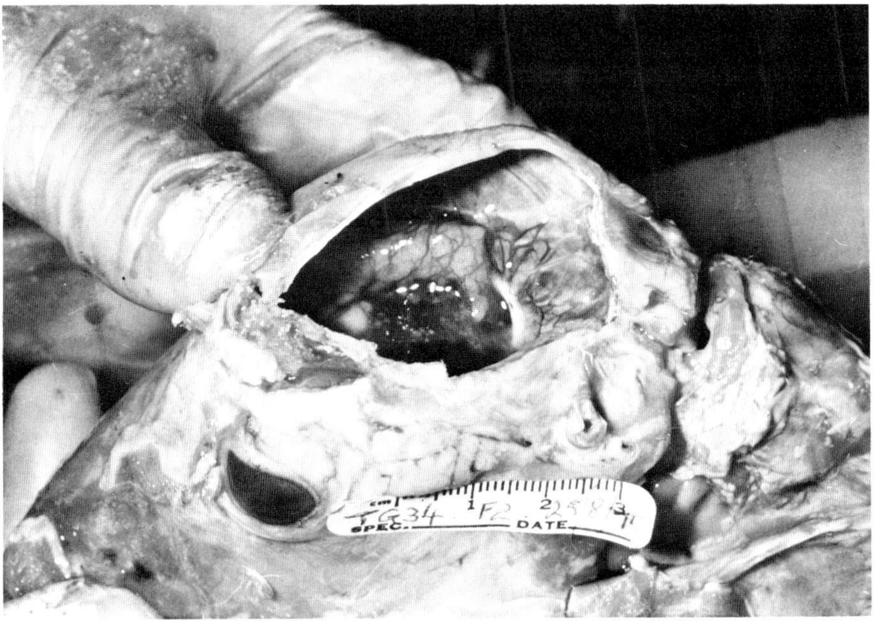

Fig. 5. Lamb showing hydranencephaly. When the calvarium was removed, most of the cerebrum was found to be replaced by fluid. The dam was infected with Akabane virus at 34 days of gestation. [Parsonson *et al.* (1975, 1977a).]

malities which included micrencephaly accompanied by a thickening of the cranial bones, with a resultant reduction in the size of the cranial cavity. Some calves had a reduction in the size of the cerebellum, which microscopically showed cavitation and collapse of the cerebellar white matter. However, serological evidence suggested that a number of these calves with micrencephaly were not infected with Akabane virus (Hartley *et al.*, 1977).

The gross and microscopic pathology of Akabane disease in lambs and kids is similar to that in calves (Figs. 4 to 6) (Nobel *et al.*, 1971; Hartley and Haughey, 1974; Parsonson *et al.*, 1975, 1977a; Kurogi *et al.*, 1977a; Narita *et al.*, 1979), with many of the pathological changes occurring concurrently in affected animals. Additional lesions were observed in the thymus of affected lambs; there was a reduced number of thymocytes in the cortex, and the medulla possessed enlarged Hassal's corpuscles.

Experimental infection studies with Akabane virus in cattle, sheep, and goats have resulted in deformed offspring with pathological lesions similar to those observed in field infections. Infection at the following different gestation stages for these species resulted in deformities: cattle, 76 to 96 days of gestation (Kurogi *et al.*, 1977b); sheep, 30 to 36 days of gestation (Parsonson *et al.*, 1975,

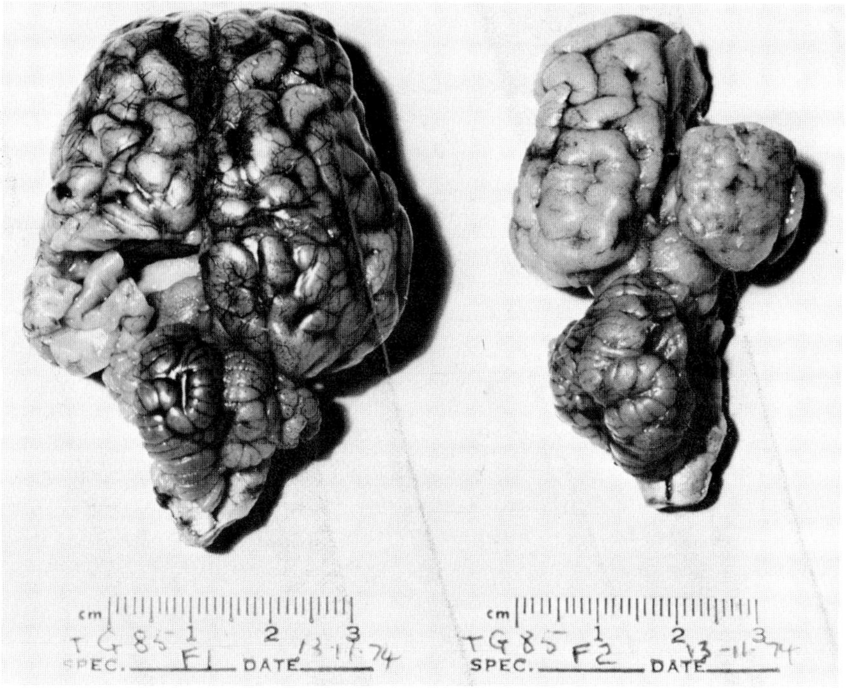

Fig. 6. The brain of a lamb showing micrencephaly (right) and the brain of its normal twin (left). Some tissue was removed for virological examination before photography. Note the marked reduction in size of the cerebrum and in the number of convolutions. The dam was infected with Akabane virus at 36 days of gestation. [Parsonson *et al.* (1975, 1977a).]

1977a); and a goat, 40 days of gestation (Kurogi *et al.*, 1977a). Some minor lesions were also observed in Japan in sheep infected between 41 and 46 days of gestation (multifocal encephalitis and porencephaly) and between 81 and 91 days of gestation (focal necrobiosis and gliosis in the white matter) (Narita *et al.*, 1979).

3. *Strain Differences*

There are data suggesting that there may be strain differences between isolates of Akabane virus. Examination of the pathogenicity of a number of isolates of Akabane virus for mice has shown that there is a significant difference between many of these isolates (Kurogi *et al.*, 1978a). Comparison of the field data from the Japanese (Omori *et al.*, 1974; Kurogi *et al.*, 1975) with data from the Australian (Hartley *et al.*, 1977; Shepherd *et al.*, 1978) epizootics of Akabane disease indicates that there may have been a higher level of abortions and stillbirths in Japan than in Australia. Further, it appears that the Japanese isolates can infect the ovine fetus over a wider gestational range (Narita *et al.*, 1979) than the limited period of the Australian isolates (Parsonson *et al.*, 1977a).

Akabane virus isolates have been grouped on the basis of cross neutralization tests (Doherty *et al.*, 1972; Metselaar and Robin, 1976; Della-Porta *et al.*, 1977) which in fact only indicates that the external viral proteins (G1 and G2) are identical or very closely related. However, as the bunyaviruses are segmented genome viruses containing three segments of single-stranded RNA, with only one segment of the RNA (medium sized) coding for the glycoproteins (Obijeski and Murphy, 1977; Gentsch and Bishop, 1979), it is possible that one or two of the three segments could differ between isolates. Thus, these isolates could contain different internal proteins, the L and N proteins coded for by the large and small RNA segments (Obijeski and Murphy, 1977). This difference could confer different virulence characteristics on the isolates. Such differences may have arisen by reassortment of the gene segments between Akabane virus isolates and other Simbu group viruses. Experimental gene reassortment has been demonstrated between two California encephalitis group bunyaviruses, snowshoe hare and La Crosse, by Gentsch *et al.* (1977). Such reassortment, however, has not been demonstrated in nature. However, Akabane virus isolates with different virulence properties warrant investigation of the possibility of gene reassortment and other properties, such as the nature of the viral glycoproteins.

E. Serology

1. Serological Tests

There are four serological tests which have been used for Akabane serology: serum neutralization (SN), hemagglutination inhibition (HI), complement fixation (CF), and agar gel diffusion precipitin (AGDP). The test most often employed is the SN test using Vero cells. All tests that have been described are microtiter SN tests, but they obviously differ in their degree of sensitivity, as significantly different amounts of virus are used in each test: 32 $TCID_{50}$ (Hartley *et al.*, 1975; Della-Porta *et al.*, 1976), 100 $TCID_{50}$ (Cybinski *et al.*, 1978), and 200 $TCID_{50}$ (Kurogi *et al.*, 1975). Because of these differences in virus levels employed, care must be taken when comparing titers between tests carried out in different laboratories. Hartley *et al.* (1975) found that 32 $TCID_{50}$ units of virus in a microtiter SN test gave titers equivalent to that found in a macrotest employing 100 $TCID_{50}$/0.1 ml.

The HI test has received more attention than the CF and AGDP tests. It was found that hemagglutination with the Simbu group viruses was both pH and NaCl molarity dependent (Goto *et al.*, 1976). The method of Clarke and Casals (1958) was modified by using a diluent with 0.4 *M* NaCl, 0.2 *M* phosphate at pH 6.0–6.2 for hemagglutination and HI. An HI test was developed for use with sera from domestic animals and gave results which correlated closely with the neutralizing antibody titers of the same sera (Goto *et al.*, 1978). An AGDP test has been developed (Gard, personal communication) in which antigens from infectious culture fluid are concentrated in a dialysis sac using polyethylene glycol (20,000 molecular weight) and then used, together with a standard reference serum, in the AGDP test. Initial trials indicated that there was a good correlation between neutralizing antibodies to Akabane virus and AGDP-positive reactions (Gard, personal communication; Della-Porta, unpublished observations), but subsequent studies indicated that the Akabane AGDP test will detect AGDP antibodies formed against other Simbu group viruses (Della-Porta and Murray, unpublished observations; St. George, personal communication). However, the AGDP test may prove useful as a simple screening test for Simbu group viruses. The CF test has principally been used for comparisons of the relationships between Simbu group viruses (Doherty *et al.*, 1972) and some preliminary studies on the CF antibody response in infected cattle (Goto *et al.*, 1978).

For diagnosis of Akabane disease, the SN test has proved most useful. It is important that a sample of serum from the dam and a presuckling sample of serum from the affected offspring be collected for the SN test. The presence of serum-neutralizing antibodies in the presuckling serum sample is an indication that infection of the fetus most likely occurred *in utero* (Omori *et al.*, 1974; Kurogi *et al.*, 1975; Hartley *et al.*, 1975, 1977). Further, elevation of immunoglobulin levels in the presuckling serum is also an indication that there was an *in utero* infection (Trainin and Meirom, 1973; Hartley and Wanner, 1974; Wanner and Husband, 1974), but lack of SN antibodies against Akabane virus in the serum of both the dam and the offspring would suggest that other infectious agents may be responsible for the deformity (Hartley *et al.*, 1977).

2. *Immune Response in Infected Animals*

Animals infected with Akabane virus rapidly develop both SN and HI antibodies (Parsonson *et al.*, 1977a; Goto *et al.*, 1978). In cattle, SN and HI antibodies can be detected between 7 and 14 days following infection (Parsonson *et al.*, 1976; Goto *et al.*, 1978), and initially these antibodies would appear to be sensitive to 2-mercaptoethanol, probably being IgM (Goto *et al.*, 1978). Field observations with sentinel cattle would indicate that the antibody response following natural and experimental infections is similar (St. George *et al.*, 1977, 1978). Significant levels of SN antibody persist in cattle for at least 2 years following infection (Parsonson *et al.*, 1976; Della-Porta, unpublished observations).

In sheep infected experimentally with Akabane virus, there was an SN antibody response similar to that observed in cattle (Parsonson *et al.*, 1976, 1977a). SN antibodies were detectable as early as 5 days after infection, which is earlier than they are detectable in cattle. There was an initial biphasic response, the initial phase probably representing the IgM response and the later response a change to IgG. Good levels of SN antibodies persisted for at least 2 years. There was no apparent difference in the SN antibody responses in sheep that produced deformed fetuses and in those that produced normal fetuses.

Fetal calves and lambs infected *in utero* with Akabane virus often developed SN antibodies against the virus (Omori *et al.*, 1974; Parsonson *et al.*, 1976, 1977a; Hartley *et al.*, 1977). Fetal lambs developed antibodies to Akabane virus between 75 and 100 days of gestation (Della-Porta *et al.*, unpublished observations). There are no data on when infected calves and goats develop antibodies. A detailed sequence study is required to determine when virus is present, when antibodies first appear in infected fetuses, and how the pathological changes occur in the development of the fetus.

Normal calves, lambs, and kids also acquire SN antibodies against Akabane virus from their dams. At the first suckling, colostrum-derived antibodies pass from the immune mothers to their offspring. In lambs, following an initial rise in SN antibodies after ingestion of colostrum, the antibody declined, to become virtually undetectable after 4 months (Murray and Della-Porta, 1976). In calves there is a similar pattern, although the SN antibodies remain at detectable levels for up to 6 months of age (Della-Porta and Murray, unpublished observations). These observations on colostrum-derived antibodies against Akabane virus emphasize the importance of obtaining presuckling serum from affected animals for the diagnosis of Akabane disease.

F. Laboratory Features and Diagnosis

1. Virus Growth Systems

Akabane virus was first isolated by intracerebral inoculation of 1- to 2-day-old suckling mice (Oya *et al.*, 1961; Doherty *et al.*, 1972). A comparison of the growth of different isolates in mice has been made by Kurogi *et al.* (1978a). Some primary isolations have also been made using baby hamster kidney (BHK-21) cells (St. George *et al.*, 1978).

The growth of the virus in a number of primary and continuous cell lines has been investigated. The virus grows in BHK-21 (Kurogi *et al.*, 1975; O'Halloran and Della-Porta, 1975; Della-Porta *et al.*, 1976), Vero (Della-Porta *et al.*, 1976), HmLu-1 (hamster kidney) (Kurogi *et al.*, 1976, 1978b), and a number of other cells including PK-15 (pig kidney), PK-81 (pig kidney), BEK-1 (bovine embryo kidney), BTR (bovine thymus), RK-13 (rabbit kidney), BSC (African green monkey kidney), LLC-MK2 (rhesus monkey kidney), and SVP (possibly derived from Vero cells) but not in primary swine kidney, primary swine testis, primary calf kidney, primary calf testis, HeLa, HEp-2, or L cells (O'Halloran and Della-Porta, 1975; Kurogi *et al.*, 1976). Growth studies in HmLu-1 cells infected at a multiplicity of about 0.1 $TCID_{50}$/cell (using uncloned virus) resulted in a virus yield of 10^7 $TCID_{50}$/0.025 ml 3 days after infection (Kurogi *et al.*, 1977c).

2. *Isolation Procedures*

Initial isolation of Akabane from field specimens or from insects has usually been accomplished by intracerebral inoculation of 1- to 2-day-old suckling mice (Doherty *et al.*, 1972; Oya *et al.*, 1961). If the mice do not develop paralysis by the tenth day after inoculation, their brains are harvested and the material is passaged, as a 1/10 dilution of a 10% suckling mouse brain preparation, by intracerebral inoculation through another set of suckling mice. Mice usually become paralyzed and die during the initial passage of Akabane virus. Some isolations have been made in BHK-21 cells (St. George *et al.*, 1978). Comparison of the tissue culture isolation system with suckling mice (O'Halloran and Della-Porta, 1975; Kurogi *et al.*, 1976) indicated that mice were more sensitive. HmLu-1, Vero, SVP, and BHK-21 cell lines were almost as sensitive as mice.

The most useful tissues for attempts to isolate the virus are the placentomas and fetal membranes, with other tissues from aborted fetuses likely to yield isolations (Della-Porta *et al.*, 1976; Kurogi *et al.*, 1976). However, it is unlikely that virus will be isolated from deformed full-term offspring because of the elimination of the virus, probably by serum-neutralizing antibodies formed in the fetus (Hartley *et al.*, 1977). In the case of full-term offspring, diagnosis of Akabane disease depends on the detection of neutralizing antibodies in precolostrum serum from the offspring and in the serum from its mother (Omori *et al.*, 1974; Hartley *et al.*, 1975, 1977).

Akabane virus has been isolated from two naturally infected calf fetuses (Kurogi *et al.*, 1976). In one case, the fetus was removed from an infected pregnant cow at 92 days of gestation and examined for pathological changes and for the presence of virus. The fetus showed signs of AG and HE. Virus was isolated from the brain, cerebral fluid, spinal cord, muscles, fetal placenta, and amnion but not from the viscera pool, intestines, amniotic fluid, or fetal blood. In the other case, virus was isolated from a fetus aborted at 134 days of gestation. Virus was isolated from the brain but not from the heart (the only tissues examined).

The virus has also been isolated from two naturally infected lamb fetuses (Della-Porta *et al.*, 1977). A number of pregnant ewes that became infected with Akabane virus were sacrificed and their fetuses examined for pathological changes, and virus isolation was attempted. One fetus taken at 89 days of gestation showed gross pathology of AG/HE and virus was isolated from the placentomas, fetal fluids, and fetal membranes but not from the lung, muscle, thymus, heart, kidney, spleen, liver, stomach contents, and cerebrospinal fluid. The other fetuses taken at 74 days of gestation had microscopic lesions of porencephaly and virus was isolated from the placentomas, fetal fluids, fetal membranes, lung, muscle, and cerebellum but not from the thymus, heart, kidney, spleen, and liver.

3. *Differential Diagnosis*

Agents other than Akabane virus can produce similar congenital deformities in cattle, sheep, and goats (Leipold *et al.*, 1972; Saperstein *et al.*, 1975). These causes may be of an infectious nature, of genetic or environmental origin, or associated with teratogenic chemicals. The deformities may also be of some unknown origin. The detection of elevated levels of specific immunoglobulins in precolostrum samples from offspring would indicate whether an infective agent was involved (Leipold *et al.*, 1972; Schultz, 1973). Diagnosis of Akabane disease can rarely be accomplished on the basis of virus isolation but depends largely on histopathology and serology.

Bluetongue virus is the most likely infectious agent to cause similar deformities. Congenital deformities in lambs have been associated with naturally occurring bluetongue virus infections based on the presence of antibody to bluetongue virus in ewes in the flock as well as with live egg-attenuated bluetongue vaccines (Shultz and Delay, 1955; Cordy and Shultz, 1961; Griner *et al.*, 1964; Young and Cordy, 1964; Osburn *et al.*, 1971a,b; Schmidt and Panciera, 1973). The pathological changes can range from HE to mild porencephaly. AG has not been reported in lambs infected *in utero* with bluetongue virus. In cattle, HE and other congenital deformities have resulted from natural infection based on the presence of antibody to bluetongue virus in the dam or experimental virus infections using egg-attenuated virus (McKercher *et al.*, 1970; Richards *et al.*, 1971; Hubbert *et al.*, 1972; Jochim *et al.*, 1974; Barnard and Pienaar, 1976). Bluetongue infections can readily be distinguished from Akabane virus infections by serology and virus isolation.

Other possible infectious causes of similar congenital deformities are Aino virus (Coverdale *et al.*, 1978; see also Section III) and possibly other Simbu group members (St. George, personal communication). Certainly not all the congenital defects in the 1974 Australian epizootic of Akabane disease were associated with fetal infection with Akabane virus (Hartley *et al.*, 1977). Attenuated Wesselsbron disease vaccine virus and its wild-type virus have been associated with ovine congenital AG/HE with the further complication of *hydrops amnii* (Coetzer and Barnard, 1977). Further, attenuated RVF vaccine virus has also been associated with AG/HE in lambs without *hydrops amnii* (Coetzer and Barnard, 1977).

Causes other than infectious agents should also be considered. Toxic substances (alkaloids) associated with lupins have been linked to AG in calves (Shupe *et al.*, 1967a,b). James *et al.* (1967) reported the appearance of crooked calves as a result of grazing on or feeding of green feed containing *Astragalus* spp. AG associated with cleft palate has been reported as an inherited congenital abnormality in Charolais neonatal calves (Leipold *et al.*, 1969, 1972). Hartley and Wanner (1974) reported cases of AG with cleft palate in 3/4 bred Charolais calves and in a 3/4 bred Devon calf, the result of an accidental father–daughter mating. The deformities associated with genetic causes usually occur as a number of isolated cases, whereas Akabane disease is often seen as major epizootics occurring in defined areas, often from winter to early summer, but not every year.

III. AINO AND OTHER SIMBU GROUP VIRUSES

The Simbu group of viruses have a wide geographical distribution (Table II), being found in all continents except Europe. Oropouche and Shuni viruses are associated with human infections. Akabane virus (see Section II) is the cause of congenital deformities in cattle, sheep, and goats. A number of other Simbu group viruses have been isolated from domestic animals (Table II) or associated with infections by serology: Aino, Douglas, Ingwavuma, Peaton, Sabo, Sathuperi, Shamonda, Shuni, and Tinaroo viruses. There is suggestive serological evidence that Aino virus may be associated with congenital bovine AG and HE (Coverdale *et al.*, 1978). There is also some serological evidence that Tinaroo virus may also be associated with *in utero* ovine infections (St. George, personal communication).

Aino virus was first isolated in Japan in 1964 from mosquitoes. One isolate was recovered from a pool of *C. tritaenorhynchus* and another from a mixed pool of *C. pipiens* and *C. pseudovishui* (Takahashi *et al.*, 1968). Subsequently, a Simbu group virus known as Samford was isolated three times from separate pools of *C. brevitarsis* (Doherty *et al.*, 1972). Aino and Samford viruses have recently been shown to be indistinguishable when compared using SN, HI, and CF tests (Miura *et al.*, 1978). Samford virus has been removed from the arbovirus catalogue (Berge, 1975) and represents Australian isolates of Aino virus. Aino virus' major host would appear to be cattle, with some infections in buffaloes and sheep. Antibodies have not been detected in man, camels, dogs, pigs, and marsupials (Doherty *et al.*, 1972; Cybinski and St. George, 1978). In Australia, the distribution of cattle which possess SN antibodies to Aino virus is similar to that for Akabane virus (Cybinski and St. George, 1978) (Fig. 2) and within the area of distribution of *C. brevitarsis*, the probable vector. However, it would appear that Aino virus is less effectively transmitted than Akabane virus, with a lower seroconversion rate in cattle.

There is some evidence that Aino virus may cause congenital HE in calves. In a Japanese study, 2 out of 20 calves with congenital AG/HE possessed SN

TABLE II
Simbu Group Viruses

Virus	Isolated from			Isolated in					
	Mosquitoes	Midges	Vertebrates	Africa	Asia	Aust.	Eur.	N. A.	S. A.
Aino	+	+			+	+			
Akabane	+	+	+(C, S)[c]	+	+	+			
Buttonwillow		+	+					+	
Douglas[a]			+(C)			+			
Faceys Paddock[a]	+					+			
Ingwavuma			+(B, P)	+	+				
Inini			+(B)						+
Kaikalur	+				+				
Manzanilla			+(OP)						+
Mermet			+(B)					+	
Nola	+			+					
Oropouche	+		+(M)						+
Peaton[a]		+	+(C)			+			
Sabo		+	+(C, G)	+					
Sango	+	+	+(C)	+					
Sathuperi	+	+	+(C)	+	+				
Shamonda		+	+(C)	+					
Shuni	+	+	+(C, M, S)	+					
Simbu	+			+					
Thimiri		+	+(B)	+	+	+			
Tinaroo[a]		+				+			
Utinga[b]									
Yaba-7[b]									

[a] Not published in the arbovirus catalogue (Berge, 1975; Karabatsos, 1978). Isolated in Australia by St. George *et al.* (personal communication). Mention of these viruses is not intended to constitute priority.

[b] Not published in the arbovirus catalogue (Berge, 1975; Karabatsos, 1978), but cited by Bishop and Shope (1979).

[c] Isolated from: B, birds; C, cattle; G, goats; M, man; OP, other primates; P, pigs; S, sheep.

antibodies to Aino virus in their precolostrum sera (Miura *et al.*, 1974). In three calves with congenital AG/HE from the New England Tablelands area of New South Wales (Fig. 2), significant levels of SN antibodies against Aino virus were detected in their precolostrum sera (Coverdale *et al.*, 1978). There was also a high level of SN antibody in the cerebrospinal fluid from one of these calves. Further, none possessed SN antibodies against Akabane virus. However, in this Australian outbreak of bovine congenital AG/HE and in the Japanese one, Akabane virus appeared to be the more important causative agent.

The effect of intracerebral inoculation of 2- to 6-month-old calves with Aino virus was studied by Moriwaki *et al.* (1977). The calves were sacrificed at between 2 and 9 days after inoculation and tissues examined for pathological changes. Lesions appeared in the central nervous system soon after inoculation and advanced to a severe disseminated nonpurulent encephalomyelitis. There was only very mild damage in the spinal cord. In contrast, calves infected with Akabane virus showed nonpurulent encephalitis in the brain stem and the spinal cord.

There have been no reported studies of experimental infections of pregnant animals with Aino virus. Hence, the actual role of this virus in congenital bovine AG/HE has yet to be definitely confirmed. There is even less evidence to associate other Simbu group viruses with congenital infections. SN antibodies against Tinaroo virus have been detected in the precolostral serum of a congenitally deformed lamb (St. George, personal communication). Akabane virus is closely related, by serological tests, to Yaba-7 and Aino virus to Shuni virus (Bishop and Shope, 1979); further investigation of these viruses is warranted. As yet, the role of Simbu group viruses, except Akabane virus, in congenital infections has to be conclusively demonstrated.

IV. RIFT VALLEY FEVER

A. History

A virus disease affecting sheep in the Rift Valley of Kenya was described by Daubney *et al.* (1931) (Fig. 7). Those working with the virus in the laboratory

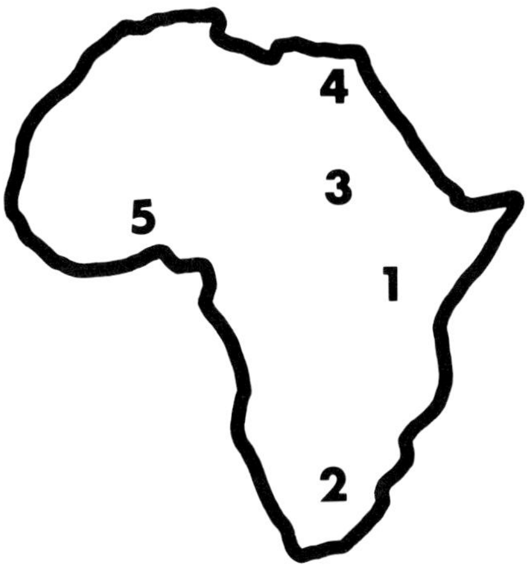

Fig. 7. Map showing the distribution of Rift Valley Fever outbreaks in Africa. (1) First recognized in Kenya in 1931; enzootic in sub-Saharan Africa. (2) South African outbreaks in 1950, 1951, 1953, and 1975. (3) Epizootics in the Sudan in 1973 and 1976. (4) Major animal and human disease in Egypt in 1977 and 1978. (5) Rift Valley Fever virus isolated from *Culicoides* in Nigeria, but no disease recognized.

developed a moderately severe febrile illness, and a similar disease was also recognized in shepherds in contact with infected sheep. Findlay (1932), in a classic study, described the pathological and histological changes produced in animals following natural or laboratory infections. The virus causes abortion and a high mortality in pregnant and newborn sheep, goats, and cattle throughout Central and South Africa (Weiss, 1957; Easterday, 1965) (Fig. 7), but a variety of wild animals, camels, buffaloes, and antelopes are also affected (Gear *et al.*, 1955). Weinbren *et al.* (1957) reported that the wild field rat *Arvicanthis abyssinicus nubilans* circulated RVF virus to high titer in its bloodstream without succumbing to the infection, and they suggested that this rodent and related species might play an important part in the spread of the infection in nature. Later studies have failed to confirm this view, and Davies (1975) pointed out that the primary vertebrate host responsible for the maintenance of RVF virus was still unknown; in his view, neither rodents nor wild bovines played any important role, nor were birds likely to be involved.

B. Epidemiology

A major epizootic of RVF occurred in South Africa (Fig. 7) in 1950 and 1951 in which an estimated 100,000 sheep and cattle died and some 20,000 human infections were recognized (Schultz, 1951; Gear *et al.*, 1955). There was a further substantial epizootic of RVF in South Africa in 1975 in which numerous human infections were recognized, at least four of which were fatal (Van Velden *et al.*, 1977). In 1977 a major extension of the geographic distribution of RVF cases occurred with the recognition of the disease in Egypt (Fig. 7). Estimates of human cases vary from 20,000 to 200,000, and at least 600 deaths were recorded in man. This represents a greatly increased severity and morbidity of the human infections as compared with previous outbreaks in Central and South Africa, and the nature of the infection was not at first recognized. Nor was the association with animal disease realized in the early stages, but on inquiry it became apparent that the human cases were associated with very large numbers of infections in sheep, goats, and camels, with a substantial mortality in all those species. A retrospective serological study carried out on animal and human sera collected for other purposes showed no evidence of RVF activity in Egypt prior to 1977, but sheep, cow, goat, and buffalo sera collected in April 1977, 6 months before the human cases were recognized, were found to have antibodies against the virus (Darwish, 1978; Meegan *et al.*, 1978). Virus isolations were made from man, sheep, a cow, a goat, a rat (*Rattus rattus*), and, for the first time ever recorded, from a horse and a camel. Following the winter of 1977–1978, further infections were recognized in animals and in man the following summer. Epidemiological evidence would incriminate mosquito transmission, with *C.*

pipiens as the most probable species. There was also strong evidence of infection of man from infected animals.

C. Control

Control of RVF virus presents many problems. An inactivated vaccine for use in man has been described by Randall *et al.* (1962, 1963, 1964), mainly for use in controlling possible laboratory infections. Inactivated vaccines have also been used in veterinary practice, as have attenuated virus strains, although sometimes when inoculated into pregnant animals they have produced abortions (Kaschula, 1953; Easterday, 1965) and into congenitally deformed lambs with AG/HE (Coetzer and Barnard, 1977). Mosquito control measures may be more effective in some instances. The possibility that the virus may persist in mosquitoes through transovarial transmission greatly complicates the whole problem of control of the disease in animals and man.

D. Diagnosis

The South African outbreaks of 1950–1951 and the Egyptian outbreaks of 1977–1978 were not recognized as RVF until several months had elapsed with the deaths of thousands of animals, and, in the Egyptian outbreaks, of many deaths in man. In both these instances, delays occurred because the disease was previously unknown in those geographical areas and the possibility of RVF was not at first considered. Kaschula (1957) has suggested that high mortality rates in lambs and calves associated with lower rates in adult sheep and cattle are indicative of RVF, and also that high abortion rates in cows and ewes provide strong supporting evidence, as do signs of liver involvement. In many animal outbreaks there may be acute febrile illnesses in men exposed to sick animals.

Laboratory diagnosis (Weiss, 1957; Easterday, 1965) may be based upon the isolation of virus from acute cases; this provides unequivocal proof of the diagnosis but requires animals (mice, hamsters, ferrets, or lambs are frequently used), cell cultures, or embryonated eggs and also carries with it the considerable risk of laboratory infections, which have been numerous and sometimes severe with this virus. Specific antibodies against RVF virus may be demonstrated by CF, HI, and SN tests or by the use of fluorescent antibody methods. For all these tests, the assistance of an experienced virological laboratory is essential.

V. BHANJA VIRUS INFECTIONS

While investigating a condition known locally as "lumbar paralysis of goats," Shah and Work (1969) isolated a virus from *Haemaphysalis intermedia* ticks collected from a paralyzed goat and named it "Bhanja virus" after the district involved. Antibodies against the virus were found in some goats and some other local animals, but the authors cautiously came to the conclusion that there was insufficient evidence to state that the virus was causally related to the paralytic syndrome. Several isolations of Bhanja virus have been made in Nigeria, mainly from *Boophilus decoloratus* ticks and *Amblyomma variegatum* ticks collected from cattle and goats, and further isolations have been made in Senegal from the blood of cattle, sheep, ground squirrel, and hedgehog, but no clear relation to animal disease was noted in West Africa. In Italy, Verani *et al.* (1970) reported the isolation of Bhanja virus from *H. punctata* ticks collected from a mountain district south of Rome where some 72% of goats were found to have Bhanja virus antibodies in the absence of any recognized disease. Under experimental laboratory conditions, Balducci *et al.* (1970) demonstrated neurotropism with Bhanja virus when they found that four out of five *M. mulatta* monkeys died following intrathecal inoculation. Across the Adriatic Sea, foci of Bhanja virus infection were found on the island of Brac and on the mainland of Yugoslavia, where many sheep and goats have antibodies against Bhanja virus (Vesenjak-Hirjan *et al.*, 1977). Following these studies in Yugoslavia, a laboratory worker contracted an infection with moderately severe neurological involvement (Calisher and Goodpasture, 1975), and at least two naturally acquired Bhanja virus infections of man have ended in death after severe neurological disease (Vesenjak-Hirjan *et al.*, 1978). The presence of Bhanja virus has also been reported from Armenia, Bulgaria, Egypt, and Somalia, but without any clear association with disease in animals or man.

The diagnosis of Bhanja virus is dependent upon laboratory tests involving either the isolation of the virus in mice or cell cultures or the demonstration of a significant rise in Bhanja virus antibodies.

VI. NAIROBI SHEEP DISEASE

An acute hemorrhagic gastroenteritis affecting sheep and goats in East Africa was described by Montgomery (1917), who presented evidence that the causal agent was a virus transmitted by *Rhipicephalus marginatus* ticks. In some outbreaks the mortality in both local and exotic sheep can be as high as 70–90%, but goats are less severely affected. In experimental studies, horses, cattle, Indian buffaloes, donkeys, pigs, dogs, rabbits, guinea pigs, white mice, and rats were unaffected (Montgomery, 1917). Daubney and Hudson, (1934) reported that the African field rat, *Arvicanthus abysinicus nubilans,* developed a viremia following experimental infection and suggested that this species might serve as a reservoir of the virus in nature. Two Blue Duikers, *Cephalohus caerulus* in a local zoo, were reported to have acquired infections with Nairobi sheep disease virus (Weinbren, 1958).

In virus-infected sheep there is an incubation period of 1–3 days, after which the temperature rises to 40°–41°C, dropping as diarrhea develops. Internal hemorrhages are common, and blood-stained exudates may be evident around the nostrils and on the hind quarters. The liver is not affected, but the gall bladder may be enlarged and occasionally hemorrhagic (Terpstra, 1969). Nephritis and myocardial degeneration have been reported by Mugera and Chema (1967).

Although the diagnosis may be suspected on clinical grounds, laboratory tests are required for a firm diagnosis. The method of choice is virus isolation in hamster kidney (BHK 21) cells followed by fluorescent antibody tests for identification; this can give a positive result in 24–48 hours (Davies *et al.*, 1977). Alternatively, suckling mice may be inoculated and sick mice examined for specific antigen using CF or fluorescent antibody techniques. During the acute febrile stage of the disease, blood is the most useful source of virus, but in the later stages or after death, spleen and mesenteric lymph nodes are more likely to yield virus than is blood. The virus is highly labile, and material for attempted virus isolation should be submitted to the laboratory on ice and preferably in phosphate-buffered saline at pH 7.2.

VII. GANJAM VIRUS INFECTIONS

A virus isolated from *Haemaphysalis intermedia* ticks collected from healthy goats in the Ganjam district, Orissa, India, was found to be different from other arboviruses and was named "Ganjam virus" (Dandawate and Shah, 1969). No disease was evident in the goats, but serological surveys revealed widespread antibodies against Ganjam virus in sera from sheep and goats collected in Orissa, Mysore, Gujar, and Kashmir (Boshell *et al.*, 1970). The virus was also isolated from the bird tick, *H. wellingtoni*, suggesting a possible avian involvement in the natural history of this virus in India (Rajagopalan *et al.*, 1970).

Laboratory studies have since revealed a close relationship, possibly amounting to identity, between Ganjam and Nairobi sheep disease viruses, with a somewhat more distant relationship between these two and Dugbe virus.

VIII. DUGBE VIRUS INFECTIONS

Dugbe virus was first isolated in Nigeria from pools of *Amblyomma variegatum* ticks collected from healthy white Fulani cattle at market in Ibadan; many isolations have been made from tick pools and cattle blood in West Africa

(Causey *et al.*, 1971; Kemp *et al.*, 1971) and from tick pools in East Africa (Tukei *et al.*, 1970), but in no instance has there been any association with disease in animals. As mentioned above, there is a serological relationship between Dugbe and Ganjam viruses, but the reaction is not one of identity.

IX. CRIMEAN-CONGO VIRUSES

These viruses are responsible for serious and sometimes fatal infections of man in Europe, Africa, and Asia, acquired through the bites of ticks, which in turn become infected by feeding upon viremic animals in normally silent foci of infection. Crimean hemorrhagic fever was first recognized in 1946 (Chumakov, 1946), and the causative virus is now known to have a complex life history involving *Hyalomma anatolicum* ticks and domestic animals as well as *H. plumbeum plumbeum* ticks and wild mammals and birds (Chumakov, 1975). In Africa, viruses which became known as "Congo viruses" were isolated from people with febrile illnesses in Central and East Africa (Simpson *et al.*, 1967; Woodall *et al.*, 1967). In West Africa, Congo viruses have been isolated from *Hyalomma, Amblyomma* and *Boophilus* ticks taken from symptom-free market animals, from the blood of similar animals, and from pooled liver and spleen of the hedgehog (Causey *et al.*, 1970).

In Asia, a virus apparently identical to Congo virus was isolated in West Pakistan from tick pools collected from a variety of wild and domestic animals, including cows, goats, camels, rodents, and the pika (*Ochotona roylei*). A related but serologically distinct virus, Hazara virus, was isolated from *Ixodes redikorzevi* ticks collected from the vole *Alticola roylei* in a subarctic habitat at an altitude of 12,000 feet in the Kaghan valley of the Hazara district (Begum *et al.*, 1970a,b).

The link between these viruses affecting animals and man in very different geographical regions and ecological situations came from Casals (1969), who demonstrated that Crimean hemorrhagic fever virus was antigenically indistinguishable from Congo virus, and these observations have been extended and confirmed by Chumakov *et al.* (1970) and Buckley (1974). The diagnosis of Congo virus infections is dependent upon laboratory studies; virus isolation attempts have been followed by a number of laboratory infections. The CF test is probably the most useful serological test.

REFERENCES

Anderson, A. A., and Campbell, C. H. (1978). *Am. J. Vet. Res.* **39,** 301–304.
Balducci, M., Verani, P., Lopes, M. C., and Nardi, F. (1970). *Acta Virol.* (*Engl. Ed.*) **14,** 237–243.
Barnard, B. J. H., and Pienaar, J. G. (1976). *Onderstepoort J. Vet. Res.* **43,** 155–158.

Begum, F., Wiseman, C. L., Jr., and Casals, J. (1970a). *Am. J. Epidemiol.* **92,** 192–194.
Begum, F., Wiseman, C. L., Jr., and Trub, R. (1970b). *Am. J. Epidemiol.* **92,** 180–191.
Berge, T. O., ed. (1975). "International Catalogue of Arboviruses Including Certain Other Viruses of Vertebrates," Publ. No. CDC 75-8301. U.S. Dep. Health, Educ., Welfare, Washington D.C.
Bishop, D. H. L., and Shope, R. E. (1979). *Compr. Virol.* **14,** 1–156.
Blood, D. C. (1956). *Aust. Vet. J.* **32,** 125–131.
Bonnor, R. B., Mylrea, P. J., and Doyle, B. J. (1961). *Aust. Vet. J.* **37,** 160.
Boshell, J., Desai, P. K., Dandawate, C. N., and Goverdhan, M. K. (1970). *Indian J. Med. Res.* **58,** 561–562.
Buckley, S. (1974). *Proc. Soc. Exp. Biol. Med.* **146,** 594–600.
Calisher, C. H., and Goodpasture, H. C. (1975). *Am. J. Trop. Med. Hyg.* **24,** 1040–1042.
Casals, J. (1969). *Proc. Soc. Exp. Biol. Med.* **131,** 233–236.
Causey, O. R., Kemp, G. E., Madbouly, M. H., and David-West, T. S. (1970). *Am. J. Trop. Med.* **19,** 846–850.
Causey, O. R., Kemp, G. E., Casals, J., Williams, R. W., and Madbouly, M. H. (1971). *Niger. J. Sci.* **5,** 41–43.
Chumakov, M. P. (1946). *Vestn. Akad. Nauk SSSR* **2,** 19–26.
Chumakov, M. P. (1975). *In* "Crimean Haemorrhagic Fever," pp. 13–45. Simferopol, USSR. (In Russ.)
Chumakov, M. P., Smirnova, S. E., and Tkachenko, E. A. (1970). *Arch. Virol.* **14,** 82–85.
Clarke, D., and Casals, J. (1958). *Am. J. Trop. Med. Hyg.* **7,** 561–573.
Coetzer, J. A. W., and Barnard, B. J. H. (1977). *Onderstepoort J. Vet. Res.* **44,** 119–126.
Cordy, D. R., and Shultz, G. (1961). *J. Neuropathol.* **20,** 554–562.
Coverdale, O. R., Cybinski, D. H., and St. George, T. D. (1978). *Aust. Vet. J.* **54,** 151–152.
Cybinski, D. H., and St. George, T. D. (1978). *Aust. Vet. J.* **54,** 371–373.
Cybinski, D. H., St. George, T. D., and Paull, N. I. (1978). *Aust. Vet. J.* **54,** 1–3.
Dandawate, C. N., and Shah, K. V. (1969). *Indian J. Med. Res.* **57,** 799–804.
Darwish, M. A. (1980). *Proc. FEMS. Symp., 6th, Brac. Yugosl.,* pp. 101–111.
Daubney, R., and Hudson, J. R. (1934). *Parisitology* **26,** 496.
Daubney, R., Hudson, J. R., and Granham, P. C. (1931). *J. Pathol. Bacteriol.* **34,** 545–579.
Davies, F. G. (1975). *J. Hyg.* **75,** 219–230.
Davies, F. G., Mungai, J. N., and Taylor, M. (1977). *Trop. Anim. Health Prod.* **9,** 75–80.
Della-Porta, A. J., Murray, M. D., and Cybinski, D. H. (1976). *Aust. Vet. J.* **52,** 496–502.
Della-Porta, A. J., O'Halloran, M. L., Parsonson, I. M., Snowdon, W. A., Murray, M. D., Hartley, W. J., and Haughey, K. J. (1977). *Aust. Vet. J.* **53,** 51–52.
Doherty, R. L., Carley, J. G., Standfast, H. A., Dyce, A. L., and Snowdon, W. A. (1972). *Aust. Vet. J.* **48,** 81–86.
Doherty, R. L., St. George, T. D., and Carley, T. D. (1973). *Aust. Vet. J.* **49,** 574–579.
Easterday, B. C. (1965). *Adv. Vet. Sci.* **10,** 65–127.
Findlay, G. M. (1932). *Trans. R. Soc. Trop. Med. Hyg.* **25,** 157–160.
Gear, J. H. S., de Meillon, B., Le Roux, A. F., Rofsky, R., Rose-Innes, R., Steyn, J. J., Cliff, W. D., and Schutz, K. H. (1955). *S. Afr. Med. J.* **29,** 514–515.
Gentsch, J. R., and Bishop, D. H. L. (1979). *J. Virol.* **30,** 767–770.
Gentsch, J., Wynne, L. R., Clewley, J. P., Shope, R. E., and Bishop, D. H. L. (1977). *J. Virol.* **24,** 893–902.
Goto, Y., Inaba, Y., Kurogi, H., Takahashi, E., Sato, K., Omori, T., Hanaki, T., Sazawa, H., and Matumoto, M. (1976). *Vet. Microbiol.* **1,** 449–458.
Goto, Y., Inaba, Y., Miura, Y., Kurogi, H., Takahashi, E., Sato, K., Omori, T., Hanaki, T., Sazawa, H., and Maturoto, M. (1978). *Vet. Microbiol.* **3,** 89–99.

Griner, L. A., McCrory, B. R., Foster, N. M., and Meyer, H. (1964). *J. Am. Vet. Med. Assoc.* **145,** 1013-1019.
Hartley, W. J., and Haughey, K. G. (1974). *Aust. Vet. J.* **50,** 55-58.
Hartley, W. J., and Wanner, R. A. (1974). *Aust. Vet. J.* **50,** 185-188.
Hartley, W. J., Wanner, R. A., Della-Porta, A. J., and Snowdon, W. A. (1975). *Aust. Vet. J.* **51,** 103-104.
Hartley, W. J., De Sarum, W. G., Della-Porta, A. J., Snowdon, W. A., and Shepherd, N. C. (1977). *Aust. Vet. J.* **53,** 319-325.
Hindmarsh, W. L. (1937). *Vet. Res. Rep., N.S.W. Dep. Agric.* **7,** 58.
Hubbert, W. T., Bryner, J. H., Estes, P. C., and Foley, J. W. (1972). *Am. J. Vet. Res.* **33,** 1879-1882.
Inaba, Y., Kurogi, H., and Omori, T. (1975). *Aust. Vet. J.* **51,** 584-585.
James, L. F., Shupe, J. L., Binns, W., and Keeler, R. F. (1967). *Am. J. Vet. Res.* **28,** 1379-1388.
Jochim, M. M., Luedke, A. H., and Chow, T. L. (1974). *Am. J. Vet. Res.* **35,** 517-522.
Kalmar, E., Peleg, B. A., and Savir, D. (1975). *Refu. Vet.* **32,** 47-54.
Karabatsos, N. (1978). *Am. J. Trop. Med. Hyg.* **27,** 371-440.
Kaschula, V. R. (1953). D.V.Sc. Thesis, Univ. of Pretoria, Pretoria.
Kaschula, V. R. (1957). *J. Am. Vet. Med. Assoc.* **131,** 219-221.
Kay, B. H., Carley, J. G., and Filippich, C. (1975). *J. Med. Entomol.* **12,** 279-283.
Kemp, G. E., Causey, O. R., Moore, D. L., and O'Connor, E. H. (1971). *Am. J. Vet. Res.* **34,** 707-710.
Konno, S., Moriwaki, M., Nakagawa, M., Uchimura, M., Kamimiyata, M., and Tojinbara, K. (1975). *Natl. Inst. Anim. Health Q.* **15,** 52-53.
Kurogi, H., Inaba, Y., Goto, Y., Muira, Y., Takahashi, H., Sato, K., Omori, T., and Matumoto, M. (1975). *Arch. Virol.* **47,** 71-84.
Kurogi, H., Inaba, Y., Takahashi, E., Sato, K., Omori, T., Muira, Y., Goto, Y., Fujiwaka, Y., Hatano, Y., Kodama, K., Fukuyama, S., Sasaki, N., and Matumoto, M. (1976). *Arch. Virol.* **51,** 67-74.
Kurogi, H., Inaba, Y., Takahaski, E., Sato, K., Goto, K., and Omori, T. (1977a). *Natl. Inst. Anim. Health Q.* **17,** 1-9.
Kurogi, H., Inaba, Y., Takahashi, E., Sato, K., Satoda, K., Goto, Y., Omori, T., and Matumoto, M. (1977b). *Infect. Immun.* **17,** 338-343.
Kurogi, H., Inaba, Y., Takahashi, E., Sato, K., Satoda, K., and Omori, T. (1977c). *Natl. Inst. Anim. Health Q.* **17,** 27-28.
Kurogi, H., Inaba, Y., Tukahashi, E., Sato, K., Akashi, H., Satoda, K., and Omori, T. (1978a). *Natl. Inst. Anim. Health Q.* **18,** 1-7.
Kurogi, H., Inaba, Y., Takahashi, E., Sato, K., Goto, Y., Satoda, K., Omori, T., and Hatakeyama, H. (1978b). *Natl. Inst. Anim. Health Q.* **18,** 97-108.
Leipold, H. W., Cates, W. F., Radostitis, D. M., and Howell, W. E. (1969). *Can. Vet. J.* **10,** 268-273.
Leipold, H. W., Dennis, S. M., and Hutson, K. (1972). *Adv. Vet. Sci. Comp. Med.* **61,** 103-150.
McKercher, D. G., Saito, J. K., and Singh, K. V. (1970). *J. Am. Vet. Med. Assoc.* **156,** 1044-1047.
Markusfeld, O., and Mayer, E. (1971). *Refu. Vet.* **28,** 51-61.
Matsuyama, T., Oya, A., Ogata, T., Kobayashi, I., Nakamura, T., Takahashi, M., and Kitaoka, M. (1960). *Jpn. J. Med. Sci. Biol.* **13,** 191-198.
Meegan, J. M., Hoogstraal, H., and Laughlin, L. W. (1980). *Proc. FEMS. Symp., 6th, Brac. Yugosl.* pp. 179-183.
Metselaar, D., and Robin, Y. (1976). *Vet. Rec.* **99,** 86.
Miura, Y., Hayashi, S., Ishihara, T., Inaba, Y., Omori, T., and Matumoto, M. (1974). *Arch. Gesamte Virusforsch.* **46,** 377-380.

Miura, Y., Inaba, Y., Goto, Y., Takahashi, E., Kurogi, H., Hayashi, S., Omori, T., and Matumoto, M. (1978). *Microbiol. Immunol.* **22,** 651-654.
Montgomery, R. E. (1917). *J. Comp. Pathol. Ther.* **30,** 28.
Moriwaki, M., Miura, Y., Hayashi, S., and Ishitani, R. (1977). *Natl. Inst. Anim. Health Q.* **17,** 95-106.
Mugera, G. M., and Chema, S. (1967). *Bull. Epizoot. Dis. Afr.* **15,** 337.
Muller, M. J. (1977). *Annu. Rep., CSIRO, Div. Anim. Health* pp. 62-63.
Muller, M. J., and Murray, M. D. (1977). *Aust. J. Zool.* **25,** 75-85.
Murray, M. D. (1975). *Aust. Vet. J.* **51,** 216-220.
Murray, M. D., and Della-Porta, A. J. (1976). *Annu. Rep., CSIRO, Div. Anim. Health* pp. 30.
Narita, M., Inui, S., and Hashiguchi, Y. (1979). *J. Comp. Pathol.* **89,** 229-240.
Nobel, T. A., Klopfer, U., and Neuman, F. (1971). *Refu. Vet.* **28,** 144-151.
Obijeski, J. F., and Murphy, F. A. (1977). *J. Gen. Virol.* **37,** 1-14.
O'Halloran, M. L., and Della-Porta, A. J. (1975). *Annu. Rep., CSIRO, Div. Anim. Health* p. 17.
Omori, T., Inaba, Y., Kurogi, H., Miura, Y., Nabuto, K., Oshashi, Y., and Matumoto, M. (1974). *Bull. Off. Int. Epizoot.* **81,** 447-458.
Osburn, B. I., Silverstein, A. M., Prendergast, R. A., Johnson, R. T., and Parshall, C. J. (1971a). *Lab. Invest.* **25,** 197-205.
Osburn, B. I., Silverstein, A. M., Prendergast, R. A., Johnson, R. T., and Parshall, C. J. (1971b). *Lab. Invest.* **25,** 206-210.
Oya, A. (1971). Cited in Berge (1975), p. 89.
Oya, A., Okuno, T., Ogata, T., Kobayashi, I., and Matsuyama, T. (1961). *Jpn. J. Med. Sci. Biol.* **14,** 101-108.
Parsonson, I. M., Della-Porta, A. J., Snowdon, W. A., and Murray, M. D. (1975). *Aust. Vet. J.* **51,** 585-586.
Parsonson, I. M., Della-Porta, A. J., and Snowdon, W. A. (1976). *Proc. Annu. Conf. Aust. Vet. Assoc.* **53,** 90-93.
Parsonson, I. M., Della-Porta, A. J., and Snowdon, W. A. (1977a). *Infect. Immun.* **15,** 254-262.
Parsonson, I. M., Snowdon, W. A., and Della-Porta, A. J. (1977b). *Annu. Rep., CSIRO, Div. Anim. Health* pp. 19.
Porterfield, J. S., Casals, J., Chumakov, M. P., Gaidamovich, S. Y., Hannoun, C., Holmes, I. H., Horzinek, M. C., Mussgay, M., Oker-Blom, N., and Russell, P. K. (1975-1976). *Intervirology* **6,** 31-24.
Rajagopalan, P. K., Sreenivasan, M. A., and Paul, S. D. (1970). *Indian J. Med. Res.* **58,** 1195-1196.
Randall, R., Gibbs, C. J., Aulisio, C. G., Binn, L. N., and Harrison, V. R. (1962). *J. Immunol.* **89,** 660-671.
Randall, R., Binn, L. N., and Harrison, V. R. (1963). *Am. J. Trop. Med. Hyg.* **12,** 611-615.
Randall, R., Binn, L. N., and Harrison, R. V. (1964). *J. Immunol.* **93,** 293-299.
Richards, W. P. C., Crenshaw, G. L., and Bushnell, R. B. (1971). *Cornell Vet.* **61,** 336-348.
St. George, T. D., Cybinski, D., and Paull, N. I. (1977). *Aust. Vet. J.* **53,** 249.
St. George, T. D., Standfast, H. A., and Cybinski, D. H. (1978). *Aust. Vet. J.* **54,** 558-561.
Saperstein, G., Leipold, H. W., and Dennis, S. M. (1975). *J. Am. Vet. Med. Assoc.* **167,** 314-322.
Schmidt, R. E., and Panciera, R. J. (1973). *J. Am. Vet. Med. Assoc.* **162,** 567-568.
Schulz, K. H. (1951). "Special Report of May, 1951." Union Dep. Health, Plague Res. Lab., Johannesburg, South Africa.
Schultz, R. D. (1973). *Cornell Vet.* **63,** 507-535.
Sellers, R. F., and Herniman, K. A. J. (1981). *Trop. Anim. Hlth. Prod.* **13,** 57-60.
Shah, K. V., and Work, T. H. (1969). *Indian J. Med. Res.* **57,** 793-798.

Shepherd, N. C., Gee, C. D., Jessop, T., Timmins, G., Carrol, S. N., and Bonner, R. B. (1978). *Aust. Vet. J.* **54,** 171-177.
Shope, R. E., Peters, C. J., and Walker, J. S. (1980). *Lancet* **i,** 886-887.
Shultz, G., and Delay, P. D. (1955). *J. Am. Vet. Med. Assoc.* **127,** 224-226.
Shupe, J. L., James, L. F., and Binns, W. (1967a). *J. Am. Vet. Med. Assoc.* **151,** 191-197.
Shupe, J. L., Binns, W., James, L. F., and Keeler, R. F. (1967b). *J. Am. Vet. Med. Assoc.* **151,** 198-203.
Simpson, D. I. H., Knight, E. M., Courtois, G., Williams, M. C., Weingren, M. P., and Kibukamusoke, J. W. (1967). *East Afr. Med. J.* **44,** 87-92.
Smithburn, K. C., Haddow, A. J., and Mahaffy, A. F. (1946). *Am. J. Trop. Med.* **26,** 189-208.
Spradbrow, P. B., and Miah, A. H. (1978). *Int. Virol.* **4,** 167.
Standfast, H. A., and Whelan, I. (1976). *Annu. Rep., CSIRO, Div. Anim. Health* pp. 33-34.
Takahashi, E., Inaba, Y., Kurogi, H., Sato, K., Goto, Y., Ito, Y., Omori, T., and Matumoto, M. (1978). *Vet. Microbiol.* **3,** 45-54.
Takahashi, K., Oya, A., Okada, T., Matsuo, R., Kuma, M., and Noguchi, H. (1968). *Jpn. J. Med. Sci. Biol.* **21,** 95-101.
Terpstra, C. (1969). Thesis, p. 172. Univ. of Utrecht. Doctor of Veterinary Medicine.
Theodoridis, A., Neville, E. M., Els, H. J., and Boshoff, S. T. (1979). *Onderstepoort J. Vet. Res.* **46,** 191-198.
Trainin, Z., and Meirom, R. (1973). *Res. Vet. Sci.* **15,** 1-7.
Tukei, P. M., Williams, M. C., Mukwaya, L. G., Henderson, B. E., Kafuko, G. W., and McCrae, A. W. R. (1970). *East Afr. Med. J.* **47,** 265-272.
Van Velden, D. J. J., Meyer, J. D., Oliver, J., Gear, H. J. S., and McIntosh, B. (1977). *S. Afr. Med. J.* **51,** 867-871.
Sellers, R. F., and Hermman, K. A. J. (1981). *Trop. Anim. Hlth. Prodn.* **13,** 57-60.
Verani, P., Balducci, M., and Lopes, M. C. (1970). *Folia Parasitol. (Prague)* **17,** 367-374.
Vesenjak-Hirjan, J., Calisher, C. H., Brudnajak, Z., Toronik, D., Skrtic, N., and Lazuck, J. S. (1977). *Am. J. Trop. Med. Hyg.* **26,** 1003-1008.
Vesenjak-Hirjan, J., Calisher, C. H., Beus, I., and Marton, E. (1980). *Proc. FEMS Symp., 6th, Brac. Yugosl.* pp. 297-301.
Wanner, R. A., and Husband, A. J. (1974). *Aust. Vet. J.* **50,** 560-562.
Watts, D. M., Pautuwatanta, A., De Foliart, G. R., Yuill, T. M., and Thompson, W. (1973). *Science (Washington, D.C.)* **182,** 1140-1141.
Weinbren, M. P. (1958). *Annu. Rep. East Afr. Virus Res. Inst.* **8,** 8.
Weinbren, M. P., and Mason, P. J. (1957). *S. Afr. Med. J.* **31,** 427-430.
Weinbren, M. P., Williams, M. C., and Haddow, A. J. (1957). *S. Afr. Med. J.* **31,** 951-957.
Weiss, K. E. (1957). *Bull. Epizoot. Dis. Afr.* **5,** 431-458.
Whittem, J. H. (1957). *J. Pathol. Bacteriol.* **73,** 375-387.
Woodall, J. P., Williams, M. C., and Simpson, D. I. H. (1967). *E. Afr. Med. J.* **44,** 93-98.
Young, J. S. (1969). *Aust. Vet. J.* **45,** 574-576.
Young, S., and Cordy, D. R. (1964). *J. Neuropathol. Exp. Neurol.* **23,** 635-659.

Part VIII

ARENAVIRIDAE

Chapter 11

Arenaviruses: Diagnosis of Infection in Wild Rodents

KARL M. JOHNSON

I.	Introduction	511
II.	Virus-Rodent Specificity	512
	A. Viruses Producing Disease in Man	512
	B. Viruses Not Producing Disease in Man	514
III.	Comparative Biology	515
	A. LCM Virus	515
	B. Lassa Virus	517
	C. Machupo and Junin Viruses	518
	D. Other Arenaviruses	519
IV.	Comparative Diagnosis	520
	A. Detection of Virus or Viral Antigens	520
	B. Detection of Immunity to Arenavirus Infection	522
V.	Conclusions	523
	References	524

I. INTRODUCTION

Ten viruses are presently classified as arenaviruses on the basis of shared structural and physicochemical properties (Fenner, 1976). These agents possess a genome composed of two segments of RNA, are membrane bound, and have no defined internal symmetry but contain a variable number of electron-dense host cell ribosomes. This unique property, which gives the round to pleomorphic 50–300-nm virion the appearance of containing granules of sand, was used to name the taxon (*arenosus* = L. sandy). Recent work, however, has demonstrated that the ribosomes are not critical for infectivity and that the viruses can

COMPARATIVE DIAGNOSIS
OF VIRAL DISEASES, VOL. IV
ISBN 0-12-429704-8

be manipulated *in vitro* to produce infectious virion devoid of ribosomes (Leong and Rawls, 1977; Bishop, personal communication).

Except for Tacaribe virus (bats), individual arenaviruses are natural parasites of one or a limited number of wild rodents and are perpetuated by intraspecific horizontal and vertical transmission from rodent to rodent without biological intervention of arthropod vectors. The basic properties, biology, and diagnosis of arenavirus infections have been described in Volume I, Chapter 17, of this series by F. A. Murphy. An extensive literature on experimental infection of the prototype arenavirus, lymphocytic choriomeningitis (LCM), in laboratory-reared mice, has been reviewed in detail elsewhere (Lehmann-Grube, 1971; Pedersen, 1979). This chapter is directed toward a more detailed description of the biology of arenaviruses in natural rodent hosts and the appropriate procedures for detecting infection in wild rodents. Emphasis is placed on those arenaviruses known to be transmitted from rodent to man: LCM, Junin, Machupo, and Lassa viruses.

II. VIRUS-RODENT SPECIFICITY

Arenaviruses are highly host-specific agents. Thus, elucidation of the important variables in the life cycle of these viruses and their role as pathogens for man rests upon careful examination of the rodent–parasite relationship, employing the natural virus host species for experimental work, and upon a comprehensive understanding of the ecology of the relevant rodent hosts. It is now clear that the experimental biology of arenaviruses must always be defined in terms of the vertebrate host selected for investigation. It also has become evident that natural virulence of a given arenavirus for man may depend not only upon the inherent pathogenicity of the agent but also upon the behavior of the natural rodent host, which determines whether or not infected rodents come into frequent contact with humans. The recognized rodent–arenavirus pairings are summarized in Table I. It is noteworthy that those agents which cause natural human infection are all found in rodent species found closely associated with man.

A. Viruses Producing Disease in Man

LCM virus is a natural parasite of *Mus musculus*. This common house mouse, which is thought to have originated in Middle Asia, is now found in close association with man on every continent. Nevertheless, LCM infection in *Mus* has not been demonstrated conclusively in Africa or Australasia. Whether this phenomenon is an artifact secondary to limited search for LCM virus on these continents or is the historic result of specific patterns of human migration from Asia and Europe in the past 5 centuries is problematic. Nevertheless, it remains true that LCM virus has almost never been found in nature in animals other than

TABLE I
Natural Hosts of Arenaviruses

Virus	Reservoir host	Geographic distribution host	Ecological distribution host	Geographic distribution virus	Major human contact	Human disease (virus)
LCM	*Mus musculus*	Worldwide	Domestic, peridomestic	World, except African, Australasia	+	+
Junin	*Calomys musculinus*, *C. laucha*, *Akodon azerae*	Argentina, Uruguay, Paraguay	Grasslands, cultivars	Argentina	+	
Machupo	*Calomys callosus*	Bolivia, Brazil, Paraguay, Peru	Grasslands forest edge, periodomestic	Bolivia	+	+
Lassa	*Mastomys natalensis*	Most of Africa	Savannah, periodomestic, domestic	West Africa	+	+
Amapari	*Oryzomys goeldii*, *Neacomys guianae*	Brazil, Peru to Venezuela	Tropical forest	Brazil	–	–
Latino	*Calomys callosus*	Bolivia, Brazil, Paraguay, Peru	Grassland forest edge, periodomestic	Bolivia, Brazil	+	–
Parana	*Oryzomys buccinatus*	Paraguay, Argentina, Brazil, Bolivia	Tropical forest–savannah border	Paraguay	–	–

M. musculus (Küpper *et al.*, 1964). In recent years, however, the virus has been transmitted to man from silently infected Syrian hamsters in Europe and North America, an epidemiological phenomenon related to the growth of the biological research industry and the associated market in hamsters as pets (Lewis *et al.*, 1965; Baum *et al.*, 1966; Ackermann *et al.*, 1972; Hotchin *et al.*, 1974; Hinman *et al.*, 1975; Biggar and Douglas, 1975).

Machupo virus, the etiological agent of Bolivian hemorrhagic fever (Johnson *et al.*, 1965), is a single host parasite of the Cricetine rodent, *Calomys callosus*. This mouse is found in the margin between grassland and riverine gallery forest in eastern Bolivia and neighboring areas of Brazil, Paraguay, Argentina, and Peru (Hershkovitz, 1959). It is quite willing to enter gardens and human dwellings, however, and in the absence of other peridomestic rodents, such as *Mus* or *Rattus*, it is the dominant rodent found in houses of the savannah region of central South America. Interestingly, Machupo virus has only been recovered from *C. callosus* in the relatively restricted region of northeastern Beni Province in Bolivia (Webb *et al.*, 1973).

Junin virus, which causes Argentine hemorrhagic fever, has been repeatedly isolated from three Cricetine rodents in Argentina: *Calomys musculinus, C. laucha,* and *Akondon azerae* (Sabattini *et al.*, 1977). None of these rodents is encountered often in houses, but all three are widely distributed in the cultivated humid pampa of east-central Argentina. *C. musculinus* achieves the highest population density and the highest virus infection rate of the three species in cultivars of maize where most human infections are acquired, but it is possible that *A. azerae* is the original rodent host of Junin virus because this species is dominant in the unmodified ecotome of the grassland pampa (deVillafañe *et al.*, 1977).

Lassa virus has so far been recovered only from the savannah rodent, *Mastomys natalensis* (Monath, 1975). This "multimammate" large mouse is widely distributed in close association with man throughout much of Africa south of the Sahara. Lassa virus, however, has been detected only in West Africa, although a closely related agent has been recovered from *M. natalensis* in Mozambique (Wulff *et al.*, 1977). The taxonomy of this rodent is under reinvestigation. Distinct forms having 32, 36, or 38 chromosomes are found in different portions of the continent, and in most localities, two of these forms exist in sympatric association (Tranier, 1974; Green *et al.*, 1978). Whether the biology of Lassa virus infection patterns differs in these apparently genetically distinct species is a question under current study.

B. Viruses Not Producing Disease in Man

With two exceptions, the remaining arenaviruses are single host parasites. Amapari virus has been isolated repeatedly from two rodents of the Amazon forest in Brazil, *Oryzomys goeldii* and *Neacomys quianae* (Pinheiro *et al.*, 1966), and Tacaribe virus was isolated in Trinidad from two species of fruit-eating bats, *Artibeus jamaicensis* and *A. lituratus* (Downs *et al.*, 1963). Most of the nonpathogenic arenaviruses have rodent hosts adapted to forest or unmodified savannah habitats where opportunity for direct contact with humans is minimal. *Calomys callosus* and *Sigmodon hispidus*, the reservoirs of Latino and Tamiami viruses, respectively, are closely linked to man-disturbed ecosystems, a fact which strongly suggests that these agents are not inherently pathogenic for humans. Nevertheless, at least two of this group of arenaviruses, Pichinde and Tacaribe, have caused infection in laboratory workers (Buchmeier *et al.*, 1974; Bishop and Casals, personal communication), and this fact warrants laboratory containment at the P-3 level when large volumes or high concentrations of nonpathogenic arenaviruses are generated.

III. COMPARATIVE BIOLOGY

The comparative biology of host-specific parasites must consider the biology of the host as carefully as that of the parasite. The only valid laboratory host model is the natural one. Thus, data available to date concerning the arenaviruses are the uneven result of several important variables: ability to colonize the host rodent; financial support available for such work, which has often been determined by the pathogenicity of a given virus for humans; and the availability of high-containment laboratory facilities for work with some of the more human virulent arenaviruses. The largest amount of information exists for the LCM–*M. musculus* pairing, although it is important to remember that the albino laboratory mice used for all of the experimental studies are quite different from wild *Mus*. Study of this model long ago elucidated a basic pattern of chronic infection with chronic excretion of virus in the urine and saliva and both horizontal and vertical intraspecific virus transmission in mice. This section is devoted to an examination of the variations on this model so far observed for different arenaviruses together with considerations leading to mouse-to-man transmission of those agents pathogenic for humans.

A. LCM Virus

The response of laboratory mice to LCM virus infection is influenced by both the mouse and virus strain and by the age of the host and route of virus administration. The classic pattern when a neurovirulent virus strain such as Armstrong is used is one of chronic, largely asymptomatic "tolerant" infection in which mice are infected by any route when less than 5 days of age. Such carrier mice infect other mice by contact and give birth to offspring which are infected as early as the oval stage of reproduction. For many years, it was felt that such mice did not produce anti-LCM antibodies; hence the primordial and celebrated concept of immune tolerance was generated. More recently it was demonstrated that such mice do, in fact, develop virus-specific antibodies which circulate in the blood complexed with virus (such animals are chronically viremic). To a variable degree, these complexes are deposited in tissues, particularly the kidney, with resultant slow development of late glomerulonephritis and a shortened life span (Hotchin, 1962; Oldstone and Dixon, 1969, 1970). Carrier mice do, however, have a virus-induced defect in specific cellular immunity, as evidenced by the ability of syngeneic LCM-immune lymphocytes to abolish the carrier state and by quantitative reduction in T-lymphocyte immune responses in carrier mice (Gilden *et al.*, 1972; Oldstone, 1973; Volkert *et al.*, 1974).

Adult mice, in contrast, when inoculated by the intracerebral route with Armstrong virus, evince a fulminant choriomeningitis and usually succumb within 6–14 days after infection. This disease is mediated by immune lymphocytes, as demonstrated by mouse survival when a variety of immunosuppressants are given, the most specific being anti-T-lymphocyte antiserum (Gledhill, 1967; Hirsch *et al.*, 1967; Cole *et al.*, 1973). When adult mice are inoculated by nonneural routes with LCM virus, the most common response is a self-limited, silent infection marked by residual permanent immunity to LCM challenge by any route (Lehmann-Grube, 1971).

Chronic LCM infection is by no means innocuous for *M. musculus*. Adult females are not as fecund as uninfected controls; infected embryos may die and be resorbed, and infected offspring grow more slowly, have higher natural mortality, and may exhibit anemia and splenomegaly not found in normal mice (Mims, 1969).

How much of this complex biology applies to natural infection of feral *M. musculus* by LCM virus? We have no direct answer because appropriate experimental studies have not been carried out. Limited field work, done in connection with the occurrence of human LCM infection, however, provides some intriguing clues. Armstrong *et al.* (1940) recovered LCM virus from 64 of 303 wild *Mus* captured in 76 different homes in Washington, D.C. Positive mice came from 34 of these dwellings, and the prevalence of active infection therein was 52%. Some of the infected mice had pathological lesions such as pleural exudates and splenomegaly. Other mice were tested for immunity to LCM virus by intracerebral virus challenge. Immunity to LCM was detected in 66% of 62 mice from homes yielding virus-positive mice but in only 11% of 47 mice from virus-negative houses. Virus-positive mice were trapped from each of the few homes where human LCM infection was documented. Rather similar data were obtained by Blumenthal *et al.* (1968) in West Germany. Thus, 139 of 356 *M. musculus* harbored LCM virus, and exactly half of the 44 putative wild colonies examined were infected. Interestingly, LCM-infected colonies yielded an average of 13.4 mice, while a comparable trap effort produced only 2.8 individuals in virus-free colonies.

A possible interpretation of these data, taken in conjunction with the experimental work from white laboratory mice, is that horizontal infection among wild *Mus* leads to abortive, epidemiologically insignificant infection with resultant immunity, but that vertical infection produces chronically infected *Mus* which excrete LCM virus into the environment. The German work also suggests that rapidly reproducing *Mus* colonies spread chronic infection by the venereal route and that resultant impairment in reproduction might account for the focal nature of LCM infection in both space and time.

B. Lassa Virus

This arenavirus is most closely related antigenically to LCM virus (Rowe *et al.*, 1970). Work on the biology of Lassa virus in *Mastomys natalensis* rodents is in its early stages and is largely unpublished. Employing colonized *Mastomys* of the 36-chromosome type from South Africa, however, Walker *et al.* (1975) showed that animals infected by the intraperitoneal route when newly born were viremic and viruric for at least 74 days. In contrast, adult animals were only transiently viremic, were intermittent shedders of small amounts of virus in urine, and developed anti-Lassa complement-fixing antibodies within 2 months of infection.

Unpublished work with wild *M. natalensis* (32 chromosomes) from a Lassa enzootic-endemic region of Sierra Leone (Johnson *et al.*, unpublished observations) has disclosed prevalence rates of virus in blood of this species of 10–20% in villages where human cases of Lassa fever occur. About 30% of such *Mastomys* have Lassa antibodies rather than viremia, and the occasional animal with both blood virus and antibody is seen.

In preliminary laboratory studies (McCormick *et al.*, unpublished observations) we found that 36-chromosome *Mastomys* from Zimbabwe developed immunofluorescent antibodies following infection as newborns or adults. Adult animals were never viruric, had only brief viremia, and did not transmit Lassa virus to susceptible cage mates, regardless of sexual pairing. Animals infected during the first 4 days of life became carriers. They were viremic and excreted up to 10^7 infectious units of Lassa virus per milliliter of urine for at least 4 months. Contact infections occurred regularly when normal mice were caged with adult carriers, but were most efficient when animals of the opposite sex were housed together. Carrier females gave birth to an equal number of offspring as compared to normal *Mastomys*, and these were all infected within the first 2 weeks of life. Whether the virus was transferred by maternal urine or milk or by intrauterine infection is not yet clear, although a single newborn from a viremic dam was found to contain large amounts of virus 2 days after birth, suggesting that true vertical transmission is possible. Other geographic and chromosomal forms of *M. natalensis* have been colonized in our laboratories and are being tested for the Lassa virus infection pattern.

C. Machupo and Junin Viruses

Machupo virus does not produce acute disease in colonized *Calomys callosus* rodents when inoculated by any route into animals of any age. The experimental biology of this infection has been discussed by Johnson *et al.* (1973) and is summarized in Table II. The striking features of this host-response pattern were the profound but partially compensated hemolytic anemia with chronic splenomegaly (extramedullary hematopoiesis) observed in "tolerant" infection, the absence of circulating neutralizing antibodies or virus–antibody complexes in such infection, the nearly complete fetal mortality observed in pregnant, chronically infected mice, and the split response observed when adult mice were infected by the intraperitoneal route. We also observed that dilution of the suckling hamster brain pool of Machupo virus used to infect adult *Calomys* influenced the pattern of response (Webb *et al.*, 1973). High dilutions of virus resulted in about 90% tolerant infections, whereas large virus doses, known to contain

TABLE II
Summary of Biology of Machupo Virus Infection in *Calomys callosus* Rodents

In newborns	In adults
No acute illness	No acute illness
Chronic viremia, viruria	Many respond as newborns
Virus in most tissues, particularly lymphoid, kidney, and salivary glands, reproductive organs	Others exhibit: Viremia, viruria, tissue virus in 30–90 days
No detectable antibodies	
No circulating complexes	Late antibody response with clearing of virus from blood, urine
Growth retardation	
	Normal reproductive capacity
Chronic hemolytic anemia with splenomegaly	No splenomegaly
Late iron-storage disease with reduced life span	Passive antibody to offspring
Major impairment of reproductive function (abortion)	Response under complex genetic control; partially virus dose dependent

defective interfering virions, led to immunocompetent infection in about two-thirds of mice. Using a median virus dose, giving about equal numbers of tolerant and competent adult responses, we also showed that the response to Machupo virus infection was at least partly under host genetic control. In a series of breeding and virus challenge experiments, we found that immunocompetent responses were non-sex-linked recessive, albeit not operating at a single classically sorting locus (Webb *et al.*, 1975). These data led us to speculate that Machupo virus exerts a significant servoregulatory brake on Calomys populations, causing venereal epizootics among genetically dominant tolerant animals when rodent numbers increase to the point where extensive intercolonial contacts occur, and in turn produce population crashes when the chronically infected cohort largely fails to reproduce.

Limited field evidence supports this general concept. Machupo virus prevalence among *Calomys* was seven times higher (35 versus 5%) in villages where human disease was or recently had been occurring, and was grossly and directly correlated with *Calomys* density. Splenomegaly was highly correlated with Machupo virus infection of *Calomys* in nature, and very few animals were ever found harboring anti-Machupo antibodies when assayed by complement-fixing (CF) neutralization, or immunofluorescent techniques.

Data reported by Sabattini *et al.* (1977) for Junin virus in both wild and experimentally infected *C. musculinus* are in general agreement with those for Machupo virus, although the reproductive influences on virus transmission were not elucidated. Chronically viremic *C. musculinus* were detected in mark and release studies, and both vertical and horizontal transmission of the agent were documented. As in the case of Machupo virus, simultaneous mating and inoculation of virus produced pregnant females which, although viremic at the time of parturition, gave birth to noninfected offspring which then acquired infection either from maternal milk or from urine.

D. Other Arenaviruses

Little definitive work has been done with most of the remaining arenaviruses. Only four isolates of Tacaribe virus were ever made, although a study by Price (1978) disclosed the presence of virus-specific neutralizing antibodies in 5 of 24 species of bats examined in Trinidad, including the vampire bat, *Desmodus rotundus*. Four of approximately 100 *Oryzomys buccinatus* rodents captured in Paraguay had Parana virus in viscera (Webb *et al.*, 1970), and Amapari virus was recovered from about 7 and 11% of *O. goeldii* and *Neacomys guianae*. About 20% of these isolates obtained from nearly 700 specimens were from blood as well as viscera, suggesting that chronic viremic infection occurs in both species (Pinheiro, personal communication). Pichindé virus was obtained from nearly 20% of several hundred *O. albigularis* rodents and in several instances repeated isolations were made from blood of trapped animals held in the laboratory (Trapido and SanMartín, 1971).

Two studies have been done with Tamiami virus in *Sigmodon hispidus* with somewhat conflicting results. Murphy *et al.* (1976) employed animals obtained from Georgia and found that newborns sustained a brief viremia, never excreted virus in the urine, and had virus in many organs, notably those of the reticuloendothelial system, for about 1 month. The advent of neutralizing antibodies was correlated with the disappearance of virus, but not with viral immunofluorescent antigen in these animals. Adult *Sigmodon* had little or no viremia or tissue virus and also developed specific antibodies. Jennings *et al.* (1970) studied Tamiami infection among wild *Sigmodon* rodents captured in the virus-type locality in Florida. They reported an 8% isolation rate, almost exclusively from viscera, over a 5-year interval and found many animals with specific CF antibodies. They also inoculated 11 wild-caught animals with Tamiami virus and observed transient but definitive viremia and excretion of virus in the urine of two animals for at least 28 days.

Eighteen strains of Latino virus were recovered from 126 *C. callosus* rodents captured in Bolivia outside the area where Machupo virus is enzootic (Webb *et al.*, 1973). Virus was cleared from tissues of adult colonized *Calomys* obtained from the Latino enzootic area within 30 days of infection and all animals exhibited immunofluorescent antibodies at this time. Following intraperitoneal inoculation of suckling *Calomys*, Latino viremia and viruria were detected, but all animals had cleared these fluids within 4 months and all had detectable antibodies. It was of interest that a minority of animals from this colony infected with Machupo virus in the first 3 days of life were still viremic and without detectable antibodies at this time. Reciprocal experiments with Machupo and Latino viruses in two *Calomys* colonies derived from their respective virus-enzootic zones also disclosed that hemolytic anemia and splenomegaly were Machupo virus specific rather than host specific.

IV. COMPARATIVE DIAGNOSIS

A. Detection of Virus or Viral Antigens

With the exception of LCM virus, very little work has been directed toward a search for the most sensitive host system for measurement of infectivity of the arenaviruses. Most of these agents were initially recognized in field laboratories where the primary interest was study of the ecology of arboviruses. Thus, combined intracerebral–intraperitoneal inoculation of mice less than 3 days of age represents the method commonly utilized for detection of most of the agents, and liver, spleen, kidney, and blood are the specimens of choice for detection of infectivity in wild rodents. All arenaviruses, with the exception of LCM, Lassa, and Latino viruses, produce encephalitis and death in suckling mice, and brain suspensions from sick animals are used as a source of virus for immunological identification.

LCM and Lassa viruses produce silent infection in suckling mice. The former agent is thus usually detected by intracerebral inoculation of adult mice. Hotchin *et al.* (1974), moreover, have shown that the sensitivity and speed of detection of LCM virus in mice can be increased by the administration of sublethal doses of bacterial endotoxin 4–7 days after inoculation of test specimens. Infected mice die within 24 hours after this procedure, whereas normal mice show no symptoms. Again, brain suspensions are prepared for identification of the isolate.

For reasons of test sensitivity and both safety and convenience in laboratory manipulation, Vero cell cultures are the current method of choice for isolation of Lassa virus from wild rodents. Most strains of this arenavirus produce cytopathic changes in Vero cells, but immunofluorescence (Cohen *et al.*, 1966; Benson and Hotchin, 1969; Wulff and Lange, 1975) and reversed passive hemagglutination (Goldwasser *et al.*, 1980) are also used to search inoculated cultures for viral antigens. Tests are regarded as definitive after 10 days of incubation.

Latino virus does not replicate in suckling mice. This arenavirus was originally recovered from rodent visceral tissues after intracerebral inoculation of suckling hamsters (Webb *et al.*, 1973). Subsequent study showed that these animals represent the only known practical host system for detection of the agent; cell cultures proved to be 4–6 dex less sensitive than baby hamsters when infected tissues or blood from rodents were assayed quantitatively.

Suckling hamsters, suckling mice, and Vero cells under agar were compared for isolation of Machupo virus from both human and *C. callosus* specimens (Jobnson *et al.*, 1967). The sensitivity of these systems varied in the order described, with hamsters best and Vero cell plaques worst. In addition, cell culture plaque assay was relatively inferior to animal inoculation when tissue homogenates as compared to whole blood or urine were assayed.

It is important to remember, whatever the host system used, that arenaviruses, particularly in tissue suspensions, often contain large amounts of defective interfering particles which inhibit replication of the virions present (Welsh *et al.*, 1975). Thus, materials should be tested at several dilutions covering a range up to at least 10^{-2} of the specimen. In the case of those arenaviruses which cause human disease, little is lost by utilizing only a 10^{-2} diluted specimen because the epidemiologically significant rodent, that is, the chronically infected, virus-shedding animal, almost always has at least this concentration of virus in the blood, viscera, and urine. An exception to this rule is the infection of colonized Syrian hamsters, which may provoke human LCM infection in laboratory workers or pet owners. Available evidence (Smadel and Wall, 1942; Hannover Larsen and Volkert, 1967) indicates that these animals sustain a semichronic (weeks) rather than a chronic, lifelong infection, thus dictating thorough quantitative testing in the search for infected animals.

If rodents chronically infected with arenaviruses have large amounts of virus in tissues, it should be possible to detect infection by direct immunochemical examination of viscera, either in tissue sections or in impression smears. Experimental work with LCM (Mims, 1970), Machupo (Johnson *et al.*, 1973), and Lassa (Walker *et al.*, 1975) suggests that this approach might be feasible employing brain, lymphoid, or reproductive tissues as sources of immunofluorescent antigen. To date, however, comparative studies have not been carried out in wild rodents.

B. Detection of Immunity to Arenavirus Infection

Although T-lymphocyte cell-mediated immunity to LCM virus has been measured in laboratory mice (Henney, 1971; Marker and Volkert, 1973), this type of assay is too cumbersome to be applied as a diagnostic method for detection of infection by arenaviruses of rodents. Similarly, the intracerebral virus challenge test which has been used in LCM infection is limited to that agent because of the unique susceptibility of mice to this virus; it also does not distinguish between animals immune after infection and antibody response from those resistant to challenge because of chronic infection.

None of the arenaviruses has been shown to possess a viral hemagglutinin. Thus, available methods for measurement of arenavirus-specific antibodies in rodent sera include CF, indirect immunofluorescence, virus neutralization, and radioimmunoassays. Details of the various methods for performance of the first three of these tests are summarized in Volume I, Chapter 17. It has been found, however, that acetone fixation of arenavirus-infected cells to glass slides for performance of the indirect fluorescent antibody test does not completely inactivate infectivity. Cell suspensions can be inactivated conveniently by exposure either to ultraviolet light (30,000 $\mu W/cm^2$) or gamma rays (1.2 million r) (Elliott and Johnson, unpublished observations) or to psoralen compounds activated by long-wave ultraviolet light (Jahrling, personal communication).

Two new methods have been tested for detection of Lassa virus antibodies in sera of *Mastomys* rodents. The first of these is reversed passive hemagglutination inhibition (RPHI) (Goldwasser *et al.*, 1980). In this procedure, anti-Lassa immune globulins are conjugated to glutaraldehyde-fixed fowl erythrocytes, and Lassa viral antigen is titrated and standardized to yield four hemagglutinating units. Antigen is incubated with dilutions of test sera for 1 hour at 37°C, and conjugated erythrocytes are then added. The test is read after 1 hour of further incubation at room temperature. This method was found to be at least as sensitive for detection of Lassa antibodies in wild rodent sera as the immunofluorescent technique and is much easier to perform, especially under field conditions. It deserves to be evaluated for other arenavirus infections.

A second technique is radioimmunoassay employing inactivated, virus-infected Vero cells as antigen, and radioactive *Staphylococcus* protein A to detect antigen–antibody binding (Richman, personal communication). Preliminary work indicates that this method yields higher titers than either immunofluorescence or RPHI and is useful in resolving doubtful, low-titer reactions observed with either of the latter methods.

The most important question with respect to serological detection of arenavirus infection in rodents is which methods are appropriate for which viruses. Much work remains to be done to elucidate this problem. LCM antibodies in wild mice are perhaps best measured by either CF or immunofluorescence, because neutralizing antibodies are present only in low titer in chronically infected carriers and the techniques utilized are very expensive to perform. Wild *C. callosus* rarely exhibit antibodies by any procedure, but newer methods should be evaluated. Both CF and neutralizing antibodies have been found in *C. musculinus* rodents infected with Junin virus (Sabattini *et al.*, 1977), and one suspects that immunofluorescence, RPHI, or radioimmunoassay might prove effective in antibody measurement. These same tests work well in the case of Lassa virus infection, and this approach is of great potential value in mapping the distribution of this agent in *Mastomys* because it appears that horizontal infection in this species uniformly results in antibody formation.

Sufficient information is not available for the other arenaviruses to make recommendations for the most appropriate methods. Neutralizing antibodies are not readily detectable after infection with Pichindé, Parana, and Latino viruses.

V. CONCLUSIONS

Methods so far evolved for recognition of arenaviruses in nature are completely empirical. The fact that at least one such agent, Latino virus, does not replicate at all in laboratory mice suggests that use of other host systems might uncover new members of this virus family. Although wild rodents may be the most promising source of new arenaviruses, the Tacaribe–bat relationship indicates that future work should be broadly based. Indeed, it may well be that there exist arenaviruses which are specific natural parasites of other mammals, including man. The diseases, if any, caused by such agents and the techniques for their recognition await future investigation.

Application of present knowledge is most significant in the case of those arenaviruses which cause human disease. Epidemiological study of wild rodents has great significance in situations, because the viruses are so host specific that infection patterns among the reservoir species are crucial to the understanding of disease occurrence in man. At present, the largest unresolved questions pertain to Lassa virus. What is its true geographic and ecological distribution in Africa?

Why are *Mastomys* more widely dispersed on the continent than the virus? Can *Mastomys* rodent control yield significant reduction in the transmission of infection to man? These and other problems can be addressed by appropriate work with rodents. Answers await only the will and the necessary resources for implementation of existing methods for measuring infection.

REFERENCES

Ackermann, R., Stille, W., Blumenthal, W., Helm, E. B., Keller, K., and Baldus, O. (1972). *Dtsch. Med. Wochenschr.* **97,** 1725.

Armstrong, C., Wallace, J. J., and Ross, L. (1940). *Public Health Rep.* **55,** 1222.

Baum, S. G., Lewis, A. M., Rowe, W. P., and Huebner, R. J. (1966). *N. Engl. J. Med.* **274,** 934.

Benson, L., and Hotchin, J. (1969). *Nature (London)* **222,** 1045.

Biggar, R. J., and Douglas, R. G. (1975). *Lancet* **1,** 856.

Blumenthal, W., Ackermann, R., and Scheid, W. (1968). *Dtsch. Med. Wochenschr.* **93,** 948.

Buchmeier, M., Adam, E., and Rawls, W. E. (1974). *Infect. Immun.* **9,** 821.

Cohen, S. M., Triandaphilli, I. E., Barlow, J. L., and Hotchin, J. (1966). *J. Immunol.* **96,** 777.

Cole, G. A., Prendergast, R. A., and Henney, C. S. (1973). *In* "Lymphocytic Choriomeningitis Virus and Other Arenaviruses" (F. Lehmann-Grube, ed.), pp. 61–71. Springer-Verlag, Berlin and New York.

deVillanfañe, G., Kravetz, F. O., Donadio, O., Percich, R., Knecher, M., Torres, M. P., and Fernández, N. (1977). *Medicina (Buenos Aires)* **37,** 128.

Downs, W. G., Anderson, C. R., Spence, L., Aitken, T. H. G., and Greenhall, A. A. (1963). *Am. J. Trop. Med. Hyg.* **12,** 640.

Fenner, F. (1976). *Intervirology* **7,** 3.

Gilden, D. H., Cole, G. A., and Nathanson, N. (1972). *J. Exp. Med.* **135,** 874.

Gledhill, A. W. (1967). *Nature (London)* **214,** 178.

Goldwasser, R. A., Elliott, L. H., and Johnson, K. M. (1980). *J. Clin. Microbiol.* **11,** 593.

Green, C. A., Gordon, D. H., and Lyons, N. F. (1978). *Am. J. Trop. Med. Hyg.* **27,** 627.

Hannover Larsen, J., and Volkert, M. (1967). *Acta Pathol. Microbiol. Scand.* **70,** 95.

Henney, C. S. (1971). *J. Immunol.* **107,** 1558.

Hershkovitz, P. (1959). *J. Mammal.* **40,** 337.

Hinman, A. R., Fraser, D. W., Douglas, R. G., Bowen, G. S., Kraus, A. L., Winkler, W. G., and Rhodes, W. W. (1975). *Am. J. Epidemiol.* **101,** 103.

Hirsch, M. S., Murphy, F. A., Russe, H. P., and Hicklin, M. D. (1967). *Proc. Soc. Exp. Biol. Med.* **125,** 980.

Hotchin, J. (1962). *Cold Spring Harbor Symp. Quant. Biol.* **27,** 479.

Hotchin, J., Sikora, E., Kinch, W., Hinman, A., and Woodall, J. (1974). *Science (Washington, D.C.)* **185,** 1173.

Jennings, W. L., Lewis, A. L., Sather, G. E., Pierce, L. V., and Bond, J. O. (1970). *Am. J. Trop. Med. Hyg.* **19,** 527.

Johnson, K. M., Wiebenga, N. H., MacKenzie, R. B., Kuns, M. L., Tauraso, N. M., Shelokov, A., Webb, P. A., Justines, G., and Beye, H. K. (1965). *Proc. Soc. Exp. Biol. Med.* **118,** 113.

Johnson, K. M., Halstead, S. B., and Cohen, S. N. (1967). *Prog. Med. Virol.* **9,** 105.

Johnson, K. M., Webb, P. A., and Justines, G. (1973). *In* "Lymphocytic Choriomeningitis Virus and Other Arenaviruses" (F. Lehmann-Grube, ed.), pp. 241–258. Springer-Verlag, Berlin and New York.

Küpper, B., Bloedhorn, H., Ackermann, R., and Scheid, W. (1964). *Zentralbl. Bakteriol., Parasitenkd., Infektionskr. Hyg., Abt. 1: Orig.* **195,** 1.
Lehmann-Grube, F. (1971). *Virol. Monogr.* **10,** 28.
Leong, W. C., and Rawls, N. E. (1977). *Virology* **81,** 174.
Lewis, A. M., Rowe, W. P., Turner, H. C., and Huebner, R. J. (1965). *Science* **150,** (*Washington, D.C.*) 363.
Marker, O., and Volkert, M. (1973). *In* "Lymphocytic Choriomeningitis Virus and Other Arenaviruses" (F. Lehmann-Grube, ed.), pp. 207–216. Springer-Verlag, Berlin and New York.
Mims, C. A. (1969). *J. Infect. Dis.* **120,** 582.
Mims, C. A. (1970). *Arch. Gesamte Virusforsch.* **30,** 67.
Monath, T. P. (1975). *Bull. W.H.O.* **52,** 577.
Murphy, F. A., Winn, W. C., Jr., Walker, D. H., Flemister, M. R., and Whitfield, S. G. (1976). *Lab. Invest.* **34,** 125.
Oldstone, M. B. A. (1973). *In* "Lymphocytic Choriomeningitis Virus and Other Arenaviruses" (F. Lehmann-Grube, ed.), pp. 185–193, Springer-Verlag, Berlin and New York.
Oldstone, M. B. A., and Dixon, F. J. (1969). *J. Exp. Med.* **129,** 483.
Oldstone, M. B. A., and Dixon, F. J. (1970). *J. Exp. Med.* **131,** 1.
Pedersen, R. (1979). *Adv. Virus Res.* **24,** 277.
Pinheiro, F. P., Shope, R. E., Paes de Andiade, A. H., Bensabeth, G., Cacios, G. V., and Casals, J. (1966). *Proc. Soc. Exp. Biol. Med.* **122,** 531.
Price, J. L. (1978). *Am. J. Trop. Med. Hyg.* **27,** 162.
Rowe, W. P., Pugh, W. E., Webb, P. A., and Peters, C. J. (1970). *J. Virol.* **5,** 289.
Sabattini, M. S., González del Ríos, L. E., Díaz, G., and Vega, V. R. (1977). *Medicina (Buenos Aires)* **37,** 149.
Smadel, J. E., and Wall, M. J. (1942). *J. Exp. Med.* **75,** 581.
Tranier, M. (1974). *Mammalia* **38,** 558.
Trapido, H., and SanMartín, C. (1971). *Am. J. Trop. Med. Hyg.* **20,** 631.
Volkert, M., Hannover Larsen, J., and Pfau, C. J. (1964). *Acta Pathol. Microbiol. Scand.* **61,** 268.
Walker, D. H., Wulff, H. T., Lange, J. V., and Murphy, F. A. (1975). *Bull. W.H.O.* **52,** 523.
Webb, P. A., Johnson, K. M., Hibbs, J. B., and Kuns, M. L. (1970). *Arch. Gesamte Virusforsch.* **32,** 379.
Webb, P. A., Johnson, K. M., Peters, C. J., and Justines, G. (1973). *In* "Lymphocytic Choriomeningitis Virus and Other Arenaviruses" (F. Lehmann-Grube, ed.), pp. 313–322. Springer-Verlag, Berlin and New York.
Webb, P. A., Justines, G., and Johnson, K. M. (1975). *Bull. W.H.O.* **52,** 493.
Welsh, R. M., Burner, P. A., Holland, J. J., Oldstone, M. B. A., Thompson, H. A., and Villarreal, L. P. (1975). *Bull. W.H.O.* **52,** 403.
Wulff, H., and Lange, J. V. (1975). *Bull. W.H.O.* **52,** 429.
Wulff, H., McIntosh, B. M., Hamner, D. B., and Johnson, K. M. (1977). *Bull. W.H.O.* **55,** 441.

Part IX

RHABDOVIRIDAE

Chapter 12

The Rhabdoviruses

WILLIAM G. WINKLER

I.	Introduction	529
II.	Morphology	530
III.	Physicochemical Properties	532
IV.	Antigenic Composition	532
V.	Host Range	533
VI.	Transmission	535
VII.	Incubation Period	536
VIII.	Pathogenesis	537
IX.	Clinical Illness	539
X.	Epidemiology	540
XI.	Diagnosis	542
	A. Virus Isolation	542
	B. Virus Propagation	544
	C. Virus Identification	544
	D. Serological Tests	545
XII.	Prevention and Control	546
	References	547

I. INTRODUCTION

The Rhabdoviridae are a family of viruses isolated from vertebrate, invertebrate, and plant hosts. Some members of the group, i.e., rabies viruses, have been known for the disease they produce in man and/or animals since the beginning of medical history. Many, however, are relatively recent additions to the Rhabdoviridae (itself a recent classification), either because they are only recently identified or because they have only recently been placed in this family as a result of newer information on their characteristics. Within the Rhabdoviridae,

ISBN 0-12-429704-8

TABLE I
Partial List of Selected Rhabdoviruses

Vertebrate and invertebrate viruses	
Genus: *Lyssavirus* (6)[a]	
Rabies	Lagos bat
Duvenhage	Mokola
Kotonkan	Obodhiang
Genus: *Vesiculovirus* (4)	
Vesicular stomatitis	Piry
Chandipura	Isfahan
No genus established (36)	
Flanders	Mt. Elgon bat
Hart Park	Bovine ephemeral fever
Egtved	Sigma
Kern Canyon	Many others
Plant and Invertebrate viruses (25)	

[a]Number of viruses in the group.

all members share, by definition, some characteristics of morphology and biochemical composition. Nevertheless, the family encompasses a diverse group of viruses with a broad host range, wide geographic distribution, and variable clinical spectrum, and are of varying significance to the microbiologists and others concerned with the nature of viruses.

As in any taxonomic system which is actively being examined, altered, and expanded, the viruses included within the Rhabdoviridae depend on who is compiling the list. It is not within the scope of this chapter to consider which viruses belong to the Rhabdovirus family. This author has accepted the 70-plus viruses identified as rhabdoviruses by the International Committee on Taxonomy of Viruses (Brown *et al.*, 1979).

Only a few of the viruses listed in the family are discussed in this chapter. Those included for discussion were selected because they are (1) representative of some characteristic of the group, (2) viruses which have been sufficiently studied to add to the knowledge of rhabdovirus characteristics, or (3) of unusual interest by virtue of their medical, veterinary, or economic importance.

A general grouping of the Rhabdoviridae is provided in Table I. This list is by no means complete, containing only 17 of the approximately 70 rhabdoviruses, but it is intended to be representative of the members of this family of viruses.

II. MORPHOLOGY

As might be expected with any group of diverse viruses having a broad host range and natural history, there is some variation in the morphology among members of the Rhabdoviridae. Morphology is especially important in the taxonomic placement of rhabdoviruses since the characteristic shape of viruses in this group is relatively easy to visualize and since many other taxonomic criteria have not yet been described for some of these viruses.

The arthropod and vertebrate viruses appear more closely related to one another morphologically than to the plant viruses. Rhabdoviruses have a basic cylindrical form which in the vertebrate and arthropod viruses is capped on one end with a cone or hemisphere, while the other end is flattened or indented; the overall configuration is a pointed rod, hence the name "rhabdovirus" (*rhabdo,* G. rod). The rhabdoviruses of plants are commonly hemispherical on both ends, giving an overall more bacilliform shape (Francki, 1973). Although there appears to be general agreement on this basic difference in plant versus animal virus morphology, exceptions with bullet-shaped virions among plant viruses and bacilliform virions among animal viruses have been reported (Howatson, 1970). There remains a question among some virologists as to whether the apparent basic difference in plant and animal virus morphology might be simply an artifact introduced by different techniques used in preparation of animal and plant virus electron micrographic specimens (Francki, 1973).

Overall size of the rhabdovirus virion is about 180 nm (range 130–380) in length by about 70 nm (range 60–95) in diameter (Brown *et al.,* 1979). Among the animal rhabdoviruses, most lyssaviruses and vesiculoviruses are about 180 nm by 70 nm; a few, including Kern Canyon, Klamath, and Barur, are shorter, approximately 132–167 nm in length; and many of the ungrouped ones, including Flanders, Hart Park, Mt. Elgon bat, and others, are longer, ranging from 218 to 230 nm in length (Murphy and Harrison, 1979).

While most animal rhabdoviruses have the characteristic bullet shape regardless of their length, several, including Kotonkan, Obodhiang, and Marco, are cone-shaped, having sides that converge rather than being parallel. In addition, variant forms of viruses which usually have a typical bullet or B morphology have been observed. Truncated or T forms of vesiculoviruses, lyssaviruses, and some unclassified rhabdoviruses have been described. These T forms appear morphologically similar to the B-shaped counterparts except that they are only about one-third the normal length (Schneider and Diringer, 1976). These T forms are also referred to as "defective-interfering" particles since they appear to play a role in inhibiting normal viral replication (Cooper and Bellett, 1959).

The outer envelope of the rhabdovirus virion is a bilayered lipid membrane embedded with projecting spikes 6–10 nm in length. The entire surface is covered with spikes except for the indented or invaginated base of the B-shaped virions, which is nude (Schneider and Diringer, 1976). The nucleocapsid within the envelope consists of a single negative-stranded RNA molecule with attached nucleoprotein arranged in a helix of about 30–35 coils whose diameter is approximately 50 nm except that the coils extending into the end caps are of smaller diameter (Howatson, 1970). In the T virions the nucleocapsid is reduced to about 8–10 coils.

III. PHYSICOCHEMICAL PROPERTIES

The major components of the rhabdovirus virion are a bilayered lipid envelope, two or three-envelope-associated proteins, the glycoprotein (G), one or two membrane (M_1 and M_2) proteins, and the nucleocapsid complex composed of the viral nucleic acid and two or three proteins including the major structural nuclear protein (N), a closely associated large protein (L), and a nonstructural protein (NS).

The bilayered lipid envelope is derived primarily from the external or internal host cell membrane, into which are inserted the viral protein components. Schneider and Diringer (1976) have postulated that for rabies virus the outer lipid layer is composed of sphingolipids, while the inner layer is made up of glycerophospholipids. Some comparisons have been made of the lipid composition in the envelopes of other lyssaviruses and vesiculoviruses, but much remains to be worked out in this area (McSharry, 1979).

Within the lipid envelope of all rhabdoviruses are inserted the G protein spikes and the membrane or matrix (M) proteins; host cell protein is not known to incorporate into the viral membrane. In the rabies virus, the G protein spikes are inserted in an interdependent hexagonal arrangement (Murphy and Harrison, 1979), whereas in many other rhabdoviruses the spikes appear to be somewhat randomly spaced and without symmetry (Murphy and Harrison, 1979). The membrane proteins are closely associated with the lipid envelope, but little is known of their exact function even though they constitute about one-third of the total viral protein mass (Bishop and Smith, 1977). Only one membrane protein has been described for many vesiculoviruses (McSharry, 1979), but two (M_1 and M_2) have been described for all of the lyssaviruses as well as some fish rhabdoviruses (McSharry, 1979).

Enclosed within the viral envelope is the helically coiled ribonucleocapsid (RNC) complex. It consists of the viral RNA and its associated major N protein. The RNC of vesicular stomatitis (VS) virus also contains two lesser proteins, the L and the NS proteins. Rabies virus contains only the L protein, and NS protein has not been identified. Relative proportions of components of the rabies virion have been determined by several workers and found to be, by weight, approximately 67% protein, 26% lipid, 4% RNA, and 3% carbohydrate (Schneider and Diringer, 1976).

IV. ANTIGENIC COMPOSITION

The major antigens of rhabdoviruses are the G protein of the membrane spike, the N protein of the RNC, and the M protein(s) of the membrane matrix.

Antibody prepared against purified complete rabies virions has virus-neutralizing capability as demonstrated by animal inoculation tests (Schneider and Diringer, 1976). Anibodies so prepared can also be quantified by plaque reduction (Sedwick and Wiktor, 1967) and rabies fluorescent focus inhibition (RFFI) tests (Smith *et al.*, 1973), two of the most reproducible and widely used serological tests. The antibodies also have weak hemagglutination inhibition properties (Halonen *et al.*, 1968) and in immunodiffusion tests (Grasset and Atanasiu, 1961) showed the development of at least two distinct precipitin lines. Sokol *et al.* (1968; 1969) also showed that antivirion serum could fix complement in the presence of purified viral antigens.

The G protein is the most important viral antigen. Inasmuch as the closely placed G protein spikes form an almost solid covering of the envelope, this is the only viral protein that comes in contact with the host cell components initially while the virion remains intact. Antibody against purified G protein is able to neutralize virus infectivity in both rabies and VS virus (Dietzschold *et al.*, 1974). G antigen is type specific and is used to differentiate between strains within a virus type. For example, among the vesiculoviruses, G antigen specificity differentiates between the various VS virus serotypes and, in the lyssaviruses, between the various rabies serotypes. Antibody prepared against G protein, in addition to having virus-neutralizing capability as measured by animal inoculation tests, also has complement-fixing (CF) ability and can be measured by the radioimmune assay (RIA) test (Dietzschold *et al.*, 1974).

The N protein of the RNC is the second major antigen. It is a group-specific antigen and is used to differentiate between the vesiculovirus group and the lyssavirus group. Antibody against purified N antigen does not have virus-neutralizing capability against intact virus but is capable of neutralizing the infectivity of purified nucleocapsids.

The M protein(s) is the third of the three major viral antigens. In studies with VS virus, antibody against purified M antigen was not capable of neutralizing virion infectivity. It does precipitate purified M antigen in double diffusion tests (Dietzschold *et al.*, 1974). It can also be assayed by complement fixation (Sokol *et al.*, 1968). The antigenic properties of M proteins in rabies virus have not been fully assessed (Schneider and Diringer, 1976).

V. HOST RANGE

Although the scope of this volume is limited to animal viruses, it would be misleading not to include a brief mention of the association between rhabdoviruses and plants. In his extensive review of plant rhabdoviruses, Francki (1973) described the plant hosts of rhabdoviruses as including many members of both the monocotyledonous and dicotyledonous plants. He described some 16 plant viruses which he putatively classified as rhabdoviruses. A number of these are known to have insect vectors. Some are of considerable economic importance.

Several authors, in reviewing virus evolution, have postulated that rhabdoviruses probably developed initially as parasites of arthropods and that the present wide distribution of these viruses as parasites of plants and vertebrates represents a signification expansion of the host range in two different directions (Sylvester and Richardson, 1970). In any case, rhabdoviruses do have a broad host range in plants, arthropods, and vertebrates, and arthropod vectors have been described for both plant and vertebrate rhabdoviruses.

A partial list of viruses and their hosts is provided in Table II. The members of the lyssavirus group all have mammalian hosts, sometimes including man. In addition to rabies, human infection with Mokola and Duvenhage viruses has been reported. Human infection with Lagos bat, Obodhiang, or Kotonkon virus has not been reported. Serological evidence of avian infection with one lyssavirus (rabies) has been reported but not confirmed (Jorgenson and Gough, 1976; Gough and Jorgenson, 1976). Experimentally, birds, amphibians, and reptiles have been infected with varying degrees of success (van Rooyen and Rhodes, 1948; Jorgenson and Gough, 1976).

Among the vesiculoviruses, only VS has been sufficiently studies to permit discussion of its host range. VS virus is a common disease of domestic ungulates.

TABLE II
Selected Rhabdoviruses and Their Natural Hosts

Virus	Man	Other mammals	Birds	Arthropods	Fish
Rabies	X	Many spp.	X[a]		
Lagos bat		Bat			
Mokola	X	Shrew			
Duvenhage	X	X[a]			
Kotonkan	X[a]	Many spp.[a]		X	
Obodhiang	X[a]			X	
Vesicular stomatitis	X	Many spp.		X	
Chandipura	X	Many spp.[a]			
Piry		Opossum			
Isfahan					
Flanders			X	X	
Hart Park			X	X	
Mt. Elgon bat		Bat		X	
Egtved					X
Kern Canyon		Bat			
Bovine ephemeral fever		Cattle			
Sigma				X	

[a] Serological evidence suggesting infection in nature, not confirmed.

In the United States it was known in the mid-nineteenth century as a horse disease. Later, cattle were recognized as an important host, and more recently swine VS disease has been identified (Karstad, 1970). Serological evidence, supported by experimental infection studies, suggests that VS infection may occur naturally in several wild animal hosts including deer, bobcats, and raccoons (Karstad *et al.*, 1956; Karstad and Hanson, 1957). Karstad (1970) has described infection in man as a result of laboratory exposure and from close contact with infected animals. Many mammalian species are susceptible to experimental infection (Andrews and Pereira, 1967).

A number of rhabdoviruses have fish hosts, one of which is listed in Table II. Egtved virus and trout hemorrhagic septicemia virus are probably the same or very closely related viruses (Howatson, 1970). As noted earlier, plant rhabdoviruses as well as strict arthropod rhabdoviruses are not within the scope of this text and are not discussed in any detail.

VI. TRANSMISSION

The routes of transmission for rhabdoviruses are as varied as their host range is broad. Among the lyssaviruses, rabies has been best studied and appears to be broadly representative of most of the group.

Rabies is clearly transmitted most often by the bite of an infected animal which introduces virus-laden saliva into the bite wound. The clinical syndrome of rabies deserves mention here as one of the more unique adaptations for virus survival. Rabies infection commonly induces bizarre behavior in the infected individual, which increases the intra- and interspecific host contact and promotes biting episodes that facilitate transmission of virus, which at this stage in clinical illness is replicating in salivary glands (Dierks *et al.*, 1969).

Rabies has also been shown transmissible through inhalation of virus-contaminated aerosols in bat caves by Constantine (1967) and Winkler (1968) and in the laboratory by Winkler *et al.* (1972). It does not appear that airborne transmission is epidemiologically important, though the potential for this to occur among terrestrial mammals in nature has been described, particularly with communal denning hosts, i.e., skunks (Verts, 1967). Oral transmission has been well documented and, as with the airborne route, its epidemiological significance is not known but appears to be minimal. Vertical transmission of rabies has been described, but its frequency and epidemiological significance remain uncertain (Martell *et al.*, 1973).

Duvenhage virus was naturally transmitted from a bat to man by a bite. This is probably the normal means of transmission for this virus, though like classic rabies, other possible routes must be considered (Meredith *et al.*, 1971).

The mode of transmission for the vesiculoviruses has not been well established, although it is presumed that arthropod vectors are one important means of spread. Hanson (1952), in reviewing the epidemiology of VS disease in the United States, postulated that biting horse flies were a probable vector. This has been supported by laboratory studies showing that horse flies, mosquitoes, and other arthropods are capable of transmitting the disease experimentally (Ferris *et al.*, 1955). VS virus has also been isolated from biting flies (Shelikov and Peralta, 1967) and mosquitoes (Sudia *et al.*, 1967) in nature. VS virus is also readily transmitted directly between infected livestock and between livestock and man by direct contact (Jonkers, 1967). Like rabies, VS apparently cannot penetrate intact skin but may enter through skin breaks or via mucous membranes. Apparently transmission via mucous membrane exposure is much more common with vesiculoviruses than with lyssaviruses.

The Sigma virus of *Drosophila* spp. flies should be mentioned, as it has been proven that vertical transmission occurs with this rhabdovirus in the arthropod host (Howatson, 1970). Transmission mechanisms for the fish viruses are not yet described.

VII. INCUBATION PERIOD

Incubation periods are affected by the usually recognized variables of inoculum dose, virus strain, site of entry (inoculation), and host variability. Even considering these factors, the incubation period for some rhabdoviruses, especially rabies and perhaps other lyssaviruses, is unduly variable.

The incubation period in rabies has been well studied. In man it is usually 3–8 weeks, although documented periods in excess of 1 year and as short as 9 days have been reported (Hattwick, 1974). No clear explanation for this extreme variability in human rabies has been offered, but animal studies suggest some possible causes.

In animals, rabies incubation periods in naturally acquired disease are commonly 2–8 weeks, though again, extremes outside of that range have been reported (Sikes, 1972).

The effect of the inoculating dose on incubation is well demonstrated in Parker's (1975) studies showing the increase in delay of onset of illness as the inoculum dose decreases; see Table III. Similar results were reported by Sikes (1972) in foxes inoculated intramuscularly (IM) with fox origin street virus, in which the mean incubation periods increased from 13 to 56 days as the inoculum dose decreased from 14,000 to 14 mouse intracerebral LD_{50}s ($MICLD_{50}$s).

The effect of virus strains on incubation periods is most clearly seen in mouse challenge experiments comparing the street and mouse-adapted fixed challenge virus standard (CVS). Fixed CVS virus which has been serially passed in one

TABLE III
Effect of Inoculum Dose with Street Rabies Virus in Experimentally Infected Skunks

No. of animals	Inoculum dose[a]	Incubation[b]	Morbidity[b]
6	100,000	20	7
6	10,000	21	7
6	1,000	41	8
5	100	47	9
2	10	80	4

[a]In weanling mouse, IC 50% lethal dose ($MICLD_{50}$); skunk origin street virus inoculated IM.
[b]In days.

host for many passages characteristically develops shorter and more constant incubation periods which are less affected by the size of the inoculum dose. With intracerebral (IC) mouse inoculation challenges, CVS characteristically has incubation periods of 5 to 7 days regardless of inoculum dose, whereas street virus incubation periods usually vary from 7 to 14 days, sometimes longer, and are commonly longer when smaller inoculum doses are used. Intramuscular challenges produce similar, if slightly more variable, results.

The effect of exposure site on incubation periods is well known with rabies. IC challenge results in shorter incubation periods as compared with IM or intraperitoneal (IP) challenges using the same inoculum dose. This has practical application in the postexposure treatment of man in that bites occurring near the central nervous system (CNS), i.e., on the face or neck, because of the shorter incubation periods, are sometimes treated more rigorously than bites occurring on more distal areas, i.e., arms or legs (World Health Organization, 1973).

In VS infections, incubation periods are somewhat shorter and less variable than with rabies. The incubation period in man is about 2–6 days and in livestock is usually 3–7 days (van Rooyen and Rhodes, 1948). Incubation periods of other vesiculoviruses have not been well studied.

VIII. PATHOGENESIS

The pathogenesis of all rhabdoviruses is incompletely understood but has been best worked out for the two major viruses in the family, rabies and VS.

Until recently, few advances had been made toward understanding the pathogenesis of rabies beyond what had been known or suspected as far back as Pasteur's time. It was evident that rabies virus had a predilection for nerve tissue, and it was accepted that by some mechanism, probably not viremia, rabies traveled to the CNS, where it replicated and then moved centripetally to other tissues, most importantly the salivary glands. In the past several years, remarkable advances have been made in elucidating the pathogenesis of rabies, advances which are likely to be significant in leading to the development of improved procedures for treatment and prevention of rabies. In his review of several years' study of experimental infections in laboratory animals, Murphy (1977) provides a comprehensive description of rabies pathogenesis as it is currently understood. It is probable that the progression of virus in the experimentally infected host as described by Murphy is identical to that which occurs in naturally infected hosts, man or animal.

Following the entry of rabies virus (via IM inoculation in these experimental studies) into the host, virus is found first in isolated myocytes, where it begins replicating and invading adjacent muscle cells, forming foci of infection. Although not proven, it has been suggested that it is at this point, prior to invasion of nerve cells, that the infectious process may be temporarily arrested and produce the longer incubation periods sometimes seen with rabies. As the infected foci expand, either immediately or following a delay, the peripheral nervous system becomes exposed to virus at the neuromuscular spindles of sensory stretch proprioceptors, at neurotendinal stretch receptors, and at neuronal endings of motor end plates. Thus, it appears possible that virus can enter the peripheral nervous system through either sensory or motor nerve routes. Additional work is needed to determine if this is actually so and if there may be a predilection for one of the other two possible routes. Murphy also observed complete intact virions in the extracellular spaces at these nerve-muscle-tendon junctions, suggesting that intercellular transfer may be accomplished by elaboration of complete virions from an infected cell, which then penetrate and infect the adjacent cell(s). He suggests this as a possible alternative to the hypothesis that early intercellular transfer is accomplished through passage of a subvirion genomic moiety. Once virus has entered the peripheral nervous system, a phase of strict neuronotropism begins in which passive virus transport occurs. Whether invasion is via sensory or motor nerve fibers, the virus apparently travels somewhat passively in the axoplasm between the axons and their myelin sheaths. A limited amount of replication (virus budding) occurs on axonal membranes.

Dean *et al.* (1963), in studies with fixed CVS virus, showed that once the rabies virus became associated with the peripheral nervous system, it traveled centripetally at the rate of about 3 mm hour. Following this, Baer *et al.* (1965) and others have shown that prompt sectioning of the nerve trunk proximal to the site of inoculation can halt virus progression. This procedure is effective when fixed virus is the inoculum only if performed within a few hours after inoculation; it may be effective for much longer periods, i.e., days, in the case of street virus inoculation (Baer *et al.*, 1968). As infection progresses, virus can be demonstrated by fluorescent microscopy in the dorsal root ganglia, then in the spinal cord, and finally in the cerebrum (Johnson, 1965; Schneider, 1969).

IX. CLINICAL ILLNESS

Infection with the lyssaviruses commonly produces neurological manifestations, as would be expected with neuronotropic viruses. Clinical rabies in man has been detailed by Hattwick (1974), who separated the clinical disease into five phases and described the major clinical symptoms of each phase, which may be summarized as follows:

1. Incubation period (2 weeks–over 1 year): asymptomatic.
2. Prodrome (2–10 days): headache, fever, nausea, vomiting, malaise, local pain or paresthesias at bite site.
3. Acute neurological phase (2–7 days): incoordination, paresis or paralysis, hydrophobia, aerophobia, confusion, hallucinations, hyperactivity.
4. Coma (0–14 days): apathy, stupor, coma.
5. Death or recovery (months for recovery): recovery with or without sequelae.

Considerable individual variation occurs. Most patients, during the acute neurological phase, have a period of marked hyperactivity (analogous to furious rabies in animals), which may be either continuous or intermittent. During this period patients are often agitated, sometimes aggressive, and commonly hyperresponsive to sensory stimuli such as touch, sound, or light. The patient may die abruptly in generalized convulsions in this period. More commonly, this excitatory period is followed by a quiescent period (dumb rabies of animals) in which the patient becomes lethargic, paralysis ensues, and death supervenes. In some patients the hyperactivity period is most dominant, whereas in others it is the quiescent period; this predominance of one or the other period may be so complete that only excitatory or quiescent behavior is observed.

An ascending paralysis has been noted in up to 20% of cases without other more common signs, and this clinical presentation is perhaps the one most easily misdiagnosed. Hydrophobia and/or aerophobia, when present, are considered almost pathognomonic for rabies.

Rabies in animals presents similarly. In animals the excitatory state is usually described as "furious" rabies and the quiescent state as "dumb" rabies. Again, either or both states may occur or either may predominate.

Vampire bat origin rabies produces a slightly different syndrome in that it is most commonly characterized by an ascending paralysis in both man and animals; this phenomenon has been recognized by persons indigenous to vampire rabies enzootic areas, and locally the disease is known as "derriengue," referring to a disorder or malfunction of the lower or rear limbs (Acha, 1967; Pawan, 1936; Watermann, 1959).

One interesting feature of animal rabies, not described in man, is altered phonation. The bark of a rabid dog or fox, or the bellow of rabid cattle, have been reported as abnormal and even characteristic for this disease; some persons who have worked with rabid livestock and carnivores suggest that they can identify individuals in a group of animals just by listening to their vocalization (Winkler, 1975). Hydrophobia is not reported as a part of the clinical picture of animal rabies, and many times rabid animals eat and drink up to the point of death.

Early reports of asymptomatic rabies carriers, especially bats, have not been substantiated (Moreno and Baer, 1980). There have also been reports of asymptomatic carrier dogs, though these have been limited to portions of Africa and Asia (Makonnen, 1972). Additional work is needed to confirm or negate these reports and to determine if the virus responsible for such illness is classic rabies or a variant virus.

Clinical disease of other lyssaviruses, i.e., Mokola and Lagos bat, are incompletely described; human illness has been associated with Mokola virus (Shope *et al.*, 1970). Experimental infections of dogs and monkeys with Lagos bat and Mokola viruses have shown that IC inoculation in some instances produces an encephalitis similar to rabies, though less severe and not necessarily fatal. IM inoculation of these viruses produced a much milder illness, often with no clinical illness at all. Duvenhage virus has not yet been thoroughly studied but produces an illness in man and animals very similar to that of rabies (Meredith *et al.*, 1971).

The vesiculoviruses affect primarily epithelial tissues. In cattle infected with VS, the prodromal illness is characterized by slightly elevated temperature, loss of appetite, and often excessive ropey salivation (Hanson, 1952). This is followed in 1-2 days by the appearance of vesicles on the gums, tongue, and lips which rupture within 24 hours, leaving shallow reddish ulcers. Barring complications of secondary bacterial infection, the eroded ulcerated areas heal rapidly, and recovery is complete within 7-10 days. While vesicles and ulcers are present, animals often eat little and salivate profusely (Karstad, 1970). In addition to oral mucous membrane lesions, vesicles and ulceration may also appear on teats and coronary bands of the feet; the latter is especially common in infected horses and mules.

Infected swine often develop lesions on the snout as well as on the feet (Henderson, 1960).

In man, VS infection is characterized by an acute influenza-like illness with fever, malaise, myalgia, headache, and occasionally nausea and vomiting (Ellis, 1964; Karstad, 1970). The illness is short, and complete recovery in 3-6 days is usual.

X. EPIDEMIOLOGY

The epidemiology of rabies virus has been well described; however, there remain some potentially significant unknowns. Rabies is normally a disease of bats and carnivores, including the domesticated dog and cat and many wild species. The virus circulates in terrestrial carnivore populations, usually erupting in epidemic proportions when first introduced into a population and eventually, in perhaps 2–10 years, establishes as an enzootic infection. Even in enzootically infected populations, localized epidemics flare up periodically, probably in response to a buildup of numbers of susceptible individuals in the population. Disease in enzootically infected areas has been known to disappear spontaneously without active intervention; no assessment has been made in such instances to determine if changes in the density of the host population may have been responsible.

Cyclic fluctuations have been described in a number of instances. Seasonal annual peaks of fox rabies in enzootically infected areas have been described by Johnston and Beauregard (1969) and Wandeler *et al.* (1974); Johnston also described 3-year cycles in fox rabies. Crandall (1975) reviewed cyclic fluctuation in arctic fox populations and regarded this as a factor in fluctuating levels of arctic fox rabies. Except for seasonal variations, cyclicity has not been well described in other species. Bigler *et al.* (1973) and McLean (1975) have described seasonal fluctuations of raccoon rabies in the southeastern United States, and Verts (1967) has described seasonality in skunk rabies. It appears that seasonal fluctuations probably relate to such factors as breeding season, young susceptibles born into the population, and annual dispersal of these younger animals. The slower cycles, i.e., 3 or more years, may be related to population buildups or to other unknown factors.

Seasonal fluctuations in bat rabies have been related to migration (Baer and Admas, 1970) in temperate zones; such fluctuations do not appear to occur in bat rabies in tropic zones.

Although rabies is regarded as an almost always fatal disease, even in severe outbreaks it appears that some portion of the infected population, perhaps 2–5%, survive with development of circulating antibody. These animals may form the nucleus for repopulation with new susceptibles and thus the maintenance of enzootic infection. Another factor in continuing maintenance of this disease is the long incubation period in some infected individuals. These animals may serve to maintain the virus in an area during the time period between decimation of one population and replacement by a new susceptible population.

A significant feature of rabies epidemiology is the apparent "compartmentalization" of the disease in nature. Although rabies may be epidemic in one host species in a given area, it seldom develops to epidemic levels in an alternate host in the same area even though alternate hosts may be present at high density levels. Only rarely, and therefore presumably with difficulty, does rabies "jump the species barrier" to become endemic in another host species. It is not uncommon, however, for single individuals in an alternate species to become infected and appear as isolated sporadic cases.

In the United States, bats are infected throughout most of their distributional range. The disease persists in these bat populations without apparently having any significant impact on terrestrial species, except for an occasional aberrant case of bat origin rabies transmitted to terrestrial mammals, including man. While the infected bat, with its great mobility, offers a seemingly excellent mechanism for spreading rabies into uninfected terrestrial mammal populations at great distances, this does not appear to happen for reasons as yet unclear. It may be that the compartmentalization effect seen between terrestrial mammals applies equally to bat–terrestrial mammal interaction.

Information on the epidemiology of other lyssaviruses is largely unavailable. Two lyssaviruses, Obodhiang and Kotonkan, are known to be insect borne, but specific data on natural transmission exist. Duvenhage virus presumably has been transmitted to man by bat bite, but little is known of the natural history of this virus.

The epidemiology of the vesiculoviruses is poorly described. Hanson (1952), in his review of VS, noted that numerous mammals and birds apparently had exposure to the virus in nature, as evidenced by the presence of antibody, but the exact means of spread and transmission remain unknown. The seasonality and geographic distribution of VS have suggested an arthropod vector, but this remains unproven.

Bats have been suggested as a possible reservoir for Cocal virus (Hanson, 1952), and various mammals have been implicated as hosts of other vesiculoviruses, but the epidemiology of most vesiculoviruses remains undetermined.

XI. DIAGNOSIS

A. Virus Isolation

Rabies and other lyssaviruses are usually isolated from brain tissue of animals (Tierkel, 1973). If brain tissue is not available, it may be necessary to use spinal cord for isolation attempts (Lee and Becker, 1972). In special studies, isolation may be attempted from saliva, salivary glands, cerebrospinal fluid, or other body tissues, but this is not recommended for primary diagnosis.

The most commonly used procedure for rabies virus isolation is IC inoculation of weanling or suckling Swiss white mice with 0.03 ml of a 5–20% suspension of brain tissue from the suspect animal. The suspension is prepared by triturating or homogenizing brain material in the selected diluent. The diluent may be any of a variety of transport-type media; usually it contains an isotonic salt solution buffered to pH 7.4–7.7, animal serum (10–50%), and bacteriostatic concentrations of antibiotics. Rabies is not an especially fastidious virus, and the diluent, buffer, and serum concentrations may vary considerably. Even plain PSS has been used successfully, though it is less satisfactory, especially if the test mateial is to be held in storage for any period of time. It is essential to be certain that the serum component of the diluent is free of rabies-neutralizing antibody, which if present would interfere with isolation attempts. Although weanling mice are most commonly used for virus isolation, suckling mice are most susceptible and give a more sensitive test; some fixed rabies viruses such as the high-egg-passage Flury strain, may not be detected when weanling mice are used and require the use of suckling mice for isolation (Koprowski, 1954). Inoculated mice are observed for clinical signs of rabies for 21 days, and animals that die are then tested by fluorescent microscopy for rabies virus. It is important, as noted by Johnson (1964) in his review of rabies diagnostic procedures, that fluorescent antibody (FA) testing be performed on test mice which succumb, since other neurotropic viruses may produce illness and death which resemble rabies.

Routes of inoculation other than IC have been used in animal isolation studies but are usually 10–100 times less sensitive. In rare instances when the test specimen is toxic to test animals by IC inoculation, other parenteral routes have been used (Winkler, 1968). Tissue culture inoculation of unknown material has been used with some success. It is most useful for detecting tissue culture-adapted virus, e.g., in testing for residual virus in inactivated tissue culture vaccines (Mitchell *et al.*, 1971). Rabies virus can be propagated successfully in embryonated chicken and duck eggs, and this technique is utilized in making human and animal rabies vaccine. It is not the choice for primary isolation of virus unless one is trying to isolate an avian embryo-adapted strain of virus (Kligler and Bernkopf, 1938; Koprowski and Cox, 1948).

Other lyssaviruses are isolated by IC mouse inoculation similar to the technique used for rabies except that in addition to brain tissue, organ pools (including heart, lung, spleen, liver, and kidney) have been used for isolation of Mokola virus by Shope *et al.* (1970) and Kemp *et al.* (1972); brain and salivary gland pools were used for primary isolation of Lagos bat virus (Boulger and Porterfield, 1958). Duvenhage virus was initially isolated from brain tissue of a fatal human case, and this is the preferred source of virus from infected animals (Meredith *et al.*, 1971). Kotonkan and Obodhiang viruses do not kill adult mice and are normally propagated by IC inoculation of suckling mice (Shope, 1975).

Among the vesiculoviruses, VS is usually isolated from vesicular fluid or tissue surrounding the vesicles (Karstad, 1970). Early isolations of VS virus were made by inoculation of vesicular fluid onto the scarified tongue of horses (Eichorn, 1917), but today inoculation of vesicular fluid into the scarified foot pad of guinea pigs is the preferred route of isolation and is particularly useful in differentiating VS from the clinically similar foot-and-mouth disease (Cotton, 1926; Olitsky *et al.*, 1927). Cocal, Piry, and Chandipura viruses are isolated in the same manner as VS virus.

B. Virus Propagation

While the lyssaviruses can obviously be propagated in the systems described for primary isolation, other techniques are commonly used in the laboratory for virus growth. For example, rabies virus is readily adapted to growth in various cell culture systems, including primary hamster kidney (Kissling, 1958), primary dog kidney (Hronovsky *et al.*, 1966), primary monkey kidney (Lang *et al.*, 1969), and primary pig kidney (Abelseth, 1964). Each of these systems has been used for producing rabies vaccine. Several cell lines are used for rabies propagation; most notable is the baby hamster kidney cell line BHK-21 (Sokol *et al.*, 1968). Two cell strains, WI-38 and MRC-5, both derived from human embryonic lung, are widely used in the laboratory and for vaccine production (Wiktor *et al.*, 1964). An excellent review of rabies virus growth in tissue culture systems has been written by Wiktor and Clark (1975). While VS virus replicates efficiently in most known mammalian and avian cell culture systems, it grows best in BHK-21 and primary chick embryo cells (McClain and Hackett, 1958). VS virus also can be propagated in several fish and mosquito cell lines (Solis and Mora, 1970; Buckley, 1969).

C. Virus Identification

Histopathology was the first specific technique used for identification of rabies virus; it is still useful in laboratories which lack more sophisticated techniques or in which rabies diagnosis is not the primary function (Lepine, 1973a). Slides of brain tissue from animals which have died of suspected rabies are prepared as impressions or smears or as thin sections, stained with an appropriate stain, usually Seller's or Mann's stain, and examined by light microscopy. Characteristic lesions of viral encephalitis, e.g., perivascular cuffing, may be present but are not specific. The presence of Negri bodies, rabies-specific intracytoplasmic neuronal inclusions, are diagnostic for rabies. Unfortunately, not all strains of rabies virus routinely produce large numbers of readily recognized Negri inclusions; this is particularly true of some of the laboratory-adapted viruses. Hence the Negri body test is regarded as less sensitive than the more recently developed FA test. The Negri body test is recommended only when the suspect animal has died or become moribund prior to sacrifice since the development of diagnostic Negri inclusions may not be complete until late in the clinical illness.

Fluorescent microscopy is the most widely used test for virus identification in the rabies diagnostic laboratory (Dean and Abelseth, 1973). Brain tissue slides are prepared as above but are stained with fluorescein-tagged antibody conjugate. Examined under the fluorescent microscope, antigen bound to tagged rabies antibody is seen with a characteristic apple green color. The FA test is more sensitive than the light microscopy test since it detects minute amounts of antigen and does not require the presence of large Negri body inclusions. For this same reason, it is the preferred test when looking at virus which may not produce large numbers of Negri inclusions. The FA test can be used effectively earlier in the clinical course of disease since it requires only the presence of antigen in the CNS and not the later pathological changes of inclusion body formation.

D. Serological Tests

The most commonly used serological test for rabies antibody determination is the mouse neutralization (MN) test (Johnson, 1973). This test involves incubation of the unknown test serum with a known quantity of rabies virus so that antibody, if present, will bind to and neutralize the virus. The mixture is then inoculated IC into test mice, and the amount of antibody present is determined on the basis of mouse survival at various dilutions as compared with mortality in control mice. While this test has been the accepted standard for many years in rabies and other serological tests, it is slow and expensive. The MN test is rapidly being replaced by the tissue culture neutralization tests, the most commonly used of which is the RFFI test developed by Smith *et al.* (1973). The RFFI test is basically the same as tne MN test, i.e., it measures neutralizing antibody, except that in this case BHK-21/S13 cells infected with antigen examined by FA microscopy for the presence of antigen are the test system rather than mouse lethality. It has the advantage of not requiring an animal colony; it is also faster and less expensive. Whether the MN or RFFI test is used, both measure neutralizing antibody and equate well with protection.

The indirect fluorescent antibody (IFA) test described by Leffingwell and Irons (1965) enjoys limited use in routine rabies serology but has limitations. It does not necessarily correlate well with neutralizing antibody, and it requires special antisera reagents for each species to be tested.

The CF test used for rabies is similar to that used for other viruses and requires no description (Kuwert, 1973). It has a major disadvantage in that the CF antibody does not necessarily correlate closely with neutralizing antibody. Nevertheless, it has some application in rabies research. Both macro- and microtechniques are used.

The RIA test is one of the newest and most sensitive tests available for rabies antibody determination (Wiktor *et al.*, 1972). Though not widely used at present, this test will likely replace older, slower, and less sensitive tests as more laboratories become equipped to perform it. The rabies RIA test is similar to other RIA tests; virus is grown in BHK-21 cells, concentrated, and iodinated. Virus is incubated with test serum, and then antispecies IgG is added; the mixture is incubated, and radioactivity of the sediment is determined with a gamma counter. The specific virus-antibody binding is determined by comparison with similarly treated control sera and is read as percent specific binding at each dilution.

The agar gel precipitin test initially described by Ouchterlony (1955) has been used for rabies for both antibody and antigen identification (Lepine, 1973b). Although some workers (Grasset and Atanasiu, 1961) found it a useful test, it suffers from lack of sensitivity and is not widely used in rabies.

The hemagglutination inhibition (HAI) test has been used on a limited basis for rabies antibody testing, as has the hemagglutination (HA) test for rabies antigen (Halonen, 1975). Because of the difficulty in removing nonspecific inhibitors, these tests are little used today. The passive hemagglutination (PHA) test has been used by several researchers (Gough and Dierks, 1971), but it has limitations in the species for which the test works and is more sensitive for IgM than IgG; hence it does not correlate well with protection (Gough and Dierks, 1975).

Commonly used serological tests for the vesiculoviruses include the MN test, in which 3-week-old mice are inoculated IC, and the CF test. The double-immuno-diffusion test has also been used for vesiculovirus serology (Murphy and Fields, 1967).

XII. PREVENTION AND CONTROL

Effective vaccines are available for prevention of rabies infection in both man and domestic animals. Vaccine is used for preexposure immunization of persons at high risk of exposure to rabies such as laboratory workers, veterinarians, wildlife biologists, and others (Morbidity, 1980). Vaccine is also used for post-exposure treatment of persons who have been bitten by rabid animals or otherwise potentially exposed to rabies. Vaccine is not effective once clinical signs of disease have developed. Several types of vaccine are available for human use, including tissue culture, avian embryo, and nerve tissue types. Almost all human rabies vaccines in use today are inactivated, though in earlier years several types of attenuated or partially inactivated vaccines were used. Specific antirabies serum or globulin is widely used with vaccine in postexposure treatment to provide immediate, though short-lived, antibody.

Animal vaccines are available for dogs, cats, and livestock. They are used for preexposure immunization; no postexposure vaccine treatment has been developed for animals (World Health Organization, 1973). These vaccines are prepared from tissue culture, avian embryo, or nerve tissue. Several live attenuated animal vaccines are available and widely used. No vaccine is available for immunization of wild animals, but current research is underway to develop an oral vaccine for immunization of wild carnivores.

No chemotherapeutic drugs have been developed for prevention or treatment of rabies. Interferon and/or interferon inducers are being evaluated as an adjuct to vaccine for prevention of rabies in postexposure treatment (Postic and Fenje, 1971; Baer and Cleary, 1972).

Control of rabies is based on vaccination of dogs and, to a lesser extent, cats. Elimination of rabies in these species is the best prevention for human rabies since most human cases result from exposure to these domestic carnivores. There are at present no good techniques for control of wildlife rabies, although intensive wildlife population reduction campaigns designed to break the cycle of transmission have been used with limited success.

There are no vaccines or specific measures for control or prevention of other lyssavirus infections. Some limited cross protection has been demonstrated between rabies and the other lyssaviruses, but much more work is needed in this area (Tignor and Shope, 1972; Tignor *et al.,* 1977).

Among the vesiculoviruses there is little available in the way of prevention and control. Even with VS, since the epidemiology is not clear, it has not been possible to design effective preventive measures. Isolation of infected animals or herds is recommended and may be of value in preventing the spread of the disease. No vaccines are commercially available for vesiculoviruses.

REFERENCES

Abelseth, M. K. (1964). *Can. Vet. J.* **5,** 84–87.

Acha, P. N. (1967). *Bull. Off. Int. Epizoot.* **67,** 343–382.

Andrews, C., and Pereira, H. G. (1967). *In* "Viruses of Vertebrates," 2nd ed., pp. 228–231. Williams & Wilkins, Baltimore, Maryland.

Baer, G. M., and Adams, D. B. (1970). *Public Health Rep.* **85,** 637–645.

Baer, G. M., and Cleary, W. F. (1972). *J. Infect. Dis.* **125,** 520–527.

Baer, G. M., Shanthaveerappa, T. R., and Bourne, G. (1965). *Bull. W.H.O.* **33,** 783–794.

Baer, G. M., Shantha, T. R., and Bourne, G. (1968). *Bull. W.H.O.* **38,** 119–125.

Bigler, W. J., McClean, R. G., and Trevino, H. A. (1973). *Am. J. Epidemiol.* **98,** 326–335.

Bishop, D. H. L., and Smith, M. S. (1977). *In* "The Molecular Biology of Animal Viruses" (D. P. Nayak, ed.), Vol. 1, pp. 167–280. Dekker, New York.

Boulger, L. R., and Porterfield, J. S. (1958). *Trans. R. Soc. Trop. Med. Hyg.* **52,** 421–424.

Brown, F., Bishop, D. H. L., and Crick, J. (1979). *Intervirology,* 1–7.

Buckley, S. M. (1969). *Proc. Soc. Exp. Biol. Med.* **131,** 625–630.

Constantine, D. G. (1967). *In* "Rabies Transmission by Air in Bat Caves," Public Health Serv. Publ. No. 1617. Natl. Cent. Dis. Control, Atlanta, Georgia.

Cooper, P. D., and Bellett, A. J. D. (1959). *J. Gen. Microbiol.* **21,** 485–497.

Cotton, W. E. (1926). *J. Am. Vet. Assoc.* **69,** 313–332.

Crandall, R. A. (1975). *In* "The Natural History of Rabies" (G. Baier, ed.), Vol. 2, pp. 23–40. Academic Press, New York.

Dean, D. J., and Abelseth, M. K. (1973). *In* "Laboratory Techniques in Rabies" (M. Kaplan and H. Koprowski, eds.), pp. 73–84. World Health Organ., Geneva.

Dean, D. J., Baer, G. M., and Thompson, W. (1963). *Bull. W.H.O.* **28,** 477–486.

Dierks, R. E., Murphy, F. A., and Harrison, A. K. (1969). *Am. J. Pathol.* **54,** 251–273.

Dietzschold, B., Schneider, L. G., and Cox, J. H. (1974). *J. Virol.* **14,** 1–7.

Eichorn, A., (1917). *Am. J. Vet. Med.* **12,** 162–164.

Ellis, E. M., and Kendall, H. E., (1964). *J.A.V.M.A.* **144,** 377–380.

Ferris, D., Hanson, R. P., Dicke, R. J., and Roberts, R. H. (1955). *J. Infect. Dis.* **96,** 184–192.

Francki, R. I. B. (1973). *Adv. Virus. Res.* **18,** 257–345.

Gough, P., and Dierks, R. E. (1971). *Bull. W.H.O.* **45,** 741–745.

Gough, P., and Dierks, R. E. (1975). *In* "The Natural History of Rabies" (G. Baer, ed.), Vol. 1, pp. 116–123. Academic Press, New York.

Gough, P., and Jorgenson, R. D. (1976). *J. Wildl. Dis.* **12,** 392–395.

Grasset, N., and Atanasiu, P. (1961). *Ann. Inst. Pasteur, Paris* **101,** 639–647.

Halonen, P. (1975). *In* "The Natural History of Rabies" (G. Baer, ed.), Vol. 1, pp. 103–113. Academic Press, New York.

Halonen, P. E., Murphy, F. A., Fields, B. N., and Reese, D. R. (1968). *Proc. Soc. Exp. Biol. Med.* **127,** 1037–1042.

Hanson, R. P. (1952). *Bacteriol. Rev.* **16,** 179–204.

Hattwick, M. A. W. (1974). *Public Health Rep.* **3,** 229–274.

Henderson, W. M., (1960). *In* "Advances in Veterinary Science," (C. A. Brandley and G. L. Jungherr, eds.), pp. 19–77. Academic Press, New York.

Howatson, A. F. (1970). *Adv. Virus Res.* **16,** 195–256.

Hronovsky, V., Benda, R., and Ciantl, J. (1966). *Acta Virol. (Engl. Ed.)* **10,** 181.

Johnson, H. N. (1964). *In* "Diagnostic Procedures for Viral and Rickettsial Diseases" (E. H. Lennette and N. J. Schmidt, eds.), 3rd ed., pp. 356–380. Am. Public Health Assoc., New York.

Johnson, H. N. (1973). *In* "Laboratory Techniques in Rabies" (M. Kaplan and H. Koprowski, eds.), pp. 94–97. World Health Organ., Geneva.

Johnson, R. T. (1965). *J. Neuropathol. Exp. Neurol.* **24,** 662–674.

Johnston, D., and Beauregard, M. (1969). *Bull. Wildl. Dis. Assoc.* **5,** 357–370.

Jonkers, A. H. (1967). *Am. J. Epidemiol.* **86,** 286–291.

Jorgenson, R. D., and Gough, P. M. (1976). *J. Wildl. Dis.* **12,** 444–447.

Karstad, L. H. (1970). *In* "Infectious Diseases of Wild Mammals (J. Davis, L. Karstad, and D. Trainer, eds.), pp. 64–67. Iowa State Univ. Press, Ames.

Karstad, L. H., and Hanson, R. P. (1957). *Am. J. Vet. Res.* **18,** 162–166.

Karstad, L. H., Adams, E. V., Hanson, R. P., and Ferris, D. H. (1956). *J. Am. Vet. Med. Assoc.* **129,** 95–97.

Kemp, G. E., Causey, O. R., Moore, R. E., Odelola, A., and Fabiyi, A.,(1972). *Am. J. Trop. Med. Hyg.* **21,** 356–359.

Kissling, R. E. (1958). *Proc. Soc. Exp. Biol. Med.* **98,** 223–225.

Kligler, I. J., and Bernkopf, H. (1938). *Proc. Soc. Exp. Biol. Med.* **39,** 212–214.

Koprowski, H. (1954). *Bull. W.H.O.* **10,** 709–724.

Koprowski, H., and Cox, H. R. (1948). *J. Immunol.* **60,** 533–554.

Kuwert, E. (1973). *In* "Laboratory Techniques in Rabies" (M. Kaplan and H. Koprowski, eds.), pp. 135–146. World Health Organ., Geneva.

Lang, R., Petermann, H. G., Branche, R., and Soulebot, J. P. (1969). *C. R. Acad. Sci.* **269,** 2287–2290.

Lee, T. K., and Becker, M. (1972). *Appl. Microbiol.* **24,** 714–716.

Leffingwell, L., and Irons, J. V. (1965). *Public Health Rep.* **80,** 999–1004.

Lepine, P. (1973a). *In* "Laboratory Techniques in Rabies" (M. Kaplan and H. Koprowski, eds.), pp. 56–72. World Health Organ., Geneva.

Lepine, P. (1973b). *In* "Laboratory Techniques in Rabies" (M. Kaplan and H. Koprowski, eds.), pp. 151–157. World Health Organ., Geneva.

McClain, M. E., and Hackett, A. J. (1958). *J. Immunol.* **80,** 356–361.

McLean, R. G. (1975). *In* "The Natural History of Rabies" (G. Baer, ed.), Vol. 2, pp. 53–77. Academic Press, New York.

McSharry, J. J. (1979). *In* "Rhabdoviruses" (D. H. L. Bishop, ed.), Vol. 1, pp. 107–117. CRC Press, Boca Raton, Florida.
Makonnen, F. (1972). *Ethiop. Med. J.* **10,** 79–86.
Martell, M. A., Ceron, F., and Alcocer, R. (1973). *J. Infect. Dis.* **127,** 291–293.
Meredith, C. D., Rossouw, A. P., and van Praag, H. (1971). *S. Afr. Med. J.* **45,** 767–770.
Mitchell, J. R., Everest, R. E., and Anderson, G. R. (1971). *Appl. Microbiol.* **22,** 600–603.
Morbidity and Mortality Weekly Report (1980). Vol. 29, No. 23, pp. 265–284. Cent. Dis. Control, Atlanta, Georgia.
Moreno, J. A., and Baer, G. M. (1980). *Am. J. Trop. Med. Hyg.* **29,** 254–259.
Murphy, F. A. (1977). *Arch. Virol.* **54,** 279–297.
Murphy, F. A., and Fields, B. N. (1967). *Virology* **33,** 625–637.
Murphy, F. A., and Harrison, A. K. (1979). *In* "Rhabdoviruses" (D. H. L. Bishop, ed.), Vol. 1, pp. 65–106. CRC Press, Boca Raton, Florida.
Olitsky, P. K., Traum, J., and Schoening, H. W. (1927). *J. Am. Vet. Med. Assoc.* **70,** 147–167.
Ouchterlony, O. (1955). *Ark. Kemi, Mineral. Geol.* **26B,** 1–98.
Parker, R. L. (1975). *In* "The Natural History of Rabies" (G. Baer, ed.), Vol. 2, pp. 41–51. Academic Press, New York.
Pawan, J. L. (1936). *Ann. Trop. Med. Parasitol.* **30,** 401–422.
Postic, B., and Fenje, P. (1971). *Appl. Microbiol.* **22,** 428–431.
Schneider, L. G. (1969). *Zentralbl. Bakteriol. Parasitenkd., Infektionskir. Hyg., Abt. 1: Orig.* **212** 1–41.
Schneider, L. G., and Diringer, H. (1976). *Curr. Top. Microbiol. Immunol.* **75,** 153–180.
Sedwick, W. D., and Wiktor, T. J. (1967). *J. Virol.* **1,** 1224–1226.
Shelikov, A. I., and Peralta, P. H. (1967). *Am. J. Epidemiol.* **86,** 149–157.
Shope, R. E. (1975). *In* "The Natural History of Rabies" (G. Baer, ed.), Vol. 1, pp. 141–152. Academic Press, New York.
Shope, R. E., Murphy, F. A., Harrison, A. K., Causey, O. R., Kemp, G. E., Simpson, D. I. H., and Moore, D. L. (1970). *J. Virol.* **6,** 690–692.
Sikes, R. K. (1972). *Am. J. Vet. Res.* **32,** 1041–1047.
Smith, J. S., Yager, P. A., and Baer, G. M. (1973). *Bull. W.H.O.* **48,** 535–541.
Sokol, F., Kuwert, E., Wiktor, T. J., Hummeler, K., and Koprowski, H. (1968). *J. Virol.* **2,** 836–849.
Sokol, F., Kuwert, E., Wiktor, T. J., and Koprowski, H. (1969). *Virology* **38,** 651–665.
Solis, J., and Mora, E. C. (1970). *Appl. Microbiol.* **19,** 1–4.
Sudia, W. D., Fields, B. N., and Calisher, C. H. (1967). *Am. J. Epidemiol.* **86,** 598–602.
Sylvester, E. S., and Richardson, J. (1970). *Virology* **42,** 1023–1025.
Tierkel, E. S. (1973). *In* "Laboratory Techniques in Rabies" (M. Kaplan and H. Koprowski, eds.), pp. 41–55. World Health Organ., Geneva.
Tignor, G. H., and Shope, R. E. (1972). *J. Infect. Dis.* **125,** 322–324.
Tignor, G. H., Murphy, F. A., Clark, H. F., Shope, R. E., Madore, P., Bauer, S. P., Buckley, S. M., and Meredith, C. D. (1977). *J. Gen. Virol.* **37,** 595–611.
van Rooyen, C. E., and Rhodes, A. J. (1948). "Virus Diseases of Man," 2nd ed. Nelson, New York.
Verts, B. J. (1967). "The Biology of the Striped Skunk," Univ. of Illinois Press, Urbana.
Wandeler, A., Wachendorfer, G., Forster, U., Krekal, H., Schale, W., Muller, J., and Steck, F. (1974). *Zentralbl. Veterinaermed., Reihe B* **21,** 735–760.
Watermann, J. A. (1959). *Caribb. Med. J.* **21,** 46–74.
Wiktor, T. J., and Clark, H. F. (1975). *In* "The Natural History of Rabies" (G. Baer, ed.), Vol. 1, pp. 155–179. Academic Press, New York.

Wiktor, T. J., Fernandez, M. V., and Koprowski, H. (1964). *J. Immunol.* **93,** 353–366.
Wiktor, T. J., Koprowski, H., and Dixon, F. J. (1972). *J. Immunol.* **109,** 464–470.
Winkler, W. G. (1968). *J. Wildl. Dis.* **4,** 37–40.
Winkler, W. G. (1975). *In* "The Natural History of Rabies" (G. Baer, ed.), Vol. 2, pp. 3–22. Academic Press, New York.
Winkler, W. G., Baker, E. F., and Hopkins, C. C. (1972). *Am. J. Epidemiol.* **95,** 267–277.
World Health Organization (1973). "Expert Committee on Rabies, 6th Report," Tech. Rep. Ser. No. 523. World Health Organ., Geneva.

Part X

RETROVIRIDAE

Chapter 13

Naturally Occurring Retroviruses of Animals and Birds

M. ESSEX AND M. WORLEY

I.	Introduction	553
II.	Classification of Retroviruses	555
III.	Principal Characteristics of Retroviruses	558
	A. Viral Genome Structure and Complexity	558
	B. Gene Products	560
	C. Replicative Cycle	567
	D. Antigenic Determinants	569
	E. Host Range and Subgroup Classification	570
	F. Defective Viruses	570
IV.	Comparative Pathobiology	572
	A. Chickens	572
	B. Cats	575
	C. Primates	582
	D. Cattle	584
V.	Detection of Retroviruses	587
VI.	Conclusions	588
	References	588

I. INTRODUCTION

Perhaps the most characteristic feature of retroviruses is the fact that the RNA genome undergoes reverse transcription through a DNA stage (Baltimore, 1970; Temin and Mizutani, 1970). Aside from this unusual attribute, which makes these agents valuable tools for the molecular geneticist, the retroviruses, or RNA tumor viruses, also represent the largest group of agents associated with the

COMPARATIVE DIAGNOSIS
OF VIRAL DISEASES, VOL. IV

ISBN 0-12-429704-8

development of cancer both under laboratory conditions and in the natural environment.

Viruses representative of this group were first detected as filterable agents associated with leukemia and fibrosarcoma in chickens about 70 years ago (Ellerman and Bang, 1908; Rous, 1911). Since that time, oncogenic retroviruses have been discovered in such divergent species as mice (Gross, 1951), cats (Jarrett *et al.*, 1964), cattle (Miller *et al.*, 1969), and subhuman primates (Theilen *et al.*, 1971). These oncogenic retraviruses are transmitted throughout their respective populations in a manner that is typical of most infectious disease agents: by horizontal transmission and/or by congenital transmission from the mother to her offspring. The retroviruses are etiologically linked primarily to leukemias and fibrosarcomas, tumors that are not necessarily associated with old age. By contrast, it appears that cancers of epithelial origin, which do appear primarily in old age, are more likely to be associated with exposure to chemical carcinogens (Cairns, 1979).

Retroviruses that cause leukemia or lymphoma in mammals are generally transmitted via saliva (Francis *et al.*, 1977; Gallo *et al.*, 1978), and those that cause leukosis in chickens are most frequently transmitted in egg albumin (Spencer *et al.*, 1976). These viruses are classified as ecotropic to indicate that they replicate efficiently in cells from the host species of origin, as distinguished from the endogenous xenotropic retroviruses, which replicate efficiently only in cells from species other than those in which they originate. Xenotropic retroviruses are genetically inherited as complete genomes which can be induced to replicate by the treatment of infected cells with certain mitogens or nucleic acid analogues. Although representative xenotropic retroviruses have been found in almost all species of birds and mammals that have been thoroughly examined, none ha, yet been associated with oncogenic activity.

The ecotropic retroviruses, on the other hand, are not genetically inherited. In some instances, closely related retroviruses may exist as endogenous xenotropic agents in two or more widely divergent species, as in the case of RD-114 virus in the domestic cat and the M-7 class of viruses of baboons (Todaro, 1978). In other instances, closely related retroviruses may exist as xenotropic endogenous agents in one species and as ecotropic oncogenic agents in a distant species, as with the gibbon lymphoma group of agents, which apparently originated as xenotropic viruses of Asian mice (Todaro, 1978). As yet, no case has been found in which closely related retroviruses act as ecotropic oncogenic agents in two or more widely divergent species.

The molecular mechanisms by which retroviruses cause leukemia and lymphoma under natural conditions remain almost completely unknown. The process by which retroviruses cause polyclonal tumors such as fibrosarcomas or acute nonlymphoid leukemias after very brief latent periods is somewhat better understood, but such agents and/or the diseases associated with them are not frequently found in nature. Such agents evolve to transduce efficiently differentiation-related genes from animal cells by recombination. The incorporation of the new gene from the cell usually coincides with the loss of a portion of the viral genome required for replication. As a result, these agents become defective for replication at the same time as they become able to transform fibroblasts *in vitro*.

The retroviruses that cause leukemia and lymphoma in outbred animals are ubiquitous, infecting most of the animals of the population at some point in their lifetime. However, only a small portion of the animals that become exposed to the virus develop persistent infections and/or clinical disease. For those that do develop disease, the induction period is both prolonged and variable.

This chapter will be devoted primarily to those retroviruses that cause disease under natural conditions in outbred animals: the avian, feline, bovine, and primate agents. The viruses that arose by "laboratory evolution" in inbred strains of mice will not be covered except to the extent that they were used as representative agents to study genome structure and function. Similarly, although endogenous nononcogenic xenotropic retroviruses have now been identified in many species, these will be covered only to the extent that they coexist in a given species along with ecotropic retroviruses. Concerning biological issues, only type C retroviruses will be covered, excluding such agents as the mouse mammary tumor virus (MMTV), a type B agent, and visna virus of sheep, a type E agent. In several other species, including reptiles, fish, guinea pigs, and horses, C-type viruses are occasionally found. Relatively little information is available on these agents, and they will not be covered in this chapter.

II. CLASSIFICATION OF RETROVIRUSES

Retroviruses are enveloped virions about 100 nm in diameter that contain a 60–70 S single-stranded RNA and an antigenically specific RNA-dependent DNA polymerase, termed "reverse transcriptase" (RT). In 1974, Retroviridae was established (Dalton *et al.*, 1974) as a family of viruses, and various members have been subdivided into six genera designated A through F, a classification based largely on morphological distinctions set forth by Bernhard (1958).

A classification of the members of the family Retroviridae is presented in Table I. Type A particles, which are 60 to 90 nm in diameter, consist of two intracellular forms, intracytoplasmic and intracisternal. The role of the intracisternal A particle (which consistutes the genus *Cisternovirus A*) is unknown; however, the intracytoplasmic A particles are considered to be precursors of either type B particles or type C particles (Sarkar and Whittington, 1977). Both forms of type A particles are double-shelled and have an electron-lucent center. Type B particles, represented by the MMTV, are associated with the genus *Oncornavirus B*. These particles bud at the cell membrane with a com-

TABLE I
Classification of the Family Retroviridae

Genus *Cisternavirus A*
Type species: Cisternavirus A murine
Genus *Oncornavirus B*
Type species: Oncornavirus B murine (murine mammary tumor virus)
Genus *Oncornavirus C*
Subgenus Oncornavirus C avian
Type species: Rous sarcoma virus
Subgenus Oncornavirus C mammal
Type species: Oncornavirus C murine (murine leukemia virus)
Genus *Oncornavirus D*
Type species: Oncornavirus D primate (Mason-Pfizer monkey virus)
Genus *Lentivirus E*
Type species: Lentivirus E sheep (visna virus)
Genus *Spumavirus F*
Type species: Spumavirus F primate (foamy virus of monkeys)

plete nucleoid to form extracellular particles, the mature form of which has an eccentric nucleoid and prominent surface spikes on its outer envelope. Members of the genus *Oncornavirus C*, more commonly termed "type C viruses," make up the majority of retroviruses, including leukemia and sarcoma viruses. These viruses bud from the plasma membrane with a crescent-shaped nucleoid (see Fig. 1). The extracellular form of the virus contains a central nucleoid.

The Mason-Pfizer monkey virus (MPMV), a type species of the genus *Oncornavirus D,* possesses both intracellular and extracellular particle types (Chopra *et al.,* 1972, 1973; Chopra and Mason, 1970). The intracellular particles are ring-shaped, measure 60 to 95 nm in diameter, and appear near the plasma membrane. The extracellular particles, measuring 100 to 120 nm in diameter, contain an electron-dense nucleoid and are free as well as attached to membrane in intercellular spaces. The latter particle contains both an outer unit membrane and a distinct membrane binding the nucleoid. MPMV can be distinguished from MMTV by the larger size of its intracytoplasmic "A" particles and the presence of knobs on the virion envelope of MPMV, in contrast to the longer spikes on the envelope of MMTV. The genus *Lentivirus E* has as its type species the visna virus of sheep. Although their size, morphology, and physicochemical properties are similar to those of oncornaviruses, lentiviruses induce chronic degenerative diseases and have not been directly implicated in the development of neoplasia.

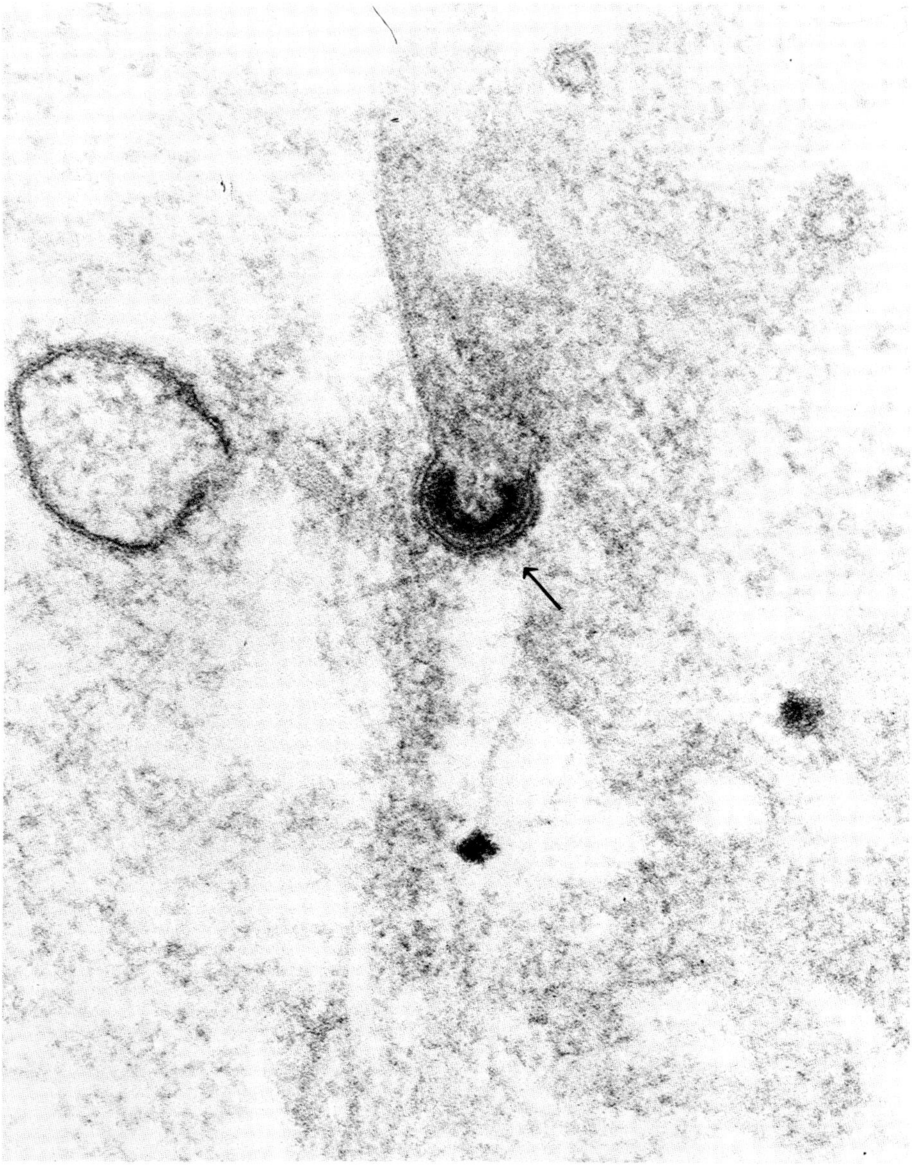

Fig. 1. C-type retrovirus budding from the plasma membrane of an infected cell. ×212,000.

The genus *Spumavirus F* contains the foamy viruses. These viruses differ slightly in their size and surface projections. Intracellular particles are ring-shaped, measuring 35 to 50 nm in diameter, and consist of an electron-opaque shell and an inner electron-lucent center. These particles are usually closely associated with either the plasma membrane or cytoplasmic vacuoles. Extracellular particles and particles within cytoplasmic vacuoles measure approximately 100 to 140 nm in diameter and consist of an electron-lucent nucleoid and an outer envelope with radiating spikes 5 to 15 nm long.

Additional criteria such as host range, species of origin, antigenicity, and pathogenicity have been used to classify retroviruses. We have chosen to discuss these topic in greater detail in more appropriate sections of this chapter.

III. PRINCIPAL CHARACTERICS OF RETROVIRUSES

Since this chapter concerns itself with the comparative aspects of leukemia, lymphoma, and fibrosarcoma, caused by retroviruses, most of the comments in this section will be restricted to the work done on type C viruses, which are the only group clearly associated with these diseases.

A. Viral Genome Structure and Complexity

The single-stranded type C viral genomic RNA has a sedimentation coefficient of about 70 S, corresponding to an estimated molecular weight of approximately 10^7 daltons (Robinson *et al.*, 1965; Duesberg, 1968; Montagnier *et al.*, 1969). Also contained in the virion are other RNA species of lower molecular weight sedimenting at 7, 5, and 4 S, respectively (Bishop *et al.*, 1970a,b; Faras *et al.*, 1973). Although none of the low-molecular-weight RNA species codes for viral products, at least some may be essential to the structure and function of the viral genome. The 7, 5, and 4 S RNA species of RNA tumor viruses are components of the host cells, and occasional 18 and 28 S host cell ribosomal RNAs are found in some virus preparations (Bishop and Varmus, 1975). The 7 S RNA is found in polyribosomal structures of uninfected cells (Erikson *et al.*, 1973; Walker *et al.*, 1974). The 5 S RNA is a part of the large subunit of normal ribosomes (Faras *et al.*, 1973). The 4 S RNA recovered from RNA tumor viruses represents a selected population of cellular tRNA species (Bonar *et al.*, 1967; Tiavnicek, 1968; Erikson, 1969; Bishop *et al.*, 1970a; Erikson and Erikson, 1970; Waters *et al.*, 1975) and can be divided into 4 S RNA species hydrogen-bonded to the 60–70 S virion RNA and the 4S RNA free inside the virus particle (Erikson and Erikson, 1971). This low-molecular-weight RNA species has been identified at tRNA trp for avian type C viruses (Dahlberg *et al.*, 1974; Harada *et al.*, 1975) and tRNA pro in the case of some but not all mammalian type C virus isolates (Peters *et al.*, 1977). Among those 4 S RNA species bound to the 60–70 S RNA is one which serves as a primer for RNA-dependent DNA synthesis *in vitro* (Canaani and Duesberg, 1972), and it is the only low-molecular-weight RNA species in the virion for which a function is known.

Denaturation of the 70 S genomic RNA leads to production of two 35 S RNA subunits (Duesberg, 1968). That the viral genome is polyploid and all 35 S subunits are similar in their sequences have been demonstrated by oliogonucleotide fingerprinting analysis using ribonuclease T1 (Billeter *et al.*, 1974; Duesberg *et al.*, 1974; Coffin and Billeter, 1976) and size measurements of infectious DNA (Hill and Hillova, 1974), as well as molecular hybridization (Baluda *et al.*, 1974). The diploidy of retroviruses is unique among the genomes of known animal viruses and may serve an essential function, such as a role in transcription of DNA from the viral genome by reverse transcriptase.

Fig. 2 represents a simplified version of the composition, structure, and topography of a haploid subunit of the retrovirus genome, using the genome of avian sarcoma virus (ASV) as a prototype; these general features probably pertain to all retroviruses, but only the genome of ASV has been characterized in detail.

Four genes have been identified in the genome of ASV: *gag,* which encodes structural proteins of the viral core; *pol,* which encodes reverse transcriptase; *env,* which encodes the glycoprotein(s) of the viral envelope; and *src,* which is responsible for neoplastic transformation of the host cell. These genes virtually account for the coding capacity of the ASV genome (Baltimore, 1974); however, genetic variants that have most or all *src* deleted from their genome ("transformation defective," or td ASV) can cause lymphoid leukosis (Biggs *et al.*, 1973); hence, the genome of ASV could contain an additional unidentified genetic determinant. The genes *gag, pol,* and *env* are all required for the replication of infectious virus and are common to all virus strains that can replicate in the absence of a helper virus; by contrast, delection of *src* has no effect on viral replication.

The genes of ASV have been ordered on the viral RNA by chemical analysis using deletion mutants and recombinant strains to identify specific segments of the genome (Joho *et al.*, 1975; Wang *et al.*, 1976a,b). Mapping of deletions by examination of heteroduplex molecules by electron microscopy has confirmed the position of *src* (Junghans *et al.*, 1977). Marker rescue experiments have

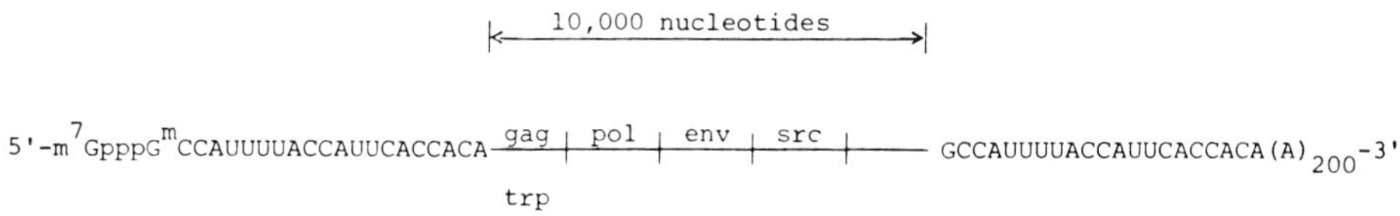

Fig. 2. Established features of the composition, structure, and topography of a haploid subunit of the retrovirus genome, using the genome of avian sarcoma virus (ASV) as a prototype.

documented the linkage between *pol* and *env* (Cooper and Castellot, 1977), and analysis of the proteins translated from viral RNA *in vitro* has substantiated the position and order of *gag* and *pol*. In addition, all strains of avian leukemia and sarcoma viruses (ALSV) share a sequence of approximately 1000 nucleotides located at the 3′ terminus of the genome and denoted "C" (for common region) (Tal *et al.*, 1977). The function of C is unknown; however, its presence in the genomes of all viable strains of ALSV, and its conservation in the face of deletions in the adjacent gene *src* (Tal *et al.*, 1977), imply that C is an essential constituent of the viral genome.

B. Gene Products

1. The Gag Gene-Coded Virion Structural Proteins

The internal structural proteins of retroviruses are a group of nonglycosylated proteins that appear to be localized within the virion envelope and whose molecular weights range from 10,000 to 30,000 daltons (see Fig. 3). There are several lines of evidence which argue that the internal structural proteins (*gag* proteins) are specified by the viral genome. First, viral proteins are synthesized upon infection of heterologous and homologous cells that were previously negative for viral proteins (Huebner *et al.*, 1964). Second, mutants that are temperature sensitive in protein processing and virion formation exist where it seems highly likely that the mutation has occurred in one of the viral proteins. Finally, it has been demonstrated that the 35 S genomic RNA extracted from virions of avian and murine retroviruses is capable of programming the synthesis of a polyprotein which contains the tryptic peptides and antigenic determinants of the *gag* proteins (von der Helm and Duesberg, 1975; Kerr *et al.*, 1976; Pawson *et al.*, 1976; Salden *et al.*, 1976).

Four major proteins which have been most extensively studied in the ALSV and in the Friend, Moloney, and Rasucher (FMR) group of murine leukemia viruses comprise the complex of internal structural proteins in retroviruses: (a) a major virion group-specific antigen present in large amounts and believed to constitute the capsid shell (avian, p27, murine p30) (Davis and Rueckert, 1972; Bolognesi *et al.*, 1973; Stromberg *et al.*, 1974); (b) a hydrophobic protein which in the murine system appears to associate with lipid (murine p15) and in the avian system may have proteolytic activity (avian p15) (von der Helm, 1977); (c) a major phosphoprotein (avian p19, murine p12) (Lai, 1976; Pal and Roy-Burman, 1975; Pal *et al.*, 1975); (d) an arginine- and lysine-rich basic protein (Herman *et al.*, 1975) that is found in association with the RNA of detergent-lysed particles (avian p12, murine p10) (Davis and Rueckert, 1972; Fleissner and Tress, 1973; Bolognesi *et al.*, 1973; Stromberg *et al.*, 1974). Neither the arrangement of the *gag* proteins in the virion nor their role in virus assembly and infection is understood yet in any detail. However, the evidence implies that the *gag* proteins

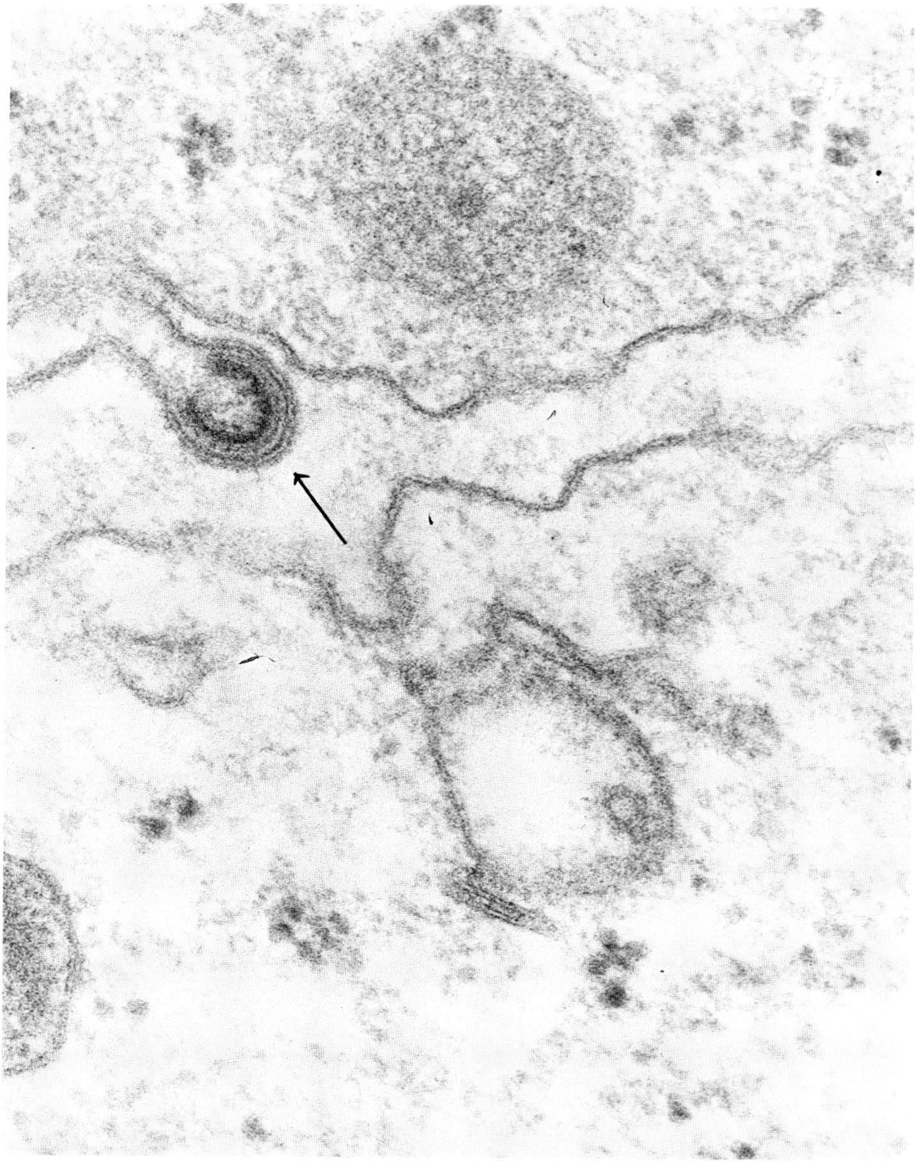

Fig. 3. Cross-sectional diagram of a mammalian C-type retrovirus particle.

are a highly conserved group (Vogt, 1977), likely to be capable of specific sets of interactions among themselves as well as with viral RNA. It is probable that the morphological differences observed between A-, B-, and C-type particles are due to differences in these proteins and/or the effects of their modification by the host cell.

Proteolytic cleavage of a polyprotein precursor appears to be the general mechanism among retroviruses for generating their *gag* proteins. Thus far such precursors have been described for avian (Vogt and Eisenman, 1973), murine (Naso *et al.*, 1975; Stephenson *et al.*, 1975b; van Zaane *et al.*, 1975), and feline (Okasinski and Velicer, 1976, 1977) viruses. Several pathways for *gag* precursor metabolism have been proposed (Eisenmann and Vogt, 1978). However, a striking variation in precursor metabolism has been demonstrated in mouse thymocytes producing AKR virus. In these cells, glycosylated forms of an uncleaved *gag* gene polyprotein precursor are observed on the outside cell surface (Snyder *et al.*, 1977; Ledbetter and Nowinski, 1977). These glycosylated *gag* gene précursors apparently correspond to the Gross cell surface antigen (GCSA) associated with endogenous Gross-type murine leukemia virus expression mouse cells (Snyder *et al.*, 1977; Ledbetter and Nowinski, 1977). The surface localization of these proteins may be of considerable biological importance in that they may provide targets for host immune responses to leukemic cells.

2. *The Pol Gene and RNA-Dependent DNA-Polymerase*

The RNA-dependent DNA-polymerase (RDDP), also known as "reverse transcriptase," has the capacity to use both polyribonucleotides and polydeoxyribonucleotides as template to synthesize complementary DNA (Baltimore, 1970; Temin and Mizutani, 1970; Baltimore and Smoler, 1971; Spiegelman *et al.*, 1970a, b; Temin and Baltimore, 1972; Verma, 1977). The purified RDDP also exhibits ribonuclease activity "RNase-H" which can selectively degrade the RNA moiety of RNA–DNA hybrids (Moelling *et al.*, 1971; Baltimore and Smoler, 1972; Keller and Crouch, 1972; Leis *et al.*, 1973). Analysis of mutants of avian and mammalian type C viruses, characterized by temperature-sensitive lesions in the RNA-dependent DNA-polymerase, DNA-dependent DNA-polymerase, and RNase-H activities has demonstrated these activities to be virus coded and essential for integration of the viral genome into the cellular DNA (Linial and Mason, 1973; Mason *et al.*, 1974; Verma *et al.*, 1974, 1976; Tronick *et al.*, 1975). Mutants in *pol* do not synthesize detectable quantities of proviral DNA (Varmus *et al.*, 1974), and neither cell transformation nor virus production ensues (Linial and Mason, 1973). Most of the viral RDDP requires a primer such as transfer RNA and some metal ions for activity (Dahlberg *et al.*, 1974; Haseltine and Baltimore, 1976; Grandgenett, 1976b). Thus, the type C viral enzyme prefers Mn^{2+} ions, while the type B and type D viruses prefer Mg^{2+} ions for their activity (Scolnick *et al.*, 1970; Howk *et al.*, 1973; Abrell and Gallo, 1973; Michalides *et al.*, 1975).

The RDDP from the murine leukemia viruses has been shown to consist of a single polypeptide of about 70,000 molecular weight (Ross *et al.*, 1971; Tronick *et al.*, 1972; Hurwitz and Leis, 1972). In contrast, the avian type C viral reverse transcriptase contains two subunits, α (70,000) and β (110,000) (Kacian *et al.*, 1971; Temin and Baltimore, 1972; Grandganett *et al.*, 1973; Gibson and Verma, 1974; Verma *et al.*, 1974). The α subunit exhibits both polymerase and nuclease activities, while the β subunit apparently enhances the binding of enzyme to the template or substrate (Verma *et al.*, 1974; Gibson and Verma, 1974; Grandganett and Green, 1974; Moelling, 1974; Grandganett, 1976a). Pulse labeling of Rauscher (R) murine leukemia virus (MuLV)-infected mouse cells has indicated that RDDP is initially synthesized in the form of a large precursor protein of about 200,000 daltons (Naso *et al.*, 1975; Arlinghaus *et al.*, 1976). Posttranslational cleavage of this high-molecular-weight precursor gives rise to an 80,000 *gag* gene-coded precursor and the viral RDDP of about 75,000 daltons (Naso *et al.*, 1975; Arlinghaus *et al.*, 1976). Similarly, ALSV-infected chick cells synthesize a polyprotein of 180,000 daltons, precipitable with anti-*gag* sera and containing tryptic peptides of $Pr76^{gag}$ (Opperman *et al.*, 1977). In both cases, the large polyprotein can be precipitated with sera raised against purified reverse transcriptase from which contaminating antibodies against *gag* proteins have been adsorbed (Jamjoom *et al.*, 1977; Opperman *et al.*, 1977). Thus revenue transcriptase appears to be synthesized in avian and murine retrovirus-infected cells in the form of a polyprotein containing the *gag* precursor covalently linked to polymerase sequences ($Pr^{gag-pol}$).

The identification and characterization of type C viruses of diverse origin has been achieved using reverse transcriptase as an antigenic marker (Aaronson *et al.*, 1971; Scolnick *et al.*, 1972a). Antisera prepared against the enzyme of a given mammalian type C virus most strongly inhibits the activity of the homologous enzyme and, to a lesser degree, enzymes of type C virus isolates of other mammalian species (Aaronson *et al.*, 1971; Scolnick *et al.*, 1972a; Parks *et al.*, 1972). However, antisera to mammalian type C viral enzymes do not inhibit the reverse transcriptase of avian type C viruses or of mammalian retroviruses that are not type C in origin (Aaronson *et al.*, 1971; Scolnick, *et al.*, 1972a). The recent development of radioimmunoassays for the reverse transcriptase of avian (Panet *et al.*, 1975; Reynolds and Stephenson, 1978) and mammalian (Krakower *et al.*, 1977) type C viruses has been described. By use of competition immunoassays for the viral reverse transcriptase, not only has it been possible to distinguish between enzymes of type C viruses of the same species (Krakower *et al.*, 1977) but application of the assays to studies of intracellular reverse transcriptase expression has led to the demonstration that translation of the type C viral genome must involve more than one initiation site (Reynolds and Stephenson, 1978).

3. *Envelope Proteins: The Env Gene*

The envelope proteins of RNA tumor viruses share many characteristics with envelope proteins of other lipid-containing viruses. They are (with one exception) glycoproteins that are localized in structures which appear as "spikes" protruding from the lipid bilayer (see Fig. 3). The spikes can be removed by proteolytic enzyme digestion without disruption of the virion but with concomitant loss of infectivity. The Bryan strain of Rous sarcoma virus (RSV), which lacks envelope glycoproteins and the spike structures, is noninfectious but can transform and replicate in cells after fusion with Sendai virus (Kawai and Hanafusa, 1973; Ogura and Friis, 1975). Thus the function of the spikes apparently is to mediate the absorption and penetration of virus particles to the susceptible host cell.

The retrovirus genome codes for the amino acid sequences of the envelope proteins. This has been most convincingly demonstrated for the avian viruses, where deletions have been found that lead to the inability of these virus mutants to synthesize the major glycoproteins (Duesberg *et al.*, 1975; Halpern *et al.*, 1976) and where spontaneous viral mutations change the properties of the glycoproteins (Zarling *et al.*, 1977). In addition, recombinants between two viruses of different subgroups have been isolated which direct the synthesis of glycoproteins with sugar moieties intermediate in character between the two parental viruses (Galehouse and Duesberg, 1978).

The two major envelope proteins of avian viruses have been termed "gp85" and "gp37" after their apparent molecular weight in SDS gel electrophoresis. Depending on the virus subgroup, 15 or 40% of the molecular weight of the gp85 is due to carbohydrate (Krantz *et al.*, 1976), which is responsible for the binding of the gp85 to lectins (Ishizaki and Bolognesi, 1976).

A third protein, p10 (Vogt *et al.*, 1975), may also be located on the envelope of avian tumor viruses in addition to the two glycoproteins. Little is known about the origin of p10 except that it is not contained in the *gag* precursor Pr76 (Vogt *et al.*, 1975) and it is reported to react with dansyl chloride when virus particles are treated with this reagent, under conditions in which the internal proteins are poorly reactive (Bolognesi *et al.*, 1973).

Both in mature virus particles and in the cell, one molecule each of gp85 and gp37 are linked by disulfide bonds (Leaminson and Halpern, 1976). Larger aggregates of the two glycoproteins apparently held together by noncovalent interactions also are observed in velocity sedimentation in sucrose gradients (Duesberg *et al.*, 1970; Leaminson and Halpern, 1976).

Antibody to gp85 precipitates a major polypeptide of molecular weight about 90,000 (gp Pr90 *env*) from cells pulse-labeled with ^{35}S methionine. Concomitant with the turnover of gp Pr90 in pulse-chase experiments of RSV-infected cells is the appearance of mature gp85 and gp37 (Moelling and Hayami, 1977; England *et al.*, 1977). Recently, the precursor-product relationship of gp Pr90 and the virion glycoproteins has been shown directly by tryptic fingerprinting. The sets of arginine containing tryptic peptides of partially deglycosylated proteins gp85 and of gp37 are distinct, and both are contained in gp Pr90 (Klemenz and Diggelmann, 1978).

The major envelope protein of the majority of mammalian type C viruses migrates at an apparent molecular weight of 70,000 on SDS gel electrophoresis. The major envelope glycoprotein of type C virus isolates of mouse (Strand and August, 1974; Hino *et al.*, 1976), woolly monkey (Hino *et al.*, 1975), baboon (Stephenson *et al.*, 1976a), and feline (Stephenson *et al.*, 1977a) origin have been isolated and studied in detail. Certain proteins, including gp70, coded for by a specific class of endogenous virus, are expressed in the mouse throughout embryonic life (Stephenson and Aaronson, 1974; Hino *et al.*, 1976). And, as a result, mice appear to develop immunological tolerance to the proteins. In contrast, expression of gp70 of other endogenous mouse type C viruses, which are subject to occasional spontaneous or chemical activation, is more tightly regulated (Stephenson *et al.*, 1975a). Since gp70 of this class of virus is not expressed during embryonic life, tolerance to antigenic determinants unique to this gp70 species fails to develop, and virus replication is thus subject to regulation by host immune surveillance mechanisms (Stephenson *et al.*, 1976b). A similar situation is seen with horizontally transmitted type C virues such as feline leukemia virus and gibbon leukemia virus.

There is accumulating evidence indicating thet gp70 expression in the mouse may in some way be linked to differentiation and development. High levels of AKR-MuLV gp70 expression have been detected in the absence of overt virus in bone marrow cells of all strains of mice examined. (McClintock *et al.*, 1977). In addition, gp70 has been found to be a constituent of the surface of normal thymocytes and to share immunological and biochemical properties with the thymocyte differentiation marker G_{IX} (Tung *et al.*, 1975; Obata *et al.*, 1975; Del Villano *et al.*, 1975).

An apparently smaller glycoprotein, gp45, has been reported to be found in variable amounts in MuLV preparations (Moroni, 1972; August *et al.*, 1974; Fleissner *et al.*, 1974; Moenning *et al.*, 1974; Ikeda *et al.*, 1975). This protein appears to be a breakdown product of gp70 (Krantz *et al.*, 1977). The ^{125}I-labeled tryptic peptide maps of both glycoproteins are similar (Elder *et al.*, 1977). Also, the content of gp45 increases at the expense of gp70 during storage of partially purified proteins (Krantz *et al.*, 1977). Finally, monospecific antisera to gp70 recognize gp45, several proteases generate a fragment the size of gp45 from purified gp70 , and the N-terminal amino acid sequences of both proteins are identical (Marquardt *et al.*, 1977; Krantz *et al.*, 1977; Charman *et al.*, 1977).

A second major MuLV envelope protein, which migrates with an apparent molecular weight of 17,000 daltons in SDS gels, has been termed "p15e" to distinguish it from the p15 internal structural protein (Ikdea *et al.*, 1975; Schafer *et al.*, 1975). There is evidence that in some instances p15e may be cleaved, giving rise to a low-molecular-weight polypeptide of about 12,000 daltons (p12e) (Arcement *et al.*, 1976). By comparing the electrophoretic mobilities of MuLV proteins under both reducing and nonreducing conditions, two groups of investigators have reported that gp70 is frequently found to be linked by disulfide bonds to a nonglycosylated protein of about 15,000 molecular weight (Leaminson *et al.*, 1977; Witte *et al.*, 1977).

Pulse-labeled MuLV (R)-infected cells contain a polypeptide precursor, gp Pr90, for the virion envelope proteins. This unstable protein is precipitable with antiserum against purified gp70 and p15e (Famulari *et al.*, 1976; Shapiro *et al.*, 1976; Naso *et al.*, 1976), and by pulse-chase experiments gives rise to cleavage products of about 70,000 and 15,000 daltons, respectively. Methionine-labeled peptide sequences analagous to those of gp70 and p15e within this precursor have been identified by tryptic digest analysis (Arcement *et al.*, 1976; Shapiro *et al.*, 1976; van Zaane *et al.*, 1976; Famulari *et al.*, 1976). Inhibition of glycosylation of the primary *env* gene product by use of 2-deoxy-D-glucose or cytochalasin B leads to formation of a 70,000-dalton nonglycosylated protein (Shapiro *et al.*, 1976) which presumably represents *env* gene translational products prior to glycosylation. The product–precursor relationships between these various *env* gene-coded proteins have been confirmed by *in vitro* protein synthesis studies (Gielkens *et al.*, 1974; van Zaane *et al.*, 1977).

4. *Src* Gene-Coded Transforming Protein(s)

Type C viral genetic sequences associated with sarcomagenic transformation have been designated *src* (Baltimore, 1974). These sequences are presumably of cellular origin and are deleted or absent from the genomes of transformation-defective type C viruses (Lai *et al.*, 1973; Scolnick *et al.*, 1974; Stehelin, *et al.*, 1976). In contrast to the extensive characterization of the *gag, pol,* and *env* gene translational products, studies of *src* gene-coded proteins have been limited. In fact, much of the evidence of an *src* gene translational product has been derived by indirect approaches. For instance, the existence of such a protein was initially based on the isolation of temperature-sensitive mutants of avian (Biguard and Vigier, 1970; Martin, 1970; Friis *et al.*, 1971; Kawai and Hanafusa, 1971; Bader and Brown, 1971) and mammalian (Scolnick *et al.*, 1972b) type C sarcoma viruses which reversibly lose expression of the transformed phenotype at their respective nonpermissive temperatures.

Direct evidence arguing for the existence of an *src* gene-coded protein has been derived by use of cell-free, *in vitro* translation systems using avian sarcoma virus. Purchio *et al.* (1977) have demonstrated the synthesis of a 60,000-dalton polypeptide in a reticulocyte cell-free translation system using subgenomic regions of the avian sarcoma virus genome as mRNA. This protein appears analagous to a 60,000-molecular-weight transformation specific protein demonstrated in ASV-transformed chicken cells and ASV-induced hamster tumor cells by immunoprecipitation of radiolabeled cell extracts with serum from tumor-bearing rabbits (Brugge and Erikson, 1977).

Attempts to detect and study sarcoma virus-specific antigens have also been undertaken in the feline system. A major advantage of this system results from the identification of a feline oncornavirus-associated cell membrane anigen designated "FOCMA" (Sliski *et al.*, 1977; Essex, 1980). While nontransformed fibroblasts of diverse mammalian species become FOCMA positive upon infection and transformation with feline sarcoma virus, the same cells remain FOCMA negative when infected with feline leukemia virus alone. The results of recent studies have shown that in both mink and rat cells nonproductively transformed by feline sarcoma virus (FeSV), an 80,000–100,000-dalton precursor polypeptide containing p15 and p12 is initially synthesized, which upon posttranslational cleavage gives rise to a 65,000-molecular-weight polypeptide lacking p15 and p12 (Stephenson *et al.*, 1977b). Feline Lymphoma cells, including those which lack detectable levels of feline leukemia virus (FeLV) structural proteins, have been shown to be FOCMA positive. These findings strongly suggest that FOCMA represents an FeSV-coded transformation specific protein.

C. Replicative Cycle

Studies on the physiology of retravirus replication have led to the conclusion that the synthesis of virus-specific DNA was required for the initiation of infection and that DNA served as the template for synthesis of progeny viral RNA (Temin, 1976). From these and other observations, Temin formulated the DNA provirus hypothesis: retroviruses transform cells and replicate through the agency of a DNA copy of the viral genome synthesized early in infection and then integrated into the genome of the host cell in a manner analagous to that of a lysogenic bacteriophage. Three major experimental observations have substantiated the DNA provirus theory:

1. Virions of retroviruses contain reverse transcriptase, the enzyme that uses viral RNA as a template to synthesize DNA (Verma, 1977), and the function of reverse transcriptase is a prerequisite for both virus replication and virus-induced cellular transformation (Mason *et al.*, 1974; Verma *et al.*, 1976).

2. The synthesis and integration of virus-specific DNA in newly infected cells have been demonstrated by molecular hybridization (Weinberg, 1977).

3. Chromosomal DNA extracted from cells infected with retroviruses can "transfect" the viral genome into premissive cells, giving rise to both viral replication and cellular transformation (Hill and Hillova, 1976).

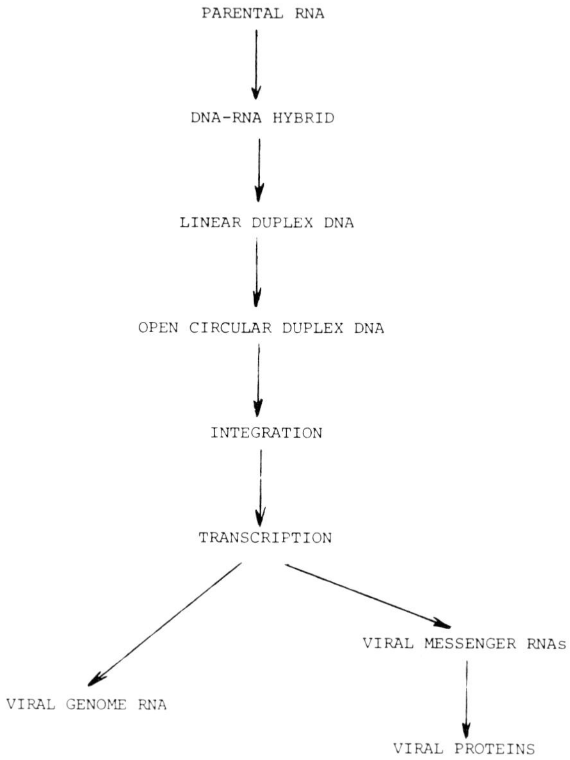

Fig. 4. The principal molecular events in the replication of retroviruses.

The principle molecular events in the replication of retroviruses are depicted in Fig. 4. Transcription of the parental viral genome by reverse transcriptase generates virus-specific DNA through a series of poorly characterized intermediates. The major stable products of synthesis are linear duplexes and closed circular duplexes (Gianni *et al.*, 1975; Guntaka *et al.*, 1975, 1976; Smotkin *et al.*, 1976). Viral DNA serves as template for the synthesis of viral RNA, generating two functionally separate pools of RNA molecules (Paskind *et al.*, 1975; Levin and Rosenak, 1976): RNA destined for encapsidation as genome and RNA destined to serve as messenger for the synthesis of viral proteins. Viral protein synthesis constitutes a minor fraction of total cellular protein synthesis and has no necessary effect on cellular metabolism; retroviruses are usually not cytopathic, and the host cell is transformed only if the infecting virus bears an appropriate transforming gene. Once established, virus production continues indefinitely and is usually limited only by the life span of the host cell.

D. Antigenic Determinants

Three major forms of antigenicity are associated with retroviruses: (a) group-specific antigens are shared by related viruses derived from a single host species; (b) type-specific antigens define the most specific serological subgroups presently identified; and (c) interspecies antigens are shared by otherwise unrelated viruses derived from different host species. Although interspecies antigens are shared between various mammalian type C viruses, there is no antigenic relatedness in either the *env* gene-coded proteins or the *gag* gene-coded proteins between avian and mammalian type C retroviruses.

The gp70s of mammalian type C virus isolates of mouse, woolly monkey, baboon, and feline origin have been isolated and studied in detail. Immunological characterization of gp70s of these diverse viruses has indicated the presence of type, group, and interspecies-specific antigenic determinants (Strand and August, 1973, 1974; Hino *et al.*, 1975, 1976; Stephenson *et al.*, 1976a). Selective removal of the carbohydrate portions of the molecule by treatment with glycosidases does not affect serological and biological properties (Bolognesi *et al.*, 1975). Competition immunoassays which measure type-specific antigenic determinants of these gp70s have been used in discriminating closely related type C virus isolates (Strand and August, 1974; Hino *et al.*, 1975, 1976; Stephenson *et al*, 1976a). Similar to the major envelope glycoprotein of mammalian type C viruses, the gp85s of avian retroviruses also possess type, group, and interspecies-specific antigenic determinants. In contrast, however, recent experiments with avian myeloblastosis virus (AMV) gp85 suggest that the carbohydrate moieties which are common to normal cell surface glycoproteins (Collins *et al.*, 1978) can influence gp85 serological reactivity *in vitro* (van Eldik *et al.*, 1978). The second component of the viral envelope in murine type C viruses, represented by p15e, has been shown to be distinct from the 15,000-dalton *gag* gene-coded protein (p15) (Schafer *et al.*, 1975). Both p15 (Strand *et al.*, 1974; Barbacid *et al.*, 1977) and p15e (Schafer *et al.*, 1975) have been shown to possess group and interspecies-specific antigenic determinants. The 12,000-dalton (p12e) cleavage product of p15e has been shown to be serologically and biochemically related to p15e (Arcement *et al.*, 1976; Karskin *et al.*, 1977). Insufficient data are available concerning the antigenic determinants possessed by the avian viral membrane spike, gp35.

The major internal polypeptide of avian (27,000 daltons) and mammalian (24,000 to 30,000 daltons depending on species) type C viruses represents the primary constituent of the core shell. This molecule possesses type, group, and interspecies-specific antigenic determinants. Therefore, antisera to feline p27 will cross-react with murine p30 but antisera to avian p27 will not recognize any of the mammalian proteins. Both murine and feline p15 as well as the analagous avian p15 possess predominantly group-specific reactivity, as previously discussed. The major phosphoproteins of avian (p19) and mammalian (p12) leukemia viruses are the most type-specific antigens of the respective viruses (Green *et al.*, 1973; Tronick *et al.*, 1973; Bolognesi *et al.*, 1975; Hunsmann *et al.*, 1976; Schwarz *et al.*, 1976). Finally, the most basic protein of avian (p12) and mammalian (p10) type C viruses possesses predominantly group-specific antigenic determinants.

E. Host Range and Subgroup Classification

Host cells can restrict the replication of retroviruses by exclusion of viral entry due to the absence of cellular receptors required for penetration of the viral genome into the host cell. Spikes protruding from the exterior of the viral envelope, composed of glycoprotein(s), are required for entry of the virus into a host cell and are responsible for the host range of the virus (Weiss, 1976). The spikes can be removed by proteolytic digestion without disruption of the virus but with concomitant loss of infectivity. Thus the function of the spikes apparently is to mediate the absorption and penetration of virus particles into the susceptible host cell. Consistent with this interference is the ability of the envelope proteins coded for by one RNA tumor virus to coat the envelope of a virus of a different strain and thereby alter the host range of the latter. At low frequency, envelope proteins can also impart their host range determinants to unrelated viruses. For example, in mixed infections, vesicular stomatitis virus (Weiss *et al.*, 1974; Weiss and Wong, 1977) or murine retroviruses (Levy, 1977; Weiss and Wong, 1977) can acquire the host range of avian retroviruses, and vice versa. The *in vitro* host range of retroviruses provides an exceptionally useful means of classification.

Both viral host range and cellular susceptibility or resistance are genetically determined. With avian leukemia or sarcoma viruses, the difference in plating efficiency between susceptible and resistant cells is usually very large, at least 10^3, often more than 10^6. Such efficient restriction has proved extremely valuable in the genetic selection for and against certain host cell variants. Therefore, the host range variants of these viruses, commonly referred to as "subgroups," constitute important genetic workers. Feline sarcoma and leukemia viruses appear to be subject to similar host range control, but this system has not yet been applied to genetic problems. The subgroups of avian leukemia and sarcoma viruses are defined by various properties of the viral envelope—antigenicity, sensitivity to interference, and host range, the last being the most practical one.

F. Defective Viruses

RNA tumor viruses which carry a mutation affecting a replicative (R) function, and therefore fail to produce infectious progeny virus, are called "replication defectives'' (rd). Two well-known rd viruses with a deletion in env function are the Bryan high-titer strain of RSV (BH RSV) and the NY8 variant of the Schmidt-Ruppin RSV of subgroup A (Sr RSV-A). In single infection with these viruses, the host cell becomes transformed, but infectious sarcoma virus is not produced (Temin, 1962, 1963; Hanafusa *et al.*, 1963; Kawai and Hanafusa, 1973). Such transformed cells do not adsorb virus-neutralizing antibody and do not elicit the production of a neutralizing antibody when injected into chickens, indicating the absence of viral surface glycoprotein from the transformed cells. If the nonproductively transformed cells are superinfected by an avian leukosis virus, the glycoprotein of the leukosis virus can be used for the maturation of infectious but still genetically defective sarcoma virus. The superinfected cells release both sarcoma and leukosis virus (Hanafusa *et al.*, 1964). The genetically defective sarcoma virus carrying the envelope glycoprotein of the leukosis helper virus is called a ''pseudotype'' (Rubin, 1965). Through its glycoprotein, the helper leukosis virus controls virus-specific surface properties of the sarcoma pseudotype; thus antigenicity, host range, and sensitivity to viral interference are identical to those of the helper virus (Vogt, 1965a,b; Hanafusa, 1965; Hanafusa and Hanafusa, 1966; Kawai and Hanafusa, 1973; Hanafusa *et al.*, 1976).

In preparations of envelope-defective avian sarcoma viruses, genetic variants with an additional defect in the polymerase can be found. These viruses lack reverse transcriptase activity and are also free of proteins which cross-react immunologically with reverse transcriptase (Hanafusa and Hanafusa, 1971; Hanafusa *et al.*, 1972; Nowinski *et al.*, 1972; Panet *et al.*, 1975).

The term ''transformation defective'' (td) is commonly used to designate nonconditional mutants of helper-independent sarcoma virus which are no longer able to transform fibroblast cultures or to induce sarcomas in the animal but still synthesize infectious progeny virus in a single infection. Defectiveness in transformation in this context refers specifically to the sarcoma-producing capacity of the virus (the *src* gene). Those td agents which have been tested still induce leukosis in the animal host and thus are not completely nononcogenic (Biggs *et al.*, 1973). Td viruses have been found in many nondefective avian sarcoma virus preparations. If the sarcoma virus has not been cloned recently, td virus usually occurs in excess of the sarcoma virus (Dougherty and Rasmussen, 1964; Hanafusa and Hanafusa, 1966; Duff and Vogt, 1969).

Genetic defectiveness in replicative and coordinate functions and the ensuing dependence on helper viruses for the production of infectious progeny are more common among mammalian than among avian sarcoma viruses. In fact, all presently known mammalian sarcoma viruses—the four strains of murine sarcoma virus (Moloney, Kirsten, Harvey, and probably Finkel-Biskis-Jinkins), the three strains of feline sarcoma (Gardner-Arnstein, McDonough, and Snyder-Theilen), and simian sarcoma virus—are unable to produce infectious progeny virus in the absence of a helper (Hartley and Rowe, 1966; Huebner *et al.*, 1966;

Aaronson and Weaver, 1971; Harvey and East, 1971; Sarma *et al.*, 1971; Levy *et al.*, 1973; Scolnick and Parks, 1973; Aaronson, 1973; Henderson *et al.*, 1974; Chan *et al.*, 1974). Rescue of defective sarcoma virus from nonproducing cell lines transformed by an mammalian sarcoma virus can be accomplished by superinfection with a wide spectrum of type C helper viruses (Huebner *et al.*, 1966; Klement *et al.*, 1969; Sarma *et al.*, 1971; Aaronson and Weaver, 1971; Scolnick *et al.*, 1972a; Aaronson, 1973; Henderson *et al.*, 1974; Levy, 1977; Weiss and Wong, 1977). There seems to be no virus-controlled or virus-specific restriction in helper activity: In principle, all mammalian and even some avian type C viruses can act as helper viruses for all mammalian sarcoma viruses. The only limitations are set by the cell, which must be susceptible to infection by the helper virus for rescue to take place.

IV. COMPARATIVE PATHOBIOLOGY

A. Chickens

Chickens develop a wide variety of diseases caused by retroviruses. From the standpoint of the poultry industry, the most important is lymphoid leukosis, because it is by far the most common. A wide variety of other neoplastic diseases of chickens are caused by retroviruses. Those agents that cause the acute leukemias, such as myeloblastosis, and the agents that cause fibrosarcomas have received intensive study by viral oncologists. A list of the diseases that may be caused in chickens by retroviruses is given in Table II.

The incubation and induction periods for the different diseases vary substantially in duration. In lymphoid leukosis, the disease usually occurs after 3–6 months of age, following initial exposure to the etiological agent either *in ovo* or at hatching. With most of the other diseases the incubation period is much shorter: only 1–4 weeks with most of the malignant solid tumors such as fibrosarcoma and 3–10 weeks with myeloblastosis, erythroblastosis, or myelocytomatosis. Only in lymphoid leukosis, however, is the estimate based on both laboratory inoculations and development of disease following natural exposure. With the other diseases, natural cases are rare and sporadic, so estimates of induction and/or incubation periods are based only on laboratory inoculations (Purchase and Burmester, 1978).

The avian retroviruses are classified into subgroups based on the genetic resistance of strains of chicken cells, on host cell receptor interference patterns, and on specificity of neutralization by antibodies (Vogt, 1970). Subgroup A represents the most common category of field isolate, but subgroup B agents are also occasionally found as field isolates. Subgroups C and D appear to exist

TABLE II
Malignant Tumors of Chickens Caused by Retroviruses

Diseases	Major pathological lesion(s)
Leukemias and lymphomas	
Long latent period	
Lymphoid leukosis	Enlargement of liver, spleen, bursa of Fabricus
Short latent period	
Myeloblastosis	Anemia, liver nodules
Erythroblastosis	Anemia, petechial hemorrhages
Myelocytomatosis	Tumors at bone surface
Solid connective tissue tumors	
Malignant	
Fibrosarcoma	Firm, diffuse infiltration of muscles and subcutis
Myxosarcoma	Soft, mucinous infiltration of muscles and subcutis
Histiocytic sarcoma	Widely distributed metastasis of several types of cells
Osteosarcoma	Infiltrative lesions of bone
Hemangiosarcoma	Vascularized infiltrative lesions
Benign	
Fibroma	Localized masses in skin or muscle
Myxoma	Localized mucinous masses
Osteopetrosis, osteoma	Localized bone growth
Hemangioma	Blood depositions on organ surfaces
Solid tumors of epithelial origin	
Nephoblastoma	Nodules in kidneys
Hepatocellular carcinoma	Nodules in liver

primarily as laboratory-derived agents. Subgroup E agents represent the genetically inherited nononcogenic avian retroviruses. Most if not all lines of chickens, as well as wild jungle fowl, have complete copies of subgroup E viruses as part of their battery of cell genes. Subgroup F and G viruses are found primarily in pheasants. Chicken cells are classified as C/A, C/B, C/O, C/AB, etc., according to whether they restrict replication only to subgroups other than A, only to subgroups other than B, are permissive to all subgroups, are permissive only to subgroups other than A and B, etc. Some subgroups will replicate in nonchicken avian cells, especially Japanese quail cells. Apart from certain strains of RSV, most avian retroviruses will not cause disease in species other than chickens. In high doses, RSV will cause tumors with low efficiency in mammals (Vogt, 1965a,b).

The major pathological lesions for the various diseases caused by avian retroviruses are summarized in Table II. Lymphoid leukosis is a disease that starts in bursal (B) lymphocytes (Payne and Rennie, 1975). The chicken usually acquires the virus before birth and maintains a state of relatively immune tolerance which allows virus replication throughout life. If first infected as juveniles or adults, most birds mount a significant virus-neutralizing antibody response which would result in elimination of the infection. The tumors begin as monoclonal follicles in the bursa (Neiman *et al.*, 1980), and subsequently a single clone of malignant cells will spread beyond the bursa to form tumors in other abdominal organs. The organs most frequently affected are the liver and spleen, but many other organs can also be involved (Purchase and Burmester, 1978). Removal of the bursa by surgical or chemical bursectomy or infection with the cytopathic infectious bursal agent can prevent the development of lymphoid leukosis, while thymectomy has no effect on disease development (Purchase and Gilmour, 1975; Purchase and Cheville, 1975).

Several field strains of avian retroviruses produce erythroblastosis. In some cases this disease is associated with the presence of many erythroblasts in the blood, while in other cases the primary characteristic is a severe anemia. Petechial hemorrhages are also usually present in mucosal surfaces of the thoracic and abdominal organs and in muscle. The clinical signs and symptoms of the anemic form of erythroblastosis are similar to those seen in myeloblastosis. In both diseases, a severe disruption occurs in the normal architecture of the bone marrow. Myeloblastosis is, however, characterized by a dramatic blood leukemia in which up to 2 million myeloblasts per mm^3 may be found in peripheral blood. Myelocytomatosis is characterized by the presence of tumors on the periosteal surfaces of bones. The tumors often occur at cartilaginous junctions. Some of the same viruses that induce myeloblastosis or myelocytomatosis, such as the BA1 strain A or MC29, also induce solid tumors, such as nephroblastoma and hepatocellular carcinoma (Purchase and Burmester, 1978). The field isolates that cause fibrosarcomas will also usually induce most of the connective tissue tumors listed in Table II, and many will also induce lymphoid leukosis, erythroblastosis, and/or myeloblastosis. Some of this broad range of pathogenesis is almost certainly due to the coexistence of etiological agents that induce different diseases. However, some cloned agents still show an ability to induce several morphologically distinct types of disease.

The avian retroviruses that cause all diseases other than lymphoid leukosis are characterized by the presence of an extra *onc* or *src* gene. In the case of the replication-competent Schmidt-Ruppin strain of RSV, this gene is positioned at the 3′ end of the *env* gene, and deletion mutants are still capable of replication in the usual fashion (Stehelin *et al.*, 1976; Wang *et al.*, 1976c; Graf and Beug, 1978).

Those viruses causing the acute leukemias or short latent period leukemias, as well as other strains of sarcoma viruses, have this *onc* gene inserted in the middle of the viral genome (Bistner *et al.*, 1977; Graf and Beug, 1978). As a result, the gene(s) that code for reverse transcriptase are usually lost, as well as the adjacent nucleotide sequences at the 3′ end of the *gag* gene and at the 5′ end of the *env* gene. This viral genome obviously then becomes incapable of self-replication and requires the presence of a replication-competent complete viral genome as helper to perpetuate itself.

All of the viral genomes that contain an *onc* or *src* insert, regardless of the position occupied by that gene, appear to have acquired that sequence from the cell, where it exists as a "regulatory" or "differentiation" gene that might otherwise be repressed. Most of these viruses thus appear to have become "supercharged" in their transformation capabilities as well as their oncogenic activities by the act of transducing a gene from normal cells where it would otherwise remain under tight control. In those systems that have been carefully examined, this gene appears to code for a protein kinase activity (Collett and Erikson, 1978) which could play an important role in a cascade of phenomena that control cell structure and metabolism. It is presumably more than a coincidence that most of the acute leukemias or short latent period avian retroviruses that have acquired this property are also capable of transforming chick fibroblasts.

None of this, however, appears to apply to the retroviruses that cause lymphoid leukosis, the disease that is by far the most common in chicken flocks. During the long latent period, it is hypothetically possible that such a defective recombinant virus may be generated, but this seems unlikely. Because the lymphoid tumors are monoclonal, unlike most short latent period tumors, it appears that a functional *onc* gene is not being efficiently transmitted from cell to cell. Rather, the lymphoid leukosis virus must be either selecting a specifically preprogrammed antigen-sensitive B cell and stimulating that clone to multiply or alternatively creating a specific but very rare insult to a random B cell, which then behaves in a malignant manner to perpetuate the clonal tumor.

B. Cats

A variety of pathological entities that occur naturally in outbred domestic cats are caused by the feline retroviruses. In the most direct sense, these include lymphoid and myeloid leukemias, various forms of lymphomas, and multicentric fibrosarcomas. Under experimental conditions, the FeSVs will also induce melanomas (McCullough *et al.*, 1972). The feline retroviruses also cause a severe immunosuppressive syndrome associated with thymic atrophy and aplastic anemia. As a result, many cats that are infected with FeLV succumb to unrelated infectious diseases that would otherwise be controlled by a fully functioning immune response (Essex *et al.*, 1975b). In fact, it appears likely that FeLV causes higher mortality in the cat population by this latter mechanism than by functioning as an oncogenic agent. A list of the diseases that are associated with infection by FeLV and FeSV is presented in Table III.

TABLE III
Diseases Associated with Infection by Feline Leukemia and Sarcoma

Diseases	Major pathological lesion(s)
Leukemias and lymphomas	
Leukemias	
Lymphoid	Lymphoblastic infiltration of bone marrow
Myeloid	Myeloblastic infiltration of bone marrow
Erythroid	Erythroblastic infiltration of bone marrow
Lymphomas	
Thymic	Thymic mass
Alimentary	Gut wall infiltration and enlargement
Multicentric	Enlargement and infiltration of lymph nodes and viscera
Undifferentiated	Localized tumors in kidney, spinal cord, skin, other locations
Solid tumors	
Fibrosarcoma	Infiltrative masses in subcutis and muscle
Melanoma	Pigmented masses in subcutis
Nonneoplastic diseases	
Aplastic anemias	Low erythrocyte levels in blood and bone marrow
Thymic atrophy	Runting, various infectious disease signs due to immunosuppression
Glomerulonephritis	Membranous depositions in kidneys

Several forms of leukemia and lymphoma occur naturally at a relatively high rate in cats: at least 40 cases per 100,000 per year (Dorn *et al.*, 1967). In lymphocytic leukemia, the primary lesions are represented by high numbers of lymphoblasts and atypical lymphoid cells in bone marrow and peripheral blood (see Fig. 5). This disease is very similar in the hematological sense to acute lymphoblastic leukemia of children, and it represents the most common form of lymphoid neoplasia in some geographical areas (Cotter and Essex, 1977). Afflicted cats usually have a severe anemia and thrombocytopenia. Aside from the lymphoid leukemia, some cats develop other forms of leukemia following infection with FeLV such as erythroleukemia, myeloid leukemia, or reticuloendotheliosis. These forms of the disease are diagnosed by routine hematological examinations of the blood and bone marrow. The lymphoid form of leukemia is apparently composed primarily of T lymphocytes (Hardy *et al.*, 1977).

Lymphoma and lymphosarcoma occurs in several forms in cats. Perhaps the most characteristic in presentation is the thymic form, where involution of the normal cellular elements is followed by replacement with neoplastic cells. Subsequently, the tumor may grow to fill much of the thoracic cavity (Fig. 6) and/or spread to the spleen and to other organs. The thoracic cavity usually contains an

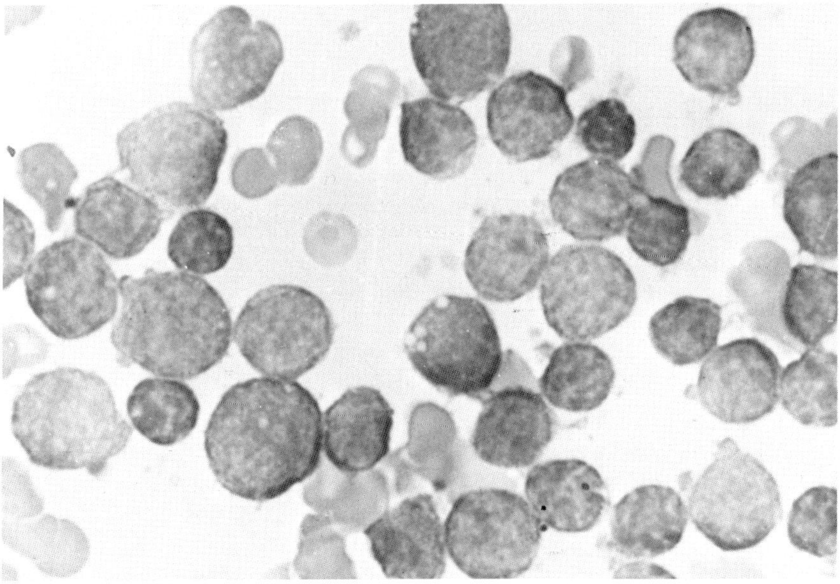

Fig. 5. A bone marrow aspirate from a cat with acute lymphoblastic leukemia. The marrow has been replaced with a population of lymphoid cells of various sizes. The predominant cell is a lymphoblast with a fine chromatin pattern, a relatively indistinct nucleolus, and a thin rim of blue cytoplasm. (Courtesy of Angell Memorial Animal Hospital, Boston, Mass.)

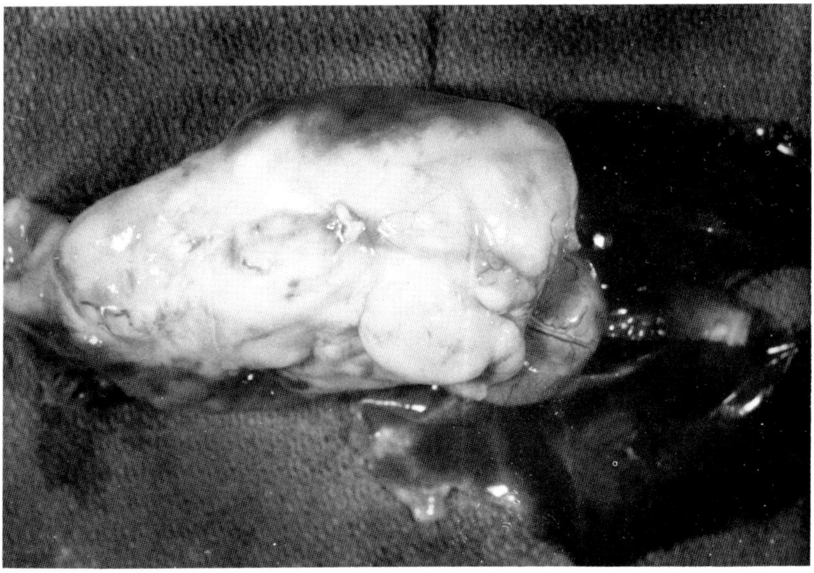

Fig. 6. Thoracic viscera from a cat with thymid lymphoma. The tumor mass encompasses the heart on the left and the lungs on the right. (Courtesy of Angell Memorial Animal Hospital, Boston, Mass.)

ascitic fluid which may contain large numbers of free leukemic lymphoblasts (Fig. 7). Cats presenting with this form of disease often have respiratory distress due to pressure on the lungs caused by the tumor and/or ascitic fluid; the disease can be readily diagnosed by the use of thoracic radiographs and the examination of thoracic fluid. This form of the disease is also made up of T lymphocytes.

Other categories of lymphoma that are seen in FeLV-infected cats are the multicentric, alimentary, and unclassified forms. The multicentric form involves the lymph nodes (Fig. 8) and other abdominal organs and may be either a T or a B form of the disease. The alimentary form apparently begins in the lymphoid follicles of the Peyer's patches lining the gut and may occur as a diffuse enlargement of the gut or as a tumor of the abdominal lymph nodes. This form of the disease appears to develop from B lymphocytes. The unclassified forms include those focal lymphomas that develop in nonlymphoid organs such as the central nervous system, the kidney capsule (Fig. 9), or the skin. The signs and symptoms associated with these forms are obviously directly related to the organ system affected, and they thus vary widely.

Although most leukemias and lymphomas of cats occur in animals that are viremic with FeLV, 25–30% occur in cats that do not contain infectious virus at the time the disease is diagnosed (Francis *et al.*, 1979a). The proportion that are virus-negative appears to be somewhat higher in the alimentary form than in the

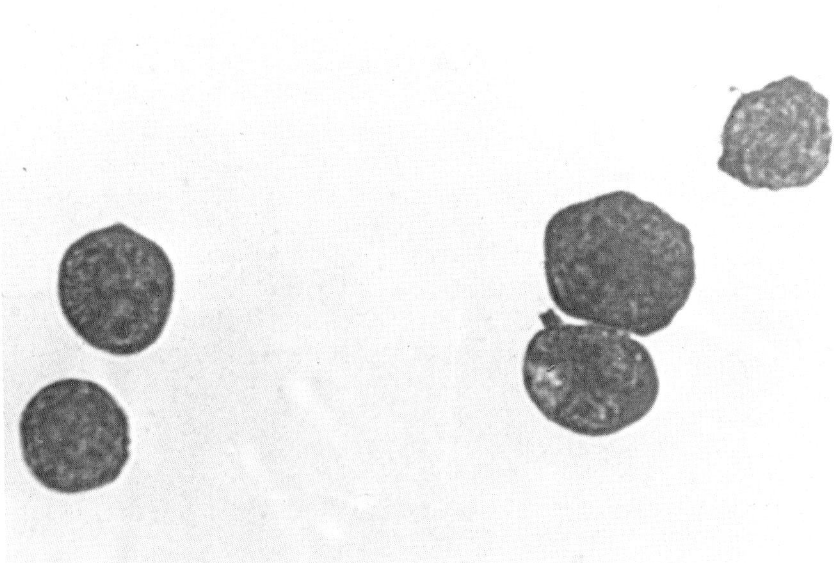

Fig. 7. Malignant lymphoblasts in pleural fluid from a cat with thoracic lymphoma. (Courtesy of Angell Memorial Animal Hospital, Boston, Mass.)

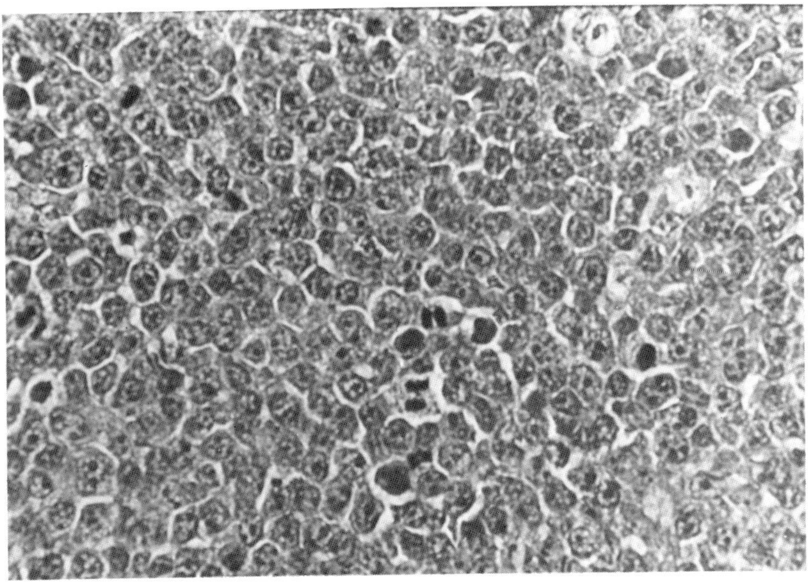

Fig. 8. Histological section of a lymphomatous lymph node. The architecture of the node has been lost, and a homogeneous population of poorly defined lymphoid cells remains. (Courtesy of Angell Memorial Animal Hospital, Boston, Mass.)

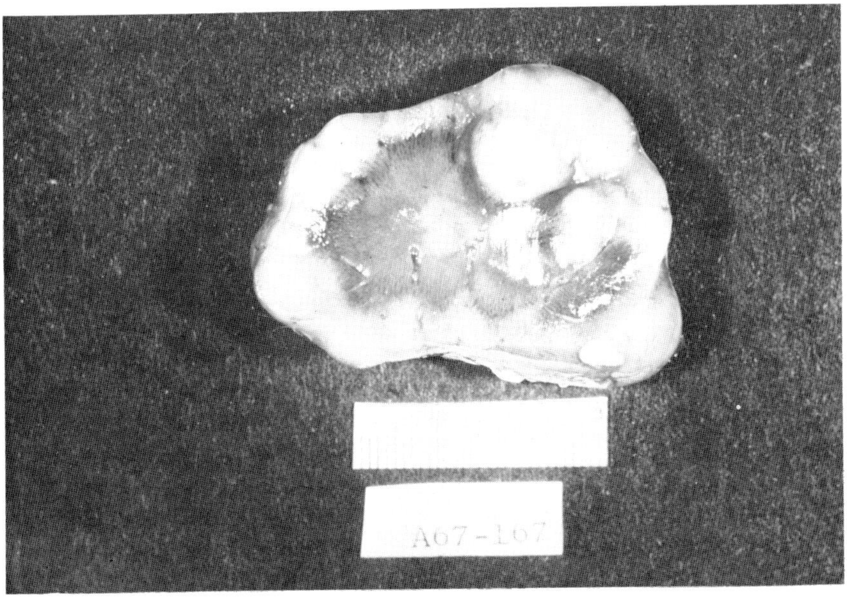

Fig. 9. Cross section of a kidney with renal lymphoma. Irregular nodules of the tumor replace much of the cortex and extend into the medulla. (Courtesy of Angell Memorial Animal Hospital, Boston, Mass.)

thymic or multicentric forms, but both virus-positive and virus-negative cases occur with all the major pathological forms of the disease. Virus-negative cases occur more frequently than virus-positive cases in older cats. However, this appears to be because of an absolute decrease in the number of cases of virus-positive leukemia in old cats since the total rate of all cases of leukemia-lymphoma is much higher in young cats and the absolute rate of virus-negative cases is at least as high in young cats as in old cats. Thus, leukemia-lymphoma in the feline species is primarily a disease of the young. The relative risk for disease development may be slightly higher for males, but as yet there is no clear evidence for increased risk for any particular breed.

The incubation or induction period for development of leukemia or lymphoma following natural exposure to FeLV is both variable and prolonged (Francis *et al.*, 1979b). It varies from about 2 or 3 months to several years after initial infection, with a mean period of 1–2 years. Most animals that become exposed to FeLV do not develop leukemia but instead rid themselves of persistent infections due to an efficient virus-neutralizing antibody response. FeLV is excreted at high levels in the saliva of persistently infected cats, and the virus is ubiquitous in the population. It is efficiently transmitted in a contagious manner, and both the disease and viremia frequently occur in household clusters (Cotter *et al.*, 1974). When persistent viremia and disease development do occur in clusters, this is probably because the animals are exposed to unusually high doses of virus at very young ages. The clusters do not appear to be related to any genetic predisposition of the cats because most contain cats with very different genetic backgrounds (Essex *et al.*, 1975c).

Recent results suggest that cases of virus-negative feline leukemia or lymphoma occur just as frequently in cats with documented previous exposure to FeLV as do cases of virus-positive feline leukemia (Hardy *et al.*, 1980). Thus, this suggests that FeLV must be considered an obvious candidate as the cause of the virus-negative cases as well as the virus-positive cases. An immunoselection hypothesis has been advanced to suggest how this might occur (Essex, 1980).

FeSV, which has a replication defective genome and is found in the presence of replication-competent helper viruses, has been isolated from several cases of naturally occurring multicentric fibrosarcomas in young cats. These agents will induce fibrosarcomas after a very short latent period when inoculated subcutaneously or intramuscularly (Snyder and Theilen, 1969). When inoculated intracutaneously or intraocularly, FeSV will induce melanomas (McCullough *et al.*, 1972).

RD-114 virus is an endogenous xenotropic agent of cats that is genetically inherited as one or more complete copies in each normal cell genome (Fischinger *et al.*, 1973; Livingston and Todaro, 1973). The virus can be activated by mitogens and various drugs, but it does not appear to play any role in tumor development. RD-114 is xenotropic, and once activated it grows efficiently only on cells of nonfeline species. It seems possible, although not yet established, that some of the RD-114 proteins may function in embryonic or fetal differentiation (Niman *et al.,* 1977).

Several subgroups of FeLV have been characterized, designated A, B, and C, based on the envelope glycoprotein antigens, interference, and host cell infectivity patterns (Sarma and Log, 1973). Subgroup A is found whenever FeLV is isolated. Some isolates also contain B with A and rarely C with A or with AB, but only A is found alone. About half of the isolates are A alone, about half are AB, and 1% or less are AC or ABC (Jarrett *et al.,* 1978). All appear to have some pathogenic activity, but no clear correlation between subgroup and pathogenicity has been established. It appears likely that different strains of FeLV can be identified within a single subgroup by RNA fingerprinting of the genome (Rosenberg and Haseltine, 1980).

All proteins coded for by FeLV and expressed as virion structural proteins are immunogenic in cats (Essex, 1980). These include the *gag* gene peptides p15, p12, p30, and p10, reverse transcriptase, and the *env* gene peptides p15e and gp70. Relatively few animals have high antibody titers to many of the proteins, but significant titers to all the proteins have been found in some nonviremic cats that are naturally exposed to the virus. As a general rule, cats that have persistent viremia have a state of antigen excess and do not have any detectable free antibodies to any of the viral proteins. At least in theory, however, a given cat might have subgroup-specific virus-neutralizing antibody actively in the presence of viremia that was confined to a different subgroup.

Another antigen that is associated with FeLV and FeSV is FOCMA, the feline oncornavirus-associated cell membrane (Essex *et al.,* 1971). FOCMA is, however, tumor and/or transformation specific, and it is not present in FeLV virions grown on FOCMA-negative cells. FOCMA is specifically expressed at the cell membrane on cells transformed by FeSV, including all strains of FeSV, but not on cells transformed by FeSV, including all strains of FeSV, but not on cells transformed by chemicals or other RNA tumor viruses (Essex *et al.,* 1980). Since this antigen can be induced in nonfeline cells by transformation with FeSV, it seems likely that it is encoded by this viral genome (Sliski *et al.,* 1977). FOCMA is also expressed on leukemia and lymphoma cells, including leukemic cells that do not release FeLV or make any of the FeLV virion structural proteins.

FOCMA is highly immunogenic in cats. It generates a humoral antibody response. The levels of antibody elicited are high in cats that resist the development of leukemia, lymphoma, or progressive fibrosarcoma, but low or undetectable in cats that develop leukemia, lymphoma malignant melanoma, or progressive fibrosarcoma (Essex, 1975). For the elicitation of FOCMA antibody, cats must either experience a generalized infection with FeLV or FeSV or be vaccinated with unfixed tumor cells. Vaccination with killed virus may elicit antiviral antibodies but not antibodies to FOCMA (Olsen *et al.,* 1977).

FOCMA antibodies exert a lytic effect on lymphoid tumor cells when in the presence of cat complement (Grant *et al.*, 1977). Most cats that become exposed to FeLV develop detectable levels of FOCMA antibodies. It appears likely that this represents an efficient immunosurveillance mechanism to prevent tumor development under natural conditions (Essex *et al.*, 1975a,d).

The role of FeLV as an immunosuppressive agent also appears to be very important under natural conditions (Cotter *et al.*, 1974). Cats that are persistently viremic with FeLV have a greatly increased risk for the development of such diseases as infectious peritonitis, septicemia, pneumonitis, and glomerulonephritis. FeLV is known to suppress the homograft rejection response and the blastogenic response normally elicited in lymphocytes by mitogens such as concanavalin A. It appears that the virion component that is most likely to be responsible for this action is either p15 or p15e (Mathes *et al.*, 1979).

C. Primates

Relatively little is known about the role of retroviruses as etiological agents of neoplastic diseases of primates. Many primate type C and D retroviruses have been isolated. They can be arbitrarily grouped into four general classes. First, a large number of endogenous inherited xenotropic retroviruses have been found in primates ranging from very primitive species to the great apes and including both Old World and New World species (Todaro, 1978; Rangan and Gallagher, 1979). None of the agents in this category have yet been shown to be oncogenic, but repeated isolations and/or visualizations of such particles have been made in placental and fetal tissues, and the possibility that such agents play a role in differentiation has been considered. Although many such viruses have been found in both the type C and type D morphological categories, the prototype agents of this class are the M7-M28 series of baboon retroviruses (Todaro *et al.*, 1973). Surprisingly, the baboon agents are closely related to the endogenous xenotropic RD-114 class of viruses found in domestic cats.

The second group of primate retroviruses is composed of several isolates of ecotropic C-type agents from gibbons (Kawakami *et al.*, 1972; Kawakami and Buckley, 1974; Gallo *et al.*, 1978). These viruses are clearly oncogenic, being associated with lymphoma and with myeloid and lymphoid leukemias. The gibbon ape leukemia viruses (GaLV) are transmitted in a horizontal manner, following excretion in saliva and urine. This class of ecotropic virus is completely distinct from any DNA sequences found in normal gibbon cells, so the agent clearly infected the gibbon population from an exogenous source. In fact, the genome of the GALVs appears to be very close to those of an endogenous retrovirus found in Asian mice (Todaro, 1978). In the sense of being composed of totally exogenous unrelated sequences, the GALVs are unlike the FeLVs or the avian lymphoid leukosis viruses. Even though whole copies of replication-competent FeLV are never found in the DNA of uninfected cat cells, most, if not all, domestic cats contain nucleotide sequences that hybridize with 30–60% of the FeLV genome (Okabe *et al.*, 1976; Koshy *et al.*, 1979). In the avian system, the exogenous subgroup A retroviruses that cause lymphoid leukosis share extensive homology with the subgroup E RAV (O) agent, which is the endogenous nononcogenic avian retrovirus that is genetically inherited as complete virus.

The third class of primate retrovirus has only a single representative, the woolly monkey sarcoma virus or the simian sarcoma virus (SSV) (Theilen *et al.*, 1971; Wolfe *et al.*, 1971). The agent is similar to FeSV or the murine sarcoma viruses in that it transforms fibroblasts, causes solid tumors that are undifferentiated fibrosarcomas, and exists as a defective genome with an excess of an associated helper virus. The helper virus, usually designated the simian sarcoma associated virus (SSAV), is serologically very similar to the GaLVs described above (Scolnick *et al.*, 1972c; Parks *et al.*, 1973). The helper virus-related nucleotide sequences in SSV are also closely related to the comparable sequences of the GaLV helper. Perhaps the most surprising aspect is why a defective transforming virus that is related to the GaLVs should arise in an New World species of monkey when the GaLVs themselves are apparently found only in an Old World species of ape, the gibbon. However, SSV was actually isolated from a woolly monkey that was maintained as a pet, and in a household where it had earlier been exposed to a gibbon that had "died of unknown causes" (Rangan and Gallagher, 1979). How the SSV actually evolved is obviously unknown, but it has been suggested that the *src* gene insert might actually be of human origin (Wong-Staal *et al.*, 1979). Upon inoculation into marmosets, SSV produces well-differentiated fibrosarcomas (Wolfe *et al.*, 1971; Wolfe and Deinhardt, 1978). Malignant progressive tumors are produced only if the animals are either inoculated at a relatively young age or given concurrent immunosuppressive therapy. Efficient replication of SSV/SSAV occurs in the inoculated animals, as opposed to the situation when RSV or FeSV is used to induce tumors in marmosets (Wolfe and Deinhardt, 1978).

The fourth class of primate retroviruses are those isolated from human tissues and/or cell cultures derived from human tissues (Gallagher and Gallo, 1975; Nooter *et al.*, 1975; Panem *et al.*, 1975; Kaplan, 1978). Since such isolations have been rare and difficult to reproduce, questions have inevitably arisen about the possibility that the putative human agents represent contaminants derived from subhuman primate stocks, especially the GaLVs and the baboon M7-M28 group. This question must be considered unresolved, however, because the GaLVs have thus far been isolated only from captive gibbons either associated with man and/or inoculated with human materials.

Since wild gibbons are not available for study or frequent importation, all studies thus far done with the GaLVs have been limited to animals that have been maintained in zoos or primate colonies. At least four distinct strains or subtypes of GaLV have been isolated. Nine cases of leukemia or lymphoma occurred in a colony of approximately 200 maintained in Thailand (De Paoli *et al.*, 1973). Four of the cases were lymphoma and five were myeloid leukemia. All apparently occurred in animals that were inoculated with either human blood containing the malarial organism (*Plasmodium falciparum*) or blood from other gibbon(s) that had themselves been inoculated with human blood. Five cases of spontaneous hematopoietic malignancies have been examined, and all have readily yielded GaLVs. These malignancies have also been induced in young gibbons in the laboratory following the inoculation of cell-free virus-containing fluid (Kawakami *et al.*, 1978).

The four strains of GaLV can be distinguished from each other by their type-specific p12 antigens, by nuclei acid hybridization, by examination of the strong-stop sequences, and by the fingerprints of the RNA genomes (Gallo *et al.*, 1978; Haseltine and Kleid, 1978; Krakower *et al.*, 1978; Reitz *et al.*, 1979). One strain was isolated from an animal with lymphoma housed in the San Francisco zoo (Snyder *et al.*, 1973). A second animal housed in the same cage also had virus-positive lymphoma. The second strain was isolated from an animal with myeloid leukemia housed in the Thailand colony. This source of virus has been used to induce subsequent cases of myeloid leukemia in inoculated animals. The third strain or type was isolated from a gibbon with lymphoma housed on Hall's Island, Bermuda. Finally, another strain of GaLV was isolated from brain tissue taken from gibbons that had been inoculated with extracts from humans having kuru.

The GaLVs are transmitted efficiently to other gibbons housed with the infected animals. Seroepidemiological studies have revealed that many healthy animals known to be exposed to the leukemic animals themselves had either circulating GaLV or antibodies to one or more of the GaLV-associated antigens (Kawakami *et al.*, 1978). The antibodies detected are directed to p30, gp70, reverse transcriptase, and an antigen found at the surface of virus-producer lymphoma cells that were established from a disease case. Goats inoculated with SSV-transformed nonproducer autocthonous cells develop antibodies that react specifically with both GaLV-producer lymphoma lines and SSV-transformed fibroblasts (Theil and Iglehart, 1980). It appears likely that this antigen is similar to FOCMA, which is a target for the immunosurveillance response in infected cats (Essex *et al.*, 1975d).

D. Cattle

Cattle develop several forms of lymphoid tumors. The most commonly observed form, designated "enzootic bovine leukosis (EBL)," is caused by a retrovirus designated "bovine leukemia virus (BLV)." The other forms are designated the "sporadic" form, the "thymic" form, and "skin leukosis." The sporadic and thymic forms occur primarily in animals under 2 years of age and are quite infrequent in occurrence. The skin and EBL forms occur primarily in older animals (Burny *et al.*, 1978).

The EBL form occurs in animals of various ages, but the peak incidence is at 4–6 years of age (Burny *et al.*, 1978). The incubation and/or induction periods can be very long. The disease clusters in certain infected herds where a majority or at least a substantial minority of the animals show evidence of infection. No breed or sex predisposition is apparent. Within a given breed, however, certain families clearly have a higher risk for development of persistent lymphocytosis, BLV infection, and EBL (Abt *et al.*, 1976). The mechanism of transmission of BLV is unknown. It is known, however, that the agent is not genetically inherited. It is either acquired as a horizontal agent (Piper *et al.*, 1975; Ferrerr *et al.*, 1976) or congenitally transmitted from the dam to progeny either in milk or across the placental barrier (Onuma and Olson, 1977). Milk, saliva, urine, and feces have all been considered as possible vehicles. Insect vectors have also been considered (Ferrerr, 1980). However, BLV is highly cell associated, and the suggestion has also been made that the agent might be transmitted in intact cells rather than in free form (Miller, 1980). Dairy cattle appear to have a higher incidence of EBL than cattle raised for beef production. This could, however, simply reflect the fact that dairy animals usually are kept alive much longer, and dairy animals are often maintained indoors under conditions of closer confinement where transmission of an agent might be facilitated.

Persistent lymphocytosis occurs in many, but not all, animals as a preleukemic syndrome (Bendixen, 1961, 1964). The major lesions seen in EBL, other than persistent lymphocytosis, are enlargement and infiltration of various tissues with malignant cells, especially the lymph nodes, muscle, gut wall, subcutis, and mammary gland. Animals may demonstrate persistent lymphocytosis for many years without developing other signs of EBL. The malignant cell itself in EBL is a B lymphocyte (Ferrerr, 1980).

The BLV, although appearing morphologically as a C-type retrovirus, does not share any major interspecies-specific antigens with other recognized mammalian retroviruses (Devare, 1980). The p24, for example, as the major core antigen of the virion, does not share common antigenic determinants with the murine, primate, or feline agents as the latter group shares such antigens with each other. Also, since the BLV is also highly cell associated, significant levels of infectious virus are rarely found free in the plasma of infected animals. Virus-neutralizing antibodies, directed primarily to the gp70 antigen, are believed to play an important role in neutralizing free virus and/or preventing virus release (Driscoll *et al.*, 1977). Once bovine leukemia cells have been cultivated *in vitro,* they often release significant amounts of BLV (Miller *et al.*, 1969). Such BLV will then infect other cells and will often grow best in fibroblasts from other species, such as sheep (van der Maaten *et al.*, 1974). The addition of substances such as mitogens to the cultured leukemia cells may aid virus release (Olson, 1977; Driscoll *et al.*, 1977). BLV causes syncitia in some cells that become infected (Diglio and Ferrerr, 1976).

BLV is completely unrelated to any endogenous sequences found in the DNA of normal bovine cells, which provides further evidence that the agent is horizontally acquired. Even normal lymphoid cells or fibroblasts from infected leukemic animals are animals are negative for DNA proviral sequences, suggesting that only the leukemic cells and/or most B cells are infected (Kettman *et al.*, 1980).

Epidemiologically, EBL has long been recognized as an infectious disease. Certain countries, such as Denmark, Germany, and the Soviet Union, have a high incidence of infected herds, while other areas, such as the British Isles, are essentially free of the disease (Olson and Baumgartener, 1975). In infected countries up to two-thirds of the herds and 15% of the animals may show evidence of infection (Baumgartener *et al.*, 1975). Within a given country there are usually preferential regions where higher rates of infection occur. The proportion of herds that are judged infected are usually classified as such on the basis of having a high proportion of animals with persistent lymphocytosis or with antibodies to the p24 or gp 70 antigen. In certain countries such as Denmark, attempts have been made to reduce the incidence of disease by the slaughter of infected herds of cattle (Bendixen, 1964). However, a significant proportion of the animals in most herds never become infected. Whether this is due to genetic resistance or other factors is unknown. Additionally, probably 5% or less of the animals that become persistently infected with BLV appear to develop EBL (Ferrerr *et al.*, 1977). This could be due to an antitumor cell surveillance mechanism, but no tumor target cell antigen has yet been identified.

Nucleic acid hybridization experiments reveals that EBL tumors are monoclonal, as are the B cell tumors of chickens (Kettman *et al.*, 1980). Provirus integration occurs only in specific locations of the host cell genome, and only one copy of proviral DNA is integrated per haploid genome. Cells that are taken at a preclinical stage of disease such as persistent lymphocytosis show a polyclonal distribution, with the BLV genome existing in different locations in different infected cells. Although the EBL tumor cell populations show a consistent integration pattern for all cells of the same tumor, different tumors show different patterns, suggesting that multiple integration sites exist that are compatible with tumor induction and/or that slightly different strains of BLV may be causing the disease in different animals (Kettman *et al.*, 1980).

Sheep also get leukosis, and under some circumstances it appears to occur preferentially in certain herds (Enke *et al.*, 1961; Paulsen *et al.*, 1973). A C-type virus has been found in some cases, and thus far it appears to be indistinguishable from BLV. BLV will infect and induce leukosis in sheep under laboratory conditions (Wittman and Urbaneck, 1969; Mammerickx *et al.*, 1976; van der Maaten and Miller, 1976), so it seems possible that the bovine agent is either BLV itself or a slight variant of BLV.

V. DETECTION OF RETROVIRUSES

Within recent years, major developments have occurred in the various methodologies useful for the analysis of viral genomes and viral proteins. Among others, these include vast improvements in ways to sequence nucleic acid strands by the Maxam–Gilbert technique, to map genomes with restriction endonucleases, to detect readily site-specific integration of whole genomes or genome fragments by the Southern blotting procedure, to isolate and characterize mRNAs by Northern blotting, to identify and characterize virus-coded peptide products by cell-free translation, and to identify specific antigenic determinants using monoclonal antibodies. A detailed review of these and other procedures would not be appropriate here, and most have been thoroughly reviewed elsewhere. Numerous other biochemical and serological procedures are routinely used for screening biological materials for retrovirus components under experimental conditions.

Known retroviruses that occur naturally in outbred species such as chickens, cattle, cats, and primates are detected primarily by serological procedures. In the case of chickens with avian retrovirus infections, a complement fixation test designated the "COFFAL test" is widely used (Sarma *et al.*, 1964). It is based on the induction of the major core antigen in chick embryo cells. Another test that is occasionally used involves nonproducer RSV-infected cells to rescue the transforming agent (Rispens *et al.*, 1970). A phenotypic mixing test which involves the use of an RAV-O pseudotype of RSV to coinfect cells containing the unknown avian leukosis virus and subsequently to plate on cells that are not permissive to RAV-O has also been described (Okazaki *et al.*, 1975).

In cats, detection of FeLV in infected animals has been done with a simple but accurate indirect immunofluorescence procedure (Hardy *et al.*, 1973). Blood films are prepared from infected animals, fixed, incubated with antiserum to FeLV p30 group-specific antigen and subsequently with the appropriate fluorescein-conjugated antiserum to cat IgG. Virus antigens are detected in phagocytic cells, lymphocytes, and platelets. Only a drop or two of blood is needed, and the dried blood films are stable for weeks or months.

In the case of bovine leukosis, several tests have also been used for screening, on both an individual and a herd basis. The major difference between the tests used for EBL and the ones used for feline leukemia or avian leukosis is that the tests for EBL are based on the detection of antibody rather than antigen. In the case of avian and feline leukemias, the persistent viremia results in an antigen excess, unlike the bovine system, where the virus is highly cell associated. An immunodiffusion test and various radioimmunoassays have been widely used for testing serum samples from cattle (Burny *et al.*, 1978).

Virus neutralization procedures, radioimmunoassays, and immunofluorescent tests have all been used for seroepidemiological screening of captive gibbons (Kawakami *et al.*, 1980).

VI. CONCLUSIONS

Viruses have been thoroughly documented as the etiological agents that cause certain forms of neoplastic diseases in several outbred species. Only viruses that can integrate into the DNA of the host cell have thus far been shown to possess oncogenic potential. Foremost among these are the herpesviruses, which have a DNA genome, and retroviruses, which contain an RNA genome but replicate through an intermediate DNA provirus stage.

Retroviruses are ubiquitous in certain species such as cats, cattle, and chickens. They replicate in many cells of the body and are transmitted in a contagious manner. In the cat, for example, the virus is primarily transmitted in saliva. Only under unusual circumstances do the infecting viruses cause disease, and when disease does result, it is usually after a prolonged and variable induction period. The events that transpire during this time period are not well understood. The class of neoplastic disease that appears to result most frequently following infection with these viruses under natural conditions is leukemia or lymphoma. Other forms of tumors are occasionally caused by retroviruses in certain species. Additionally, some of these agents also may be pathogenic to their hosts by causing a variety of nonneoplastic diseases. These include chronic degenerative diseases such as skeletal paralysis, aplastic anemias, and various secondary diseases resulting from thymic atrophy and immunosuppression.

During the last 5–10 years, a vast amount of basic information has been obtained on the molecular biology of retroviruses, including events related to host cell transformation. Most of this information has been generated using highly selected laboratory strains of retroviruses. Although leukemia and lymphoma represent the major diseases caused by these agents in nature, we still know very little about the molecular mechanisms involved in this process. Hopefully this information will be forthcoming in the next 5–10 years.

REFERENCES

Aaronson, S. A. (1973). *Virology* **52**, 562–567.

Aaronson, S. A., and Weaver, C. A. (1971). *J. Gen. Virol.* **13**, 245–251.

Aaronson, S. A., Parks, W. P., Scolnick, E. M., and Todaro, G. J. (1971). *Proc. Natl. Acad. Sci. U.S.A.* **68**, 920–924.

Abrell, J. W., and Gallo, R. C. (1973). *J. Virol.* **12**, 431–439.

Abt, D. A., Marshak, R. R., Ferrerr, J. F., Piper, C. E., and Bhatt, D. M. (1976). *Vet. Microbiol.* **1**, 287–300.

Arcement, L. J., Karshin, W. L., Naso, R. B., Jamjoom, G., and Arlinghaus, R. (1976). *Virology* **69**, 763–774.

Arlinghaus, R. B., Naso, R. B., Jamjoom, G. A., Arcement, L. J., and Karskin, W. L. (1976). *In* "Animal Virology" (D. Baltimore, A. S. Huang, and C. F. Fox, eds.), pp. 689–716. Academic Press, New York.

August, J. T., Bolognesi, D. P., Fleissner, E., Gilden, R. V., and Nowinski, R. C. (1974). *Virology* **60,** 595-601.
Bader, J. P., and Brown, N. R. (1971). *Nature (London)* **234,** 11-12.
Baltimore, D. (1970). *Nature (London)* **226,** 1209-1211.
Baltimore, D. (1974). *Cold Spring Harbor Symp. Quant. Biol.* **39,** 1187-1200.
Baltimore, D., and Smoler, D. F. (1971). *Proc. Natl. Acad. Sci. U.S.A.* **68,** 1507-1511.
Baltimore, D., and Smoler, D. (1972). *J. Biol. Chem.* **247,** 7282-7287.
Baluda, M. A., Shoyab, M., Markham, P. D., Evan R. M., and Drohan, W. N. (1974). *Cold Spring Harbor Symp. Quant. Biol.* **39,** 869-874.
Barbacid, M., Stephenson, J. R., and Aaronson, S. A. (1977). *Cell* **10,** 641-648.
Baumgartner, L. E., Olson, C., Miller, J. M., and Van der Maaten, M. J. (1975). *J. Am. Vet. Med. Assoc.* **169,** 1189-1191.
Bendixen, H. J. (1961). *Mod. Vet. Pract.* **42,** 33.
Bendixen, H. J. (1964). *Bull. Off. Int. Epizoot.* **62,** 675-700.
Bernhard, W. (1958). *Cancer Res.* **18,** 491-509.
Biggs, P. M., Milne, B. S., Graf, T., and Bauer, H. (1973). *J. Gen, Virol.* **18,** 399-403.
Billeter, M. A., Parsons, J. T., and Coffin, J. M. (1974). *Proc. Natl. Acad. Sci. U.S.A.* **71,** 3560-3564.
Biquard, J. M., and Vigier, P. (1970). *C. R. Acad. Sci.* **271,** 2430-2433.
Bishop, J. M., and Varmus, H. (1975). *In* "Cancer" (F. F. Becker, ed.), Vol. 2, pp. 3-48. Plenum, New York.
Bishop, J. M., Levinson, W. E., Quintrell, N., Sullivan, D., Fanshier, L., and Jackson, J. (1970a). *Virology* **42,** 182-188.
Bishop, J. M., Levinson, W. E., Sullivan, D., Fashier, L., Quintrell, N., and Jackson, J. (1970b). *Virology* **42,** 727-736.
Bistner, K., Hayman, M. J., and Vogt, P. K. (1977). *Virology* **82,** 431-448.
Bolognesi, D. P., Luftig, R., and Shaper, J. H. (1973). *Virology* **56,** 549-564.
Bolognesi, D. P., Collins, J. J., Leis, J. P., Moenning, V., Schafer, W., and Atkinson, P. H. (1975). *J. Virol.* **16,** 1453-1463.
Bonar, R. A., Sverak, L., Bolognesi, D. P., Langlois, A. J., Beard, D., and Beard, J. W. (1967). *Cancer Res.* **27,** 1138-1143.
Brugge, J. S., and Erikson, R. L. (1977). *Nature (London)* **269,** 346-348.
Burney, A., Bex, F., Chantrenne, H., Cleuter, Y., DeKogel, D. Ghysdael, J., Kettman, R., Le Clercq, M., Leunen, J., Mammerickx, M., and Portetelle, D. (1978). *Adv. Cancer Res.* **28,** 252-311.
Cairns, J. (1979). "Cancer, Science, and Society." Freeman, San Francisco, California.
Canaani, E., and Duesberg, P. (1972). *J. Virol.* **10,** 23-31.
Canaani, E., Duesberg, P., and Dina, D. (1977). *Proc. Natl. Acad. Sci. U.S.A.* **74,** 29-33.
Chan, E. W., Schiop-Stansly, P. E., and O'Connor, T. E. (1974). *J. Natl. Cancer Inst.* **52,** 469-473.
Charman, H. P., Marquardt, H., Gilden, R. V., and Grosjlan, S. (1977). *Virology* **83,** 163-170.
Chopra, H. C., and Mason, M. M. (1970). *Cancer Res.* **30,** 2081-2086.
Chopra, H. C., Hooks, J. J., Walling, M. J., and Gibbs, C. J., Jr. (1972). *J. Natl. Cancer Inst.* **48,** 451-463.
Chopra, H. C., Ebert, P., Woodside, N., Kyedar, J., Albert, S., and Brennan, M. (1973). *Nature (London), New Biol.* **243,** 159-160.
Coffin, J. M., and Billeter, M. A. (1976). *J. Mol. Biol.* **100,** 293-318.
Collett, M. S., and Erikson, R. L. (1978). *Proc. Natl. Acad. Sci. U.S.A.* **75,** 2021-2025.
Collins, J. J., Montelaro, R. C., Denny, T. P., Ishizaki, R., Langlois, A. J., and Bolognesi, D. P. (1978). *Virology* **86,** 205-217.

Cooper, G. M., and Castellot, S. B. (1977). *J. Virol.* **22,** 300–307.
Cotter, S. M., and and Essex, M. (1977). *Am. J. Pathol.* **87,** 265–268.
Cotter, S. M., Essex, M., and Hardy, W. D., Jr. (1974). *Cancer Res.* **34,** 1061–1069.
Dahlberg, J. E., Sawyer, R. C., Taylor, J. M., Faras, A. J., Levinson, W. E., Goodman, H. M., and Bishop, J. M. (1974). *J. Virol.* **13,** 1126–1133.
Dalton, A. J., Melnick, J. L., Bauer, H., Beaudreau, G., Bentveljen, P., Bolognesi, D., Gallo, R., Graffi, A., Haguenau, F., Heston, W., Huebner, R., Todaro, G., and Heine, U. I. (1974). *Intervirology* **4,** 201–206.
Davis, N. L., and Rueckert, R. R. (1972). *J. Virol.* **10,** 1010–1020.
Del Villano, B. C., Nave, B., Croker, B. P., Lerner, R. A., and Dixon, F. J. (1975). *J. Exp. Med.* **141,** 172–187.
De Paoli A., Johnson, D. O., and Noll, W. W. (1973). *J. Am. Vet. Med. Assoc.* **163,** 624–628.
Devare, S. (1980). *In* "Viruses in Naturally Occurring Cancers" (M. Essex, G. J. Todaro, and H. zur Hausen, eds.), pp. 943–952. Cold Spring Harbor Press, Cold Spring Harbor, New York.
Diglio, C. A., and Ferrerr, J. F. (1976). *Cancer Res.* **36,** 1056–1067.
Dorn, C. R., Taylor, D. O. N., and Hibbard, H. H. (1967). *Am. J. Vet. Res.* **28,** 993–1001.
Dougherty, R. M., and Rasmussen, R. (1964). *Natl. Cancer Inst. Monogr.* **17,** 337–341.
Driscoll, D. N., Baumgartener, L. E., and Olson, C. (1977). *J. Natl. Cancer Inst.* **58,** 1513–1515.
Duesberg, P. H. (1968). *Proc. Natl. Acad. Sci. U.S.A.* **60,** 1511–1518.
Duesberg, P. H., Martin, G. S., and Vogt, P. K. (1970). *Virology* **41,** 631–646.
Duesberg, P. H., Vogt, P. K., Beemon, K., and Lai, M. (1974). *Cold Spring Harbor Symp. Quant. Biol.* **39,** 847–857.
Duesberg, P. H. Karvai, S., Want, L.-H., Vogt, P. K., Murphy, H. M., and Hanafusa, H. (1975). *Proc. Natl. Acad. Sci. U.S.A.* **72,** 1569–1573.
Duff, R. G., and Vogt, P. K. (1969). *Virology* **39,** 18–22.
Eisenman, R. N., and Vogt, V. M. (1978). *Biochim. Biophys. Acta* **473,** 187–239.
Elder, J. H., Jensen, F. C., Bryant, M. L., and Lerner, R. A. (1977). *Nature (London)* **267,** 23–28.
Ellerman, V., and Bang, O. (1908). *Zentralbl. Bakteriol., Parasitenkd. Infektionskr. Hyg.* **46,** 595–609.
England, J. M., Bolognesi, D. P., Dietzschold, B., and Halpern, M. S. (1977). *Virology* **76,** 437–439.
Enke, K. H., Jungwitz, H., and Rössger, M. (1961). *Dtsch. Tieraerztl. Wochenschr.* **68,** 359–364.
Erikson, E., and Erikson, R. L. (1970). *J. Mol. Biol.* **52,** 387–392.
Erikson, E., and Erikson, R. L. (1971). *J. Virol.* **8,** 254–260.
Erikson, E., Erikson, R. L., Hendry, B., and Pace, N. R. (1973). *Virology* **53,** 40–45.
Erikson, R. L. (1969). *Virology* **37,** 124–129.
Essex, M. (1975). *Adv. Cancer Res.* **21,** 175–248.
Essex, M. (1980). *In* "Viral Oncology" (G. Klein, ed.), pp. 205–209. Raven, New York.
Essex, M., Klein, G., Snyder, S. P., and Harrold, J. B. (1971). *Int. J. Cancer* **8,** 384–390.
Essex, M., Cotter, S. M., Hardy, W. D., Jr., Hess, P. W., Jarrett, W., Jarrett, O., Mackey, L., Laird, H., Perryman, L., Olsen, R. G., and Yohn, D. S. (1975a). *J. Natl. Cancer Inst.* **55,** 463–467.
Essex, M., Hardy, W. D., Jr., Cotter, S. M., Jakowski, R. M., and Sliski, A. (1975b). *Infect. Immun.* **11,** 470–475.
Essex, M., Jakowski, R. M., Hardy, W. D., Jr., Cotter, S. M., Hess, P., and Sliski, A. (1975c). *J. Natl. Cancer Inst.* **54,** 637–641.
Essex, M., Sliski, A. H., Cotter, S. M., Jakowski, R. M., and Hardy, W. D., Jr. (1975d). *Science (Washington, D.C.)* **190,** 790–792.
Essex, M., Sliski, A. H., Worley, M., Grant, C. K., Snyder, H. W., Jr., Hardy, W. D., Jr., and Chen, L. B. (1980). *In* "Viruses in Naturally Occurring Cancers" (M. Essex, G. J. Todaro,

and H. zur Hausen, eds.), pp. 589–602. Cold Spring Harbor Press, Cold Spring Harbor, New York.

Famulari, N. G., Buckhagen, D. L., Klenk, H. D., and Fleissner, E. (1976). *J. Virol.* **20,** 501–506.

Faras, A. J., Gaigpin, A. C., Levinson, W. E., Bishop, J. M., and Goodman, H. M. (1973). *J. Virol.* **12,** 334–338.

Ferrerr, J. (1980). *In* "Viruses in Naturally Occurring Cancers" (M. Essex, G. J. Todaro, and H. zur Hausen, eds.), pp. 887–900. Cold Spring Harbor Press, Cold Spring Harbor, New York.

Ferrerr, J. F., Baliga, V., Diglio, C., Graves, D., Kenyon, S. J., McDonald, H., Piper, C., and Wuu, K. (1976). *Vet. Microbiol.* **1,** 159–184.

Ferrerr, J. F., Piper, C. E., and Baliga, V. (1977). *In* "Bovine Leukosis: Various Methods of Molecular Virology" (A. Burny, ed.), pp. 323–336. Comm. Eur. Communities, Luxembourg.

Fischinger, P. J., Peebles, P. T., Nomura, S., and Haapala, D. K. (1973). *J. Virol.* **11,** 978–985.

Fleissner, E., and Tress, E. (1973). *J. Virol.* **12,** 1612–1615.

Fleissner, E., Skeda, H., Tung, J. S., Vitetta, E., Tress, E., Hardy, W. D., Jr., Stockert, E., Boyse, E. A., Pincus, T., and O'Donnell, P. (1974). *Cold Spring Harbor Symp. Quant. Biol.* **39,** 1057–1066.

Francis, D. P., Essex, M., and Hardy, W. D., Jr. (1977). *Nature (London)* **269,** 252–254.

Francis, D. P., Cotter, S. M., Hardy, W. D., Jr., and Essex, M. (1979a). *Cancer Res.* **39,** 3866–3870.

Francis, D. P., Essex, M., Cotter, S. M., Jakowski, R. M., and Hardy, W. D., Jr. (1979b). *Leuk. Res.* **3,** 435–441.

Friis, R. R., Toyoshima, K., and Vogt, P. K. (1971). *Virology* **43,** 375–389.

Galehouse, D. M., and Duesberg, P. H. (1978). *J. Virol.* **25,** 86–96.

Gallagher, R. E., and Gallo, R. C. (1975). *Science (Washington, D.C.)* **187,** 350–353.

Gallo, R. C., Gallagher, R. E., Wong-Staal, F., Aoki, T., Markham, P. D., Shetlas, H., Ruscetti F. R., Valerio, M., Walling, M. J., O'Keefe, R. T., Saxinger, W. C., Smith, R. G., Gillespie, D. H., and Reitz, M. S. (1978). *Virology* **84,** 359–373.

Gianni, A. M., Smotkin, D., and Weinberg, R. A. (1975). *Proc. Natl. Acad. Sci. U.S.A.* **72,** 447–451.

Gibson, W., and Verma, I. M. (1974). *Proc. Natl. Acad. Sci. U.S.A.* **71,** 4991–4994.

Gielkens, A. L. J., Salden, M. H. L., and Bloemendal, H. (1974). *Proc. Natl. Acad. Sci. U.S.A.* **71,** 1093–1097.

Graf, T., and Beug, H. (1978). *Biochim. Biophys. Acta* **516,** 269–299.

Grandgenett, D. P. (1976a). *J. Virol.* **17,** 950–961.

Grandgenett, D. P. (1976b). *In* "Animal Virology" (D. Baltimore, A. S. Huang, and C. F. Fox, eds.), pp. 215–226. Academic Press, New York.

Grandgenett, D. P., and Green, M. (1974). *J. Biol. Chem.* **249,** 5148–5152.

Grandgenett, D. P., Gerard, G. F., and Green, M. (1973). *Proc. Natl. Acad. Sci. U.S.A.* **70,** 230–234.

Grant, C. K., DeBoer, D. J., Essex, M., Worley, M. B., and Higgins, J. (1977). *J. Immunol.* **119,** 401–408.

Green, R. W., Bolognesi, D. P., Schafer, W., Pfister, L., Hunsmann, G., and deNoronha, F. (1973). *Virology* **56,** 565–579.

Gross, L. (1951). *Proc. Soc. Exp. Biol. Med.* **76,** 27–32.

Guntaka, R. V., Mahy, B. W. J., Bishop, J. M., and Varmus, H. E. (1975). *Nature (London)* **253,** 507–511.

Guntaka, R. V., Richards, O. C., Shank, P. R., Kung, H. J., Davidson, N., Fritsch, E., Bishop, J. M., and Varmus, H. E. (1976). *J. Mol. Biol.* **106,** 337–357.

Halpern, M. S., Bolognesi, D. P., and Friis, R. (1976). *J. Virol.* **18,** 504–510.

Hanafusa, H. (1965). *Virology* **25,** 248–252.
Hanafusa, H., and Hanafusa, T. (1966). *Proc. Natl. Acad. Sci. U.S.A.* **55,** 532–536.
Hanafusa, H., and Hanafusa, T. (1971). *Virology* **43,** 313–317.
Hanafusa, H., Hanafusa, T., and Rubin, H. (1964). *Proc. Natl. Acad. Sci. U.S.A.* **51,** 41–46.
Hanafusa, H., Baltimore, D., Smoler, D., Watson, K. F., Yaniv, A., and Spiegelman, S. (1972). *Science (Washington, D.C.)* **177,** 1188–1192.
Hanafusa, T., Hanafusa, H., and Rubin, H. (1963). *Proc. Natl. Acad. Sci. U.S.A.* **49,** 572–576.
Hanafusa, T., Hanafusa, H., Metroka, C. E., Hayward, W. S., Rettenmier, C. W., Sawyer, R. C., Doughterty, R. M., and Di Stephano, H. S. (1976). *Proc. Natl. Acad. Sci. U.S.A.* **73,** 1333–1337.
Harada, F., Sawyer, R. C., and Dahlberg, J. E. (1975). *J. Biol. Chem.* **250,** 3487–3497.
Hardy, W. D., Jr., Old, L. J., Hess, P. W., Essex, M., and Cotter, S. M. (1973). *Nature (London)* **244,** 266–269.
Hardy, W. D., Jr., Zuckerman, E. E., MacEwen, E. G., Hayes, A. A., and Essex, M. (1977). *Nature (London)* **270,** 249–251.
Hardy, W. D., Jr., Zuckerman, E. E., McClelland, A. J., Snyder, H. W., Jr., Essex, M., and Francis, D. P. (1980). *In* "Viruses in Naturally Occurring Cancers" (M. Essex, G. J. Todaro, and H. zur Hausen, eds.), pp. 677–698. Cold Spring Harbor Press, Cold Spring Harbor, New York.
Hartley, J. W., and Row, W. P. (1966). *Proc. Natl. Acad. Sci. U.S.A.* **55,** 780–785.
Harvey, J. J., and East, J. (1971). *Int. Rev. Exp. Pathol.* **10,** 265–272.
Haseltine, W. A., and Baltimore, D. (1976). *In* "Animal Virology" (D. Baltimore, A. S. Huang, and C. F. Fox, eds.), pp. 175–213. Academic Press, New York.
Haseltine, W. A., and Kleid, D. G. (1978). *Nature (London)* **278,** 358–363.
Henderson, I. C., Lieber, M. M., and Todaro, G. J. (1974). *Virology* **60,** 282–287.
Herman, A. C., Green, R. W., Bolognesi, D. P., and Vanaman, T. C. (1975). *Virology* **64,** 339–348.
Hill, M., and Hillova, J. (1974). *Biochim. Biophys. Acta* **355,** 7–48.
Hill, M., and Hillova, J. (1976). *Adv. Cancer Res.* **23,** 237–298.
Hino, S., Stephenson, J. R., and Aaronson, S. A. (1975). *J. Immunol.* **115,** 922–927.
Hino, S., Stephenson, J. R., and Aaronson, S. A. (1976). *J. Virol.* **18,** 933–941.
Howk, R. A., Rye, L. A., Killeen, L. A., Scolnick, E. M., and Parks, W. P. (1973). *Proc. Natl. Acad. Sci. U.S.A.* **70,** 2117–2121.
Huebner, R. J., Armstrong, D., Okuyan, M., Sarma, P. S., and Turner, H. C. (1964). *Proc. Natl. Acad. Sci. U.S.A.* **51,** 741–750.
Huebner, R. J., Hartley, J. W., Rowe, W. P., Lane, W. T., and Capps, W. I. (1966). *Proc. Natl. Acad. Sci. U.S.A.* **56,** 1164–1169.
Hunsmann, G., Claviez, M., Moenning, V., Schwartz, H., and Schafer, W. (1976). *Virology* **69,** 157–168.
Hurwitz, J., and Leis, J. P. (1972). *J. Virol.* **9,** 116–129.
Ikeda, H., Hardy, W. D., Jr., Tress, E., and Fleissner, E. (1975). *J. Virol.* **16,** 53–61.
Ishizaki, R., and Bolognesi, D. P. (1976). *J. Virol.* **17,** 132–139.
Jamjoom, G. A., Naso, R. B., and Arlinghaus, R. (1977). *Virology* **78,** 11–34.
Jarrett, O., Hardy, W. D., Jr., Golder, M. C., and Hay, D. (1978). *Int. J. Cancer* **21,** 334–337.
Jarrett, W. F. H., Martin, W. B., Crighton, G. W., Dalton, R. G., and Stewart, M. F. (1964). *Nature (London)* **202,** 566–567.
Joho, R. H., Billeter, M. A., and Weissmann, C. (1975). *Proc. Natl. Acad. Sci. U.S.A.* **72,** 4772–4776.

Junghans, R. P., Hu, S., Knight, C. A., and Davidson, N. (1977). *Proc. Natl. Acad. Sci. U.S.A.* **74,** 477–481.

Kacian, D. L., Watson, K. F., Burny, A., and Spiegelman, S. (1971). *Biochim. Biophys. Acta* **266,** 365–383.

Kaplan, H. S. (1978). *Leuk. Res.* **2,** 253–272.

Karskin, W. L., Arecement, L. J., Naso, R. B., and Arlinghaus, R. B. (1977). *J. Virol.* **23,** 787–798.

Kawai, S., and Hanafusa, H. (1971). *Virology* **46,** 470–479.

Kawai, S., and Hanafusa, H. (1973). *Proc. Natl. Acad. Sci. U.S.A.* **70,** 3493–3497.

Kawakami, T. G. (1980). *In* "Viruses in Naturally Occurring Cancers" (M. Essex, G. J. Todaro, and H. zur Hausen, eds.), pp. 719–728. Cold Spring Harbor Press, Cold Spring Harbor, New York.

Kawakami, T. G., and Buckley, P. M. (1974). *Transplant. Proc.* **6,** 193–196.

Kawakami, T. G., Huff, S., Buckley, P. M., Dungworth, D., Snyder, S., and Gilden, R. (1972). *Nature (London), New Biol.* **235,** 170–171.

Kawakami, T. G., Sun, L., and McDowell, T. S. (1978). *In* "Advances in Comparative Leukemia Research 1977" (P. Bentvelzen, J. Hilgers, and D. S. Yohn, eds.), pp. 33–36. Elsevier, Amsterdam.

Keller, W., and Crouch, R. (1972). *Proc. Natl. Acad. Sci. U.S.A.* **69,** 3360–3364.

Kerr, I. M., Olshevsky, U., Lodish, H. F., and Baltimore, D. (1976). *J. Virol.* **18,** 627–635.

Kettman, R., Meunier-Rotival, M., Marbaix, G., Cortadas, J., Cleuter, Y., Mammerickx, M., Burny, A., and Bernardi, G. (1980). *In* "Viruses in Naturally Occurring Cancers" (M. Essex, G. J. Todaro, and H. zur Hausen, eds.), pp. 927–942. Cold Spring Harbor Press, Cold Spring Harbor, New York.

Klement, V., Hartley, J., Rowe, W. P., and Huebner, R. J. (1969). *J. Natl. Cancer Inst.* **43,** 925–930.

Klemenz, R., and Diggelman, H. (1978). *Virology* **85,** 63–74.

Koshy, R., Wong-Staal, F., Gallo, R. C., Hardy, W. D., Jr., and Essex, M. (1979). *Virology* **99,** 135–144.

Krakower, J. M., Barbacid, M., and Aaronson, S. A. (1977). *J. Virol.* **22,** 331–339.

Krakower, J. M., Tronick, S. R., Gallagher, R. E., Gallo, R. C., and Aaronson, S. A. (1978). *Int. J. Cancer* **22,** 715–720.

Krantz, M. J., Lee, V. C., and Hung, P. P. (1976). *Arch. Biochem. Biophys.* **174,** 66–73.

Krantz, M. J., Strand, M., and August, J. T. (1977). *J. Virol.* **22,** 804–815.

Lai, M. M. C. (1976). *Virology* **74,** 287–301.

Lai, M. M. C., Duesberg, P. H., Horst, J., and Vogt, P. K. (1973). *Proc. Natl. Acad. Sci. U.S.A.* **70,** 2266–2270.

Leaminson, R. N., and Halpern, M. S. (1976). *J. Virol.* **18,** 956–986.

Leaminson, R. N., Shander, M. H. M., and Halpern, M. S. (1977). *Virology* **76,** 437–439.

Ledbetter, J., and Nowinski, R. C. (1977). *J. Virol.* **23,** 315–322.

Leis, J. P., Berkower, I., and Hurwitz, J. (1973). *Proc. Natl. Acad. Sci. U.S.A.* **70,** 466–470.

Levin, J. G., and Rosenak, M. J. (1976). *Proc. Natl. Acad. Sci. U.S.A.* **73,** 1154–1158.

Levy, J. A. (1977). *Virology* **77,** 811–825.

Levy, J. A., Hartley, J. W., Rowe, W. P., and Huebner, R. J. (1973). *J. Natl. Cancer Inst.* **51,** 525–529.

Linial, M., and Mason, W. S. (1973). *Virology* **53,** 258–273.

Livingston, D. M., and Todaro, G. J. (1973). *Virology* **53,** 142–151.

McClintock, P. R., Ihle, J. N., and Joseph, D. R. (1977). *J. Exp. Med.* **146,** 422–434.

McCullough, B., Schaller, J., Shadduck, J. A., and Yohn, D. S. (1972). *J. Natl. Cancer Inst.* **48,** 1893–1896.

Mammerickx, M., Dekegel, D., Burny, A., and Portetelle, D. (1976). *Vet. Microbiol.* **1,** 347–356.

Manson, W. S., Friis, R. R., Linial, M., and Vogt, P. K. (1974). *Virology* **61,** 559–574.

Marquardt, H., Gilden, R. V., and Brosglen, S. (1977). *Biochemistry* **16,** 710–717.

Martin, G. S. (1970). *Nature (London)* **227,** 1021–1023.

Mathes, L. W., Olsen, R. G., Hebebrand, L. C., Hoover, E. A., Schaller, J. P., Adams, P. W., and Nichols, W. S. (1979). *Cancer Res.* **39,** 950–955.

Michalides, R., Schlom, J., Dahlberg, J., and Peck, K. (1975). *J. Virol.* **16,** 1039–1050.

Miller, J., and Van der Maaten, M. J. (1980). *In* "Viruses in Naturally Occurring Cancers" (M. Essex, G. J. Todaro, and H. zur Hausen, eds.), pp. 901–910. Cold Spring Harbor Press, Cold Spring Harbor, New York.

Miller, J. M., Miller, L. D., Olson, C., and Gillette, K. G. (1969). *J. Natl. Cancer Inst.* **43,** 1297–1305.

Moelling, K. (1974). *Cold Spring Harbor Symp. Quant. Biol.* **39,** 969–973.

Moelling, K., and Hayami, M. (1977). *J. Virol.* **22,** 598–607.

Moelling, K., Bolognesi, D. P., Bauer, H., Busen, Plassmann, H. W., and Hausen, P. (1971). *Nature (London), New Biol.* **234,** 240–243.

Moenning, V., Frank, H., Hunsmann, G., Schneider, I., and Schafer, W. (1974). *Virology* **61,** 100–111.

Montagnier, L., Golde, A., and Vigier, P. (1969). *J. Gen. Virol.* **4,** 449–452.

Moroni, C. (1972). *Virology* **47,** 1–7.

Naso, R. B., Arcement, L. J., and Arlinghaus, R. B. (1975). *Cell* **4,** 31–36.

Naso, R. B., Arcement, L. J., Karskin, W. L., Jarnjoom, G. A., and Arlinghaus, R. B. (1976). *Proc. Natl. Acad. Sci. U.S.A.* **73,** 2326–2330.

Neiman, P., Payne, L. N., Weiss, R., Jordan, L., and Eisenman, R. N. (1980). *In* "Viruses in Naturally Occurring Cancers" (M. Essex, G. J. Todaro, and H. zur Hausen, eds.), pp. 519–528. Cold Spring Harbor Press, Cold Spring Harbor, New York.

Niman, H. L., Gardner, M. B., Stephenson, J. R., and Roy-Burman, P. (1977). *J. Virol.* **23,** 578–586.

Nooter, K., Aarssen, A. M., Bentvelzen, P., DeGroot, F. G., and Van Pelt, F. G. (1975). *Nature (London)* **256,** 595–597.

Nowinski, R. C., Watson, K. F., Yaniv, A., and Spiegelman, S. (1972). *J. Virol.* **10,** 959–963.

Obata, Y., Ikeda, H., Stockert, E., and Boyse, E. A. (1975). *J. Exp. Med.* **141,** 188–197.

Ogura, H., and Friis, R. (1975). *J. Virol.* **16,** 443–446.

Okabe, H., Twiddy, E., Gilden, R. V., Hatanaka, M., Hoover, E. A., and Olsen, R. G. (1976). *Virology* **69,** 798–801.

Okasinski, G. F., and Velicer, L. F. (1976). *J. Virol.* **20,** 96–106.

Okasinski, G. F., and Velicer, L. F. (1977). *J. Virol.* **22,** 74–85.

Okazaki, W., Purchase, H. G., and Burmester, B. R. (1975). *Avian Dis.* **19,** 311–317.

Olsen, R. G., Hoover, E. A., Schaller, J. P., Mathes, L. E., and Wolff, L. H. (1977). *Cancer Res.* **37,** 2082–2085.

Olson, C. (1977). *In* "Bovine Leucosis: Various Methods of Molecular Virology" (A. Burny, ed.), pp. 244–245. Comm. Eur. Communities, Luxembourg.

Olson, C., and Baumgartener, L. E. (1975). *Bovine Pract.* **10,** 15–22.

Onuma, M. and Olson, C. (1977). *In* "Bovine Leucosis: Various Methods of Molecular Virology" (A. Burny, ed.), pp. 95–118. Comm. Eur. Communities, Luxembourg.

Opperman, H., Bishop, J. M., Varmus, H. E., and Levintow, L. (1977). *Cell* **12,** 993–1005.

Pal, B. K., and Roy-Burman, P. (1975). *J. Virol.* **15,** 540-549.

Pal, B. K., McAllister, R. M., Gardner, M. B., and Roy-Burman, P. (1975). *J. Virol.* **16,** 123-131.

Panem, S., Prochownik, E. V., Reale, F. R., and Kirsten, W. H. (1975). *Science (Washington, D.C.)* **189,** 297-299.

Panet, A., Baltimore, D., and Hanafusa, T. (1975). *J. Virol.* **16,** 146-152.

Parks, W. P., Scolnick, E. M., Ross, J., Todaro, G. J., and Aaronson, S. A. (1972). *J. Virol.* **9,** 110-115.

Parks, W. P., Scolnick, E. M., Noon, M. C., Watson, C. S., and Kawakami, T. G. (1973). *Int. J. Cancer* **12,** 129-137.

Paskind, M. P., Weinberg, R. A., and Baltimore, D. (1975). *Virology* **67,** 242-248.

Pawson, T., Martin, G. S., and Smith, A. E. (1976). *J. Virol.* **19,** 950-967.

Payne, L. N., and Rennie, M. T. (1975). *Vet. Rec.* **96,** 454-456.

Peters, G., Harada, F., Dahlberg, J. E., Panet, A., Haseltine, W. A., and Baltimore, D. (1977). *J. Virol.* **21,** 1031-1041.

Piper, G. E., Abt, D. A., Ferrerr, J. F., and Marshak, R. R. (1975). *Cancer Res.* **35,** 2714-2716.

Purchase, H. G., and Burmester, B. R. (1978). *In* "Diseases of Poultry" (M. F. Hofstad, B. W. Calnek, W. M. Reid, and H. W. Yoder, Jr., eds.), pp. 418-468. Iowa State Univ. Press, Ames.

Purchase, H. G., and Cheville, N. F. (1975). *Avian Pathol.* **4,** 239-245.

Purchase, H. G., and Gilmour, D. G. (1975). *J. Natl. Cancer Inst.* **55,** 851-855.

Purchio, A. F., Erikson, E., and Erikson, R. L. (1977). *Proc. Natl. Acad. Sci. U.S.A.* **74,** 4661-4665.

Rangon, S. R. S., and Gallagher, R. E. (1979). *Adv. Virus Res.* **24,** 1-123.

Reitz, M. S., Wong-Staal, F., Haseltine, W. A., Kleid, D. G., Trainor, C. D., Gallagher, R. E., and Gallo, R. C. (1979). *J. Virol.* **29,** 395-400.

Reynolds, R. K., and Stephenson, J. R. (1978). *Virology* **81,** 328-340.

Rispens, B. H., Long, P. A., Okazaki, W., and Burmester, B. R. (1970). *Avian Dis.* **14,** 738-751.

Robinson, W. S., Pitkanen, A., and Rubin, H. (1965). *Proc. Natl. Acad. Sci. U.S.A.* **54,** 137-144.

Rosenberg, Z., and Haseltine, W. A. (1980). *Virology* **102,** 240-244.

Ross, J., Scolnick, E. M., Todaro, G. J., and Aaronson, S. A. (1971). *Nature (London), New Biol.* **231,** 163-167.

Rous, P. (1911). *J. Exp. Med.* **13,** 397-411.

Rubin, H. (1965). *Virology* **26,** 270-275.

Salden, M., Asselbergs, F., and Bloemenda, H. (1976). *Nature (London)* **259,** 696-699.

Sarkar, N. H., and Whittington, E. S. (1977). *Virology* **81,** 91-106.

Sarma, P. S., and Log, T. (1973). *Virology* **54,** 160-170.

Sarma, P. S., Turner, H. C., and Huebner, R. J. (1964). *Virology* **23,** 313-321.

Sarma, P. S., Baskar, J. F., Gilden, R. V., Gardner, M. B., and Huebner, R. J. (1971). *Proc. Soc. Exp. Biol. Med.* **137,** 1333-1337.

Schafer, W., Hunsmann, G., Moenning, V., De Noronha, F., Bolognesi, D. P., Green, R. W., and Huper, G. (1975). *Virology* **63,** 49-59.

Schwarz, H., Hunsmann, G., Moennig, V., and Schafer, W. (1976). *Virology* **69,** 169-178.

Scolnick, E. M., and Parks, W. P. (1973). *Int. J. Cancer* **12,** 138-142.

Scolnick, E. M., Rands, E., Aaronson, S. A., and Todaro, G. J. (1970). *Proc. Natl. Acad. Sci. U.S.A.* **67,** 1989-1996.

Scolnick, E. M., Parks, W. P., Todaro, G. J., and Aaronson, S. A. (1972a). *Nature (London), New Biol.* **235,** 35-40.

Scolnick, E. M., Stephenson, J. R., and Aaronson, S. A. (1972b). *J. Virol.* **10,** 653-657.

Scolnick, E. M., Parks, W. P., Todaro, G. J., and Aaronson, S. A. (1972c). *Nature (London), New Biol.* **235,** 1-6.
Scolnick, E. M., Maryak, J. M., and Parks, W. P. (1974). *J. Virol.* **14,** 1435-1444.
Shapiro, S. Z., Strand, M., and August J. (1976). *J. Mol. Biol.* **107,** 459-477.
Sliski, A. H., Essex, M., Meyer, C., and Todaro, G. J. (1977). *Science (Washington, D.C.)* **196,** 1339-1340.
Smotkin, D., Yoshimura, F., and Weinberg, R. A. (1976). *J. Virol.* **20,** 621-626.
Snyder, H. W., Stockert, E., and Fleissner, E. (1977). *J. Virol.* **23,** 302-314.
Snyder, S. P., and Theilen, G. H. (1969). *Nature (London)* **221,** 1074-1075.
Snyder, S. P., Dungworth, D. L., Kawakami, T. G., Callaway, E., and Lan, D. T.-L. (1973). *J. Natl. Cancer Inst.* **51,** 89-96.
Spencer, J. L., Crittenden, L. B., Burmester, B. R., Romero, C., and Witter, R. L. (1976). *Avian Pathol.* **5,** 221-226.
Spiegleman, S., Burny, A., Das, M. R., Keydar, J., Schlom, J., Travnicek, M., and Watson, K. (1970a). *Nature (London)* **227,** 1029-1031.
Spiegleman, S., Burny, A., Das, M. R., Keydar, J., Schlom, J., Travnicek, M., and Watson, K. (1970b). *Nature (London)* **228,** 430-432.
Stehelin, D., Varmus, H. E., Bishop, J. M., and Vogt, P. K. (1976). *Nature (London)* **260,** 170-175.
Stephenson, J. R., Reynolds, R. K., Tronick, S. R., and Aaronson, S. A. (1975a). *Virology* **67,** 404-414.
Stephenson, J. R., Tronick, S. R., and Aaronson, S. A. (1975b). *Cell* **6,** 543-548.
Stephenson, J. R., Hino, S., Garrett, E. S., and Aaronson, S. A. (1976a). *Nature (London)* **261,** 609-611.
Stephenson, J. R., Peters, R. L., Hino, S., Donahoe, R. M., Long. L. K., Aaronson, S. A., and Kelloff, G. J. (1976b). *J. Virol.* **19,** 890-898.
Stephenson, J. R., Essex, M., Hino, S., Hardy, W. D., Jr., and Aaronson, S. A. (1977a). *Proc. Natl. Acad. Sci. U.S.A.* **74,** 1219-1223.
Stephenson, J. R., Khan, A. S., Sliski, A. H., and Essex, M. (1977b). *Proc. Natl. Acad. Sci. U.S.A.* **74,** 5608-5612.
Strand, M., and August, J. T. (1973). *J. Biol. Chem.* **248,** 5627-5633.
Strand, M., and August, J. T. (1974). *J. Virol.* **13,** 171-180.
Strand, M., Wilsnack, R., and August, J. T. (1974). *J. Virol.* **14,** 1575-1583.
Stromberg, K., Hurley, N. E., Davis, N. L., Rueckert, R. R., and Fleissner, E. (1974). *J. Virol.* **13,** 513-528.
Tal, J., Kung, H.-J., Varmus, H. E., and Bishop, J. M. (1977). *Virology* **79,** 183-187.
Temin, H. M. (1962). *Cold Spring Harbor Symp. Quant. Biol.* **27,** 407-415.
Temin, H. M. (1963). *Virology* **20,** 235-240.
Temin, H. M. (1976). *Science* **192,** 1075-1080.
Temin, H., and Baltimore, D. (1972). *Adv. Virus Res.* **17,** 129-186.
Temin, H., and Mizutani, S. (1970). *Nature (London)* **226,** 1211-1213.
Theil, H. J., and Iglehart, D., Matthews, T. J., Butchko, A. W., and Bolognesi, D. P. (1980). *In* "Viruses in Naturally Occurring Cancers" (M. Essex, G. J. Todaro, and H. zur Hansen, eds.), pp. 847-856. Cold Spring Harbor Press, Cold Spring Harbor, New York.
Theilen, G. H., Gould, D., Fowler, M., and Dungworth, D. L. (1971). *J. Natl. Cancer Inst.* **47,** 881-889.
Tiavnicek, M. (1968). *Biochim. Biophys. Acta* **166,** 757-783.
Todaro, G. J. (1978). *Br. J. Cancer* **37,** 139-158.
Todaro, G. J., Tevethia, S. S., and Melnick, J. L. (1973). *Intervirology* **1,** 399-404.
Tronick, S. R., Scolnick, E. M., and Packs, W. P. (1972). *J. Virol.* **10,** 885-888.

Tronick, S. R., Stephenson, J. R., and Aaronson, S. A. (1973). *Virology* **54,** 199–206.
Tronick, S. R., Stephenson, J. R., Verma, I. M., and Aaronson, S. A. (1975). *J. Virol.* **16,** 1479–1482.
Tung, J., Vitetta, E. S., Fleissner, E., and Boyse, E. A. (1975). *J. Exp. Med.* **141,** 198–205.
van der Maaten, M. J., and Miller, J. M. (1976). *Bibl. Haematol.* **43,** 377–379.
van der Maaten, M. J., Miller, J. M., and Boothe, A. D. (1974). *J. Natl. Cancer Inst.* **52,** 491–497.
van Eldik, L. J., Palson, J. C., Green, R. W., and Smith, R. E. (1978). *Virology* **86,** 193–200.
van Zaane, D., Gielkens, A. L. J., Deeker-Michielson, M. J. A., and Bloemers, H. P J. (1975). *Virology* **67,** 544–552.
van Zaane, D., Dekker-Michielsen, M. J. A., and Bloemers, H. P. J. (1976). *Virology* **75,** 113–129.
van Zaane, D., Gielkens, A. L. J., Hesselink, W. G., and Gloemers, H. P. J. (1977). *Proc. Natl. Acad. Sci. U.S.A.* **74,** 1855–1859.
Varmus, H. E., Guntaka, R. V., Heasley, S., Fan, W. J., and Bishop, J. M. (1974). *Proc. Natl. Acad. Sci. U.S.A.* **71,** 3874–3878.
Verma, I. M. (1977). *Biochim. Biophys. Acta* **473,** 1–34.
Verma, I. M., Mason, W. S., Drost, S. D., and Baltimore, D. (1974). *Nature (London)* **251,** 27–31.
Verma, I. M., Varmus, H. E., and Hunter, E. (1976). *Virology* **74,** 16–29.
Vogt, P. K. (1965a). *Adv. Virus Res.* **11,** 293–385.
Vogt, P. K. (1965b). *Virology* **25,** 237–242.
Vogt, P. K. (1970). *Bibl. Haematol.* **36,** 153–167.
Vogt, P. K. (1977). *Compr. Virol.* **9,** 341–456.
Vogt, V. M., and Eisenman, R. (1973). *Proc. Natl. Acad. Sci. U.S.A.* **70,** 1734–1738.
Vogt, V. M., Eisenman, R., and Diggelmann, H. (1975). *J. Mol. Biol.* **96,** 471–493.
von der Helm, K. (1977). *Proc. Natl. Acad. Sci. U.S.A.* **74,** 911–915.
von der Helm, K., and Duesberg, P. H. (1975). *Proc. Natl. Acad. Sci. U.S.A.* **72,** 614–618.
Walker, T. A., Pace, N. R., Rickson, R. L., Eickson, and Behr, F. (1974). *Proc. Natl. Acad. Sci. U.S.A.* **71,** 3390–3398.
Wang, L.-H., Duesberg, P. H., Kawai, S., and Hanafusa, H. (1976a). *Proc. Natl. Acad. Sci. U.S.A.* **73,** 447–451.
Wang, L.-H., Galehouse, D., Mellon, P., Duesberg, P., Mason, W. S., and Vogt, P. K. (1976b). *Proc. Natl. Acad. Sci. U.S.A.* **73,** 3952–3956.
Wang, L.-H., Duesberg, P. H., Mellon, P., and Vogt, P. K. (1976c). *Proc. Natl. Acad. Sci. U.S.A.* **73,** 1073–1077.
Waters, L. C., Mullin, B. C., Bailiff, E. G., and Popp, R. A. (1975). *J. Virol.* **16,** 1608–1612.
Weinberg, R. A. (1977). *Biochim. Biophys. Acta* **473,** 39–56.
Weiss, R. A. (1976). *In* "Receptors for RNA Tumor Viruses" (R. F. Bears, Jr. and E. G. Bassett, eds.), Int. Symp. Ser., No. 9, pp. 237–325. Raven, New York.
Weiss, R. A., and Wong, A. L. (1977). *Virology* **76,** 826–834.
Weiss, R. A., Boettinger, D., and Love, D. N. (1974). *Cold Spring Harbor Symp. Quant. Biol.* **39,** 913–918.
Witte, O. N., Tsukamoto-Adley, A., and Weissman, I. L. (1977). *Virology* **76,** 539–533.
Wittman, W., and Urbaneck, D. (1969). *Arch. Exp. Vet. Med.* **23,** 709–713.
Wolfe, L. G., and Deinhardt, F. (1978), *Primates Med.* **10,** 96–118.
Wolfe, L. G., Deinhardt, F., Theilen, G., Rabin, H., Kawakami, T., and Bustad, L. (1971). *J. Natl. Cancer Inst.* **47,** 1115–1120.
Wong-Staal, F., Josephs, S., Dalla Favera, R., and Gallo, R. (1979). *In* "Modern Trends in Human Leukemia III" (R. Neth, R. C. Gallo, P. H. Hofschneider, and K. Mannweiler, eds.), pp. 553–560. Springer-Verlag, Berlin and New York.
Zarling, D. A., Moser, A. G., and Temin, H. M. (1977). *J. Virol.* **21,** 105–112.

Chapter 14

Spumavirinae: Foamy Virus Group Infections: Comparative Aspects and Diagnosis

JOHN J. HOOKS and BARBARA DETRICK-HOOKS

I. Introduction 599
II. Description of the Virion 601
A. Morphology and Size 601
B. Physicochemical Properties 602
C. Antigenic Composition 605
III. Biological Features 607
A. Growth of Virus in Tissue Culture 607
B. Infection in the Natural Host 608
C. Infection in Experimental Hosts 609
IV. Association with Human Disease 610
V. Immunity 611
VI. Epidemiology 612
VII. Laboratory Diagnosis 613
A. Viral Isolation Methods 613
B. Serological Techniques 613
VIII. Concluding Remarks 616
References 616

I. INTRODUCTION

Over 25 years have passed since foamy viruses were first recognized as contaminants in primary monkey kidney cell cultures (Enders and Peebles, 1954). During that time, these viruses have captured the attention of investigators with

COMPARATIVE DIAGNOSIS
OF VIRAL DISEASES, VOL. IV

ISBN 0-12-429704-8

TABLE I
Spumavirinae (Foamy Viruses)

Name	Number of serotypes
Hamster foamy virus (HFV)	1
Feline syncytial virus (FSV)	1
Bovine syncytial virus (BSV)	1
Simian foamy virus (SFV)	9
Human foamy virus	?

widely diverse interests. Their presence in primary cell cultures continues to plague producers of polio virus vaccine, diagnostic laboratories, and investigators preparing virus stock pools for research (Hsiung, 1968; Hull, 1968). Furthermore, the mechanisms by which these reverse transcriptase-containing viruses can produce a lifelong persistent infection continue to create enthusiastic interest (Hooks and Gibbs, 1975). Now, however, by far the greater interest in these viruses concerns their possible involvement in human diseases, especially nasopharyngeal carcinoma (Achong *et al.*, 1971; Young *et al.*, 1973; Epstein *et al.*, 1974; Loh *et al.*, 1977; Achong and Epstein, 1978; Nemo *et al.*, 1978; Brown *et al.*, 1978; Cameron *et al.*, 1978).

The Spumavirinae or foamy viruses are one of the three subfamilies of the Retroviridae (Kurstak, 1977). These viruses have been found in subhuman primates, cows, cats, hamsters, and man (Table I) (reviewed in Hooks and Gibbs, 1975). Like the other retroviruses, the foamy viruses are RNA viruses which contain a reverse transcriptase and mature by budding from cytoplasmic membranes. However, unlike the other retroviruses, which induce transformation, tumor production, or clinical disease, the foamy viruses have not unequivocally demonstrated these properties.

One of the fascinating and characteristic properties of the foamy viruses is their ability to persist in the infected host for long periods of time in the presence of circulating antibody (Hooks and Gibbs, 1975). Studies suggest that their ability to persist may depend on the depression of the cell-mediated immune (CMI) response in conjunction with the presence of a reverse transcriptase in the virion and infectious viral DNA in infected cells (Parks *et al.*, 1971; Chiswell and Pringle, 1977, 1978; Chu and Hooks, 1977; Hooks and Detrick-Hooks, 1979).

It is our aim in this chapter to review the major features of the foamy viruses and foamy virus infections. Three general features of foamy virus infections which will be stressed are the ubiquity of the viruses, the persistence of the viruses in the absence of apparent disease, and finally, the comparative virology and diagnosis of the viruses.

II. DESCRIPTION OF THE VIRION

A. Morphology and Size

The foamy viruses are spherical 90- to 140-nm particles which have a buoyant density in sucrose ranging from 1.15 to 1.18 gm/cm^3 (Parks *et al.*, 1971; Hooks *et al.*, 1973; Hooks and Gibbs, 1975; Brown *et al.*, 1978).

Morphologically, the foamy viruses resemble the other retroviruses (Dalton *et al.*, 1974; Seman and Domochowski, 1977). The extracellular virion is composed of an electron-lucent core and a core shell which together are referred to as the ''nucleoid'' (Fig. 1). The nucleoid measures 35 to 50 nm in diameter and is surrounded by an envelope which is separated from the nucleoid by an electon-lucent space. The envelope contains numerous radiating spikes which are 5 to 15 nm long with tip-to-tip spacing of 5 to 10 nm (Clarke *et al.*, 1967, 1969a,b;

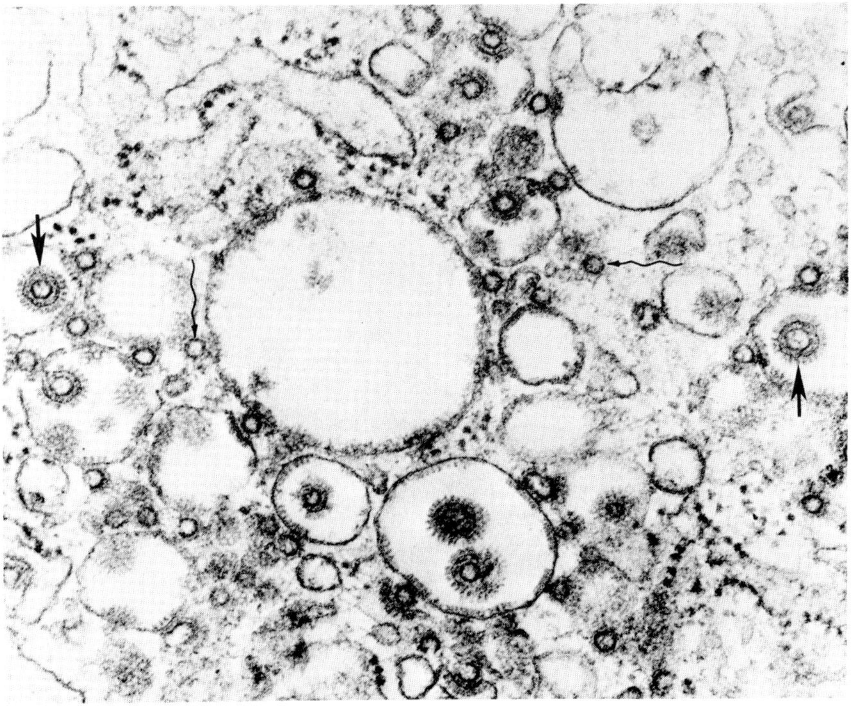

Fig. 1. Simian foamy virus type 6 infected HEK cells. Nucleoids (wavy arrows) are seen budding into cytoplasmic vacuoles, and complete virions (straight arrows) are seen within the vacuoles. × 120,000. (From Hooks *et al.*, 1972.)

Clarke and Attridge, 1968; Malmquist *et al.*, 1969; Clarke and McFerran, 1970; Dermott *et al.*, 1971; Chopra *et al.*, 1972; Hooks *et al.*, 1972, 1973).

Within the infected cell, the virus is first seen as a nucleoid in the cytoplasm. This nucleoid acquires its envelope by one of two processes. Usually, the nucleoid buds into cytoplasmic vacuoles or buds from the plasma membrane. The viral envelope is derived from the cell membrane. When the budding process is completed, the nucleoid of the virion maintains its ring-shaped electron-lucent center (Malmquist *et al.*, 1969; Dermott *et al.*, 1971; Chopra *et al.*, 1972) (Fig. 2). The second mechanism by which the nucleoid acquires its envelope has been seen infrequently and only with simian foamy virus type 8 and bovine syncytial virus (Boothe *et al.*, 1970; Hooks *et al.*, 1973). The intracellular nucleoid is found in close association with tubular membrane profiles which surround the nucleoid to form a complete virion (Fig. 3). Under these circumstances, the virus does not bud from the cell membrane, nor do the viral envelopes contain the typical spikes.

Despite morphological and maturation similarities between the foamy viruses and the other retroviruses, the foamy viruses maintain their ring-shaped structure with an electron-lucent center. Thus the core of the foamy virus particle does not change during the maturation process. This, however, is not the case with the other retroviruses, for they undergo an extracellular maturation in which the core of the nucleoprotein is rearranged and hence appears considerably different from the original intracellular particle.

B. Physicochemical Properties

1. Stability

Foamy virus particles contain lipid, as evidenced by their sensitivity to chloroform and ether (Enders and Peebles, 1954: Rustigian *et al.*, 1955; Johnston, 1961, 1971; Stiles *et al.*, 1964; Riggs *et al.*, 1969; Hooks *et al.*, 1972, 1973). The thermalability of the virus is shown by the loss of 90% of its infectivity in a 24-hour period at 37°C and the complete loss of infectivity in a 30-minute period at 56°C (Hooks and Gibbs, 1975; Hruska and Takemoto, 1975). In contrast, virus infectivity is extremely resistant to inactivation by ultraviolet light, which is a characteristic of the retroviruses (Parks and Todaro, 1972). The foamy viruses also retain their infectivity after freezing and thawing and lyophilization.

2. Nucleic Acid

The inhibition of viral replication by 5-bromodeoxyuridine provided early evidence that the nucleic acid was RNA (Johnston, 1971; Hooks *et al.*, 1972; Parks and Todaro, 1972). The presence of RNA in the foamy virus particle was

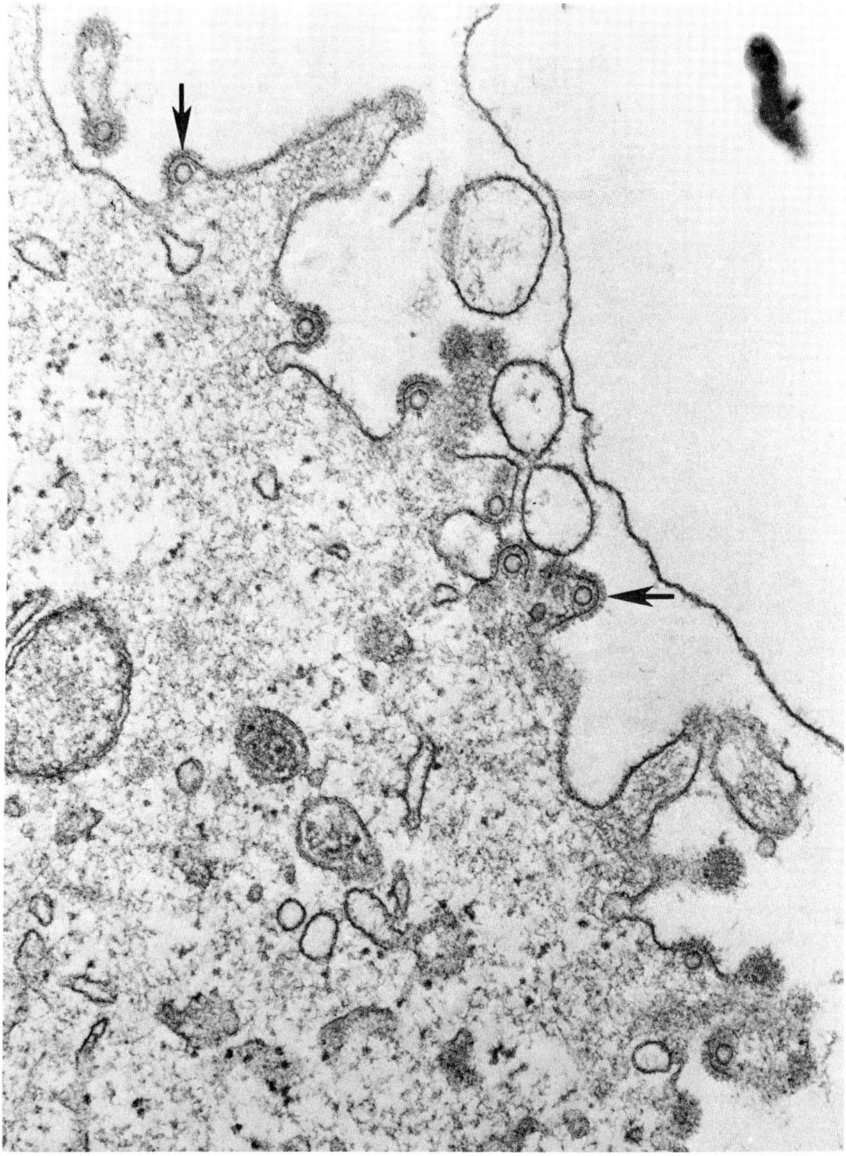

Fig. 2. Simian foamy virus type 6 infected HEK cells. Ring-shaped nucleoids are seen budding from the plasma membrane. × 90,000. (From Hooks *et al.*, 1972.)

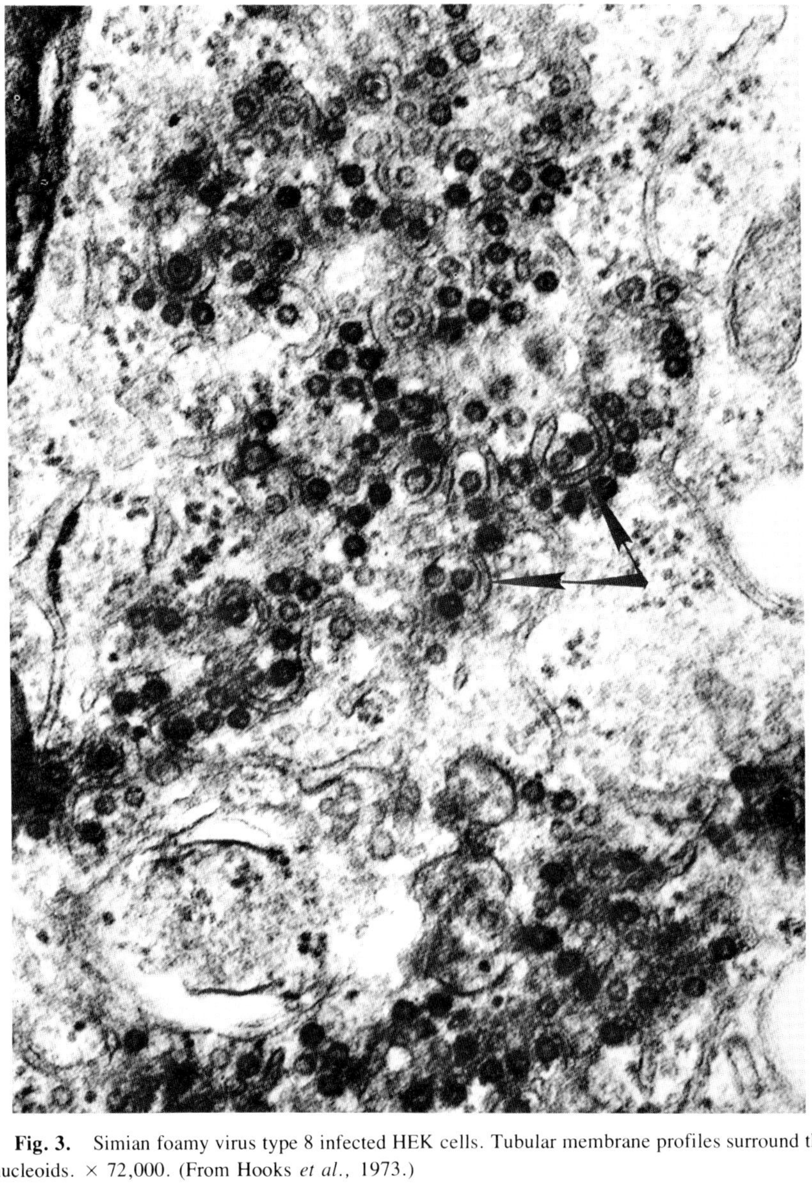

Fig. 3. Simian foamy virus type 8 infected HEK cells. Tubular membrane profiles surround the nucleoids. × 72,000. (From Hooks *et al.*, 1973.)

further substantiated by the incorporation of [^{3}H]uridine (Parks *et al.*, 1971; Parks and Todaro, 1972). However, treatment of infected cells with actinomycin D inhibited the incorporation of [^{3}H]uridine and decreased virus yield. Hruska and Takemoto (1975) isolated the hamster foamy virus RNA and showed that it consisted of several species: 62, 40, 28–30, and 18–20 S. The 60 S RNA probably represents most of the viral RNA and, like the C-type virus RNA, this RNA is composed of multiple segments held together by hydrogen bonds.

3. Proteins

Fortified with the knowledge that foamy viruses had properties similar to the properties of the RNA tumor viruses, Parks and co-workers (1971) looked for, and demonstrated the presence of, a reverse transcriptase. These and further studies substantiated the finding that foamy viruses have RNA-dependent DNA polymerase activity with enzymatic properties similar to those of retroviruses (Parks *et al.*, 1971; Scolnick *et al.*, 1972; Hooks *et al.*, 1973; Hruska and Takemoto, 1975; Rabin *et al.*, 1976; Liu *et al.*, 1977; Brown *et al.*, 1978; Chiswell and Pringle, 1979). The polymerase activity exhibited a marked preference for poly(rA)·oligo(dT) but could also utilize poly(rC)·oligo(dG). The cation requirement for the enzyme reaction favored Mn^{++} ions, which is also observed with most of the C-type retroviruses. In contrast, the enzyme activity for mouse mammary tumor virus, monkey Mason-Pfizer virus, visna virus, and bovine leukemia virus is greater with magnesium. Liu and co-workers (1977) partially purified the enzyme and showed that it has a molecular weight of 80,000.

The physicochemical properties of the other foamy virus protein components have not been investigated.

C. Antigenic Composition

Cats, cows, and hamsters each possess their own distinct foamy virus serotype (Malmquist *et al.*, 1969; Riggs *et al.*, 1969; Hruska and Takemoto, 1975). In contrast, six virus serotypes have been isolated from monkeys and two virus serotypes have been isolated from apes (Rustigian *et al.*, 1955; Johnston, 1961, 1971; Stiles *et al.*, 1964; Rogers *et al.*, 1967; Hooks *et al.*, 1972, 1973). Two additional simian serotypes appear to be new and are presently being characterized (Hooks and Gibbs, 1975; Barahona *et al.*, 1976). The distribution of these virus serotypes in primate species is shown in Table II. During the past decade, there have been several reports of foamy virus isolations from human tissues. To date, these isolates have been shown to be closely related antigenically to the known simian or ape foamy viruses and are discussed in detail in Section IV.

Within the foamy virus group, the simian, bovine, feline, and hamster viruses

TABLE II
Simian Foamy Virus Isolations

Primate species	Virus serotype isolated
Pro simians	
Galago	5
Old World Primates	
Rhesus (*Macaca mulatta*)	1, 2, 3
Cynomolgus (*M. fascicularis*)	1, 2
Formosan rock macaque (*M. cyclopsis*)	1, 2
Bonnet (*M. radiata*)	not identified
Pigtailed macaque (*M. nemestrina*)	1
Vervet (*Cercopithecus pygerythrus*)	1
Grivet (*C. aethiops*)	1, 2, 3
Mangabey (*Cercocebus* spp.)	2
Baboon (*Papio* spp.)	1, 2, 3
New World Primates	
Squirrel (*Saimiri* spp.)	4
Spider (*Ateles* spp.)	8
Capuchin (*Cebus* spp.)	new type
Red Uakari (*Cacajao rubicundus*)	new type
Apes	
Chimpanzee (Pan)	6, 7
Humans	1, 6

do not cross-react by neutralization. There is, however, cross reactivity among the simian foamy viruses, indicating the presence of some common antigens. For example, serotypes 1 through 6 and 8 are neutralized only by their homologous antisera. SFV-7 has a low level of cross reactivity with antisera prepared against SFV-2 (Hooks *et al.*, 1972). However, when immunofluorescent (FA), complement fixation (CF), or hemagglutination inhibition (HAI) assays are used to analyze the antigenic relationships, the simian foamy viruses demonstrate cross reactions (Stiles, 1968, Peries and Todaro, 1977; Nemo *et al.*, 1978; Brown *et al.*, 1978). These findings suggest that the simian foamy viruses may have one or more proteins in common.

Within the Retroviridae, the foamy viruses (Spumavirinae) are antigenically distinct from the members of the other two subfamilies, Oncornavirinae and Lentivirinae (Johnston, 1974; Hooks and Gibbs, 1975; Hruska and Takemoto, 1975). Furthermore, foamy virus reverse transcriptase is antigenically distinct from the reverse transcriptase of other retroviruses (Scolnick *et al.*, 1970; Parks *et al.*, 1971; Liu *et al.*, 1977). In addition, the foamy viruses are also antigenically distinct from the other RNA and DNA viruses (Johnston, 1974; Hooks and Gibbs, 1975; Hruska and Takemoto, 1975).

III. BIOLOGICAL FEATURES

A. Growth of Virus in Tissue Culture

The foamy viruses can induce both a productive and a persistent infection in cell cultures. The productive infection is the more frequent outcome of the virus-cell interaction and takes place in numerous cell lines from a variety of mammalian hosts. However, cell multiplication is required for virus replication (Parks and Todaro, 1972; Hooks *et al.*, 1976; Shroyer and Shalaby, 1978). The viruses propagate in both epithelial and fibroblastic cells of human, monkey, rabbit, cow, pig, rat, hamster, and chicken origin (Hooks and Gibbs, 1975). Persistent virus infections have been reported with SFV-1 in Hep-2 cells and BHK-21 cells (Clarke *et al.*, 1970). Unlike the other retroviruses, foamy viruses have not been shown to induce cellular transformation.

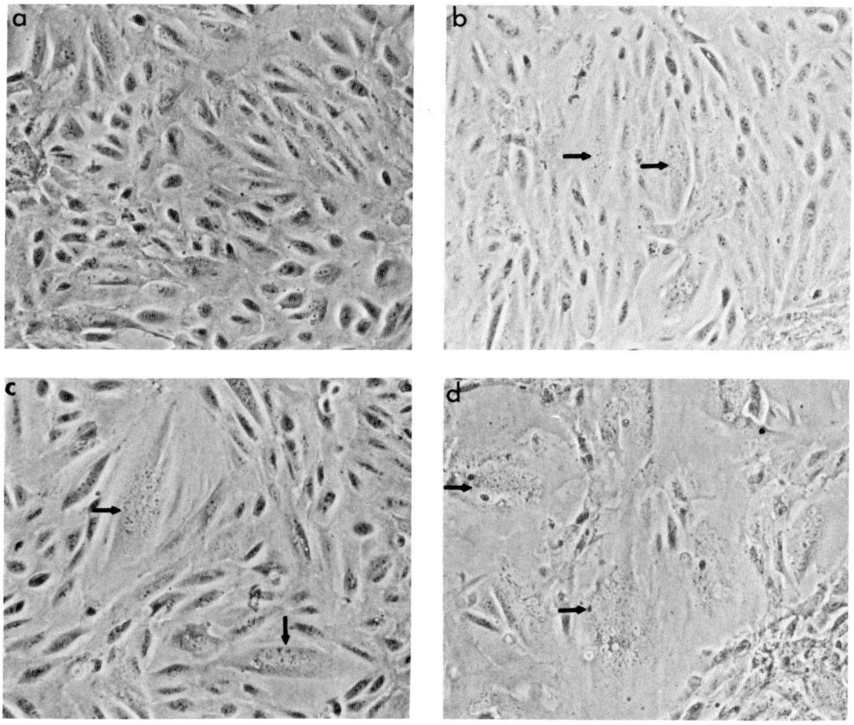

Fig. 4. Development of foamy virus (SFV-8)-induced CPE in HEK cells. (a) Normal HEK cells; (b) early foci of multinucleation (day 3); (c) gradual increase in the number of nuclei within the syncytia (day 4); (d) numerous areas of multinucleation (day 6).

The cytopathic effect (CPE) induced by foamy viruses is characterized by the formation of a vacuolated, foamy syncytia without inclusion bodies. Initially cytopathic changes consist of small areas of multinucleation (Fig. 4 b, c). The number of nuclei increases until large areas of multinucleation are seen (Fig. 4d). The vacuolated, foamy effect is usually not observed until late in the infection, after a majority of the cells in the monolayer have grossly visible CPE. This type of cytopathology is not unique to the foamy viruses. In fact, similar cellular changes can be seen in cells infected with measles virus, mumps virus, respiratory syncytial virus, and some strains of herpes simplex virus and herpes zoster virus.

The time of the appearance of CPE varies with the virus strain, titer, passage history, and type of cells used to propagate the virus. Most of the virus strains will induce CPE within 4 to 5 days, and virus yields of 10^3 to 10^5 mean tissue culture infective dose ($TCID_{50}$)/ml are readily attainable.

A detailed description of the virus attachment, penetration, multiplication, assembly, and release has been reviewed and will not be discussed in detail (Hooks and Gibbs, 1975). However, there are some distinguishing features of foamy virus replication which will be briefly mentioned. Virus replication takes place in both the nucleus and the cytoplasm. In fact, nuclear immunofluorescence is a general characteristic of the foamy viruses which differentiates them from the other retroviruses (Carski, 1960; Malmquist *et al.*, 1969; Fleming and Clarke, 1970; Hooks *et al.*, 1972; Parks and Todaro, 1972). Another important feature of foamy virus replication is that, like the other members of the Retroviridae, infectious viral DNA is produced in infected cells (Chiswell and Pringle, 1977, 1978; Chu and Hooks, 1977).

B. Infection in the Natural Host

The foamy viruses have been isolated from numerous tissues from both normal and clinically ill apes, monkeys, cows, cats, and hamsters (Hooks and Gibbs, 1975). To date, these viruses have not been shown to induce clinical disease in their natural host or in experimental animals. Inoculation of the foamy viruses into seronegative natural hosts results in seroconversion in the absence of clinical disease (Ruckle, 1958a,b; Malmquist *et al.*, 1969; McKissick and Lamont, 1970; Swank and Hsiung, 1975; Feldman *et al.*, 1975). In addition, the inoculation of these viruses into laboratory animals such as rabbits, newborn and adult mice and hamsters, guinea pigs, day-old chicks, and embryonating hens' eggs has not yet resulted in clinical disease (Rustigian *et al.*, 1955; Johnston, 1961; Plummer, 1962; Stiles *et al.*, 1964; McKissick and Lamont, 1970; Johnston, 1971; Hooks *et al.*, 1972, 1973).

The distribution of the virus within the host is widespread. For example, SFV 6 and 7 have been isolated from the following chimpanzee tissues grown *in vitro:* brain, spinal cord, sympathetic ganglia, spleen, thymus, kidney, lymph node, salivary gland, and lung (Rogers *et al.*, 1967; Hooks *et al.*, 1972). The Simian foamy viruses are not found in serum or feces and are infrequently detected in urine (Hooks and Gibbs, 1975; Feldman *et al.*, 1975). Similar widespread distribution patterns are seen in cats and cows (Malmquist *et al.*, 1969; Gaskin and Gillespie, 1972; Gillespie and Scott, 1973). A number of investigators have looked for and found foamy viruses in peripheral blood leukocytes (Malmquist *et al.*, 1969; Gillespie and Scott, 1973; Hooks and Gibbs, 1975; Heberling and Kalter, 1975; Feldman *et al.*, 1975; Barahona *et al.*, 1976; Shroyer and Shalaby, 1978; Hooks and Detrick-Hooks, 1979; Rhodes-Feuillette *et al.*, 1979). The presence or persistence of the virus within the leukocytes may account for the widespread distribution of the virus within the host.

Foamy virus infections in the natural host are characterized by persistence of the virus in the presence of neutralizing antibody. Several lines of evidence suggest that once the animal is infected with the virus, the virus remains for a long period of time, probably for the life of the host (Hooks and Gibbs, 1975; Swank and Hsiung, 1975). Johnston (1971) repeatedly isolated foamy virus from monkey throat swabs over a 10-week interval. During a 16-week period, virus was repeatedly isolated from experimentally infected juvenile rhesus monkeys (Feldman *et al.*, 1975). High neutralizing antibody levels were also detected in chimpanzees which were followed for 8 years (Hooks and Gibbs, 1975).

C. Infection in Experimental Hosts

Recent studies have shown that experimental infection of rabbits with foamy virus results in a persistent infection which closely resembles the infection observed in the natural hosts (Johnston, 1974; Swank and Hsiung, 1975; Hooks and Detrick-Hooks, 1979). The virus can persist in the presence of neutralizing antibody and is recovered from numerous tissues and peripheral blood leukocytes in the absence of clinical disease.

Since virus infections of the reticuloendothelial system can alter the host's immune response, the experimental foamy virus infection in rabbits was used to determine if foamy viruses alter immune responses (Woodruff and Woodruff, 1975; Hooks and Detrick-Hooks, 1979). The humoral immune response to sheep red blood cells was not affected in foamy virus-infected rabbits. In contrast, leukocytes from rabbits with a foamy virus infection showed a depressed CMI response as determined by [^{3}H]thymidine uptake and immune interferon induction (Hooks and Detrick-Hooks, 1979). This depression was transient and lasted for a 2-week period. Since CMI aids in elimination of infection, a depression of CMI responses early in infection may aid the foamy virus in establishing a persistent infection. This is the first demonstration that a foamy virus can adversely affect a persistently infected host.

Following immunosuppression, primary or reactivation infections are quite common, especially with viruses belonging to the herpesvirus group (Armstrong *et al.*, 1976). Early in the course of a persistent foamy virus infection, one of the foamy virus-infected rabbits developed a herpesvirus infection. The ability of a foamy virus infection to increase susceptibility to infection or reactivation is of interest, especially since it is common to isolate a number of different viruses and parasites from the natural host with a persistent foamy virus infection (Hooks and Gibbs, 1975).

IV. ASSOCIATION WITH HUMAN DISEASE

During the past decade there have been three reports of human foamy virus isolations. Achong and co-workers (1971) found a virus morphologically similar to the foamy viruses in a cell line derived from a nasopharyngeal carcinoma of a Kenyan African. Additional studies demonstrated that the human isolate had the morphological and biological characteristics of foamy viruses (Epstein *et al.*, 1974; Loh *et al.*, 1977; Achong and Epstein, 1978). There are some discrepancies in the immunological relationship of this isolate to the known foamy viruses. Epstein and co-workers have characterized this isolate as a new, immunologically distinct member of the foamy virus group (Epstein *et al.*, 1974). In contrast, others have shown that by neutralization, FA, and CF assays, the human isolate is immunologically indistinguishable from the known chimpanzee virus, SFV-6 (Nemo *et al.*, 1978; Brown *et al.*, 1978; Hooks, unpublished observations). Seroepidemiological studies on human populations have indicated that the virus is not widespread in most populations. Brown *et al.* (1978) have failed to find neutralizing antibody to the SFV-6 human isolate in sera from a total of 256 humans. Included in this sera collection was sera from 25 North Americans with nasopharyngeal carcinoma, sera from 45 normal Africans, and sera from 25 Africans with Burkitt's lymphoma. Achong and Epstein (1978), using an indirect immunofluorescence assay, also failed to find antibody to the human isolate in sera from normal and tumor patients from Tunisia, Singapore, and Great Britain, as well as sera from eight African Burkitt's lymphoma patients. In contrast, when Achong and Epstein (1978) studied sera from humans in Kenya, they found antibody to the isolate by immunofluorescence. This antibody was detected in 17 of 97 sera tested. These included 25% (10/42) of the nasopharyngeal carcinoma patients, 25% (4/16) of patients with other tumors of the oronasopharynx, and 17% (2/12) of patients with tumors of other parts of the body. Recently, Loh and co-workers (1980) found an antibody to this virus isolate in 7% of individuals from nine Pacific-island territories. They suggest that this virus is ubiquitous in the Pacific communities.

These studies suggest that a foamy virus which is immunologically closely related to SFV-6 was isolated from a cell line derived from a patient with nasopharyngeal carcinoma. Seroepidemiological studies indicate that a high percentage of nasopharyngeal carcinoma patients from Kenya have antibody to the isolate. Additional studies indicate that antibody to the virus is dispersed in Pacific-island populations.

The second report of a foamy virus from human tissues involved the isolation of SFV-1 from peripheral blood cells from a patient with leukemia (Young *et al.*, 1973). A preliminary report describes the isolation of a foamy virus from the brain of a patient with dialysis encephalopathy (Cameron *et al.*, 1978).

It is reasonable to assume that the foamy virus group may have a human counterpart which could have antigens cross-reacting with the known foamy viruses. However, confirming evidence is not clear-cut because antibody has not been demonstrated in patients who harbor the virus and the isolated viruses are immunologically closely related to monkey and ape viruses (SFV-1 and SFV-6).

V. IMMUNITY

Both the natural and experimental hosts acquire circulating antibody 1 to 4 weeks after a foamy virus infection (Hooks and Gibbs, 1975; Feldman *et al.*, 1975; Swank and Hsiung, 1975; Hooks and Detrick-Hooks, 1979). This is followed by a long-lasting production of antibody and the continued presence of the virus. The humoral immune response can be measured by neutralization, CF, FA, HAI, and precipitation assays.

The mechanisms by which foamy viruses can avert elimination by the humoral immune responses have been investigated (Hooks *et al.*, 1976). The relative importance of the specific immunological mechanism used by the host to protect itself against a viral infection depends, at least in part, upon the route by which the virus spreads from one cell to another. Viruses can spread by one or more of three different routes: extracellularly from infected cells to nearby or distant uninfected cells (type 1 spread); directly from infected to contiguous uninfected cells as a result of cell fusion or viral budding (type 2 spread); or from parent to progeny cells during cell division (type 3 spread) (Notkins, 1974; Hooks *et al.*, 1976). In general, it appears that neutralizing antibody can stop type 1 spread but not type 2 or 3 spread. Foamy viruses can spread by the type 1 and 2 routes and in all likelihood by the type 3 route. Neutralizing antibody and antibody-mediated cell lysis cannot stop foamy virus spread (Hooks *et al.*, 1976).

A CMI response to the foamy viruses has not been demonstrated (Hooks and Detrick-Hooks, 1979). This may well be due to the low antigenicity of foamy virus pools which are used for *in vitro* CMI assays. It has been shown, however, that foamy virus infections *in vivo* can depress the CMI response (Hooks and Detrick-Hooks, 1979).

VI. EPIDEMIOLOGY

Both horizontal and vertical transmission of foamy viruses have been described. Horizontal transmission most likely occurs by the respiratory route (Hooks and Gibbs, 1975; Swank and Hsiung, 1975). Foamy viruses are routinely detected in throat and pharyngeal swabs. Furthermore, monkeys without antibody to foamy viruses frequently develop antibody 1 to 3 months after entry into a primate colony (Ruckle, 1958a,b). Alternatively, the uninfected susceptible host can become infected *in utero* (Hacket and Manning, 1971; Hooks and Gibbs, 1975).

The distribution of antibody to the foamy viruses is relatively species specific and widespread within the natural host population. In addition, numerous investigators have shown that there is a correlation between the presence of circulating antibody and the ability to isolate the virus (Ruckle, 1958a,b; Carski, 1960; Johnston, 1961, 1971; Stiles *et al.*, 1964; Hooks *et al.*, 1972, 1973; Gaskin and Gillespie, 1972; Shroyer and Shalaby, 1978). Therefore, the presence of circulating antibody not only demonstrates the existence of a previous infection but may also be indicative of an ongoing persistent infection.

Sera collected from animals at the time of capture have been used in some studies to determine virus distribution in nature. Ruckle (1958a,b) found that 70% of the cynomolgus monkeys that arrived at a holding facility had neutralizing antibody to SFV-1. Sera from 5 of 24 baboons bled after capture contained neutralizing antibody to SFV-3 (Kalter and Heberling, 1975). Hooks and coworkers (1972) found that sera collected from 16 chimpanzees in the African bush contained neutralizing antibody to either or both SFV-6 and SFV-7.

The presence of feline syncytial virus (FSV) and bovine syncytial virus (BSV) also appears to be widespread in cat and cow populations, respectively. The presence of precipitating antibody to FSV was found in 50 of 180 cat sera tested, while antibody to BSV was detected in 139 of 462 cow sera (Malmquist *et al.*, 1969; Gaskin and Gillespie, 1972).

The presence of antibody in human sera to the human foamy virus isolate has already been discussed in Section IV. As was noted, this isolate was immunologically indistinguishable from the chimpanzee virus, SFV-6. Human sera have also been tested for neutralizing antibody to the simian foamy viruses, and with one exception, no antibody was found (Hooks and Gibbs, 1975; Kalter and Heberling, 1975; Nemo *et al.*, 1978; Brown *et al.*, 1978). Antibody to feline and bovine foamy viruses has not yet been investigated in human populations.

VII. LABORATORY DIAGNOSIS

A. Viral Isolation Methods

The only method for isolating or cultivating foamy viruses is by inoculation of tissue culture. These viruses will replicate and produce a CPE in a variety of human and animal cell cultures. However, the cells must be multiplying for the virus to replicate. Therefore, cell cultures should be inoculated when the cultures are less than 50% confluent. Primary and first subcultures of human embryo kidney, rabbit kidney, or rat kidney cell cultures are appropriate for use in isolation procedures. Primary cultures from monkeys, cows, and cats should not be used because of the likelihood of these cultures containing their own foamy virus. Furthermore, rapidly growing cell lines, such as Vero and HeLa cells, may not be the best choice for some isolates since the cells may replicate faster than the virus and the CPE is not readily detectable.

The best source of virus from living animals is throat swabs or peripheral blood leukocytes. Virus can also be isolated from numerous tissues, but kidney and spleen are probably better candidates. The viruses are not readily isolated from 10% tissue suspensions; rather, explantation or co-cultivation techniques should be used. The presence of foamy virus is indicated by the typical CPE, which initially consists of small areas of multinucleation. With some of the isolates, this can proceed to involve most of the cell cultures, and the cultures will contain numerous areas of vacuolated multinucleated syncytia (Fig. 5a,b). On the other hand, some isolates reveal themselves only as one or two small areas of multinucleation. Rapid serial passage of both cells and supernatant fluids is required to establish these isolates in culture. In addition to the development of CPE, the presence of foamy viruses is also indicated by FA and electron microscopy (EM).

B. Serological Techniques

Identification of the virus cannot be based on CPE or EM but must depend upon serological tests. The serological assays which have been used are neutralization, CF, FA, HAI, and precipitation. Neutralization is required for identification because of the cross reactivity detected when CF, FA, or HAI assays are used. Reference reagent antisera to the simian foamy viruses may be obtained from Dr. Richard Heberling, Southwest Foundation for Research and Education, World Health Organization Collaborating Center for Reference and Research in Simian Viruses, San Antonio, Texas, 78284, USA.

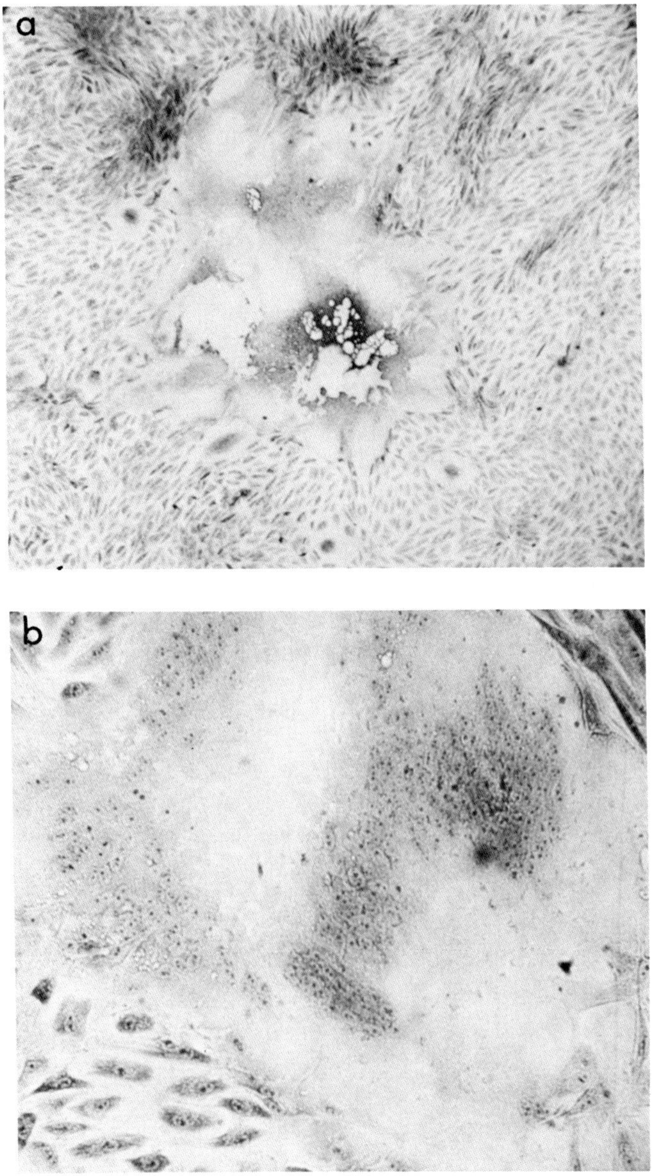

Fig. 5. Foamy virus (SFV-6)-induced CPE in rabbit kidney cells (PRK). (a) Foci of vacuolated multinucleation; (b) higher magnification of syncytia showing extensive multinucleation.

1. Neutralization

Neutralization assays for the detection of foamy virus antibodies depend upon the ability of antisera to inhibit foamy virus CPE or plaque formation (Hooks *et al.*, 1972, 1976; Parks and Todaro, 1972). We have found that the microneutralization assay is rapid and as sensitive as the inhibition of plaque formation. The infectivity assay in microplates consists of adding a 0.1 ml volume of PRK cells (1×10^5 cells/ml) to the microplate wells. Within the next 3 hours, serial tenfold dilutions of virus in a 0.1 ml volume are added to the wells in quadruplicate. The cultures are incubated at 36°C in a 5% CO_2 atmosphere, and the medium is changed every 3 to 4 days. For most virus stocks, the infectivity, as determined by development of CPE, can be read within 10 days (Hooks and Detrick-Hooks, 1979). For the neutralization assay, 100 $TCID_{50}$ of virus is mixed with an equal volume of serial twofold dilutions of heat-inactivated sera. The mixture, along with virus controls, is incubated at room temperature for 1 hour. The virus-antisera mixture and virus controls are each inoculated in four PRK cell cultures in microwells in a 0.1 ml volume. Inhibition of foamy virus CPE after a 7- to 10-day period is recorded as positive neutralization.

2. Fluorescent Antibody Staining

FA is performed using standard techniques. Sera are precipitated with ammonium sulfate. Immunoglobulin G (IgG) fractions are separated on a Sephadex G-200 column and labeled with fluorescein isothiocyanate (Carski, 1960; Hooks *et al.*, 1972, 1976; Brown *et al.*, 1978). Excess fluorescein is removed by passing the conjugated globulin through a G-25 fine Sephadex column.

Infected cells grown on coverslips are washed with phosphate-buffered saline (PBS) and fixed with acetone for 5 minutes at room temperature. The cells are then incubated with fluorescein-labeled antiviral IgG at 37°C for 20 min. The coverslips are washed with PBS, mounted on slides, and observed through a fluorescent microscope.

Viral antigens are initially observed in the nuclei of infected cells approximately 20 hours after inoculation. Within a few hours after nuclear staining is observed, cytoplasmic fluorescence is noted around the nucleus. As the infection progresses, nuclear fluorescence is diminished and cytoplasmic staining becomes more granular.

3. Complement Fixation

CF testing can be carried out in microtiter systems using standarized techniques (Stiles, 1968; Brown *et al.*, 1978). The preparation of antigen is probably the limiting factor in the usefulness of CF testing. Because of the low yield of virus, it is possible to obtain foamy virus antigens suitable for CF reagents for only SFV-1, 2, 3, 6, and 7.

4. Hemagglutination Inhibition

Peries and Todaro (1977) succeeded in preparing SFV-1 stocks containing 10^{12} virus particles by propagating the virus in a clone of canine thymus cell (Fcf 2th, Naval Biomedical Research Laboratories). Using these preparations, they demonstrated virus hemagglutinating activity with guinea pig red blood cells. A standard microtiter assay system is used with a PBS or veronal-gelatine solution at pH 6.0 to 8.2 as the diluent. The reaction takes place at both room temperature and at 37°C. Using this system, HI by specific anti-foamy virus antibody was demonstrated. Cross reactivity was observed between SFV-1 antigen and antisera prepared against SFV-2, 4, 5, 6, and 7.

VIII. CONCLUDING REMARKS

The wide distribution of the foamy viruses and their ability to establish persistent infections probably contribute to their survival in nature. Nevertheless, the ubiquity of these viruses can create problems in diagnostic and research studies. Investigators should be cognizant of their presence both *in vivo* and *in vitro*.

However, the isolation and serological identification of foamy viruses can be an extremely difficult task. Many of the primary foamy virus isolates do not replicate well in tissue culture. To facilitate isolations, the investigator must use explantation and co-cultivation techniques with viable cells rather than 10% suspensions, dividing indicator cells rather than stationary cells, and subculturing procedures. With the use of these techniques, more viruses belonging to this group may be discovered.

A better awareness of the presence of foamy viruses and the comparative diagnostic procedures used for isolations should lead to a better understanding of comparative virology and the association of these viruses with human disease.

REFERENCES

Achong, B. G., and Epstein, M. A. (1978). *J. Gen. Virol.* **40,** 175.

Achong, B. G., Mansell, P. W. A., Epstein, M. A., and Clifford, P. (1971). *J. Natl. Cancer Inst.* **46,** 299.

Armstrong, J. A., Evans, A. S., Rao, N., and Ho, M. (1976). *Infect. Immun.* **14,** 970.

Barahona, H., Garcia, F. G., Melendez, L. V., King, N. W., Ingalls, J. K., and Daniel, M. D. (1976). *J. Med. Primatol.* **5,** 253.

Boothe, A. D., Van Der Maaten, M. J., and Malmquist, W. A. (1970). *Arch. Gesamte Virusforsch.* **31,** 373.

Brown, P., Nemo, G., and Gajdusek, D. C. (1978). *J. Infect. Dis.* **137,** 421.

Cameron, K. R., Birchall, S. M., and Moses, M. A. (1978). *Lancet* **ii,** 796.

Carski, T. (1960). *J. Immunol.* **84,** 426.

Chiswell, D. J., and Pringle, C. R. (1977). *J. Gen. Virol.* **36,** 551.
Chiswell, D. J., and Pringle, C. R. (1978). *Virology* **90,** 344.
Chiswell, D. J., and Pringle, C. R. (1979). *J. Gen. Virol.* **43,** 429.
Chopra, H. C., Hooks, J. J., Walling, M. J., and Gibbs, C. J., Jr. (1972). *J. Natl. Cancer Inst.* **48,** 451.
Chu, C. T., and Hooks, J. J. (1977). *Chin. J. Microbiol.* **10,** 91.
Clarke, J. K., and Attridge, J. T. (1968). *J. Gen. Virol.* **3,** 185.
Clarke, J. K., and McFerran, J. B. (1970). *J. Gen. Virol.* **9,** 155.
Clarke, J. K., Attridge. J. T., Dane, D. S., and Briggs, M. (1967). *J. Gen. Virol.* **1,** 565.
Clarke, J. K., Attridge, J. T., and Gay, F. W. (1969a). *J. Gen. Virol.* **4,** 183.
Clarke, J. K., Gay, F. W., and Attridge, J. T. (1969b). *J. Gen. Virol.* **3,** 358.
Clarke, J. K., Samuels, J., Dermott, E., and Gay, F. W. (1970). *J. Virol.* **5,** 624.
Dalton, A. J., Melnick, J. L., Bauer, H., Beaudreau, G., Bentvelzen, P., Bolognesi, D., Gallo, R., Graffi, A., Haguenu, F., Heston, W., Huebner, R., Todaro, G., and Heine, U. I. (1974). *Intervirology* **4,** 201.
Dermott, E., Clarke, J. K., and Samuels, J. (1971). *J. Gen. Virol.* **12,** 105.
DiGiacomo, R. F., Hooks, J. J., Sulima, M. P., Gibbs, C. J., Jr., and Gajdusek, D. C. (1977). *J. Am. Vet. Med. Assoc.* **171,** 859.
Enders, J., and Peebles, T. (1954). *Proc. Soc. Biol. Med.* **86,** 277.
Epstein, M. A., Achong, B. G., and Ball, G. (1974). *J. Natl. Cancer Inst.* **53,** 681.
Feldman, M. D., Dunnick, N. R., Barry, D. W., and Parkman, P. D. (1975). *J. Med. Primatol.* **4,** 287.
Fleming, W. A., and Clarke, J. K. (1970). *J. Gen. Virol.* **6,** 277.
Gaskin, J. M., and Gillespie, J. H. (1972). *Am. J. Vet. Res.* **34,** 245.
Gillespie, J. H., and Scott, F. W. (1973). *Adv. Vet. Sci. Comp. Med.* **17,** 164.
Hacket, A. J., and Manning, J. S. (1971). *J.A.V.M.A.* **158,** 948.
Heberling, R. L., and Kalter, S. S. (1975). *Amer. J. Epidemiol.* **102,** 35.
Hooks, J. J., and Detrick-Hooks, B. (1979). *J. Gen. Virol.* **44,** 383.
Hooks, J. J., and Gibbs, C. J., Jr. (1975). *Bacteriol. Rev.* **39,** 169.
Hooks, J. J., Gibbs, C. J., Jr., Cutchins, E. C., Rogers, N. G., Lampert, P., and Gajdusek, D. C. (1972). *Arch. Gesamte Virusforsch.* **38,** 38.
Hooks, J. J., Gibbs, C. J., Jr., Chou, S., Howk, R., Lewis, M., and Gajdusek, D. C. (1973). *Infect. Immun.* **8,** 804.
Hooks, J. J., Burns, W., Hayashi, K., Geis, S., and Notkins, A. L. (1976). *Infect. Immun.* **14,** 1172.
Hruska, J. F., and Takemoto, K. K. (1975). *J. Natl. Cancer Inst.* **54,** 601.
Hsiung, G. D. (1968). *Bacteriol. Rev.* **32,** 185.
Hull, R. (1968). *Virol. Monogr.* **2.**
Johnston, P. (1961). *J. Infect. Dis.* **109,** 1.
Johnston, P. (1971). *Infect. Immun.* **3,** 793.
Johnston, P. (1974). *Lab. Anim. Sci.* **24,** 159.
Kalter, S. S., and Heberling, R. L. (1971). *Bacteriol. Rev.* **35,** 310.
Kurstak, E. (1977). *In* "Comparative Diagnosis of Viral Diseases" (E. Kurstak and C. Kurstak, eds.), Vol. 1, pp. 1–22. Academic Press, New York.
Liu, W. T., Natori, T., Chang, K. S. S., and Wu, A. M. (1977). *Arch. Virol.* **55,** 187.
Loh, P. C., Achong, B. C., and Epstein, M. A. (1977). *Intervirology* **8,** 204.
Loh, P. C., Matsuura, F., and Mizumoto, C. (1980). *Intervirol.* **13,** 87.
McKissick, G. E., and Lamont, P. H. (1970). *J. Virol.* **5,** 247.
Malmquist, W. A., Van Der Maaten, M. J., and Boothe, A. D. (1969). *Cancer Res.* **29,** 188.
Nemo, G., Brown, P., Gibbs, C. J., Jr., and Gajdusek, D. C. (1978). *Infect. Immun.* **20,** 69.

Notkins, A. L. (1974). *Cell. Immunol.* **11,** 478.
Parks, W., and Todaro, G. (1972). *Virology* **47,** 673.
Parks, W., Todaro, G., Scolnick, E., and Aaronson, S. (1971). *Nature (London)* **229,** 258.
Peries, J., and Todaro, G. (1977). *J. Gen. Virol.* **34,** 195.
Plummer, G. (1962). *J. Gen. Microbiol.* **29,** 703.
Rabin, H., Neubaurer, R. H., Woodside, N. J., Cicmance, J. L., Wallen, W. C., Lapin, B. A., Agrba, V. A., Yakoleva, L. A., and Chuvirou, G. N. (1976). *J. Med. Primatol.* **5,** 13.
Rhodes-Feuillette, A.. Fortuna, S., Lesneret, J., Dubouch, P., and Peries, J. (1979). *J. Med. Primatol.* **8,** 308.
Riggs, J. L., Oshire, L. S., Taylor, D. O. N., and Lennette, E. H. (1969). *Nature (London)* **222,** 1190.
Rogers, N., Basnight, M., Gibbs, C. J., Jr., and Gajdusek, D. C. (1967). *Nature (London)* **216,** 446.
Ruckle, G. (1958a). *Arch. Gesamte Virusforsch.* **8,** 139.
Ruckle, G. (195Ob). *Arch. Gesamte Virusforsch.* **8,** 167.
Rustigan, R., Johnston, P., and Reihart, H. (1955). *Proc. Soc. Exp. Biol. Med.* **88,** 8.
Scolnick, E., Rands, E., Aaronson, S. A., and Todaro, G. (1970). *Proc. Natl. Acad. Sci. U.S.A.* **67,** 1789.
Scolnick, E., Parks, W., Todaro, G., and Aaronson, S. (1972). *Nature (London), New Biol.* **235,** 35.
Seman, G., and Dmochowski, L. (1977). *In* "Comparative Diagnosis of Viral Diseases" (E. Kurstak and C. Kurstak, eds.), Vol. 2, pp. 111-205. Academic Press, New York.
Shroyer, E. L., and Shalaby, M. R. (1978). *Am. J. Vet. Res.* **39,** 555.
Stiles, G. (1968). *Proc. Soc. Exp. Biol. Med.* **127,** 225.
Stiles, G. E., Bittle, J. L., and Cabasso, V. J. (1964). *Nature (London)* **201,** 1350.
Swank, N. S., and Hsiung, G. D. (1975). *Infect. Immun.* **12,** 470.
Woodruff, J. F., and Woodruff, J. J. (1975). *In* "Viral Immunology and Immunopathology" (A. L. Notkins, eds), pp. 393-418. Academic Press, New York.
Young, D., Samuels, J., and Clarke, J. K. (1973). *Arch. Gesamte Virusforsch.* **42,** 228.

Chapter 15

Lentivirinae: Maedi/Visna Virus Group Infections

M. BRAHIC and A. T. HAASE

I.	Introduction	620
II.	Description of Viruses	620
	A. Structural Composition of Lentiviruses and Organization	620
	B. Replication	620
	C. Comparative Aspects	624
III.	Description of Diseases	626
	A. Symptomatology	626
	B. Pathology	627
	C. Immune Response	629
	D. Antigenic Variation	629
	E. Differential Diagnosis	630
	F. Host Range	630
	G. Epidemiology	631
	H. Methods of Disease Control	631
IV.	Methods	632
	A. Basic Experimental Protocol	632
	B. Comments on Specific Procedures	633
V.	Pathogenesis	637
	A. Visna Life Cycle in the Animal	637
	B. Restriction of Proviral DNA Expression *in Vivo*	638
	C. Slowness—Persistence	639
	D. Unresolved Questions in Pathogenesis	639
VI.	Conclusion	640
	References	641

ISBN 0-12-429704-8

I. INTRODUCTION

The late Bjorn Sigurdsson (1954) introduced the term "slow infections" to draw attention to the unusually long incubation period and symptomatic course of several diseases that appeared in epidemic form in Iceland in the period 1930–1950. These diseases include visna and maedi, and the Icelandic form of scrapie (rida). Visna and maedi are inflammatory conditions of the central nervous system (CNS) and lungs caused by viruses with conventional properties, whereas scrapie is caused by an agent with unconventional properties and is a degenerative noninflammatory disease of the CNS (Hunter, 1972). In this chapter we describe the viruses that cause visna and maedi, and the diseases they produce, taking these lentiviruses as the prototypes of the other kinds of conventional viruses that can cause slow infections with predominantly inflammatory pathology (Johnson *et al.*, 1974; Townsend *et al.*, 1975). In addition, we discuss the methodology relevant to addressing the novel issues of pathogenesis raised by the lentiviruses.

II. DESCRIPTION OF VIRUSES

The viruses that cause visna and maedi, and two other agents that cause similar diseases of the lungs and CNS of sheep, progressive pneumonia virus (PPV) (Kennedy *et al.*, 1968), and zwoegerziekte virus (De Boer, 1975) comprise a subfamily of Retroviridae designated "Lentivirinae" (from *lentus*, L. slow). The other subfamilies of Retroviridae are the RNA tumor viruses (Oncovirinae), and the foamy viruses (Spumavirinae) (Kurstak, 1978).

A. Structural Composition of Lentiviruses and Organization

The structural properties of lentiviruses that form the bases for their inclusion in the family Retroviridae are summarized in Table I. The salient taxonomic property that distinguishes all retroviruses is the virion-associated reverse transcriptase that transfers information from the RNA genome of the virus to a DNA intermediate in the cell (Baltimore, 1970; Temin and Mizutani, 1970). Other aspects of structure shared for the most part with other retroviruses (see Bolognesi, 1974, for a review) are listed in the table. Differences between lentiviruses and other retroviruses will be taken up in detail following Section B.

B. Replication

1. Single Growth Cycle in Vitro

Visna virus (taken as the prototype of lentiviruses) will infect cells derived from many species of vertebrates (Thormar and Sigurdardottir, 1962), but with

TABLE I
Structural Characteristics of Lentiviruses

Morphology[a]

Spherical enveloped virions, 80–120 nm in diameter with a 40-nm central electron-dense core; surface bears 8-nm knoblike projections.

Physical and chemical properties

Isopycnic density: 1.15–1.16 gm/cm^3 in sucrose; Isoelectric point: 3.8[b]

Sedimentation: 600 S in sucrose.[c]

Probable composition: 60% protein, 35% lipid, 3% carbohydrate, 2% RNA.[d]

Inactivation: relatively resistant to ultraviolet irradiation; infectivity abolished by lipid solvents, periodate, phenol, trypsin, ribonuclease, formaldehyde, and low pH (less than 4.2).

Thermal stability: infectivity preserved for months in the presence of serum at −50°C; relatively stable at 0–4°C; infectivity destroyed at 56°C; at 37°C one-half the infectivity is lost in 6–8 hours, 90% in 20–30 hours.[e]

Proteins: structural organization and antigenicity[f]

Four structural polypeptides designated gp135, p30, p16, and p14 (see August *et al.*, 1974, for nomenclature) comprise 90% of the protein mass of the virus.[f]

gp135 is the glycoprotein of the knoblike projections emanating from the virion surface and elicits type-specific neutralizing antibody.[g]

p30 and p14 are located in the virion core;[h] p30 is the major group-specific antigen, with antigenic determinants shared by visna, maedi, PPV, and zwoegerziekte virus.[i]

Enzymes[j]

RNA-directed DNA polymerase or reverse transcriptase; dimer with subunits of molecular weight 68,000; located in the virion core.

RNAs

Virion nucleic acid is single-stranded RNA with extensive secondary structure; sediments at 60–70 S in sucrose gradients. On denaturation, 2–3 subunits sedimenting at 35 S are released.[k]

Base composition (%): C: 16, A: 36, G: 26, U: 22.[l]

Each subunit contains essentially identical information; i.e., the genome is polyploid.[m]

Subunits have a molecular weight of 3.6 × 10^6, are of plus strand polarity, and have a poly(A) tract at the 3′ end.[n]

Virions also contain ribosomal and low-molecular-weight RNAs derived from the cell. One species of tRNA binds to the 70 S genome and functions as the primer for DNA synthesis.[o]

[a]Thormar and Cruickshank (1965); Coward *et al.* (1970); Chippaux-Hyppolite *et al.* (1972); Takemoto *et al.* (1973). [b]Haase and Baringer (1974). [c]Stone *et al.* (1971b). [d]Haase (1975); August *et al.* (1977). [e]Thormar (1961b, 1965). [f]Mountcastle *et al.* (1972); Haase and Baringer (1974); Lin and Thormar (1974); Lin (1977). [g]Mountcastle *et al.* (1972); Lin and Thormar (1979); Scott *et al.* (1979); Bruns and Frenzel (1979). [h]Lin (1977). [i]Stowring *et al.* (1979). [j]Lin and Thormar (1970, 1972); Stone *et al.* (1971a); Lin *et al.* (1973); Haase *et al.* (1974b); Scolnick and Parks (1974); Lin and Papini (1978). [k]Harter *et al.* (1969, 1971); Brahic *et al.* (1971, 1973); Lin and Thormar (1971); Friedmann *et al.* (1974); Haase *et al.* (1974a). [l]Vigne *et al.* (1977). [m]Beemon *et al.* (1976); Vigne *et al.* (1977, 1978). [n]Gillespie *et al.* (1973); Vigne *et al.* (1977). [o]Haase *et al.* (1974a); Faras *et al.* (unpublished observations).

few exceptions (Harter *et al.*, 1968) it replicates efficiently only in ovine cells; the highest yields (50–100 PFU/cell) are obtained in cultures established from sheep choroid plexus (SCP) (Sigurdsson *et al.*, 1960). In a single-step multiplication cycle, the period of latency lasts about 24 hours and is followed by a phase of exponential growth concluded by 72–96 hours (Thormar, 1963). In this period of virus production, the cells exhibit cytopathic effects (CPE) of three kinds (Fig. 1): formation of round refractile cells, of stellate cells, and of polykaryocytes. The last also can be induced at high multiplicities by virions inactivated by ultraviolet irradiation; i.e., visna virus, like paramyxoviruses, can fuse cells from without (Harter and Choppin, 1967b). Because of the viral-induced CPE, the culture undergoes complete degeneration by the completion of the life cycle. The typical CPE of lentiviruses provides a basis for virus identification and quantal assay, by end point dilution or plaque methods (Harter and Choppin, 1967a; Harter, 1969; Haase and Levinson, 1973; Trowbridge, 1974).

2. *Biochemical Events*

Visna virus enters the cell by fusion (Chippaux-Hyppolite *et al.*, 1972), and its RNA is released into the cytoplasm within the first hour of infection (Brahic *et al.*, 1977). Reverse transcription of the RNA is initiated in the cytoplasm and completed in the nucleus (Harris *et al.*, in press). The completed viral DNA in the nucleus consists of linear duplex molecules equivalent to a transcript of a subunit of viral RNA and to rare circular forms. Many of the linear molecules have a gap in the plus strand (Harris *et al.*, in press). A variable proportion of viral DNA is associated with high-molecular-weight cellular DNA (Haase and Varmus, 1973; Haase *et al.*, 1976; Clements *et al.*, 1979) but there is no evidence as yet that establishes covalent linkage of viral and host cell sequences.

Viral DNA synthesis continues throughout the life cycle of the virus to reach 200–300 copies per cell (Haase, 1975; Traynor and Haase, 1977; Harris *et al.*, in press) representing about a tenfold amplification over input RNA. The mechanisms involved in this amplification, the source of templates, and the import of this extensive synthesis of DNA for the viral life cycle are as yet unclear. It seems likely that much of the DNA synthesized late in infection is the result of super infection and is superfluous for production of viral RNA and progeny virus. The early phase of amplification in the first 20 hours of infection is by contrast necessary for a full yield of viral RNA and progeny virus (Haase *et al.*, in preparation).

Viral RNA synthesis is initiated asynchronously in the nucleus of infected cells as early as 5–7 hours after infection (Haase *et al.*, in preparation) and proceeds exponentially to reach levels of several thousand copies of RNA per cell (Brahic *et al.*, 1977). The aggregations of subunits to form the 70 S complex begins during virus maturation and is completed extracellularly (Brahic and

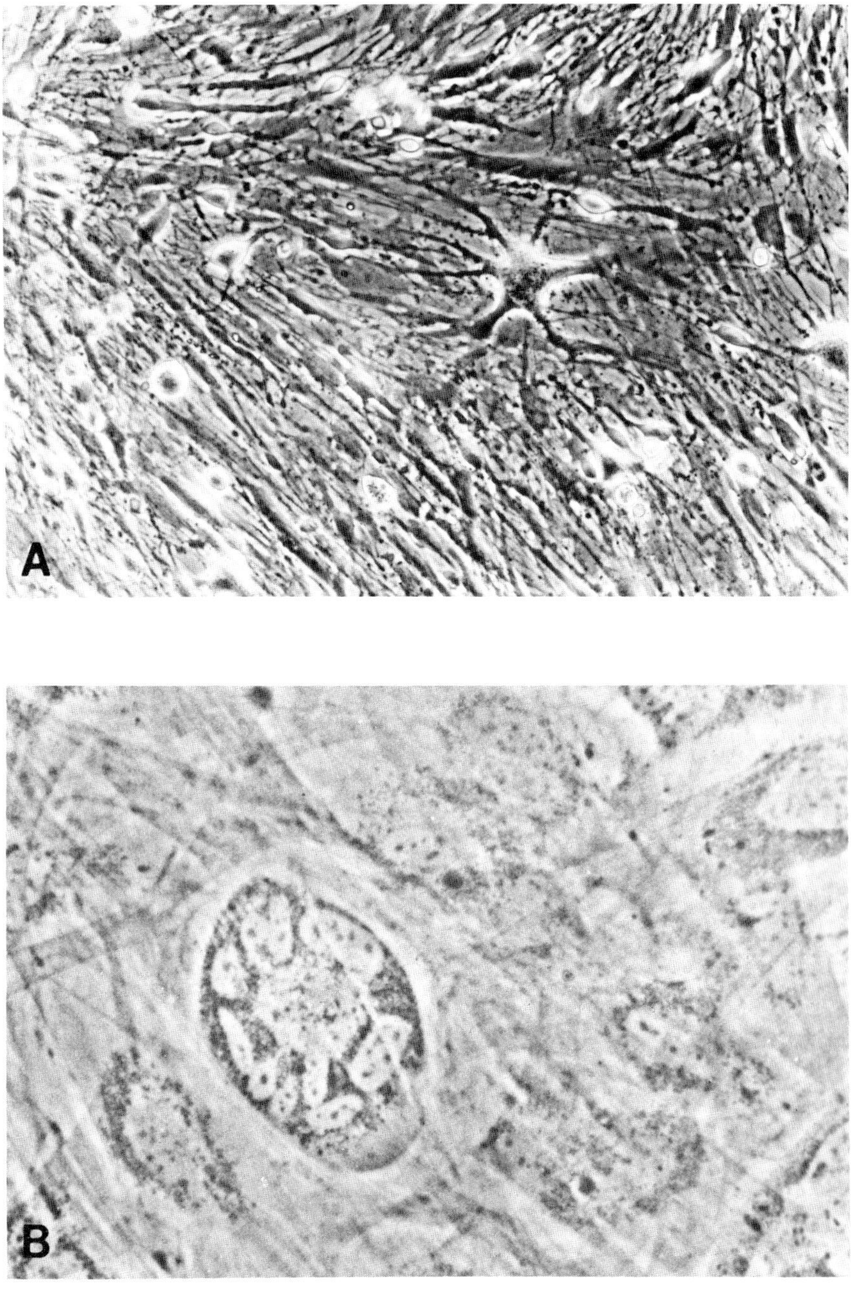

Fig. 1. Cytopathic effects of Lentivirinae. Phase micrographs of SCP infected with visna virus (A) Round refractile degenerating cells and stellate cell. Original magnification × 100. (B) Multinucleated giant cells. × 400.

Vigne, 1975). There are three classes of mRNA with sedimentation coefficients of 35 S, 28 S, and 21 S (Filippi *et al.*, 1979). By analogy to other retroviruses the 35 S and 28 S mRNAs may code for the major core polypeptides and reverse transcriptase, and envelope glycoprotein respectively (S. R. Weiss *et al.*, 1977); as is the case with other retroviruses, synthesis of polypeptides takes place in the cytoplasm (Harter *et al.*, 1967; Thormar, 1969), and virion polypeptides are generated by proteolytic cleavage of larger precursors, (Vogt and Eisenman, 1973; Gielkens *et al.*, 1976; Vigne *et al.*, unpublished observations), although the details of the cleavage of the primary translational products are not yet known. In the final step in the life cycle virions are formed by budding from the plasma membrane of the infected cell (Fig. 2) (Thormar, 1961a; Chippaux-Hyppolite *et al.*, 1972; Dubois-Dalcq *et al.*, 1976).

C. Comparative Aspects

1. Lentiviruses Vis-à-Vis Other Retroviruses

The structure and replication of lentiviruses and the other retroviruses are generally similar, but there are a number of characteristics that taken together distinguish the lentiviruses as a separate subfamily.

a. Lytic Effects Versus Transformation and Tumors. Lentiviruses as a group are associated *in vivo* with persistent infections and slowly evolving pathology of the lungs and CNS of sheep; *in vitro* these viruses fuse and kill cells. It is unlikely that these viruses transform cells (one early report of transformation has not been confirmed) (Takemoto and Stone, 1971), and they do not cause tumors in animals. No nucleotide sequence homology can be detected by hybridization between lentiviruses and various members of the Oncornavirus subfamily (Harter *et al.*, 1973; Quintrell *et al.*, 1974; Stehelin *et al.*, 1976).

b. Virion Proteins and Morphogenesis. The lentivirus type- and group-specific antigens do not share determinants with the group-specific antigen of other retroviruses, including isolated examples (Stowring *et al.*, 1979) of retroviruses that are associated with slow infections (equine infectious anemia virus) or cause cell fusion (bovine syncytial virus) or CPE (reticuloendotheliosis virus). The polypeptides of lentiviruses also differ somewhat in size and number from oncornaviruses. The glycopeptide gp135 is apparently larger, although this may reflect only the extent of glycosylation, and there is one report that the glycopeptide has a molecular weight of 70,000 (Bruns and Frenzel, 1979). There are three core polypeptides in lentiviruses and four in oncornaviruses. Lentiviruses do not have a second internal membrane either, since the nucleocapsid is immediately apposed to the virion envelope in the formation of the virion bud.

Replication. In contrast to most other retroviruses, there are no detectable endogenous lentiviruses or viral genes in uninfected cells (Haase and Varmus,

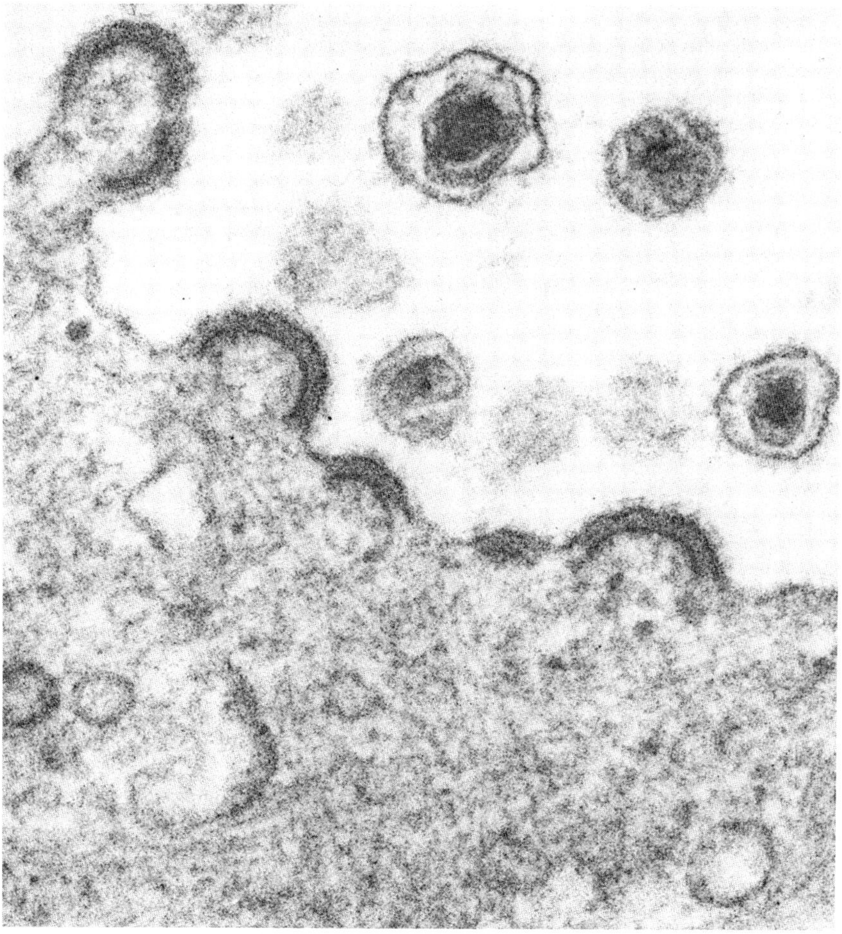

Fig. 2. Morphology and morphogenesis of Lentivirinae. Electron micrograph of an SCP cell infected with visna virus. Various stages in the formation of budding particles and extracellular particles with a central electron-dense core are evident. × 167,000. (Electron micrograph provided by Dr. J. Richard Baringer.)

1973); replication does not require cellular DNA synthesis or mitosis (confluent cultures are fully permissive for the growth of visna virus); mature forms of viral DNA are found essentially only in the nucleus; and the extent of amplification is unusual, although not wholly without precedent (Keshet and Temin, 1979).

2. Distinctions between Lentiviruses

The lentiviruses are known now to be so closely related that it is most reasonable to consider them to be strains of a single virus type (Gudnadottir, 1974). The viruses isolated in visna, maedi, progressive pneumonia (PP), and zwoegerziekte produce identical CPE *in vitro* and pathology *in vivo*. Thus animals with lesions in the CNS of visna will also have pulmonary pathology, and conversely, animals with maedi regularly evidence histopathological changes in the CNS as well as the lungs (Gudnadottir, 1974). At the molecular level the nucleotide sequences of the viruses are highly related, whether measured by hybridization (Harter *et al.*, 1973), by mapping the RNAs or peptides of the major antigens (Scott *et al.*, 1979), or by immunological assays (M. J. Weiss *et al.*, 1977; Stowring *et al.*, 1979). There are minor differences between strains in the glycopeptide gp135, demonstrable in neutralization assays or peptide maps (Scott *et al.*, 1979). These distinctions, however, are subtle and occur between strains of viruses isolated in visna (Gudnadottir, 1974), just as between visna viruses and isolates of maedi viruses (Thormar and Helgadottir, 1965) or PPV (Takemoto *et al.*, 1971). The designation "visna virus," for example, is therefore largely historical, and refers to the site of isolation of virus or predominant pathology in an individual animal.

III. DESCRIPTION OF DISEASES

The following description of the symptoms and pathology of visna and maedi is drawn from the original description by Sigurdsson and co-workers in Iceland (Sigurdsson *et al.*, 1957; Sigurdsson and Palsson, 1958) and the more recent work of others on experimental infections (Narayan *et al.*, 1974; Petursson *et al.*, 1976; Palsson *et al.*, 1977). No distinction is made between natural and experimental infection since both lead to the same observations. Although visna and maedi are described separately, as noted above, this distinction is somewhat arbitrary, as it is clear that they are two manifestations of the same viral infection (Gudnadottir, 1974).

A. Symptomatology

1. Visna

Visna affects both sexes and, in the field, is very infrequent in animals under 2 years of age. The onset of the disease is insidious. The first signs consist of a slight aberration of gait, especially of the hind quarters, fine trembling of the lips, unnatural tilting of the head and, in rare instances, blindness. The symptoms gradually progress to paresis or even total paralysis. Fever is absent. If left unattended, animals die of inanition, hence the name "visna," which means wasting in Icelandic. When helped to obtain food and water, the animal can survive for remarkably long periods of time. The duration of this protracted clinical phase is highly variable. Some animals die in a few weeks, but most survive 1 to 2 years. The incubation period of experimentally infected sheep is also extremely irregular, ranging from 2 months to over 10 years.

2. *Maedi*

Like visna, maedi is found in both sexes and only in adults. Early signs consist of a slow progressive loss of condition accompanied by dyspnea. The respiratory rate can reach 80–120 per minute. With time the respiration becomes more and more difficult; breathing requires the use of accessory muscles, and is accompanied by characteristic rhythmic jerks of the head. Sometimes there is dry coughing; no appreciable amount of fluid is present in the respiratory tract, and there is no nasal discharge. As with visna, the incubation period and clinical phase can vary from a few months to several years. The animal often dies from an acute terminal bacterial pneumonia.

B. Pathology

1. *Visna*

The major pathological change characteristic of visna is destruction of tissue in inflammatory foci that are found principally in the neuroparenchyma bordering the ventricles, in choroid plexus, and in the meninges (Sigurdsson *et al.*, 1962; Petursson *et al.*, 1976). In the periventricular lesions, the inflammatory infiltrate begins beneath the ependymal cells lining the ventricle and extends into the deeper gray and white matter. The inflammatory cells are often arrayed as a cuff surrounding blood vessels or are collected in distinct foci. Infiltration of the meninges is usually more diffuse and less marked. The choroid plexus is constantly and more intensely involved to the point where it may resemble a lymph node with organized follicles and active germinal centers (Fig. 3). The typical pleocytosis in visna is the result of shedding of inflammatory cells into the cerebrospinal fluid (CSF). In animals inoculated intracerebrally, cell counts of 1000–15,000/mm^3 are usually observed. This pleocytosis is sustained for the duration of the infection at lower levels of 50–200 cells/mm^3 (Griffin *et al.*, 1978). The magnitude of the cell counts in the CSF, and the changes in the course of disease vary considerably from animal to animal.

The inflammatory infiltrates consist mainly of three cell types (Georgsson *et al.*, 1977). Lymphocytes predominate, ranging in morphology from small to large blastlike cells. Macrophages are the next most common, and plasma cells are also present, especially in the choroid plexus. They exhibit all transitional stages from the beginning of differentiation to mature plasma cells with granular endoplasmic reticulum.

The destruction of cells in inflammatory foci probably involves all elements, with secondary demyelination as a consequence (Sigurdsson *et al.*, 1962). The nature of the demyelination was reexamined recently using electron microscopy.

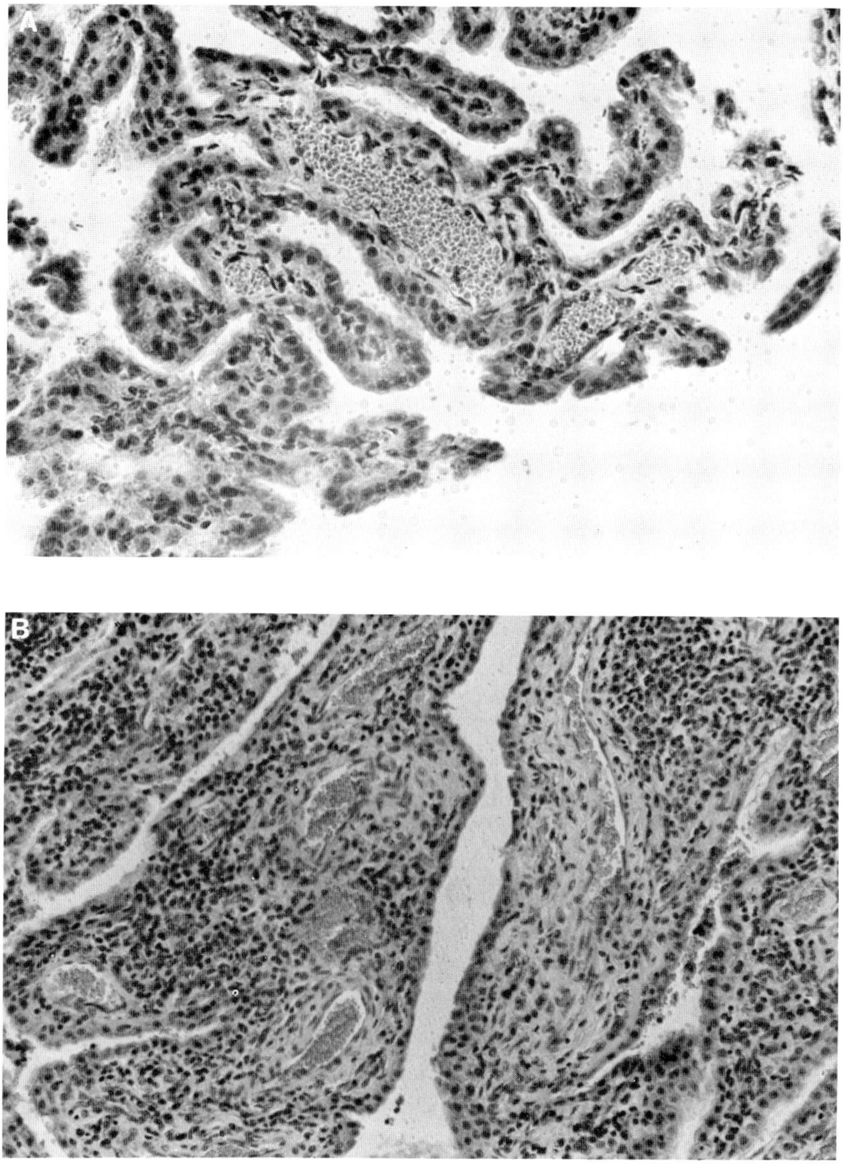

Fig. 3. Pathological changes in the choroid plexus. (A) Normal choroid plexus. (B) Choroid plexus from an animal infected with visna virus, exhibiting dense infiltration by lymphocytes and other inflammatory cells. Sections stained with hemotoxylin and eosin. × 160.

One month after inoculation, myelin breakdown was found to be minimal and, when present, was of the secondary type. Macrophages were observed in contact with myelinated axons of normal appearance and did not contain myelin breakdown products (Georgsson *et al.*, 1977). In contrast, focal lesions of primary demyelination, with conserved axons, were observed in the CNS of long term infected sheep presenting paralysis (Georgsson, Nathanson, Petursson, personal communication).

2. *Maedi*

Maedi is an interstitial pneumonia. The most remarkable change at the gross level in advanced cases is a two- to threefold increase in the weight of the lung. The histopathological lesion consists of thickening of the interalveolar septa caused by infiltration by lymphocytes, monocytes, and macrophages in varying proportions. The thickening may be so pronounced as to obliterate the alveola completely. In these areas there may be epithelialization of alveoli. In addition, lymphoid accumulations, with the formation of follicles and germinal centers analogous to those described in visna, are scattered throughout the lung parenchyma (Palsson, 1976).

C. Immune Response

The immune response in visna in many respects is not very different from that seen in more conventional types of infection. Complement-fixing (CF) antibodies are the first to be detected. They usually appear a few weeks after inoculation, rise to a maximum titer in a few months, and stay at about the same level throughout the course of the disease (Gudnadottir and Kristinsdottir, 1967). Neutralizing antibodies appear later (between 1 and 4 months postinoculation), reach a maximum at about 1 year, and stay at this level for the rest of the animal's life (Gudnadottir and Palsson, 1965; Gudnadottir, 1974; Petursson *et al.*, 1976). The neutralizing antibody titer can be remarkably elevated (>1/2000). Virus is also neutralized in serum and CSF by nonimmunological factors (Thormar *et al.*, 1979). PBL and cells present in the CSF can be stimulated to incorporate [^{3}H]thymidine by contact with virus-infected homologous mammary tissue cells. This cellular immune response appears only a few days after inoculation, peaks at about 2 weeks, and decreases to control level in about 1 month (Griffin *et al.*, 1978). No virus-specific cytotoxic T-cell activity has been demonstrated in visna-infected animals.

D. Antigenic Variation

Neutralizing antibody recognizes type-specific antigenic determinants of the virion glycoprotein; neutralization tests are therefore useful in distinguishing different strains of Lentivirinae (Scott *et al.*. 1979). Neutralization tests have shown interestingly that variations in virus strains occur during infection in individual animals (Gudnadottir, 1974; Narayan *et al.*, 1977). Thus if an animal is inoculated with plaque-purified strain 1514 of visna virus, after many months of infection new strains of virus can be isolated from PBL that are no longer neutralized by antibody that neutralizes the inoculum strain. Longitudinal studies have shown that in time, neutralizing antibodies are produced to the new strains that appear in a given animal (Narayan *et al.*, 1978). However, the new strains do not successively replace parental virus; both the inoculum strain and variant strains can be isolated from PBL simultaneously, and the inoculum as well as the new strains replicate *in vitro* with equivalent efficiency. The emergent strains of virus have been shown to arise by point mutations in the gene coding for the envelope glycoprotein (Scott *et al.*, 1979; Clements *et al.*, 1980).

E. Differential Diagnosis

None of the clinical signs of maedi are pathognomonic. A pneumonia with an unusually protracted course occurring in the adult is suggestive, but a definite diagnosis can be obtained only at autopsy. The great increase in weight of the lung, and the diffuse, compact, homogeneous lesions are characteristics of maedi. The histological examination will usually confirm the diagnosis.

In its early stages, visna can easily be confused with other CNS diseases such as abscesses, trauma, or parasitic lesions. Later, the progressive, protracted course of the disease, the absence of fever, and the pleocytosis in the CSF are in favor of visna. Tremor of the head, grinding of teeth, or itching, which are characteristic of scrapie, are always absent in visna. Histological examination of the CNS, particularly choroid plexus, is helpful in establishing the diagnosis (Palsson, 1976).

Virological methods of diagnosis include isolation of virus and serological studies. Virus can be isolated from the CNS, lung, spleen, PBL, and CSF (rarely in disease). Because of the limitation in virus, replication *in vivo* (see Section V, A), tissue explantation, and blind passage are often required to isolate virus. The most useful serological test in visna/maedi is complement fixation, since CF antibody appears early in disease, is maintained at detectible levels throughout the disease, and is cross-reactive between all strains of virus. Neutralization tests are of much less value in diagnosis because they appear much later in disease and because of the strain specificity of neutralizing antibody.

F. Host Range

Maedi and visna have been described only in sheep and goats (Palsson, 1976). Transmission to other species has been attempted in the laboratory. All attempts have failed, even after inoculation with large doses of virus and observation of the animal for long periods of time (Thormar, 1976). Some breeds of sheep are probably more susceptible to the agent than others. The Icelandic breed, a primitive, highly inbred sheep which has lived in almost complete isolation for more than a thousand years, seems to be particularly susceptible to the agent.

G. Epidemiology

Visna/maedi-related viruses are probably endemic in sheep throughout the world. Well-documented cases of progressive interstitial pneumonia similar to maedi have been reported from Canada, Denmark, France, Germany, Greece, Holland, India, Norway, Rumania, South Africa, the United States, and the Soviet Union (Palsson, 1976).

The epidemiology of visna/maedi has been extensively documented by the Icelandic authors (Palsson, 1976). The agent was introduced into Iceland in 1933 by 20 apparently healthy sheep which had been imported from Halle, Germany. The animals were kept in quarantine for 2 months, then sent to 14 farms in various parts of the island. Two foci of maedi appeared a few years later in two widely separated areas, and by the late 1930s, it became apparent that this new contagious lung disease had spread to a large part of the country. Visna was first recognized in the early 1940s in flocks where maedi had caused losses for some years. From the outset, it was noted that visna often developed in animals already afflicted with maedi. Transmission of the disease within a flock occurred during the winter when the animals were kept indoors in close contact with each other. On the other hand, during the summer when they roamed freely on common pasture, the communicability of the disease was found to be very low. The spread of the agent from flock to flock very likely occurred in the fall when sheep were gathered in large numbers and sorted out into their original flocks. Spread without direct contact between animals was extremely rare. Transmission presumably occurs by the respiratory route by virus disseminated by the dry cough. Young lambs may be infected via the milk, since virus is excreted in the milk, and lambs raised on infected ewes often develop disease at a young age. Transplacentral transmission, if it occurs, is rare (De Boer *et al.,* 1978).

H. Methods of Disease Control

No vaccine has been developed to date that satisfactorily prevents disease, and no therapeutic agents are available, although isatin-β-thiosemicarbasone and phosphonoformate inhibit replication of visna *in vitro* (Haase and Levinson, 1973; Sundquist and Larner, 1979).

Maedi and visna were eventually eradicated from Iceland, where they had a major economic impact, through a massive slaughtering and restocking program conducted between 1944 and 1954. Every animal in a flock where a case of maedi or visna had been recognized was killed. Restocking was done with animals from a remote western part of the island where the diseases never appeared. Maedi reappeared in a few flocks between 1952 and 1965; in each case the entire flock was destroyed. By 1965 it was estimated that 100,000 sheep had died of maedi or visna and 650,000 sheep had been killed in the eradication program (Palsson, 1976).

IV. METHODS

The persistence of lentiviruses in the face of the host immune response, and the slow evolution of the infections they cause, are two important issues in virus pathogenesis. In our view, these issues can be resolved satisfactorily only by methods suitable for quantitative analysis of virus replication in the complex setting of tissues in the infected animal. Accordingly, we will restrict our discussion to methods that provide visual and quantitative information about virus replication at the single-cell level. We believe these methods, with some modification, will have general applicability to problems of animal virus pathogenesis in other systems. The details of methods that have been utilized with more conventional objectives, such as virus diagnosis and classification, are cited appropriately in the text.

A. Basic Experimental Protocol

The design of experiments that quantitatively assess synthesis of virus nucleic acid and proteins in single cells is illustrated in a flow diagram (Fig. 4). SCP cultures are infected *in vitro* under conditions where infection is initiated synchronously, and where replication is fully permissive. As infection proceeds, cells are collected by trypsinization and divided into two portions. One portion is lysed; the amount of virus protein is determined by radioimmunoassay (RIA), or the average number of copies per cell of viral DNA or RNA is measured by hybridization in solution. Cells in the other aliquot are deposited on slides and fixed.

In the case of proteins the number of cells with detectable antigen is determined by immunofluorescence (IFA) employing sera which are specific for a given virion polypeptide. By comparison with the content of that polypeptide in the entire culture, it can be shown that a cell scored as positive by immunofluorescence contains 1 pg of viral antigen per cell (10^7 molecules of p30 per cell). For example, 72 hours after infection virtually all the cells are stained by anti-p30 (Fig. 5); of this ~ 1% of the total cell protein is p30; since there are about 100 pg of protein per cell, 1 pg of p30 per cell gives detectable fluorescence. At earlier

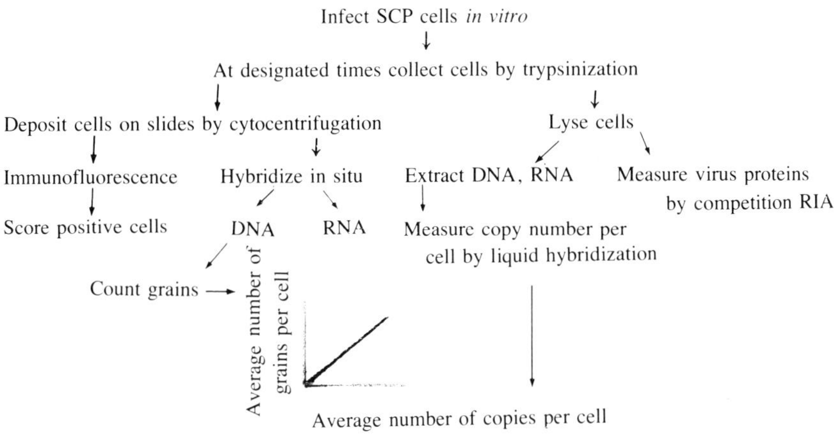

[a] SCP cell cultures are infected at a moi of 3 PFU/cell. After adsorption of virus for 2 hours at 4°C, infection is initiated synchronously by addition of warm medium. Cells are collected at specified times after infection by trypsinization, and the average content of protein or nucleic acid per cell is determined in one aliquot as described in the text. The remaining cells are sedimented onto slides where, after fixation, the content of antigen or nucleic acid in individual cells is assayed by immunofluorescence, or by hybridization in situ.

Fig. 4. Flow diagram of procedures to quantitate the synthesis of viral macromolecules in single cells.

times there is a comparable relationship between the content of p30 per cell and the fraction of cells that gives positive fluorescence, as would be expected if these cells account for most of the antigen production by the culture. Thus cells in infected tissues that stain positively with anti-p30 serum must contain about 10^7 p30 per cell.

In the case of nucleic acids, the cells fixed on slides are hybridized *in situ* to a [^{3}H]labeled probe under conditions where viral DNA or RNA will be detected (Fig. 6). After radioautographic exposure, the number of grains per cell is enumerated, and the number of grains per cell per minute of exposure is plotted against the average number of genome copies per cell (from liquid hybridization). From this relationship one can determine the number of genome copies of viral nucleic acid in a given cell in tissue sections.

B. Comments on Specific Procedures

1. Viral Proteins

The procedure to detect viral antigens by immunofluorescence and by radioimmunoassay are well-established and are described in detail elsewhere (Haase *et al.*, 1977; Stowring *et al.*, 1979).

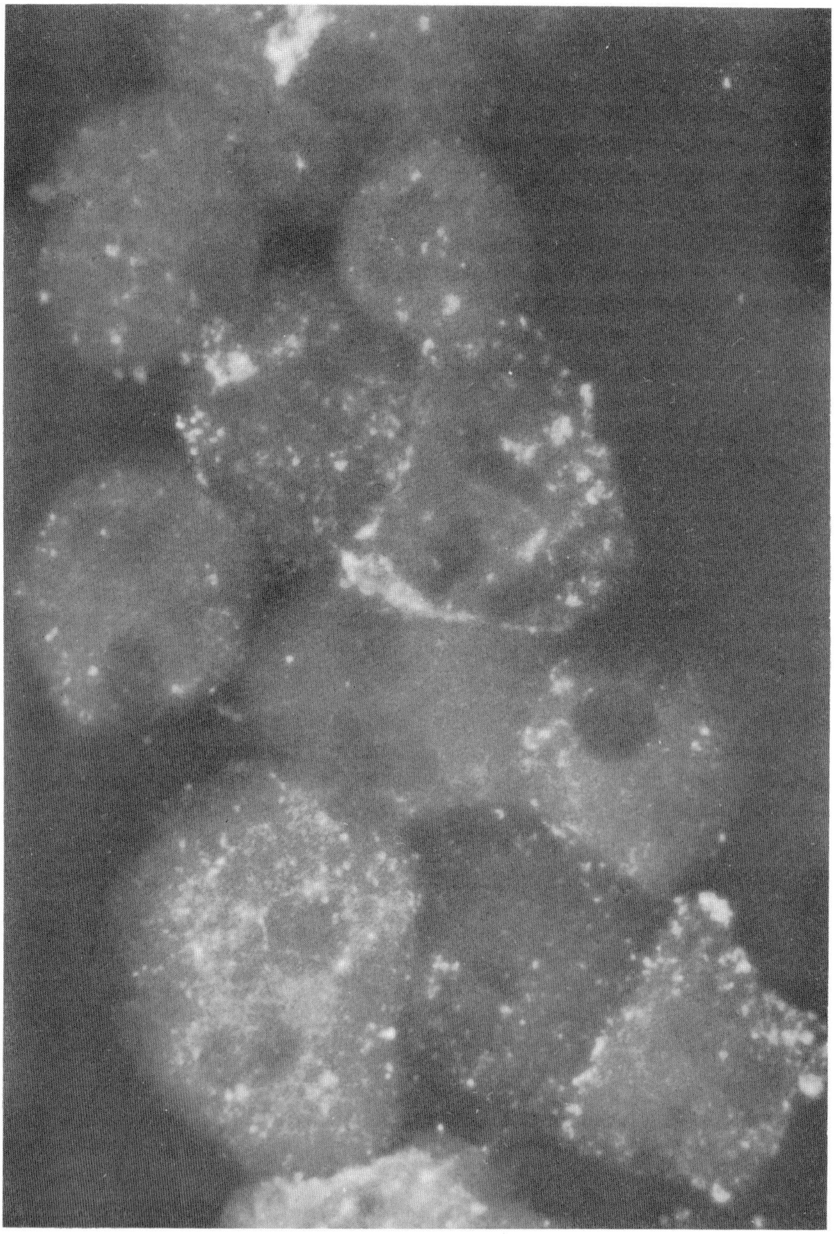

Fig. 5. Immunofluorescence. SCP cells infected for 48 hours with visna virus were collected by trypsinization, sedimented onto glass slides, fixed with periodate-polylysin-paraformaldehyde, and stained with goat antivisna p30 and fluorescein-conjugated rabbit antigoat IgG. × 250.

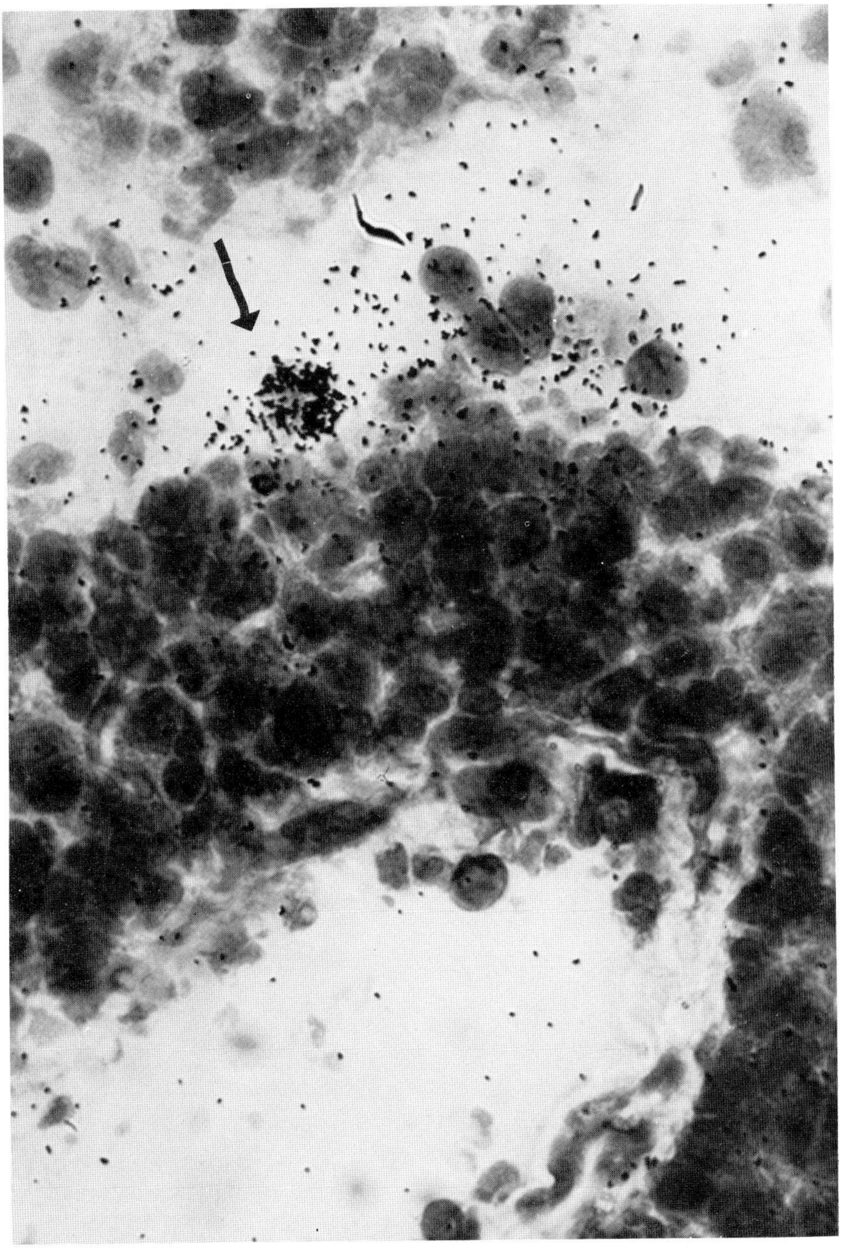

Fig. 6. *In situ* hybridization. A 10-μm section was cut from frozen SCP from an animal infected with visna virus. After hybridization to [^{3}H]cDNA, the slide was washed, coated with emulsion, and exposed for 4 days. The slide was developed and stained with Giesma. The arrow points to a cell containing viral RNA displaced by an inflammatory infiltrate surrounding a blood vessel. × 160.

2. Viral Nucleic Acids—Molecular Hybridization

Various molecular hybridization techniques have been used extensively for the study of the replication mechanisms of visna virus in tissue culture and more recently for the study of the genetic regulative mechanisms responsible for the persistence of the infection in the animal and the slowness of the disease. We will briefly describe the principles and major usefulness of the techniques. For specific experimental details, the reader is referred to the original publications.

These techniques require the synthesis of radioactive, viral-specific, nucleic acid "probes." Two different types of probes can be prepared.

The first type is cDNA made *in vitro* using the viral 70 S RNA as template. This probe is synthesized usually in the presence of actinomycin D in a reaction in which purified virus provides both reverse transcriptase and viral 70 S RNA. It is possible to synthesize cDNA of a length exceeding 1000 nucleotides by conducting the reaction at 45°C and by using deoxyribonucleotide triphosphates at a concentration of 0.1 m*M*. This cDNA hybridizes stoichiometrically to viral 70 S RNA, demonstrating that it is a faithful copy of its template (Brahic and Haase, 1978). By using [^{3}H]dTTP with a specific activity of $\simeq$90 Ci/m*M*, the cDNA is labeled to a specific activity of $\simeq 2.5 \times 10^{8}$ cpm/μg. This probe can be used to detect both proviral DNA and viral RNA. Alternatively, cDNA can be synthesized in a reaction containing purified viral 70 S RNA, avian myeloblastosis virus DNA polymerase, and "random primers," prepared by digesting calf thymus DNA with pancreatic DNase (Taylor *et al.*, 1976).

The second probe is double-stranded DNA synthesized *in vitro* in a reaction which does not contain actinomycin D (Haase and Varmus, 1973).

3. Liquid Hybridization

Liquid hybridization methods are used to detect and quantitate viral nucleic acids. The methods are well-established and are described in detail elsewhere for both proviral DNA (Haase and Varmus, 1973) and viral RNA (Brahic *et al.*, 1977).

4. In Situ Hybridization

This technique was originally described by Gall and Pardue (1971) for the detection of specific DNA sequences in histological preparations. In this method, trypsinized cells or cryostat sections are fixed on a microscope slide and reacted with radioactive cDNA under conditions which allow the formation of hybrid molecules. In the case of cells infected with visna virus, if the slides are treated with ribonuclease and denatured, the reaction detects proviral DNA, when they are hybridized without any treatment, the assay detects viral RNA. This type of assay is potentially extremely useful to detect and locate proviral DNA or its transcription products in tissue sections. Unfortunately, the efficiency of the standard *in situ* hybridization methods is relatively low (on the order of 10%), which limits its use to the detection of large viral genomes present in relatively large numbers per cell. We recently improved the efficiency of the assay to nearly 100% (i.e., saturation of the viral sequences with cDNA is achieved). Briefly, this is obtained by increasing the diffusion of the cDNA through the specimen, by hybridizing at an optimum temperature in cDNA excess, and by stabilizing hybrids during autoradiography. This assay is also quantitative; a calibration curve has been constructed which relates the number of grains over a cell to the number of viral genomes present in the cell. By using [^{3}H]cDNA with a specific activity of 2×10^8 dpm/μg, one copy of viral RNA genome (molecular weight 3×10^6) gives rise to four grains above background after 3 weeks of autoradiographic exposure. The complete description of the assay for viral RNA has been published (Brahic and Haase, 1978). An example of this assay is shown in Fig. 5. In order to detect proviral DNA, the technique must be modified as follows: (i) The specimen is fixed in methanol-acetone (1:2) for 4 minutes at −20°C. (ii) After pretreatments with HCl, heat, and proteinase K, the preparation is treated for 30 min. at 37°C with a mixture of 100 μg/ml of pancreatic RNase and 10 units/ml of T1 RNase in $2 \times$ SSC, followed by extensive washing in $2 \times$ SSC and dehydration in graded ethanol. (iii) The DNA is postfixed in 5% paraformaldehyde for 2 hours at room temperature, followed by washing in $2 \times$ SSC and dehydration in graded ethanol. (iv) Proviral DNA is denatured with 95% formaldehyde in $0.1 \times$ SSC at 65°C for 15 minutes, followed by quenching in ice water and dehydration in graded ethanol. (v) Dextran sulfate (10% final concentration) is included in the hybridization medium and the cDNA concentration is reduced to 0.06 ng/μl to reduce background for low copy numbers (one to ten copies of visna DNA per cell). The sensitivity is comparable to that achieved for the detection of viral RNA.

V. PATHOGENESIS

Since the original studies of Sigurdsson, visna has been considered a paradigm of slow infection. From this standpoint, the pathogenesis of this disease is of great interest. In this section we review what is known of the life cycle of the virus in the animal (*in vivo*), including some new information obtained at the molecular level. This will be a point of departure for a reconstruction of the events responsible for the characteristic slowness and persistence of the infection. We will review the information pertinent to two unresolved questions of pathogenesis, namely, the mechanism of spread of the infection and the nature of the tissue lesions.

A. Visna Life Cycle in the Animal

The first clue to the mechanism of persistence and slowness was uncovered by virus isolation studies carried out on experimentally infected animals. Following intracerebral inoculation, the virus can be consistently reisolated from various tissues. These include, in order of decreasing frequency, choroid plexus, brain, spinal cord, lung, spleen, lymph nodes, bone marrow, and peripheral leukocytes. This distribution of organs and the frequency of isolation do not vary greatly with the time of infection. Virus can be reisolated from the choroid plexus 2 weeks after inoculation and is still present in this organ years later (Petursson *et al.*, 1976; Narayan *et al.*, 1977). There are several remarkable aspects of these isolation studies: (i) there is a clear affinity of the virus for the brain, lung, and hematopoietic system. (ii) In all these organs the virus is predominantly cell associated. This is particularly clear in the blood. No virus is present in the plasma, but virus can be reisolated from purified peripheral blood leukocytes. One notable exception is CSF, where low amounts of free virus can be found in the first weeks of disease. (iii) In all these organs at any time during the course of the disease, the amount of virus present is minimal. The amount of virus present in a tissue homogenate is generally too low to be titered. In many instances, tissues have to be explanted to increase the chances of isolation. Electron microscopic examination of tissues lends further support to the concept that viral replication is greatly restricted *in vivo*. Virus particles are never observed even in the most involved sites, such as the choroid plexus (Georgsson *et al.*, 1977).

B. Restriction of Proviral DNA Expression *in Vivo*

In sheep cells cultivated *in vitro*, visna virus undergoes an acute lytic cycle in which the yield of viral progeny is high. The principal events of this life cycle are (i) transfer of genetic information to a DNA copy of the genome (proviral DNA), (ii) transcription of proviral DNA into viral mRNAs and progeny genomic viral RNA, and (iii) synthesis of viral proteins and assembly and release of new virions through a budding process.

In sharp contrast viral replication is severely restricted *in vivo*. The first clue to the mechanism of restriction was uncovered by studies which showed that proviral DNA can be detected by *in situ* hybridization in a significant fraction of the cells in the choroid plexus, but that only an occasional cell contains the major viral polypeptide p30 as detected by immunofluorescence (Haase *et al.*, 1977). More recently, quantitative analysis of viral nucleic acids synthesis *in vivo*, using *in situ* hybridization, has shown that, although the amount of proviral DNA per cell in the choroid plexus is relatively high (60 to 70 copies per cell), the amount of viral RNA is about two orders of magnitude lower than during a permissive viral replication cycle *in vitro* (Brahic *et al.*, in press). These results demonstrate that choroid plexus cells *in vivo* impose a restriction on proviral DNA expression at the transcriptional level.

C. Slowness—Persistence

This restriction of virus gene expression provides an explanation for the persistence of the infection and the slowness of the disease. The virus persists in the face of a vigorous immune response because most infected cells are not producing viral antigens and are therefore undetectable by the immune surveillance mechanisms. The slowness of the disease reflects the rate of accumulation of tissue lesions which, in turn, develop at a tempo set by the production and spread of the virus. This is minimal and corresponds to the continuous presence of a very small fraction of cells in which the expression of proviral DNA is spontaneously induced.

Accordingly, the questions of slowness and persistence can be redefined as questions about the mechanism of restriction of proviral DNA expression. This is the subject of ongoing research, but results already available indicate that the immune response and interferon do not play a role in this restriction. Indeed, fetal thymectomy (Narayan *et al.*, 1977) or a vigorous immunosuppressive regimen (Nathanson *et al.*, 1976) do not increase the level of virus expression, and visna virus is exceptionally resistant to interferon (Carroll *et al.*, 1978). On the other hand, the results obtained by *in situ* hybridization show that the restriction is mediated by a block at the transcriptional level of gene expression. The mechanism of this transcriptional block is unknown at the present time. It could involve a classic repressor; other possibilities include a gene dosage effect and regulatory controls imposed by the integration of proviral DNA into the host chromosome.

D. Unresolved Questions in Pathogenesis

Two important questions about the pathogenesis of lentivirus infections have not been resolved as yet at a fundamental level: how virus is disseminated, and how tissue is destroyed.

The problem of virus dissemination is posed differently in natural infections and in experimental infections, but this difference may only be apparent. In natural infections, the question is how virus can cause a general infection, and progressive destruction of tissue in the CNS, when it is introduced via the respiratory tract, presumably in small amounts, and only small amounts of virus are replicated in the animal. In experimental infections, the question is how virus is disseminated widely and spreads within the CNS after the appearance of the host immune response, since these defense mechanisms should interdict the spread of extracellular virus, destroy virus-producing cells, and convert the infection to a latent state, or at least confine the spread to cell to cell. If a site is discovered where virus can replicate efficiently in the initial phase of infection, both problems would be resolved. In such a postulated permissive acute phase, virus would be produced in high titer and generally disseminated. Later intermittent activation of virus gene expression in the CNS and elsewhere would account for the apparent ubiquity of infection and progressive spread.

Other explanations for spread include antigenic variation and transport of virus in PBL. Although antigenic variants that are not neutralized by preexisting antibody could account for spread, the expected pattern of each variant succeeding the parental type and the preceding strain has not been observed in visna (Narayan *et al.*, 1978). Certainly lentiviruses could spread by the familiar mechanism of association with circulating cellular elements since it is known already that virus is associated with PBL (Petursson *et al.*, 1976). Virus circulating in PBL as proviral DNA with a low frequency of expression would represent a reasonable explanation for the mode of diffusion of infection.

It has been hypothesized that the tissue lesions in visna are immunologically mediated because of the analogous histopathological appearance of the viral lesions and the lesions of acute experimental allergic encephalomyelitis (Panitch *et al.*, 1976). This hypothesis has been tested by submitting sheep to an immunosuppressive regimen capable of preventing experimental allergic encephalomyelitis. The coordinate suppression of the inflammatory infiltration of the CNS and lack of tissue destruction are in accord with the postulated immune mechanism of tissue destruction. The nature of the antigens that may be involved is unknown.

VI. CONCLUSION

The Lentivirinae are closely related strains that comprise a subfamily of retroviruses that cause lytic infections *in vitro* and slow, persistent infections of the lungs and CNS of sheep. As a group, they are a paradigm of slow infections caused by conventional agents in which the pathological changes are inflammatory and destructive. At the molecular level, the documented alterations in virus gene expression in the infected cells in the animal already suggest that the lentiviruses will be a fruitful system for analysis of fundamental problems in animal virus pathogenesis. The methodology which the authors feel to be appropriate for this task is presented in this chapter.

ACKNOWLEDGMENTS

We thank Harriet Lukes for typing the manuscript. Work in the authors' laboratory is supported by grants from the United States Public Health Service and the American Cancer Society and is project MRIS 3367 in the Veterans Administration. Dr. Haase is a medical investigator of the Veterans Administration. We thank Caroline Mota for the graphic work in the chapter.

REFERENCES

August, J. T., Bolognesi, D. P., Fleissner, E., Gilden, R. V., and Nowinski, R. C. (1974). *Virology* **60,** 595.

August, M. J., Harter, D. H., and Compans, R. W. (1977). *J. Virol.* **22,** 832.

Baltimore, D. (1970). *Nature (London)* **226,** 1209.

Beemon, K. L., Faras, A. J., Haase, A. T., Duesberg, P. H., and Maisel, J. E. (1976). *J. Virol.* **17,** 525.

Bolognesi, D. P. (1974). *Adv. Virus Res.* **19,** 315.

Brahic, M., and Haase, A. T. (1978). *Proc. Natl. Acad. Sci. U.S.A.* **75,** 6125.

Brahic, M., and Vigne, R. (1975). *J. Virol.* **15,** 1222.

Brahic, M., Tamalet, J., and Chippaux-Hyppolite, C. (1971). *C. R. Acad. Sci., Ser. D* **272,** 2115.

Brahic, M., Tamalet, J., Filippi, P., and Delbecchi, L. (1973). *Biochimie* **55,** 885.

Brahic, M., Filippi, P., Vigne, R., and Haase, A. T. (1977). *J. Virol.* **24,** 74.

Brahic, M., Stowring, L., Ventura, P., and Haase, A. T. *Nature (London)*, in press.

Bruns, M., and Frenzel, B. (1979). *Virology* **97,** 207.

Carroll, D., Ventura, P., Haase, A., Rinaldo, C. R., Jr., Overall, J. C., Jr., and Glasgow, L. A. (1978). *J. Infect. Dis.* **138,** 614.

Chippaux-Hyppolite, C., Taranger, C., Tamalet, J., Pautrat, G., and Brahic, M. (1972). *Ann. Inst. Pasteur, Paris* **123,** 409.

Clements, J. E., Narayan, O., Griffin, D. E., and Johnson, R. T. (1979). *Virology* **93,** 377.

Clements, J. E., Pederson, F. S., Narayan, O., and Hazeltine, W. A. (1980). *Proc. Natl. Acad. Sci. U.S.A.* **77,** 4454.

Coward, J. E., Harter, D. H., and Morgan, C. (1970). *Virology* **40,** 1030.

De Boer, G. F. (1975). *Res. Vet. Sci.* **18,** 15.

De Boer, G. F., Terpstra, C., and Houwers, D. J. (1978). *Bull. Off. Int. Epiz.* **89,** 487.

Dubois-Dalcq, M., Reese, T. S., and Narayan, O. (1976). *Virology* **74,** 520.

Filippi, P., Brahic, M., Vigne, R,, and Tamalet, J. (1979). *J. Virol.* **31,** 25.

Friedmann, A., Coward, J. E., Harter, D. H., Lipset, J. S., and Morgan, C. (1974). *J. Gen. Virol.* **25,** 93.

Gall, J. G., and Pardue, M. L. (1971). *In* "Nucleic Acids," Part D (L. Grossman and K. Moldave, eds.), Methods in Enzymology, Vol. 21, p. 470. Academic Press, New York.

Georgsson, G., Palsson, P. A., Panitch, H., Nathanson, N., and Petursson, G. (1977). *Acta Neuropathol.* **37,** 127.

Gielkens, A. L. J., Van Zaane, D., Bloemers, H. P. J., and Bloemendal, H. (1976). *Proc. Natl. Acad. Sci. U.S.A.* **73,** 356.

Gillespie, D., Takemoto, K., Robert, M., and Gallo, R. C. (1973). *Science (Washington, D.C.)* **179,** 1328.

Griffin, D. E., Narayan, O., and Adams, R. J. (1978). *J. Infect. Dis.* **138,** 340.

Gudnadottir, M. (1974). *Prog. Med. Virol.* **18,** 336.

Gudnadottir, M., and Kristinsdottir, K. (1967). *J. Immunol.* **98,** 663.

Gudnadottir, M., and Palsson, P. A. (1965). *J. Immunol.* **95,** 1116.

Haase, A. T. (1975). *Curr. Top. Microbiol. Immunol.* **72,** 101.

Haase, A. T., and Baringer, J. R. (1974). *Virology* **57,** 238.

Haase, A. T., and Levinson, W. (1973). *Biochem. Biophys. Res. Commun.* **51,** 875.

Haase, A. T., and Varmus, H. E. (1973). *Nature (London), New Biol.* **245,** 237.

Haase, A. T., Garapin, A. C., Faras, A. J., Taylor, J. M., and Bishop, J. M. (1974a). *Virology* **57,** 259.

Haase, A. T., Garapin, A. C., Faras, A. J., Varmus, H. E., and Bishop, J. M. (1974b). *Virology* **57,** 251.

Haase, A. T., Traynor, B. L., Ventura, P. E., and Alling, D. W. (1976). *Virology* **70,** 65.
Haase, A. T., Stowring, L., Narayan, O., Griffin, D., and Price, D. (1977). *Science* (*Washington, D.C.*) **195,** 175.
Haase, A. T., Brahic, M., Carroll, D., Scott, J., Stowring, L., Traynor, B., and Ventura, P. (1978). *In* "Persistent Viruses: ICN-UCLA Symposia on Molecular and Cellular Biology" (J. G. Stevens, G. J. Todaro, and C. F. Fox, eds.), Vol. 11, pp. 643-654. Academic Press, New York.
Harris, J. D., Scott, J. V., Traynor, B., Brahic, M., Stowring, L., Ventura, P., Haase, A. T., and Peluso, R. *Virology,* in press.
Harter, D. H. (1969). *J. Gen. Virol.* **5,** 157.
Harter, D. H., and Choppin, P. W. (1967a). *Virology* **31,** 176.
Harter, D. H., and Choppin, P. W. (1967b). *Virology* **31,** 279.
Harter, D. H., Hsu, K. C., and Rose, H. M. (1967). *J. Virol.* **1,** 1265.
Harter, D. H., Hsu, K. C., and Rose, H. M. (1968). *Proc. Soc. Exp. Biol. Med.* **129,** 295.
Harter, D. H., Rosenkranz, H. S., and Rose, H. M. (1969). *Proc. Soc. Exp, Med.* **131,** 927.
Harter, D. H., Schlom, J., and Spiegelman, H. (1971). *Biochim. Biophys. Acta* **240,** 435.
Harter, D. H., Axel, R., Burny, A., Subhash, G., Schlom, J., and Spiegelman, S. (1973). *Virology* **52,** 287.
Hunter, G. D. (1972). *J. Infect. Dis.* **125,** 427.
Johnson, K. P., Byington, D. P., and Gaddis, L. (1974). *Adv. Neurol.* **6,** 77.
Kennedy, R. C., Eklund, C. M., Lopez, C., and Hadlow, W. J. (1968). *Virology* **35,** 483.
Keshet, E., and Temin, H. M. (1979). *J. Virol.* **31,** 376.
Kurstak, E. (1978). *In* "Comparative Diagnosis of Viral Diseases" (E. Kurstak and C. Kurstak, eds.), Vol. 1, pp. 1-22. Academic Press, New York.
Lin, F. H. (1977). *J. Virol.* **25,** 207.
Lin, F. H., and Papini. M. (1978). *Biochim. Biophys. Acta* **561,** 383.
Lin, F. H., and Thormar, H. A. (1970). *J. Virol.* **6,** 702.
Lin, F. H., and Thormar, H. A. (1971). *J. Virol.* **7,** 582.
Lin, F. H., and Thormar, H. (1972). *J. Virol.* **10,** 228.
Lin, F. H., and Thormar, H. (1974). *J. Virol.* **14,** 782.
Lin, F. H., and Thormar, H. (1979). *J. Virol.* **29,** 536.
Lin, F. H., Genovese, M., and Thormar, H. (1973). *Prep. Biochem.* **3,** 525.
Mountcastle, W. E., Harter, D. H., and Choppin, P. W. (1972). *Virology* **47,** 542.
Narayan, O., Silverstein, A. M., Price, D., and Johnson, R. T. (1974). *Science* (*Washington, D.C.*) **183,** 1202.
Narayan, O., Griffin, D. E., and Silverstein, A. M. (1977). *J. Infect. Dis.* **135,** 800.
Narayan, O., Griffin, D. E., and Clements, J. E. (1978). *J. Gen. Virol.* **41,** 343.
Nathanson, N., Panitch, H., Palsson, P. A., Petursson, G., and Georgsson, G. (1976). *Lab. Invest.* **35,** 444.
Palsson, P. A. (1976). *In* "Slow Virus Diseases of Animals and Man" (R. H. Kimberlin, ed.), pp. 17-43. North-Holland Publ., Amsterdam.
Palsson, P. A., Georgsson, G., Petursson, G., and Nathanson, N. (1977). *Acta Vet. Scand.* **18,** 122.
Panitch, H., Petursson, G., Georgsson, G., Palsson, P. A., and Nathanson, N. (1976). *Lab. Invest.* **35,** 452.
Petursson, G., Nathanson, N., Georgsson, G., Panitch, H., and Palsson, P. A. (1976). *Lab. Invest.* **35,** 402.
Quintrell, N., Varmus, H. E., Bishop, J. M., Nicholson, M. O., and McAllister, R. M. (1974). *Virology* **58,** 569.
Scolnick, E. M., and Parks, W. P. (1974). *Virology* **59,** 168.
Scott, J. V., Stowring, L., Haase, A. T., Narayan, O., and Vigne, R. (1979). *Cell* **18,** 321.

Sigurdsson, B. (1954). *Br. Vet. J.* **110,** 341.
Sigurdsson, B., and Palsson, P. A. (1958). *J. Exp. Pathol.* **39,** 519.
Sigurdsson, B., Palsson, P. A., and Grimsson, H. (1957). *J. Neuropathol. Exp. Neurol.* **16,** 389.
Sigurdsson, B., Thormar, H., and Palsson, P. A. (1960). *Arch. Gesamte Virusforsch.* **10,** 368.
Sigurdsson, B., Palsson, P. A., and Van Bogaert, L. (1962). *Acta Neuropathol.* **1,** 343.
Stehelin, D., Guntaka, R. V., Varmus, H. E., and Bishop, J. M. (1976). *J. Mol. Biol.* **101,** 349.
Stone, L. B., Scolnick, E., Takemoto, K. K., and Aaronson, S. A. (1971a). *Nature (London)* **229,** 257.
Stone, L. B., Takemoto, K. K., and Martin, M. A. (1971b). *J. Virol.* **8,** 573.
Stowring, L., Haase, A. T., and Charman, H. P. (1979). *J. Virol.* **29,** 523.
Sundquist, B., and Larner, E. (1979). *J. Virol.* **30,** 847.
Takemoto, K. K., and Stone, L. B. (1971). *J. Virol.* **7,** 770.
Takemoto, K. K., Mattern, C. F. T., Stone, L. B., Coe, J. E., and Lavelle, G. (1971). *J. Virol.* **7,** 301.
Takemoto, K. K., Aoki, T., Garon, C., and Sturm, M. M. (1973). *J. Natl. Cancer Inst.* **50,** 543.
Taylor, J. M., Illmensee, R., and Summers, J. (1976). *Biochim. Biophys. Acta* **442,** 324.
Temin, H. M., and Mizutani, S. (1970). *Nature (London)* **226,** 1211.
Thormar, H. (1961a). *Virology* **14,** 463.
Thormar, H. (1961b). *Arch. Gesamte Virusforsch.* **10,** 501.
Thormar, H. (1963). *Virology* **19,** 273.
Thormar, H. (1965). *Res. Vet. Sci.* **6,** 117.
Thormar, H. (1969). *Acta Pathol. Microbiol. Scand.* **75,** 296.
Thormar, H. (1976). *In* "Slow Virus Diseases of Animals and Man" (R. H. Kimberlin, ed.), pp. 97–114. North-Holland Publ., Amsterdam.
Thormar, H., and Cruickshank, J. G. (1965). *Virology* **25,** 145.
Thormar, H., and Helgadottir, H. (1965). *Res. Vet. Sci.* **6,** 456.
Thormar, H., and Sigurdardottir, B. (1962). *Acta Pathol. Microbiol. Scand.* **55,** 180.
Thormar, H., Wisniewski, H. M., and Lin, F. H. (1979). *Nature (London)* **279,** 245.
Townsend, J. J., Baringer, J. R., Wolinsky, J. S., Malamud, N., Mednick, J. P., Panitch, H. S., Scott, R. A. T., Oshira, L. S., and Cremer, N. E. (1975). *N. Engl. J. Med.* **292,** 990.
Traynor, B. L., and Haase, A. T. (1977). *Abstr. Annu. Meet. Am. Soc. Microbiol.* **S362,** 339.
Trowbridge, R. S. (1974). *Appl. Microbiol.* **28,** 366.
Vigne, R., Brahic, M., Filippi, P., and Tamalet, J. (1977). *J. Virol.* **21.** 386.
Vigne, R., Filippi, P., Brahic, M., and Tamalet. J. (1978). *J. Virol.* **28,** 543.
Vogt, V. M., and Eisenman, R. (1973). *Proc. Natl. Acad. Sci. U.S.A.* **70,** 1734.
Weiss, M. J., Gulati, S. C., Harter, D. H., Sweet, R. W., Spiegelman, S., and Lopez, C. (1975). *J. Gen. Virol.* **29,** 335.
Weiss, M. J., Zeelong, E. P., Sweet, R. W., Harter, D. H., and Spiegelman, S. (1977). *Virology* **76,** 851.
Weiss, S. R., Varmus, H. E., and Bishop, J. M. (1977). *Cell* **12,** 983.

Part XI

UNCLASSIFIED VIRUSES

Chapter 16

Equine Infectious Anemia

LEROY COGGINS

I. Introduction 647
II. Characteristics of EIA Virus 648
A. Morphology 648
B. Physicochemical Properties 649
C. Antigenic Composition 650
III. Comparative Biology 650
A. Clinical Features in the Horse 650
B. Experimental Host 650
C. Pathogenesis 651
IV. Immunity 652
V. Epizootiology 653
VI. Comparative Diagnosis 655
A. Differential Diagnosis 655
B. Virus Isolation 655
C. Serological Procedures 656
VII. Prevention and Control 656
References 657

I. INTRODUCTION

Equine infectious anemia (EIA) was first recognized and described as a clinical entity in horses in France (Ligneé, 1843) and has since been reported from nearly all countries of the world (Dreguss and Lombard, 1954). Infected horses show a wide spectrum of clinical signs, many of which are immunologically mediated. It has been known for some time that the causative virus persists in the infected horses indefinitely (Stein *et al.*, 1955), but the realization that the majority of carriers show little or no clinical signs came only after the develop-

ISBN 0-12-429704-8

ment and widespread use of the agar gel immunodiffusion test for EIA (Coggins and Norcross, 1970).

The virus persists in the blood of infected horses and is readily transmitted by transfusions of whole blood. The disease occurs naturally in areas where congregated horses are exposed to large populations of blood-sucking insects.

Clinical EIA has resulted in economic losses, especially in the horse racing industry. Recurrences of illness in chronically infected horses were all too common at racetracks before the development of an accurate diagnostic test capable of detecting the inapparent carrier. This diagnostic test, along with better knowledge about transmission of EIA, has made effective control of the infection possible.

The purpose of this chapter is to review the current knowledge of this intriguing persistent virus infection and its associated disease. The finding that EIA virus is a retrovirus (Charman *et al.*, 1976) has renewed the biomedical significance of the disease as a model for the study of mechanisms of viral persistence and the pathogenesis of immunologically mediated lesions, especially those shared with the RNA tumor viruses.

II. CHARACTERISTICS OF EIA VIRUS

A. Morphology

The virions of EIA virus were first observed by electron microscopy on thin sections of infected horse leukocyte cultures (Ito *et al.*, 1969; Tajima *et al.*. 1969); later, viral particles were seen in equine kidney and dermal cells infected with EIA virus (Kono and Yoshino, 1974; Matheka *et al.*, 1976; Weiland *et al.*, 1977) and in purified preparations of virus from infected horse serum (Nakajima *et al.*, 1974). It is of interest to note that, as yet, EIA virus has not been seen in tissues from infected horses by electron microscopy even though intensive searches have been made.

The viral particles resemble the mammalian C-type particles seen with RNA tumor viruses and are formed by a process of budding from the plasma membrane into extracellular spaces or into membrane-bounded cytoplasmic vesicles. Prominent electron-dense crescents, which are destined to become mature virions, form just beneath the surface membrane with no intervening space and sometimes extend over a distance of several particles. The size of the virions varies greatly, from 80–180 nm in diameter, but most of the particles are about 100 nm in diameter (Fig. 1).

Surface projections or knobs arranged over the surface of the virus particle have been reported (Weiland *et al.*, 1977). These knobs appear to be very easily destroyed and have been seen only in carefully preserved preparations. In con-

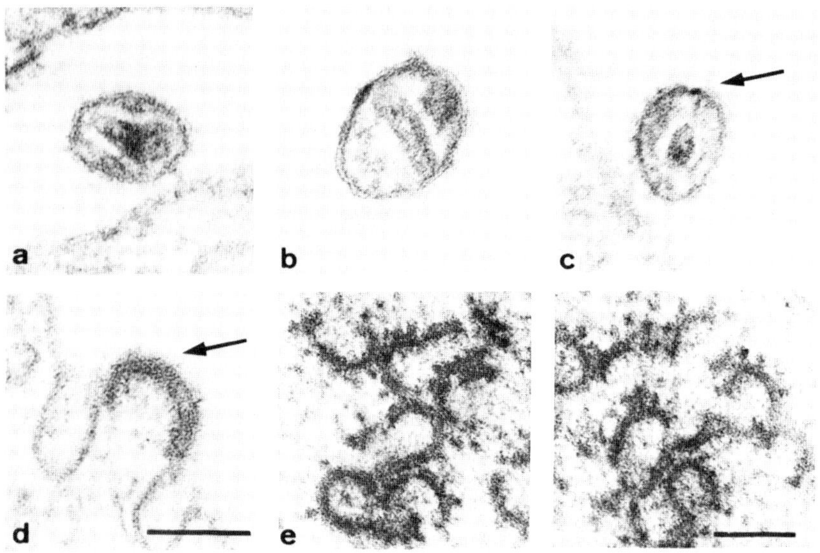

Fig. 1. Ultrathin sections of EIA virus. Extracellular particles of different diameters with (a) conical and (b) tubular cores or with (c) cores transversely cut. (d) Particle in the stage of budding. Surface projections are visible (arrows in c, d). (e) Intracytoplasmic structures in infected cells. Bar: 100 nm. (Reprinted with permission from Weiland *et al.*, 1977.)

trast to the typical morphology of the C-type particles, EIA virus particles often contain conical or tubular cores and a lump of electron-dense amorphous material between their loose envelopes and internal cores. This characteristic internal arrangement most closely resembles that of visna virus, a lentivirus or type E retrovirus (Gonda *et al.*, 1978). Intracytoplasmic structures which may be precursor particles have also been described (Matheka *et al.*, 1976; Weiland *et al.*, 1977; Gonda *et al.*, 1978) and are similar to those reported for other retroviruses (Takemoto *et al.*, 1971).

B. Physicochemical Properties

EIA virus has all the physicochemical characteristics of a retrovirus. Its buoyant density is 1.162 (Matheka *et al.*, 1976). It is sensitive to ether (Nakajima and Obara, 1964) and contains a 70 S molecular weight RNA genome composed of two 34 S subunits of single-stranded RNA (Cheevers *et al.*, 1977). It is relatively resistant to ultraviolet irradiation (Nakajima *et al.*, 1973), has a DNA-dependent stage in its growth cycle (Kono *et al.*, 1970a), and possesses a reverse transcriptase (Nakajima *et al.*, 1970; Charman *et al.*, 1976; Archer *et al.*, 1977; Cheevers *et al.*, 1977). In addition, specific proviral DNA has been demonstrated in equine cells infected *in vitro* with EIA virus (Rice *et al.*, 1978).

C. Antigenic Composition

The structural polypeptides of EIA virus have not been well delineated yet, but preliminary studies indicate that there are at least three nonglycosylated polypeptides with molecular weights of 25,000, 14,000, and 11,000, and two glycoproteins of 80,000 and 40,000 (Ishizaki *et al.*, 1978; Cheevers *et al.*, 1978). The 25,000 molecular weight polypeptide, which is the major viral protein, has been used in the agar gel immunodiffusion test in the diagnosis of EIA infection (Norcross and Coggins. 1971; Coggins *et al.*, 1978). Serum neutralization tests have demonstrated a number of different strains of EIA virus (Kono, 1973).

High-titered preparations of EIA virus containing about $10^{7.5}$ $TCID_{50}$/ml have hemagglutinated guinea pig erythrocytes with titers ranging from 16 to 32 units/0.05 ml. The hemagglutinin has been shown to be associated with the virus particles and has been inhibited specifically by sera from horses recovering from EIA infection. Neuraminidase did not destroy the hemagglutinin receptors on the erythrocytes (Sentsui and Kono, 1976).

III. COMPARATIVE BIOLOGY

A. Clinical Features in the Horse

The prominent signs of EIA are fever, anemia, anorexia, rapid loss of weight, ventral edema, and terminal depression (Dreguss and Lombard, 1954). Horses with severe, acute clinical disease may die as early as 2 to 3 weeks after infection, but the majority of infected horses recover to become healthy-appearing carriers after a few repeated episodes of clinical disease.

The incubation period after inoculation of a virulent strain of virus is usually between 5 and 30 days but may be as long as 90 days, especially when blood from a field case of EIA has been injected. The recurring episodes of fever have a surprisingly regular pattern, occurring about every 2 weeks (Dreguss and Lombard, 1954) during the first few months of infection. After this period of time, recurrences of clinical illness become much less frequent and may be absent for long periods after the first year of infection (Kono, 1973).

The clinical signs of anemia and edema may not be evident during the initial period of pyrexia but become more pronounced with succeeding episodes of disease. Horses with early severe symptoms are more apt to die than horses with milder or no signs of illness.

B. Experimental Host

Species of the genus *Equidae* are the only susceptible hosts for EIA virus and, thus far, EIA virus has been found to replicate only in cell cultures of equine origin. Ordinary laboratory animals are not susceptible to infection.

The unmodified EIA virus from field cases propagates in macrophage cultures prepared from equine blood, spleen, or bone marrow. The virus was grown first in primary horse bone marrow cultures and horse leukocyte cultures by Kobayashi (1961a,b). A cytopathic effect was seen only at high multiplicities of infection. End point dilutions of a virus titration showed no effects from the virus, and infection was determined after a subpassage in cultures and measurement of the viral antigen in a complement fixation test. The horse leukocyte culture system remains the only *in vitro* assay for the wild virus and is plagued with a number of problems which contribute to poor reproducibility. The donor of both leukocytes and serum for the cultures must be chosen carefully in order to obtain cultures that are satisfactory for virus propagation. In addition, the persistent infection of leukocytes from most horses with equine herpesvirus type 2 virus precludes the use of their cells.

Viral antigen in tissue culture has been demonstrated by immunofluorescence in the cytoplasm of infected cells (Crawford *et al.*, 1971; Ushimi *et al.*, 1972), but this technique has not been satisfactory for the determination of the end point in a virus assay. Apparently, a threshold of virus ($\geq 10^3$ $TCID_{50}$) is necessary before viral antigen becomes detectable by immunofluorescence. In addition, immunofluorescent viral antigen has been noted to modulate with time in infected cultured cells.

C. Pathogenesis

Once EIA virus gains entry into the body, it is taken up by macrophages in which it is able to replicate. Viral antigen has been demonstrated by immunofluorescence in many tissues, predominantly in the spleen, lymph nodes, and cells thought to be macrophages in the liver. Other types of cells apparently are refractory to this virus (McGuire *et al.*, 1971b). The highest titers of virus are found in the blood and tissue concurrent with the appearance of the initial pyrexia. These titers then decrease but rise again in association with periods of recurrent fever (Kono *et al.*, 1971).

In some asymptomatic animals, infectious virus cannot be demonstrated in the serum or whole blood, but the infection can be transmitted with washed leukocytes (Coggins and Kemen, 1976). Presumably, in these cases, the virus is present in some restricted form, perhaps as integrated proviral DNA, and becomes expressed only after the leukocytes are washed free of serum and injected into another horse. The lack of a satisfactory *in vitro* virus assay has prevented detailed virological studies of such restricted infections. Data show, however, that a horse once infected remains infected throughout its lifetime.

A marked decrease in platelets usually occurs in concert with each fever peak, the significance of which is not fully understood. Because the platelet counts return to normal levels rapidly as the fever dissipates, it is assumed that the platelets are clumping or aggregating in the tissues rather than being destroyed. Thrombi composed of platelets block smaller vessels and result in ventral edema and colic. It is not known if viral antigen is involved specifically in this platelet aggregation.

Anemia, another characteristic feature of EIA, results from two general mechanisms (Henson and McGuire, 1974): decreased hematopoiesis and complement-mediated hemolysis. Anemia is not seen prior to the appearance of precipitating antibody but then may become very severe and is often the primary cause of death in the late stages of the infection.

Circulating infectious virus–antibody complexes have been demonstrated in the sera of infected horses and are thought to be responsible for the occurrence of subclinical glomerulonephritis in EIA-infected horses (Henson and McGuire, 1974). However, sera of many carrier horses are not infectious even though they may contain high levels of precipitating antibody, and their blood readily infects other horses.

The most severe lesions and the largest accumulation of mononuclear cells in liver and lymphoid tissues are seen in horses that experience frequent exacerbations of clinical disease. Animals that have been asymptomatic for several months or years may have no detectable lesions. The accumulation of lymphocytes may be the result of chronic antigenic stimulation by viral antigen which is released during the recurring periods of viral replication and clinical illness. The type of lymphocyte involved in these chronic lesions has not been determined.

IV. IMMUNITY

The inability of EIA-infected horses to clear virus from their blood and tissues, even though they have recovered from the clincal symptoms and remain healthy-appearing for years, has been well documented (Schalk and Roderick, 1923; Stein *et al.*, 1955; Coggins *et al.*, 1972). Because the blood of infected horses remained infectious, early workers thought the immune response to EIA virus was lacking or severely impaired. More recently, a number of different tests have demonstrated the presence of anti-EIA antibody and a cell-mediated immune response to EIA virus (Henson and McGuire, 1974; Kono *et al.*, 1978).

Precipitating (Coggins *et al.*, 1972) and complement-fixing antibody (Kono and Kobayashi, 1966) are produced within 2 to 4 weeks after infection. The precipitating antibodies remain indefinitely (Coggins *et al.*, 1972), but tests for complement-fixing antibody are inhibitied often by non-complement-fixing IgG(T) immunoglobulin (McGuire *et al.*, 1971a), making the detection of complement-fixing antibody unreliable as an indicator of infection. Neutralization antibody appears later, at approximately 40 days after inoculation, and persists throughout the life of the horse. Such antibody is strain specific compared to precipitating antibody, which is group specific and is not satisfactory for use in diagnosis of EIA.

Horses that have recovered from a particular strains of EIA virus are usually refractory to further challenge with the homologous virus. Such horses, however, are hypersensitive to challenge with the heterologous virus, rapidly developing a severe clinical reaction, and may die within a few days. Sera from recovered horses have neutralizing antibody which protects against low levels (about 3 logs) of homologous virus (Coggins and Kemen, 1976). The neutralizing effect was overcome by using high levels of virus (5–6 logs). Thus, neutralizing antibody response is not highly effective against EIA virus and is delayed in appearance following infection.

The cellular immune processes in EIA have been studied (Banks and Henson, 1973; Kono *et al.*, 1978; Abid, 1978) to determine if the response is depressed in EIA infection. Results showed that lymphoblastogenesis to nonspecific antigens was not reduced during EIA infection in horses, but there was a temporary virus-specific depression of lymphocyte blastogenesis during clinical recrudescence of illness. Cell-mediated immune responses returned to normal with clinical recovery. Thus it seems that the inability of the horse to eliminate EIA virus may be due to the weak and ineffective neutralizing antibody response in association with a depression of the cell-mediated immune responses during the period of rapid virus replication.

Kono *et el.* (1970b) have proposed a mechanism of virus mutation or antigen modulation to explain viral persistence in EIA. Their data indicate that EIA virus is able to shift its surface antigens sufficiently to escape neutralization in the serum. This idea is used to explain the occurrence of periodic attacks of clinical disease and fever but does not suggest why these episodes tend to cease after a period of time. Antigenic variation in conjunction with a delayed, weak neutralizing antibody response resulting from the fragility of the viral surface antigens may be responsible for the inability of the horse to cope satisfactorily with EIA virus.

V. EPIZOOTIOLOGY

Transmission of EIA infection occurs by three principal means. In each of these, the potential for transmission appears to be greatest if the infected donor shows clinical signs of EIA, at which time it has higher levels of virus in its blood and tissues (Kono, 1973).

Infection can occur when the virus crosses the placenta to infect the fetus. The mechanism responsible for passage of virus across the barrier is not known. In one study, all mares with clinical signs of EIA during gestation infected their fetuses (Kemen and Coggins, 1972). Thus it appears that high levels of virus in the blood are needed before the virus crosses the placenta. The infected fetus may be aborted or born alive as a carrier. A fetus aborted at 6 months was found to have both EIA virus and precipitating antibody. The newborn foal may be infected from ingestion of colostrum, most probably containing virus-infected leukocytes.

Natural transmission of EIA virus may occur also from insects feeding on infected horses and then on susceptible horses. Since the virus does not multiply within the insects, the transmission is purely mechanical. Horse flies are important vectors because of the amount of blood carried on their mouth parts and the tendency for them to be interrupted during feeding by the horse in response to their painful bites. Horse flies readily transmit EIA from horses with high titers of virus in their blood.

Finally, EIA infection may result from the transfer of infective blood on needles and instruments. Biologicals may become contaminated in this manner, and EIA virus may survive in these products at refrigerator temperatures for several months. Of course, blood transfusion is an ideal way of transferring the virus to recipient animals.

Thus, it is assumed that EIA infection is maintained in nature by mare-to-foal transmission either *in utero* or via colostrum, then by insect transmission from the acutely infected foals to susceptible horses which are pastured in close proximity. Needle transmission no doubt increases the incidence of EIA infection, but it does not account for the continued spread of infection on breeding farms after the use of contaminated needles is abandoned.

The incubation period for EIA has been observed to be as short as 5 days and as long as 90 days. The time seems to depend on virus dose, virus strain, and individual host response to the virus. With the very long incubation periods, antibody may be produced several weeks before clinical signs become evident. Experimental data indicate that the longest time between inoculation and antibody production is 45 days (Coggins *et al.*, 1972). Thus, the longer incubation periods probably represent clinical recrudescence of illness rather than legitimate incubation periods. The original illnesses may have been too mild to be detected. The scanty data on viremia show that the highest titer of virus in the blood is found during the first febrile episode. The levels decrease thereafter as the fever subsides and increase again with the reappearance of clinical signs. The virus titer, however, never again reaches the level of the initial viremia and clinical illness (Kono, 1973).

Horses have been reported to carry EIA virus in their blood for as long as 18 years (Stein *et al.*, 1955). A majority of these carriers remained healthy-appearing for many years and appeared to be less important in the transmission of EIA than clinically affected horses. However, some of these horses became clinically ill again, at which time they were able to transmit EIA virus.

VI. COMPARATIVE DIAGNOSIS

A. Differential Diagnosis

As a clinical entity, EIA often can be diagnosed on the basis of its history and clinical signs. A number of other equine diseases, however, may mimic EIA, piroplasmosis probably being the most common one in areas where this infection occurs. Typically, the affected horse has recurring episodes of fever, depression, anemia, loss of weight, and dependent edema of the abdomen, sheath, and limbs. Bouts of illness are repeated about every 2 weeks for the first several months after infection; then sickness may recur only whenever the horse is stressed.

B. Virus Isolation

As yet, there is no satisfactory way to assay wild EIA virus in tissue culture. Although the virus can be cultivated in leukocyte cultures, the technique has many disadvantages and is little used. Inoculation of a susceptible horse remains the most reliable means of detecting EIA virus in the blood of an animal.

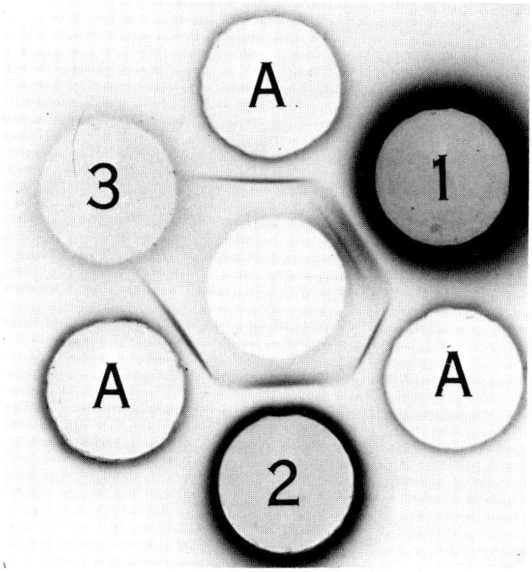

Fig. 2. Agar gel immunodiffusion test used in the diagnosis of EIA infection. Viral antigen is placed in the center well and antibody-positive control serum is placed in wells marked *A*. Wells marked 1 and 2 contain sera from EIA-infected horses, and well 3 contains serum from a non-infected horse.

C. Serological Procedures

A number of serological tests have been used in the diagnosis of EIA infection. The complement fixation test, one of the first attempted, has not been satisfactory for the diagnosis of EIA because IgG(T) in the horse does not fix complement. The complement fixation test alone cannot be used to diagnose EIA infection (McGuire *et al.*, 1971a). The serum neutralization test has two disadvantages that limit its use as a diagnostic test. Neutralizing antibody is strain specific, and since there are many strains of EIA virus, a number of tests would be required. Secondly, neutralizing antibody is produced late in the infection, usually sometime after 45 days, which may be long after the clinical disease.

Although a number of serological tests have been developed in recent years, the agar gel immunodiffusion test remains the most useful test in the diagnosis of EIA (Fig. 2). It accurately and reliably detects antibody to EIA virus which is directly correlated with viremia in the blood. Except for maternally acquired antibody in the young foal, all horses found to have precipitating antibody have had EIA virus in their blood as well (Coggins *et al.*, 1972). Maternal antibody in the uninfected foal wanes after about 6 months of age.

VII. PREVENTION AND CONTROL

Thus far, active immunization against EIA virus has not been successful. Relatively nonpathogenic EIA viruses have been isolated from horses infected under field conditions (Coggins *et al.*, 1972), and the virus has been attenuated by passage in tissue culture systems (Kono *et al.*, 1970b). None of these viruses have, however, provided adequate protection against a challenge with other more virulent viruses and, perhaps a more serious defect, these viruses have produced persistent virus carriers. Thus to date, it has not been possible to devise a safe and effective vaccine which would prevent or cure EIA infection. In addition, all treatments aimed at eliminating the virus from the carrier horse have been ineffective.

The control of EIA has centered on the interruption of the virus transmission from infected horses, since they are the only known reservoir of EIA virus. If infected horses are detected and controlled or destroyed, then the spread of this infection can be stopped and the infection, most likely, eradicated rapidly from a population of horses.

Although the number of horseflies and other blood-sucking insects can be reduced on a farm or racetrack by good sanitary measures and use of insecticides, it is usually not feasible to reduce these vectors to the level necessary to prevent all possible transmission. Thus the most practical approach to EIA control boils down to the recognition of all infected horses and the isolation or destruction of them to remove all possibility of transmission. Success in such control programs has been surprisingly easy to accomplish (Coggins and Auchnie, 1977). Freedom from EIA can be maintained by testing all incoming horses.

REFERENCES

Abid, H. N. (1978). Ph.D. Thesis, Cornell Univ., Ithaca, New York.

Archer, B. G., Crawford, T. B., McGuire, T. C., and Frazier, M. E. (1977). *J. Virol.* **22,** 489-497.

Banks, K. L., and Henson, J. B. (1973). *Infect. Immun.* **8,** 679-682.

Charman, H. P., Bladen, S., Gilden, R. V., and Coggins, L. (1976). *J. Virol.* **19,** 1073-1079.

Cheevers, W. P., Archer, B. G., and Crawford, T. B. (1977). *J. Virol.* **24,** 489-497.

Cheevers, W. P., Ackley, C. M., and Crawford, T. B. (1978). *J. Virol.* **28,** 997-1001.

Coggins, L., and Auchnie, J. A. (1977). *J. Am. Vet. Med. Assoc.* **170,** 1299-1301.

Coggins, L., and Kemen, M. (1976). *Int. Conf. Equine Infect. Dis., 4th* pp. 14-22.

Coggins, L., and Norcross, N. L. (1970). *Cornell Vet.* **60,** 330-335.

Coggins, L., Norcross, N. L., and Nusbaum, S. R. (1972). *A. J. Vet. Res.* **33,** 11-18.

Coggins, L., Matheka, H. D., and Charman, H. P. (1978). *J. Equine Med. Surg., Suppl.* **1,** 351-358.

Crawford, T. B., McGuire, T. C., and Henson, J. B. (1971). *Arch. Gesamte Virusforsch.* **34,** 332-339.

Dreguss, M. N., and Lombard, L. S. (1954). "Experimental Studies in Equine Infectious Anemia." Univ. of Pennsylvania Press, Philadelphia.

Gonda, M. A., Charman, H. P., Walker, J. L., and Coggins, L. (1978). *Am. J. Vet. Res.* **39,** 731-740.

Henson, J. B., and McGuire, T. C. (1974). *Prog. Med. Virol.* **18,** 143-159.

Ishizaki, R., Green, R. W., and Bolognesi, D. P. (1978). *Intervirology* **9,** 286-294.

Ito, Y., Kono, Y., and Kobayashi, K. (1969). *Arch. Gesamte Virusforsch.* **28,** 411-416.

Kemen, M. J., and Coggins, L. (1972). *J. Am. Vet. Med. Assoc.* **161,** 496-499.

Kobayashi, K. (1961a). *Virus* **11,** 240-256.

Kobayashi, K. (1961b). *Virus* **11,** 189-201.

Kono, Y. (1973). *Proc. Int. Conf. Equine Infect. Dis., 3rd* pp. 175-186.

Kono, Y., and Kobayashi, K. (1966). *Natl. Inst. Anim. Health Q.* **6,** 204-207.

Kono, Y., and Yoshino, T. (1974). *Natl. Inst. Anim. Health Q.* **14,** 155-162.

Kono, Y., Yoshino, T., and Fukunaga, Y. (1970a). *Arch. Gesamte Virusforsch.* **30,** 252-256.

Kono, Y., Kobayashi, K., and Fukunaga, Y. (1970b). *Natl. Inst. Anim. Health Q.* **10,** 113-122.

Kono, Y., Kobayashi, K., and Fukunaga, Y. (1971). *Natl. Inst. Anim. Health Q.* **11,** 11-20.

Kono, Y., Sentsui, H., and Murakami, Y. (1978). *J. Equine Med. Surg., Suppl.* **1,** 363-374.

Ligneé, M. (1843). *Rec. Med. Vet.* **20,** 30.

McGuire, T. C., von Hoosier, G. L., and Henson, J. B. (1971a). *J. Immunol.* **107,** 1738-1744.

McGuire, T. C., Crawford, T. B., and Henson, J. B. (1971b). *Am. J. Pathol.* **62,** 283-291.

Metheka, H. D., Coggins, L., Shively, J. N., and Norcross, N. L. (1976). *Arch. Virol.* **51,** 107-114.

Nakajima, H., and Obara, J. (1964). *Natl. Inst. Anim. Health Q.* **4,** 129-134.
Nakajima, H., and Obara, J. (1964). *Natl. Inst. Anim. Health Q.* **4,** 129-134.
Nakajima, H., Tanaka, S., and Ushimi, C. (1970). *Arch. Gesamte Virusforsch.* **30,** 273-280.

Nakajima, H., Mizuno, Y., Yasuda, K., and Ushimi, C. (1973). *Arch. Gesamte Virusforsch.* **41,** 135-137.

Nakajima, H., Yoshino, T., and Ushimi, C. (1974). *Infect. Immun.* **10,** 667.

Norcross, N. L., and Coggins, L. (1971). *Infect. Immun.* **4,** 528-531.

Rice, N. R., Simek, S., Ryder, O. A., and Coggins, L. (1978). *J. Virol.* **26,** 577–583.
Schalk, A. M., and Roderick, K. M. (1923). *N.D. Agric. Exp. Stn., Bull.* No. 168.
Sentsui, H., and Kono, Y. (1976). *Infect. Immun.* **14,** 325–331.
Stein, C. D., Mott, L. O., and Gates, D. W. (1955). *J. Am. Vet. Med. Assoc.* **126,** 277–287.
Tajima, M., Nakajima, H., and Ito, Y. (1969). *J. Virol.* **4,** 521–527.
Takemoto, K. K., Mattern, C. F. T., Stone, L. B., Coe, J. E., and Lavelle, G. (1971). *J. Virol.* **7,** 301–308.
Ushimi, C., Henson, J. B., and Gorham, J. R. (1972). *Infect. Immun.* **5,** 890–895.
Weiland, F., Matheka, H. D., Coggins, L., and Hartner, D. (1977). *Arch. Virol.* **55,** 335–340.

Chapter 17

Astroviruses in Diarrhea of Young Animals and Children

DAVID R. SNODGRASS

I. Introduction ... 659
II. Description of Astroviruses ... 660
A. Morphology ... 660
B. Physicochemical Properties ... 660
III. Comparative Biology and Pathogenesis ... 664
A. Pathogenesis in Animals ... 664
B. Pathogenesis in Man ... 665
C. Tissue Culture ... 665
IV. Serology ... 667
V. Epidemiology ... 667
VI. Laboratory Diagnosis ... 667
A. Demonstration of Virus in Feces ... 667
B. Demonstration of Virus in Gut Sections ... 668
C. Demonstration of a Serological Response ... 668
References ... 669

I. INTRODUCTION

The name "astrovirus" was first used by Madeley and Cosgrove in 1975 to describe a "small round virus" observed by electron microscopy in stools from babies in Scotland with gastroenteritis. They believed that this virus was morphologically distinct from various other small viruses observed in stools. The virus particles were circular in outline, with a surface configuration of a five- or six-pointed star; hence the suggested name "astrovirus." Since this initial description, viruses fulfilling these morphological criteria have also been reported

COMPARATIVE DIAGNOSIS
OF VIRAL DISEASES, VOL. IV

ISBN 0-12-429704-8

from diarrheic children in England (Kurtz *et al.*, 1977; Ashley *et al.*, 1978) and Australia (Schnagl *et al.*, 1978). Viruses indistinguishable in appearance have also been identified in feces from calves (Woode and Bridger, 1978) and lambs (Snodgrass and Gray, 1977). Confirmation that the astroviruses were in fact vertebrate viruses, and not bacteriophages or cell debris, was first obtained by transmission of lamb astrovirus to gnotobiotic lambs (Snodgrass and Gray, 1977).

There is no official approval for the name "astrovirus" from the International Committee on Taxonomy of Viruses, and indeed insufficient information as yet exists to enable these viruses to be classified. However, it has found favor with those actively working with neonatal diarrhea in man and animals, and so is used in this chapter in preference to more clumsy alternatives.

II. DESCRIPTION OF ASTROVIRUSES

A. Morphology

No differences have been detected in the appearance of astroviruses in feces of children, lambs, or calves. After negative contrast staining with either 1% potassium phosphotungstic acid (pH 7.0) or 1% ammonium molybdate (pH 5.3), the viruses appear circular in outline and 28–30 nm in diameter (Madeley and Cosgrove, 1975; Snodgrass and Gray, 1977; Woode and Bridger, 1978). Empty particles, which have a rim 3–4 nm thick, are rarely seen (Woode and Bridger, 1978). A five- or six-pointed stellate configuration is apparent on some particles, being seen on about 10% of the human astroviruses stained with potassium phosphotungstate (Madeley and Cosgrove, 1975). However, after staining with ammonium molybdate, some surface structure can be seen on nearly all particles, with recognizable star shapes on many of them (Fig. 1). The lamb astrovirus is usually observed in feces as large aggregates, with occasional bridging structures between particles. The human astrovirus can occur in quasi-crystalline arrays, with a 6.5-nm gap between adjacent particles (Madeley and Cosgrove, 1975).

Examination of thin sections of infected lamb small intestine shows astroviruses in villus epithelial cells (Fig. 2) and occasionally in subepithelial macrophages. The viruses are seen in viroplasm (Fig. 3) or crystalline arrays (Fig. 4) within the cytoplasm of infected cells, and occasionally within membranes or in vacuoles (Fig. 5). Hollow-cored particles are sometimes evident. The mean diameter of 103 of these particles is 24.8 ± 0.6 nm.

B. Physicochemical Properties

Experiments to characterize lamb astrovirus are incomplete, but some information has already been obtained (A. J. Herring, personal communication). The

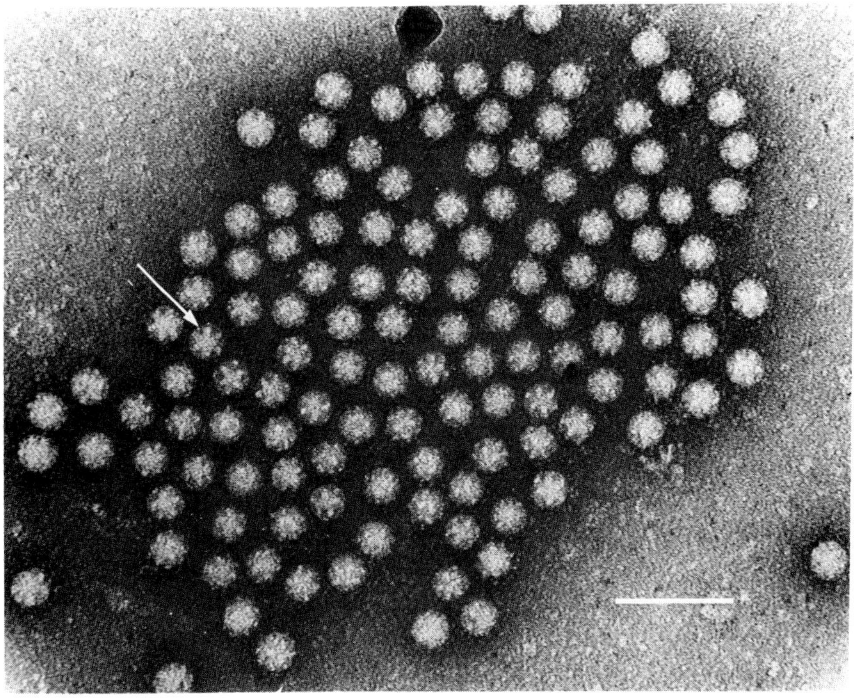

Fig. 1. Lamb astrovirus particles in intestinal content of experimentally infected gnotobiotic lamb. The arrow indicates a group of particles showing a star-like surface structure. The bar represents 100 nm. Stained with ammonium molybdate. (Reproduced by permission of *Archives of Virology*.)

work is limited by the fact that the astrovirus is present in feces and intestinal contents largely in aggregates, which cannot be disrupted by ultrasound, and also by the fact that the only method of virus assay is electron microscopic visualization. The best preparations have been made from scrapings of small intestinal mucosa of infected lambs. In cesium chloride density gradients, virus particles are seen most often at a density of 1.38 gm/ml. In the single experiment so far in which a substantial preparation of astrovirus has been achieved, RNA extraction yielded a single-stranded RNA molecule with a molecular weight 2.7×10^6 and with a sedimentation coefficient of 35 S on sucrose gradients. This molecule contains a poly(A) tract.

The virus is stable to the following range of treatments as determined by electron microscopy: organic solvents—chloroform, Arcton 113; high salt concentrations—2 *M* NaCl, 2 *M* CsCl; detergents—1% SDS, 1% Sarcosyl, 1% Triton X100; enzymes—trypsin. The virus is sensitive to 3 *M* urea in PBS after 30 minutes at 37°C.

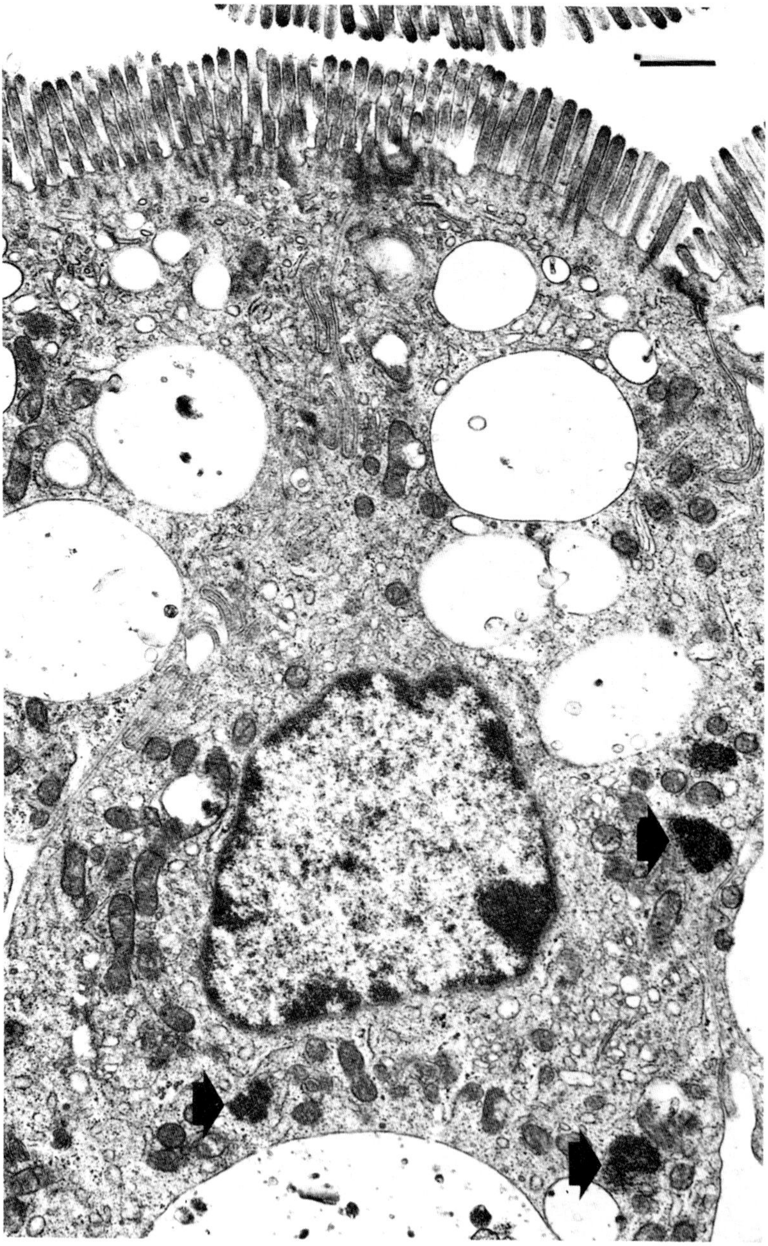

Fig. 2. Villus epithelial cell containing astrovirus (arrowed). Bar respresents 1 μm. Figs. 2, 3, 4, and 5 stained with lead citrate and uranyl acetate.

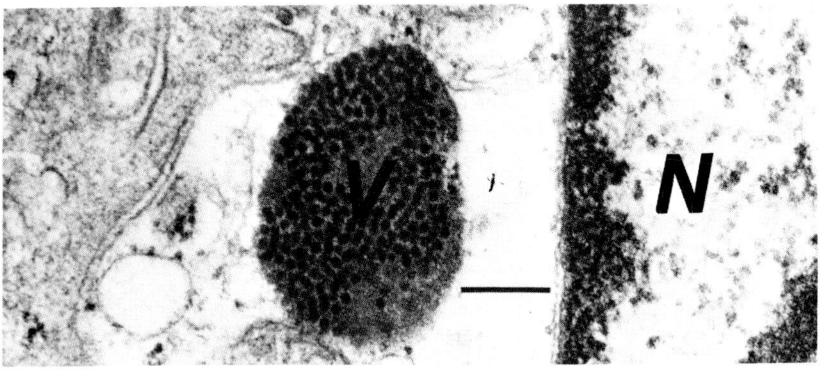

Fig. 3. Viroplasm (V) adjacent to nucleus (N) of epithelial cell. The bar represents 200 nm.

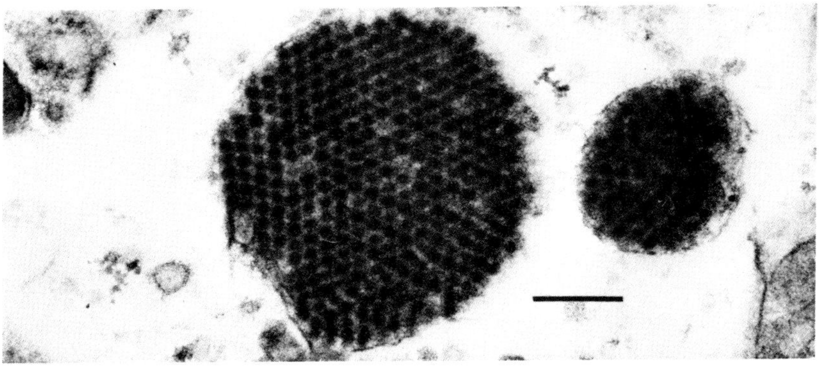

Fig. 4. Crystalline array of astrovirus in epithelial cell. The bar represents 200 nm.

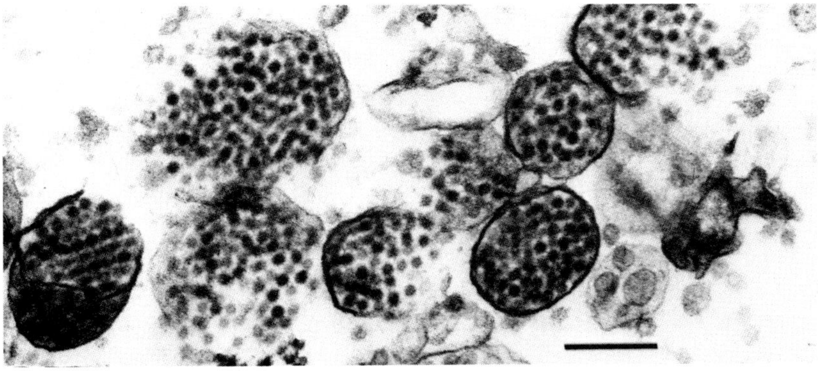

Fig. 5. Membrane-bound astrovirus in cytoplasmic vacuole. The bar represents 200 nm.

III. COMPARATIVE BIOLOGY AND PATHOGENESIS

A. Pathogenesis in Animals

Astrovirus infections have been produced experimentally in gnotobiotic lambs (Snodgrass and Gray, 1977) and calves (Woode and Bridger, 1978) by oral inoculation of bacteria-free fecal filtrates (0.22 μm) of intestinal contents containing astrovirus. However, pathogenesis has so far been studied only in lambs (Snodgrass *et al.*, 1979). A mild diarrhea occurred in lambs after an incubation

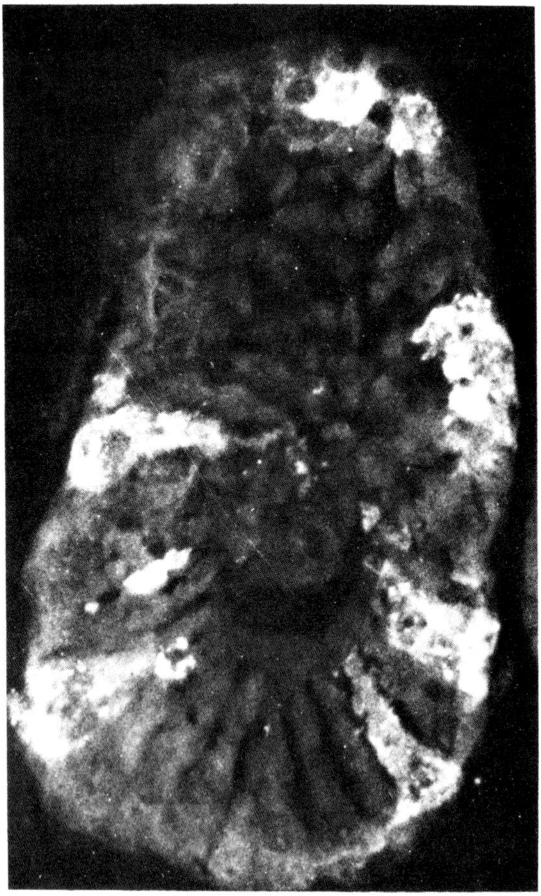

Fig. 6. Section of lamb small intestine stained by immunofluorescence with lamb antiserum to lamb astrovirus. Virus antigen is present in epithelial cells on villi. (Reproduced by permission of *Archives of Virology*.)

period of about 48 hours and lasted 1–2 days. Virus excretion in feces was detected at a time approximately coincident with the onset of diarrhea. By use of immunofluorescence on tissue sections, the virus was found to multiply only in the small intestine, predominantly in the epithelial cells on the distal parts of the villi (Fig. 6). There was also evidence of occasional infection of subepithelial cells. The greatest numbers of infected enterocytes were found during the incubation period, and only scattered fluorescent cells were seen after diarrhea had commenced.

Histological damage was confined to the middle and posterior small intestine. Villi were shorter and more spatulate than in control lambs and lined with a crenated epithelium, which contained some cuboidal cells. The villous lamina propria contained infiltrates of macrophages, lymphocytes, and neutrophils. Occasional intracytoplasmic inclusions could be seen, which were shown by electron microscopy to consist of large numbers of astrovirus particles. Measurements of villus height and crypt depth throughout the infection confirmed the partial villus atrophy, villus height at the time of onset of diarrhea being only half of that in control lambs. Crypt hypertrophy developed subsequent to the villus atrophy and was still marked after clinical recovery.

Electron microscopic examination of thin sections of small intestine confirmed that some villus epithelial cells were cuboidal and also demonstrated infected necrotic epithelial cells sloughing into the gut lumen (Fig. 7).

Obviously no detailed pathogenetic mechanism can be postulated on this limited evidence. However, the demonstration of epithelial cell infection and subsequent destruction is similar to that of other viral enterites, particularly rotaviral and coronaviral infections (Pensaert *et al.*, 1970; Mebus *et al.*, 1975; Snodgrass *et al.*, 1977), and such damage is likely to interfere with the normal digestive and absorptive functions of these cells.

B. Pathogenesis in Man

Studies with human astrovirus infections in adult volunteers have shown the ability of the virus to infect the gastrointestinal tract after oral inoculation with bacteria-free fecal filtrates (Kurtz *et al.*, 1979). Symptoms produced were mild, with diarrhea and vomiting in only 1 of 17 adults and mild constitutional symptoms in several of the others.

C. Tissue Culture

Attempts to adapt lamb astrovirus to ovine embryo kidney cell cultures by routine techniques have failed. Application of techniques that have been found helpful for rotavirus propagation, i.e., trypsin treatment of both virus and cultures (Theil *et al.*, 1978), centrifugation of cell cultures after inoculation (Banat-

potassium phosphotungstic acid (pH. 7.0), and examined in the electron microscope at a magnification of 40,000. We have compared this examination of crude stool suspension with results obtained by clarification and ultracentrifugation and found both to be equally sensitive. Techniques for detecting astroviruses in calves and children have usually employed concentration procedures (Kurtz *et al.*, 1977; Madeley *et al.*, 1977; Ashley *et al.*, 1978; Woode and Bridger, 1978). The particles have to be distinguished from other small viruses present in feces, particularly calici-like viruses (Madeley and Cosgrove, 1976). The stellate configuration of astroviruses is usually distinct from the large surface hollows of the calicivirus, and in particular the calicivirus "Star of David" configuration with a central hollow is unmistakable. However, where very few particles are present, it may be impossible to be certain of their identity.

The ability of the human astrovirus to produce nonproductive infections in cell culture that can be detected by immunofluorescence can also be utilized as a diagnostic method provided a specific immune astrovirus serum without antibody to rotavirus is available.

The future application of modern techniques suitable for examining large numbers of samples, such as enzyme-linked immunosorbent assay, will depend on the production of high-titer specific antisera, while as yet only convalescent sera are available.

B. Demonstration of Virus in Gut Sections

Immunofluorescent staining of cryostat sections of small intestine obtained at biopsy or necropsy can be used to demonstrate astrovirus antigen in epithelial cells. However, the absence of infected cells is not a conclusive negative finding, as the highest rate of cell infection is present during the incubation period (Snodgrass *et al.*, 1979).

C. Demonstration of a Serological Response

Serological diagnosis of astrovirus infections has been used only in human infections, and the methods available are either immunofluorescence or immune electron microscopy (Kurtz *et al.*, 1977, 1979; Ashley *et al.*, 1978).

ACKNOWLEDGMENTS

I would like to thank my colleagues K. W. Angus, E. W. Gray, and A. J. Herring for their cooperation throughout this work.

REFERENCES

Ashley, C. R., Caul, E. O., and Paver, W. K. (1978). *J. Clin. Pathol.* **31,** 939-943.

Banatvala, J. E., Totterdell, B., Chrystie, I. L., and Woode, G. N. (1975). *Lancet* **2,** 821.

Kurtz, J. B., Lee, T. W., and Pickering, D. (1977). *J. Clin. Pathol.* **30,** 948-952.

Kurtz, J. B., Lee, T. W., Craig, J. W., and Reed, S. E. (1979). *J. Med. Virol.* **3,** 221-230.

Lee, T. W., and Kurtz, J. B. (1977). *Lancet* **2,** 406.

Madeley, C. R., and Cosgrove, B. P. (1975). *Lancet* **2,** 451-452.

Madeley, C. R., and Cosgrove, B. P. (1976). *Lancet* **1,** 199-200.

Madeley, C. R., Cosgrove, B. P., Bell, E. J., and Fallon, R. J. (1977). *J. Hyg.* **78,** 261-273.

Mebus, C. A., Newman, L. E., and Stair, E. L. (1975). *Am. J. Vet. Res.* **36,** 1719-1725.

Pensaert, M., Haelterman, E. O., and Burnstein, T. (1970). *Arch. Gesamte Virusforsch.* **31,** 321-334.

Schnagl, R. D., Holmes, I. H., and Mackay-Scollay, E. M. (1978). *Med. J. Aust.* **1,** 304-307.

Snodgrass, D. R., and Gray, E. W. (1977). *Arch. Virol.* **55,** 287-291.

Snodgrass, D. R., Angus, K. W., and Gray, E. W. (1977). *Arch. Virol.* **55,** 263-274.

Snodgrass, D. R., Angus, K. W., Gray, E. W., Menzies, J. D., and Paul, G. (1979). *Arch. Virol.* **60,** 217-226.

Theil, K. W., Bohl, E. H., and Saif, L. J. (1978). *J. Am. Vet. Med. Assoc.* **173,** 548-551.

Woode, G. N., and Bridger, J. C. (1978). *J. Med. Microbiol.* **11,** 441-452.

Index

A

Abortion
  Akabane virus, 488, 489
  bovine enterovirus, 28, 29
  foot-and-mouth disease, 26
  togavirus, 391–394
N-Acetylethyleneamine, viral inactivator, 53
Adenovirus, canine
  with canine SV-5 virus, 202
  differentiation from canine distemper, 269
AE, *see* Avian encephalomyelitis
Aedes
  Ross River virus, 412
  Wesselsbron virus, 410
Aedes aegypti, Israel turkey meningoencephalitis, 410
Aedes caballus, Middelburg virus, 412
Aedes sollicitans, eastern equine encephalitis virus, 369
Aedes vexans, Akabane virus, 486
Aerophobia, 539
African horse sickness virus, 449
  characterization, 80
  control, 80, 81
  morphology, 69
  pathogenesis, 80
  transmission, 80, 81
Agar gel diffusion test
  enterovirus encephalomyelitis, 49
  vesicular disease, 44, 45
Agar gel diffusion precipitin test
  Akabane virus, 493
  rinderpest, 280
Agar gel immunodiffusion test, equine infectious anemia virus, 655, 656
Agar gel precipitin test
  coronavirus, 320
  rabies virus, 546
  togavirus, 464, 465
Aino virus, *see also* Bunyaviridae
  classification, 481
  differentiation, Akabane virus, 496
  pathology, 499
  transmission, 497
Akabane virus, 482–497
  abortion, 488, 489
  antibody, 493, 494
  arthrogryposis, 488, 489
  Australia, 484, 485, 491, 492
  classification, 481
  clinical signs, 487, 488
  diagnosis, 494–497
  differentiation, 496, 497
  epidemiology, 483–486
  growth system, 494, 495
  history, 482
  host range, 486
  hydranencephaly, 487–490
  immune response, 493, 494
  isolation, 495, 496
  Israel, 485
  Japan, 485, 491, 492
  micrencephaly, 490, 491
  morphology, 482
  pathology, 488–491
  physicochemical properties, 482
  protein, 482, 483
  serology, 492, 493
  strain difference, 491, 492
  transmission, 486, 487
    transplacental, 486
  vector, 484, 486, 487
    feeding preference, 486
Alphavirus, *see also* specific virus; Togavirus
  amino acid composition, 339, 340

antibody induction, 349
antigen, 344–349
character, 349
reactivity, 342
buoyant density, 335
carbohydrate, 340–343
chemical organization, 335–344
classification, serology, 348
envelope, 343
etiology, 333, 334
genetic difference, 351
genome, 344
geographic distribution, 333
glycoprotein
antigenic characteristic, 349
organization, 342
heat stability, 350
hemagglutination, 342
hemolytic activity, 340
host range, cell culture, 350
immunization, 345
infectivity, 342
lipid binding, 340
morphology, 335–344
neutralization testing, 344–347
nucleocapsid, 343, 344
oligonucleotide fingerprint, 350, 351
phosphate content, 340
phospholipid content, 343
physicochemical properties, 335
plaque size, 349
protein
molecular weight, 339
structural, 337–339
antigenic specificity, 345–349
ribonucleic acid, 344
Semliki forest virus complex, 345
surface charge, 350
variation, 349–355
Venezuelan equine encephalitis complex, 345
virulence, 349–351
Western equine encephalitis complex, 345
Amapari virus, *see also* Arenavirus
host range, 513, 514
isolate, 519
Amblyomma variegatum
Bhanja virus, 502
Dugbe virus, 501
Anemia
equine infectious anemia virus, 652
retrovirus, 574, 576
Anopheles, Getah virus, 413
Anopheles funestrus, Akabane virus, 486
Anorexia
canine distemper virus, 259
rotavirus, 118
Antibody, specific neutralizing, enterovirus, 38, 39
Aphthovirus, *see also* Foot-and-mouth disease virus; Picornavirus
distinguishing feature, 8
host range, 17
immunity, 36, 37
Arboviral equine encephalitis, diagnosis, 413
Arbovirus, Reoviridae, 68
Arenaviridae, *see* Arenavirus
Arenavirus, 511–524, *see also* specific virus
classification, 511, 512
diagnosis, 520–523
host range, 513
human infection, 512–514
immunity, 522, 523
infection, 515–520
isolation, 520–522
nonhuman, 514, 515
rodent specificity, 512–515
transmission, 515–520
Argentine hemorrhagic fever, 514
Arthritis, reovirus infection, 96
Arthrogryposis
Aino virus, 497, 499
Akabane virus, 488, 489
Arthropod-borne virus; *see also* specific arthropod
Bunyaviridae, 481
orbivirus, 68, 72, 83, 88–92
rhabdovirus, 534
togavirus, 331–426
Asian virus, *see* Influenza virus
Astroviridae, *see* Astrovirus
Astrovirus, 659–668
classification, 660
comparative biology, 664–666
diagnosis, 667, 668
in diarrhea, 659–668
epidemiology, 667
in feces, 667, 668
in intestinal section, 661–664, 666, 668

morphology, 661-663
pathogenesis
animal, 664, 665
human, 665
physicochemical properties, 661, 662
serological response, 667, 668
tissue culture, 665, 667
ASV, *see* Avian sarcoma virus
Aura virus, *see also* Alphavirus; Togavirus
encephalitis, 380
epizootiology, 412
Avian encephalomyelitis, diagnosis, 50
Avian infectious bronchitis virus, *see also* Coronavirus
classification, 302
clinical signs, 308
isolation, 316
serological testing, 320, 321
serotype, 304
vaccine, 308, 322
Avian myeloblastosis virus, 569
Avian sarcoma virus, *see also* Retrovirus
genome, 559

B

Bangor isolate, *see* Yucaipa virus
BDV, *see* Bovine diarrhea virus
Bebaru virus, 345
Bee chronic paralysis virus, 449
Bhanja virus, 501, 502, *see also* Bunyaviridae
classification, 481
diagnosis, 502
human, 502
vector, 501, 502
Blind staggers, 396
Bluecomb, 97, 98, *see also* Turkey enteritis coronavirus
Bluetongue virus, 81-85, *see also* Orbivirus
antigenic spectrum, 81
clinical signs, 82-84
control, 84, 85
differentiation
Akabane virus, 496
peste des petits ruminants, 286
ecology, 84
electron micrograph, 79
Eubenangee subgroup, 83
latency, 84
morphology, 69
ribonucleic acid, 70
temperature-sensitive mutant, 82
transmission, 81, 84
vaccine, 81, 82
Boophilus decoloratus, Bhanja virus, 502
Bolivian hemorrhagic fever, 514
Bordatella bronchiseptica, with canine SV-5 virus, 202
Border disease, 450, 453
cytopathic effect, 459
immunofluorescence, 464
Bovine diarrhea virus, *see also* Togavirus
classification, 442
clinical signs, 452
epidemiology, 456, 457
immunity, 453
immunofluorescence, 463, 464
morphology, 443, 444
pathology, 453
susceptibility
cell lines, 462
cultured cells, 461
transmission, 453
vaccine, 457
virus neutralization, 466
Bovine enterovirus, *see also* Enterovirus
clinical signs, 26-29
critical age, 27
description, 6
diagnosis, 47, 48
effect on reproduction, 28, 29
in enteric disease, 27
hyperimmune serum, 48
isolation, 48
pathogenesis, 26-29
in respiratory disease, 27, 28
serotype, 15, 16
in winter dysentery, 27
Bovine ephemeral fever, *see also* Rhabdovirus
host, 534
transmission, 81
Bovine leukemia virus, 584-586
Bovine respiratory syncytial virus, 221-226, *see also* Pneumovirus; Respiratory syncytial virus
antibody, 226
clinical signs, 223, 224
diagnosis, 224-226
experimental, 224

history, 221, 222
host tissue range, 222
immunity, 226
isolation, 224, 225
natural, 223
physicochemical properties, 222, 223
propagation, 223
serology, 225, 226
vaccination, 226
Bovine rhinovirus, *see also* Rhinovirus
clinical signs, 34
cytopathic effect, 46, 47
diagnosis, 46, 47
pathogenesis, 34
Bovine syncytial virus, *see* Spumavirinae
BTV, *see* Bluetongue virus
Bunyamwera virus, 480
Bunyaviridae, 479–504, *see also* specific virus
classification, 480, 481
disease, 481
human infection, 500–502, 504
morphology, 480
ribonucleic acid, 492
serology, 480, 481
transmission, 481
Bunyavirus, see Bunyaviridae
Burkitt's lymphoma, 610

C

Calf diarrhea coronavirus, 307, *see also* Coronavirus
isolation, 316
serology, 321
vaccine, 323
Calicivirus, *see also* specific virus
antigenic structure, 16, 17
chemical properties, 12, 13
control, 39
cytopathic effect, 52
description, 7
diagnosis, 51, 52
heat stability, 12
host range, 19
immunity, 39, 40
molecular weight, 12
morphology, 9, 11
physicochemical properties, 9
protein coat, 12
Cancer, retrovirus-caused, 554
Canine coronavirus, 307, *see also* Coronavirus
differentiation from canine distemper, 269
isolation, 316
serological testing, 321
Canine distemper virus, 236, 259–272, *see also* Morbillivirus
antigen identification, 274
antigenic relationship
measles virus, 242, 243
rinderpest virus, 243, 244
biological variation, 263, 264
cerebrospinal fluid, 272
clinical signs, 259, 260
control, 268, 269
cytopathic effect, 247–249
demyelinating encephalomyelitis, 240, 261, 264
diagnosis, 269, 272
epizootiology, 271
host range, 261, 263
immunity, 264–267
isolation, 271
lymphocyte blastogenesis, 271
measles virus inhibition, 275, 276
pathogenesis, 260, 261
pathology, 270
persistence in brain, 260, 261
route of infection, 259
seasonal prevalence, 267
serology, 271, 272
strain, 263, 264
transmission, 267
vaccination, 243, 244, 264, 265, 268, 269
virion, 238, 239
Cardiovirus, *see also* Picornavirus
description, 6, 7
diagnosis, 51
distinguishing feature, 8
epizootiology, 41
host range, 18
immunity, 38
Catarrhal vaginitis, role of bovine enterovirus, 28, 29
CCV, *see* Canine coronavirus
CDV, *see* Canine distemper virus
CDCV, *see* Calf diarrhea coronavirus
Cell fusion inhibition, 247
Cell survivor technique, 249
Cerebrospinal meningitis, 396
Challenge virus standard, 536, 537

Changuinola virus, 90, 91
Chikungunya virus, 345
CIEP, *see* Counterimmunoelectrophoresis
Cisternovirus A, 555
Cloacal pasting, in reovirus infection, 96
Clone, influenza virus, 155
Cocal virus, 542, *see also* Rhabdovirus
COFFAL test, 587
Colorado tick fever, 85–87
  diagnosis, 87
  ecology, 85, 86
  morphology, 69
  pathogenesis, 86, 87
  transmission, 85, 86
Complement fixation
  Akabane virus, 493
  coronavirus, 319
  equine infectious anemia, 656
  rabies virus, 545
  rinderpest, 280
  rotavirus, 126
  Spumavirinae, 615
  togavirus, 416, 466
  vesicular disease, 44
Congenital defect, Akabane disease, 487–491
Congo virus, *see* Crimean–Congo virus
Contagious pustular dermatitis, differentiation from peste des petits ruminants, 286
Coronaviridae, *see* Coronavirus
Coronavirus, 301–323, *see also* specific virus
  agar gel precipitin test, 320
  antibody detection, 318–322
  antigen
    detection, 313–315
    in tissue extract, 316, 317
  antigenic composition, 304, 305
  cat, 309
  cattle, 307
  cell culture-immunostaining, 318
  characteristics, 302–304
  chicken, 308
  clinical signs, 305–311
  complement fixation, 319
  diagnosis, 305, 306, 311–322
  dog, 307
  economic importance, 302
  electron microscopy, 311, 312
  fluorescent antibody test, 313, 314
  hemadsorption-elution-hemagglutination assay, 315
  hemagglutination, 318
  host, 303
    laboratory, 315–317
  human, 308, 310
  identification, 315–318
  immune electron microscopy, 312
  immunoperoxidase technique, 314
  indirect fluorescent antibody test, 319, 320
  indirect hemagglutination test, 320
  intestinal infection, 305–308
  isolation, 315–318
  mice, 307, 309, 310
  parrot, 310
  pathology, 305–311
  plaque formation, 317
  prevention, 322, 323
  rabbit, 310, 311
  rat, 309
  replication, detection, 317, 318
  respiratory-pharyngeal form, 314
  respiratory infection, 308–311
  serology, 318, 319
  single radial hemolysis test, 320
  swine, 306–309
  turkey, 308
  virus neutralization test, 319
Corriparta virus, 90, 91
Counterimmunoelectrophoresis, in rotavirus diagnosis, 126
Coxsackie B5 virus, human, 19
Crimean-Congo virus, 504, *see also* Bunyaviridae
  classification, 481
Crimean hemorrhagic fever, 504
Cross-neutralization test, flavivirus, 362, 363
CTF, *see* Colorado tick fever
Culex
  Aino virus, 497
  Getah virus, 413
  Japanese encephalitis virus, 409, 410
  Palyam virus, 89
  West Nile virus, 411
Culex annulirostris, Murray valley encephalitis virus, 412
Culex melanura, eastern equine encephalitis virus, 399
Culex pipiens, Rift valley fever virus, 500, 501
Culex tarsalis, western equine encephalitis virus, 401
Culex tritaeniorhynchus
  Akabane virus, 486
  Japanese encephalitis virus, 378

Culicoides
bluetongue virus, 81, 82, 84
Palyam group, 90
Culicoides brevitarsis
Akabane virus, 482–486
bovine ephemeral fever, 81
Culicoides pallidipennis, African horse sickness, 80, 81
Culiseta melanura, eastern equine encephalitis virus, 397
Cytoplasm, rotavirus replication, 112

D

D'Aguilar virus, 74–78
Dehydration
canine distemper virus, 259
rotavirus, 118, 123, 124
Demyelinating encephalomyelitis, 260, 261, 264
Demyelination, visna virus, 627, 629
Dengue virus, *see also* Flavivirus; Togavirus
nucleotide composition, 359
Dengue-2 virus, 357
Deoxyribonucleic acid
proviral
equine infectious anemia virus, 649
Lentivirinae, 636–638
retrovirus, 553, 567, 568, 586
synthesis, Lentivirinae, 622
Depression, caused by rotavirus, 119
Dermacentor adersoni, Colorado tick fever, 85, 86
Derriengue, 539
Diabetes
experimental, reovirus induced, 96
model, foot-and-mouth disease, 26
Diarrhea
coronavirus, 305
rinderpest, 275
rotavirus, 105–107, 118, 119, 135
swine enterovirus, 33, 34
Diethylaminoethyl dextran, coronavirus isolation, 316
Disaccharase level, rotavirus infection, 123, 124
Disinfectant, picornavirus, 12
Diploidy, retrovirus, 559
DNA, *see* Deoxyribonucleic acid
Duck virus hepatitis, diagnosis, 50
Dugbe virus, 503, 504, *see also* Bunyaviridae
classification, 481
Duovirus, *see* Rotavirus
Duvenhage virus, *see also* Rhabdovirus
clinical signs, 540
host, 534
isolation, 543
transmission, 535, 542

E

EAV, *see* Equine arteritis virus
Eastern equine encephalitis virus, *see also* Alphavirus; Togavirus
antibody, 370
antigenic variance, 355
bird
domestic, 373–375, 383, 384
wild, 399
bovine, 377
chicken, 375, 376, 383, 384
chrololopic relationships, 397
chukar partridge, 375, 384
climatological factor, 398
clinical signs, 367–380
duck, 383
geographic distribution, 395–397
heart involvement, 384
human, 396, 397
infection
pattern, 370
source, 370, 375–377, 399
isolation, 413, 414
after death, 383
occurrence, 395–397
overwintering mechanism, 400
pathogenicity, 369
pathology, 380–384
Pekin duck, 375
pheasant, 375
clinical signs, 374
reptile, 399
seasonal distribution, 397, 398
swine, 375–377
turkey, 375, 383
transmission, 369, 375, 397
cycle, 398–400
vaccine, 421
vector, 398, 399
viremia, 370

EDIM, *see* Epizootic diarrhea of infant mice
EE, *see* Enterovirus encephalomyelitis
EEE, *see* Eastern equine encephalitis virus
Egtved virus, 535
EHD, *see* Epizootic hemorrhagic disease of deer
Ehrlichiosis, differentiation from canine distemper, 269
EIA, *see* Equine infectious anemia virus
Electron microscopy
  coronavirus, 311, 312
  rotavirus, 124, 125
ELISA, *see* Enzyme-linked immunosorbent assay
EM, *see* Electron microscopy
Embryonic death, swine enterovirus, 32, 33
Enamel hypoplasia, canine distemper virus, 260
Encephalomyelitis
  measles complication, 250
  togavirus-caused, 331-426
Endoplasmic reticulum, rotavirus association, 110
Enteric disease, bovine, enterovirus, 27
Enteritis, reovirus infection, 96
Enterovirus, *see also* Bovine enterovirus; Picornavirus; Simian enterovirus; Swine enterovirus
  antigenic structure, 15
  description, 6
  distinguishing feature, 8
  epizootiology, 41, 42
  host range, 18, 19
  immunity, 38, 39
Enterovirus encephalomyelitis, *see also* Swine enterovirus
  clinical signs, 30, 31
  description, 6
  pathogenesis, 30, 31
Env, 559, 564-566
Enzootic bovine leukosis, 584-586
Enzyme-linked immunosorbent assay
  rotavirus, 126, 127
  togavirus, 419, 467
Eperythrozoon coccoides
  with lactic dehydrogenase virus, 454
  with murine hepatitis virus, 312
Epidemic polyarthritis, 412
Epidemic tremor, 373
Epizootic diarrhea of infant mice virus, *see also* Rotavirus
  characteristics, 106
  morphology, 110
  physicochemical properties, 113
Epizootic hemorrhagic disease of deer virus, 449
Epizootic pneumonia in seals, 178, *see also* Influenza virus
Equine arteritis virus, *see also* Togavirus
  antigenic relationship, lactic dehydrogenase virus of mice, 448
  classification, 442
  clinical signs, 453, 454
  epidemiology, 457
  immunofluorescence, 464
  morphology, 444, 445
  pathology, 454
  susceptibility
    cell lines, 463
    cultured cells, 462
  vaccine, 457
Equine infectious anemia virus, 647-657, *see also* Retrovirus
  agar gel immunodiffusion test, 655, 656
  antigen, 650
  clinical signs, 650
  control, 656, 657
  culture, 651
  diagnosis, 655, 656
  epizootiology, 653, 654
  host, 650
  hypersensitivity, 653
  immunity, 652, 653
  immunofluorescence, 651
  incubation, 654
  isolation, 655
  morphology, 648, 649
  pathogenesis, 651, 652
  persistance, 653
  physicochemical properties, 649
  platelet involvement, 651, 652
  proviral DNA, 649
  replication, 651
  reverse transcriptase, 649
  ribonucleic acid, 649
  serology, 656
  vaccine, 656
Equine rhinovirus, *see also* Rhinovirus
  clinical signs, 35
  description, 7
  diagnosis, 47
  host range, 18
Equine-1 virus, *see* Influenza virus, equine

Equine-2 virus, *see* Influenza virus, equine
Erythroblastosis, 572, 574
Erythrocyte, viral replication, 86, 87
Escherichia coli
with avian infectious bronchitis, 308
differentiation, coronavirus infection, 305, 306
with rotavirus, 124
Ethyleneamine, viral inactivator, 53
Exanthema subitem, differentiation from measles, 257

F

FC, *see* Feline calicivirus
Febrile exanthem, Getah virus, 394, 395
Feline calicivirus, *see also* Calicivirus
epizootiology, 43
pathogenesis, 35, 36
vaccine, 40, 54
Feline infectious peritonitis virus, 309, *see also* Coronavirus
isolation, 316
serological testing, 321
Feline syncytial virus, *see* Spumavirinae
Fetal mummification, swine enterovirus, 32, 33
Fibrosarcoma, retrovirus-caused, 554, 572, 575–582
FIPV, *see* Feline infectious peritonitis virus
Flavivirus, 356–367, *see also* specific virus; Togavirus
antigen, 357
composition, 361–365
determinant, 364
nonstructural, 364
relationship, 449
variation, 365, 366
biological variation, 365, 366
chemical organization, 359–361
complex-reactive determinant, 364
cross-neutralization, 362, 363
density, 356, 357
envelope, 357, 358
etiology, 333, 334
genome, 358, 359
geographic distribution, 333
glycoprotein, 361
group-reactive determinant, 364
host range, 367
morphology, 359–361
nucleocapsid, 361, 364
physicochemical properties, 356–359
soluble complement-fixing antigen, 357, 365
structural protein, 356, 357
type-specific determinant, 364
Fluorescent antibody test
avian encephalomyelitis, 51
coronavirus, 313, 314
duck virus hepatitis, 50
rabies virus, 543–545
Spumavirinae, 615
FMD, *see* Foot-and-mouth disease virus
Foamy virus, *see* Spumavirinae
Foot-and-mouth disease virus, *see also* Aphthovirus; Picornavirus
antibody, physicochemical change, 37
carrier, 41
clinical signs, 20–26
description, 5, 6
diagnosis, 43–46
differentiation
peste des petits ruminants, 286
rinderpest, 279
economic importance, 4–6
electron micrograph, 10
epizootiology, 40, 41
group reactivity, 14
immunogenicity, 13, 14
pathogenesis, 20–26
serotype, 14, 15
vaccine, 52, 53
Forage poisoning, 396
Fowl plague, *see* Influenza A virus
Frankel technique, 52

G

Gag, 559, 560, 562
Ganjam virus, 503, *see also* Bunyaviridae
classification, 479
Garbage, role in swine vesicular disease, 42
Gastroenteritis
acute infectious, rotavirus-caused, 105, 106
seasonal incidence, 135
GET, *see* Getah virus
Getah virus, *see also* Alphavirus; Togavirus
epizootiology, 412, 413
febrile exanthem, 394, 395

Gibbon ape leukemia virus, 582–584
Glomerulonephritis
equine infectious anemia virus, 652
retrovirus, 576
Goat pox, differentiation, peste des petits ruminants, 286
Gorman grouping, 71

H

Haemophilus influenza suis, with influenza virus, 169
Haemophysalis intermedia
Bhanja virus, 501
Ganjam virus, 503
Hairy shaker disease, 450, 453
Hamster foamy virus, *see* Spumavirinae
Hard pad disease, 260
Hazara virus, 504
HCV, *see* Hog cholera virus; Human coronavirus
Heat stability
calicivirus, 12
picornavirus, 11
HECV, *see* Human enteric coronavirus
HEHA, *see* Hemadsorption-elution-hemagglutination assay
HEV, *see* Hemagglutinating encephalomyelitis virus
Hemadsorption-elution-hemagglutination assay, coronavirus, 315
Hemagglutinating encephalomyelitis virus, 308, 309, *see also* Coronavirus
isolation, 316
serological testing, 322
Hemagglutinating virus of Japan, *see* Sendai virus
Hemagglutination inhibition
Akabane virus, 492, 493
rabies virus, 546
rinderpest, 281
Spumavirinae, 616
togavirus, 415, 416
Hemagglutinin
influenza virus, 156–161, 176
Herpesvirus, 588
canine, with canine SV-5 virus, 202
Herpes simplex virus, with Japanese encephalitis virus, 385
HI, *see* Hemagglutination inhibition
Highlands J virus, 345
Hog cholera virus, *see also* Togavirus
antibody, 452
classification, 442
clinical signs, 451, 452
epidemiology, 455, 456
immunofluorescence, 463
morphology, 443
multiplication, 451
pathology, 451
susceptibility
cell lines, 461
cultured cells, 460
transmission, 455
vaccine, 456
virulence, 451
virus neutralization, 465, 466
Hong Kong virus, *see* Influenza virus
Human coronavirus, *see also* Coronavirus
isolation, 316
serology, 304, 321
Human enteric coronavirus, *see also* Coronavirus
isolation, 316
serology, 321
Human foamy virus, *see* Spumavirinae
Hyalomma, Crimean-Congo virus, 504
Hydranencephaly
Aino virus, 497–499
Akabane virus, 487–489
Hydrophobia, 539
Hyperactivity, 539
H1N1 virus, *see* Influenza virus
H2N2 virus, *see* Influenza virus
H3N2 virus, *see* Influenza virus

I

IBDV, *see* Infectious bursal disease virus
IBV, *see* Avian infectious bronchitis virus
IDD, *see* Immuno-double-diffusion test
IEM, *see* Immunoelectron microscopy
IF, *see* Immunofluorescence
IFF, *see* Indirect immunofluorescence technique
IGV, *see* Infantile gastroenteritis virus
Immune system, effect of lactic dehydrogenase virus, 454

Immuno-double-diffusion test, influenza virus, 157, 158, 176
Immuno-electro-osmophoresis, togavirus, 467
Immunoelectron microscopy
  coronavirus, 312
  rotavirus, 124, 125
Immunofluorescence
  canine distemper virus, 270
  coronavirus, 313
  equine infectious anemia virus, 651
  Lentivirinae, 632, 633
  measles, 257, 258
  nuclear, Spumavirinae, 608
  retrovirus, 587
  rotavirus, 125
  togavirus, 463, 464
Immunoperoxidase technique
  coronavirus, 314
  rotavirus, 127, 132, 133
Immunosuppression, retrovirus, 575, 582
Indirect fluorescent antibody test
  coronavirus, 319, 320
  rabies virus, 545
Indirect hemagglutination test, coronavirus, 320
Indirect immunofluorescence technique
  rotavirus, 125
  togavirus, 467
Infant mortality, bovine enterovirus, 28, 29
Infantile gastroenteritis virus, *see also* Rotavirus
  discovery, 107
  morphology, 110
  physicochemical properties, 113
Infectious bursal disease virus, 97, 98
Infectious canine hepatitis, differentiation from canine distemper, 273
Infectious leukoencephalomyelitis, 450, *see also* Togavirus
Infectious myocarditis of goslings, 97, 98
Infectious pancreatic necrosis virus, 97, 98
Infertility
  in bovine enterovirus, 28, 29
  in swine enterovirus, 32, 33
Influenza virus, 151–180, *see also* specific virus
  antigen, 155–157
  antigenic drift, 165
  antigenic relationship, 157, 158
  antigenic similarity, 161
  Asian, 178–180
  avian, 172–177
    antigenic character, 176
    classification, 157, 158
    host range, 173–175
    isolation, 173, 176
    pathology, 173–175
    replication, 173
    source, human virus, 178–180
    subtype, 166, 167
    transmission, 176
    virulence, 176
  classification
    antigenic, 157–169
    serological, 156
  clone, 155
  coding capacity, 154, 156, 161
  comparative tryptic peptide analysis, 161
  composition, 154–157
  diagnosis, 180
  equine, 171, 172
    antigenic drift, 171
    classification, 157, 158
    relation to avian, 171, 172
    subtype, 165
    vaccine, 172
  genome, 154, 155
  hemagglutinating property, 153
  hemagglutinin, 156–161, 176
    cleavage, 157
  history, 153, 154, 169
  Hong Kong, 178–180
  human, 152, 153, 157, 158
    nonhuman source, 152, 178–180
    relation to swine, 169, 170
    subtype, 164
  H1, 161
  H1N1, 178–180
  H2N2, 178–180
  H3N2, 178–180
  immunity, 156, 157
  immuno-double-diffusion test, 157, 158, 176
  monoclonal antibody, 162
  neuraminidase, 154–161, 176
  nomenclature, 157–169
  nucleocapsid, 155, 156
  pathogenicity, 157
  polypeptide, 155–157
  ribonucleic acid, 154, 155, 161
  seal, 178
  serology, 180
  structure, 154–157
  subtype, 157–168
  swine, 169–171
    classification, 157, 158

Hong Kong, 170
relation to human, 169, 170, 179
reservoir, 170
source for human virus, 178–180
subtype, 165
transmission, 153
whale, 178
Influenza A virus, *see also* Influenza virus
classification, 157, 162, 165
reservoir, 152
subtype, 163–165
Influenza B virus, 153, 165, 169, *see also* Influenza virus
Influenza C virus, 153, 165, 169, *see also* Influenza virus
Inoculation, intradermal-lingual, foot-and-mouth disease, 46
IP technique, *see* Immunoperoxidase technique
IPNV, *see* Infectious pancreatic necrosis virus
Isatin-β-thiosemicarbasone, 631
Israel turkey meningoencephalitis virus, *see also* Flavivirus; Togavirus
clinical signs, 379
epizootiology, 410
pathology, 385
IT, *see* Israel turkey meningoencephalitis virus
Ixodes ricinus, louping ill virus, 410

J

J virus, 210, 211, *see also* Paramyxovirus
Japanese encephalitis virus, *see also* Flavivirus; Togavirus
clinical signs, 378
epizootiology, 408–410
equine, 378
pathology, 384
histopathology, 394
host, 392, 408, 409
human, 424
nucleotide composition, 359
overwinter maintenance, 409
seasonal, 408
spermatogenesis disorder, 394
swine, 378
clinical signs, 391–394
pathology, 391–394
transmission, 378, 408, 409
vaccine, 424, 425
vector, 409, 410
viremia, 392
JE, *see* Japanese encephalitis virus
Junin virus, *see also* Arenavirus
host range, 513, 514
infection, 519

K

Kata, 236, *see also* Peste des petits ruminants virus
Kemerovo virus, 87, 88
Kennel cough complex, 202
Koplik's spots, 251
Kotonkan virus, *see also* Rhabdovirus
host, 534
isolation, 537
transmission, 542
Kunitachi virus, 221
Kunjin virus, 357

L

Lactase, receptor, rotavirus, 116, 134
Lactic dehydrogenase virus of mice, *see also* Togavirus
antigenic relationship, equine arteritis virus, 448
classification, 442
diagnosis, 460–462
with *Eperythrozoon coccoides,* 454
epidemiology, 457, 458
in immune system, 454
immunofluorescence, 464
morphology, 444–446
pathogenesis, 454
in tumor growth, 454
Lactogenic immunity, rotavirus, 138, 139
Lactose, dehydrating effect, 124
Lagos bat virus, *see also* Rhabdovirus
clinical signs, 540
host, 534
isolation, 533
Langat virus, 359
Lassa virus, *see also* Arenavirus
antibody, 517
carrier, 517
experimental, 517

host range, 513, 514
isolation, 522
natural, 517, 518
radioimmunoassay, 523
reversed passive hemagglutination inhibition, 522
Latino virus, *see also* Arenavirus
host range, 513, 515
infection, 520
isolation, 521
LCM, *see* Lymphocytic choriomeningitis virus
LDV, *see* Lactic dehydrogenase virus of mice
Lentivirinae, 619-640, *see also* specific virus
antigen, 624
antigenic variation, 629, 630
control, 631, 632
cytopathic effect, 622, 623
deoxyribonucleic acid
proviral, 636-638
synthesis, 622
diagnosis, 630
dissemination, 639, 640
distinction, 625, 626
enzyme, 621
host range, 630, 631
hybridization
in situ, 636, 637
liquid, 608
molecular, 635, 636
identification, 622
immune response, 629
infection, persistance, 639
lytic effect, 624
morphogenesis, 624, 625
morphology, 621, 625
nucleic acid, 633
pathogenesis, 637-640
pathology, 627-629
physicochemical properties, 621
polypeptide synthesis, 624
protein, 621
quantitative analysis, 632-637
replication, 620, 622, 624, 625
reverse transcriptase, 620
reverse transcription, 622
ribonucleic acid, 621
symptomatology, 626, 627
transcriptional block, 639
transmission, 631
Lentivirus, *see* Lentivirinae
Lentivirus E, 556, *see also* Lentivirinae; Visna virus
Leptospirosis, differentiation from canine distemper virus, 269
Lesion, neurological
enterovirus encephalomyelitis, 31
swine vesicular disease, 32
Lethal intestinal virus of infant mice, 307, 310, *see also* Coronavirus
Leukemia, retrovirus-caused, 554, 574-582
Leukopenia
canine distemper virus, 259
measles, 251
Venezuelan equine encephalitis virus, 371
LI, *see* Louping ill virus
Litter size, effect of swine enterovirus, 33
LIVIM, *see* Lethal intestinal virus of infant mice
Louping ill virus, *see also* Flavivirus; Togavirus
clinical signs, 378, 379
epizootiology, 410
pathology, 384, 385
red grouse, 384, 385
with tick-borne fever, 410
Lumbar paralysis of goats, 501
Lymphocyte blastogenesis, 272
Lymphocytic choriomeningitis virus, 449, *see also* Arenavirus
antibody, 516
host range, 512-514
immunity, 522, 523
infection, 515-517
isolation, 522
Lymphoma, retrovirus-caused, 554, 576-582
Lymphoid leukosis, retrovirus-caused, 554, 559, 572-575
Lyssavirus, 530, *see also* Rhabdovirus; specific virus

M

Machupo virus, *see also* Arenavirus
antibody, 518
clinical signs, 518, 520
genetic control, 519
host range, 513, 514
isolation, 521
Maedi virus, 619-640, *see also* Lentivirinae
clinical signs, 627

diagnosis, 630
pathology, 629
Malignant catarrhal fever, differentiation from rinderpest, 280
Marine mammal, virus, 7
Mason-Pfizer monkey virus, 556
Mayaro virus, 345
Measles virus, 236, *see also* Morbillivirus
adaptational process, 258
antigenic relationship
canine distemper virus, 242, 243
rinderpest virus, 244
complication, 252, 253
control, 156, 157
diagnosis, 257-259
differentiation
exanthema subitem, 257
rubella, 257
epidemiology, 255, 256
human, 251-259
immunity, 254, 255
immunofluorescence, 257, 258
isolate, 253, 254
isolation, 258
effect on lymphocyte, 254, 255
prevention, 156, 157
route of infection, 251
serology, 258, 259
skin lesion, pathology, 257
symptomatology, 255, 256
transmission, 255
in tuberculosis, 254
vaccination, 156, 157
Medipest virus, *see* Morbillivirus
MHV, *see* Murine hepatitis virus
Micrencephaly, in Akabane virus, 490, 491
MID, *see* Middleburg virus
Middleburg virus, *see also* Alphavirus; Togavirus
abortion, 394
epizootiology, 412
MN, *see* Mouse neutralization test
Mokola virus, *see also* Rhabdovirus
clinical signs, 540
host, 534
isolation, 543
Morbillivirus, 235-287, *see also* Paramyxovirus; specific virus
antigenicity, comparative, 242-245
biological properties, 242-251
buoyant density, 237, 238
cell culture, 245-251
cell fusion inhibition, 247
cytopathic effect, 247-249
growth cycle, 246
heat stability, 241
isolation, 245
morphology, 237
nucleocapsid, 238, 239
passaged virus, 245, 246
persistant infection, 249-251
pH stability, 241
physicochemical properties, 237-242
plaquing, 249
polypeptides, 240, 241
replication, 246, 247
resistance
chemical agents, 241, 242
radiation, 241
ribonucleic acid, 239, 240
subgenomic particle, 250
temperature-sensitive mutant, 250
virion structure, 237-241
Mosquito, *see also* specific genus
bluetongue virus, 82
eastern equine encephalitis virus, 369
Venezuelan equine encephalitis virus, 404, 405
Mouse neutralization test, rabies virus, 545
Mucambo virus, 345
Mucosal disease virus, *see also* Togavirus
classification, 442
differentiation from rinderpest, 279, 280
Mudfever, *see* Turkey enteritis coronavirus
Multiple sclerosis
with canine distemper virus, 263
with measles virus, 236
Mumps virus, *see also* Paramyxovirus
relation to canine SV-5 virus, 201
Murine virus, *see* Cardiovirus
Murine hepatitis virus, 307, 309, 310, *see also* Coronavirus
isolation, 316
serological testing, 321
serotype, 304
Murine leukemia virus, *see also* Retrovirus
protein, structural, 560
Murine mammary tumor virus, 556
Murine parainfluenza type 1 virus, *see* Sendai virus

Murray valley encephalitis virus, *see also* Flavivirus; Togavirus
  clinical signs, 379, 380
  epizootiology, 412
  nucleotide composition, 359
  structural protein, 357, 358
MV, *see* Measles virus
MVE, *see* Murray valley encephalitis virus
Mycoplasma
  with avian infectious bronchitis, 308
  with bovine rhinovirus, 34
  with canine SV-5 virus, 202
  with parainfluenza-3 virus, 198
Myeloblastosis, 572, 574
Myelocytomatosis, 572, 574
Myocarditis, in reovirus infection, 96, 97

N

NA, *see* Neuraminidase
Nairobi sheep disease virus, *see also* Bunyaviridae
  classification, 481
  clinical signs, 503
  diagnosis, 503
  transmission, 502
Nairovirus, 480, 481, *see also* Bunyaviridae; specific virus
Nariva virus, 210, *see also* Paramyxovirus
Nasopharyngeal carcinoma, Spumavirinae, 600, 610, 611
NCDF, *see* Neonatal calf diarrhea virus
Ndumu virus, *see* Alphavirus; Togavirus
NDV, *see* Newcastle disease virus
Necrotic stomatitis, oral, in peste des petits ruminants, 282, 283
Neonatal calf diarrhea virus, 107, *see also* Rotavirus
  morphology, 110
  physicochemical properties, 113
Nephrosis, in reovirus infection, 96
Nervous signs, in canine distemper, 260
Neu test, *see* Neutralization test
Neuraminidase, in influenza virus, 155–161, 176
Neutralization test
  alphavirus, 344, 345
  equine rhinovirus, 47
  Lentivirinae, 630
  Spumavirinae, 615
  togavirus, 416–418
Newcastle disease virus, *see also* Paramyxovirus
  antibody, natural, 218
  antigenic type, 212
  Beach's form, 214
  Beudette's form, 215
  biotyping system, 212, 213
  in chicken embryo, 213, 214
  cytopathic effect, 214
  diagnosis, 217
  Doyle's form, 214
  history, 211
  Hitchner's form, 215
  host range, 216
  immunity, 217, 218
  isolation, 217
  lentogenic strain, 212
  mesogenic strain, 212
  morphology, 190, 192–194
  plaque, 214
  properties, 211–214
  serology, 217
  strain differentiation, 212, 213
  symptoms, 214, 215
  transmission, 215
  vaccination, 218
  velogenic strain, 212, 213
Nucleoid, 601, 602

O

O agent, *see also* Rotavirus
  characteristics, 106
  morphology, 110
  physicochemical properties, 113
Obodhiang virus, *see also* Rhabdovirus
  host, 534
  isolation, 543
  transmission, 542
ODE, *see* Old dog encephalitis
Old dog encephalitis, 260, 261
Oligonucleotide fingerprint, alphavirus, 350, 351
Onc, 574, 575
Oncogenic virus, 553–588
Oncornavirus B, 555, 556
Oncornavirus C, 556
Oncornavirus D, 556

O'nyong'nyong virus, 345
Oral lesion, rinderpest, 273, 274
Orbivirus, 69-94, *see also* Reoviridae
  acid sensitivity, 70
  antigenic grouping, 69-73
  characterization, 69-73
    problem, 92-94
  classification, 70-72
  diagnosis, 92-94
  distinguished from reovirus, 79
  genetic reassortment, 93
  Gorman grouping, 71, 72
  host, vertebrate, 72
  morphology, 73-80
  ribonucleic acid, 69
  serotype, 72
  temperature-sensitive mutant, 93
  tubule, 80
  ungrouped, 90, 92
  vector, 72
  virus particle layer, 71
Orphan virus, 6
Orthomyxoviridae, 149-184, *see also* Influenza virus
Orthomyxovirus, *see* Influenza virus
Otitis media, measles complication, 252

P

PAGE, *see* Polyacrylamide gel electrophoresis
Palyam virus, 89, 90
Pancreatic enzyme, in rotavirus infection, 134
Parainfluenza virus
  from budgerigar, 220, 221
  canine, differentiation from canine distemper, 269
Parainfluenza-1 virus, human, 205, *see also* Sendai virus
Parainfluenza-2 virus, *see* SV-5 virus
Parainfluenza-3 virus, 195-200, *see also* Paramyxovirus
  antibody, natural, 200
  clinical signs, 198
  cytopathic effect, 196, 197
  diagnosis, 199, 200
  experimental, 198, 199
  hemagglutination, 196, 197
  histopathology, 199
  history, 195, 196
  host tissue range, 196
  immunity, 200
  isolation, 199, 200
  morphology, 191-194
  natural, 197-199
  respiratory disease complex, 198
  serology, 196, 200
  vaccine, 200
Parainfluenza-4 virus, *see* Paramyxovirus
Paralysis
  enterovirus encephalomyelitis, 30, 31
  rabies, 539
  visna virus, 626
Paramyxoviridae, *see* Paramyxovirus
Paramyxovirus, 185-287, *see also* specific virus
  antigenic relationship, 205
  avian, 211-221
  classification, 188, 189
  diagnosis, 208-210
  experimental, 208-210
  genome, 189
  hemagglutination, 205
  history, 204
  host cell range, 204, 205
  immunity, 210
  isolation, 208, 209
  morphology, 188, 189, 194
  natural, 205, 206
  nucleocapsid, 194
  pathology, 207, 208
  physicochemical properties, 198, 199
  replication, 195
  rodent, 204-211
  serology, 209, 210
  viral envelope, 194
Parana virus, 519, *see also* Arenavirus
  host range, 513
Parvovirus, canine, 269
Passive hemagglutination, togavirus, 467
Pasteurella haemolytica, with parainfluenza-3 virus, 198, 199
Pericarditis, role of swine enterovirus, 33, 34
Peromyscus agent, 210
Peromyscus virus, *see* Paramyxovirus
Persistent infection, morbillivirus, 249-251
Peste des petits ruminants virus, 236, 281-287, *see also* Morbillivirus
  antigen
    identification, 287
    relationship, 244, 245
  clinical signs, 282-284
  control, 285, 286

cytopathic effect, 248
diagnosis, 286, 287
epizootiology, 285
host range, 284
immunity, 285
isolation, 287
pathology, 286
prevention, 285, 286
route of infection, 281, 282
serology, 287
strain, 284, 285
transmission, 285
Pestivirus, 442, *see also* Togavirus; specific virus
Pheasant, eastern equine encephalitis virus, 373-375, 383
Phlebovirus, 480, 481, *see also* Bunyaviridae; Rift valley fever virus
Phonation, altered, 539
Phosphonoformate, 631
Pichindé virus, 519, 520, *see also* Arenavirus
host range, 515
Picornaviridae, *see* Picornavirus
Picornavirus, 1-55, *see also* specific virus
antigenic structure, 13-17
buoyant density, 11
chemical properties, 12, 13
classification, 4, 5
clinical signs, 19-36
description, 8-17
diagnosis, 43-52
distinguishing feature, 8
epizootiology, 40-43
heat stability, 11
host range, 17-19
immunity, 36-40
molecular weight, 11, 12
morphology, 8-11
pathogenesis, 17-34
pH sensitivity, 11
physicochemical properties, 9
prevention, 52-54
protein coat, 12
site of infection, 20
Pixuna virus, 345
Placenta, site of bovine enterovirus infection, 29
Plaque-neutralization test
bluetongue virus, 83, 84
bovine enterovirus, 48
togavirus, 417
vesicular disease, 45
Plasmodium falciparum, 584
Platelet count, equine infectious anemia virus, 651
PMV, *see* Paramyxovirus
Pneumonia
maedi virus, 627, 629
measles complication, 256
swine enterovirus, 33, 34
Pneumonia virus of mice, *see also* Pneumovirus
antibody production, 228
clinical signs, 228
history, 226
host tissue range, 226, 227
immunity, 228
pathology, 227
physicochemical properties, 227
Pneumovirus, 187-228, *see also* Paramyxovirus; specific virus
morphology, 195
nucleocapsid, 195
physicochemical properties, 195
viral envelope, 195
Pol, 559, 562, 563
Polioencephalomyelitis, *see also* Enterovirus encephalomyelitis
lactic dehydrogenase virus of mice, 454
Pollution, reovirus marker, 95
Polyacrylamide gel electrophoresis, virus classification, 71, 73
PPRV, *see* Peste des petits ruminants virus
Porcine enterovirus, serotype, 16
Progressive pneumonia virus, 620, 626, *see also* Lentivirinae
Propagation, *in vitro,* rotavirus, 128-134
Protein kinase, 575
Pseudoreplica technique, rotavirus, 124, 125
Pulmonary edema, African horse sickness, 80
PVM, *see* Pneumonia virus of mice

R

Rabies fluorescent focus inhibition, 533, 545
Rabies virus, *see also* Rhabdovirus
compartmentalization, 541
cyclic fluctuation, 541
epidemiology, 540-542

host range, 534
identification, 544, 545
incubation, 536, 537
isolation, 542, 543
pathogenesis, 537, 538
physicochemical properties, 532
propagation, 544
replication, 538
transmission, 535
vaccine, 542, 546, 547
vampire bat, 539
Rabbit infectious cardiomyopathy, 311
Radioimmunoassay
Lassa virus, 523
rabies virus, 545
solid-phase, rotavirus, 126
togavirus, 418, 419, 467
Rash, measles, 252
Rat coronavirus, 309, *see also* Coronavirus
isolation, 316
serological testing, 322
RCV, *see* Rat coronavirus
RDDP, *see* Reverse transcriptase
RD-114 virus, 580
Replication, viral
in erythrocyte, 86, 87
tubule, 79, 80
Reproduction
effect of bovine enterovirus, 28, 29
effect of swine enterovirus, 32, 33
Reoviridae, 65–147, *see also* specific virus
classification, 68, 69, 73
orbivirus infection, 67–99
provisional, 97–99
reovirus infection, 67–99
ribonucleic acid, 69
Reovirus, 94–97, *see also* Reoviridae
acid sensitivity, 70
avian, 96, 97
with canine SV-5 virus, 206
diagnosis, 95
distinguished from orbivirus, 79
genetic reassortment, 93, 95
mammalian, 94, 95, 97
morphology, 69
neurovirulence, 94, 95
ribonucleic acid, 95, 96
temperature-sensitive mutant, 95
transmission, 94, 95
virus particle layers, 71
Respiratory disease
enterovirus, 27, 28
reovirus, 96
Respiratory syncytial virus, *see also* Bovine respiratory syncytial virus; Pneumovirus
morphology, 190–192
Retroviridae, 553–632, *see also* Lentivirinae; Retrovirus; Spumavirinae
Retrovirus, 553–588, *see also* specific virus
A particle, 555, 556
adsorption, 564
antigenic determinant, 569, 570
antigenic marker, 563
avian
clinical signs, 568
subgroup, 572, 573
B particle, 555, 556
C particle, 556
cattle, 584–586
chicken, 572–575
classification, 555–558
coating ability, 570
COFFAL test, 587
defective, 570–572
detection, 587
diploidy, 559
ecotropic, 554
env, 559, 564–566
feline, 575–582
antigen, 581
incubation, 580
subgroup, 581
feline oncornavirus-associated cell membrane, 567, 581, 582
gag, 559, 560, 562
gene product, 560–567
genome
structure, 508–560
translation, 563
glycoprotein, 559, 564–566
helper virus, 571, 572
homology, 582, 583
host range, 570
immune response, target, 562
immunofluorescence, 587
immunosuppression, 575, 582
inheritance, 554
molecular mechanism, 554, 555
morphology, 555–558, 561, 562
neoplastic transformation, 559, 566, 567, 575

onc, 574, 575
pathobiology, 572–586
pol, 559, 562, 563
primate, 582–584
protein
precursor, 562
structural, 559, 560, 562
protein kinase, 575
radioimmunoassay, 563
replication, 559, 567, 568
defective, 570, 571
regulation, 557
restriction, 570
reverse transcriptase, 555, 559, 562, 563, 567, 574
reverse transcription, 553, 559
ribonucleic acid, 555, 558
spike, 564, 570
src, 559, 566, 567, 583
extra, 574, 575
transformation defective, 559, 571
transmission, 554, 582, 584, 588
tryptic fingerprinting, 565
vaccine, 581
virus release, 586
xenotropic, 554
Reverse transcriptase
equine infectious anemia virus, 649
Lentivirinae, 620
retrovirus, 555, 559, 562, 563, 567, 574
Spumavirinae, 600, 605, 606
Reverse transcription
Lentivirinae, 622
retrovirus, 553, 559
Reversed passive hemagglutination inhibition, Lassa virus, 522
RFFI, *see* Rabies fluorescent focus inhibition
Rhabdoviridae, *see* Rhabdovirus
Rhabdovirus, 529–547, *see also* specific virus
animal, 531, 533–535
antigenic composition, 532, 533
classification, 529–531
clinical signs, 539, 540
control, 546, 547
diagnosis, 542–546
epidemiology, 540–542
host range, 533–535
identification, 544, 545
incubation period, 536
isolation, 542, 543
morphology, 530, 531
pathogenesis, 537, 538
physicochemical properties, 532
plant, 531, 533
prevention, 546, 547
propagation, 544
replication inhibition, 531
ribonucleic acid, 531
serological testing, 535, 546
T form, 531
transmission, 535, 536
vector, 533
Rhinovirus, *see also* Bovine rhinovirus; Equine rhinovirus; Picornavirus
antigenic structure, 16
description, 7
distinguishing feature, 8
epizootiology, 41
host range, 17, 18
immunity, 37, 38
Rhipicephalus marginatus, Nairobi sheep disease virus, 502
RIA, *see* Radioimmunoassay
Ribonuclease, 562
Ribonucleic acid
double-stranded
reovirus, 95, 96
rotavirus, 114, 115
messenger, 568
single-stranded
Bunyaviridae, 492
equine infectious anemia virus, 649
flavivirus, 358, 359
influenza virus, 154, 155
Lentivirinae, 621
morbillivirus, 239, 240
retrovirus, 555, 558
rhabdovirus, 531
Spumavirinae, 602, 605
togavirus, 447, 448
transfer, 558
Ribosome, arenavirus, 511
Rice dwarf virus, morphology, 69
Rift valley fever virus, 480, 499–501, *see also* Bunyaviridae
classification, 481
clinical signs, 500
control, 501
diagnosis, 501

differentiation from Akabane virus, 496
distribution, 499
epidemiology, 500, 501
history, 499, 500
vector, 500, 501
Rinderpest virus, 236, 272–281, *see also* Morbillivirus
antigen identification, 280
antigenic relationship
canine distemper virus, 243
measles virus, 244
clinical signs, 273–274
control, 279
convalescent period, 275
diagnosis, 279–281
differentiation from peste des petits ruminants, 286
in domestic animals, 275, 276
epizootiology, 278, 279
host range, 275, 276
immunity, 277, 278
incubation period, 273
isolation, 284
in laboratory animals, 276
mucosal period, 273–275
pathology, 280
prevention, 279
prodromal period, 273
recovery mechanism, 278
route of infection, 272, 273
serology, 280, 281
strain, 277
transmission, 278
vaccine, 277, 278
in wild animals, 276
RNA, *see* Ribonucleic acid
RNA-dependent DNA polymerase, *see* Reverse transcriptase
RNA polymerase, rotavirus, 116, 117
RNA tumor virus, *see* Retrovirus
Rodent, cardiovirus host, 18, 41
Ross river virus, *see also* Alphavirus; Togavirus
clinical signs, 379
epizootiology, 412
muscle disease, 379
Rotavirus, 105–141, *see also* Reoviridae; specific virus
age relationship, 136, 137
antigen
group-specific, 118
location, 115, 116
species-specific, 118
antigenic relationship, 117, 118
in calf, 118, 119
classification, 107, 108
clinical features, 118–120
control, 139, 140
cytopathic effect, 129, 130
dehydrating effect, 123, 124
diagnosis, 124–128
differentiation
canine distemper, 269
coronavirus, 305
epidemiology, 135–137
genetic diversity, 136, 137
growth *in vitro,* 134
host cell receptor, 116
immunity, 137–139
immunofluorescence, 131
incidence, 108
intestinal effect, 122, 123
in lamb, 119
in man, 119, 120
morphogenesis, 108–112, 130
pancreatic enzyme, role, 134
particle/infectivity ratio, 114
particle type, 109–111
passive protection, 138
pathogenesis, 120–124
pathology, 120–124
physicochemical properties, 112–117
in pig, 119
prevention, 139, 140
propagation *in vitro,* 130–134
protein composition, 115, 116
ribonucleic acid, 114, 115, 117
seasonal incidence, 135
serial passage, 134
serotype, 117
site of infection, 120–122
species difference, 135, 136
subclinical infection, 120
substitute antigen, 128
transmission, 136, 137
treatment, 119, 139, 140
tubule formation, 112, 130
vaccination, 117, 139
virus particle layers, 71
D-xylose malabsorption, 124
Rous sarcoma virus, 556

RR, *see* Ross river virus
RSV, *see* Bovine respiratory syncytial virus
Rubella virus, 442
 antigenic relationship, 450
 differentiation from measles, 257
Rubivirus, 442
RV, *see* Rinderpest virus
RVF, *see* Rift valley fever virus

S

St. Louis encephalitis virus, *see also* Flavivirus; Togavirus
 composition, 356
 geographic distribution, 366
 morphology, 360
 nucleotide composition, 359
 oligonucleotide map, 366
 phospholipid content, 358
 strain separation, 365, 366
 virulence, 366
San Miguel sea lion virus, *see also* Calicivirus
 clinical signs, 35
 electron micrograph, 10
Scrapie, 620
SDAV, *see* Sialodacryoadenitis virus
Semliki forest virus, *see also* Alphavirus; Togavirus
 clinical signs, 380
 epizootiology, 411
 morphology, 336
Semliki forest virus complex, 345
Sendai virus, *see also* Parainfluenza-1 virus; Paramyxovirus
 diagnosis, 208–210
 experimental, 206–208
 hemagglutination, 205
 host cell range, 204, 205
 host factors, 207
 immunity, 210
 infection pattern, 205, 206
 isolation, 208, 209
 mouse antibody production test, 209
 natural, 205, 206
 pathology, 207, 208
 relation to human parainfluenza-1 virus, 205
 serology, 209, 210
 study, 204
Serum neutralization test
 Akabane virus, 492, 493
 alphavirus, 346, 347
 equine infectious anemia virus, 656
 vesicular disease, 45
SF, *see* Semliki forest virus
Sheep pox, differentiation from peste des petits ruminants, 286
SHFV, *see* Simian hemorrhagic fever virus
Shipping fever, *see* Parainfluenza-3 virus
Sialodacryoadenitis virus, 309, *see also* Coronavirus
 isolation, 316
 serological testing, 322
Sigma virus, *see also* Rhabdovirus
 host, 534
 transmission, 536
Simbu group virus, 497–499, *see also* Bunyaviridae; specific virus
Simian enterovirus, diagnosis, 50
Simian foamy virus, *see* Spumavirinae
Simian hemorrhagic fever virus, *see also* Togavirus
 classification, 442
 clinical signs, 455
 electrophoresis, 449
 epidemiology, 458, 459
 morphology, 446
 pathology, 455
Simian sarcoma virus, 583
Simian virus SA.11, *see also* Rotavirus
 characteristics, 106
 morphology, 110
 physicochemical properties, 113
SIN, *see* Sindbis virus
Sindbis virus, *see also* Alphavirus; Togavirus
 carbohydrate composition, 341
Single radial hemolysis test, coronavirus, 320
SLE, *see* St. Louis encephalitis virus
SMEDI virus, 32, 33
 diagnosis, 49
SMS, *see* San Miguel sea lion virus
Spumavirinae, 599–616, *see also* specific virus
 antibody, 612
 antigenic composition, 604, 605
 characteristics, 600
 complement fixation, 615
 cytopathic effect, 608, 613, 614
 diagnosis, 613–616

epidemiology, 612
fluorescent antibody test, 615
hemagglutination inhibition, 616
human, 610, 611
immunity, 611, 612
immunosuppression, 609, 610
infection
experimental, 609, 610
natural, 608, 609
isolation, 613
morphology, 601, 602
nasopharyngeal carcinoma, 600, 610, 611
neutralization, 615
nuclear immunofluorescence, 608
nucleic acid, 602, 603
persistance, 609
physicochemical properties, 602–605
protein, 605
replication, 608
reverse transcriptase, 600, 606
serology, 613
spread, 611
stability, 602
tissue culture, 607, 608
Spumavirus F, 558
Src, 559, 566, 567, 583
extra, 574, 575
SSPE, *see* Subacute sclerosing panencephalitis
Stillbirth
bovine enterovirus, 28, 29
togavirus, 391–394
Subacute sclerosing panencephalitis
clinical signs, 252, 253
with measles virus, 236, 249, 255
SVD, *see* Swine vesicular disease
SV-5 virus, *see also* Paramyxovirus
canine
antigenic relationship, 202
clinical signs, 202
diagnosis, 203
experimental, 202, 203
hemagglutination, 201
history, 201
host cell range, 201
immunity, 203, 204
isolation, 203
natural, 202
serology, 203
vaccination, 203, 204
simian, 202
Swine enterovirus, *see also* Enterovirus; Enterovirus encephalomyelitis; Swine vesicular disease
clinical signs, 29–34
diagnosis, 48, 49
pathogenesis, 29–34
in reproduction, 32, 33
Swine fever virus, 442, *see also* Togavirus
Swine vesicular disease, *see also* Swine enterovirus
clinical signs, 31, 32
diagnosis, 43–46
differentiation, foot-and-mouth disease, 32
epizootiology, 42
relation to coxsackie virus, 19
vaccine, 52
Syncytial virus of rabbits, 97, 98

T

Tacaribe virus, *see also* Arenavirus
antibody, 519
host range, 515
Talfan disease, *see* Enterovirus encephalomyelitis
Tamiami virus, 520, *see also* Arenavirus host range, 516
TECV, *see* Turkey enteritis coronavirus
Teschen disease, *see* Enterovirus encephalomyelitis
TGEV, *see* Transmissible gastroenteritis virus
Thymic deficiency, with murine hepatitis virus, 310
Thymocyte, with retrovirus, 565
Tick, *see also* specific genus
Colorado tick fever, 85, 86
Kemorovo virus, 87
Tick-borne encephalitis virus, 359
Togaviridae, *see* Togavirus
Togavirus, 331–468, *see also* specific virus
abortion-causing, 391–394
agar gel precipitation, 464, 465
age effect, 385, 386
antibody detection, 465–467
antigen
composition, 335–367
relationship, 449–451
structure, 449–451

arthropod-borne, 331–426
biology, comparative, 367–394
blood-brain barrier, 388
budding, 337
cell culture, 414
classification, 442, 443, 446, 449
clinical signs, 332, 367
complement fixation, 416, 466
control, 421–426
diagnosis, 413–421, 459–467
enzyme-linked immunosorbent assay, 419, 467
encephalitis-causing, 367–391
epidemiology, 455–459
epizootiology, 395–413
geographic distribution, 332, 333
glycoprotein structure, 337
hemagglutination, 449
hemagglutination inhibition, 415, 416
host defense mechanism, 385, 387
host range, 355
human, 334, 369, 394, 396, 397, 401–404, 424
identification, 459–465
immune response
alteration, 390
pathogenic role, 389, 391
immunoelectro-osmophoresis, 467
immunofluorescence, 463, 464
indirect fluorescent antibody test, 419
indirect immunofluorescence, 467
infectiousness, 447
interferon, effect of, 386, 387
isolation, 413–415, 459–463
morphology, 335–367, 443–446
muscle involvement, 388
neuropathology, 367
neutralization test, 416–418
non-arbo, 441–468
passive hemagglutination, 467
pathogenesis, 367–394, 451–455
genetic determinant, 390, 391
host factor, 385
pathological lesion, 388, 389
pathology, 451–455
physicochemical properties, 335–367, 446–449
plaque-reduction test, 417
prevention, 421–426
radioimmunoassay, 418, 419, 467
replication
enhancement, 390
rate, 387
site, 388
ribonucleic acid, 447, 448
ribonucleoprotein, 447
serological test, 419–421
stillbirth, 391–394
structural polypeptide, 448, 449
vector control, 425, 426
virulence difference, 386, 387
virus neutralization, 465, 466
Toxocara canis, with Japanese encephalitis virus, 385
Transmissible enteritis, *see* Turkey enteritis coronavirus
Transmissible gastroenteritis virus, 306, 307, *see also* Coronavirus
clinical signs, 305
isolation, 316
serological testing, 321, 322
vaccine, 322
Transmission, cross-placental, rotavirus, 136
Tropical sprue, 308
Trout hemorrhagic septicemia virus, 535
Trypsin, in rotavirus infection, 134
Tuberculosis, effect of measles virus, 254, 255
Tumor
growth, effect of lactic dehydrogenase virus, 454
malignant, retrovirus-caused, 573
Turkey enteritis coronavirus, 308, *see also* Bluecomb; Coronavirus
isolation, 316
serological testing, 322
Turkey parainfluenza virus, 220, *see also* Paramyxovirus
Type 1 hemadsorption virus, *see* Parainfluenza-3 virus

U

Uukuvirus, 480, 481, *see also* Bunyaviridae
Una virus, *see also* Alphavirus; Togavirus
encephalitis, 380
epizootiology, 412

V

Vaccine
avian infectious bronchitis virus, 308
bluetongue virus, 81, 82
bovine diarrhea virus, 457
bovine respiratory syncytial virus, 226
canine distemper virus, 243, 244, 264, 265, 268, 269
canine SV-5 virus, 203, 204
coronavirus, 322, 323
eastern equine encephalitis virus, 421
equine arteritis virus, 457
feline calicivirus, 40, 54
foot-and-mouth disease, 50, 51
adjuvant, 53
hog cholera virus, 456
interference by infectious bursal disease virus, 98
Japanese encephalitis virus, 424, 425
measles virus, 156, 157
morphology, 556
Newcastle disease virus, 218
parainfluenza-3 virus, 200
peste des petits ruminants, 285
retrovirus, 581
Rift valley fever virus, 501
rinderpest virus, 277, 278
rotavirus, 117
Sendai virus, 210
swine enterovirus, 53, 54
swine vesicular disease, 52
Venezuelan equine encephalitis virus, 422–424
Wesselsbron disease, 393
western equine encephalitis virus, 421
VEE, *see* Venezuelan equine encephalitis virus
Venezuelan equine encephalitis virus, *see also* Alphavirus; Togavirus
abortion, 393, 394
biological variation, 351
bovine, 377
clinical signs, 367–380
distribution, 403, 404
dog, 377, 378
extraneural tissue involvement, 382, 383
hematological abnormality, 371, 373
host, 405–408
human, 403, 404
infection
experimental, 372
pattern, 371
source, 371, 377, 378
interepidemic maintenance, 406, 407
isolation, 414, 415
after death, 383
occurrence, 403, 404
pathogenicity, 369
pathology, 380–383
subtype, geographic distribution, 352–354
swine, 377
transmission cycle, 404–408
vaccine, 415, 422–424
improperly inactivated, 407
interference by heterologous antibody, 423
vector, 404, 405
viremia, 371
Venezuelan equine encephalitis complex, 345
VES, *see* Vesicular exanthema of swine
Vesicle
foot-and-mouth disease, 25
swine vesicular disease, 29, 30
Vesicular disease, *see* Foot-and-mouth disease virus; Swine vesicular disease; Vesicular exanthema of swine; Vesicular stomatitis virus
Vesicular exanthema of swine, *see also* Calicivirus
clinical signs, 35
control, 54
diagnosis, 43–46
epizootiology, 42
Vesicular stomatitis virus, *see also* Rhabdovirus; Vesiculovirus
clinical signs, 540
diagnosis, 43–46
epidemiology, 542
host range, 534, 535
incubation, 537
isolation, 543
propagation, 544
protein, 532
transmission, 536
Vesiculovirus, 530, *see also* Rhabdovirus; specific virus
prevention, 547
serological test, 546
VIA antigen, 14

Viremia
African horse sickness, 80
Colorado tick fever, 86
Virology, mammalian, history, 5
Virus, *see* specific virus
Virus neutralization test
coronavirus, 319
enterovirus encephalomyelitis, 49
simian enterovirus, 50
togavirus, 465, 466
vesicular disease, 45
Visna virus, 619–640, *see also* Lentivirinae
clinical signs, 626, 627
diagnosis, 630
growth cycle, 620, 622
life cycle, 637, 638
pathology, 627–629
replication, restriction, 638
VS, *see* Vesicular stomatitis virus

W

Wasting disease, coronavirus, 308
Wallal virus, 90, 91
Warrego virus, 90, 91
WEE, *see* Western Equine encephalitis virus
Wesselsbron virus, *see also* Flavivirus; Togavirus
clinical signs, 392, 393
differentiation from Akabane virus, 496
epizootiology, 410, 411
maintenance cycle, 411
pathology, 392, 393
vaccine, 393
vector, 410, 411
West Nile virus, *see also* Flavivirus; Togavirus
antigenic relation, 365
clinical signs, 379
epizootiology, 411
pathology, 385
vector, 411
Western equine encephalitis virus, *see also* Alphavirus; Togavirus
bird, domestic, 373–375, 383, 384
bovine, 377
chicken, 375
climatological factor, 401, 402
clinical signs, 367–380
geographical distribution, 400, 401
host, 402
human, 401, 402
infection, source, 371, 375
isolation, 413, 414
after death, 383
occurrence, 400, 401
overwintering mechanism, 403
pathogenicity, 369
pathology, 380–384
reptile, 402, 403
swine, 377
transmission cycle, 401–403
transplacental infection, 394
vaccine, 421
vector, 401
Western equine encephalitis complex, 345
Whataroa virus, 345
Winter dysentery, enterovirus role, 27
WN, *see* West Nile virus
WSL, *see* Wesselsbron virus

X

D-Xylose malabsorption, 124

Y

Yucaipa virus, *see also* Paramyxovirus
clinical signs, 219
history, 219
host range, 219
properties, 219
related isolates, 219, 220
Y62-33 virus, 345

Z

Zwoegerziekte virus, 620, 626, *see also* Lentivirinae

0121 isolate, *see* Yucaipa virus